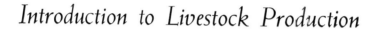

Introduction to Livestock Production

A SERIES OF BOOKS IN *Agricultural Science*

Animal Science Editors: G. W. SALISBURY AND E. W. CRAMPTON

Plant and Soil Science Editors: IVER JOHNSON AND M. B. RUSSELL

Introduction to Livestock Production

INCLUDING

DAIRY AND POULTRY

Second Edition

H. H. Cole, Editor

University of California, Davis

W. H. Freeman & Company

San Francisco

PRINTED IN THE UNITED STATES OF AMERICA

LIBRARY OF CONGRESS CATALOG NUMBER: 66–16377
STANDARD BOOK NUMBER: 7167 0812–4

5 6 7 8 9 10

Preface to the Second Edition

The objective of the second edition of *Introduction to Livestock Production,* like that of the first, is to provide students with a comprehensive view of the livestock industry and to show how the science of biology is utilized in modern livestock management practice.

The authors and editor have been highly gratified by the wide acceptance of the first edition, and believe that this edition is an improvement upon it. For example, in order to emphasize that knowledge from the sciences is being put to practical use in solving livestock problems, more applied biology has been included in the livestock management chapters. This change has made it possible to reorganize the nutrition section in order to include such topics as the role of nutrients in the animal body, pathways in nutrient utilization, estimation of the nutritional value of feeds, and the nutrient requirements for various physiological functions. Certain parts of the text give the student a glimpse of some of the metabolic complexities. Though these processes will not be completely understood without additional study, we think that the discussions provided here will show the beginning student why the animal science curriculum includes training in chemistry, physics, and biology. It would, for example, be absurd to expect beginning students to memorize or to comprehend fully the metabolic pathways described in Chapter 29; yet, in our opinion, it is important for them to know that these pathways exist.

Although the broad scope of our subject matter has prevented us from giving the text a more international point of view, it is encouraging to know that Dr. Jaime E. Escobar considered the first edition to be of sufficient value for students in Spanish-speaking countries to warrant publication of a Spanish translation.

In this second edition more attention has been given to interchapter correlation of the subject matter, which is always a problem with a text by many contributing authors. However, I take the view that any lack of closely knit organization is more than compensated for by the fact that each author can speak with the authority gained from research and from having taught the topic on which he is writing. To give but one example: Who could speak with the authority of Dr. Ralph Phillips on the subject, "The Livestock Industry: Its Scope and Potential"? Dr. Phillips, who has devoted the past 19 years to problems of agriculture throughout the world, has held prominent positions with the Food and Agricultural Organization (FAO) of the United Nations and with the Foreign Agricultural Service of the United States Department of Agriculture.

To my colleagues who have given freely of their time and advice, I am most grateful. I also wish to thank Miss Mary Bigelow, Mrs. Vivian Baker, and Mrs. Carlene Blaylock for their devoted services in matters pertaining to editing.

> "But natural and experimental history is so varied and diffuse, that it confounds and distracts the understanding unless it be fixed and exhibited in due order."
>
> FRANCIS BACON, *Novum Organum*

January 1966   H. H. COLE

Preface to the First Edition

During the past quarter century the livestock industry has been emerging from "a way of life" to a sound business enterprise based on advances made both in the biological sciences—nutrition, genetics, physiology—and in the sciences concerned with disease control, as well as upon accepted economic principles. Though one may have nostalgic recollections of the good old days, as do the editor and the authors of this volume, it behooves us to acquaint the potential livestock producers of tomorrow with this new and exciting era of the animal, dairy, and poultry husbandman. Large livestock enterprises have become commonplace. Many feed yards, for example, finish over 50,000 head of cattle annually. Yes, this means fewer livestock managers, but at the same time it means more positions of great responsibility comparable to those available in industry. These changes demand a new type of leadership based upon a sound understanding of basic biological principles and an understanding of modern business methods.

In the past, livestock producers raised animals of their choice and sold them for what they could get at the market place. In order to succeed, the livestock producer of today, and of tomorrow, must raise animals that meet consumer specifications and which will retail at a price the consumer is willing to pay. This concept of producing a high quality, standardized product has been developing over a considerable period. For example, as early as 1926, a group of beef producers was organized for the purpose of establishing a federal program for identification of high quality beef. This initial stimulus led to the development of federal meat grading. The introduction of standardized meat, milk, and egg products in the market has played an important part in creating the high consumer acceptability of these products. Much more can and must be done. Furthermore, the economy of production must be emphasized if animal products are to compete successfully with plant products. Though the preference for animal products is high, the consumer does consider price as has been dramatically illustrated in the competition between butter and margarines.

What are the livestock, poultry, and dairy industries doing about improving the quality and economy of their products? The current practices of the poultry industry serve as an outstanding example of the way in which modern concepts of the biological sciences have been used to produce quality products economically; nutritionists, geneticists, physiologists, and disease-control specialists have all been recruited for the task. Other phases of the livestock industry are on the move: the use of the hormone diethyl-stilbestrol to increase the rate and economy of gain in cattle and sheep; the use of crossbreeding to increase the number of pigs weaned; the use of artificial insemination to hasten improvement through breeding; and the development of drugs and vaccines for disease control are but a few examples of the ways in which advances in the biological sciences are being utilized to increase productivity.

This is the story that unfolds in the succeeding pages; this together with an overall picture of the livestock industry. This book, designed for beginning students interested in some phase of livestock production, is the product of 40 authors, all of whom are accepted leaders in specialized fields. The student has an opportunity, therefore, to become acquainted early with points of view all too frequently reserved for advanced courses. Our aim has been to develop these concepts in a manner understandable to the high school graduate and to arouse in the student early in his training a keen desire to delve deeper into the intriguing possibilities of the further application of the sciences in improving livestock production. Though this text is designed primarily as a beginner's text, livestock producers, I am sure, will find much of interest. In many instances, advances relating to production of one species have not been adapted to others. In reading the manuscript, I have been fascinated by the inherent possibilities of such progress.

Some, no doubt, will decry the lack of emphasis on livestock expositions. In the past, these shows have had a tremendous influence in improving livestock. Somehow they have been left behind. Let us hope that they will rise from the ashes of nurse cows, emphasis on fancy points unrelated to productivity, and subtle means of hiding defects and develop programs which will demonstrate how modern methods can be used in improving our herds and flocks.

The editor is indebted to the authors for their patience in accepting suggested changes; to many colleagues for reviewing different portions of the manuscript; to W. H. Freeman and Company for their cooperation and advice; and to Mrs. June Law, Mrs. Vivian Baker, and Miss Joan Popovich for their assistance in preparing the index and in other matters relating to editing.

> Come shepherd, let us make an honorable retreat;
> Though not with bag and baggage, yet with scrip and scrippage.
> SHAKESPEARE, *As You Like It*

March 1962   H. H. COLE

Contributors

F. N. ANDREWS, *Purdue University, Lafayette, Indiana.*

J. M. BELL, *University of Saskatchewan, Saskatoon, Canada.*

O. G. BENTLEY, *University of Illinois, Urbana, Illinois.*

V. R. BOHMAN, *University of Nevada, Reno, Nevada.*

G. ALVIN CARPENTER, *University of California, Berkeley, California.*

D. W. CASSARD, *University of Nevada, Reno, Nevada.*

M. T. CLEGG, *University of California, Davis, California.*

H. H. COLE, *University of California, Davis, California.*

HARRY W. COLVIN, JR., *University of California, Davis, California.*

P. T. CUPPS, *University of California, Davis, California.*

L. B. DARRAH, *Cornell University, Ithaca, New York.*

RALPH M. DURHAM, *Texas Technological College, Lubbock, Texas.*

R. H. DUTT, *University of Kentucky, Lexington, Kentucky.*

HAROLD GOSS, *University of California, Davis, California.*

DAVID HALLETT, *United States Department of Agriculture, Washington, D.C.*

LORIN E. HARRIS, *Utah State University, Logan, Utah.*

LLOYD HENDERSON, *3300 Ameno Drive, Lafayette, California.*

HAROLD F. HINTZ, *University of California, Davis, California.*

K. R. JOHNSON, *University of Idaho, Moscow, Idaho.*

R. G. LOY, *University of California, Davis, California.*

HAROLD P. LUNDGREN, *United States Department of Agriculture, Albany, California.*

S. W. MEAD, *University of California, Davis, California.*

J. C. MILLER, *Oregon State College, Corvallis, Oregon.*

C. E. MURPHEY, *United States Department of Agriculture, Washington, D.C.*

A. V. NALBANDOV, *University of Illinois, Urbana, Illinois.*

J. E. NELLOR, *Michigan State University, East Lansing, Michigan.*

A. W. NORDSKOG, *Iowa State University, Ames, Iowa.*

JOHN W. OSEBOLD, *University of California, Davis, California.*

A. M. PEARSON, *Michigan State University, East Lansing, Michigan.*

RALPH W. PHILLIPS, *United States Department of Agriculture, Washington, D.C.*

JOHN C. PIERCE, *United States Department of Agriculture, Washington, D.C.*

G. A. RICHARDSON, *Oregon State College, Corvallis, Oregon.*

W. C. ROLLINS, *University of California, Davis, California.*

JEAN TWOMBLY SNOOK, *Cornell University, Ithaca, New York.*

G. M. SPURLOCK, *University of California, Davis, California.*

W. J. STADELMAN, *Purdue University, Lafayette, Indiana.*

D. F. STEPHENS, *United States Department of Agriculture, El Reno, Oklahoma.*

H. H. STONAKER, *Colorado State University, Fort Collins, Colorado.*

CLAIR E. TERRILL, *United States Department of Agriculture, Beltsville, Maryland.*

E. J. WARWICK, *United States Department of Agriculture, Beltsville, Maryland.*

W. O. WILSON, *University of California, Davis, California.*

Contents

Section IV. Physiological Mechanisms and Livestock Production

Section V. Nutrition and Livestock Production

Section VI. Livestock Management

Section VII. Classification, Grading, and Marketing of Livestock and Their Products

Section VIII. Livestock Diseases

Chapter 1

The Livestock Industry: Its Scope and Potential

And the man increased exceedingly, and had much cattle, and maid-servants, and menservants, and camels, and asses.

Genesis 30:43

The world's barnyard is large indeed. It comprises over six billion animals and birds—approximately twice the number of people in the world.

Livestock and poultry have major places in the economies of practically all countries. They produce the meat, milk, and eggs that are important parts of the diet of most peoples. In many countries, much of the power for farm work and for transport of farm products to the market place is still provided by cattle, horses, donkeys, mules, camels, llamas, water buffaloes, and yaks. Wool, mohair, hides and skins, and the numerous by-products of livestock-processing industries are used by man for many purposes. The manure produced by livestock makes an important contribution to the maintenance of soil fertility.

We must examine the scope of the livestock industry from several viewpoints in order to appreciate fully the contributions of farm animals and poultry to human welfare. Data on numbers of livestock and poultry give some indication of the importance of the industry, but global and even national figures are difficult to comprehend. For this reason, data will be summarized also in terms of the number of animals or birds per person in the human population. Although these numbers are important, the amounts of usable products actually harvested are of greater significance. Thus we shall also examine figures for global and national production of animal products. Here, too, the significance of these products for human welfare may be more readily grasped if products harvested are expressed in amounts per person. Efficiency of production is also important. It will become more so as competition between foods from animal and plant sources increases; hence, some data that indicate variations in levels of efficiency will also be examined.

There are wide variations in the manner in which man utilizes his farm animals and poultry in different parts of the world. There are also wide variations in the productive capacities of different types and breeds. Some of these variations will become apparent as we consider the data on numbers of farm animals and poultry, the amounts of products harvested from them, and the levels of productivity per animal or bird in the livestock and poultry populations. From this background of variation, some concept may be formed of the yet unexploited potential of the industry. It is also necessary to recognize both the important role of animal protein in human nutrition, and the relation of the increase in the human population to future demands for animal products and to the availability of land for various types of agricultural production.

Before delving into the present and potential contributions of the livestock industry to human welfare, we should look back to the time when man had no domesticated animals and birds; when he had to depend upon his prowess as a hunter for his supply of meat, hides, and skins; and when his food other than meat had to be obtained from wild sources or from such domesticated plants as he could cultivate without the aid of animal power.

1-1. THE BEGINNINGS OF LIVESTOCK PRODUCTION

The domestication of livestock was one of man's major steps in his struggle for mastery over his environment. Previously, the domestication of crop plants, particularly the cereals, had given man a dependable source of food, and the development of these crops over the centuries greatly increased the number of people the earth could support. The domestication of food-producing animals not only brought man a more dependable source of animal protein than the produce of the hunt, but also further expanded the number of people the earth could support, since those animals provided a means of converting into human food many plants that are not edible by man (Fig. 1-1). The domestication of the horse put a source of speed and power at man's disposal—a source that was not excelled until the invention of steam and internal-combustion engines. Only a few other achievements in man's history, such as the mastery of fire, the invention of the wheel, the harnessing of power from electricity, and the accomplishment of nuclear fission, rank with the domestication of crops and livestock in giving man the

Fig. 1-1. Simmental cattle graze on alpine pastures in Switzerland. These pasture lands are poorly adapted for producing plant foods for human consumption. The sturdy cattle are used for the production of meat and milk and as a source of power on Swiss farms and in several neighboring countries.

means with which to achieve mastery over his environment and to utilize more fully the resources of the earth he lives on.

Even though the domestication of livestock took place in prehistoric times, it was in reality a recent event in the long history of man. Reed (1959) points out that, in spite of a prolific literature, the central problems concerning the origins and early history of animal domestication remain unsolved. However, on the basis of evidence that he and others have accumulated, Reed concluded that goats were probably domesticated 8,000 to 9,000 years ago, and cattle, sheep, and pigs some time thereafter. There is fairly definite proof that cattle had been domesticated by about 3,000 to 4,000 B.C., or about 5,000 to 6,000 years ago.

Domestication was centered in the village-farming communities of the hilly, grassy, open-forested flanks of the Near Eastern mountain ranges (Reed, 1959), that is, in the Palestinian, Lebanese, and Zagros mountains. From this primary center, the village-farming way of life diffused in all directions, spreading cereal agriculture and the rearing of basic, domestic food-producing animals. In Egypt, Thessaly, Baluchistan, the Indus Valley, and probably even in China, or at least northern China, village-farming life began later and seems to have received a cultural stimulus from southwestern Asia. Whether domesticated animals were actually taken to these other regions or whether only the idea was communicated, we do not know. Each probably occurred as man learned to make use of domesticated animals for meat and, eventually, milk, wool, and draft power.

Although its center was in the Near East, domestication must have actually taken place at many points and under a variety of circumstances. The turkey had been domesticated in southern Mexico before the North American continent was discovered and occupied by Europeans. Such species as the reindeer and the camel's relatives—the llama and the alpaca—must have been domesticated in or near their present habitats. But this is not the place for an extensive discussion of domestication. The objects here are twofold: to emphasize the relatively recent occurrence of domestication, and to give some indication of the benefits accruing to man as a result of its occurrence.

The potential values of domestication were probably little appreciated in preliterate cultures when the process first began. However, as man attained knowledge and understanding, he gradually adapted his birds and animals to his needs. Thus, long before there was any real understanding of the science of genetics, productive animals and birds had been developed, and wide variations in form and function had resulted from the process of selection.

1-2. SIZE OF THE LIVESTOCK INDUSTRY

Numbers of the most important types of farm animals and poultry in the world and in the United States are given in Table 1-1. Most of the data were taken from the tabulations of the Food and Agriculture Organization of the United Nations (FAO, 1964a). Since many countries do not have reliable statistics, some of the figures that contributed to the world totals in Table 1-1 are based on informed guesses. The figures for poultry are probably the least reliable. FAO does not publish world totals for poultry, as it does for large animals, although it does publish the figures for many countries. Hence, the totals for chickens, ducks, geese, and turkeys are merely the totals for those countries listed by FAO, and thus undoubtedly somewhat below world totals. Data for several countries include only total figures for all poultry; these figures are included in the total number of chickens, thus inflating somewhat the data for chickens and deflating those for ducks, geese, and turkeys. For horses, asses, and mules, the United States published only a single figure covering all three in recent years, and even that tally has now been discontinued. Hence, the figure given is not current, but it is the latest available. In spite of these and a few other limitations, the data in Table 1-1 give a good indication of the size of each major segment of the livestock industry, and of the size of the entire industry. The numbers for livestock total 3,086,800,000; those for poultry, 3,297,583,000. There are, of

TABLE 1-1.	*Approximate numbers of the most important types of farm animals and poultry in the world and in the United States.*

Types of Farm Animals and Poultry	Approximate Numbers	
	World	United States
Farm animals		
Cattle	952,900,000	100,002,000
Water buffaloes	101,800,000	—
Sheep	998,300,000	31,320,000
Goats	352,400,000	3,647,000
Pigs	550,600,000	57,000,000
Horses	64,700,000	⎫
Asses	40,400,000	⎬ 3,089,000
Mules	15,300,000	⎭
Camels	10,400,000	—
Poultry		
Chickens	3,189,050,000*	368,452,000
Ducks	58,737,000	12,139,000
Geese	18,392,000	1,152,000
Turkeys	31,404,000	6,488,000

Note: In general, the data represent the situation as of 1962; however, many of the figures that contributed to the world totals are for earlier years.
* This figure includes a number of national totals wherein all types of poultry are reported together.

course, seasonal variations in the numbers of animals in every country, and this is particularly true of poultry.

The ratios of livestock and poultry populations to human populations vary a great deal from country to country. Since it is hardly possible here to show the many contrasts that exist between countries, comparisons are limited to the average world statistics on the one hand and the statistics in the United States on the other.

Suppose the farm animals and poultry of the world were divided equally among all the people. How many animals and birds would each person have in his private barnyard? If all those in the United States were divided among the people in this country, how many animals and birds would each person have? The answers are given in graphic form in Figure 1-2. Thus, for each of the 3,114,000,000 people in the world in 1962, there would have been just over $\frac{1}{3}$ head of cattle (including water buffaloes), just under $\frac{1}{3}$ head of sheep, just over $\frac{1}{10}$ of a goat, somewhat less than $\frac{1}{6}$ of a pig, and about $\frac{1}{20}$ of a horse (including asses, mules, and camels). For poultry, there would have been about one chicken per person, and about $\frac{1}{25}$ of a bird of other types. Figure 1-2 also shows that the United States has fewer sheep, goats, and horses (including asses and mules) per person than the world as a whole,

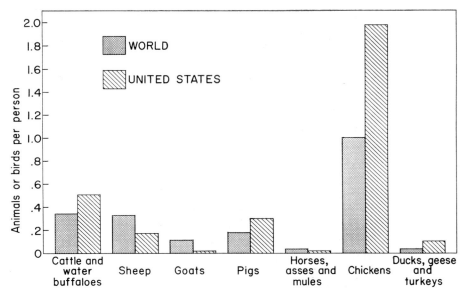

Fig. 1-2. The relative size of the livestock and poultry populations of the world and of the United States in relation to the size of the human populations.

but that there are more cattle, pigs, and poultry in the United States for each person than in the world as a whole.

1-3. THE HARVEST OF LIVESTOCK PRODUCTS

Some groups of people in the world still assess the value of their livestock in terms of numbers. But it is the amount of usable products harvested that actually determines the value of an animal or a bird to its owner and to the human race. The size of the annual harvest of some of the most important livestock and poultry products is given in Table 1-2 (data from FAO, 1964a). In addition to showing the total production of various products in the world and in the United States, this table shows the approximate amount of each product produced each year per person. It will be seen that, with the exceptions of mutton and lamb and of milk other than that from cows, the per capita production of food from animals is much higher in the United States than in the world as a whole.

Although the production of meat, eggs, and milk is high in the United States in relation to the number of people, other countries rank higher in some products. For example, in 1962 the production of beef and veal per person was 87.4 lb. in the United States, 253.7 lb. in New Zealand, and 189.3

| TABLE 1-2. | *Products harvested from farm animals and poultry in the world and in the United States, in terms of total production and production per person in the population.* |

Type of Product	Total Production (metric tons)		Production per Person (lb.)	
	World	United States	World	United States
Beef and veal	30,400,000	7,398,000	21.5	87.4
Mutton and lamb	5,900,000	367,000	4.2	4.3
Pork	29,500,000	5,371,000	20.9	63.5
Poultry meat	7,343,700	4,124,900	5.2	48.7
Eggs (hen)	14,000,000	3,798,800	9.9	44.9
Milk				
Cow	324,200,000	57,119,000	229.5	674.9
Water buffalo	15,700,000	—	11.1	—
Goat	8,200,000	—	5.8	—
Sheep	6,800,000	—	4.8	—
Wool				
Grease	2,568,000	135,900	1.8	1.6
Clean	1,474,000	59,700	1.0	.7

Note: In general, these figures reflect the situation as of 1962, when the population of the world was 3,114,000,000 and that of the United States was 186,591,000.

lb. in Australia. At the other end of the scale, the countries of the Far East produced only 2.3 lb. of beef per person. Production of milk per person was 674.9 lb. in the United States, 4,802.2 lb. in New Zealand, 1,393.8 lb. in Australia, 1,358.4 lb. in the Netherlands, and 1,038.8 lb. in Canada. The countries of the Far East were again at the other end of the scale, with production of only 68.8 lb. per person per year, even when milk from water buffaloes, goats, and sheep was included. The foregoing figures have not been adjusted to account for exports or imports; they merely indicate levels of domestic production in relation to the sizes of the human populations.

Water buffaloes, goats, and sheep make important contributions to the milk supply in certain parts of the world; in fact, these three groups of animals taken together produce 8.6% or somewhat more than 1/12 of the world's milk supply. In India the Murrah and related breeds of water buffaloes produce more milk than do cattle. Other types of water buffaloes, such as some of those in Ceylon (Fig. 1-4) and in southern China (Phillips, 1945; Phillips, Johnson, and Moyer, 1945), are used almost entirely for work. Some breeds of sheep produce high yields of milk, particularly if there has been some selection in order to take advantage of high milk-yielding potentials. For example, Finci (1957) reports that 22,519 Awassi ewes in 109 flocks in Israel averaged 615 lb. of milk during the 1955–56 season (Fig. 1-5). The highest yield for one ewe was about 1,950 lb.

Fig. 1-3. An Indian woman prepares chapatties. The food supply comes mostly from plant sources, but the urn in lower left corner contains goat's milk that will be passed around during the meal. Although cattle are the main producers of milk, such other species as the goat, sheep, yak, and water buffalo are important producers in various parts of the world. [United Nations photo.]

Power is an important part of the harvest from livestock each year (Figs. 1-6 and 1-7). Much of the power on farms in the more highly developed countries is now supplied by tractors or other mechanical equipment, but most of the power used on farms in other parts of the world is still supplied by animals. Although work production is more difficult to measure than food and fiber production, estimates derived (Phillips, 1950–51) from data published by Acock in 1950 indicate that, at that time, 86.4% of the draft power in the world's agriculture was provided by animals (Fig. 1-8). In the United Kingdom, North America, and Oceania, animals provided only 22.7%, 24.5%, and 62.5% of the power, respectively, whereas in the U.S.S.R. the figure was 78.7%. Averages for other regions were Europe, 85.6%; Africa, 98.3%; Near East, 98.9%; Latin America, 99.1%; Far East, 99.9%. The more advanced countries have moved much further in the mechanization of agriculture since these data were tabulated. Some additional mechanization has taken place in other areas also, but there has been relatively little in most of the countries that in 1950 depended primarily upon animals for power in agriculture.

Another general measure of the size of the harvest of livestock and poultry products is the amount produced per animal or per bird in the livestock and poultry populations. Summaries of such measures for the major livestock and poultry products in the world and in the United States are shown in Table 1-3. The productivity of livestock and poultry of the United States is generally well above the world average according to these generalized measures. However, when compared with individual countries, the United States does not rank first in every category. For example, the milk produced per animal in the cattle population in 1962 was 4,198.4 lb. in the Netherlands, 3,369.2 lb. in Denmark, 1,808.7 lb. in New Zealand, and 1,259.2 lb. in the United States. Beef and veal production in these same countries, per animal in the cattle populations, was 164.8 lb. in Denmark, 152.5 lb. in the Netherlands, 95.6 lb. in New Zealand, and 163.1 lb. in the United States. In contrast with these levels of productivity, cattle in the countries of the Far East as a whole yielded only 132.3 lb. of milk and 9.5 lb. of beef and veal per animal in the population.

Such generalized estimates of the productivity of livestock and poultry populations must be taken for what they are. Although they are useful for broad comparisons, they do not reflect the productivity of specialized groups

Fig. 1-4. Water buffaloes provide power for the preparation of rice fields in Ceylon. Some breeds and types of water buffaloes are used primarily for milk production, others for work. [FAO photo.]

Fig. 1-5. An Awassi ewe. Although sheep are used primarily for meat and wool production, they are an important source of milk in some parts of the world. The Awassi, a Near Eastern breed, is being further developed as a milking breed in Israel. [Finci, 1957.]

of animals. They do not take into account such factors as the relative portions of the cattle populations in different countries that may be used for milk or for meat production, or the large numbers of bullocks in countries where cattle supply much of the draft power (Figs. 1-9 and 1-10). To make this point perfectly clear, let us examine the production of dairy cows in the United States in relation to the figure of 1,259.2 lb. of milk given in Table 1-3. According to the United States Department of Agriculture (1963), there were 17,086,000 milk cows on farms in the United States in 1962, and their average yield of milk was 7,370 lb. During the same year 2,006,534 cows on test in Dairy Herd Improvement Associations had average milk yields of 11,032 lb.

1-4. ANIMAL PRODUCTS IN HUMAN NUTRITION

Before leaving the subject of products of the livestock industry and turning to methods and problems of their production, we should delve a bit further into the role of these products in human welfare and particularly into their role in human nutrition.

**TABLE
1-3.** | *Products harvested annually per animal or bird in the livestock and poultry populations.*

	Production per Animal or Bird in Population (*lb.*)	
Product	*World*	*United States*
Beef and veal	70.3	163.1
Mutton and lamb	13.0	25.8
Pork	118.1	207.7
Poultry meat	4.9	23.4
Eggs (hen)	9.7	22.7
Milk		
Cow	750.1	1,259.2
Water buffalo	340.0	—
Goat	51.3	—
Sheep	15.0	—
Wool		
Grease	5.7	9.6
Clean	3.2	4.2

Note: These are very generalized indications of levels of productivity, derived from livestock statistics and from data on amounts of various products produced annually.

Fig. 1-6. Threshing grain by treading in Pakistan. In many parts of the world cattle still provide much of the power for work on farms. [FAO photo.]

Fig. 1-7. Zebu oxen being hitched to a cart loaded with rice straw in India. Owing to scarcity of other materials, the straw is used for livestock feed as well as for thatching roofs and for fuel. [United Nations photo.]

People who live in the United States and other countries having high animal productivity accept as the normal dietary pattern one that contains substantial amounts of animal protein. The housewife builds her menus around meat and other animal products. But the diet of the majority of the world's people is based primarily upon plant products and often upon the produce of a single crop plant such as rice, wheat, or cassava. The amount of food available is also inadequate in many areas. This problem was emphasized in the *Third World Food Survey* (FAO, 1963), which showed that as much as half the people in the world suffer from hunger, malnutrition, or both.

The nature of the problem may be seen in Figure 1-11 where data on calorie, total protein, and animal protein supplies in 49 countries are charted (FAO, 1964b). In these countries calorie supplies vary from a high of 3,510 per person per day in New Zealand to a low of 1,800 in the Philippines. Total protein supplies vary from a high of 112 grams per person per day (also in New Zealand) to a low of 42 in both Ceylon and the Philippines.

For animal protein, New Zealand also leads with a high of 77 grams per person per day, while India (No. 44) has the lowest level, 6 grams, followed by Ceylon and Pakistan with 7 grams each, and Jordan with 9 grams.

In general, the countries that are low in calorie supplies are also low in protein supplies. In turn those that are low in total protein are also low in animal protein. However, there are notable exceptions, particularly Turkey, Yugoslavia, and to a lesser extent Greece. These countries, although having limited supplies of animal protein, have substantial supplies of protein from plant sources.

In view of the critical importance of protein in nutrition, data on protein supplies are charted separately in Figure 1-12. This chart is based on data for 43 countries (FAO, 1964b): because certain data are for years different from those on which Figure 1-11 is based, different levels may be shown for the same countries. In this chart it will be seen that, although supplies of animal protein are generally high in those countries that have an abundance of total protein, and low where total protein is low, there are some important exceptions. For example, Yugoslavia, Greece, and Turkey had more than 90 grams of total protein per day, yet had only 26, 27, and 15 grams, respectively, of animal protein. Similarly, Syria and the United Arab Republic had more than 75 grams of total protein per person per day, yet only 17 and 13

Fig. 1-8. A Mongolian horse in Chinghai Province, China. Although the draft horse has been replaced largely by tractors in some countries, horses are still used extensively for draft, for riding, and as pack animals in many parts of the world. [Photo by Ralph W. Phillips.]

Fig. 1-9. A Chiana or Chianina bull. These large, active, fast-growing cattle are native to central Italy, where they are used for work, for meat, and to some extent for milk production.

grams, respectively, of animal protein. On the other hand, Sweden had only 61 grams of total protein per person per day but, of this amount, 52 grams was animal protein.

In order that we may pursue the subject of animal protein supplies one step further, variations in the levels of these supplies, and in their composition, are charted in Figure 1-13. This chart is based upon FAO (1964b) data for the same 43 countries included in Figure 1-12. It shows how much of the animal protein in each country comes from each of four major types of products: meat and poultry, eggs, fish, and milk and milk products. For example, in the United States where there were 65 grams of animal protein supplies per person per day when these data were compiled, 32 grams were from meat and poultry, 25 grams from milk and milk products, 5 grams from eggs, and 3 grams from fish. Careful examination of the chart will reveal wide variations among countries, in both amounts and proportions of animal protein from different kinds of animal products.

1-5. AN INDUSTRY OF DIVERSITY

Although farm animals and poultry are important in practically every country, their uses differ considerably from country to country. In the foregoing discussion it has been possible only to indicate some of the extreme

Fig. 1-10. A Charollais bull. This beef breed, which is native to France, is now attracting attention in the United States. This animal was photographed at six years of age, when it weighed approximately 2700 lbs.

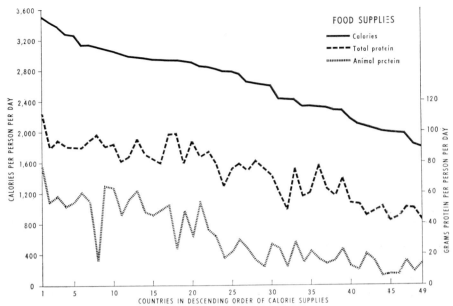

Fig. 1-11. Variations in supplies of calories, total protein, and animal protein in 49 countries. Countries are arranged from left to right, in descending order of calorie supplies, as follows: 1—New Zealand; 2— Ireland; 3—Denmark; 4—Switzerland; 5—United Kingdom; 6—Australia; 7—Finland; 8—Turkey; 9—United States; 10—Canada; 11— Netherlands; 12—Sweden; 13—Uruguay; 14—Austria; 15—Belgium-Luxembourg; 16—Germany, Federal Republic; 17—France; 18—Yugoslavia; 19—Norway; 20—Greece; 21—Argentina; 22—Israel; 23—South Africa; 24—Brazil; 25—Spain; 26—Italy; 27—Mexico; 28—Southern Rhodesia; 29—United Arab Republic; 30—Portugal; 31—Paraguay; 32—Mauritius; 33—Chile; 34—Honduras; 35—Venezuela; 36—Syria; 37—Lebanon; 38— China (Taiwan); 39—Japan; 40—Peru; 41—Libya; 42—Colombia; 43— Surinam; 44—India; 45—Ceylon; 46—Pakistan; 47—Ecuador; 48—Jordan; 49—Philippines.

variations in productivity and to hint at a few of the many reasons behind these variations (Fig. 1-14).

Farm animals and poultry are kept under conditions ranging from those of the high Andes and the Himalayas to those of the low, steaming tropics; from the far north to the equator; from the lush farm lands of the Corn Belt to the semiarid ranges of the Navajo reservation. Environment is an important factor in determining the carrying capacity of the land, the kinds of animals and birds that may be used to best advantage, and the amounts of produce that may be expected from them (Phillips, 1948; 1956; 1958; 1964).

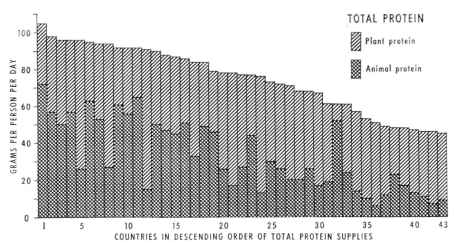

Fig. 1-12. Variations in supplies of total protein and animal protein in 43 countries. Countries are arranged from left to right in descending order of total protein supplies, as follows: 1—New Zealand; 2—Argentina; 3—France; 4—Ireland; 5—Yugoslavia; 6—Canada; 7—Finland; 8—Greece; 9—Australia; 10—Denmark; 11—United States; 12—Turkey; 13—Switzerland; 14—Belgium-Luxembourg; 15—Austria; 16—United Kingdom; 17—Israel; 18—Norway; 19—Germany, Federal Republic; 20—Italy; 21—Syria; 22—Chile; 23—Netherlands; 24—United Arab Republic; 25—South Africa; 26—Portugal; 27—Spain; 28—Mexico; 29—Paraguay; 30—Japan; 31—Brazil; 32—Sweden; 33—Venezuela; 34—China (Taiwan); 35—Libya; 36—India; 37—Peru; 38—Colombia; 39—Surinam; 40—Philippines; 41—Mauritius; 42—Pakistan; 43—Ceylon.

The kinds of animals or birds available as breeding stock; the systems of farming or ranching; the skills and knowledge of the farmers or ranchers; the degree to which diseases and parasites have been brought under control; the extent to which owners have access to new knowledge; the educational and cultural backgrounds of owners, which affect their capacity or their willingness to adopt new methods; the economic incentives that stimulate the achievement of higher levels of production—all these and other factors help to determine levels of productivity (Figs. 1-16 and 1-17).

The Masai tribesman—who herds his cattle in the dry, hot tribal areas of East Africa; who measures his wealth in terms of numbers of animals rather than in terms of their capacity to produce meat or milk; who on occasion uses some of his cattle as the purchase price for a bride; who long ago learned to puncture the jugular vein to obtain blood, which he drinks fresh for food, and thus retains his live animals as basic wealth and as a hedge against unusually hard times; and whose animals would have no appeal to specialized producers of beef or milk—is still very much in the cattle business. But the development of his business is limited by many factors, and the degree of productivity of his enterprise stands in sharp contrast to that achieved in the specialized beef and dairy industries of many advanced countries.

The millions of chickens that scavenge for a living—a few to each farm or village household in many parts of the world, and from which the owners harvest perhaps sixty small eggs per hen during a year—make small but important contributions to the supply of animal protein for human consumption (Fig. 1-18). This part of the world's poultry industry contrasts sharply with the highly specialized chicken industry in the United States,

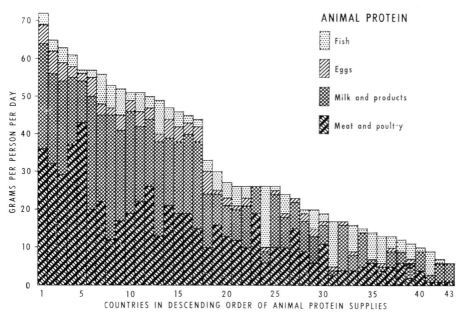

Fig. 1-13. Variations in the level of animal protein supplies, and in the composition of those supplies, in 43 countries. Countries are arranged from left to right in descending order of animal protein supplies, as follows: 1—New Zealand; 2—United States; 3—Canada; 4—Australia; 5—Argentina; 6—Ireland; 7—Denmark; 8—Finland; 9—Sweden; 10—Switzerland; 11—United Kingdom; 12—France; 13—Norway; 14—Belgium-Luxembourg; 15—Germany, Federal Republic; 16—Austria; 17—Netherlands; 18—Israel; 19—South Africa; 20—Greece; 21—Chile; 22—Italy; 23—Paraguay; 24—Portugal; 25—Yugoslavia; 26—Venezuela; 27—Colombia; 28—Mexico; 29—Spain; 30—Brazil; 31—Japan; 32—Syria; 33—Surinam; 34—Turkey; 35—China (Taiwan); 36—Philippines; 37—United Arab Republic; 38—Peru; 39—Mauritius; 40—Libya; 41—Ceylon; 42—Pakistan; 43—India.

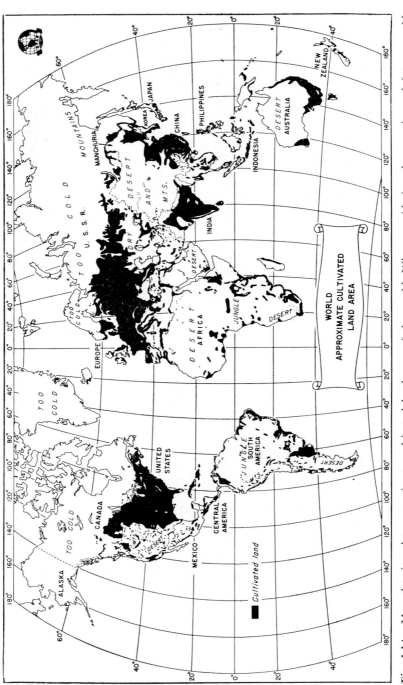

Fig. 1-14. Map showing the approximate cultivated land area of the world. When considering the questions of future world population and how those people are to be fed, we must recall that only 28.3% of the world's surface is land. The land surface in turn is only about 10.7% arable. A further 19.2% is permanent grassland, 30% is forested, and 40% is built-on or waste land or is used for other purposes. Some of the grasslands and forested lands are potentially arable, but livestock provide the only currently available means of harvesting a crop from much of the earth's land surface.

Fig. 1-15. Churning butter from whole yak milk in the Himalayan high-lands of Nepal. Yaks and yak-cattle hybrids make important contributions to the milk and butter supply in the Tibetan and Mongolian highlands of Asia. [Phillips, Tolstoy, and Johnson, 1946. FAO photo by S. Siegenthaler.]

an industry with centralized breeding farms and hatcheries, breeding systems planned and supervised by highly trained geneticists, feeding systems based on the latest results of nutritional research, highly organized processing and marketing systems, and a ready demand for its products based on a level of economic development at which people can afford to buy substantial amounts of eggs and chicken meat.

Equally sharp is the contrast between the system of pork production in China (Phillips and Hsu, 1944; Phillips, Johnson, and Moyer, 1945) and those practiced in the Corn Belt of the United States or the system utilized for the production of bacon-type hogs in Denmark. In both the latter systems, the farmers have access to carefully bred stock, generally practice modern and efficient methods of feeding and management, have access to feeds from which they can provide balanced rations, and either specialize in pork production or fit it into a general farm operation as an important phase of the farm business. In both countries there is a ready market for pork products. Pork is a highly favored article of diet among the Chinese (Fig. 1-19). However, the economic level is such that pork, for most people, is a luxury, a feast food to be enjoyed only on special occasions and then in

Fig. 1-16. Ankole cattle of East Africa. This unusual-looking breed is among the many distinct types found in Africa that make important contributions to man's food supply by transforming roughage into meat. [Photo by Uganda Department of Information.]

relatively small amounts. The demands made on the land for the production of crops for human consumption are so high that the production of a crop such as corn, primarily for the feeding of pigs, is not a practical possibility. In the 1940's pigs in many areas were weaned at one or two months of age, were raised afterward primarily on roughage and farm refuse, and were kept as an important source of manure. This period of "stretching up" lasted anywhere from six months to two years, and during this time a pig could change hands several times. Eventually the animals were put on a fattening ration for two or three months; often high-quality by-products from mills, oil presses, or breweries were used; the animals were then sold for slaughter. Such a system of production results in animals with a minimum of muscle development and a great deal of fat, in contrast with animals that are adequately fed on a balanced ration, high in concentrates, from the time they begin to take feed other than their dam's milk until the time of slaughter.

It is quite natural, in the study of livestock production, to concentrate upon the breeds, the methods of feeding and management, and the market requirements and methods of marketing that prevail in our own country, and to overlook livestock production's inter-relations with other aspects of United States and world agriculture. However, the livestock and poultry industries in the United States are but segments of the nation's agriculture. This, in turn, is but a part of the world picture. Improved communications, high-speed transportation, more travel and exchange of information, greater attention to international trade, and other factors are constantly bringing countries and peoples closer and closer together. As a result, it is becoming increasingly difficult, if not impossible, for either the livestock and poultry industries or the entire agricultural industry of a country to develop in isolation. Thus the scope and potential of our own livestock and poultry industries can be best understood and appreciated if considered against a world background (Figs. 1-20, 1-21 and 1-22). Moreover, there are important lessons to be learned from the experience of other countries, and on occasion we may find it profitable to draw upon the sources of germ plasm available in those countries (Phillips, 1961). In the development of our livestock industry we have depended primarily upon the breeds our forefathers brought with them from the British Isles and adjacent parts of northwestern Europe. When new stock was desired, we have returned to the same sources. For our poultry breeds, we have ranged somewhat farther afield. But we have made use of relatively few of the sources of germ plasm in the world. For example,

Fig. 1-17. Boran steers in Kenya. This is one of the several African breeds containing a high percentage of zebu blood that are well adapted to tropical and subtropical grazing conditions in Africa. [Photo by Kenya Department of Information.]

Fig. 1-18. Vaccination of chickens against Newcastle disease in Thailand. The large basket contains birds awaiting their turn. In contrast with the situation in countries where typical production units have large numbers of chickens kept under conditions of intensive management, many chickens (as well as other poultry) are kept in small groups—a few to a farm or a village household—in many parts of the world, and these birds are important sources of animal protein, particularly in the underdeveloped parts of the world. [FAO photo by S. Bunnag.]

our zebu cattle, which we call Brahmans, and which have come from India rather than the primary sources of our other breeds of cattle, are derived from primarily one breed, the Kankrej, with presumably some small admixtures of blood from two or three other zebu breeds. In India and Pakistan

Fig. 1-19. A fat hog being delivered to market by wheelbarrow in Chengtu, China. Pork is a favorite food of the Chinese. Production is carried out in association with intensive rice culture in the south, and with wheat and other cereal culture in the north. Every effort is made to avoid loss of weight in transit to market. [Photo by Ralph W. Phillips.]

Fig. 1-20. An Indian villager herds a sow into an underground hog house. This type of housing is used to protect the animals from the extreme summer heat of the plains of central India. Although swine make very important contributions to man's food supply in many parts of the world, pork is used only to a limited extent in India. [Photo by Ralph W. Phillips.]

there are six major groups and at least 28 more or less distinct breeds of zebus (Phillips, 1944; Joshi and Phillips, 1953). In Africa there are eight major groups of cattle, comprising some 33 distinct types and breeds. Many of these contain considerable amounts of zebu blood, and some are essentially pure zebu (Joshi, McLaughlin, and Phillips, 1957).

1-6. AN INDUSTRY OF CHANGE

The livestock industry is not only diverse, but subject to change. It has shared in the changes brought about by the agricultural revolution that has taken place in many countries during the present century; it has changed markedly in the United States in recent decades; and it will probably continue to change. The rapid increase in human population that is now taking place throughout the world is certain to have its effects upon the pattern of crop and livestock production.

The agricultural revolution of recent decades is well advanced in the

Fig. 1-21. Merino "hoggets" in the yards at Glentanner Station in New Zealand. Ben Ohau Range, in the background, rises to 8,000 feet. These sheep are part of a flock of eight to nine thousand, mostly Merinos, that graze the foothill country on this station. Their fine wool stands in sharp contrast with the wool of the Barbary sheep, shown in Fig. 1-22, which is used primarily for carpets.

United States and in some of the other highly developed countries. Less-developed countries are striving to bring about similar revolutions in their agriculture. This is a process that will continue for many years. It is having marked effects on cropping and livestock production systems, and on both rural and urban peoples.

In 1910, 34.9% of the population of the United States lived on farms. By 1962 the proportion had dropped to 7.7%. This in itself is an indication of the marked increase in the amount of food and other agricultural products that one farmer can produce. This increase is, in part, a result of the farmers growing use of mechanical power and equipment, upon which he has become more and more dependent. In 1910 there were 24,211,000 horses and mules on farms in the United States. The number increased to a peak of 26,723,000 in 1918, then gradually declined. By 1960, the last year in which the number of horses and mules was recorded, it had dropped to 3,089,000, or only about

11.5% of the number in 1918. This is but one indication of the extent to which farmers have altered their methods to take advantage of modern technical and scientific developments and to meet competition. There has also been a remarkable shift in the productivity per worker. In 1820 one farm worker produced enough to meet the needs of four persons. By 1940 the number had increased to 11 persons; by 1958, to 23. The productivity per worker is greater on larger farms because they can afford more mechanization. The number of farms in the U.S. has dropped from 5,647,800 in 1950 to an estimated 3,286,230 in 1966, and the average size has increased from 215 to 315 acres. (U.S.D.A. Statistical Reporting Service, Aug. 1966).

Within this overall pattern of agricultural change, there have been important changes in the livestock industry in the United States. The southeastern states, with their traditional emphasis on cotton and tobacco, now have a well-established broiler industry. Florida, long known for its citrus fruit and vegetable production, is now among the major cattle-producing states. Cattle feeding is expanding in the Range states and along the Pacific coast. In a study of changing sources of farm output, Durost and Barton (1960) have shown that the relative importance of livestock production, in

Fig. 1-22. Barbary sheep in Libya. In many parts of the world sheep make important contributions to man's food and fiber supplies, utilizing much land that is unsuited to intensive agriculture. [FAO photo.]

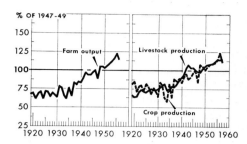

Fig. 1-23. Changes in farm output in the east north central region of the United States (Ohio, Indiana, Illinois, Michigan, and Wisconsin), a region in which changes in farm output from livestock and crop production have tended to parallel each other rather closely in recent years. [Durost and Barton, 1960.]

relation to crop production, had increased during the past decade in the New England, Middle Atlantic, South Atlantic, East South Central, West South Central, and, to a lesser degree, the Pacific regions. In other regions— East North Central, West North Central, and Mountain—changes in crop and livestock production had roughly paralleled each other. Graphs showing changes in the East North Central and South Atlantic regions are shown in Figures 1-23 and 1-24. In addition to such overall changes in farm output, changes in consumer demand have resulted in changes in animal type and in the feeding and management of livestock and poultry. Over the past several decades, there has been a decided shift from the lard type to the leaner type of hog. Beef steers go to market at a younger age than they did at the beginning of the century. Turkeys are smaller and more compact. Emphasis upon a high level of fat in milk has decreased, and skim milk is now a regular article of stock in the grocery store.

1-7. THE IMPACT OF HUMAN POPULATION ON THE LIVESTOCK INDUSTRY

If we accept the emergence of *Homo habilis* in East Africa as a starting point, then somewhat more than 1,800,000 years have been required for man to build his numbers up to the present level—3,114,000,000 in 1962. Yet, by the year 2000, or in less than 40 years, that number will have nearly doubled, and perhaps even more than doubled.

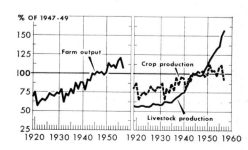

Fig. 1-24. Changes in farm output in the South Atlantic region of the United States (Delaware, Maryland, Virginia, West Virginia, North Carolina, South Carolina, Georgia, and Florida), a region in which farm output from livestock has increased rapidly, in relation to output from crop production, in recent years. [Durost and Barton, 1960.]

According to the United Nations (1964), there will be 7,410,000,000 people in the world at the turn of the century if recent trends continue. However, recognizing that some changes in rate of increase will probably take place, and taking various factors into account, the United Nations has made high, medium, and low projections, according to which the population may be 5,296,000,000, 5,965,000,000, or 6,828,000,000 by 2000 A.D. For purposes of discussion here, the medium estimate, 6,000,000,000 people in round numbers, has been taken as a probable level.

This reflects a magnitude of increase completely unparalleled in man's history (Fig. 1-25), and the size of the increase predicted for the decades just ahead is such that it will influence man's way of life through its effects upon land use pattern, production and distribution, space for living and recreation, and in many other ways. Inevitably, it will affect the livestock industry.

FAO (1963) concluded that, even to sustain the world's population at its present unsatisfactory dietary level, world food supplies would need to be increased by 36% by 1975, and 123% by the year 2000. If, in addition, the nutritional level is to be reasonably improved, world food supplies will have to be increased by 51% by 1975 and 174% by 2000. These estimates were based on modest goals per person per day, for 1975 and 2000, respectively, as follows: kcal—2,550 and 2,600; total-protein—75 and 79 grams animal protein—31 and 35 grams. In terms of types of food, if these improved nutritional goals are to be reached in 1975 and 2000, respectively, supplies of cereals will have to be increased by 35% and 110%, pulses by 85% and 225%, and animal products by 60% and 210%. The necessary increases may well be even greater because of inequality of distribution among countries.

Since the population will be increasing more rapidly in the less-developed, low-calorie countries than in the world as a whole, food supplies in those countries will need to be increased more rapidly. Food supplies in the low-calorie countries will need to be increased by 79% by 1975 and 293% by 2000 if an improved level of nutrition is to be achieved. Cereals will need to be increased by 40% and 130%, pulses by 100% and 275%, and animal products by 120% and 485%, by 1975 and 2000, respectively. These estimates are based on the following even more modest target figures for 1975 and 2000, respectively: kcal—2,350 and 2,450; total protein—69 and 74 grams; animal protein—15 and 21 grams.

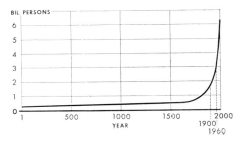

Fig. 1-25. The growth of the world's population from 1 A.D. and the anticipated growth to the end of the twentieth century. [Brown, 1963.]

Let us now look at the land upon which all of this food (except fishery products) must be produced. Since settled agriculture began, the amount of agricultural land has increased as the population increased. However, in the densely populated countries that now encounter dietary difficulties, little land remains that can be brought under cultivation at a reasonable cost. In the less densely populated countries there are substantial areas that are potentially arable. Exactly how much of this land could be brought under cultivation economically is not known and the amount would, in fact, change with changing circumstances.

Kellogg (1964) cites an estimate by Orvedal of 6,589 million acres of currently arable plus potentially arable land. The potentially arable portion amounts to about 84% of the land that is currently classified as arable or under tree crops (FAO, 1964a). This is not surplus acreage that is available, merely for the taking, for crop production. Many areas are inaccessible or for other reasons cannot be developed economically at this time. Large areas are under permanent meadows or pastures, or are forested. To plow these pastures or clear the forests would adversely affect the supplies of animal and forest products, and the need for both types of products is increasing as the population increases, just as is the need for food crops.

Nevertheless, new lands will be brought under cultivation by plowing grasslands and clearing forested lands, as well as by irrigating more desert and semidesert lands, and creating more polders. Ways will no doubt be found to turn large areas of seemingly useless tropical forests, such as those in the Amazon and Congo basins, into productive farm land. Some of the cleared forest lands will no doubt be used for grazing, thus offsetting in part the plowing up of grasslands. But forests may make their own inroads on arable and grass lands, if intensive plantation forestry expands to meet rising needs for lumber and other forest products. Increasing demands for dwelling space, factories, highways, and airports will also reduce the amounts of land available for agricultural purposes.

Thus, any attempt to assess future land availabilities in precise terms is beset by the existence of many imponderables. However, the nature of the problem is indicated in Figure 1-26, which shows the changes in amounts of land available per person, at 20-year intervals from 1920 to 2000, assuming a fixed land use pattern as of about 1962 (FAO, 1964a), and using United Nations (1964) medium projections of world population in 1980 and 2000. During this period of only 80 years, the total land available per person will have dropped from 18.5 to 5.6 acres per person, and the amount of land in arable and tree crops will have dropped from 1.9 to 0.6 acres.

Regardless of efforts to bring new land under cultivation, the inexorable increase of population pressure on the land is such that the amount of land available per person for food production must shrink substantially below present levels. For example, the arable land available per person in 1960, as shown in Figure 1-26, was 1.2 acres per person. Even if all the additional

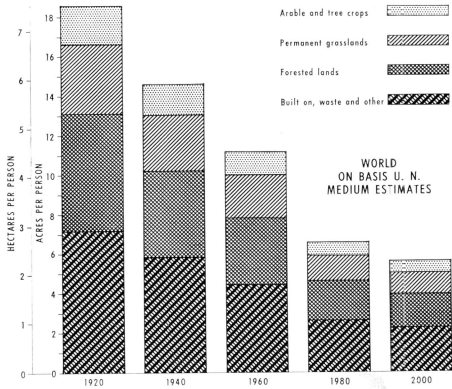

Fig. 1-26. Changes in amounts of land available per person in the world for various purposes from 1920 to 2000 A.D., in relation to the land use pattern in 1962 or the latest prior year for which data were available (FAO, 1964a), and assuming population increases through 1980 and 2000 according to the U.N. medium estimates. [United Nations, 1964.]

potentially arable land mentioned earlier, which amounts to just over 3,000 million acres, could conceivably be brought under cultivation by the turn of this century, the amount of arable land per person would be somewhat less than in 1960, and the amounts of pasture and forested lands projected for the year 2000 would then be substantially lower than the levels of about 1.1 and 1.7 shown in Figure 1-26, since portions of these lands would have been converted to crop production.

To put the problem another way, the land available per person for all purposes in 2000 A.D. (5.61 acres) will be only slightly greater than the amount of arable and permanent grassland combined that was available per person in 1920 (5.42 acres).

The effect of the rapid increase in population upon the amount of land available per person is being felt most acutely in the most densely populated

regions of the world, and the problems ahead will be greatest in those regions. However, the more sparsely populated regions are also feeling, and
will continue to feel, the effects, as may be seen from Figure 1-27. Changes
in the amounts of land available per person in northern North America
from 1920 to 2000 are portrayed in this figure, based on the amount of land
available for various purposes in 1962, and on the assumption that the
United Nations medium projections of population numbers for 1980 and
2000 are correct, rather than the high or low estimates based on somewhat
different assumptions. Since 81.6% of the arable land and 92.4% of the
permanent grassland in this region lies in the United States, it follows that

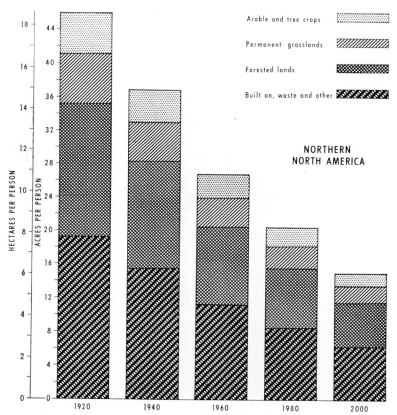

Fig. 1-27. Changes in amounts of land available per person in northern
North America for various purposes from 1920 to 2000 A.D., in relation to
the land use pattern in 1962 or the latest prior year for which data were
available, (FAO, 1964a) and assuming population increases through 1980
and 2000 according to the U.N. medium estimates (United Nations, 1964).
Northern North America includes the United States, Canada, Bermuda,
Greenland, and St. Pierre and Miquelon.

the trends shown in Figure 1-27 reflect rather accurately the changes in the amount of land available for food production that have taken place or may be anticipated in the United States.

If we assume that the present or similar rates of population increase continue beyond the year 2000, both in northern North America and in the world as a whole, it is evident that man soon will be facing really serious problems in meeting his needs for food. Although population masses would probably continue to concentrate in cities and towns, thus leaving substantial rural areas available for food production, the relative amounts of such land per person for food production certainly would decrease below either present amounts or those projected for 2000 A.D., and competition between crop and livestock production would increase steadily. Soon most of the land suitable for crop production that had not been used for buildings, highways and airfields, would be used for crops for direct human consumption. There would be little space for the production of feed for intensive livestock production, and man would be forced to subsist largely on foods from plant sources. If the population upsurge continued unabated, these plant foods would have to be supplemented by yeasts grown in culture tanks, algae grown in sewage disposal holding tanks, and perhaps seaweed. The catch of fish, particularly from marine sources, could be expected to increase, but not at a rate sufficiently rapid to replace other animal proteins or to keep pace with a continually rising population. In addition to the problem of producing enough food of sufficiently high quality to meet the needs of a continually rising population, man would also encounter the ever-increasing adverse effect of overpopulation upon economic development. Thus, the economic capacity of the average person to purchase the more expensive foods of animal origin might well fall behind his need for those foods.

1-8. ANSWERS TO THE DILEMMA

Hopefully, the series of events postulated in the previous paragraph will not take place. Some alternatives to a continuing population explosion are equally grim—a large segment of the population could be wiped out by an atomic war or famine and disease could decimate people living in the more populous areas. A happier trend of events would be for man to take over population control in a sensible manner. Rational methods are available, and research is under way on others. Available methods are in wide use in some countries; in other countries, programs are in progress (Stycos, 1964). It is to be hoped that these programs will be expanded, and that the now rapidly rising population curve will level off at a reasonable density—one that will allow reasonable space for living, and room for enough livestock production to insure each person a good supply of animal protein.

However, before measures to control the population upsurge can be effective and lead to a leveling off and reasonable stabilization of population, there will be increases like those envisioned in the United Nations projections. Thus, livestock production must be more efficient if its products are to compete with plant foods in the human diet. There must be improvement both in the economy of production and in the quality of the products.

Such improvements will be essential if those peoples who now enjoy an abundance of animal products are to continue to have such a supply available to them. In the densely populated, less developed parts of the world, the problem is of a different nature. As we have seen, the inhabitants of many of these countries now have only small supplies of animal products per person. Will these peoples continue to depend primarily upon food from plant sources, or—as is being attempted in Japan (Yang, 1962)—can systems of farm management be devised to fit more animal production into the agricultural systems of the densely populated countries? The following average percentages of the gross energy in feed that, when eaten by various kinds of farm animals, is converted into human food (Morrison, 1956), are worth noting: pork, 20%; milk (from dairy cows), 15%; eggs, 7%; poultry meat, 5%; beef, 4%; lamb, 4%. These estimates were based on somewhat better than average production, under good feeding practices.

If the present level of animal products per person is to be maintained, will the increase in the total supplies of animal products be achieved primarily by increases in numbers of animals? Doubling the numbers of livestock and poultry by the end of this century, in order to meet the needs of a human population that will have approximately doubled in the same period, does not appear to be a feasible solution for the world as a whole. Rather, much of the increase must come from improvements in efficiency of production, and through better breeding, feeding, and management.

The solutions to these complex problems may be quite different in different countries. Some countries having relatively small human populations and large areas suited to livestock production may concentrate even more than they do now on livestock and poultry production for domestic consumption and export. Other countries, where the populations are already very dense and where the land is already overcrowded, may find it necessary to depend even more on plant products and to seek their essential protein supplies from plant sources. Between these two extremes many combinations are possible. But the object here is not to attempt to provide solutions for the many problems facing livestock and poultry producers, or even to suggest specific solutions that may be applicable in the decades ahead. Rather, the object is to indicate something of the challenge that lies ahead for the producers of livestock and poultry, and for those who serve these industries.

The producer of livestock or poultry, in this world of agricultural change, must draw increasingly upon knowledge from many sources if he is to put on the market the kinds of products consumers want, and if he is to breed, feed, and manage his animals or birds in such ways that they will produce efficiently and bring him a reasonable profit. To produce high quality products that will not cost too much more than plant products, he must have an understanding of animal genetics, physiology and nutrition, and the complex problems of feeding, management, and marketing. Butter affords an excellent example of an animal product that has declined in popularity because of the disparity between its price and that of substitutes manufactured from plant products.

Today, meat products, milk, and eggs are unsurpassed in consumer preference, but the livestock producer cannot rest on his laurels. He must, by individual and cooperative effort, improve still further the quality of his products and simultaneously increase the efficiency with which they are produced, in order that they will not be priced out of the market. He must understand the laws of supply and demand, and the overall economic situation in his own and other countries, so that he may assess short- and long-term demands for his products, as well as possible shifts in consumer preferences and the conditions of production.

It is against this background that we undertake the study of the various aspects of livestock and poultry production.

REFERENCES AND SELECTED READINGS

References marked with an asterisk are of general interest.

* Acock, A. M., 1950. *Progress and Economic Problems in Farm Mechanization.* Food and Agriculture Organization of the United Nations, Rome, Italy. 88 pp.

* Brown, Lester R., 1963. *Man, Land & Food.* Foreign Agricultural Economic Report No. 11, U.S. Department of Agriculture, Washington, D.C. 153 pp.

Durost, Donald D., and Glen T. Barton, 1960. *Changing Sources of Farm Output.* A.R.S. Product-Research Report No. 36, U.S. Department of Agriculture, Washington, D.C. 57 pp.

* FAO, 1963. *Third World Food Survey.* F.F.H.C. Basic Study No. 11, Food and Agriculture Organization of the United Nations, Rome, Italy. 102 pp.

*————, 1964a. *Production Yearbook, 1963.* Food and Agriculture Organization of the United Nations, Rome, Italy. 503 pp.

*————, 1964b. *State of Food and Agriculture, 1964.* Food and Agriculture Organization of the United Nations, Rome, Italy. 240 pp.

Finci, M., 1957. *The Improvement of the Awassi Breed of Sheep in Israel.* Bull. Research Council of Israel, Section B, Biology and Geology, 68, No. 1-2: 106 pp.

*Joshi, N. R., E. A. McLaughlin, and R. W. Phillips, 1957. *Types and Breeds of African Cattle.* Agricultural Studies No. 37, Food and Agriculture Organization of the United Nations, Rome, Italy. 297 pp., illus.

*Joshi, N. R., and R. W. Phillips, 1953. *Zebu Cattle of India and Pakistan.* Agricultural Studies No. 19, Food and Agriculture Organization of the United Nations, Rome, Italy. 256 pp., illus.

Kellogg, Charles E., 1964. Potentials for food production. *Farmer's World.* The Yearbook of Agriculture, 1964 Department of Agriculture, Washington, D.C., pp. 57–69.

*Morrison, F. B., 1956. *Feeds and Feeding.* Morrison, Ithaca.

Phillips, R. W., 1944. The cattle of India. *J. Heredity*, 35:273–288.

————, 1945. The water buffalo of India. *J. Heredity*, 36:71–76.

*————, 1948. *Breeding Livestock Adapted to Unfavorable Environments*. Agricultural Studies No. 1, Food and Agriculture Organization of the United Nations, Rome, Italy. 182 pp., illus.

————, 1950–51. Expansion of livestock production in relation to human needs. *Nutrition Abstracts and Reviews*. 21:241–256.

*———— (Editor), 1956. *Recent Developments Affecting Livestock Production in the Americas*. Agricultural Development Paper No. 55, Food and Agriculture Organization of the United Nations, Rome, Italy. 181 pp., illus.

————, 1958. The effects of climate on animals and their performance: an introduction to a symposium. *J. Heredity*, 49:47–51.

*————, 1961. Untapped sources of animal germ plasm. *Germ Plasm Resources*, pp. 43–75. Ed. Ralph E. Hodgson. American Association for the Advancement of Science, Washington, D.C.

*————, 1963. Animal products in the diets of present and future world populations. *J. Animal Science*, 22:251–262.

*————, 1964. Animal agriculture in the emerging nations. In *Agricultural Sciences for the Developing Nations*, pp. 15-32. Ed. A. H. Moseman. Publication No. 76. American Association for the Advancement of Science, Washington, D.C.

————, and T. Y. Hsu, 1944. Chinese swine and their performance compared with modern and crosses between Chinese and modern breeds. *J. Heredity*, 35:365–379.

*Phillips, R. W., R. G. Johnson, and R. T. Moyer, 1945. *The Livestock of China*. U.S. Department of State Publication 2249, Far Eastern Series 9, U.S. Department of State, Washington, D.C. 174 pp., illus.

Phillips, R. W., I. A. Tolstoy, and R. G. Johnson, 1946. Yaks and yak-cattle hybrids in Asia. *J. Heredity*, 37:162–170, 206–215.

Reed, C. A., 1959. Animal domestication in the prehistoric Near East. *Science*, 130:1629–1639.

Stycos, J. Mayone, 1964. The outlook for world population. *Science*, 146:1435–1440.

* United Nations, 1958. *The Future Growth of World Populations*. United Nations, New York. 75 pp.

*————, 1964. *Provisional Report on World Population Prospects, As Assessed in 1963*. United Nations, New York. 316 pp.

USDA, 1963. *Agricultural Statistics*, 1963. U.S. Department of Agriculture, Washington, D.C. 635 pp.

Yang, W. Y., 1962. *Farm Development in Japan*. Agricultural Development Paper No. 76, Food and Agriculture Organization of the United Nations, Rome, Italy. 76 pp., illus.

Section I

Livestock Products

Chapter 2

Meats

And who abstains from meat that is not gaunt?
SHAKESPEARE, *Richard II*

2-1. INTRODUCTION

The term meat, as used in this discussion, will refer to all parts of the dressed carcass, whether beef, pork, veal, lamb, or mutton. Meat not only consists of muscular tissue, bones, and fat, but also includes the edible glands and organs removed at slaughter.

The sale of meat animals provides the largest single source of farm income, as shown in Fig. 2-1. Meat currently provides 30 cents of every dollar of farm income.

2-2. MEAT CONSUMPTION

The per capita meat consumption[1] in the United States was 169 lb. in 1963; this is an increase of 21 lb. above the average for the 1947–49 base period. USDA predictions for 1964 indicate a further increase to 172 lb. per capita. The consumption in percent is shown for each species in Fig. 2-2. More than 94% of all meat consumed was beef and pork; veal, lamb, and mutton consumption totaled less than 6%. Compared to the 1947–49 base period, the consumption of beef increased from 44% to 55% of the total, and pork declined from 46% to less than 39%. The proportions of veal, lamb, and mutton have also declined during the same period. Lard consumption has declined from more than 12 lb. during the base period to less than

[1] Poultry meats are excluded.

Fig. 2-1. The source of the farmer's dollar for 1963. [From the American Meat Institute, Chicago, Ill.]

6.5 lb. in 1963, with predictions indicating a continued decrease for 1964.

There is great variation in the per capita meat consumption among the countries of the world. In 1963 the per capita consumption was highest in New Zealand with 240 lb. and lowest in Japan with 13 lb. However, consumption in Japan is up from less than 1 lb. per capita in 1946. The other leading meat consuming countries in the world in lb. per capita are as follows: Australia—219 lb.; Argentina—216 lb.; Uruguay—203 lb.; United States—169 lb.; Canada—140 lb.; United Kingdom—139 lb.; France—132 lb.; Denmark—130 lb.; Belgium and Luxembourg—123 lb.; West Germany —117 lb. It is of interest to note that the USSR (Russia) consumed only 73 lb. per capita.

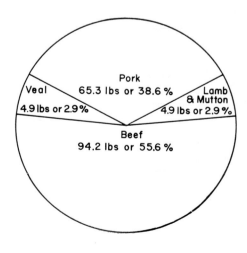

Fig. 2-2. Amounts and proportions of various kinds of meat consumed in the United States in 1963. [Data from National Food Situation, NFS-109, August 1964, published by Economic Research Service, USDA.]

2-3. MEAT AS A FOOD

The importance of meat in the diet is well known. Research has shown meat to be an excellent source of most B-complex vitamins, especially niacin. Lean pork contains relatively large amounts of thiamin. The greatest contribution of meat to the diet is its protein content, which is important chiefly because it supplies all of the essential amino acids in adequate amounts.

Because of similarities in composition and in nutritive value, meat from swine, sheep, and cattle is often grouped with poultry and fish in evaluations of the realtive importance of various classes of foods. The data of Clark et al. (1947) are presented in this manner (Fig. 2-3). Meat not only is a good source of B-vitamins and amino acids, but also supplies large amounts of iron. Liver is an outstanding source of this nutrient. Of particular interest to a weight-conscious public is the fact that lean meat supplies a relatively low number of calories.

The digestibility of edible meat—normally well over 95%—is sometimes used as a standard for comparison with other foods. Frequently, one hears that fat meat, particularly pork, is indigestible. There is no basis for this claim, although fat is digested more slowly.

The meat industry should be aware of the problem of atherosclerosis (hardening of the arteries), which is the leading cause of death in the United States today. Meat and, more particularly, animal fats have been claimed to increase the blood cholesterol level, which many physicians believe is responsible for atherosclerosis. In spite of the fact that some of the foremost lipid researchers have cautioned against major changes in the fatty acid composition of the diet (Kummerow, 1964), people have unquestionably altered their eating habits as a result of statements in the popular press and advertisements. However, cholesterol is synthesized by the body even in the absence of

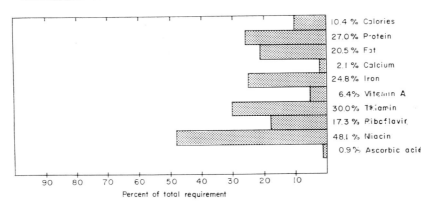

Fig. 2-3. Percentage of total dietary requirement for each nutrient supplied by meat, poultry, and fish

dietary sources. The best information indicates that the dietary level of cholesterol is only one of a number of contributory factors that cause atherosclerosis (Maddox, 1960). Our present knowledge indicates that hardening of the arteries can best be avoided by adequate exercise and avoiding obesity. Except for a reduction in total calories, no major dietary changes are generally being recommended.

2-4. COMPOSITION OF MEAT

The American Meat Institute Foundation (1960) gives the average composition of the edible portion of fresh meat cuts as 17% protein, 20% fat, 62% moisture, and 1% ash. Since wide variability in composition exists, an average value serves only as a general guide.

The percentage of separable lean in a carcass or cut varies widely and is inversely related to fat content (Fig. 2-4). It is interesting to note that the percentage of bone and tendon declines directly with muscle.

The protein of the carcass is mostly in two tissues, connective tissue and muscle. In meat, muscle tissue contains the largest amount and the highest nutritive value of protein. The muscle proteins can be subdivided into structural or stroma proteins, contractile proteins, and sarcoplasmic or water-soluble proteins. The last two classes contain numerous enzymes, including

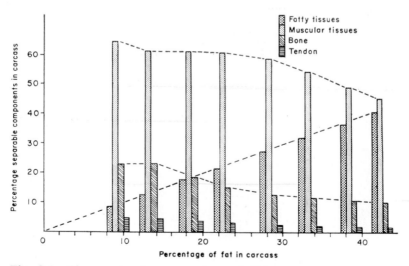

Fig. 2-4. Changes in the proportions of different components as the carcass changes in fatness. [From E. H. Callow, "Comparative Studies of meat II. The changes in the carcass during growth and fattening, and their relation to the chemical composition of the fatty and muscular tissues." *J. Agr. Sci.* 38:174, 1948.]

many glycolytic and oxidative enzymes in the sarcoplasmic fraction and in the myosin of the contractile group. Myosin, which comprises about 38% of all muscle proteins, is the most abundant protein in the body; in addition to its role in muscle contraction, it serves as an adenosine triphosphate splitting enzyme, ATPase. Thus, meat is composed of an unusually high concentration of enzymes.

2-5. WHAT THE CONSUMER WANTS IN MEAT

All members of the meat team—the producer, the packer, and the retailer —should keep in mind that they are in business to serve the consumer. Consequently, they should know why the consumer eats meat and what he desires for maximum satisfaction, and then gear production, processing, and distribution systems toward best meeting consumer desires. The popularity of meat in this country depends largely upon its savory characteristics and, to a lesser extent, upon its high nutritive value. Tenderness, flavor, juiciness, leanness, and attractiveness, which includes color and firmness, appear to be the main criteria the consumer considers.

Tenderness. According to consumer studies, tenderness is the most important attribute for acceptance of beef. Tenderness is less critical for veal, lamb, and pork because there is less variability from animal to animal. Although tenderness can be altered by cooking, aging, and enzyme treatment, only the pre-slaughter factors will be considered at this point. Many characteristics have been suggested as possible indicators of tenderness, such as conformation, maturity, finish, marbling, and muscle structure.

CONFORMATION. Recent studies by Ramsey and coworkers (1963) indicate that body type or conformation is unrelated to tenderness. These investigators found that Jersey steers of inferior beef type produced, on the average, more tender steaks than Hereford or Angus steers of good beef type. Differences in the tenderness of steaks from Hereford, Angus, Brahman-British cross, Santa Gertrudis and Holstein steers were not significant. Steaks from Brahman steers were rated the least tender of all breeds studied.

MATURITY. A number of researchers have shown that younger, more immature beef is more tender. The determination of maturity in the retail market is a difficult task. In the carcass, maturity is estimated by the redness and degree of ossification of the spinous process. Since about 3 to 5% of the old cows that are marketed produce tender meat (Goeser, 1964), it is obvious that maturity is not the only factor that determines tenderness. Nevertheless, tender meat is most likely to be obtained from cattle, hogs, and lambs if they are marketed when they are relatively immature.

FINISH OR DEGREE OF FATNESS. Several workers have reported that fatter hogs are more tender than lean hogs, but this distinction is not clear-cut. Some evidence points to a similar distinction for beef. It is doubtful whether additional external finish beyond the amounts needed to permit distribution to retailers without surface drying or slime growth is advantageous for tenderness.

MARBLING. Until recently, the amount and distribution of intramuscular fat, commonly called marbling, has been quite generally accepted as an indicator of tenderness by the meat trade, although evidence showing that there is little correlation between marbling and tenderness was submitted about 30 years ago (Hostetler et al., 1936). Since then, a number of workers have shown that marbling accounts for only about 10% of the tenderness variation in beef. Similar values are not available for pork, but the evidence suggests that marbling may be more closely related to tenderness than it is in beef (Batcher and Dawson, 1960; Harrington and Pearson, 1961).

MUSCLE STRUCTURE. Muscle structure and tenderness may be related. Jourbert (1956) found considerable variation in the muscle fiber diameter of different breeds of cattle; however, there seems to be little relationship between tenderness and muscle-fiber diameter (Hiner et al., 1953). Although Adams et al. (1960) found an inverse relationship between collagen (white connective tissue) and tenderness, other factors appear to be more important. With the electron microscope and other new tools for studying the structure of meat, the true basis for tenderness may come to light.

EFFECT OF EXERCISE. Mitchell and Hamilton (1933) demonstrated that, contrary to the popular view, steers exercised for 131 days were slightly more tender and had less collagen in their muscles than similar unexercised animals. This work was later confirmed by Bull and Rusk (1942). Thus, moderate exercise over a considerable length of time does not appear to adversely affect meat tenderness.

Flavor. Flavor is undoubtedly an important factor in consumer acceptance, yet personal likes and dislikes vary widely. Generally, flavor becomes more pronounced as an animal matures, as evidenced by the full flavor of beef in comparison with the blandness of veal. Like that of most other foods, meat flavor is a blend of odor and taste. Available evidence indicates that the components determining meat flavor are water soluble (Kramlich and Pearson, 1958) and are basically the same for all meats (Hornstein and Crowe, 1960; 1963). Although fat does not contribute directly to the basic meaty flavor, it probably causes the characteristic differences in flavor and odor of cooked meat from animals of different species (Hornstein and Crowe, 1960; 1963).

Off-flavors in meats are often the result of careless handling, but this discussion will be concerned only with problems related to production. Sex odor, or boar odor, in pork has been the basis for condemnation of boars and stags by the Meat Inspection Division of the USDA. However, the meat from boars is not unhealthful, and both federal and state meat inspectors are presently allowing boar carcasses to be used for processing purposes.

Research has shown that boar odor is localized in the unsaponifiable fraction of the fat and is not readily volatilized below the boiling point (Craig et al., 1962). Thus, boar meat can be successfully incorporated into processed meats that are eaten without prior heating (Williams et al., 1963). Regulations have been changed to allow the incorporation of boar and stag meat into sausage, and, as a result, the price spread between boars and barrows has narrowed.

In sheep, mutton flavor has unquestionably contributed to the low consumption of lamb and mutton in the U.S. However, milk-fed lamb has a bland flavor, and educating the public to this fact would probably do much toward increasing the popularity of this excellent product. Beef flavor problems have been largely caused by off-flavors coming from certain feeds, such as wild onions. Removal of most feed flavor can be accomplished simply by taking the animals off offending feeds for about 10 days prior to slaughter.

Juiciness. Certainly juiciness adds to the overall acceptability of meat, yet the separation of true juiciness from other palatability factors is difficult. Taste panel studies indicate that marbling contributes more to juiciness than to tenderness. On the other hand, neither moisture content nor press fluid have been found to be closely associated with juiciness. In general, it appears that fat covering and both inter- and intra-muscular fat tend to increase juiciness.

Leanness. Numerous consumer studies based upon visual appraisal have shown that the housewife will select lean meat when given a choice between cuts differing widely in fatness. However, organoleptic studies of beef have shown that fatness improves acceptability. It takes more feed to put on a pound of fat than a pound of lean. Thus, for economy, beef producers should concentrate on developing strains of cattle yielding highly palatable meat with a minimum of fat.

Attractiveness. Since the advent of the self-service market, the retailer and meat packer have become increasingly aware of the importance of meat appearance. Color has probably been the most important single factor in attractiveness; thus, two-toned pork and dark or discolored beef and lamb sell for a discount.

Hall et al. (1944) studied dark-cutter beef, which fails to brighten on cutting and exposure to air. The color is normally bright at pH 5.6 or below; at 5.7 it becomes shady or dull; at 6.5 or above it is dark. The ultimate pH of the meat depends on the amount of lactic acid produced after death, which in turn depends upon the amount of glycogen present at the time of death. By preventing undue excitement, by feeding in transit, and by avoiding undue exposure to inclement weather in marketing, the incidence of dark-cutter beef can be greatly reduced.

Pale, watery, and two-toned pork has also been shown to be related to the glycogen content of different muscles at the time of slaughter (Briskey et al., 1959). A uniformly dark product, which does not appear to be objectionable to the consumer, can be produced by using exercise or other means to deplete glycogen levels prior to slaughter. However, the incidence of spoilage increases as the pH is raised (Ingram, 1948); thus, this method of achieving a desirable meat color may not be feasible.

2-6. PROCESSING PRACTICES AND THEIR IMPACT

Today, many housewives hold full-time jobs outside the home, and consequently the time available for housework has been reduced. This is but one example of the kinds of social changes that have altered consumer demand. In addition, other new developments promise to bring about drastic revisions in many trade and processing practices; these revisions have had, or promise to have, a marked influence upon the entire meat and livestock business.

The Super Market. The advent of self-service meats has influenced the development of the meat-type hog and is causing a revision in requirements for beef carcasses. Self-service markets, relying upon meat items as major drawing cards, recognize the importance of providing customer satisfaction. The consumer is offered a wide selection of cuts varying in price and in quality, and in making his own choices from among these items, he reflects his preferences to the packer and producer. High quality meats, attractively packaged and displayed, have increased gross meat sales at the super market.

Tenderization. Naturally occurring meat enzymes tenderize meat during aging, but rapid aging is a recent innovation. The use of ultraviolet lights to control microbial growth at elevated temperatures has made possible faster aging of entire beef carcasses. By speeding up the breakdown of connective tissue, aging can be accomplished in 2 to 4 days at 60 to 75°F. Though an increase in tenderization can be achieved, this method is not adapted for lower-grade carcasses that lack sufficient fat covering. However, recent studies indicate that the concurrent use of antibiotics will allow higher holding temperatures and will make it possible for lower-grade meat to be aged in 24 hours or less.

Among the new developments, the one having the greatest potential for revolutionizing the production, processing, and distribution of meats today is a tenderizing method that utilizes exogenous enzymes. Goeser et al. (1960) make use of pre-slaughter injection of a proteolytic enzyme preparation into the vascular system of the animal. Apparently this enzyme preparation does not cause shock when administered to the animal and does not become active until the temperature is raised during cooking. This method presumably tenderizes low-grade beef, as well as the cheaper cuts from the front quarter.

The post-slaughter injection method has also been used for tenderizing both carcasses and cuts. However, this method poses greater difficulties in making the injection than does the live-animal procedure. Nonetheless, large quantities of low-grade meat are now being dipped or injected with enzyme solutions for tenderization.

Sausage Production. According to estimates, one-twelfth of all meat is consumed as sausage. Sausage production has shown in increase for several years. This can be attributed to an increased demand for convenience items, such as lunch meats and frankfurters. Increased sausage consumption has provided a good market for boneless meat from low-grade beef and heavy hogs.

2-7. THE MEAT ANIMAL AND ITS PRODUCTS

The yield of an animal, or the dressing percentage, is defined as follows:

$$\text{Dressing } \% = \frac{\text{Carcass weight}}{\text{Live weight}} \times 100.$$

Carcass yield varies considerably, depending upon a large number of factors. Weighing conditions of both the animal and the carcass will have considerable effect upon the yield. Such factors as the amount of fill (water and/or feed) before weighing of the live animal and the length of time the carcass is held in storage before it is weighed are determinants. Thus, dressing percentage has little meaning unless the conditions of weighing are accurately described. The method of dressing, the inherent size of the digestive tract and other vital organs, and the amount of finish also affect dressing percentage.

Cattle. Dressing percentage for different grades of breef are shown in Table 2-1. With the exception of the commercial grade, dressing percentage declines directly with grade, but there is considerable overlap between adjacent grades. The carcasses of many fat cows are graded as commercial, because their age prevents them from being graded as U.S. Good or higher. That is, many animals within this grade have finish enough for the higher grades, but are too old to qualify. Fat animals in this grade, therefore, have a high dressing percentage.

RELATIVE AMOUNTS AND VALUES OF PARTS. A sketch of a beef animal hanging from the rail is shown in Fig. 2-5. The carcass, which usually constitutes only 50 to 60% of the live weight, accounts for about 90% of the live animal's value (Table 2-2). The hide is the most valuable by-product, composing 6 to 11% of the weight and about 5% of the total live value. Other by-products account for only a small amount of each animal's total value. The

TABLE 2-1. | *Dressing percentage of slaughter cattle for various U.S. grades.*

Grade	Dressing Percentage*
Prime	62–66
Choice	58–62
Good	56–58
Standard	52–56
Commercial	56–60
Utility	45–52
Cutter and Canner	40–48

* A normal range is given, but much larger variation occurs within any given grade.

round, loin, and rib make up 27 to 29% of the live animal's weight, but because they are preferred for steaks and oven roasts, they represent 62 to 67% of the total value of the carcass (Table 2-3). The chuck is the other major contributor, making up about 13 to 16% of the weight and accounting for 25 to 29% of the value. Thus, approximately 90% of the total carcass value of the animal is obtained from the loin, round, rib, and chuck, which represent only 39 to 45% of the live weight. Approximately 17 to 18% of the live weight consists of the kidney knob, brisket, plate, flank, and foreshank, but these parts account for only 10% of the total value of the live

TABLE 2-2. | *Weight and value of the carcass and noncarcass parts of cattle expressed as percentages of the weight and value of the live animal.*

Part	Live Weight	Live Value
Hide	6.0–11.0	4.6
Feet	1.3–1.8	0.1
Heart	0.3–0.5	0.3
Caul fat	0.0–3.0	0.3
Liver	0.9–1.2	0.8
Lungs	0.9–1.3	0.2
Kidneys	0.4–0.6	0.2
Head		
Tongue	0.2–0.4	0.3
Head and cheek meat	0.3–0.5	0.4
Head bones	2.3–2.5	0.2
Remainder of viscera	17.0–28.0	2.6
Carcass, dressed*	50.0–62.0	89.7

* Based on U.S. Good carcass with no allowance for costs of slaughtering.

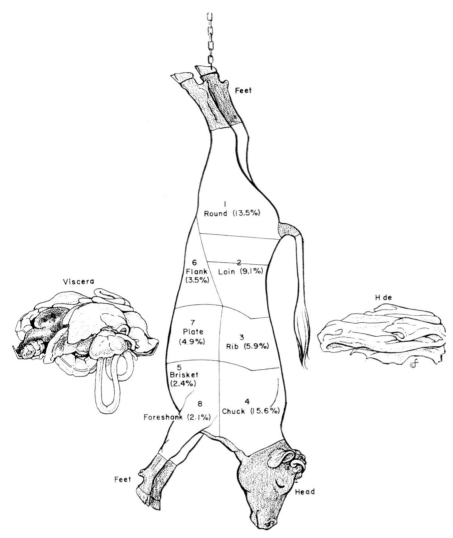

Fig. 2-5. This live intact beef hanging on rail shows wholesale cuts and percentages of each on a live basis, U.S. Choice Grade.

animal. The hindquarter makes up only 25 to 30% of the live weight, yet it constitutes about 52 to 57% of the value.

Helser and coworkers (1930) reported on the effect of fattening on the yield of various wholesale cuts. Carcass yield and the percentages of loin, flank, kidney knob, and plate all increased markedly with higher levels of finish; the percentage of rib increased slightly. On the other hand, the percentages of chuck and foreshank were not greatly influenced by finish.

TABLE 2-3.	*Weight and value of forequarters, hindquarters, and high-priced cuts, expressed as percentages of the weight and value, respectively, of live animals in the various grades.*

	High-priced Cuts*		Forequarter		Hindquarter	
Grade	Weight	Value	Weight	Value	Weight	Value
Prime	28.6	65.3	32.5	44.9	29.6	55 1
Choice	28.5	66.7	30.9	43.8	28.3	56.1
Good	28.6	66.5	28.5	43.3	28.0	56.7
Standard	27.7	63.5	27.4	45.8	26.8	54.1
Commercial	28.0	59.5	30.3	48.1	27.4	51.7
Utility	26.7	62.2	24.6	45.5	25.4	54.6

* High-priced cuts include the loin, rib, and round (rump on).

The percentage of round declined with increased fatness. Because it is necessary to trim off excessive fat, the percent of the carcass represented by retail cuts of roasts and steaks may actually decline with increased fattening. (Fig. 2-6).

STEER VERSUS HEIFER BEEF. Beef heifers of the English breeds fatten at an earlier age and at a lighter weight than steers; if kept in the feed lot much

Fig. 2-6a. A good meaty heifer. Live wt.—915 lb.; carcass wt.—525 lb.; carcass grade—choice No. 3; dressing %—57.4; fat trim—21.9%; bone—7.7%; lean trim—15.4%; lb. of carcass in roasts and steaks—281.9.

Fig. 2-6b. A fat wasty heifer. Live wt.—940 lb.; carcass weight—580 lb.; carcass grade—choice No. 5; dressing %—61.6; fat trim—30.3%; bone—7.3%; lean trim—14.3%; lb. of carcass in roasts and steaks—271.2. Note that, because of its high percentage of fat trim, this heifer with a high dressing percentage and with a higher body weight actually had a lower yield of pounds of roasts and steaks than did its mate shown in *a*. [Courtesy R. E. Rust, Dept. of Animal Science, Iowa State University, Ames, Iowa.]

beyond 900 lb. live weight, they become excessively fat and take on many of the carcass attributes of older cows. In addition to being discounted for overfatness, heifers sell for less than steers of equal quality and grade. This discrimination does not appear to be justified, since most carcass-cutting tests have shown that if heifers are slaughtered at lighter weights, their carcasses are equal to those of steers of similar grades and quality. The trade practice of penalizing heifers is no doubt an outgrowth of the lowered dressing percentage associated with pregnancy.

TYPE AND YIELD. The claim is frequently made that "beef-type" cattle produce carcasses yielding a greater proportion of high-priced cuts. The work of Wilson and Curtis (1893) at Iowa has often been cited to support this claim, yet their data show little difference between beef-type and dairy-type steers. This work was done more than 70 years ago. More recent work (Butler, 1957; Butler et al., 1956; Stonaker et al., 1952; Willey et al., 1951) indicates that there are no major differences in wholesale cut-out percentages between comprest and conventional-type beef cattle or between cattle containing various proportions of Hereford and Brahman stock. Similarly, a comparison of the percentages of front and hind quarters in cattle of different types has revealed little difference (Stroble et al., 1951; Stonaker et al., 1952).

Table 2-1 demonstrates that one of the major causes of variability in the value of live cattle is variation in carcass yield, or dressing percentage. In general, beef-type cattle have a higher dressing percentage than dairy-type cattle handled identically. Otherwise, there is little advantage in yield for beef-type over dairy-type cattle (Branaman et al., 1962).

Hogs. Table 2-4 gives the percentages of the different wholesale cuts by weight and value on the basis of live weight for three grades of pork. Since the grade standards are based chiefly on backfat thickness, the differences noted depend primarily upon differences in fatness. The lean cuts and, to a lesser extent, the primal cuts decline as fatness increases, and the percentage of fat trim increases.

AMOUNTS AND VALUES OF PARTS. The lean cuts make up 39% of the weight but 73% of the value of a U.S. No. 1 pig, as compared to 37% of the weight and 72% of the value for a U.S. No. 3 pig. The percentage of fat trim increases from 17% of the weight and 5% of the value for the U.S. No. 1

TABLE 2-4. | *Weight and value of wholesale pork cuts expressed as percentages of the weight and value of live hogs in the various grades.*

	U.S. No. 1		U.S. No. 2		U.S. No. 3	
Wholesale Cut	*Weight*	*Value*	*Weight*	*Value*	*Weight*	*Value*
Ham, skinned	14.7	30.3	13.7	28.8	13.0	28.0
Loin	11.8	24.9	11.5	24.7	11.1	24.5
New York shoulder†	12.8	18.0	13.0	18.6	12.9	19.0
Belly	10.0	14.1	10.1	14.5	9.7	14.3
Lean trim	1.8	1.8	1.6	1.6	1.4	1.5
Fat trim plus leaf fat	16.6	5.0	19.2	5.9	21.4	6.7
Spareribs	1.9	3.5	1.9	3.6	1.9	3.7
Jowls	2.0	1.1	2.0	1.1	2.0	1.2
Neckbones	1.3	0.6	1.3	0.6	1.3	0.6
Feet	1.9	0.6	1.9	0.6	1.9	0.6
Lean cuts**	39.3	73.2	38.2	72.1	37.0	71.5
Primal cuts‡	49.3	87.3	48.3	86.6	46.7	85.8
Dressing %	74.24		74.74		75.05	
Av. backfat thickness (in.)	1.42		1.72		1.96	

Source: Data taken from hogs slaughtered at Michigan Agricultural Experiment Station, Meats Laboratory.
† New York shoulder is essentially the same as the Boston butt plus picnic, but is left in one piece.
** Lean cuts include loin, ham, and New York shoulder.
‡ The term primal cuts means major cuts from value standpoint and includes lean cuts plus belly.

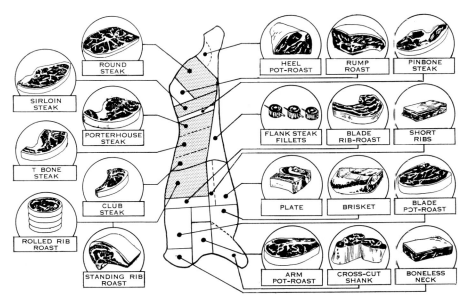

Fig. 2-7. Location of retail cuts of a beef carcass. [Courtesy National Livestock and Meat Board.]

pig, to 21% of the weight and 7% of the value for the U.S. No. 3 pig. The increase in the percentage of fat trim tends to compensate for the decline in the percentage of lean cuts. As a result of the increased value of the fat trim and an increase in dressing percentage as fatness increases, the spread in value between grades becomes narrower, often causing farmers to comment that packers do not pay for quality. Nevertheless, pork consumption has gone down in recent years, owing to consumer resistance to fat pork; thus, it behooves the producer to market leaner hogs.

LEAN OR PRIMAL CUTS. Any improvement of pork carcasses brought about by a progeny-testing program must necessarily be based upon proper methods of evaluating the progeny of the sires being proven. One of the controversies has been whether the most rapid progress could be made by determining the percent of the carcass composed of lean cuts or the percent composed of primal cuts. Some have used primal cuts that include the belly, a fat cut, in addition to the lean cuts because the belly, from which bacon is derived, is an important part of the carcass. However, if leanness of the carcass is the trait to be emphasized, the weight of the lean cuts should be used.

LEANNESS AND DRESSING PERCENTAGE. On the average, fat hogs have a higher dressing percentage than lean ones; however, this correlation is relatively low. A truly meaty hog may have a high dressing percentage, and a hog that is lean but lacks muscling may have a low carcass yield. Thus,

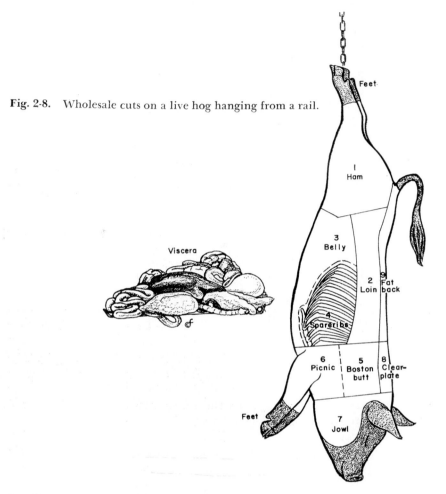

Fig. 2-8. Wholesale cuts on a live hog hanging from a rail.

breeders must reject sires of low-yield hogs deficient in muscling or meatiness.

LENGTH OF CARCASS. A few years ago, length of carcass was given great emphasis in progeny evaluation programs and in carcass judging. More recent evidence shows that the relationship between length of carcass and the percentage of lean or primal cuts is slight, although positive. Present evidence indicates that the use of length in swine carcass evaluation is of questionable value.

Lamb and Mutton. Generally, the carcass yield of U.S. Prime and Choice grade lambs is 50 to 54%. Variations outside this range are due to the weight of the fleece and to weighing conditions. U.S. Good lambs usually dress 47 to 50%. Dressing percentage for the lower grades tends to decline directly as grades decrease.

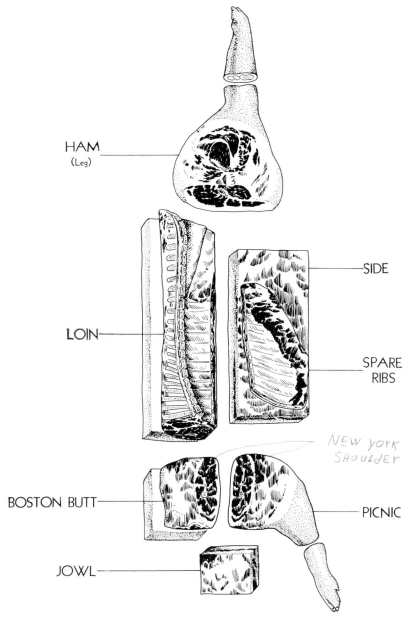

HAM
(Leg)

SIDE

LOIN

SPARE
RIBS

NEW YORK
SHOULDER

BOSTON BUTT

PICNIC

JOWL

Fig. 2-9. Location of wholesale cuts of pork. Note that the spareribs are removed from the overlying flesh leaving the side or clear belly for processing of bacon. [Courtesy National Livestock and Meat Board.]

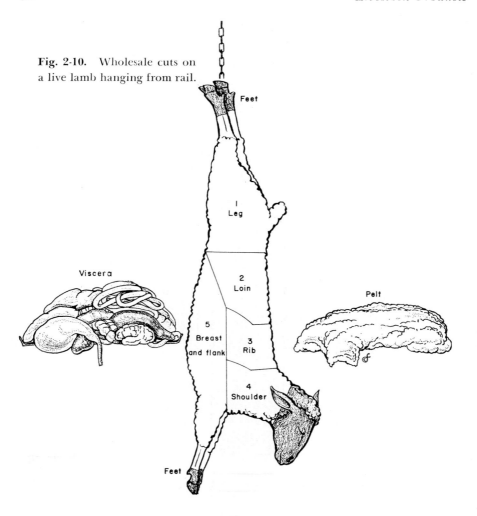

Fig. 2-10. Wholesale cuts on a live lamb hanging from rail.

The value of lamb and mutton by-products is more variable than for the other species. The variation is due to differences in quality of the pelt and to the fluctuations in wool prices. Many lambs are sold after shearing, when the pelt is of little value, but others are sold in full fleece. Since the full fleece often weighs 6 to 8 lb. or more, great variability in value occurs. The amount of wool also has a marked influence on the dressing percentage.

THE PERCENTAGE AND VALUE OF CUTS. Although the method of cutting lamb differs from that used for beef, the differences in value between desirable and less desirable cuts are probably very much alike (Table 2-5). Although the leg, loin, and rack make up only a little more than 29% of the live

TABLE 2-5. *Weight and value of wholesale cuts of lamb expressed as percentages of the weight and value of the live animal.**

Wholesale Cut	Weight	Value
Leg (long cut)	16.7	43.2
Loin	7.3	22.2
Hotel rack	5.2	10.4
Shoulder	13.3	6.7
Breast and shank	9.5	17.4

* Based on a dressing percentage of 52.

weight, they constitute about 76% of the total value of the carcass. The wholesale cuts for lamb are shown in Fig. 2-10.

MUTTON VERSUS LAMB. In the United States any sheep carcass that fails to show the break joint is classified as mutton (see Fig. 37-12). Consequently, there is a wide variation in fatness, weight, conformation, flavor, and other characteristics of mutton carcasses. The trade has generally paid a premium for lamb and has severely discounted mutton, regardless of age. Old crop lambs (sold at about a year of age after being fattened on irrigated pasture

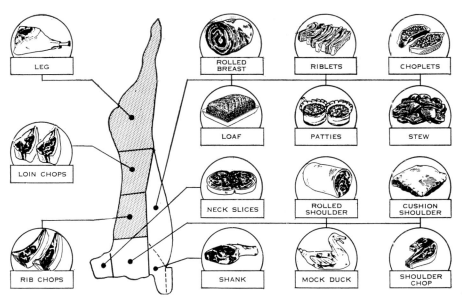

Fig. 2-11. Location of retail cuts of lamb. Wholesale cuts are indicated by heavy lines, but are not labeled. [Modified from figure supplied through courtesy of National Livestock and Meat Board.]

or in dry lot) sell at considerably lower prices than milk-fed early spring lambs.

INFLUENCE OF CONFORMATION. Ljundahl (1942) and Henneman (1942) found that long-legged rangy lambs have a higher percentage of leg than do compact short-legged lambs. On the other hand, the shorter-legged lambs have a greater percentage of loin and rack, with no great difference in shoulder. Thus, it appears that conformation has little effect on the yields of higher priced cuts.

Veal and Calf. The major difference between beef, calf, and veal is a matter of maturity or age. In general terms, veal is less than 3 months of age, calf is less than 12 months, and beef, more than 12 months at the time of slaughter. In comparison with beef, both veal and calf are lighter colored, have less fat, have less flavor and firmness, and are more watery. Since calf is older than veal, its flesh more closely resembles beef. Owing to seasonal production, veal and calf have not been promoted, and demand has remained relatively low. The term "baby beef" has been used in the past to denote cattle finished on full feed at 12 to 18 months of age. However, the use of this term has become less common. Since many calves go directly to the feed lot for fattening at weaning time, much of the beef consumed today would have been classed as "baby beef" twenty years ago.

By-products. A major contributing factor to the expansion of the meat packing industry in the United States has been the efficient utilization of by-products. Custom butchering has largely replaced farm slaughtering. In the past, the margin of profit of the meat packer had been based mainly upon efficient handling and marketing of by-products, such as hides, tallow, animal feeds, fertilizer, and natural sausage casings. Recently, however, there has been a marked decline in the prices received for by-products.

The relative values of hides, tallow, and animal feeds have declined in the past several years. Lard, which at one time sold for more per pound than the packer paid for live hogs, today sells for less. These changes have resulted mostly from competition from other products. Lard has competition from vegetable shortenings and cooking oils. Leather has been partially replaced by synthetic leather products, which can be produced more cheaply. Most inedible tallow was formerly used to manufacture soap, but synthetic detergents have now taken over a large part of the market. Consequently, tallow has so declined in price that its inclusion in mixed poultry and swine rations has become feasible. The development of the soybean oil meal industry has resulted in a cheap and competitive product for meat scraps. As a result of this decline in the values of by-products, export trade in hides, tallow, lard, and animal feeds has greatly expanded.

Table 2-6 shows the relative changes in the prices of hides and live steers. Although the number of cattle slaughtered has increased in recent years, the

TABLE 2-6. *The relative prices of 100 pounds of hides and of live steers from 1910 to 1963.*

Years	Heavy Native Steer Hides	Average Price All Steers	Price Hides/ Price Steers (%)
1910–1914	$17.19	$ 7.57	227
1915–1919	30.55	11.93	256
1920–1924	18.90	9.76	194
1925–1929	18.03	11.76	154
1930–1934	9.71	7.58	128
1935–1939	13.49	9.94	111
1940–1944	14.70	13.26	111
1945–1949	22.31	23.57	95
1950–1954	19.88	29.06	67
1955–1960	13.40	24.28	55
1963	13.72	25.93	53

Source: Data supplied by American Meat Institute Foundation, Chicago, Illinois.

total income from hides has steadily decreased. If such trends in the values of all by-products were to continue, the costs of slaughtering would necessarily need to be charged against the carcass. The end result would be an increase in meat prices without any direct benefit—and with possible loss— to the producer.

2-8. EFFECTS OF FEEDING AND MANAGEMENT UPON THE CARCASS

Feeding and management have marked effects upon the carcass and its rate of development. Figure 2-12 shows the influences of early and late maturity and/or the level of nutrition upon various body parts. A high plane of nutrition and early maturity have the same gross effect: more rapid growth of all body parts. Conversely, a low plane of nutrition during the growth phase of an animal's life lengthens the duration necessary for the various developmental changes to take place. The end result of such nutritional restriction on the early maturing animal is to change the growth pattern to that of the late maturing animal (McMeekan, 1940, 1941). In spite of changes in the growth pattern, the order of development for the various body parts is essentially unchanged. Thus, different tissues attain their maximum growth in a definite order with nervous tissue first, followed by bone, muscle, and fat. Fat deposition follows a definite pattern also, with development of mesenteric fat early in life, followed by kidney fat, subcutaneous fat, and finally marbling (intramuscular fat).

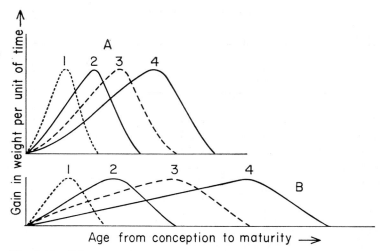

Fig. 2-12. The rate of weight increase of different body parts and tissues in sheep, showing their order of development as affected by early and late maturity and/or level of nutrition. A—early maturity or high plane of nutrition, B—late maturity or low plane of nutrition. Curves: (1) head, brain, cannon, kidney fat; (2) neck, bone, tibia-fibula, intermuscular fat; (3) thorax, muscle, femur, subcutaneous fat; (4) loin, fat, pelvis, marbling fat. [Reprinted from J. Hammond. "Progress in the physiology of farm animals." Butterworths Scientific Publications, London. Vol. 2, p. 437. 1955.]

The correlation between stage of development and the relative proportions of bone, muscle, and fat is pertinent. If the animal is slaughtered at the correct stage of growth, it should be possible to obtain the most nearly desirable combination of tissues, that is, fat, bone, and muscle. However, the effect of plane of nutrition upon the edibility of the meat produced has not been thoroughly investigated. Recent research on the effect of delayed growth by USDA researchers (Winchester and Howe, 1955; Winchester and Ellis, 1957) indicates that dietary restriction does not seem to influence either tenderness or juiciness of the meat produced. This work has since been verified by Yeates (1964) in Australia.

2-9. RECOGNITION OF CARCASS TRAITS IN LIVE ANIMALS

The improvement of meat animals could be more easily achieved if superior carcasses could be recognized in the live animal. Toward this achievement, several methods for determining fat content of the live animal have been proposed.

Live Probe. The live probe method of measuring backfat thickness on hogs was developed by Hazel and Kline (1952). A steel ruler is inserted through a skin incision. The method is a good indicator of lean cuts of pork and has been widely used by the swine industry. It has been criticized on the basis that it measures fatness directly and leanness indirectly. The method is not effective in sheep or cattle.

Lean Meter. This technique (Andrews and Whaley 1954) permits measurement of subcutaneous fat thickness by the use of a needle that indicates differences in electrical conductivity for fat and lean tissues. It appears to offer little or no advantage over the live probe method and is more expensive.

Thermistor Probe. The thermistor probe measures subcutaneous fat thickness by showing temperature changes as the needle passes from fat to lean tissues (Warren et al., 1959). The method has appeared to be useful in measuring fatness in cattle.

Ultrasonics. Ultrasound has been used extensively by the steel industry for the detection of flaws and, more recently, by the medical profession for locating and studying abnormal tissues. Basically, sound waves are generated and sent out through the tissues; whenever a change in density occurs, a reflection is returned and amplified on an oscilloscope. Temple and co-workers (1956) used the instrument to measure fat thickness in cattle. Since then, it has been used successfully in lambs and hogs.

Stouffer (1959) developed a method of plotting the cross-sectional area of the rib-eye muscle from ultrasonic reflections and angular readings in hogs, lambs, and cattle. Even though the method is effective, its usefulness is questionable, since Cole et al. (1960) found rib-eye area to have a low predictive value for total carcass lean.

Live-Animal Measurement. Extensive measurements of live animals have been made and related to carcass traits, but, in general, they have not proven useful. It is questionable whether further investigation on live measurements is warranted.

Potassium-40. The naturally occurring radioisotope K^{40} is found in all living cells and is concentrated largely in the muscle tissues. However, Kirton et al. (1961) found that both bone and fatty tissue contribute appreciable gamma radioactivity, and that K^{40} counting does not indicate only muscle tissue. Recently, Lawrie and Pomeroy (1963) and Gillett et al. (1964) demonstrated that the potassium content of different muscles of the pig may differ by more than 30%; this would indicate a further limitation in accuracy for predicting muscle tissue from K^{40}. Because background radioactivity interferes, special equipment for shielding and counting is needed. Thus,

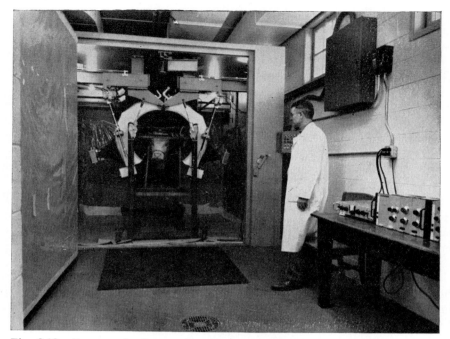

Fig. 2-13. Large animal potassium-40 counter at University of Illinois with animal in position for counting. Note the heavy lead shielding necessary to reduce background radiation. [Courtesy G. S. Smith, Department of Animal Science, University of Illinois, Urbana, Illinois.]

the cost of satisfactory counters and equipment may often be prohibitive. Although the K^{40} method has been shown to be useful for measuring sizable differences in fatness, it is not sufficiently accurate to predict differences of a smaller magnitude (Kirton and Pearson, 1963).

Antipyrine and Other Body-water Diluents. A method that involves injecting antipyrine (or one of several other compounds, such as urea) and noting its dilution has been investigated as a means of predicting fatness or leanness. However, the reports on its accuracy are conflicting.

Specific Gravity. Specific gravity, or relative density, has long been recognized as a good measure of leanness. Carcass density can be accurately determined by underwater weighing, but this procedure obviously cannot be used with live animals. It has been proposed that specific gravity may be determined by measuring body volume on the basis of air displacement (Liuzzo et al., 1958) or by helium dilution (Siri, 1956). Although these methods appear sound, mechanical problems have prevented the desired accuracy.

Creatinine Excretion. The creatinine coefficient (mg of creatinine excreted in 24 hr per kg of body weight) is theoretically proportional to leanness. Studies of farm animals have failed to show the method to be sufficiently accurate for practical improvement programs (Saffle et al., 1958).

Conformation. According to information reported by Pierce (1957) and unpublished data of Carroll and Clegg, animals of good conformation yield a higher proportion of retail cuts than animals of poor conformation. For example, Carroll and Clegg have data on a limited number of steers showing that the percent of trimmed retail shortloin, sirloin, top round, bottom round, and knuckle accounts for 66.3% of the hindquarter weight for Herefords as compared to 61.6% for Holsteins. Although this difference would not be expected, in view of the similarity in percentage of wholesale cuts between cattle differing in conformation, the differences were statistically significant.

To summarize the information on recognition of carcass traits in the live animal, we might say that although some advances in estimating fat content have been made, the problem remains, for the most part, unsolved.

REFERENCES AND SELECTED READINGS

References marked with an asterisk are of general interest.

Adams, R., D. L. Harrison, and J. L. Hall, 1960. Comparison of enzyme and Waring blender methods for determination of collagen in beef. *Agr. and Food Chem.,* 8:229.

*American Meat Institute Foundation, 1960. *The Science of Meat and Meat Products.* Freeman, San Francisco.

Andrews, F. N., and R. M. Whaley, 1954. *A Method for the Measurement of Subcutaneous Fat and Muscular Tissues in the Live Animal.* Purdue University Press.

Batcher, O. M., and E. H. Dawson, 1960. Consumer quality of raw and cooked pork. *Food Tech.,* 14:69.

Branaman, G. A., A. M. Pearson, W. T. Magee, R. M. Griswold, and G. A. Brown, 1962. Comparison of the cutability and eatability of beef- and dairy-type cattle. *J. Animal Sci.,* 21:321.

Briskey, E. J., R. W. Bray, W. G. Hoekstra, R. H. Grummer, and P. H. Phillips, 1959. The effect of various levels of exercise in altering the chemical and physical characteristics of certain pork ham muscles. *J. Animal Sci.,* 18:153.

*Bull, S., 1951. *Meat for the Table.* McGraw-Hill, New York.

———, and H. P. Rusk, 1942. *Effect of Exercise on Quality of Beef.* Ill. Agr. Expt. Sta. Bull. 488.

Butler, O. D., 1957. The relation of conformation to carcass traits. *J. Animal Sci.,* 16:227.

———, B. L. Warwick, and T. C. Cartwright, 1956. Slaughter and carcass characteristics of short-fed yearling, Hereford, and Brahman X Hereford steers. *J. Animal Sci.,* 15:93.

Clark, F., B. Friend, and M. C. Burk, 1947. *The Nutritive Value of Per Capita Food Supply.* USDA Misc. Publ. 616.

Cole, J. W., L. E. Orme, and C. M. Kincaid, 1960. Relationship of loin eye area, separable lean of various beef cuts, and carcass measurements to total carcass lean in beef. *J. Animal Sci.,* 19:89.

Craig, H. B., A. M. Pearson, and N. B. Webb, 1962. Fractionation of the component(s) responsible for sex odor/flavor in pork. *J. Food Sci.,* 27:29.

Gillett, T. A., A. M. Pearson and A. H. Kirton, 1964. Variation in potassium and sodium in muscles of the pig. *J. Animal Sci.,* 24:177.

Goeser, P. A., 1964. *Personal Communication.* Research Laboratories, Swift and Company.

―――, H. F. Bernholdt, M. Hogan, and G. E. Brissey, 1960. Tendered meat through antemortem vascular injection of proteolytic enzymes. *Food Tech.,* 14:41 (Abstract).

Hall, J. L., C. E. Latschar, and D. L. Mackintosh, 1944. *Quality of Beef.* Part IV: *Characteristics of Dark Cutting Beef. Survey and Preliminary Investigation.* Kansas Agr. Expt. Sta. Tech. Bull. 58.

Harrington, G., and A. M. Pearson, 1962. The chew count as a measure of tenderness in pork loins with varying degrees of marbling. *J. Food Sci.,* 27:106.

Hazel, L. N., and E. A. Kline, 1952. Mechanical measurement of fatness and carcass value on live hogs. *J. Animal Sci.,* 11:313.

Helser, M. D., P. M. Nelson, and B. Lowe, 1930. *Influence of the Animal's Age upon the Quality and Palatability of Beef.* Iowa Agr. Expt. Sta. Bull. 272.

Henneman, H. A., 1942. *The Relationship of Rate of Growth in Lambs to Body Measurements and Carcass Value.* Mich. State University, M.S. Thesis.

Hiner, R. H., O. G. Hankin, H. S. Sloane, C. R. Fellers, and E. E. Anderson, 1953. Fiber diameter in relation to tenderness of beef muscle. *Food Research,* 18:364.

*Hinman, R. B., and R. B. Harris, 1942. *The Story of Meat.* Swift and Company, Chicago.

Hornstein, I., and P. F. Crowe, 1960. Meat flavor chemistry: Studies on beef and pork. *J. Agr. and Food Chem.,* 8:494.

―――, 1963. Meat flavor: Lamb. *J. Agr. and Food Chem.,* 11:147.

Hostetler, E. H., J. E. Foster, and O. G. Hankins, 1936. *Production and Quality of Meat from Native and Grade Yearling Cattle.* N.C. Agr. Expt. Sta. Bull. 63.

Ingram, M., 1948. Fatigue musculaire, *p*H et prolifération bactérienne dans la viande. *Ann. Inst. Pasteur.,* 75:139.

Jourbert, D. M., 1956. An analysis of factors influencing post-natal growth and development of the muscle fiber. *J. Agr. Sci.,* 47:59.

Kirton, A. H., A. M. Pearson, R. H. Nelson, E. C. Anderson, and R. L. Schuch, 1961. Use of naturally occurring potassium-40 to determine carcass composition of live sheep. *J. Animal Sci.,* 20:635.

Kirton, A. H., and A. M. Pearson, 1963. Relationships between potassium content and body composition. Annals N.Y. Acad. Sci., 110:221.

Kramlich, W. E., and A. M. Pearson, 1958. Some preliminary studies on meat flavor. *Food Tech.,* 23:567.

Kummerow, F. A., 1964. The role of polyunsaturated fatty acids in nutrition. *Food Tech.,* 18 (6): 49.

Lawrie, R. A., and R. W. Pomeroy, 1963. Sodium and potassium in pig muscle. *J. Agr. Sci.,* 61:409.

Liuzzo, J. A., E. P. Reineke, and A. M. Pearson, 1958. Determination of specific gravity by air displacement. *J. Animal Sci.,* 17:513.

Ljundahl, W. A. *Significant Factors in the Determination of Carcass Quality in Lamb.* Mich. State Univ. M.S. Thesis (1942).

Maddox, G., 1960. Effects of fats on heart disease. *Food and Nutr. News,* 31(7):1.

McMeekan, C. P., 1940. Growth and development of the pig with special reference to carcass quality characters. *J. Agr. Sci.,* 30:276, 387, 511.

―――, 1941. Growth and development of the pig with special reference to carcass quality characters. *J. Agr. Sci.,* 31:1.

Mitchell, H. H., and T. S. Hamilton, 1933. The effect of long-continued muscular exercise upon the chemical composition of the muscles and other tissues of beef cattle. *J. Agr. Research*, 46:917.

Pierce, J. C., 1957. The influence of conformation, finish, and carcass weight on the percentage yield of wholesale and retail cuts of beef. *Proc. Recip. Meat Conf.*, 10:119.

Ramsey, C. B., J. W. Cole, B. H. Meyer and R. S. Temple, 1963. Effects of type and breed of British, Zebu and dairy cattle on production, palatability and composition. II. Palatability differences and cooking losses as determined by laboratory and family panels. *J. Animal Sci.*, 22:1001.

Saffle, R. L., L. E. Orme, D. E. Sutton, D. E. Ullrey, and A. M. Pearson, 1958. A comparison of urinary and blood serum creatinine with live probe as measures of leanness for live swine. *J. Animal Sci.*, 17:480.

Siri, W. E., 1956. Apparatus for measuring human body volume. *Rev. Sci. Instr.*, 27:729.

Stonaker, H. H., M. H. Hazaleus, and S. S. Wheeler, 1952. Feedlot and carcass characteristics of individually fed comprest and conventional type Hereford steers. *J. Animal Sci.*, 11:15.

Stouffer, J. R., 1959. Status of the application of ultrasonics in meat animal evaluation. *Proc. Recip. Meat Conf.*, 12:161.

Stroble, C. P., C. B. Roubicek, and N. W. Hilston, 1951. Carcass studies of steer

progeny. *Proc. West. Sect., Am. Soc. Animal Prod.*, 11:155.

Temple, R. S., H. H. Stonaker, D. Howry, G. Posakony, and M. H. Hazaleus, 1956. Ultrasonic and conductivity methods for estimating fat thickness in live cattle. *Proc. West. Sect., Am. Soc. Animal Prod.*, p. 70.

Warren, R. B., V. H. Arthaud, C. H. Adams, and R. M. Koch, 1959. Thermistor thermometer for estimating fat thickness on live beef cattle. *J. Animal Sci.*, 18:1469.

Willey, N. B., O. D. Butler, J. K. Riggs, J. H. Jones, and P. J. Lyerly, 1951. The influence of type on feedlot performance and killing qualities of Hereford steers. *J. Animal Sci.*, 10:195.

Williams, L. D., A. M. Pearson, and N. B. Webb, 1963. Incidence of sex odor in boars, sows, barrows and gilts. *J. Animal Sci.*, 22:166.

Wilson, J., and C. F. Curtis, 1893. *Steer Feeding.* Iowa Agr. Exp. Sta. Bull. 20.

Winchester, C. F., and N. R. Ellis, 1957. *Delayed Growth of Beef Cattle.* USDA Tech. Bul. 1159.

Winchester, C. F., and P. E. Howe, 1955. *Relative Effects of Continuous and Interrupted Growth on Beef Steers.* USDA Tech. Bul. 1108.

Yeates, N. T. M., 1964. Starvation changes and subsequent recovery of adult beef muscle. *J. Agr. Sci.*, 62:267.

*Ziegler, P., 1954. *The Meat We Eat.* 5th Ed. Interstate, Danville.

Chapter 3

Milk and Milk Products

He asked water and she gave him milk; she brought forth butter as a lordly dish.

Judges 5:25

3-1. INTRODUCTION

The early history of dairying is somewhat obscure, but archeological excavations indicate that men of the Old Stone Age were the first to domesticate the cow, and that, in prehistoric times, milk and butter were common human foods. The selection of cows on the basis of milk production is credited to the nomadic Aryans of Central Asia.[1]

During the following centuries, the recognition of the nutritional importance of milk and milk products increased until in 1956 they were established as one of the four groups of foods contributing the essentials of an adequate diet. In the United States, dairy foods account for about 14% of the consumer's food budget.

As the cow population expanded throughout the world and as milk production became a specialized phase of agriculture, methods were developed for concentrating and preserving milk nutrients for future use, barter, or international trade. Dairying progressed from an art to a science, drawing upon genetics, chemistry, physics, bacteriology, heat and refrigeration engineering, business, and other branches of knowledge, in order to improve dairy commodities, create new ones, and profitably distribute them in both domestic and world commerce.

[1] The excellent summary of the historical background and origin of the present dairy industry by Rusoff (1955) is very interesting.

In the United States, one of the important dairy countries, the total milk production in 1963 was approximately 124.8 billion lb., of which the producers sold approximately 117.8 billion lb. Approximately 47 5% of the latter was used for fluid purposes; the remainder was used in the manufacture of approximately 1.4 billion lb. of butter, 1.6 billion lb. of cheese, 1.4 billion lb. of cottage cheese, 1 billion gallons of frozen dairy products, and almost 2 billion lb. of nonfat dry milk solids. An estimate of the cash farm income from dairying in 1964 (exclusive of the milk used on the farm and the income from the sales of cows, heifers, and calves from dairy herds—approximately 1.6 billion dollars in 1962) exceeds 4.7 billion dollars.

It is noteworthy that, although the per capita consumption of cream has declined from 10.3 lb. in 1953 to 8.3 lb. in 1963, the per capita consumption of other fluid dairy products appears to have stabilized at approximately 300 lb. The consumption of butter reached an all-time low of about 6.7 lb. per capita in 1963. On the other hand, the consumption of cheese, not including cottage cheese, has shown a steady increase over the years to about 9.4 lb. per capita in 1963. The use of cottage cheese appears to have plateaued at approximately 4.5 lb. per capita. Ice cream consumption remains at approximately 18 lb. per capita, and that of ice milk has increased from approximately 2.1 lb. in 1953 to approximately 5.5 lb. per capita in 1963. Nutritional science, aided by industry-supported research, education, and promotion, and coupled with a healthy economy, is encouraging consumer purchasing of dairy products.[2]

3-2. THE PRODUCTION OF MILK

Although milk from nearly every species of domestic mammal has affected man's economy, the modern agriculturalist is primarily interested in the dairy cow, the buffalo, the goat, the ewe, and, to a degree, the mare and the sow as sources of milk. The dairy industry has evolved around the first three especially, and of these the dairy cow is most important in countries where dairying is of major agricultural significance. Since milks from these industrially important mammals differ mainly in percentage composition, and therefore in some physical properties (see Table 3-1), the term "milk," as used in the following pages, will refer to cow's milk.

Milk is the white fluid, exclusive of the colostrum, secreted by the mammary glands of mammals. The composition and quantity are dependent upon several physiological and environmental factors (see Chapter 23). For example, the first-drawn, middle, and last milks differ in composition and quantity; mastitis and certain physiological disturbances also influence milk secretion. Therefore, natural milk is generally considered as the complete secretion of a normal mammary gland of a healthy cow. Other factors

---

[2] A well-documented treatment of the nutritional value of milk has been published (Rusoff, 1964).

TABLE 3-1. *Characteristic composition of milks of important mammals, in percentage of total weight.*

Mammal	Water	Total Solids	Fat	Protein	Lactose	Ash
Human	87.8	12.2	3.8	1.2	7.0	0.2
Cow	87.3	12.7	3.9	3.3	4.8	0.7
Goat	87.6	12.4	3.7	3.3	4.7	0.7
Water buffalo	76.8	23.2	12.5	6.0	3.8	0.9
Ewe	81.6	18.4	6.5	6.3	4.8	0.8
Sow	82.4	17.6	5.3	6.3	5.0	1.0
Mare	90.2	9.8	1.2	2.3	5.9	0.4

—age, stage of lactation, month of lactation, etc.—also are taken into consideration in production testing (see Chapter 32). From the analyses of a series of monthly one-day composites, throughout the lactation period, it is possible to determine the lactation yield and percentage composition of the total milk of the individual cow for the specific lactation. The production records of many individual cows have been used to estimate the average composition of individual breed milks. It should be recognized, however, that because of the wide variation in the composition of milk between individuals within herds and between herds within breeds, breed averages serve only to characterize the composition of breed milks and may not be representative of a specific sample of milk.

Table 3-2 was compiled from the analyses of individual cow samples taken at approximately six-week intervals throughout the year from 827 cows in commercial herds in widely separated parts of Oregon. Milks from abnormal udders were excluded. The data represent 141 samples from 1 Ayrshire herd, 523 from 2 Brown Swiss herds, 812 from 5 Guernsey herds, 1137 from 6 Holstein herds and 1122 from 9 Jersey herds. The breed means show the characteristic composition of breed milks, and the herd means illustrate the differences between herd milks within breeds. They do not reveal the extent of deviation from the means of either the individual cow or herd samples. A statistical analysis of the data has been made by von Krosigk (1959, 1960).

Interrelationships Among Major Milk Constituents. Positive correlations exist between the percentages of fat and total solids, fat and protein, fat and solids-not-fat, protein and total solids, and protein and solids-not-fat. In view of the growing recognition of the nutritional and economic values of the solids-not-fat and of protein, and the existence of a practical test for fat, much research has centered on the possibility of finding a simple mathematical equation with which to calculate the solids-not-fat content from the fat content with an accuracy sufficient for market purposes. The con-

TABLE 3-2. | *Mean composition of milk from individual herds of five dairy breeds in Oregon, in percentage of total weight.*

Breed	Fat	Solids-not-fat	Protein	Lactose	Ash	Protein in Total Solids	Energy/lb. (kilocal)
Ayrshire	4.21	8.97	3.34	4.95	0.68	25.3	349
Brown Swiss	4.12	9.43	3.69	4.99	0.75	27.2	355
	3.84	9.34	3.58	5.01	0.75	27.1	341
Guernsey	4.76	9.37	3.53	5.12	0.72	25.0	380
	5.10	9.08	3.58	4.79	0.71	25.2	390
	4.79	9.54	3.64	5.17	0.73	25.4	385
	5.06	9.41	3.71	4.97	0.73	25.6	395
	5.06	9.52	3.61	5.18	0.73	24.8	396
Holstein	3.56	8.61	3.08	4.87	0.66	25.3	314
	3.58	8.47	3.02	4.79	0.66	25.1	312
	3.68	8.78	3.08	5.03	0.67	24.7	322
	3.67	8.82	3.10	5.04	0.68	24.8	322
	3.65	8.78	3.32	4.79	0.67	26.7	323
	3.34	8.51	3.07	4.78	0.66	25.9	303
Jersey	5.58	9.64	3.84	5.09	0.71	25.2	430
	6.52	9.90	4.15	5.03	0.72	25.3	468
	5.23	9.49	3.88	4.91	0.70	26.4	405
	6.15	9.82	4.01	5.09	0.72	25.1	450
	5.38	9.71	3.94	5.06	0.71	26.1	416
	5.32	9.58	3.92	4.95	0.71	26.3	411
	5.62	9.91	3.93	5.26	0.72	25.3	429
	5.51	9.70	3.92	5.07	0.71	25.4	421
	5.16	9.63	3.92	5.00	0.71	26.5	405
BREED MEANS							
Ayrshire	4.21	8.97	3.34	4.95	0.68	25.3	349
Brown Swiss	3.97	9.37	3.63	4.99	0.75	27.2	347
Guernsey	4.99	9.44	3.64	5.07	0.73	25.2	383
Holstein	3.59	8.68	3.11	4.90	0.67	25.4	316
Jersey	5.58	9.70	3.95	5.04	0.71	25.8	424

sensus now is that relationships between fat and solids-not-fat and between fat and protein contents are not linear, and that equations derived from one set of data may not satisfy the data from another source.

As illustrated in Fig. 3-1, the linear regression equations for the various breeds represented in Table 3-2 differ with regard to the regression coeffi

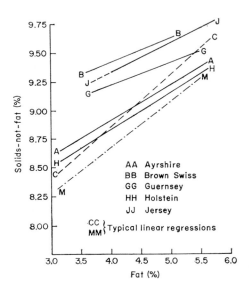

AA Ayrshire
BB Brown Swiss
GG Guernsey
HH Holstein
JJ Jersey

CC }
MM } Typical linear regressions

Fig. 3-1. The relationships between the percentages of fat and of solids-not-fat in milks of different breeds.

cients and especially to the Y intercepts. It is obvious that a single linear equation, as represented by curves C-C or M-M, is inadequate for estimating the solids-not-fat percentage from the percentage of fat in producer milks. The errors resulting from this indirect method for estimating one major component of milk from a measured content of another are too great for even within-breed milks to be of practical value (Erb et al., 1963). Individual breed equations relating solids-not-fat to protein percentages appear to be more accurate.

Some Constancies in Milk Composition. The ratio of the percentage of protein to that of the total solids in herd milks varies only within narrow limits, regardless of breed or fat percentage; Brown Swiss milk, however, is characteristically high. In the Oregon study referred to previously, the mean protein content of the yearly milk yield of the 23 herds was 25.64% of the total solids. The percent of total solids may be estimated by multiplying the protein percentage by the factor 3.9.

The energy value of natural milk is closely related to its fat percentage and may be calculated from the equation $E = 133 + 52f$, where E = kilocalories per pound of milk and f = percent fat.

Energy Corrected Milk. Since the feed energy required by the cow for lactation is proportional to the milk energy yield, and since the cow's ability to produce milk may be measured in terms of milk energy yield, the yield of milk from the individual cow or herd is frequently expressed in terms of the energy value of a reference milk. Overman and coworkers

(1933) chose milk containing 4% fat as the reference milk and, from an equation similar to the foregoing, developed and introduced the Fat Corrected Milk (4%) formula: FCM = 15F + 0.4M, in which FCM = pounds of milk equivalent in energy to the reference milk, M = pounds of milk, and F = pounds of fat in the milk being compared. To illustrate the use of this formula, let us take an example of 100 pounds of 3.5% milk: FCM = (15 × 3.5) + (0.4 × 100) = 52.5 + 40 = 92.5, the number of pounds of reference milk (4%) equivalent to 100 pounds of 3.5% milk. This relation may also be expressed as FCM = M(0.15f + 0.4), where f = percent fat. It is perhaps unfortunate that the term Energy Corrected Milk (ECM) was not used. It might also be suggested that the formula ECM = 15.25F + 0.39M appears to be more representative of the data from which Table 3-2 was compiled than the simplified formula. Students of dairy husbandry might be interested in pursuing the idea suggested by Gaines and Overman (1938), namely, that since the energy yield of milk tends to be a simple multiple of the protein yield (kilocalories per pound of milk = 102.6 × percent protein), more significance might be attached to the protein-energy relationship than to the fat-energy relationship.

The relationship between the energy value and the protein percentage of the herd milks illustrated in Table 3-2 may be expressed by this equation: Energy, kilocalories per pound = −75.73 + 126.78 × protein percent. If, for example, a 3.3% protein milk is considered as the reference, other milks may be energy corrected to it on a protein (3.3%) basis by the formula ECM = −0.221M + 37P, where M = pounds of milk and P = pounds of protein. The correlation between protein and energy is not as high as that between fat and energy, but it may prove significant in studies pertaining to milk production and to the biological and genetic factors in milk secretion.

3-3. CHEMICAL COMPOSITION OF MILK

The variation in the quantitative gross composition of milk due to environmental and genetic factors has been considered, in the previous section. These variations are important to producers, milk dealers, and, to a degree, to nutritionists, but the composition of milk, in order to be fully appreciated, must be considered in its entirety. In the compilation made by Macy, Kelly, and Sloan (1953), representative values for more than 100 components of human, goat, and cow milk are tabulated and summarized. In Table 3-3, many of these data for cow's milk have been condensed and, in some instances, slightly revised in light of recent findings; enzymes are also listed.

It must be recognized that the data in Table 3-3 do not represent average values. The values may be expected to fall within the range of values characteristic of normal milk secreted by healthy cows. Much information of particular interest to the nutritionist has been omitted. For example, the

TABLE | *Partial evaluation of representative cow's milk.*
3-3. |

Constituent	Per 100 ml	Per qt
Water (g)	87.3	826.12
Total solids (g)	12.7	120.18
Fat (g)	3.7	35.01
Solids-not-fat (g)	9.0	85.17
Total protein (g)	3.3	31.23
Caseins (g)	2.6	24.60
β-lactoglobulins (g)	0.34	3.22
α-lactalbumin (g)	0.06	0.57
Other albumin and globulin types (g)	0.3	0.28
Nonprotein nitrogenous compounds (g)	0.02	0.19
Lactose (g)	4.96	46.94
Ash (g)	0.72	6.81
Calcium (mg)	122	1,182.9
Phosphorus (mg)	96	908.4
Magnesium (mg)	12	113.6
Potassium (mg)	138	1,305.9
Sodium (mg)	58	548.9
Chlorine (mg)	103	974.7
Sulfur (mg)	30	283.9
Vitamins		
Fat soluble		
Vitamin A (μg)	34	321.2
Carotenoids (μg)	38	359.6
Vitamin D (U.S.P. units)	2.36	22.3
Vitamin E (mg)	0.06	0.57
Vitamin K (Dam-Glavind units)	100	946.3
Water soluble		
Ascorbic acid (mg)	1.6	15.1
Biotin (μg)	3.5	33.1
Choline (mg)	13	123.02
Folic acid (μg)	0.23	2.2
Inositol (mg)	13	123
Nicotinic acid (μg)	85	840.36
Pantothenic acid (μg)	350	3,312.05
Pyridoxine (μg)	48	454
Riboflavin (μg)	157	1,485.7
Thiamine (μg)	42	397
Vitamin B_{12} (μg)	0.56	5.3
Other accessories		
Phospholipids (as lecithin) (g)	0.057	0.539
Cholesterol (g)	0.014	0.1325
Energy		
Combustible (kcal)	72	681
Physiological (kcal)	66	625
Enzymes of known significance		
Catalase, peroxidase, xanthine oxidase, phosphatases, lipases, proteases		

Source: Revised from Macy, Kelly, and Sloan (1953).

mere summation of fat and protein values fails to indicate either the wide distribution of fatty acids (saturated and unsaturated) in milk fat or the completeness of the proteins in dispensable and indispensable amino acids. The original publication of Macy et al. (1953) and the report by Rusoff (1964) describe milk as a well-balanced composition of nutrients. This consideration is the basis for much of the opposition by nutritionists and members of the medical profession to fortification and unregulated modification of milk.

3-4. THE PHYSICAL PROPERTIES OF MILK

The complexity of its chemical composition suggests that milk possesses interesting physical properties and leads one to wonder how it can be a stable system. It is supersaturated with milk fat, proteins, and calcium phosphate salts.

Structure. As indicated in Table 3-4, milk is a colloidal system[3] in which an aqueous solution of mineral salts, lactose, and some of the serum proteins constitutes the continuous phase. Calcium caseinate, calcium phosphate, and

TABLE 3-4. | *Structural composition of milk.*

Constituent	Average diameter of globules (mμ)*	State	Visibility
Fat			
As secreted	2500–3000	Emulsion	⎫ Readily visible
In homogenized milk	250	Emulsion	⎭ under microscope
Calcium caseinate	5–100	Colloidal	⎫
Calcium phosphate	5–100	Colloidal	⎪ Visible under
Lactoalbumins and lactoglobulins	5–15	Molecular but having	⎬ ultramicroscope
(whey proteins)		colloidal properties	⎭
Lactose	—	Molecular solution	⎫
Mineral salts	—	Ionic solution	⎬ Invisible

* 1 mμ = about 4×10^{-8} inch.

probably calcium citrate, in various degrees of association, form a dispersed phase; the milk fat forms a more coarsely suspended phase. The dispersions are illustrated in Fig. 3-2. The system is stable as secreted, but the emulsified

[3] A colloidal system is heterogeneous and consists of a continuous phase and at least one dispersed phase. The dimensions of the dispersed particles are larger than ions and simple molecules, but are small enough to combat the force of gravity. The stability of the system is dependent upon Brownian movement and electrostatic repulsion forces within the dispersed phase.

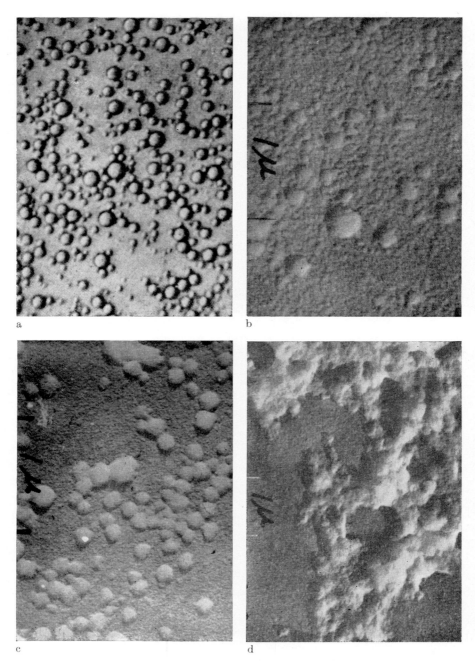

Fig. 3-2. Microphotograph (*a*) and electron microphotographs (*b*), (*c*), (*d*), of coarse and colloidal dispersions in milk. (*a*) Fat globules, approximately 600 X; (*b*) dry nonfat milk, approximately 60,000 X; (*c*) calcium caseinate-calcium phosphate, approximately 60,000 X; (*d*) acid whey proteins (coagulated), approximately 28,000 X. [Taken from the PhD thesis of S. H. Dalal, Oregon State College, 1953.]

fat may be separated as cream, either by natural sedimentation or by centrifugation; the calcium complex, as well as the serum proteins, may be isolated by ultrafiltration.

The Milk Fat Emulsion. Although the fat is coarsely dispersed by colloidality standards, it is present in milk in tiny spheres emulsified in an aqueous medium. The emulsion is stabilized by a rather complex material that confers properties characteristic of colloidal systems to it. The composition of the so called fat globule "membrane" is not entirely understood, but it may be regarded as consisting mainly of a phospholipid-protein complex that serves as a bridge between the fat and aqueous phases. Although the bridge is not securely anchored at either end, it fulfills its primary purpose of keeping the fat emulsified before and during nursing by the young. It serves also to protect the fat against oxidation and the action of milk lipase. However, its fragility is such that care must be taken in handling milk by both producer and processor if rancidity, cream plug, and oxidation flavors are to be avoided.

Cream, the emulsified fat concentrated in milk plasma, if properly handled, is a stable product and will withstand pasteurization, freezing, and careful drying.

Milk Plasma. When the fat is removed from milk as cream, the part remaining is known as separated milk, skim milk, nonfat milk, or milk plasma. As previously indicated, it consists of proteins and calcium phosphate colloidally dispersed in milk serum, the latter being a water solution of lactose, mineral salts, and molecularly dispersed serum proteins. The physical properties of the plasma mainly determine those of the whole milk.

With the exception of butter, the manufacture of dairy products, including the specialties, is based on an understanding of the physical and chemical properties not only of the milk plasma as a whole but also of its major components. For example, the milk plasma is stable under heating and drying conditions, thus permitting the manufacture of pasteurized skim milk, condensed skim milk, and nonfat dry milk. On the other hand, the addition of a very small quantity of rennet to warm milk or milk plasma renders the system so unstable that a clot quickly forms, which later shrinks, exudes whey, and becomes the curd—the raw material for most cheeses. When lactose-fermenting organisms are allowed to develop acidity in milk plasma, the increased concentration of cations results in the coagulation of the main protein—casein. This is the basic reaction in the manufacture of cottage cheese. The quality of such dairy products is dependent upon precise scientific control of the reactions involved.

Useful Physical Properties

SPECIFIC GRAVITY. Owing to the wide variation in percentage composition, the density of milk varies considerably. A value of 1.032 for the specific gravity (15°C/15°C) is considered sufficiently representative for practical purposes. From the assumption that the composition of milk fat and the relationship between the lactose, proteins, and mineral salts of the plasma are constant and, consequently, that definite values for the specific gravities of the fat and the plasma solids may be accepted, many formulas have been recommended for calculating the total solids content of milk from the fat percentage and the lactometer or densitometer readings under specific temperature conditions. The protein content, especially the caseins, of the plasma solids has been shown to vary without a compensating variation in the lactose and mineral salts components. The resulting variation in the density of the plasma solids prohibits the devising of a standard formula for precise calculations of total solids. For the same reasons, multiple regression equations derived from the density, fat, and nonfat measurements for one source of milk may prove inaccurate for milks from other sources. The lactometer or densitometer, however, is a useful tool for indicating gross watering of herd milk and for production testing of milk from individual animals for solids-not-fat.

ELECTRICAL CONDUCTANCE. The capacity of milk to conduct an electric current is dependent upon the concentration of certain ions, especially the chloride ion. This property has been applied in the detection of mastitis, but its usefulness is limited, owing to the variation in the conductance of normal milks.

SURFACE TENSION. The presence in milk of hydrophilic proteins and phospholipides results in a lower surface tension than that of water. A figure of 52 dynes/cm at 25°C may be considered representative. A progressive lowering of surface tension is a fairly reliable indication of lipase action and of the development of rancidity in cooled milk.

FREEZING POINT. The freezing point of milk is below that of water; thus the addition of water to milk is manifested by a rise in the freezing point. An exact method for determining the freezing point has been developed and accepted by the Association of Official Agricultural Chemists (1960) as the official method for detecting the addition of water to milk. The presence of added water is indicated if the freezing point of the milk is above −0.530°C. The 3% tolerance formerly allowed when −0.550°C was the standard mean freezing point of milk is no longer accepted. The limitations of this method are pointed out in a review report by Shipe (1959).

3-5. ENZYMES OF MILK

During milk secretion, the fat, protein, and sugar are simultaneously and continuously formed from dissimilar materials provided by the blood. The complex chemical reactions are catalyzed by efficient individual enzymes and enzyme systems, and it is not surprising that these catalytic agents find their way into milk. Those of recognized significance to the producer and processor are listed in Table 3-3.

Catalase. The catalase content varies probably more than any other enzyme and is increased during udder disturbances, which result in increased leucocyte counts. A high catalase test is not necessarily indicative of mastitis.

Oxidases. The peroxidase content is fairly constant, and since the enzyme is not inactivated below high heat-treatment temperatures, application of this knowledge is sometimes made to detect overheating or to regulate heat treatment. The content of xanthine oxidase is quite variable, but its presence is significant. It is closely associated with the fat globules and, being fairly resistant to heat, may conceivably play an important role in the development of flavor defects of an oxidative nature.

Proteases. It is an accepted fact that at least one protease (galactase), not of bacterial origin, is present. For the extensive conversion of simple nitrogenous compounds to the complex proteins of milk during milk secretion, it must be assumed that an efficient protease system is involved. It is surprising, therefore, that proteases are not among the prominent enzymes of milk. Galactase is considered to play a small part in the ripening of cheese made from raw milk.

Phosphatases. Of the phosphatases present in milk, the greatest significance is attached to the alkaline class. Its concentration in normal milk is fairly constant, and its activity is greatly reduced under the conditions of pasteurization. The phosphatase test, therefore, has been found to be the most reliable for assuring proper pasteurization, providing that precise methods of testing are followed, and the results properly interpreted.

Lipases. Milk is able to catalyze the hydrolysis of a wide range of esters and fats. The very extensive literature on the subject was reviewed by Herrington (1954). However, the results of current research may reveal much about the lipolytic activity of milk. So far, it is recognized that milk probably contains at least two true lipases, which, through their action on the complex triglycerides of milk fat, release fatty acids, some of which confer odors and flavors that are undesirable in most dairy products but which are desirable in certain types of cheese. Normally, the emulsified fat as secreted is protected against lipolysis. However, if the milk is subjected to undue agitation in the milking process, or if the cooled milk is warmed and cooled with even mild agitation, the natural protection is weakened, and rancidity is likely to occur. The milk from cows in advanced lactation occasionally becomes rancid on mere cooling.

3-6. MILK PRODUCTS

Milk is a unique fluid, not merely because of its chemical composition and physical properties, but also because it provides the materials for a wide variety of foods. Milk as secreted is already a prepared food, manufactured by the cow from raw materials, many of which are nonedible to humans. Dairy products are designed to provide concentrated components, tasty vari-ations, or readily preserved commodities.

Creams, Butter, Milk Fat, Buttermilk. The fat in milk may be concen-trated by centrifugation to form light, medium, heavy, or plastic cream; by churning the cream, the fat may be further concentrated to form butter, with buttermilk as a by-product. From cream or butter, the fat may be isolated as practically pure milk fat. No chemical reactions are involved in any of these processes.

The lipid material in milk is an extremely complex mixture of organic compounds consisting mainly of true fats (triglycerides), phospholipides, and sterols; it contains vitamin A, carotenoids, vitamin E, and traces of vitamins D and K. The triglycerides contain a variety of saturated fatty acids ranging mainly from the C_4 to C_{26} carbon lengths and of unsaturated fatty acids of lengths from C_{10} to C_{20}, including di-, tri-, and tetra, etc., polyunsaturates. This variety, coupled with a coefficient of digestibility of 97, is responsible for the unique nutritional and culinary quality of milk lipids and the acceptable flavor and aroma of milk, cream, sweet buttermilk, and other food products containing these lipids. Milk lipids satisfy hunger, both at mealtime and between meals, and help to meet the fatty acid and vitamin needs of the body.

Cheeses. The cheese of early Biblical times was probably little more than coagulated milk proteins (curd) admixed with fat and whey. As man learned to harness and control the activities of bacteria, molds, enzymes, and hydro-gen ions, many different varieties of cheese were developed. Today, of the 800 or more named cheeses, 400 have been described, of which all but about 30 have commercial significance either as staple foods or as the gourmet's delight. For convenience, they may be classified as very hard (Romano), hard (Cheddar, Swiss), semisoft (Brick, Trappist, Blue) and soft (Camembert, Neufchatel, Cottage, Mysost).

The moisture and fat (and indirectly the protein) contents of cheeses are so strictly regulated that the fat content of the common varieties of natural cheeses, excluding cottage, ranges from 28 to 32%, the protein from 22 to 27%, and the lactose from 1.5 to 2.0%. The calcium and phosphorus contents vary, however, only from approximately 0.3 to 0.9% calcium and from 0.2 to 0.7% phosphorus, largely because of the variation in the pH and the action of rennet during processing. The common varieties may be expected to contain about 6,000 international units of vitamin A, from 2 to 3 mg of riboflavin, small amounts of thiamin and niacin, and from 3 to 5 mg of iron per pound.

The nutritive values of the common varieties are not markedly different. The consumer's choice is made on the basis of flavor and aroma. These develop during the aging and ripening process and are largely dependent upon the type and action of the organisms, or group of organisms, in the starter culture. Scientists have yet to find either a single compound or a mixture of compounds that duplicates cheese flavor and aroma. Diacetyl appears to be a common component of all cheese flavors.

Uncreamed cottage cheese is a cultured, but not a ripened cheese. It provides approximately 1.4 g of fat, 77 g of protein, 12 g of milk sugar, 400 mg of calcium, 790 mg of phosphorus, 1.4 mg of iron, 55 international units of vitamin A, and 1.4 mg of riboflavin per pound. The dressing for creamed cottage cheese is usually cultured with a flavor-producing organism (or organisms) and gives the cottage cheese a better taste than the uncreamed product. The normal minimum fat content is 4%.

Casein.[4] Approximately 80% of the total protein of milk consists of colloidally dispersed calcium caseinate. It can be precipitated as calcium paracaseinate by the addition of rennet to skim milk; this precipitate may then be processed into cheese or commercial "rennet casein." The latter, at one time, was extensively used in plastic manufacture. More commonly, the caseinate is converted to casein by acidifying skim milk under carefully controlled conditions. The precipitated casein may be processed into cottage cheese and edible casein products, or dried for industrial use. The expanding markets for nonfat milk as a food have so increased the market quotation for casein that the industrial use of casein has become somewhat restricted to paper coating, glue, and paints, even though research has developed methods for its use in the manufacture of casein wool, paint brush bristles, and other fibers. It is fortunate that the necessity of utilizing this high-quality protein for industrial use is rapidly declining.

Whey Proteins. Approximately 50% of the milk solids of the original milk are retained in the serum, or whey, remaining after the removal of the casein. These consist of the serum proteins, lactose, mineral salts, and water-soluble vitamins. The major portion of the proteins may be recovered by heating and filtering the acidified whey, or by less simple means using electrokinetic principles.

---

[4] This refers to whole casein, which actually consists of a complex association of three main caseins that may be further dissociated or fractionated.

Milk Sugar. The mineral salts that remain in the filtrate may be removed by ion-exchange treatment. Lactose, the main solute in the effluent, may be isolated by crystallization and drying methods, if purity is desired. Besides being low in sweetness, lactose has distinctive nutritional qualities (National Dairy Council, 1962).

Dried and Condensed Products. As pointed out previously, milk, although a complicated system, has remarkably stable physical properties. The individual components possess good chemical stability. Thus, by employing sound engineering principles, milk, cream, skim milk, and whey may be condensed, dried, or frozen, without chemical treatment of any kind. The products are physically tailored according to the ultimate use. The instantizing process for dry milk solids, for example, induces mainly physical rather than chemical changes.

Cultured Products. Milk, separated milk, and whey are good culture media for several types of bacteria and yeasts. By judicious selection of cultures and careful control of their activity, a wide variety of dairy products has been developed. These products range from the relatively simple acid and alcoholic fermented beverages, such as cultured buttermilk and Kefir, to the semisolid products, such as yogurt and certain soft cheeses. The characteristic flavors and textures of these products result from the chemical changes in the lactose, proteins, and fat caused by the microorganisms.

Frozen Dairy Products. Unlike almost all other dairy products, ice cream, sherbets, and ice milks are compounded dairy products. But, except for the added fruits, nuts, eggs, chocolate, flavorings, sugar, stabilizers, and emulsifiers, all ingredients are of milk origin.

For the student who desires to look further into the production, processing, or distribution of milk and milk products, there is available a good selection of textbooks dealing with specific dairy commodities. If he is interested in developing new dairy products or in salvaging the components of milk that are not now completely utilized as human or animal foods, he will find the book by Whittier and Webb (1950) very helpful. It is well to recognize, however, that the dairy foods market is very strictly regulated for the protection of the consumer and that federal and state standards (Agricultural Marketing Service, 1959) have been established for the minimum composition of nearly 100 milk and non-milkfat products.

Although tables showing the composition of dairy foods are readily available, it should be recognized that the values are based on milks of an average fat content of approximately 3.9%, and though acceptable for most nutritional purposes, they are not indicative of the wide variation likely to exist in the composition of dairy products. The compiled values shown in Table 3-5 may be used to calculate the yield and composition of products

TABLE 3-5. | *Percentages, by weight, of solids-not-fat, protein, and milk sugar contained in natural (unstandardized) milks and their skim milks and creams.*

	Fat Test of Milk (%)						
Constituent	3.0	3.5	4.0	4.5	5.0	5.5	6.0
Solids-not-fat (SNF)							
in milk	8.43	8.60	8.84	9.20	9.50	9.62	9.74
in skim	8.69	8.91	9.21	9.63	10.00	10.18	10.36
in cream (40%)	5.21	5.35	5.53	5.78	6.00	6.11	6.22
Protein							
in milk	3.02	3.14	3.33	3.60	3.83	3.95	4.01
in skim	3.12	3.25	3.47	3.75	4.03	4.15	4.27
in cream (40%)	1.87	1.95	2.08	2.25	2.42	2.49	2.56
Lactose							
in milk	4.74	4.78	4.82	4.89	4.94	4.98	4.99
in skim	4.90	4.95	5.02	5.09	5.20	5.27	5.31
in cream (40%)	2.84	2.87	2.89	2.95	2.97	2.98	2.99

$$\% \text{ SNF in skim} = \frac{100}{100 - \text{fat } \% \text{ in whole milk}} \times \% \text{ SNF in whole milk}$$

$$\% \text{ SNF in cream} = \frac{100 - \text{fat } \% \text{ in cream}}{100} \times \text{SNF } \% \text{ in skim}$$

$$\% \text{ constituent in skim} = \frac{\% \text{ constituent in whole milk} \times \text{water } \% \text{ in skim}}{\% \text{ water in whole milk}}$$

$$\% \text{ constituent in cream} = \frac{\% \text{ constituent in whole milk} \times \text{water } \% \text{ in cream}}{\% \text{ water in whole milk}}$$

made from milks of different fat content and to determine the results of standardization, by the addition or removal of cream, the addition or removal of skim milk, or the addition of nonfat solids of such milks to predetermined fat, solids-not-fat, or protein contents. When it is desired to estimate the yield of skim milk from a supply of whole milk, the following equation is useful:

$$W = \frac{C - M}{C - S} \times 100$$

where C = cream test (% fat)
M = milk test (% fat)
S = skim test (% fat)
W = pounds of skim per 100 pounds of whole milk.

REFERENCES AND SELECTED READINGS

References marked with an asterisk are of general interest.

Agricultural Marketing Service, USDA, 1959. *Federal and State Standards for the Composition of Milk Products (and Certain Non-milkfat Products).* Agr. Handbook no. 51.

———, 1961. *Cheese Buying Guide for Consumers.* Marketing Bulletin no. 17.

———, 1964. *A Summary of Laws and Regulations Affecting the Cheese Industry.* Agr. Handbook no. 265.

Association of Official Agricultural Chemists, 1960. *Official Methods of Analysis.* 9th Ed. Washington, D.C., pp. 192–196.

Carskadon, T. R., 1946. The juice of life. *Esquire,* Aug., pp. 44–45.

Erb, R. E., U. S. Ashworth, L. J. Manus, and N. S. Golding, 1963. Estimating solids-not-fat and protein in milk from different sources using percentage milk fat, protein, and solids-not-fat. *Jour. Dairy Sci.,* 46: 1217–1227.

Federal Reserve Bank of Philadelphia, 1955. The cow: a source of wealth, of health, a ward of the state. *Business Review,* April, May, June.

Gaines, W. L., and O. R. Overman, 1938. Interrelations of milk-fat, milk-protein and milk-energy yield. *J. Dairy Sci.,* 21(6):211–274.

Hannay, E. E., 1928. *Dairying and Civilization.* California Dairy Council, San Francisco.

Herrington, B. L., 1954. Lipase, a review. *J. Dairy Sci.,* 37(7):770–789.

Hoard's Dairyman, 1937. Milk through the centuries. 82 (17):479, 501.

Jenness, R., and S. Patton, 1959. *Principles of Dairy Chemistry.* Wiley, New York.

Kon, S. K., and A. T. Cowie, 1961. *Milk: The Mammary Gland and Its Secretion,* Vol. II. Academic Press, New York and London.

Krosigk, C. M. von, 1959. *Genetic Influences on the Composition of Cow's Milk.* Ph.D. thesis, Iowa State University, Ames, Iowa.

———, G. A. Richardson, and J. O. Young, 1960. Factors influencing the composition of individual samples of cow's milk. *Jour. Dairy Sci.,* 43:877.

Krosigk, C. M., J. O. Young, and G. A. Richardson, 1960. The composition of cow's milk. *Jour. Dairy Sci.,* 43:877.

National Dairy Council, 1962. The nutritional significance of lactose. *Dairy Council Digest,* 33, no. 5, Sept.–Oct.

Landman, D., 1955. Cheese and the man with the tasty tongue. *True; the man's magazine,* 35(212):22–25, 72–76.

Macy, I. G., H. J. Kelly, and R. E. Sloan, 1953. *The Composition of Milks.* Natl. Acad. of Sci., Natl. Research Council, Pub. 254. Washington, D.C.

*Milk Industry Foundation, 1964. *Milk Facts.* Washington, D.C.

*National Dairy Council, 1962. *How Americans use their dairy foods,* Chicago.

———, 1963. The nonfat composition of milk. *Dairy Council Digest,* 34, no. 3, May–June.

Overman, O. R., and W. L. Gaines, 1933. Milk energy formulas for various breeds of cattle. *J. Agr. Research,* 46(12):1109–1120.

———, 1948. Linearity of regression of milk energy on fat percentage. *J. Animal Sci.,* 7:55–59.

Page, L., and E. F. Phipard, 1956. *Essentials of an Adequate Diet.* Agr. Inf. Bull. 160, USDA.

*Pirtle, T. R., 1926. *History of Dairying.* Mojonnier Bros. Co., Chicago.

Regan, J. C., 1940. Milk. *Bull. Medical Soc. of the County of Kings, N.Y.,* 19(7):1–8.

*Rogers, Associates of, 1935. *Fundamentals of Dairy Science,* 2nd Ed. Reinhold, New York.

Rusoff, L. L., 1955. The miracle of milk. *Jour. Dairy Sci.,* 38:1057–1068.

————, 1964. *The Role of Milk in Modern Nutrition.* Borden's Review of Nutrition Research, New York. Vol. 25, No. 2-3, 17-49.

*Sanders, G. P., 1953. *Cheese Varieties and Descriptions.* Agr. Handbook No. 54, USDA.

Shipe, W. F., 1959. The freezing point of milk. A review. *Jour. Dairy Sci.,* 42:1745–1762.

Watt, B. K., and A. L. Merrill, 1963. *Composition of Foods.* Agr. Handbook No. 8, Agr. Res. Service, USDA.

Webster, G., 1959. Nature's strangest chemical. *Codfish, Cats and Civilization.* Doubleday, Garden City, pp. 144–160.

Whittier, E. O., and B. H. Webb, 1950. *Byproducts from milk.* Reinhold, New York.

Chapter 4

Wool and Mohair

We must cut our coat according to our cloth, and adapt ourselves to changing circumstances.

DEAN W. R. INGE, *Anglican Prelate (1860-1954)*

4-1. INTRODUCTION

Those accustomed to handling bulk wool are able to judge its important qualities, such as fineness, length, crimp, "soundness," and color, by feel and appearance. In fact, most wool is bought and sold on the basis of such judgments. Although this information does certainly go a long way toward predicting the behavior of wool during processing and the attractiveness and serviceability of manufactured wool products, there is need for refinement of these judgments. More precise and quantitative information on the wool fibers as a raw material is needed for modification and for manufacture.

Wool is still highly prized as a textile fiber because of its rich appearance, softness, and suitability for tailoring. Garments made of wool are comfortable. The fabrics are relatively soil resistant and easy to clean. They are flame resistant. Because of this combination of desirable qualities—not found in any other textile fiber—wool maintains today a market close to 13% of the 3 billion pounds of all fabrics used annually for apparel goods in the United States. The finest fabrics used for men's and women's clothing are made from wool. America is spending close to 30 billion dollars annually for apparel. The markets for textile fibers are growing along with our population, which is expected to reach 225 million by 1975. Wool's share in tomorrow's markets as always, will be determined by the qualities it offers for the price.

But even though wool continues to be in great demand for apparel, its popularity is being seriously challenged by the man-made fibers. Were it not for our increasing population and rising standard of living, the demand for wool would be diminishing quickly. The products manufactured from these man-made fibers have taken an important share of wool's markets by promoting ease of care and certain other properties, such as shrinkage resistance, quickness and smoothness of drying, wrinkle resistance, moth resistance, abrasion resistance, and durability.

To meet this competition most effectively, it is essential that there be made available as complete a picture as possible of the fiber and its behavior, including fiber damage, distortion, and weakening under the diverse conditions that occur in processing and in use. It is necessary, for example, to know not only the distribution of fiber lengths, diameters, and crimp, but also such information as fiber surface characteristics, chemical stability, stability to heat and light, color characteristics, moisture content, strength, and elastic properties (Fig. 4-1). Such information enables the mill and the scientist to devise better ways to modify and manipulate wool in their endeavor to adapt it to changing circumstances. This is the basis for research aimed to help the wool industry keep abreast of the trend in modern manufacture, which demands greater efficiency and new and superior products made with the highest degree of uniformity.

In the several wool research laboratories throughout the world, scientists are using highly specialized tools to gain the needed information. For example, they are using the optical microscope, the electron microscope, the X-ray diffraction camera, the electron spin resonance spectrometer, the nuclear magnetic resonance spectrometer, the ultraviolet and infrared spectrometers, and specialized test equipment to measure such mechanical properties as strength and elasticity of fiber, yarn, and fabric. Aided by such information, other scientists are using chemical means to modify wool structure, by introducing new chemical groups to the molecules of wool, by attaching resins to the surface, and by blocking chemical centers that cause wool to break down. As a result of all this work, there are being manufactured wools that have new desired properties, such as easy-care performance of fabrics.

This discussion is intended as a glimpse into what the scientists are learning and doing about wool. In general, the information being obtained on wool applies also to mohair, the fiber from the Angora goat. Mohair is similar to wool in origin, structure, composition, stability, and behavior. A main difference is the greater surface smoothness of mohair, which is responsible for its relatively high luster.

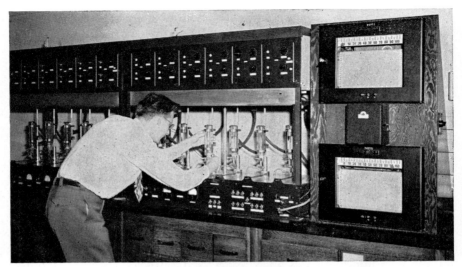

Fig. 4-1. The fiber test instrument is used to study the stress-strain behavior of the wool fibers. It records the force required to stretch a fiber. This force is related ultimately to the structure of the various fiber components. When fibers are tested in this instrument, information is plotted electrically. The information obtained (see lower part of Fig. 4-8) is a load-elongation autograph of the wool fiber. It shows that, as you apply force to stretch a wool fiber, it elongates, first of all very slowly. Finally, it suddenly begins to yield and then elongates more rapidly. It slows down somewhat at around 30% stretch. The important thing is that the fiber recovers its original length if the force is removed. This is one of the characteristics that distinguishes wool from most other fibers. From a comparison of these autographs of wool fibers, some untreated and some purposely modified with chemicals, important information is derived, leading to a better understanding of how wool behaves in processing and in use.

4-2. GENERATION AND MORPHOLOGY OF WOOL FIBERS

Wool is the end product of a remarkable process that occurs in the tubelike sac—the follicle—located in the outer layers of the sheep's skin (Fig. 4-2). The basic materials for the wool fiber are individual pre-wool cells, which are generated continuously in the bulblike base of the follicle. Pre-wool cells are similar to other living cells in that they are spherical in shape and contain a nucleus suspended in fluid protoplasm. The formation of new cells forces the older ones outward through the follicle channel. During this passage, the cells die; some of them elongate in shape, some of them flatten; the nuclei disintegrate, and the protoplasm becomes fibrous; finally, the assembly of modified cells is fused to form a continuous, complex filament—the wool fiber. Although complex in morphology, the greater part of the fiber is made up of insoluble, sulfur-containing proteins, the so-called keratins. The molecules of these proteins differ significantly in size, composition, and properties from one part of the fiber to another. Most of the molecules are threadlike and coiled.

Similar threadlike keratin molecules are the building blocks from which hair, feathers, nails, skin, horns, and hoofs are made. In these substances, too, the chemical composition varies, but the differences are relatively small compared with the wide differences in their physical structures. The manner in which the keratin molecules are cemented and hooked together accounts for the wide differences in the structures of these materials. Similarly, it is true that, by modifying the manner in which the yarns are made and woven, the texture and the performance characteristics of wool can be varied; nevertheless, the unique properties of wool fabrics, such as tailorability, comfort, softness, appearance, and ease of recovery from elongation depend upon specific properties of the keratin building blocks.

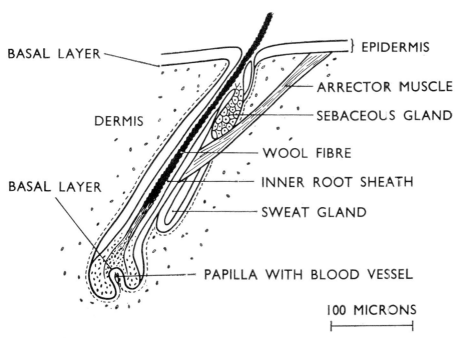

Fig. 4-2. The wool follicle. The wool fiber is formed from the epithelial cells of the follicle.

4-3. WOOL'S MOLECULAR ARCHITECTURE
AND SPECIAL PROPERTIES

Felting of Wool. If we explore wool's structure under an ordinary microscope, observing the outside of the fiber and cut sections of the fiber, we see that wool consists of two distinct components, the outer sheath (the cuticle) and the inner core (the cortex). The scales or cuticle surround the spindle-shaped cortical cells (Fig. 4-3). One kind of cortical cell makes up the hard segment called para cortex; another kind makes up the soft segment, the ortho cortex. The difference in elasticity of the adjoining segments is responsible for crimp in wool. The more elastic layer of ortho cells lies on the outside of the crimp wave. The overlapping scales are responsible for wool's ability to felt. In some wools, especially those from less well-bred sheep, and in most hair fibers, there is a third component—the medulla.

The cuticle is essentially a system of close-packed, flattened cells, commonly called scales. They overlap one another, like tiles on a roof. Because the scale edges are rather well defined, the surface of wool exhibits much greater friction when the fibers are moved in a direction against the scales than when moved in the opposite direction. The net effect is a tendency of wool fibers to entangle when they are moved, especially when wet and soft. The entanglements are called felt. Felting can occur on the sheep's back, where it appears as cots. The felting of wool fibers occurs in yarns and fabrics, as well as in unprocessed wools. Felting is one of the main causes of shrinkage when wool fabrics are laundered. Although such shrinkage is

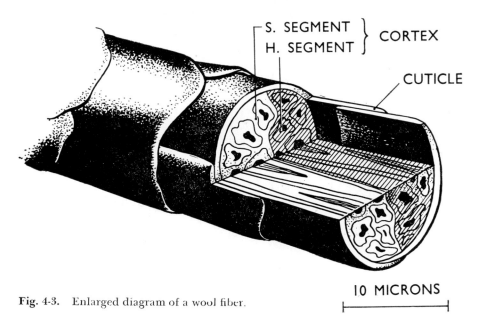

Fig. 4-3.   Enlarged diagram of a wool fiber.

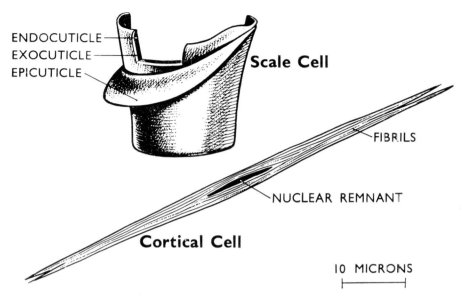

ENDOCUTICLE
EXOCUTICLE
EPICUTICLE

Scale Cell

FIBRILS

NUCLEAR REMNANT

Cortical Cell

10 MICRONS

Fig. 4-4. Wool fiber components. The scale itself is shown. It is made up of at least three distinct structures, the epicuticle, exocuticle, and endo-cuticle.

a very undesirable property of wool, felting can be a useful property in finishing wool fabrics after weaving. If desired, a controlled degree of felting can be applied during finishing, to impart a "body" to the fabric. Such effects are not obtainable in synthetic fabrics. The controlled felting of wool fabrics is called fulling.

Moisture Absorption and Water-shedding Properties. When examined under the electron microscope, another feature of the fiber cuticle is seen, namely, the presence, on the outside surface, of a very thin membrane—the epicuticle (Fig. 4-4). This membrane is protein; yet its composition and structure are such that it repels water, whereas the rest of wool's protein is capable of absorbing relatively large amounts of moisture. Paradoxically, this moisture can pass in and out of the epicuticle. Wool's ability to "breathe" and hold moisture contributes to the comfort in wearing wool garments. And, certainly, the relative ease with which it sheds water is also important.

Morphological Basis for Elastic Properties and Softness. In some fibers the cortex of the wool fiber constitutes as much as 90% of the wool sub-stance. It is responsible for wool's elastic behavior and its crimp. The cortex is essentially a thread of fused, needle-shaped cells. Because of their shape, they are commonly called spindle cells and frequently contain remnants of nuclei. These cells can be separated from wool by treatment with acid or with proteolytic enzymes such as papain (Fig. 4-5). Each cell consists of a bundle of fibrils; the fibrils, in turn, are made up of smaller fibrils, called microfibrils; and the microfibrils consist of bundles of the threadlike keratin molecules. The fibrous elements are embedded in an amorphous cementing material, which has a higher sulfur content and is smaller in molecular size than the threadlike molecules of the fibrils and microfibrils.

The cortex of crimped wool fibers has a bilateral structure; that is, it consists of two distinct regions in which the cortical cells have different elastic properties. The cells of these regions are called ortho cells and para cells (Fig. 4-6). The manner in which these two kinds of cells are distributed determines the degree and character of the fiber crimp. Crimp in wool is important because it contributes to the softness of the fiber and fabrics, and it probably contributes to wool's spinnability.

Fig. 4-5. Spindle cells isolated from wool exhibit special optical properties (birefringence) in polarized light because the thread molecules that compose the cells lie in one direction. The ortho and para cells are not distinguishable in this picture.

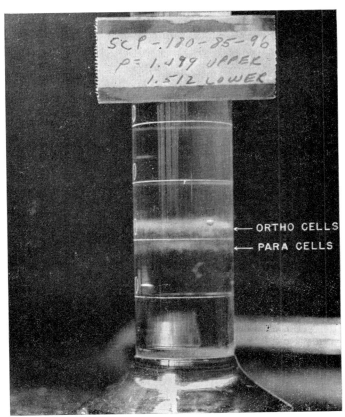

Fig. 4-6. Separation of ortho and para spindle cells is made possible by placing the mixture in a liquid column of varying density, which causes the cells to fall into layers according to their density.

Some wool fibers, and most hair fibers, contain a third main component —the medulla. The medulla consists of a group of air-filled cells and, when present, is found at the center of fibers. Kemp is a kind of hair normally produced by some sheep. It is an extreme form of medullated fiber in which the interior is almost all medulla. Kemp fibers are often flattened and exhibit a shallow crimp. These fibers are highly undesirable in wool because they are chalky in appearance and weak in strength, and exhibit poor dye receptivity. In most well-bred sheep, kemp fibers are virtually absent.

Molecular Basis for Wool Fiber Behavior. Although microscopic methods reveal important information about wool's cuticle, cortex, and medulla, other methods are required to determine the composition, structure, and behavior of the keratin molecules that make up these structural units.

The X-ray diffraction camera shows that some of the threadlike keratin

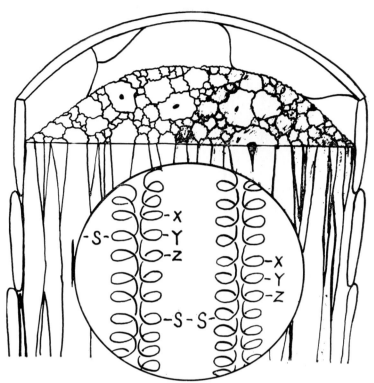

Fig. 4-7. The heart of the wool fiber, shown schematically. In the circle diagram in the center is an illustration of the coiled, threadlike keratin molecules that make up the fibrillar structures in the cortex. The coils are cemented together, back to back, by hydrogen bridges. Groups of coils are tied together by two sulfur atoms of a disulfide bridge. It is interesting to compare natural rubber with wool. Rubber is similarly made up of coiled thread molecules, but the natural rubber that comes from the tree is a sticky material, of no use to man. About a hundred years ago, Goodyear discovered that addition of sulfur would improve the qualities of rubber and make a material that had elastic characteristics. This process became known as vulcanization, a good example of chemical modification of a natural substance. Over the years, the vulcanization process has become more clearly understood. We now have rubber materials with a wide range of properties, depending upon the amount of sulfur that ties together the chains of atoms. The sulfur atoms in the heart of the wool fiber determine its quality in the same way. If the two sulfur atoms are separated (and they split very easily), wool no longer is useful.

molecules are coiled and that they uncoil when wool is stretched. Moreover, when the extending force is released, the threads recoil. This uncoiling and coiling is the basis for the relatively long-range elasticity exhibited by wool. The ability of wool garments to "hang out" wrinkles relates to the recovery of the molecules from uncoiling.

In some chemical environments, the coiled thread molecules in wool become disrupted and the threads collapse into random shapes. The result is a shortening of the fiber. Fibers can be made to shorten as much as 40% by such treatments. This shortening phenomenon is called supercontraction. It results when the chemical environment causes cleavage of chemical bridges that maintain the coiled structure. Two kinds of bridges are recognized: the disulfide bridges |—S—S—| in which two sulfur atoms hold the thread molecules together, and the hydrogen bonds |—H $\cdots$ O=|, which connect hydrogen atoms anchored on one molecule to neighboring atoms, such as oxygen, anchored on another molecule (Figs. 4-7 and 4-8). When wool is exposed to extreme conditions in a solution that facilitates breaking the

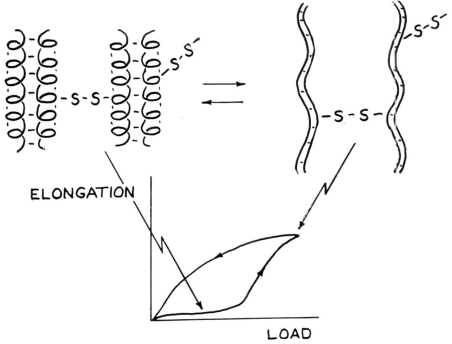

Fig. 4-8. Wool's mechanical behavior in relation to molecular structure. Sulfur atoms keep wool from pulling apart even though the coiled thread molecules are stretched out. The sulfur atoms also help in recovery of the fiber when the force is removed, as illustrated in the load-elongation record shown in this figure.

maximum number of disulfide and hydrogen bonds, the fiber not only supercontracts, but falls apart, thus freeing the keratin protein molecules from one another, which then dissolve in the solution.

Using solutions of the dissolved wool protein, scientists are determining the sizes, shapes, and electrical charges of the molecules and their tendency to interact with themselves and with other molecules. To do this, several kinds of specialized equipment are required. One such piece of equipment is the ultracentrifuge, in which the protein solutions are whirled at speeds up to 60,000 revolutions per minute (Fig. 4-9). The high centrifugal force causes the larger molecules to separate from the smaller ones in a manner similar to the separation of cream from milk. By the use of special optical methods, the rates at which these molecules separate in the solutions are measured. From such measurements, the weights and shapes of the molecules are calculated. The thread molecules of wool differ from those of silk which has no sulfur, yet both are protein. The thread molecules of wool and silk are different from the thread molecules of cotton, which is cellulose and not protein. Because the cellulose molecules are not as flexible, cotton does not exhibit the extensibility of wool.

Similar optical measurements are made with electrophoresis equipment, in which the protein solutions are subjected to electrical fields that separate molecules of different electrical charges.

Fig. 4-9. The ultracentrifuge is one of the instruments used to separate and study the keratin molecules isolated from wool.

Altogether, the information on sizes, shapes, charges, and other properties of molecules, including the interactions of the molecules with themselves and with molecules entering wool, is providing an understanding of the differences in the elastic behavior and other properties observed among wools.

In other studies, keratin protein molecules are degraded into their constituent amino acids by boiling them in acid solution. So far, eighteen amino acids have been identified as building blocks of the wool proteins. Chromatographic techniques are being used to determine the number, kinds, and distribution of the amino acids in the keratins that have been isolated from different parts of the wool fiber.

It is known that the amino acids lysine, histidine, and arginine contribute to wool's affinity for acid dyes; other amino acids, aspartic acid and glutamic acid, contribute to wool's ability to bind basic dyes. Moreover, moisture binding in wool is similarly related to the number, kinds, and distribution of specific amino acids. Important supplementary information on the different degrees of moisture binding in wool is being obtained from studies using nuclear magnetic resonance—a new and powerful physical tool that makes it possible to determine the effects of molecular structure on the mobility of the hydrogen atoms of water.

4-4. IMPROVING WOOL PRODUCTS THROUGH THE MODIFICATION OF PHYSICAL AND MOLECULAR STRUCTURE

Modification of the structure of wool is reducing the particular advantages of the man-made fibers over wool, and the threat these fibers have posed to wool markets is being diminished by the superior wools that are being developed. Scientists are acquiring new knowledge of wool, new techniques for treating and finishing wool, and new and cheaper chemicals. They are finding ways to incorporate superior and durable properties into wool with little treatment. Because only minimum treatment is necessary, they are able to achieve these objectives with minimum sacrifice of wool's natural desirable qualities.

Scientists in laboratories throughout the world have made important progress in the development of wool fabrics that dry more smoothly and have durable pleats and creases. Among the new types of all-wool fabrics that have appeared are foam-backed wool and stretch materials.

Research on wool is under way to explain fiber yellowing and to find effective ways to prevent it. Yellowing of wool fibers can be caused by a number of agents and conditions. It occurs on the sheep's back, in processing, and in use by the consumer. In the fleece, yellow stains are caused by normal urine and fecal pigments, as well as by the end products in urine of phenothiazine, which is used for control of parasites. Fleece yellowing also results from stains caused by pigments from bacteria and fungi growing in the fleece. Similar stains can occur in wool products. Alkali and light can also cause yellowing. Most of these stains cannot be removed by washing and are difficult to remove by present bleaching methods. Bleaching, at best, is costly and damaging to fiber quality.

Each kind of yellowing involves either the chemical interaction of wool protein with a colored substance or the generation in the fiber of a colored chemical group that becomes anchored to the wool protein. As the chemical mechanisms of discolorations become better understood, the efforts to devise effective and economical ways to minimize or overcome these reactions is facilitated. The chemical blocking of amino acid tyrosine residues in wool sometimes inhibits yellowing by light. Work to develop even more effective treatments is in progress.

Important progress has been made toward practical control of yellowing

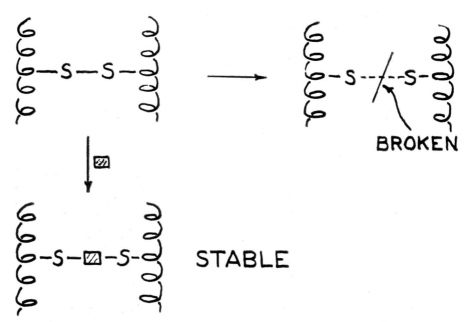

Fig. 4-10. How chemical modification can improve wool's stability to alkaline solutions. The alkali splits the two sulfur atoms that tie together the coils of threadlike molecules in the heart of the cells of these fibers. When we place a fabric in alkali, first it becomes harshened and weakened, and then, on longer contact with the alkali, falls apart. If the fiber is modified by inserting a chemical between these two sulfur atoms (as illustrated at the bottom of the figure), the fiber looks and behaves the same as normal wool, but is stable to alkali. The modification of chemically reduced wool with bismaleimides does this.

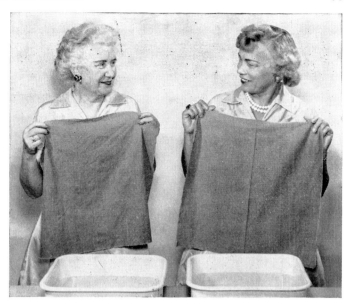

Fig. 4-11. Laboratory comparisons of new treatments for easy-care wool fabrics. The two fabrics shown have just been wetted in warm soapy water. The fabric on the right emerges smooth and with a durable pleat. The control (untreated) fabric on the left emerges mussy and a crease previously present is missing. The treated fabric is shrink-resistant to repeated home laundering.

of damp or wet shorn wool. When the wool is sheared the fleece often is damp, or even wet, and it is always dirty. Damp and dirty, it is baled and kept for perhaps some time. Under these conditions it yellows and eventually weakens. Scientists at the USDA's Wool and Mohair Laboratory at Albany, California find that this undesirable yellowing is caused by a combination of chemical and microbial action and can be prevented or minimized if the wool is dusted with paraformaldehyde or formaldehyde prior to baling. If the aldehyde is equivalent to about 1% of the weight of the wool, it is sufficient to prevent the temperature rise within the bale and to eliminate most of the yellowing.

This is not to suggest that aldehyde treatment makes it unnecessary to observe care and good practice in the baling and storage of fleece wool. However, it is inevitable that some wool will be sheared and baled when wet and dirty; under these adverse conditions, the aldehyde treatment should be of practical value to the sheep raiser, the wool buyer, and the processor.

Special emphasis is being given to finding improved treatments to prevent the shrinkage of wool. Some treatments involve the use of protein-degrading agents such as chlorine, which break down the keratin protein.

Handled correctly, they reduce the sharpness of the scale edges and reduce the differential-friction effect and, hence, the felting shrinkage. The one problem with these agents is that they also decrease fiber strength. A new treatment for control of laundering shrinkage, which does not decrease fiber strength, has been developed by the USDA's Wool and Mohair Laboratory. It is called the WURLAN treatment. It is designed primarily to inhibit shrinkage in machine laundering and performs exceptionally well. Moreover, the treated fabrics are more resistant to pilling and abrasive wear.

WURLAN-treated fabrics have come into large scale commercial use as "machine launderable" men's sport shirts, and women's and children's wear. (The name WURLAN however does not appear on any wool product on the market, because manufacturers prefer to use their own labels.)

The WURLAN treatment employs the same chemicals used to make the man-made fibers. These chemicals react to form ultrathin resin films sur-

Fig. 4-12. Pilot plant for mechanical processing of wool. This room houses equipment for experimental studies of the operations of processing scoured wool into worsted fabrics. The operations include carding, drawing, combing, gilling, roving, spinning, twisting, slashing, beaming, and weaving. With this machinery, research is carried out on the processing of natural and chemically modified wools.

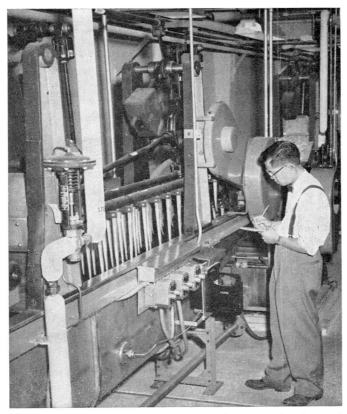

Fig. 4-13. Experimental scouring of raw wool, the first operation in wool processing, involves removal of grease and dirt by washing in an aqueous detergent solution. The first bowl of the scouring train is shown here. In this bowl, warm-water soaking loosens dirt so that it drops through a screen into a settling basin. Chemical agents in the next two bowls remove grease and suint (dry perspiration). The final bowl rinses the wool. Between bowls the wool passes between high-pressure rolls. A "stock dryer" returns the wool to its normal moisture content before it is processed further. One of the aims of scouring studies is to find ways to minimize fiber felting during scouring; another aim is to find better ways of treating scouring-waste liquors.

rounding and chemically grafted to each wool fiber so that they remain permanent through washing and drycleaning. Even though the total weight added to the wool by the film is 1% or less, the resistance to felting shrinkage on laundering of the treated wools is comparable with that of blends that contain at least 50 percent man-made fibers. Obviously the small amount of chemicals in the new man-modified wools is a distinct advantage for wool.

Since the films anchored to the wool are so thin, there is relatively little change in wool properties. For example, the moisture uptake of treated wools is essentially the same as that of normal wool.

New wool products are being developed from wools that are chemically modified to improve their resistance to acids, alkalies, and bleaches. Alkali, for example, splits the pairs of sulfur atoms that tie together the coils of threadlike molecules. Wool fabric is harshened and weakened and falls apart by exposure to alkali. But when the sulfur bridges are reinforced with an appropriate chemical inserted between the sulfur atoms, the treated fabric looks and behaves like normal wool, but is stable in the presence of alkali (Fig. 4-10). Faster and more effective methods of dyeing wool are now possible with the use of special chemicals as dye-assisting agents, which facilitate the penetration of wool by the dye.

Significant advances have been made toward fashioning durable pleats and creases in wool garments. In principle, the setting of pleats or creases

Fig. 4-14. Pilot-scale processing study of easy-care wool fabrics. The fabric is first impregnated with resin by leading it through a trough containing the liquid and then squeezing the cloth between high-pressure rolls. After this treatment it is led through a drying chamber to heat-set the treating chemical. Similar studies are made on variously treated wools to obtain information of direct use to the wool industry.

is the same as the setting of waves in hair. Chemicals are used to expedite rearrangements in fiber structure. During this rearrangement, the disulfide bridges are opened and new bonds are formed. Durable pleats and creases can be set into goods that have received treatment for shrink resistance as well as into untreated goods (Fig. 4-11).

Research is under way to evaluate numerous new and cheaper chemicals and a very large number of so-called textile auxiliaries that are potentially capable of improving the processing of wools and of altering the appearance, feel, and serviceability of fabrics. The number of new and improved wool textiles that may be obtained by chemical modification is very great.

Many possibilities for modification of wool and mohair remain to be explored. Chemical treatments of wool fibers, yarns, and fabrics are resulting in the creation of products that have new and improved performance characteristics while retaining the useful fibrous form of the original materials.

The chemical treatment of wool and mohair thus makes them more adaptable to changing circumstances by minimizing fiber damage in processing and by tailoring these fibers to particular uses. It then enables them to maintain their present markets, to regain those lost to synthetics, and to fulfill new uses as needs arise. The modification of wool and mohair gives great promise of increasing the utilization of these fibers by extending their usefulness (Figs. 4-12, 4-13, and 4-14).

REFERENCES AND SELECTED READINGS

Alexander, P., and R. F. Hudson, 1954. *Wool and its Chemistry and Physics.* Reinhold, New York.

Bergen, W. von, and H. R. Mauersberger, 1948. *American wool handbook.* Textile Book Publishers, Inc., New York.

Harris, M., 1954. *Handbook of Textile Fibers.* Harris Research Laboratories, Inc., Washington, D.C.

Kaswell, E. R., 1953. *Textile Fibers, Yarns, and Fabrics.* Reinhold, New York.

Matthews, J. M., 1947. *Textile Fibers, their Physical, Microscopical, and Chemical Properties.* Wiley, New York.

Meredith, R., 1956. *The Mechanical Properties of Textile Fibers.* North-Holland Pub. Co., Amsterdam.

Moncrieff, R. W., 1954. *Wool Shrinkage and its Prevention.* Chemical Pub. Co., Inc., New York.

Neurath, H., and K. Bailey, 1954. *The Proteins.* Vol. II, Chapter 23. Academic, New York.

Preston, J. M., 1953. *Fibre Science.* The Textile Institute, Manchester, England.

Proceedings of the First International Wool Textile Research Conference, Australia, 1955. Commonwealth Scientific and Industrial Research Organization, Melbourne, Australia.

Proceedings of the Second Quinquennial International Wool Textile Research Conference, England, 1960. The Textile Institute, Manchester, England.

Review of the Textile Progress (Annual Review). The Textile Institute, Manchester, England, and The Society of Dyers and Colourists, Bradford, England.

Speel, H. C., and E. W. K. Schwarz, 1957. *Textile Chemicals and Auxiliaries.* Reinhold, New York.

Chapter 5

Eggs and Poultry Meats

It's as full of good-nature as an egg's full of meat.
RICHARD B. SHERIDAN, *A Trip to Scarborough* (1777)

Until quite recently, the production of poultry meat or eggs was primarily a small operation and was a minor source of income on most farms. However, specialized poultry farms have now come into being, largely as a result of improvements in technology. The poultry industry in 1963 brought a gross income to American agriculture in excess of 3.3 billion dollars. As late as 1950, a sizable percentage of the chicken meat produced was a by-product of the egg industry, but by 1960, the poultry industry had become highly specialized, owing to the development of units for the production of eggs or of chicken, turkey, or duck meat. Seldom are any two phases of the industry found on a single farm.

5-1. THE EGG INDUSTRY

The egg industry in 1963 had a gross income of 1.92 billion dollars from the sale of eggs and, at the end of the laying year, the hens. This was about 56% of the total gross income of the poultry industry. Historically, egg production was concentrated in the north central states. The abundant feed supply was one of the primary reasons for the poultry concentration in that area. Recently, egg production has moved nearer to centers of population as indicated by data in Table 5-1. The production of eggs for freezing and drying is largely confined to the central states. Market egg-production centers are found along both coasts and in recent years have been expanding rapidly in the southeastern states.

TABLE 5-1. | *Ranking of states according to 1963 gross income (in thousands of dollars) from eggs and chickens.*

Rank	State	Value	Rank	State	Value
1	California	201,356	26	Tennessee	30,586
2	Georgia	117,253	27	Washington	30,314
3	Pennsylvania	106,026	28	Kentucky	27,952
4	Iowa	91,428	29	Massachusetts	24,873
5	North Carolina	84,842	30	Kansas	22,314
6	Texas	79,516	31	Louisiana	18,695
7	Alabama	71,831	32	Oregon	17,755
8	Ohio	70,023	33	Oklahoma	14,508
9	Minnesota	66,439	34	New Hampshire	13,919
10	Mississippi	65,533	35	West Virginia	11,800
11	Indiana	61,772	36	Maryland	9,841
12	Arkansas	61,303	37	Utah	8,769
13	New York	60,033	38	North Dakota	8,585
14	New Jersey	59,545	39	Hawaii	8,277
15	Illinois	51,503	40	Colorado	7,927
16	Wisconsin	46,368	41	Idaho	7,317
17	Virginia	44,812	42	Montana	6,369
18	Florida	42,275	43	Vermont	5,837
19	Maine	39,030	44	Delaware	5,548
20	Missouri	38,383	45	Arizona	5,350
21	South Carolina	36,766	46	New Mexico	4,795
22	Michigan	36,613	47	Rhode Island	3,499
23	Nebraska	31,794	48	Wyoming	1,896
24	South Dakota	31 221	49	Alaska	404
25	Connecticut	31,137	50	Nevada	358
				Total:	1,924,290

Source: Agricultural Marketing Service, USDA.

Chemical Composition and Nutritive Value of Eggs. The egg is composed of three principal parts—shell, albumen or white, and yolk. Each of these major parts is subdivided further as shown in Fig. 5-1. The shell and its membranes serve as a package for the edible portions. The package is not perfect and can be penetrated by gases, water, and bacteria; for this reason, care must be exercised to prevent contamination of the contents of the egg or excessive loss of moisture.

The shell of an egg is composed of about 94% calcium carbonate, 1% calcium phosphate and magnesium carbonate, and 4% protein. The two shell membranes, which are composed primarily of protein fibers, form the inner lining of the shell.

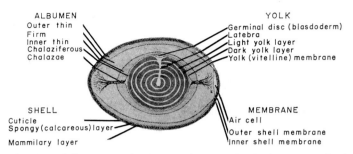

ALBUMEN
Outer thin
Firm
Inner thin
Chalaziferous
Chalazae

YOLK
Germinal disc (blasdoderm)
Latebra
Light yolk layer
Dark yolk layer
Yolk (vitelline) membrane

SHELL
Cuticle
Spongy (calcareous) layer
Mammilary layer

MEMBRANE
Air cell
Outer shell membrane
Inner shell membrane

Fig. 5-1. The parts of an egg. [Source: Agricultural Marketing Service, USDA Handbook No. 75.]

The egg albumen, or white, as observed in fresh eggs, consists of four layers. The layer nearest the shell membrane is the outer thin white. The next layer is the outer thick white. Inside the thick white is the inner thin white; next to the yolk is the inner thick white, which is sometimes referred to as the chalaziferous layer. The chalaza are connected to the inner thick white, which adheres closely to the yolk or vitelline membrane. The chalaza serve as anchors to hold the yolk near the center of the egg. The albumen contains about 87% water, 12% protein, about 1% carbohydrate, and less than 1% minerals. The proteins of the albumen are ovalbumin, ovomucin, conalbumin, ovomucoid, ovoglobulins, lysozyme, and avidin.

The egg yolk is composed of about 48% water, 17.5% protein, 32.5% fat, and 2% minerals. The total chemical composition of the egg is greatly influenced by the individual percentage compositions of the shell, albumen and yolk. The exact percentage that each contributes to the total weight of the egg varies with the age of the eggs, the handling conditions, the strain of chicken, the age and, to a lesser extent, the diet of the bird. Average values are albumen, 60%; yolk, 30%; and shell, 10%.

The high nutritive value of the whole egg is well known. Dried whole eggs are often used as the standard against which other foods are rated. Eggs are an excellent source of amino acids, unsaturated fatty acids, essential vitamins, and minerals.

The exact amount of each of these nutrient groups can be varied widely by modifying the diet of the hen. A detailed review of the studies on the nutritive value of eggs was published by Everson and Souders (1957).

Shell Eggs. Most of the eggs produced are marketed in the shell. Since the shell is not removed until just prior to use, the handling practices that prevail on the farm and on the way to market exert a major influence on egg quality. Grade specifications require that all eggs have clean shells. With good management of a laying flock, a majority of all eggs can be produced with clean shells. In wet weather, or with poor management, when the percentage of eggs requiring cleaning will be high, all eggs are usually cleaned. The cleaning of dirty shells is one of the egg industry's major problems.

When the egg is laid, its fluid contents completely fill the shell. As the egg cools, the contents shrink slightly. This results in the formation of an air space (called the air cell) between the two shell membranes, usually in the large end of the egg. The air cell continues to increase in size as moisture is lost from the egg through the porous shell.

Microbiological Problems. The egg has several natural barriers to microbiological contamination. The first defense is the shell. As long as the egg is dry, bacteria cannot penetrate the shell. However, eggs are moist when laid and are frequently cleaned in water. In addition, moisture condenses on eggs as they are moved from cool areas to warm, humid areas. If bacteria get through the pores of the shell, the membranes serve as another defense. The third line of defense is the albumen, which contains the proteins lysozyme, avidin, and conalbumin. Lysozyme protects the egg by breaking down the bacterial cell wall, and avidin inhibits bacterial growth by tying up the vitamin biotin, which the organisms require for growth. Conalbumin also functions as an inhibitor, by combining with any iron present.

Eggs are normally sterile at the time of production. They are exposed to microorganisms that exist in the nest bedding, on the dust in the air, or on the wires of cages. When eggs are cleaned, the protein material sealing the pores is largely removed. As long as the egg is dry, this presents no problem, but if the eggs are washed, or if eggs are allowed to sweat during marketing, bacteria may penetrate the shell.

Since one of the most emphasized consumer requirements is that the eggs be clean, many different models of egg washers are available. These can be divided into three classes based on method of cleaning: immersion, jet spray, or brush type. Some models incorporate more than one of these methods for dirt and stain removal. In the use of any type of washer, great care must be taken to maintain strict sanitation and to use recommended sanitizers with the cleaners. Eggs can be washed without setting up conditions that will result in rotting; however, according to most reports, bacterial spoilage (either green or black rot) can be traced to improper cleaning methods. Much research has been devoted to egg cleaning, most of which is summarized in a report by Kahlenberg et al. (1952).

Chemical and Physical Factors Related to Shell Eggs. Egg quality is highest at the time the egg is laid. The decrease in quality is a result of chemical reactions. The most commonly observed changes in quality are a thinning of the outer thick albumen and a flattening of the yolk. These changes proceed at varying rates, depending on the temperature and the rate of respiration of the egg. A fresh egg has a pH of about 7.6 when laid.

This increases to about 8.3 in 24 hours and to about 9.2 after 3 weeks. The higher the pH, the more rapidly the albumen thins. The pH rise is largely attributable to loss of carbon dioxide from the egg. The rate can be reduced by using mineral oil to partially seal the pores of the shell.

One physical quality-factor that causes much concern is shell strength. As egg handling becomes more and more mechanized, the strength of the shell becomes an increasingly important factor. The extent of mechanization varies widely among plants. A view of extensive mechanization is given in Fig. 5-2. With this equipment eggs are handled mechanically from the point when they are candled for interior quality until they are sorted for size into the designated container, either a carton for one dozen, or a 30-egg flat.

Two of the factors that determine consumer satisfaction with eggs are flavor and yolk color. Flavor can be altered during storage, particularly if the eggs are stored with odoriferous products, such as apples, cabbages, onions, or gasoline. Yolk color is influenced mostly by the diet of the hen. A feed containing large amounts of highly pigmented ingredients can cause dark orange yolks. Most consumers prefer a relatively light yellow yolk.

Fig. 5-2. Mechanization in egg handling and cartoning. [Courtesy of Food Machinery and Chemical Company, Packaging Division, Riverside, Calif.]

Egg Products. Eggs that do not meet the requirements of the shell egg trade are frequently sold to egg breakers. In times past, the egg breakers relied on farm-flock production to supply their needs. There are indications that, because of the decrease in farm flocks the commercial egg breakers will have to breed flocks to produce specifically for their purposes.

Egg products come in a variety of forms for many applications. The list of egg products, with their specifications and principal uses, shown in Table 5-2 was compiled by Koudele and Heinsohn (1960).

LIQUID AND FROZEN EGG PRODUCTS. A few eggs are sold as liquid nonfrozen products; these are of little importance on the total market, although there is sometimes a local demand for small quantities. The major problems encountered in producing liquid and frozen eggs are sanitation and making a complete separation from the yolks. Egg breaking is accomplished either by hand or by mechanical equipment.

Quality of these products is determined by testing the performance of samples. Albumen is tested to determine the volume of a whipped quantity. Yolk or whole-egg quality is frequently rated on the basis of the deepness of the orange-yellow color. Wholesomeness is indicated by a low bacterial count.

EGG SOLIDS. The production problems of the frozen-egg business also occur in the production of egg solids. Handling of the product during the drying operation creates additional problems; thus special techniques and precautions must be used to insure the yield of a high quality product. Many egg solids are exported, and the importers in other countries demand a product free of *Salmonella sp.*[1] Pasteurization of all egg products for interstate shipment was made mandatory by USDA, effective June 1, 1966.

The name Egg Solids was officially adopted by egg processors for their dried products in 1952. The technological improvements made in the methods used in the production of dried eggs in the late 1940's have made it possible to obtain stable, flavorful egg solids such as albumen, yolk or whole egg.

NEW EGG PRODUCTS. Members of the egg industry are attempting to increase the per capita egg consumption by developing new egg products. The entire food industry is continuing to offer more heat-and-serve and ready-to-eat items. The new egg-rich products are following the same pattern. Among the items suggested as new products are a frozen drink made from orange-egg concentrate, a frozen French toast, a frozen fried egg, and a frozen egg Cantonese. Some convenience egg products now available that will probably gain national distribution are freeze-dried scrambled eggs, a refrigerated

[1] *Salmonella* is a genus of bacteria, many species of which cause severe gastrointestinal disturbances, when they are eaten.

TABLE | *Egg products and their uses.*
5-2. |

Egg Product	Specifications	Principal Uses
Frozen whole eggs	A mixture of whites and yolks in natural proportions, with no additives; contains a minimum of 25.5% egg solids.	In cakes, milk pies, cookies, sweet doughs, and other pastries.
Frozen, fortified whole eggs	Whole eggs to which extra yolks and sugar, salt, or syrup have been added. Made according to packers' own formulas.	Same as for whole eggs
Standard frozen whites	Whites with a minimum of 11.5% egg solids, and fat content not over 0.03%.	In angel food and white cakes, meringues and icings; in candy making.
Quick-whipping frozen whites	Whites specially processed prior to freezing, to produce quicker whipping than regular whites.	In angel food cakes.
Frozen plain yolks	Yolks with a minimum of 45% egg solids and no additives. "Dark Yolk" must show No. 4 or No. 5 NEPA* color.	In egg noodles, which by law must contain $5\frac{1}{2}$% egg solids; also in baby foods.
Frozen sugared yolks	Yolks with a minimum of 43% egg solids, containing 10% sugar.	In cakes, milk pies, sweet goods, and other pastries; in French ice cream; also in baby foods.
Frozen salted yolks	Yolks with a minimum of 43% egg solids. containing 10% salt.	In mayonnaise.
Whole-egg solids	Whole eggs in natural proportions, dried by a spray drier; in powdered form, with 2–4% moisture.	Limited use in cake mixes. Used for human relief feeding.
Glucose-free whole-egg solids	Same as whole egg solids, except that glucose has been removed from the liquid before drying.	Used mostly in school lunch programs.
Fortified whole-egg solids	Whole eggs to which extra yolks and sugar, salt, or syrup have been added. (These additives help the dried product retain its lifting ability in cake making) Made according to packers' own formulas.	In cakes
Flake albumen solids	Albumen dried on pans in cabinet driers; in flake form, with 12–14% moisture.	In candy making.
Standard powdered albumen solids	Flake albumen ground to a fine powder.	In confections, meringue powders, and cake icings; sometimes mixed with spray dried albumen for making angel food cake mixes; also exported.
Spray-dried albumen solids	Whites with fat content not over 0.03%, dried in the spray drier; a fine powder, with 5–8% moisture. Sometimes it is blended with powdered flake albumen.	In angel food cakes; in angel food cake mixes; also exported.
Standard yolk solids	Yolks with 45% solids dried in the spray drier; in powdered form, with 3–5% moisture.	In doughnut and other cake mixes; in sweet doughs and Danish pastry; also in noodles.
Glucose-free yolk solids	Same as standard yolk solids except that glucose has been removed from the liquid before drying.	Same as for standard yolk solids.

* National Egg Products Association.

hard-cooked egg roll, and an egg salad. New egg products are being developed at a constantly increasing rate, but most are still getting only local or regional distribution.

5-2. THE POULTRY MEAT INDUSTRY

The per capita consumption of poultry meat increased rapidly between 1940 and 1960. The actual increase was from 14.1 lb. per person in 1940 to more than 30.0 lb. per person in 1963. There are numerous reasons for this dramatic increase. Several trade organizations waged outstanding promotion campaigns. Poultry breeders improved the market types of meat birds, and the nutritionists formulated more efficient diets. Improvements in management, preventive medicine, and medication were also important factors. The overall effects on production efficiency of broilers can best be shown by comparting the 1940 standard expectations of 5% mortality, the provision of 5 lb. of feed per lb. of meat (live weight), and a 3 lb. bird at 12 to 14 weeks of age with the practical results of 1960. In 1960, broiler growers expected less than 2% mortality, the provision of 2.5 or less lb. of feed per lb. of meat (live weight), and a 3.5 lb bird in about 9 weeks. The improvement in the turkey industry has been comparable. In 1940, poultry processing consisted of killing the birds and handpicking them, either with or without dipping them in scalding water to loosen the feathers. Very few birds were eviscerated in processing plants. The evisceration was left for the butcher or for the housewife. Today, almost all poultry is full eviscerated at the processing plant, and much of it is cut up to be sold as parts.

Fryers or Broilers. Broilers or fryers are young, usually 8- to 10-week-old chickens. The commercial broiler-production centers have shifted over the years. The three states around Chesapeake Bay (Delaware, Maryland, and Virginia) are known as the Delmarva area, which was the first concentrated broiler-production center. Recently, the Southeastern states, principally Georgia and Alabama, have taken the lead in total production. The value of the broiler industry, as determined by gross income, for each state is listed in Table 5-3.

As shown in Table 5-3, much of the poultry is produced at points quite distant from the markets. Chickens cannot be shipped alive economically, so the slaughtering is done close to the site of production, and the birds are shipped completely eviscerated and ready to cook. Processing plants range in capacity from those that process a few hundred chickens a day to others that handle 15,000 chickens per hour. All plants that ship interstate maintain government inspection. Each chicken is individually inspected for wholesomeness, and the federal inspectors see that plant sanitation is maintained.

The processing procedure is similar in all plants. The live chickens are delivered to the plant in crates. The birds are hung by both feet on shackles

TABLE | *Ranking of states according to 1963 gross income (in thousands of dollars) from*
5-3. | *broilers.*

Rank	State	Value	Rank	State	Value
1	Georgia	168,799	26	Oklahoma	5,155
2	Arkansas	117,478	27	Oregon	4,977
3	Alabama	104,748	28	Florida	4,448
4	North Carolina	99,561	29	New York	3,671
5	Maryland	77,393	30	New Jersey	3,502
6	Mississippi	70,242	31	Idaho	3,167
7	Texas	62,432	32	New Hampshire	2,371
8	Delaware	61,704	33	Michigan	2,122
9	Maine	42,140	34	Hawaii	1,873
10	California	37,700	35	Iowa	1,728
11	Pennsylvania	25,223	36	Illinois	1,227
12	Virginia	20,195	37	Utah	970
13	Tennessee	16,771	38	Rhode Island	783
14	Missouri	14,677	39	Kansas	758
15	Indiana	13,633	40	Nebraska	478
16	Louisiana	12,987	41	Vermont	384
17	Washington	11,385	42	Colorado	323
18	West Virginia	10,022	43	Arizona	323
19	South Carolina	9,741	44	North Dakota	—
20	Wisconsin	9,566	45	South Dakota	—
21	Connecticut	7,463	46	Montana	—
22	Minnesota	6,793	47	Wyoming	—
23	Kentucky	6,136	48	New Mexico	—
24	Ohio	5,291	49	Nevada	—
25	Massachusetts	5,263	50	Alaska	—
				Total:	1,058,280

Source: Agricultural Marketing Service, USDA.

attached to a conveyor that takes the chickens to the killing area, where
the throat is cut to allow complete bleeding. The chickens are allowed to
bleed until dead before they are moved into a scalder. The water temperature and the scalding time vary, depending on the equipment used. A scalder
operates at 123°F for a soft-scald. Other commonly used temperatures are
126 to 130°F for a semi-scald, 138 to 142°F for a sub-scald, and, in a few
plants, 150°F or more for a hard-scald. The scalding time is reduced as the
higher temperatures are used. The exact scalding procedure is important in
that excessive scalding reduces tenderness of the muscle tissue (Shannon et
al., 1957) and insufficient scalding makes it very difficult to remove the
feathers. In some plants, the birds are taken through the pickers on the

shackles. Such equipment is shown in Fig. 5-3. After the birds are picked, they are taken from the shackles on the killing, scalding, and picking line and are placed on the eviscerating line. In plants that use a different type of picking equipment, the birds are removed from the shackles prior to picking.

It is necessary to carefully control the picking time and the force exerted by the rubber-finger pickers, since excessive picking results in a less tender muscle tissue. However, it is necessary to remove all feathers and protruding pinfeathers in order to market the birds.

After broilers are picked, they are shackled to the eviscerating line. On this line, a trained government inspector examines each bird for wholesomeness, by looking at the internal organs and the carcass. After inspection, the inedible portions—the head, feet, digestive system (except for the gizzard), and reproductive and excretory systems—are discarded. The heart and liver are cleaned and returned to the packaged bird as edible giblets along with the gizzard. As soon as evisceration is complete, the birds are placed in chilling equipment or in tanks. Poultry meat should be chilled to a temperature of 35 to 40°F as rapidly as possibly. Until 1958, the chilling

Fig. 5-3. Mechanical removal of feathers from chickens. [Courtesy of Gordon Johnson Company, Kansas City, Mo.]

was done by placing the birds in tanks of ice and water. The water was usually agitated by bubbling air through it or by using a circulating pump. In 1958, mechanical cooling equipment was introduced and was rapidly adopted by the industry. Figure 5-4 shows one type of mechanical chilling equipment.

Ice-Packed or Fresh, Dry-Packed Fryers. More than 80% of the broilers or fryers marketed are sold as ice-packed fresh birds. Such birds are shipped long distances. Wire-bound boxes that hold 24 chickens and about 25 lb. of ice are used for most shipments. The shelf-life of the fresh meat is usually determined by the control of dehydration, discoloration, and microbiological spoilage. As long as the birds are kept in ice, there is little chance that dehydration will take place. Microbiological spoilage and discoloration are caused by numerous factors. These are discussed in detail by Dawson and Stadelman (1960).

The most important single factor for prevention of bacterial spoilage in the processing and packaging plant is sanitation. If the poultry is packed in ice or in a dry package with relatively few bacteria or fungi present, it will take longer for them to build up to a number that will cause spoilage. One

Fig. 5-4. The discharge end of a continuous chiller for poultry meat. [Courtesy of Morris and Associates, Raleigh, N.C.]

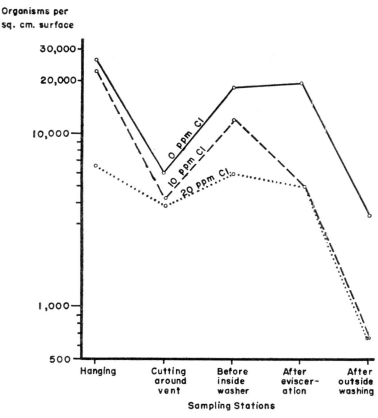

Organisms per sq. cm. surface

Fig. 5-5. The effect of in-plant chlorination of water on bacterial populations. [Source: Dawson and Stadelman, 1960.]

common procedure for improving sanitation is the use of chlorinated water containing about 20 ppm of active chlorine for flushing surfaces of equipment that come in contact with the birds. The value of chlorination in a poultry processing plant is discussed by Gunderson et al. (1954). Figure 5-5 shows the effectiveness of chlorine in an eviscerating room.

A second important factor is the temperature maintained during storage of poultry. At a temperature of 32°F, the shelf-life was 18 days; at 37°F, the bacterial spoilage occurred after 11 days; and at 47°F, spoilage occurred after only 6 days. These data were reported by Shannon and Stadelman (1957). Other effective factors are packaging and the use of antibiotics. (Dawson and Stadelman, 1960).

Much consideration is being given to proper packaging of fresh fryers. Most birds are sold as cut-up birds or as whole, eviscerated birds. The whole birds are usually packaged in polyethylene or cellophane bags. In some stores that offer specials, the birds are displayed on crushed ice, without packaging. The cut-up birds are packed in fiber boats and are overwrapped with various types of plastic films. The most satisfactory films for over-wrapping can be heat sealed. In some areas, large volumes of fryers are sold as fresh parts. This type of product requires packaging similar to that for cut-up poultry.

Roasting Chicken. Roasters are 11- to 16-week-old chickens that are mar-keted at weights of 4 to 8 lbs. It is estimated that about 8% of the broiler chickens are grown to roaster size. The management problems with roaster production are greater than with fryers. One of the major problems is the development of breast blisters, which cause the roaster to be downgraded. The quality of roasters is frequently improved by the use of estrogenic ma-terials during the last six weeks prior to marketing.

Frozen Chicken. Frozen fryers may be purchased in all forms mentioned for fresh fryers. Packages of frozen fryer parts are one of the most popular poultry items. Discoloration or darkening of the long bones and the meat adjacent to them is one of the biggest problems in marketing frozen young chickens. With mature chickens, or even with roasters of 13 weeks of age, this is not a problem, since the bones of older birds do not discolor.

Roaster chickens and stewing hens are usually packaged in flexible film bags. The film must have low vapor transfer, have good strength at low temperatures, and be transparent. Other desirable features are low gas transmission and heat shrinkability or stretchability, which permit tight fits to be attained and maintained.

Frozen chicken keeps well at temperatures of less than 0°F. The effects of time and temperature of frozen storage on the quality of chicken were investigated by Klose et al. (1959). They found that storage temperatures of less than 0°F were necessary to maintain high quality frozen poultry meat.

Precooked Chicken. Among the first precooked frozen foods offered in retail stores were chicken pies. Since this beginning, many other dishes have appeared. Fried chicken as a frozen "warm-and-serve" item is one of the most popular.

The precooked products add to the industry's problems. One of the big-gest problems is to control the flavor so that, when served, the fried chicken or chicken pie will have a "just cooked" flavor. According to Hanson et al. (1959), storage conditions have a big influence upon flavor.

Turkeys. Turkey production is a highly specialized business. The distri-bution of production is shown in Table 5-4.

The turkey industry has gone through several phases. First, production was greatly affected by the disease, blackhead, before the discovery of the

TABLE 5-4. | *Ranking of states according to 1963 gross income (in thousands of dollars) from turkeys.*

Rank	State	Value	Rank	State	Value
1	California	64,073	26	Kentucky	2,666
2	Minnesota	54,624	27	Alabama	2,245
3	Iowa	33,084	28	Massachusetts	2,088
4	Missouri	22,084	29	South Carolina	1,957
5	Texas	18,859	30	Washington	1,600
6	Wisconsin	16,900	31	Connecticut	1,417
7	Ohio	14,766	32	Delaware	1,356
8	Utah	14,471	33	Maryland	1,099
9	Indiana	13,559	34	Montana	1,097
10	Virginia	13,430	35	Idaho	1,097
11	North Carolina	11,563	36	Arizona	994
12	Arkansas	11,307	37	New Jersey	917
13	Colorado	9,858	38	Mississippi	777
14	Pennsylvania	7,501	39	Florida	648
15	Oregon	7,419	40	Tennessee	416
16	Oklahoma	5,652	41	New Hampshire	404
17	Michigan	5,121	42	Louisiana	155
18	Illinois	4,880	43	Maine	147
19	Georgia	4,818	44	Rhode Island	124
20	Nebraska	4,019	45	Vermont	115
21	North Dakota	3,369	46	New Mexico	41
22	West Virginia	3,007	47	Wyoming	14
23	South Dakota	2,995	48	Nevada	—
24	New York	2,902	49	Alaska	—
25	Kansas	2,798	50	Hawaii	—
				Total:	373,336

Source: Agricultural Marketing Service, USDA.

life cycle of the causative organism. Range rotation and the use of medicants have significantly reduced losses from this disease. Then, in 1939, the Broad-Breasted Bronze turkey was introduced at the Seventh World Poultry Congress in Cleveland, Ohio, giving the industry a meatier and more efficient bird. During the following decade, the large bronze birds were by far the most popular variety. The small, family-size varieties were the next to be in demand, the Beltsville Small White turkey being the most popular. Interest has recently centered around a large white variety, whose strains offer the advantages of white pin feathers and a high efficiency of feed conversion. These make good small roasters if slaughtered and processed at 12 to 14 weeks of age. One of the major problems facing the turkey producers is the high cost of poults, which is due to low egg production and poor hatchability. However, research and improved breeding techniques are making progress toward the solution to this problem.

Processing Turkeys. The processing equipment for turkeys is similar to that for chickens. The speed of handling turkeys is lower than that with which chickens are handled. The large plants process as many as 2,000 turkeys per hour. A turkey eviscerating line is shown in Fig. 5-6.

An electric shock is frequently used to reduce struggling of turkeys during slaughter. The turkeys are stunned with high voltage, low amperage currents just before their throats are cut. This humane slaughter treatment reduces the number of bruises on the wings that would otherwise result from struggling at slaughter.

More than 75% of all turkeys processed are frozen. The meat must be adequately aged before it is frozen in order to assure tenderness. The turkeys must be held at temperatures of 45°F or less for about 20 hours. They are usually packaged and frozen on the day following slaughter.

Turkeys are frozen either in an air blast freezer (at −30°F) or in a liquid freezer, in which a solution of propylene glycol or sodium chloride is used as the freezing medium. The solutions are usually maintained at about 0°F. Before they are frozen, dressed turkeys are packaged in one of the several flexible plastic bags available.

Shelf-Life of Frozen Turkeys. Properly packaged turkey meat keeps well when stored at temperatures of less than 0°F. Turkey fat is more susceptible

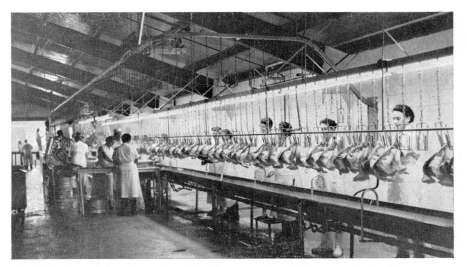

Fig. 5-6. A turkey eviscerating line in operation. [Courtesy of Gordon Johnson Company, Kansas City, Mo.]

to become rancid than chicken fat, since it contains very little natural antioxidants. If special attention is devoted to packaging, thus preventing surface dehydration and "freezer burn," and if proper temperatures are maintained, turkey meat can be stored for 8 months with no detectable rancidity.

A large quantity of turkey meat is used in frozen turkey pies and in turkey rolls. These products are still relatively new, but they are already established as having consumer appeal. Frozen, prestuffed, ready-to-cook turkeys are another product that becomes more widely accepted each year.

Ducks and Geese. The duck production centers of the United States are located on Long Island, New York, and in an area of northern Indiana and southern Michigan. The meat-type Pekin duck is one of the fastest growing of all domestic animals. Many commercially produced ducks weigh as much as 7 lb. at 8 weeks of age.

The largest goose farm in the United States is in New Mexico. Geese are also grown extensively in the north central states. In many localities, geese are grown to be used for weeding row crops. They have been used success-fully for weeding cotton and strawberries.

The large amount of wax in the feathers and the tight feathering make feather removal from ducks and geese difficult. For a cleaner picking job, a process known as wax picking is frequently used. The birds are scalded and partially picked with conventional poultry equipment. They are then dipped into a special hot wax at about 185°F. This is allowed to cool and is then peeled off by hand, removing all down and feathers.

Most ducks and geese are packaged in flexible film bags and are frozen for wholesale or retail distribution.

Other Poultry Meats. There are specialized production centers for Cornish game hens, pheasants, and other game birds. The Cornish game hen is a young broiler-type chicken that is processed at about 5 weeks of age to yield a 1 lb. eviscerated bird.

Game birds are processed, in most instances, with much hand labor. The birds are usually sold as gift packages or to a specialized institutional trade.

By-products. Because the margins of profit available to the egg or poultry processing organizations are constantly decreasing, it is becoming essential to develop uses for what have hitherto been considered waste products. Some progress has been made, but much more attention will have to be devoted to by-product utilization in the future.

Some examples of utilization of waste products from egg processing plants are the developments of technical grade albumen for uses other than food and of animal food products—mink, cat and dog foods—from leakers and blood spot eggs. The poultry processing industry is utilizing visceral waste as frozen mink food and hydrolyzed feathers as a dog food ingredient, and it has succeeded in increasing the number of edible parts by selling feet for human consumption.

REFERENCES AND SELECTED READINGS

References marked with an asterisk are of general interest.

Dawson, L. E., and W. J. Stadelman, 1960. *Microorganisms and their Control on Fresh Poultry Meat.* North Central Regional Publication 112. Michigan State Univ. Tech. Bull. 278.

Everson, Gladys J., and Helen I. Souders, 1957. Composition and nutritive importance of eggs. *J. Am. Dietetic Assoc.,* 33:1244–1254.

*Gunderson, M. F., M. W. McFadden, and T. S. Kyle, 1954. *The Bacteriology of Commercial Poultry Processing.* Burgess, Minneapolis.

Hanson, H. L., L. R. Fletcher, and H. Lineweaver, 1959. Time-temperature tolerance of frozen foods. XVII. Frozen fried chicken. *Food Technol.,* 12:221–224.

Kahlenberg, O. J., J. E. Gorman, H. E. Goresline, M. A. Howe, Jr., E. R. Baush, 1952. *A Study of the Washing and Storage of Dirty Shell Eggs.* USDA Circular No. 911.

Klose, A. A., M. F. Pool, A. A. Campbell, and H. L. Hanson, 1959. Time-temperature tolerance of frozen foods. XIX. Ready-to-cook cut-up chicken. *Food Technol.,* 13:477–484.

*Koudele, Joe W., and E. C. Heinsohn, 1960. *The Egg Products Industry of the United States.* Part I. *Historical Highlights,* 1900–1959. North Central Regional Publication 108. Kansas State Univ. Bull. 423.

Shannon, G., W. W. Marion, and W. J. Stadelman, 1957. Effect of temperature and time of scalding on the tenderness of breast meat of chicken. *Food Technol.,* 11:284–286.

Shannon, G. W., and W. J. Stadelman, 1957. The efficacy of chlortetracycline at several temperatures in controlling spoilage of poultry meat. *Poultry Sci.,* 36:121–123.

Section II

Types and Breeds
of Livestock

Chapter 6

Breeds of Beef Cattle

6-1. HISTORICAL ASPECTS OF BREED INTRODUCTION AND DEVELOPMENT

Stocks Introduced by Early Settlers. Many of the earliest explorers and settlers of the New World brought cattle. The first to do so was Columbus, on his second voyage in 1493 to the West Indies.

Cattle entered what is now the United States by two general routes. Spanish explorers, settlers, and missionaries introduced cattle to the Southwest, to Florida, and to other areas of southern North America. These cattle multiplied and became the bases of the herds of Texas Longhorns (which were of great importance to the nation's cattle industry during the years immediately after the Civil War), of the range herds in a great part of the West, and of the "native" range cattle of Florida and other Southern coastal areas. Settlers from the British Isles and other Northern European countries brought cattle from their homelands into the eastern United States and subsequently took them westward as settlement proceeded across the nation.

Most of the early importations were from the British Isles, but were made before specialized beef breeds as we know them today had been developed. Most were valued more for their milk production and their usefulness as draft animals than for meat. Indeed, ordinances were passed in some New England areas making it a criminal offense to slaughter an animal for beef before it had passed the age of usefulness for other purposes. A specialized beef industry did develop during the colonial period in some eastern areas—notably the Piedmont section of North and South Carolina and Virginia (Towne and Wentworth, 1955).

After the importations by early settlers in the seventeenth century, few additional importations were made for a period of more than 100 years. Historical records, though scanty, indicate that little effort was made to develop improved breeds from the cattle originally imported from Spain or Northern Europe. Writing of cattle in the eastern sections of the nation, Allen (1868) said, "As immigration proceeded from the eastern coast to the interior, their neat cattle went with the people, intermixing still more in their new and scattered localities, until they became an indefinite compound of all their original breeds, and composing, as we now find them, a multitude of all possible sorts, colors, shapes and sizes. Thus our 'native cattle' as we call them, have no distinctive character, or quality, although in some of the States, as a stock, they are better than in others." He gave cattle of Spanish descent even less approbation in saying: "As for the Texas cattle, we do not name them as an economical beast at all. We have only described them to be shunned."

Early American pioneers were presumably too busy subduing a new land and wresting a living from it to pay attention to improvement of beef cattle. When, later in our history, such improvement began to be considered, well-defined breeds were available in other countries, and efforts centered on importing these breeds and grading up native stocks to them. It is useless to speculate now whether native cattle in certain areas may have had virtues that would have been worth preserving. They were not preserved and have contributed very little to the nation's present stocks.

Several experiments were conducted during the first third of the twentieth century in which the performance of "native" cattle was compared with that of purebred beef animals and of purebred X native crosses. In all cases the purebreds and their crosses produced animals more nearly meeting market demands, and the production efficiencies of most were equal to or superior to those of the natives (Cullison, 1940). Thus the decision to import and grade up to purebreds was apparently a sound one, although only a few of the many types of natives were used in the comparisons.

Introduction of Purebreds. Shorthorns brought to this country from England in 1783, 1791, 1812, and 1817, are believed to have been of the breed destined to be an important beef producer. Large numbers were imported during the general period between 1820 and 1850. Many if not most of these Shorthorns were what are today termed "dual-purpose," but they dominated the beef industry of the United States for much of the nineteenth century and their offspring were referred to as the big 'farmers' cow" used for both meat and milk production in farming areas of the country. Bulls of these strains were also used on the western ranges but rather quickly gave way to the Hereford when sizable numbers of this breed became available during the latter part of the century.

During the nineteenth century, in both Britain and the United States, unfortunate pedigree fads and speculation in pedigrees occurred in the Shorthorn breed. These culminated in an 1873 sale in New York in which

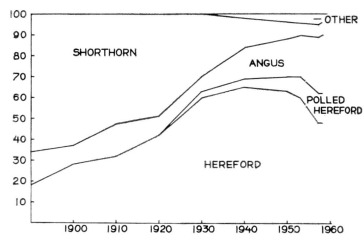

Fig. 6-1. Registrations, 1890–1959, expressed as percentages of total beef cattle registrations accounted for by each breed in each year. Polled Shorthorns are included with Shorthorns. [Graph courtesy of Dr. H. H. Stonaker, Colorado State University.]

109 head of Bates-bred "straight Duchess" cattle averaged $3504 and a cow sold for $40,500. Shortly thereafter the boom collapsed.

In the meantime, Scotch breeders, of whom the most prominent was Amos Cruickshank, had been improving the breed for beef purposes through selection for shorter legs, broader backs, earlier maturity, greater feeding capacity, and thicker and smoother fleshing. Scotch cattle were imported to the United States in numbers in the latter quarter of the nineteenth century and quickly assumed a dominant role in the production of beef by Shorthorns. Breeders have continued to rely heavily on stock imported from Scotland except for brief periods of the breed's history in America.

The first Herefords were imported in 1817 by the statesman Henry Clay of Kentucky, but this English breed did not find its way to this country in appreciable numbers until the general period between 1850 and 1890—especially toward the end of this period. The most prominent pioneer herd in the United States was that of Gudgell and Simpson of Independence, Missouri, which operated from 1877 to 1916; many animals were imported by this firm. Following the dispersal of the herd in 1916, particularly after about 1935, a fad for cattle of Gudgell and Simpson breeding developed; cattle that traced in all lines of their pedigrees to this herd (termed "airtight" pedigrees) were in demand.

Angus cattle first came to the United States in 1873, when George Grant of Victoria, Kansas, imported four bulls. Large enough numbers were brought here during the next twenty years to establish the breed firmly. The

Fig. 6-2. Herefords on a western pasture. This breed has been the most popular on western ranges for many years. [Photo courtesy Soil Conservation Service, USDA.]

Angus attained its earliest popularity in the corn belt states, and this area is still its stronghold. In recent years adoption of the breed has been rapid in other areas, including the western ranges, where it is found in appreciable numbers.

Other British breeds have been introduced from time to time: the Galloway (1853), the West Highland (1893), the Long-Horns (1817), the Sussex

Fig. 6-3. Angus steers in an Iowa feedlot. The Middle West has long been the area of this breed's greatest popularity. [Photo courtesy American Angus Association.]

Fig. 6-4. Polled Hereford heifers on a coastal Bermuda grass pasture in Georgia, an area where the breed is popular. [Photo courtesy Georgia Coastal Plain Experiment Station, Tifton, Ga.]

(1884), and perhaps others. These have either disappeared or have to date been maintained in such small numbers that they have had little or no impact on the cattle industry.

The Charolais, a French breed, is first known to have been imported in 1936 from a herd in Mexico; small numbers have been imported since. Most have come from Mexico because disease problems in France have prevented direct imports. Numbers are being increased through top crossing on cattle of other breeds. Those with five top crosses are considered purebred. The breed is handicapped in the United States by its limited numbers, but it has attracted considerable attention, and cattlemen will follow its future progress with interest.

The only non-European type of beef cattle imported to the United States, and currently of any importance here, is the Zebu. Animals of this type are of the *Bos indicus* species. In the United States they are called "Brahmans." The American Brahman breed represents an amalgamation of several breeds imported from India. Brahman cattle are distinctive in appearance, being characterized by a hump over the shoulders, large drooping ears, and loose skin in the dewlap and navel regions. First known importations were in 1849, 1854, and 1885, but the importations that established the breed were in 1905, 1906, and the early 1920's. Only limited numbers were imported and the breed was built up through top crossing on cattle of other breeds. Animals with four top crosses are registered as purebred.

The Brahman has had its greatest usefulness in cross combinations with the British breeds and with unimproved cattle, and for this reason it probably exerts a greater influence on the nation's cattle industry than would be indicated by registration figures.

Breeds Developed in America. In many parts of the southern and southwestern United States, neither the Brahman nor European breeds are ideal for beef production. The averages of extensive crossbreeding experiments in these areas show conclusively that crosses of Brahman bulls on cows of British breeding will result in calves 25 to 30 pounds heavier at a normal weaning age than British calves; crossbred cows will wean calves 60 to 80 pounds heavier. Postweaning growth rates under winter feeding conditions have not exceeded those of British steers and gains have usually been less efficient, but growth on summer pastures has usually been superior. Few objective carcass data are available, but observation and some objective evidence indicate that the carcasses of crossbreeds are superior to those of Brahmans in finish, conformation, and palatability of lean tissue, and they approach but do not equal British types in these characters. Crossbreds on the average have dressing percentages that are higher than those of the purebred British types and that approach those of the Brahman.

The superior growth rates of the crosses are an excellent example of heterosis or "hybrid vigor," since they have exceeded both parental types (Kincaid, 1962).

In order to eliminate the disadvantages of a continual crossbreeding program, several new breeds have been developed with two objectives: stabilizing as much as possible the vigor, production qualities, and reason-

Fig. 6-5. Purebred Brahman cattle in a pasture near the Texas-Louisiana border. The picture illustrates some of the colors common in the breed. [Photo courtesy American Brahman Breeder's Association.]

Fig. 6-6. A group of Santa Gertrudis yearling heifers, a breed developed in America by the King Ranch, Kingsville, Texas, from a Brahman-Shorthorn foundation. It is usually considered to be the first strictly American breed of cattle. [Photo courtesy Santa Gertrudis Breeders International, P.O. Box 1340, Kingsville, Texas.]

ably good carcass qualities of the Brahman-European crossbred types, and selecting for further improvement in closed populations based on crossbred foundations. The new strains or breeds developed thus far include the Santa Gertrudis, Beefmaster, Brangus, and Charbray. As a group the cattle of these breeds are larger than those of the British breeds; they have only a trace of the hump of the Brahman, but they show definite evidence of Brahman ancestry in ear shape, loose skin in the dewlap and sheath areas, and type of hair coat.

Trends in Breed Popularity. No census data are available to show conclusively the proportions of different breeds of all beef cattle on the nation's farms and ranches. Data on numbers registered in a recent year (Table 6-1) give a general idea of relative importance but are far from precise estimates of distribution in the total population. Figure 6-1 shows trends in numbers registered in various breeds for the first 60 years of this century. In the early 1900's the Shorthorn was the most numerous breed, but it was long ago exceeded in number by the Hereford. In more recent years the Angus has become very popular. The polled (hornless) strains of both Herefords and Shorthorns, established during the early years of the present century, have substantially increased in popularity. In 1964 registrations of the polled strains constituted approximately one-third of the total registrations of each breed.

The data vividly illustrate the fact that breeds change in popularity. In the long run these changes probably depend upon how profitable the breeds

Fig. 6-7. Shorthorn cows on pasture. This breed is the largest in body size of the British breeds. [Courtesy American Shorthorn Association.]

TABLE | *Total 1964 registrations and leading states in registrations by breed.*
6-1.

Breed	Total Registrations	Leading States
All Hereford*	493,160	Texas, Oklahoma, Nebraska, Kansas, Montana
Polled Hereford*	175,045	Texas, Missouri, Illinois, Kentucky, Tennessee
Angus	372,737	Missouri, Texas, Iowa, Oklahoma, Illinois
Shorthorn	29,605	Iowa, Illinois, Kansas, Nebraska, Missouri
Polled Shorthorn	11,243	Illinois, Iowa, Nebraska, Missouri, Kansas
Brahman†	12,382	Texas, Florida, Louisiana
Santa Gertrudis**	14,337	Texas, Florida, Louisiana, Arkansas, Oklahoma
Charolais	4,060	—
Charbray	4,621	—
Brangus	6,548	—
Galloway	2,250	—
Scotch Highland	896	—

* All Herefords, both horned and polled, are registered by the American Hereford Association, which does not report horned and polled types separately. Subtracting polled registrations as reported by the American Polled Hereford Association leaves a total of 318,115 as apparent registrations of horned Herefords.

† 1963 figures.

** Number given is for animals classified rather than registered in the conventional sense.

are in a given era, although skillful breed promotion and effective selling efforts are no doubt factors of importance, especially for short periods.

6-2. HISTORY AND CHARACTERISTICS OF BEEF CATTLE BREEDS

Most of the older breeds (Table 6-2) were the result of development by one or more breeders in an area of a kind of cattle that was useful and that had characteristics distinguishing it from other strains or breeds. Most breeds were developed from mixtures of various native strains and came to be recognized as breeds only after they had achieved at least some degree of popularity outside the area of origin.

Studies of the genetic histories of the Shorthorn (McPhee and Wright, 1925), Angus (Stonaker, 1943 and Sprague et al., 1961), and Hereford (Willham, 1937) all show that rather intensive inbreeding was practiced during the formative stages. Inbreeding was apparently employed intentionally at times in order to intensify desired traits, but it probably also occurred somewhat automatically because numbers were limited and because the best foundation animals were closely related. Whatever the reasons, a few foundation animals in each breed are often closely related to all the other animals in the breed. In general, much less inbreeding has been practiced by modern breeders. In one of the newer American breeds, the Santa Gertrudis (Rhoad, 1955), similar procedures including inbreeding were followed during the formative stages.

The more recently developed American breeds (Table 6-3) are all based on crossbred foundations, with the crosses made for the purpose of founding a breed—usually after exploratory crosses had suggested the desirability of the particular cross.

Breed differences are of two kinds. The readily apparent differences in color, color pattern, and presence or absence of horns serve as breed trademarks and may also have economic value in certain cases. Tables 6-2 and 6-3 summarize information on breed differences of this kind.

Differences between breeds in traits such as fertility, mothering ability, efficiency of gains, or carcass desirability are potentially more important to the person making selection of a breed or breeds for his own operation. Adaptability, or the ability to achieve maximum productivity in a given environment, is also of prime importance.

Research information has given us evidence of some breed differences in such traits, while long-time observations have indicated the probable existence of other such differences. However, no completely objective comparisons of different breeds have been made in which each breed was systematically sampled and representatives observed under similar conditions for productivity. Again, a great deal of evidence and observation shows immense variation in productive characteristics within breeds. Thus,

TABLE 6-2. | *Beef cattle breeds imported to the United States or developed from imported breeds.*

Breed	Origin		Importation to U.S.		Breed Standards	
	Country	Approx. Date	First	In Large Numbers	Color	Polled or Horned
Aberdeen Angus	Scotland	Late 1700's	1873	1878–1900	Black	Polled
Red Angus	United States	Assn. est. 1954			Red	Polled
Charolais	France	Before 1775	1936		Cream to light wheat	Horned
Hereford	Western England	Middle 1700's	1817	1850–1890	Red with white face	Horned
Polled Hereford	United States	Early 1900's			Red with white face	Polled
Shorthorn	Northeast England	Late 1700's	1783	1820–1850	Red, white, or roan	Horned
Polled Shorthorn	United States	Early 1900's			Red, white, or roan	Polled
American Brahman	India	Amer. assn. est. 1924	1849	1905–1925	Varies; steel gray most common, but reds and darker shades occur	Horned

Foundation Stocks column:
- Aberdeen Angus: Stocks native to area
- Red Angus: Red segregates from black Angus herds
- Charolais: Stocks native to area
- Hereford: Stocks native to area
- Polled Hereford: Polled mutants or "sports" from horned herds
- Shorthorn: Stocks native to area
- Polled Shorthorn: Polled mutants or "sports" from horned herds
- American Brahman: Amalgamation of several breeds from India

TABLE 6-3. *Beef cattle breeds developed in the United States from crossbred foundations.*

Breed	Origin			Breed Standards		Date Registry Association Founded
	Founder	Approximate Date	Foundation Stocks	Color	Polled or Horned	
Santa Gertrudis	King Ranch, Kingsville, Texas	First crosses 1910. Intensive efforts to form breed started in 1918	$\frac{5}{8}$ Shorthorn, $\frac{3}{8}$ Brahman ancestry	Cherry red	Horned	1951
Beefmaster	Lasater Ranch, Falfurrias, Texas	1908, efforts intensified in 1930's	Brahman, Shorthorn, and Hereford breeds*	Varies; is not a factor in selection	Horned	1961
Brangus	Clear Creek Ranch, Welch, Oklahoma	1942	$\frac{5}{8}$ Angus, $\frac{3}{8}$ Brahman ancestry	Black	Polled	1949
Red Brangus	Paleface Ranch, Spicewood, Texas	1946	Angus and Brahman†	Red	Polled	1956
Charbray	Several ranches in Texas	Late 1930's	$\frac{1}{8}$ to $\frac{1}{4}$ Brahman and $\frac{7}{8}$ to $\frac{3}{4}$ Charolais	Light cream or wheat	Horned	1940

* Developed by mass selection in multiple sire herds. Percentages of each breed in ancestry not known.
† No fixed proportion of ancestry of two breeds required, but cattle must be of type intermediate between the two parent breeds.

in spite of the fact that there may be some real differences in breed averages, variation within breed is greater. The selection of a breed then becomes of less importance than selection of productive animals within the breed used.

6-3. ECONOMIC AND ESTHETIC IMPORTANCE OF DIFFERENCES BETWEEN BEEF BREEDS

Color. A uniform color in a group of cattle provides an optical illusion of uniformity of type and conformation. A distinctive color or color pattern characteristic of a breed forms a trademark by which it is readily identified; this may serve as an aid in popularizing the breed. It is for these reasons that a majority of breeds have adopted standard colors or color patterns or have limited the number of acceptable colors to a very few. If, however, the color or pattern adopted as standard is difficult or impossible to fix, breeders are forced to practice selection for it and to cull off-color animals regardless of their other merits.

The color of the Hereford may be taken as an example. The distinctive color pattern of this breed—red body with white face, white underline, and other white markings—is of unknown origin, although there are cattle of other breeds on the continent of Europe with somewhat similar markings. The white face is dominant, and the fact that calves of any cross between the Hereford and another breed exhibit it to varying degrees may well have been a factor in the popularization of the Hereford breed. The exact color pattern desired has proved difficult to fix, and many animals have had too much white along the back (linebacks) or too much red on the neck (red necks); in fact, linebacks and red necks still occur occasionally. A genetic study (Stanley et al., 1958) indicates that the preferred amount of white "is the result of an intermediate genetic situation and that it is not likely that selection of breeding stock for this trait will fix the color pattern for this desired intermediate." Also, in the Hereford breed some breeders have preferred a dark and some a light shade of red. Studies at several experiment stations have shown no relationship between shade of color and productivity. To the extent that selection is made for color, selection pressure for traits of greater economic value is reduced.

The black color of the Angus has been a valuable trademark. Black shows a variable degree of dominance in crosses with other breeds. It is dominant over red and in crosses with red Shorthorns, solid black offspring result. In crosses with Herefords, the offspring are black-bodied, with white faces and some white on other extremities. Crosses with white Shorthorns are usually blue roan (a mixture of black and white hairs), with white spotting sometimes occurring. Brahman crosses are usually solid black, and the offspring of crosses with the Charolais are usually a smoky, dark cream hue.

From the beginning of recorded history of the Angus breed, the predominant color has been black, but a gene for red color is present in low frequency as a recessive and occasional red calves are born. Complete elimination of a recessive gene such as this is very difficult. Reds breed true for color when intermated, and breeders in both Great Britain and the United States have assembled red herds. In the United States a breed association was organized in 1954, and the Red Angus is now an established breed. So far as is known the characteristics of the Red Angus were initially the same as those of the parent breed except for color. Different selection criteria in the two breeds could make them markedly different in the future.

Color is related to productivity in some climates. In hot areas with intense sunlight, light coat colors absorb less heat from the sun than darker ones. They thus aid in maintaining normal body temperatures.

White udders may lead to "snow burn" if cows calve in the spring before snow is gone. Pigment around the eyes in white-faced cattle reduces the incidence of cancer of the eyelid.

Horned versus Polled. Cattlemen have long disagreed on the question of whether beef cattle should be horned or polled. Horns make market animals more difficult to handle and also result in bruises. For this reason most market cattle of horned breeds are dehorned. If performed when the cattle are young, this operation is not difficult, and if properly done involves little danger of loss. However, the operation constitutes additional work and there is always the danger of infection or insect injury, particularly in areas where the screwworm is present.

Advocates of the horned type believe that these cattle have advantages in production. The principal advantage claimed is that in multiple sire herds, especially under range conditions, horned bulls scatter and breed their cows, while polled ones tend to gather in groups and spend their time fighting, so that lowered calf crops result. It has also been observed that many polled bulls have a tendency to allow the penis or sheath to protrude, with a resulting possibility of injury and temporary or permanent sterility. It is not known whether this tendency is another effect of the gene for polled cattle or whether it is mere coincidence that it occurs in the polled type.

Some breeders feel that horned cows are better mothers under certain conditions, especially in areas where predatory animals are prevalent.

No studies are known to have been made on any of the foregoing ideas. It is quite possible that economic considerations may favor one of the two kinds under some environmental and management situations, while the second type might be favored in other cases.

In cattle of British origin the hereditary factor for polledness behaves in most cases as a simple dominant, although there are exceptions indicating a more complicated mode of inheritance.

Angus cattle are polled and are apparently homozygous for the gene for polledness, since in crosses with all breeds, except the Brahman or those based on Brahman crossbred foundations, only polled offspring are produced. In crosses of the latter type variable percentages of the calves have horns or heavy scurs, thus further indicating that the inheritance of horns is not a simple one-factor pair situation.

Present-day Polled Herefords and Polled Shorthorns descend entirely from polled mutant or "sport" animals that have occurred in the parent horned breeds. The formation of these breeds (or perhaps it is better to term them strains within breeds) is an interesting example of applied genetics. The Polled Hereford was established by Warren Gammon of Iowa, who in 1901 purchased four polled bulls and seven cows he had located by circularizing all Hereford breeders. Other mutants have been utilized as they have appeared from time to time in horned herds. Horned blood was infused during the formative stages of the breed, both to increase numbers of cattle and to improve the average quality. In the establishment of the breed it was necessary to use all of the few polled animals available, regardless of other merits. Infusion of horned blood has continued to the present time.

Gestation Length and Birth Weight. Gestation periods and birth weights of calves of the same breed vary considerably; for example, the average gestation length in Herefords in different studies has been reported to be from 279 to 286 days (see review by Brakel et al., 1952). It therefore does not seem advisable to attempt to give breed averages for either characteristic. However, trends in within-herd comparisons conclusively indicate that real breed differences exist.

In purebred Angus cattle, gestation length averages three to eight days less and calves on the average are five to eight pounds lighter at birth than the Hereford or Shorthorn. Both traits tend to be transmitted in crosses. Calves from Hereford cows but sired by Angus bulls *on the average* have shorter gestation periods and are lighter at birth than if sired by Hereford bulls (Gerlaugh et al., 1951).

Observations of the above tendencies led to the hypothesis that it would be desirable to breed first-calf heifers of other breeds to Angus bulls, obtain calves with lower birth weights, and thus reduce calving difficulties. Chambers et al. (1954a) studied the question and found that while on the average there was less difficulty in calving of Hereford heifers bred to Angus bulls, there was a great deal of difference between bulls; some Angus bulls sired calves that were bigger and gave more calving trouble than those sired by some Hereford bulls. To minimize calving troubles in first-calf heifers they recommended using small, fine-boned bulls, with choice of breed a secondary consideration.

Purebred Brahman cattle on the average have gestation periods a few days longer than any British breed, and crossing Brahman bulls on British cows tends to result in longer gestations (Wheat and Riggs, 1958) and heavier birth weights (Godbey et al., 1959) than pure British matings.

Milking and Mothering Ability. As discussed more fully in Chapter 14, weaning weight of calves is a very important factor in profitable beef production. Comprehensive comparisons of British breeds have not been made, but scattered data from several experiment stations and on-the-farm testing programs strongly suggest that Angus cows wean heavier calves than Herefords. The data on present-day Shorthorns is uncertain.

In the southern states the Brahman cow and Brahman-British crossbreds wean calves considerably heavier than British types (Kincaid, 1962). Cattle of the newer breeds based on Brahman-British foundations have also been shown to raise heavy calves. The superiority of cows with some Brahman background is one of the principal reasons for the popularity of this type in many southern areas.

Post Weaning Growth. Comprehensive comparisons of the British breeds for ability to gain rapidly and efficiently after weaning, either in the feed lot or on pasture, have not been made. Available data suggest some *average* superiority of the Hereford, followed closely by the Shorthorn and Angus. The Charolais and its crosses have been shown in several tests to grow faster than British types. Rapid growth is one of the traits that have stimulated interest in the breed. Brahman crossbreds and breeds based on them have generally shown superior gaining ability under Southern pasture conditions.

Meat Quality. For a period of many years the Angus breed has enjoyed an unequaled record in carcass competition at leading shows. The winning record apparently depends largely upon the high average ability of animals of this breed to marble well without excessive external fat. Recent summaries of data from all entries in carcass contests and from experiment stations where limited breed comparisons have been made indicate some superiority of Angus carcasses in marbling, rib eye area, and carcass grade. But the margin over other breeds is small, and there is a great deal of overlap.

The Brahman and, to a lesser degree, its crosses have been shown to produce less tender lean than the British breeds but also to have more variability, suggesting the possibility of more rapid improvement (Cartwright et al., 1958). Brahman and Brahman crossbreds usually dress 1 to 3 perceent higher than British-type cattle.

Charolais crosses with the British breeds in experiments of limited scope have demonstrated their ability to produce carcasses with high lean content and apparent good eating quality, as judged by tenderness ratings (Damon et al., 1960; Warwick, 1960). These carcasses, however, have less marbling

than those of British cattle and in several comparisons have rated lower, by
U.S. government grade standards.

Adaptation to Range. Although one breed, the Hereford, has dominated
the range cattle industry for many years, all the breeds are being raised
successfully under range conditions. Since no comprehensive comparisons
have been made, it is uncertain whether the Hereford's adaptation to range
conditions is truly superior, or whether its popularity rests on other factors.

The Hereford originated in an area of west central England where cattle
were produced on grass, with little or no supplementary feeding. It has
been surmised that the environment in its native home may have caused
selection of types especially well adapted to range and pasture management.
It is generally recognized that the ability to thrive under the rigors of the
American range, to survive during periods of adversity such as severe winters,
and to maintain satisfactory fertility records under many conditions are
characteristics that have contributed to the breed's popularity.

During recent years the Angus has been increasing in popularity among
range operators. To what extent this breed will replace the Hereford cannot
be predicted at this time.

Heat Adaptation. Cattle of the British breeds were developed in areas of
temperate climate and most of them do not have a thermoregulatory mechanism good enough to maintain normal body temperatures in a hot environment. The Brahman and its crosses are superior in their ability to tolerate
heat (Rhoad, 1940). This characteristic is thought to be one of the reasons
Brahmans and their crosses are able to make more rapid summer pasture and
feedlot gains in many areas (Cartwright, 1955).

Problems of beef production under subtropical conditions are complicated,
however; in addition to heat tolerance, there are such factors as quality of
pasture forage and insect and parasite infestations. Having animals with
ability to withstand heat does not automatically solve all beef production
problems in these areas.

Hereditary Defects in Beef Cattle. Many hereditary defects have long
been known in both beef and dairy cattle (see Rice et al., 1957, and Gilmore,
1950, for listings). Few defects have been frequent enough to constitute
major problems, although they have occasionally been serious enough in
some breeds to result in extensive culling of animals related to those producing defective animals. However, until about 1940, it had been generally
assumed that such defects were of little importance to industry. Events since
then have made it necessary to modify this view.

During the late 1930's and 1940's a small extra-low-set, compact type of
Hereford, generally known as "Comprest"—and apparently due to a semi-
dominant mutant gene resulting in an extreme type of dwarf when homozy-

gous (Stonaker, 1954)—appeared in the breed and for a time was popular in show rings. Its practical usefulness was so much in question that research on this type was conducted at several experiment stations. This research, as well as the experience of practical cattlemen, showed conclusively that these cattle had little or no carcass superiority and that they had several serious production difficulties—including low weaning weights, slow growth, and production of dwarfs when intermated (Stonaker et al., 1952; Chambers et al., 1954b; Stonaker, 1954). They very quickly lost their popularity. A type known as "compact" in the Shorthorn breed had a similar history. These types were easy to eliminate since animals carrying the gene were phenotypically recognizable and could be culled.

At about the same time, or a little later, a recessive hereditary type of dwarfism (Johnson et al., 1950), popularly known as "snorter," became a problem among Hereford and Angus cattle and possibly occurred in other breeds. Dwarfism of this type has undoubtedly caused more turmoil among purebred beef cattle breeders than any other factor. Research has clearly established that "snorter" dwarfism is inherited as a recessive. Why it occurs frequently enough to constitute a problem is not conclusively established. However, research at several stations indicates that *on the average* the gene for dwarfism may have some effect in the heterozygous condition and that normal animals carrying the gene in the recessive or hidden form may be slightly shorter in body, head, and legs (Marlowe, 1964). These effects are not definite enough to be diagnostic for the presence or absence of the gene in individual animals, but they may be great enough so that if selection were to be based on these characteristics, carriers would be favored in a greater than chance proportion in the selection of breeding stock, with consequent increase in frequency of the gene. The alternative view is that dwarfism increased in frequency because several prominent animals happened to be carriers and the popularity of their progeny resulted in wide dissemination of the gene.

Much research has been done in attempts to find anatomical or physiological methods of identifying normal carriers of the dwarf gene, but to date no successful method has been found (Bovard, 1960). Breeders have used "clean pedigree" animals—animals with no known carriers in their pedigrees—to reduce the incidence of dwarfism. This has resulted in culling many otherwise desirable animals who had one or more known carriers in their pedigrees. Some progeny testing of bulls to locate noncarriers has been done by breeding them to groups of known carrier cows.

Other types of dwarfism are also known (Bovard, 1960) but none are believed to have been frequent enough to present major problems. There is disagreement among research workers on the genetic relationships of dwarfs of these types to the snorter.

Occurrence of another category of defects depends partly upon hereditary tendencies toward susceptibility and partly upon environmental effects.

Cancer eye is an example of this kind (Anderson, 1960), and it has been a particular problem in the Hereford breed. Selection against cancer eye is difficult since it occurs most frequently in older animals, which may already have several progeny in a herd or dispersed among other herds. The fact that it is partially hereditary, however, indicates that culling the affected animals and their progeny and other close relatives should reduce its frequency. Uterine prolapse is a defect whose elimination is beset with many of the same problems (Woodward and Quesenberry, 1956). Maintaining intensive selection on traits of this kind would greatly reduce the selection intensity which could otherwise be put on other traits of economic importance.

The problem of animals that have long sheaths or habitually allow the penis to dangle from the sheath has been referred to earlier in connection with the polled gene. These conditions are not limited to polled cattle. The Brahman breed and many of the newer breeds based on Brahman crossbred foundations have these problems. No studies are known to have been made on the heritability of these conditions, but observation suggests hereditary differences. If these exist, selection for improvement should be effective.

Some beef animals have dispositions that make them difficult to handle under normal management conditions. All breeds have animals of this kind, but they are apparently more frequent in the Brahman and related breeds and in the Angus. Again, no scientific studies have been made on heritability of differences in temperament, but observation suggests that it is desirable to select docile animals for breeding purposes.

6-4. SUMMARY

The historical material on the beef cattle breeds used in the United States and the discussion of the popularity and characteristics of the various breeds make it apparent that all breeds have both strong points and weaknesses and that there is no one best breed for all conditions. Although real differences exist between breed averages, there is much hereditary variation within all breeds. Selection of inherently productive animals within a breed is likely to be of as much or more importance than selection of a breed.

REFERENCES AND SELECTED READINGS

References marked with an asterisk are of general interest.

*Allen, Lewis F., 1868. *American Cattle, Their History, Breeding and Management.* Orange Judd Co., New York.

Anderson, David E., 1960. Studies on bovine ocular squamous carcinoma ("cancer eye"): V. Genetic aspects. *J. Heredity* 51(2):51–58.

Bovard, K. P., 1960. Hereditary dwarfism in beef cattle. *Anim. Breeding Abstracts,* 28(3):223–237.

Brakel, W. J., D. C. Rife, and S. M. Salisbury, 1952. Factors associated with the duration of gestation in dairy cattle. *J. Dairy Sci.*, 35(3):179–194.

*Briggs, Hilton M., 1958. *Modern Breeds of Livestock*. Revised ed. Macmillan, New York.

Cartwright, T. C., 1955. Responses of beef cattle to high ambient temperatures. *J. Animal Sci.*, 14(2):350–362.

———, O. D. Butler, and Sylvia Cover, 1958. Influence of sires on tenderness of beef. *Proc. 10th Res. Conf.*, Amer. Meat Instit. Found., pp. 75–79.

Chambers, Doyle, J. A. Whatley, Jr., and W. D. Campbell, 1954a. A study of the calving performance of two-year-old Hereford heifers. Okla. Agr. Exp. Sta. Misc. Pub. MP-34, pp. 39–43.

Chambers, Doyle, J. A. Whatley, Jr., and D. F. Stephens, 1954b. Growth and reproductive performance of large- and small-type Hereford cattle. Okla. Agr. Exp. Sta. Misc. Pub. MP-34, pp. 50–54.

Cullison, A. E., 1940. *The Influence of Breeding on the Performance of Beef Calves Produced in Mississippi*. Miss. Agr. Exp. Sta. Bull. 347.

Damon, R. A., Jr., R. M. Crown, C. B. Singletary, and S. E. McCraine, 1960. Carcass characteristics of purebred and crossbred beef steers in the gulf coast region. *J. Animal Sci.*, 19(3):820–844.

Gerlaugh, Paul, L. E. Kunkle, and D. C. Rife, 1951. *Crossbreeding Beef Cattle*. Ohio Agr. Exp. Sta. Research Bull. 703.

Gilmore, Lester O., 1950. Inherited nonlethal anatomical characters in cattle—a review. *J. Dairy Sci.*, 33(3):147–165.

Godbey, E. G., W. C. Godley, L. V. Starkey, and E. D. Kyzer, 1959. Braham × British and British × British Matings for the Production of Fat Calves. S. Carolina Agr. Exp. Sta. Bull. 468.

Johnson, L. E., G. S. Harshfield, and W. McCone, 1950. Dwarfism, an hereditary defect in beef cattle. *J. Heredity*, 41:177–181.

Kincaid, C. M., 1962. Breed crosses with beef cattle in the South. *Southern Cooperative Series Bull*. 81.

Marlowe, Thomas J., 1964. Evidence of selection for the snorter dwarf gene in cattle. *J. Animal Sci.*, 23(2):454–460.

McPhee, H. C., and S. Wright, 1925. Mendelian analysis of the pure breeds of livestock. III. The Shorthorn. *J. Heredity*, 16:205–215.

*Ornduff, D. R., 1957. *The Hereford in America*. Printed privately by the author.

*Rhoad, A. O., 1940. A method of assaying genetic differences in the adaptability of cattle to tropical and subtropical climates. *Empire J. Exp. Agr.*, 8:190–198.

*———, 1955. Procedures used in developing the Santa Gertrudis breed. In *Breeding Beef Cattle for Unfavorable Environments*. Univ. of Texas Press, Austin, Tex., pp. 203–210.

*Rice, V. A., F. N. Andrews, E. J. Warwick, and J. E. Legates, 1957. *Breeding and Improvement of Farm Animals*. 5th ed. McGraw-Hill, New York.

*Sanders, A. H., 1918. *Short-horn Cattle*. Sanders Pub. Co.

*———, 1928. *A history of Aberdeen-Angus cattle*. New Breeders' Gazette, Chicago.

Sprague, J. I., Jr., W. T. Magee, and R. H. Nelson, 1961. A pedigree analysis of Aberdeen Angus cattle. *J. Heredity* 52(3):129–132.

Stanley, M. E., D. Chambers, and D. E. Anderson, 1958. *Inheritance of Color Pattern and Shade of Hair Color in Hereford Cattle*. Okla. Agr. Exp. Sta. Misc. Pub. MP-51, pp. 50–54.

Stonaker, H. H., 1943. The breeding structure of the Aberdeen Angus. *J. Heredity*, 34:323–328.

———, 1954. Dwarfism in beef cattle. *Proc. West. Sect. Amer. Soc. Animal Prod.*, 5:239–242.

———, M. H. Hazaleus, and S. S. Wheeler, 1952. Feedlot and carcass characteristics of individually fed comprest and conventional type Hereford steers. *J. Animal Sci.*, 11:17–25.

*Towne, C. W., and E. N. Wentworth, 1955. *Cattle and Men*. University of Oklahoma Press, Norman, Okla.

*Warwick, E. J., 1958. Fifty years of progress in breeding beef cattle. *J. Animal Sci.,* 17(4):922–943.

*———, 1960. Genetic aspects of production efficiency in beef cattle. Proc. of Conf. "Beef for Tomorrow," Natl. Aca. Sci., Natl. Research Council Pub. 751, pp. 82–92.

Wheat, J. D., and J. K. Riggs, 1958. Heritability and repeatability of gestation

length in beef cattle. *J. Animal Sci.,* 17(1):249–253.

Willham, O. S., 1937. Genetic History of Hereford Cattle in the United States. *J. Heredity,* 28:283–294.

Woodward, R. R., and J. R. Quesenberry, 1956. A study of vaginal and uterine prolapse in Hereford cattle. *J. Animal Sci.,* 15(1):119–124.

Chapter 7

Breeds of Dairy Cattle

"Those who survey the work done in this department will arrive at the conviction that among all of the experiments made, not one has been carried out to such an extent and in such a way as to make it possible to determine the number of different forms under which the offspring appear, or to arrange these forms with certainty according to their separate generations, or definitely to ascertain their statistical relations."

GREGOR MENDEL

7-1. INTRODUCTION

Dairy cattle are well adapted to the types of agriculture practiced in the United States. They are found in every state. According to the latest estimates, dairy cows are found on about 50% of the farms; many of the small herds are on farms that produce other agricultural products. There is a marked tendency for dairy herds to be concentrated around densely populated areas, where they furnish a supply of fluid milk.

Approximately one-third of the cattle in the United States are kept for the production of milk. Of these, approximately 80% belong to the major dairy breeds, 15% belong to the dual-purpose and minor dairy breeds, and 5% are beef cattle or "native" cattle lacking the characteristics of any particular breed. As our agriculture becomes more specialized, a greater percentage of the cattle kept for milk production will probably belong to the specialized dairy breeds. Consumer preferences—combined with high costs for labor, equipment, land, and other items used in the production of milk—have created an economic situation conducive to the production of a

high volume of milk of a composition conforming to the market requirements.

Taxonomically, cattle belong to the class *Mammalia,* the order *Arterio-
dactyla* (even-toed hoofed animals), and the suborder *Pecora,* which includes
deer and giraffes. This suborder is characterized by a specialized four-
compartmented stomach, which enables these animals to digest and utilize
roughages. Cattle belong to the family Bovidae, which includes sheep, goats,
and antelopes, and are characterized taxonomically by hollow horns that
contain a bony core and do not shed. Cattle are further grouped in the
genus *Bos,* which differentiates them from the other genera of this family.
Our domestic cattle belong to two species within this genus. One is the
species of European origin, *Bos taurus,* also called *Bos typicus primigenius.*
These animals have a flat forehead and poll with the horns growing from
the junction of the lateral and posterior borders of the skull, a small dewlap,
and a lowing cry. The second species—*Bos indicus,* commonly called "zebu"
or "Indian cattle"—have the large hump over the withers, unlike the cattle
of European origin, a large dewlap, large drooping ears, and a grunting
voice. Most of the dairy cattle in the United States belong to *Bos typicus,*
but animals of the Red Sindhi breed, belonging to the *Bos indicus* species,
are being used experimentally in crossbreeding for increased heat tolerance
in the southeastern United States.

The immediate ancestor of our major breeds of dairy and beef cattle
is thought to be the great ox of Europe, *Bos primigenius.* Some authorities
consider that another distinct species, *Bos brachycerus,* also called the Celtic
Ox or Celtic Shorthorn, was the ancestor of some of our modern breeds.
Bos primigenius became extinct during the seventeenth century; the last
survivors were found in the forests of north central Europe. The semiwild
"White Park Cattle" found in England are considered to be descendants
of the *Bos brachycerus* species.

Whether our breeds arose from one species or two has little bearing upon
the improvement of our modern breeds but it is historically interesting. Information on the relationship of the "White Park Cattle" of England to
our modern breeds could be obtained by comparing their respective "blood
groups." The blood groups are based upon antigens in the red blood corpuscles and agglutinins in the blood sera, which, when mixed with blood
from other animals, cause clumping and hemolysis of the red cells. Their
presence in the blood of an animal is controlled by inheritance, and the
finding of an identical blood group in two different animals indicates that
these two animals are genetically alike for this factor; that is, both animals
have the same genetic factors responsible for this characteristic and are
therefore related. If some identical blood groups should be found in these
semiwild cattle and our modern breeds it would indicate either that our
modern breeds descended from these animals or that both had common an-

cestors; the number of similar groups would give an estimate of the extent of their relationship.

7-2. IMPORTATION OF CATTLE

The early importation of cattle into the United States is discussed in Chapter 6. Although the cattle imported by the early settlers were probably representative of the kinds found in the different areas from which the colonists migrated, they were interbred and soon lost their identity. It is not surprising that the early settlers knew nothing of breeding, because the mechanisms governing reproduction and inheritance had not been discovered. Hamm discovered the spermatozoon in 1677, and Von Baer discovered the egg in 1827. Spallenzani showed in 1780 that the spermatozoa were necessary for the genesis of the young, but the actual process of fertilization was not understood until about 1875. Mendel discovered the fundamental laws governing inheritance in 1866, but his discoveries were ignored when first published, and not rediscovered until 1900. Thus the modern concepts of breeds and breeding are of recent origin and early animal improvement was based primarily on observations of men who were breeding the animals for various uses.

Fig. 7-1. Ayrshire cow, Neshaminy Miss Phett. She produced 166,941 lb. of milk and 7,646 lb. of fat in 10 lactations. [Courtesy Ayrshire Breeders Association.]

TABLE 7-1. | *Origin, importation, and characteristics of the major breeds of American dairy cattle.*

	Ayrshire	*Brown Swiss*	*Guernsey*	*Holstein*	*Jersey*
Country of origin	Scotland	Switzerland	Island of Guernsey	Holland	Island of Jersey
First importation to U.S.*	1822	1869	1830	1852	1850
Breed association formed†					
European	1877	1911	1814	1873	1833
U.S.	1875	1880	1877	1871	1868
Herdbook established†					
European	1877	1911	1822	1873	1866
U.S.	1875	1889	1877	1885	1868
Body size (lb.)**					
Female	1,200	1,400	1,100	1,500	1,000
Male	1,850	2,000	1,700	2,200	1,500
Av. gestation period (days)	277.9	289.7	283.9	278.8	278.8
Desirable color markings	Red, mahogany brown, or combination of these with white	Solid brown	Fawn, with white markings, yellow skin	Black and white	Fawn, with or without white markings

* Earlier importations occurred but available records indicate these later importations eventually formed the nucleus for the present-day breeds.
† Approximate dates.
** Minimum mature weight in milking or breeding condition. Purebred Dairy Cattle Association.

The second wave of importation of dairy cattle into the United States began about the middle of the nineteenth century (Table 7-1). The cattle imported at this time were probably representative of the improved cattle in their respective areas, and following importation they were so mated as to maintain their identity. Closely associated with these importations was the formation of the breed associations and the establishment of herd books. Some individual breeders kept private herd books for pedigrees. The breed organizations were established primarily to record pedigrees, to keep the breeds "pure," and to protect the importers from unscrupulous dealers. The present breed organizations continue to register cattle and keep pedigrees, but they perform other functions, which are discussed in greater detail in Section 7-6.

The country of origin and the approximate foundation dates of the modern breeds of dairy cattle are given in Table 7-1. It is of interest to note that some of the herd books were established earlier in the United States than in the country of origin.

7-3. BREEDS

A breed may be defined as a group of animals related by descent and developed for a special function. Thus, dairy cattle breeds are breeds developed primarily for milk production. In the United States, cattle kept primarily for milk production belong to the Ayrshire, Brown Swiss, Guernsey, Holstein, and Jersey breeds. Figures 7-1 to 7-5 are photographs of cows of these breeds. In addition, two dual-purpose breeds, the Milking Shorthorn and Red Polled, are kept for milk, but their numbers are small in comparison with the previously mentioned dairy breeds. The number of animals registered in each of the different breeds for the year 1964 is as follows: Ayrshire, 16,631; Brown Swiss, 17,292; Guernsey, 75,332; Holstein, 262,419; Jersey, 40,885; Milking Shorthorn, 8,327; a total of 395,783 head. This represents a drop of about 15% from the 1959 figure. If we assume that the above figures are correct, and that the average life of the cow in the herd is five years, the estimated number of registered dairy cattle in the United States

Fig. 7-2. Brown Swiss cow, Lee's Hills Keeper's Raven. She produced 34,851 lb. of milk, 1,579.3 lb. of fat, 3 times milking in 365 days. [Courtesy the Brown Swiss Breeders Association.]

should be about 1,978,915. The latest estimate of the total number of cattle kept for dairy purposes is about 33,000,000; thus registered cattle constitute approximately $6\frac{1}{2}\%$ of the total number of dairy cows. Data on the actual numbers of cows belonging to the different breeds are not available, but the Holstein-Friesian Association estimates that Holsteins, both purebred and grade, account for approximately 60% of the dairy cattle in the United States.

Cattle were kept for dairy purposes long before the modern breeds developed. For example, in Holland butter production from cow's milk became so important that specialized buildings, "butterhouses," were constructed for butter storage and distribution before 1288. Students of breed history do not agree on the origin of the different breeds of dairy cattle, but animals conforming to our present idea of a specialized breed probably existed in certains areas for various periods before they were formally recognized as a distinct breed. The consensus of opinion is that most of the modern breeds were probably formed from crossbred foundation stock—that is, the "native" animals were crossed with animals from other areas. This was probably followed by selection within the group for animals with the desired characteristics, but the methods and precision of the selection cannot be determined. In the eighteenth century, Bakewell was one of the first men to

Fig. 7-3. Guernsey cow, Lush Acres Hermes' Quest. She produced 14,651 lb. of milk and 900 lb. of fat in 305 days, 2 times milking. [Courtesy American Guernsey Cattle Club.]

Fig. 7-4. Holstein cow, Princess Breezewood R A Patsy. She produced 36,821 lb. of milk and 1,866 lb. of fat in 365 days, 2 times milking. [Courtesy Holstein Friesian Association of America.]

apply modern methods to the improvement of animals. Although he worked with beef cattle, his ideas—"Like begets like" and "Breed the best to the best"—were probably applied to the early selection of cattle for dairy purposes and played a role in the selection of the animals that eventually became the dairy breeds. Laws prohibiting the importation of cattle to the islands of Guernsey and Jersey probably were important in the formation of these two breeds as we know them today, because these laws prevented crossing of the island cattle with others so that the uniformity in traits of the cattle on each island was maintained.

7-4. BREED CHARACTERISTICS

Inherited characteristics form the bases on which the individual breeds are distinguished. Those traits that are easily identified and inherited in a simple manner characterize the different breeds. Color markings fall into this category and are the trademarks of many breeds. As already mentioned, the origins of our modern breeds are not known but the information available suggests that the circumstances of origin of the various breeds were quite similar. Foundation animals from which the breeds arose were apparently crossed with animals from other areas. These crosses with widely differing animals increased the number of different genes in the foundation stocks

Fig. 7-5. Jersey cow, Marlu Milady's Fashion with a lifetime production record of 199,084 lb. of milk and 9,908 lb. of fat. [Courtesy of the American Jersey Cattle Club].

and formed a wide genetic basis from which animals with the desired characteristics could be selected. Selection of animals that were superior for a given productive characteristic and breeding them together tended to concentrate the genes governing this trait, with resulting improvement. This selection automatically caused them to become more alike genetically than nonselected animals; that is, more homozygous than animals bred randomly. As a result of the formation of the breed societies, which record the ancestry of animals, the matings of more closely related animals could be controlled, and the process of standardization of the animals within the group accelerated. When the genes controlling characteristics are few in number and their inheritance is simple, the animals become alike very quickly because a larger proportion of the animals become more homozygous for the characteristic. However, when a large number of genes control the characteristic, the standardization of a group of animals is more difficult and the process requires a longer time. All of our present-day breeds are still heterozygous or "mixed" for the genes controlling size and productive characteristics.[1]

---

[1] For a more thorough explanation of genetics and inheritance, see Chapter 12, *Introduction to Modes of Inheritance.*

Blood Types. The general heterozygosity of the different breeds has been demonstrated clearly by the blood typing of cattle during recent years. Neimann-Sørensen et al. (1956) reported 47 alleles of the B group blood type in three Danish breeds of cattle. In the five American dairy breeds, 89 alleles have been reported for the same system. On the other hand, the Jersey breed in Denmark has the same B alleles, except for two, that are found in Jerseys in the United States. Thus the Jersey breed in Denmark is more closely related to the Jersey breed in the United States than to other breeds of cattle in Denmark. Stormont (1958) compared the frequencies of alleles between Guernseys, Herefords, Holsteins, Jerseys, and Shorthorns. He found differences in frequencies of the alleles between the breeds and between strains within the breeds. Because genetic differences are present in individuals within the present-day breeds, selection for higher production is effective. If a breed should become completely homozygous, then improvement by selection would not be possible within the breed because all of the animals would be alike. Many genes control most of the productive traits and their random distribution in the offspring suggests that the production of a homozygous group or breed of animals is unlikely. If, however, some easily identified traits, for example, certain blood groups, can be associated with a productive characteristic, more rapid selection for that productive characteristic would be possible and selection would be more effective.

The many factors that determine blood types are also important in the identification of the parents of an individual. If the blood types of the calf, its dam, and the probable sires are obtained, it is almost always possible to identify the sire accurately, because only one of the individuals in question can transmit all the factors producing the specific blood type found in the calf. A notable exception will occur if the unidentified sires are identical twins.

Body Size. A wide variation in size exists between breeds. The minimum weights for cows and bulls of the five major dairy breeds, as published in the Unified Score Card of the Purebred Dairy Cattle Association, are listed in Table 7-1. Average weights for animals of the various breeds are not available but minimum weights have been suggested by the Purebred Dairy Cattle Association (Table 7-1). In recent years the animals in all breeds have tended to be larger. Many of the animals holding the production records for their respective breeds are among the larger animals of the breed.

Composition of Milk. As indicated in Table 3-2, the milk composition of each breed differs. In general, the milk from the Guernsey and Jersey breeds is high in fat and solids-not-fat, that from the Ayrshire and Brown Swiss is intermediate, and that from Holsteins is lowest. The variation between breeds is greatest for percentage fat, less for percentage solids-not-fat, and least for percentage ash. There is a good correlation between the percentage fat and the percentage solids-not-fat, so that in general the breeds and individuals whose milks have a high content of fat also contain greater amounts of other solids in the milk. Individual cows within a breed, however, show marked differences in the percentage of each constituent. Within the Holstein breed, for example, milk from individual cows ranges from 2.6 to 6.0% fat, 2.4 to 6.5% protein, 3.9 to 5.7% lactose, 0.56 to 0.87% ash, and 10.7 to 17.6% total solids (Overman et al., 1939). In spite of the wide individual differences, the breed differences in milk composition are real and are characteristic of the breed. The composition of milk is genetically determined to a certain degree, but very little information is available concerning the manner in which it is inherited. In the future, selection of breeding animals will probably be based on the solids-not-fat content of the milk as well as total milk and fat production.

There is a breed difference in the ability to convert carotene to vitamin A. Holsteins, Ayrshires, and Brown Swiss convert more carotene to vitamin A than Guernseys and Jerseys. The yellow color of the milk from the Guernsey and Jersey is due to the secretion of carotene instead of vitamin A into the milk. On similar rations that are adequate in carotene, the vitamin A value (carotene plus vitamin A) per unit of fat is the same in the different breeds, but a greater portion of the vitamin A activity is present as carotene in the milk with the yellow color. The carotene and vitamin A are secreted in conjunction with the fat; therefore the total amount secreted per unit of milk is greater in the milk with a high fat content even though the concentration per unit of fat is the same. The carotene is also stored in the body fat and secreted by the glands in the skin, giving the fat and glandular secretions a yellow color.

Factors other than breed affect the composition of milk. For example, high environmental temperatures decrease the solids-not-fat in milk, and finely ground hay, if fed as the only source of roughage, depresses the percentage of milk fat. Therefore, selection for changes in milk composition should be conducted on animals under similar environmental conditions and should be based on samples collected over the entire lactation period.

Quantity of Milk. Breeds also vary in the total amount of milk and fat produced. An estimate of the potential of the better animals within the different breeds may be obtained from the production of the daughters of proved bulls used in artificial breeding organizations (Table 7-2).

As shown in Table 7-2, the greatest difference between the breeds with respect to their milk is the amount of milk they produce. Because the breeds producing less milk generally have a higher content of fat, fat production per lactation shows less breed variation. A cow's milk-producing ability, when measured under standard conditions, depends on her maximum daily yield at the peak of lactation and ability to maintain this yield. Normally,

TABLE 7-2. | *Production of daughters of proved bulls used in artificial insemination studs. 2X-305 day M.E. Records.*

Breed	No. of Sires Proved	Milk (lb.)	Fat (%)	Fat (lb.)
Ayrshire	48	10,650	4.2	448
Brown Swiss	75	11,554	4.2	482
Guernsey	202	9,337	4.9	459
Holstein	457	13,542	3.7	503
Jersey	162	8,921	5.4	477
Shorthorn	17	8,853	3.9	349
Red Poll	1	11,387	3.7	427
Red Dane	1	9,992	4.5	447

Source: Dairy Herd Improvement Letter, ARS 44-58, July 1959.

daily production of milk rises for four to six weeks following parturition and then declines slowly until the cow is dried off. Persistency is the term used to describe the ability of the cow to continue to produce at a high level throughout her lactation and is very important as far as total production is concerned. Cows with the same productive rate at the peak of lactation may vary as much as 50% in total production per lactation because of differences in persistency. The few data available indicate very little difference in persistency between the Holstein, Jersey, and Guernsey breeds, with more variability between offspring of sires within each breed than between the breeds.

Color. Coat color is one trait that distinguishes the different breeds. In comparison with productive characteristics, its inheritance is simple, and it affords an easy method for identifying the breed of an animal. The preferred colors for the different breeds are listed in Table 7-1. Many breeds have color standards for registration of animals. Color markings that prevent the registry of Holsteins are solid black, solid white, black in the switch, black from the hoof to the knee or hock, black encircling the leg touching the hoof head, black belly, black and white intermixed to give color other than black and white, and red and white. These rigid color standards are hard to maintain and result in the loss of a few valuable productive animals to the purebred breeders. Black is dominant to red, but in spite of continued selection against red, red and white animals are occasionally born from black and white parents. In these parents the genetic factor for red is present but is masked by the black. Brown Swiss cows with any white or off-color markings above the underside of the belly or with a white core in the switch do not meet the color standards of the breed, and these markings must be recorded at the time of registration; white or off-colored spots, pink noses, and light streaks on the side of the face are objectionable. Black or brindle colors are objectionable in Ayrshires, and a golden yellow skin pigmentation and clear muzzle are favored in Guernseys. As discussed in an earlier section, the yellow skin and fat found in the Guernsey and Jersey breeds are due to the storage of carotene in the fat and fatlike materials. It is related to their lesser ability to convert carotene to vitamin A, which suggests that these breeds' requirement for this vitamin might be higher than that of other breeds. Some data suggest they do have a higher requirement but it does not appear to be much greater; critical experiments to establish how much difference have not been performed.

Heat Tolerance. Preliminary data suggest that differences in heat tolerance—that is, the ability to produce milk when the environmental temperature is high—exist between the popular dairy breeds. However, very few animals within each breed have been tested and no definite conclusions can be made about whether these differences are associated with breed or with individuals within the breed. The limited data available indicate that the Brown Swiss is more heat-tolerant than the Jersey, and the Jersey slightly more than the Holstein. However, many factors—level of milk production, body size, surface area, and the number and activity of sweat glands—affect the heat tolerance of the individual animal. Many animals will have to be tested to establish if a true breed difference exists.

Reproductive Efficiency. There are indications that breeds may differ in reproductive efficiency, but experimental evidence is not available to verify this possibility.

Milk Fever. According to three different investigators (Metzger, 1936; Henderson, 1938; and Hibbs, 1946), Jerseys are more susceptible to milk fever than the other breeds. The incidence of milk fever in the different breeds as reported by Henderson is Ayrshire 6.0%, Brown Swiss 15.3%, Guernsey 8.6%, Holstein 5.6%, and Jersey 29.2%. These percentages are based upon the total number of calvings, exclusive of the first calvings in a large herd. The number of cows in each breed is not given and the levels of production of milk by the animals having milk fever and those not having milk fever are not reported. The author states that the high-producing cows are more susceptible. These data indicate a breed difference in the susceptibility to milk fever. Other factors play a role in causing milk fever but relative importance to breed differences in susceptibility has not been evaluated.

Summary. The present-day dairy breeds differ in blood type, size, color, milk composition, and total milk production. Data also suggest that heat tolerance, breeding efficiency, and susceptibility to milk fever are breed characteristics. However, more information on larger numbers of animals is needed to establish these as breed characteristics. Most of these differences are controlled genetically but many of the traits are also modified by environment. The breeds are relatively homozygous for those breed characteristics that are controlled by a few genes that are easily identified and inherited in a simple manner. However, they are still quite heterozygous for such characteristics as milk production and milk composition, which are determined by a large number of factors that may be inherited in a complex way. Many of the productive characteristics are also modified by environmental factors; this complicates the selection for the genetic factors controlling them.

7-5. TYPE

One of the earliest references to the relationship between type and production is found in Youatt's book, *Cattle, Their Breeds, Management and Diseases,* writen in 1838. He quotes an earlier description of the Suffolk cow: "A clean throat, with little dewlap, a snake head, thin short legs, the ribs springing well from the center of the back, the carcase large, the belly heavy, the backbone ridged, the chine thin and hollow, the loin narrow, the udder square, large, loose, and creased when empty, the milk veins remarkably large and rising in knotted puffs; and this so general, that I scarcely ever saw a famous milker that did not possess this point, a general habit of leanness, hip bones high and ill covered, and scarcely any part of the carcase so formed and covered as to please the eye that is accustomed to fat beasts of the finer breeds." In many ways the above fits the description of dairy type as described by the Unified Score Card. Understandingly, these animals were not pleasing to the eye accustomed to selecting cattle primarily on the basis of their merit for meat.

Type may be defined as the presumed relationship between the animal's body conformation and its ability to perform a given function. For example, in cattle we speak of dairy type and beef type for animals whose primary function is milk production and meat production respectively. An animal's type may be expressed by a numerical score, determined by its degree of conformity to the "ideal type" animal. In dairy cattle the ideal type and method for scoring have been standardized and published by the Purebred Dairy Cattle Association. Except for size and color markings there is a striking similarity in the general conformation of the ideal type animal in the breeds of American dairy cattle. The score card for dairy cows is shown in Fig. 7-6. In order for an individual to be able to compute a correct type score for a cow, it is essential that he know the parts of the animal, as shown in the same figure.

Although type is not closely related to the production of dairy cattle, it is important to breeders. For example, a weakly attached udder in a high-producing cow may break down and shorten her productive life. Pendulous udders are more subject to injury. Crooked feet and legs may restrict the

DAIRY COW UNIFIED SCORE CARD

Copyrighted by The Purebred Dairy Cattle Association, 1943. Revised, and Copyrighted 1957
Approved — The American Dairy Science Association, 1957

Breed characteristics should be considered in the application of this score card	Perfect Score

Order of observation

1. GENERAL APPEARANCE — 30

(Attractive individuality with, feminity, vigor, stretch, scale, harmonious blending of all parts, and impressive style and carriage. All parts of a cow should be considered in evaluating a cow's general appearance) — 10

    BREED CHARACTERISTICS — (see reverse side)
    HEAD — clean cut, proportionate to body; broad muzzle with large, open nostrils; strong jaws; large, bright eyes; forehead, broad and moderately dished; bridge of nose straight; ears medium size and alertly carried

    SHOULDER BLADES — set smoothly and tightly against the body — 10
    BACK — straight and strong; loin, broad and nearly level
    RUMP — long, wide and nearly level from **HOOK BONES** to **PIN BONES**; clean cut and free from patchiness; **THURLS**, high and wide apart; **TAIL HEAD**, set level with backline and free from coarseness; **TAIL**, slender

    LEGS AND FEET — bone flat and strong, pasterns short and strong, hocks cleanly moulded. **FEET**, — 10 short, compact and well rounded with deep heel and level sole. **FORE LEGS**, medium in length, straight, wide apart, and squarely placed. **HIND LEGS**, nearly perpendicular from hock to pastern, from the side view, and straight from the rear view

2. DAIRY CHARACTER — 20

(Evidence of milking ability, angularity, and general openness, without weakness; freedom from coarseness, giving due regard to period of lactation)

    NECK — long, lean, and blending smoothly into shoulders; clean cut throat, dewlap, and brisket — 20 **WITHERS**, sharp. **RIBS**, wide apart, rib bones wide, flat, and long. **FLANKS**, deep and refined. **THIGHS**, incurving to flat, and wide apart from the rear view, providing ample room for the udder and its rear attachment. **SKIN**, loose, and pliable

3. BODY CAPACITY — 20

(Relatively large in proportion to size of animal, providing ample capacity, strength, and vigor)
    BARREL — strongly supported, long and deep; ribs highly and widely sprung; depth and width of barrel — 10 tending to increase toward rear
    HEART GIRTH — large and deep, with well sprung fore ribs blending into the shoulders; full crops; — 10 full at elbows; wide chest floor

4. MAMMARY SYSTEM — 30

(A strongly attached, well balanced, capacious udder of fine texture indicating heavy production and a long period of usefulness)
    UDDER — symmetrical, moderately long, wide and deep, strongly attached, showing moderate cleavage — 10 between halves, no quartering on sides; soft, pliable, and well collapsed after milking; quarters evenly balanced
    FORE UDDER — moderate length, uniform width from front to rear and strongly attached — 6
    REAR UDDER — high, wide, slightly rounded, fairly uniform width from top to floor, and strongly — 7 attached
    TEATS — uniform size, of medium length and diameter, cylindrical, squarely placed under each quarter, — 5 plumb, and well spaced from side and rear views
    MAMMARY VEINS — large, long, tortuous, branching — 2
    "Because of the natural undeveloped mammary system in heifer calves and yearlings, less emphasis is placed on mammary system and more on general appearance, dairy character, and body capacity. A slight to serious discrimination applies to overdeveloped, fatty udders in heifer calves and yearlings."

Subscores are not used in breed type classification.   **TOTAL** — **100**

PARTS OF A DAIRY COW

Fig. 7-6. Dairy Cow Unified Score Card. [Courtesy Purebred Dairy Cattle Association.]

EVALUATION OF DEFECTS

In a show ring, disqualification means that the animal is not eligible to win a prize. Any disqualified animal is not eligible to be shown in the group classes. In slight to serious discrimination, the degree of seriousness shall be determined by the judge.

EYES
1. Total blindness: *Disqualification.*
2. Blindness in one eye: *Slight discrimination.*
3. Cross-eyes: *Slight discrimination.*

WRY FACE
Slight to serious discrimination.

CROPPED EARS
Slight discrimination.

PARROT JAW
Slight to serious discrimination.

SHOULDERS
Winged: *Slight to serious discrimination.*

TAIL SETTING
1. Wry tail or other abnormal tail settings: *Slight to serious discrimination.*

LEGS AND FEET
1. Lameness — apparently permanent and interfering with normal function: *Disqualification.*
— apparently temporary and not affecting normal function: *Slight discrimination.*

2. Bucked knees: *Slight to serious discrimination.*
3. Evidence of arthritis, crampy hind leg: *Serious discrimination.*
4. Boggy hocks: *Slight to serious discrimination.*

ABSENCE OF HORNS
No discrimination.

LACK OF SIZE
Slight to serious discrimination.

UDDER
1. Blind quarter: *Disqualification.*
2. Abnormal milk (bloody, clotted, watery): *Possible disqualification.*
3. Udder definitely broken away in attachment: *Serious discrimination.*
4. A weak udder attachment: *Slight to serious discrimination.*
5. One or more light quarters, hard spots in udder, obstruction in teat (spider): *Slight to serious discrimination.*
6. Side leak: *Slight discrimination.*

DRY COWS
Among cows of apparently equal merit: *Give strong preference to cows in milk.*

FREEMARTIN HEIFERS
Disqualification unless proved pregnant.

OVERCONDITIONED
Slight to serious discrimination.

TEMPORARY OR MINOR INJURIES
Blemishes or injuries of a temporary character not affecting animal's usefulness: *Slight discrimination.*

EVIDENCE OF SHARP PRACTICE
1. Animals showing signs of having been operated upon or tampered with for the purpose of concealing faults in conformation, or with intent to deceive relative to the animal's soundness: *Disqualification.*
2. Uncalved heifers showing evidence of having been milked: *Serious discrimination.*

Fig. 7-6. Dairy Cow Unified Score Card (continued).

animal's ability to move about. The breed associations have a type classification program in which the cows of the individual breeder are compared with the breed ideal. When animals are classified by this method the components of type are scored individually and can be used as a basis for the selection of future breeding animals to strengthen weak points prevalent in the breeding herd. If a breeder's herd scores low on udder attachments or on feet and legs the breeder should select his next herd sire from families within the breed that are outstanding in these respects. Comparative judging, as practiced at Dairy Cattle Shows, is based indirectly on type. Type also has an economic value to the breeder. When he is selling breeding animals of similar productive capacities, the better type animals are in greater demand and usually sell for a better price.

7-6. BREED ASSOCIATIONS

Many of the dairy cattle breed associations were formed in the last quarter of the nineteenth century. They were organized for the recording of animals that were distinctive in their breeding and for protecting the importers of cattle from unethical practices by some of the exporters. This protection was accomplished by requiring pedigrees of the animals, and the original herd books were primarily used for recording pedigrees. The present associations perform several functions for their members. They are responsible for the registration of animals, they promote their breed, they conduct production testing and type classification programs. Most of them have selective registration programs.

Several testing programs are supervised by the breed organizations. The broad aspects of the testing programs are the same for all the breeds, but they differ in specifics. These programs are discussed in Chapter 32.

The breed associations also conduct the type classification programs. In addition, most of them have selective registration programs based on production records and type classification. Any animal that is the offspring of registered parents may be registered by the breed association if it is normal and meets the color standards.

Selective registration or breed recognition programs are designed to give recognition to outstanding animals within the breed. Programs for the different breeds, although not identical, are similar, and generally they recognize outstanding production, superior type, and superior transmitting ability for production and type. For example, the Holstein-Friesian Association of America offers the following selective registration programs: gold medal dam, silver medal 'type' sire, silver medal 'production' sire, and gold medal sire.

To qualify as a gold medal dam, a cow must meet the following requirements:

1) She must have produced 100,000 lb. of milk or have averaged 12,500 lb. of milk and 500 lb. of fat for her recorded lactations.
2) She must have scored at least 80 points as a two- or three-year-old, 81 points as a four- or five-year-old, or 82 points at 6 years or older.
3) She must have at least three progeny tested for production and three classified for type.

The progeny must qualify as follows:

1) Daughters
   Each must have same qualifications as outlined under the first two categories for the cow.
2) Sons
   a) Each must have an official proof (10 or more daughter-dam pairs) certifying that the average of all his daughters is at least 13,000 lb. of milk and 525 lb. of fat, or have been recognized as a silver medal production sire.
   b) Each must have been recognized as a silver medal type sire or have scored at least 82 when classified, regardless of age.

A cow may classify as a gold medal dam on the records of her progeny, if she has no record. The same daughters do not have to qualify both for production and type.

A silver medal "type" sire must have at least 10 daughters meeting specific type requirements; a silver medal "production" sire must have at least 10 daughters meeting specific production requirements. If a sire qualifies for both type and production he becomes a gold medal proved sire.

Details of other breed selective programs may be obtained from the yearbooks published by the breed organizations, or from the breed organizations.

With the continued improvement in the productive characteristics of grade cattle, selective registration and continued selection for better production in the purebred cattle will be more important and the selection standards will be higher. For example, the lactations completed under the HIR testing program of the Holstein breed during 1959 averaged 13,612 lb. of milk, 3.69% fat, and 502 lb. of fat on a 305-day, 2 ×, mature equivalent basis. These averages are practically identical with those reported for grade and purebred progeny of Holstein bulls used in artificial insemination (Table 7-2). Both of these groups are selected, but with the continued use of bulls that transmit high production, the population of cows that are both grade and purebred will increase in productivity, and fewer bulls will have the ability to transmit increased milk and fat production above the average of the breed.

REFERENCES AND SELECTED READINGS

References marked with an asterisk are of general interest.

*Bowling, G. A., 1942. The introduction of cattle into Colonial North America. *J. Dairy Sci.*, 25:129–154.

*Caldwell, W. H., 1941. *The Guernsey.* American Guernsey Cattle Club, Peterborough, N.H.

*Gow, R. N., 1936. *The Jersey.* American Jersey Cattle Club, New York.

Henderson, J. A., 1938. Observations on reproduction and associated conditions in a herd of dairy cattle. *Cornell Vet.*, 28: 173–195.

Hibbs, J. W., W. E. Krauss, C. F. Monroe, and T. S. Sutton, 1946. Studies on milk fever in dairy cows. I. The possible role of vitamin D in milk fever. *J. Dairy Sci.*, 29:617–624.

Metzger, H. J., and H. B. Morrison, 1936. The occurrence of milk fever in the Kentucky Station over a period of 20 years. *Proc. Am. Soc. Animal Prod.*, pp. 48–52.

Neimann-Sørensen, A., P. H. Sørensen, E. Anderson, and J. Moustgaard, 1956. Danish investigations on blood groups of cattle and pigs. *Proceedings of Seventh International Congress of Animal Husbandry*, Madrid, pp. 87–111.

Overman, O. R., O. F. Garrett, K. E. Wright, and E. P. Sanman, 1939. *Composition of Brown Swiss Milk with Summary of Data on Composition of Milk from Cows of Other Dairy Breeds.* Illinois Agr. Expt. Sta. Bull. 457.

*Prentice, E. P., 1942. *American Dairy Cattle.* Harper, New York.

*Prescott, F. S., and F. T. Price, 1930. *Holstein Friesian History.* Holstein-Friesian World Inc., Lacona.

*Sanders, A. H., 1925. The taurine world. *National Geographic Magazine.* 48:591–710.

Stormont, C., 1958. On the applications of blood groups in animal breeding. *Proc. Tenth International Genetics Congress*, 1:206–224.

*Youatt, W., 1838. *Cattle, Their Breeds, Management and Diseases.* Baldwin and Cradock, p. 174.

Chapter 8

Breeds of Swine

8-1. INTRODUCTION

Swine have long been important animals on American farms. They have traditionally been called the "mortgage lifters" in the corn belt. Swine have gone through more type changes than any other class of livestock. However, the center of swine production has remained constant during this century: the corn belt has served as the hub of the industry.

The listing below shows the leading states in hog production during the period from 1958 to 1962. Iowa produces more hogs yearly than the next two states combined, in spite of the fact that Illinois occasionally produces more corn than Iowa.

No. on farms[1]	*State*
12,467,000	Iowa
7,191,000	Illinois
4,872,000	Indiana
3,849,000	Missouri
3,523,000	Minnesota
2,608,000	Ohio
2,452,000	Nebraska
1,877,000	Wisconsin
1,654,000	Georgia
1,395,000	North Carolina

[1] From USDA Agricultural Statistics, 1964

The per capita consumption of pork has declined somewhat in relation to that of beef and poultry. This decrease in popularity of pork products has been the principal incentive to the swine breeders to improve their breeds; some have made great effort, others have lagged.

About 1880, according to Vaughn (1941), there began a trend toward what was called the "cob-roller" type. This was a small-framed, quick-fattening type further characterized by small litter size and a slow rate of gain. Because commercial swine producers found it to be uneconomical, purebred breeders shifted about 1908 to the other extreme of very tall "stilt-like" hogs. Ranginess was stressed to the point that judges used their canes to measure height. Great arch of back was also encouraged. These animals did not reach a proper degree of finish until they weighed well over 200 lb. In the 1930's, therefore, another change in emphasis toward a smaller type took place, but at first no special attention was given to back-fat thickness and percentage of lean. Since World War II, all the breed associations have initiated programs to develop swine that would have a minimum of waste and a maximum percentage of the lean cuts—ham, loin, picnic, and Boston butt.

Swine have a rapid rate of reproduction and a short generation interval. It has therefore been easy to make marked changes in type within the breeds; these changes have tended to follow the demands for fat or meat. Thus, during World War I the great demand for fats stimulated production of fat hogs. The current trend is toward a lean or "meat type" hog. The discussion here will center principally on the current state of the breeds. There is little to be gained from studying the history of swine breeds, other than learning some things not to do.

Swine breeds at one time were classified into "lard" or "bacon" breeds. Most of the American breeds were classified as lard hogs, and the British and Scandinavian breeds as bacon hogs. The basis for this classification was that American hogs were an important source of fat, and the English and Scandinavian breeds were used in the production of Wiltshire type bacon, made from the whole side of the hog. The current meat-type hog presumably grows somewhat faster and has a larger ham than representatives of the bacon breeds. Some of the new breeds represent an attempt to borrow the more desirable traits from both bacon and lard breeds.

Currently several of the American breeds are considered to be meaty and efficient. The degree to which they are held in esteem by packers tends to be related to the improvement efforts that the respective breed associations have made. Concerted work toward improving breeds of swine has now become a principal function of swine record associations, which in the past were largely just recording and promotion agents.

Any statement characterizing a breed of swine is likely to be out of date in a few years. Individual herds also may deviate sharply from the breed average. The performance characterizations that follow are based on the current state of the breeds, as exactly as the writer can determine them from recent testing station and experiment station results.

Some associations will point to their Production Registry data as an indication that their litter size is good. However, litters of less than eight are not submitted for Production Registry. Hence any estimate based on these records is biased, since the small litters are not considered.

8-2. POPULAR MODERN BREEDS AND THEIR TRAITS

A list of some of the more popular breeds of swine and a few of their traits are given in Table 8-1. Though the figures listed for the traits are

TABLE 8-1. | *Some popular breeds of swine and their traits.*

Breed	Origin	Color	Litter Size	No. of Pigs Raised to 154 Days[†]	Rate of Gain	Carcass Merit
Hampshire	United States	Black, white belt	8.66*		Average**	Excellent (heavy loin)
Duroc	United States	Red	9.78* 9.31†	6.92	138.9‡ Excellent**	Average
Poland China	United States	Black, white points	7.98* 7.84†	5.82	Below average** 132.6‡	Good (heavy ham)
Berkshire	England	Black, white points	7.74*		Average**	Average
Spotted Swine	United States	Black and white spotted	8.70†	6.92	139.0‡	Good
Yorkshire	England	White	10.75*		Fair**	Average (long loin, small eye)
American Landrace	Denmark	White	9.74*		Excellent**	Average
Tamworth	England	Red	7.43*		Below average**	Poor (lack muscling)
Chester White	United States	White	9.33* 8.78†	6.56	136.1‡ Average**	Poor (too fat)

* Lush, J. L. and A. E. Molln, 1942. *Litter size and weight as permanent characteristics*, USDA Technical Bul. 836.
† Bradford, G. E., A. B. Chapman, and R. H. Grummer, 1953. Performance of hogs of different breeds and from straightbred and crossbred dams on Wisconsin farms. *J. Animal Sci.*, 12:582–590.
** Hazel, L. N., 1959. Quoted by Dean C. Wolf in *Farm Journal*, May, 1959.
‡ Weight at 154 days as reported by Bradford *et al.*, cited above.

the averages of fairly large groups, they may not represent true averages of the breeds. Note, however, the very close agreement of the data on litter size as reported by Lush and Molln (1942) with the data of Bradford et al. (1953).

Color. Color is one of several characteristics used in identifying the various breeds. It is considered to be of such importance that in the Hampshire, aberrations in color—including width of the white belt—prohibit approximately 25% of the animals from registration. Spotted Swine should have no more than 80% nor less than 20% white on body. The color red is recessive to black or white in crosses. Hence the Duroc cross is not as identifiable by color at the market place as some other breeds, such as the Hampshire. The ability of Hampshires to transmit their color pattern to their offspring has sometimes caused some undesirable crossbred hogs to have the Hampshire color; this has tended to lower the opinion of this breed. In the Chester White, American Landrace, and Yorkshire, white is dominant to black or red, except in the cross with Hampshires, where the body is often white with a bluish colored belt. Producers sometimes discriminate against white hogs in areas where sunburn is a problem.

Face Length. The deviation from the regular-shaped head of swine is most prevalent in the Berkshires and Yorkshires. Extreme dish of face in Berkshires has limited their increase, since commercial hog producers have found that these short-faced hogs have difficulty in eating from self-feeders and drinking from automatic waterers. Breeders have partially corrected this abnormality, but it is still one factor limiting the popularity of the breed. The tendency for pug noses has been disadvantageous to Yorkshires, too. Hampshire and Tamworth hogs have longer snouts than other American breeds.

Body Size. Growth rate data (Table 8-2) indicate that the breeds are quite close together, with the exception of the Duroc, whose growth rate is still outstanding. Similarly, length is comparable among breeds other than the Spotted Poland, Poland China, and Duroc, which are shorter.

Meatiness. Meatiness is the term used to distinguish the animals that have a high proportion of muscle to fat. The ideal meat-type hog has larger hams and greater cross section area of loin than the conventional bacon-type.

Selection has been effective in improving the meatiness of all of the major breeds (Table 8-2). The Duroc is still somewhat smaller in loin eye size and in cutout, but it has been greatly improved. As selection plateaus for each of the breeds, differences tend to become smaller.

In recent years Yorkshire breeders have imported many English Large Whites, which are more rugged, faster gaining pigs than the Canadian

| TABLE 8-2. | *Information on the rate and efficiency of gain and on carcass characteristics of several breeds of swine.* |

| | Pen Average | | Boar Backfat Average Probe (inches) | Barrow Carcass Data | | | |
| | | | | Length (inches) | Backfat (inches) | H & L† (%) | Loin Eye (sq. inches) |
Breed	Gain (lb.)	Eff.* (lb.)					
Hampshire	1.90	280	.89	30.0	1.27	40.5	4.40
Duroc	2.13	269	1.02	29.4	1.40	38.6	3.93
Landrace	2.00	281	1.00	29.6	1.60	39.1	4.02
Poland China	1.91	283	.95	29.3	1.39	39.5	4.16
Spotted Swine	1.93	271	.90	29.0	1.40	40.6	4.28
Yorkshire	1.89	277	1.00	30.2	1.44	39.2	4.38
1956 All Breed Average	1.95	303	1.31	28.9	1.61	33.2	3.50

Source: From 1963 Fall Summary of Iowa Swine Testing Station by Breeds.
* Pounds of feed per 100 lb. of gain.
† Percent of carcass composed of ham and loin.

Yorkshires, but have a tendency to get fatter. Breeders are making a great effort to improve Yorkshire meatiness, and since meatiness is highly heritable, this improvement should be fairly easy to accomplish (Table 8-2). The Danish Landraces, like the Yorkshires, are used primarily for producing Wiltshire sides. Hence they have been selected for leanness and length, rather than for heavy hams and big loins. The carcass yields a good percentage of lean cuts. In the early stages of the development of the meat-type hog, many thought the Tamworth was the answer because of its lean look and great length. Iowa Testing Station results indicate, however, that the Tamworth is among the fattest of the breeds.

Fig. 8-1. Hampshire boar, Superman, bred by C. T. Keen and Son, Le Grande, Iowa. He was one of the boars tested in the first test at the Iowa Station; several littermates made outstanding records, and his sire, Automation, is a leading sire in the breed.

Fig. 8-2. Duroc boar, Kayward Constructor, bred by LaVerne Reitz, Gilman, Ill., owned by Kayward Farm, Iowa City, Iowa. This boar made an outstanding record in one of the Illinois testing stations.

Cutout percent has been increasing recently. Animals in the first Iowa test in 1956 yielded an average of 32.3% closely trimmed ham and loin; today that average is 39%, with a range from 37% to as high as 44%, but few carcasses are found at the lower end of the scale.

Specific data on carcass traits of swine breeds are given in Table 8-2. The figures are averages of 300 boars and 100 barrows and may not accurately reflect averages for the breeds as a whole. This table shows that much improvement has been made in carcass traits and that some change has occurred in efficiency during a seven-year period. It further shows that the differences between breeds are not great at the present time.

Prolificness and Reproduction. The prolificness of Yorkshires and Durocs is considered to be superior to that of the other breeds. Iowa tests indicate that the Duroc is the only American breed that has successfully competed with crossbreds in this trait. The chief drawback is incidence of blind or inverted nipples and lack of enough nipples for good commercial production of litters of 10 to 12. The improvement of prolificness has been emphasized in breeding programs. According to experimental work, the Yorkshire breed farrows the largest litters. Bradford et al. (1953) found the Spotted

Fig. 8-3. Poland China boar, Famous, a famous certified meat sire (CMS), owned by Harvey Richardson, Elmore City, Okla. Member of high-testing litter in Ohio Testing Station. His progeny have qualified him for superior meat sire rating.

Fig. 8-4. Spotted Swine boar, named L. & M., bred by Carver and Byers, Nevada, Iowa, owned by Bill Hemm, Sheffield, Iowa. This boar was tested in the Iowa Swine Testing Station and subsequently made a remarkable record of siring both high testing and show pigs.

Swine to compare favorably with the Durocs on the basis of number of young raised (Table 8-1).

The Hampshire, the Chester White, and the American Landrace are considered average to good for prolificness. Sows of the Chester White and Hampshire breeds are good mothers. The Chester White sows have excellent underlines. The Landrace produces very prolific sows when crossed with American breeds.

Poor litter size has apparently contributed to the decline of the Poland China in popularity. Although tests indicate that crossbred gilts (young crossbred sows) carrying half Poland genes perform satisfactorily as mothers, the stigma of poor litter size still clings to the breed. At present a great effort is being made to improve the litter size by breed-sponsored herd improvement programs. The Poland has been extensively used for crossing with the Scandinavian and British breeds. Such crossings, popular in the corn belt, seem to produce a combination of meatiness, prolificness, and mothering ability in the crossbred female. Like the Poland China, the Berkshire has long been considered by commercial producers to be poor in prolificness, especially of gilt litters.

Practically all commercially raised hogs are crossbred. The principal function of the breeds is to supply boars for use in crossbreeding programs. Thus it is important that breeders consider the highly heritable traits in their

Fig. 8-5. Chester white boar, bred by Burton Lofgren, St. James, Minn., owned by Parkison and Rodebaugh, Rensselaer, Ind. This boar's littermates made outstanding cutout records in the Minnesota Station.

Fig. 8-6. Berkshire boar, Locust Creek Hallmark Flash CMS., bred by John Randall Tucker, Newton, Mo., owned by Elmer Monson, Blair, Neb. This boar made a good record in the on-farm test conducted by the association.

selection programs, since selection for traits of low heritability would avail them little. Crossbreeding has the most beneficial effect on those traits with low heritability. There are wide differences in litter size, not only between breeds, but also between individuals within a breed. Iowa swine records indicate that about seven pigs per sow are now being raised throughout the state. Production registry has a minimum standard requirement of eight raised. There is a wide range in this trait; sows have been known to raise fifteen pigs, and, of course, some raise only one.

Rate and Economy of Gain. Gains may vary from 1 lb. or less per day after weaning to more than 2.5 lb. per day. Efficiencies as good as 240 lb. feed per 100 lb. gain have been reported. There have been a few testing station boars that weighed 200 lb. at 105 to 110 days of age.

The Duroc was long popular with the commercial producer because of its ruggedness and gaining ability. Iowa tests indicate that it is the only American breed that has successfully competed with crossbreeds in rate and economy of gain. Lack of meatiness, however, curtailed its popularity somewhat, but this deficiency is being overcome (Table 8-2).

In growth rate and efficiency the Hampshires and Berkshires are average. The Yorkshire has not been popular with the commercial producers because of its poor gaining ability and lack of ruggedness under farm conditions. However, due to the breeding of imported English Large White Yorkshires

Fig. 8-7. CMS Yorkshire boar, Crab Tree Toastmaster, bred by Cerny Brothers, Dorchester, Neb., owned by C. J. Cooper and Sons, Hartley, Iowa. This is a certified meat sire based on the performance of his offspring.

Fig. 8-8. Landrace boar, Rob Boe Samson, bred by Robert Boesch and Son, Woodburn, Ind. He was a high record boar at the National Landrace Testing Station.

with the American Yorkshire, recent Iowa Testing Station results have shown the breed to be good in rate of gain and excellent in efficiency. Chester White records in testing stations indicate that the breed has been average to poor in rate and economy of gain.

Two breeds that differ greatly in rate and economy of gain are the Landrace and Tamworth. Long selected under "hothouse" conditions, the Landrace has not proved rugged enough to withstand the environment provided by many American farms. Testing station results at Iowa have shown it to be excellent in rate and efficiency of gain in the summer and average to good in the winter. In contrast, the Tamworth is a rugged hog capable of surviving rigorous conditions. The Tamworth Breed Association Testing Station data indicate that the breed is average to poor in gain and efficiency.

8-3. OTHER BREEDS

A few other breeds need to be mentioned. Most of these were developed at state and federal experiment stations. Among them are the Minnesota Nos. 1, 2, and 3; Beltsville No. 1; and Palouse. The Wessex Saddleback, which has been imported recently, is being produced in some areas. A breed called LaCombe has been developed from a crossbred Chester White and Landrace foundation. None of these breeds seem to have the overall excellence to replace the established breeds, which experiments indicate are doing

Fig. 8-9. Tamworth boar, Retta Rays Pard Echo, bred by Retta Ray Farm, Bowling Green, Ky. This boar was tested at the National Tamworth Testing Station.

a satisfactory job. Establishing a new breed is difficult. Perhaps the principal result of the development of the new breeds has been the resulting competition, which has induced improvement of the already existing breeds.

REFERENCES AND SELECTED READINGS

Reference marked with an asterisk is of general interest.

Bradford, G. E., A. B. Chapman, and R. H. Grummer, 1953. Performance of hogs of different breeds and from straightbred and crossbred dams on Wisconsin farms. *J. Animal Sci.,* 12:582–590.

Hazel, L. N., 1959. (Quoted by Dean Wolf in *Farm Journal,* May, 1959.)

Iowa Swine Testing Station, 1968. *Biannual Reports.*

Lush, J. L., and A. E. Molln, 1942. *Litter Size and Weight as Permanent Characteristics.* USDA Tech. Bull. 836.

*Vaughn, H. W., 1941. *Types and Market Classes of Livestock.* College Book Co., Columbus.

Chapter 9

Breeds of Sheep
and Goats

Ewes yearly by twining, rich masters do make:
The lambs of such twinners for breeders go take.
W. YOUATT, *Sheep (1837)*

9-1. DOMESTICATION AND EARLY HISTORY

Sheep and goats were probably among the earliest animals to be domesticated—some six to eight thousand years ago, as shown by evidence reviewed by Reed (1959). The tractability of sheep, which can be observed in wild forms today, and the versatility of their products, including meat, milk, wool, and skins, were no doubt dominant factors leading to their early husbandry by man. Sheep are used even for work and sport in rare cases.

Breeds of sheep have been developed in relatively isolated geographic areas where adaptation to peculiar environmental factors has been dominant. Thus the Merino of Spanish origin thrives in hot, dry climates and the herding or flocking instinct is well developed. These sheep have been accustomed through many centuries to long trails, and those that failed to stay with the flock probably did not leave descendants. On the other hand, the British breeds—usually kept in enclosed pastures in a more northern climate—do not band together as tightly as the Merino and also, like wild types, have retained the tendency to breed at a time for lambing to occur when spring temperatures and plant growth are favorable for survival. The fat-tailed and fat-rumped sheep in the deserts of Asia store fat during the lush season that can be drawn on during the often long periods when plant growth is dormant. Fine-wool Merinos were developed in a dry climate; the coarse- and long-wool types, such as the Lincoln and Romney, in a more humid climate.

Breeds of sheep have also resulted from selection for specialization to fill the need for meat or for a particular kind of wool. Some breeds kept primarily for meat, such as the Blackhead Persian or the Wiltshire Horn, are called wolless although they produce some wool fibers that are shed. Some native breeds in Central and Southern Europe, such as the German East Friesian or the Italian Langhe, are kept primarily for milk; the world-famous Roquefort cheese is made from ewe's milk. The Navajo type, developed from unimproved Spanish sheep, produces a coarse hairy outer coat from which Navajo rugs and blankets are made.

The wild Rocky Mountain Bighorn Sheep has long been present in North America but it has never been domesticated; all of our present domestic breeds of sheep or their ancestors have been imported. Domesticated sheep were first introduced on the American Continent by the Spanish conquerors (Carmen et al., 1892); Columbus brought sheep on his second voyage in 1493. Sheep were brought into the English Colonies almost as soon as they were settled, beginning with Jamestown in 1609.

Important importations of Spanish Merino sheep into the United States began about 1801. Earlier importations probably came to the Southwest by way of Mexico. Others came from France, Silesia, Sweden, and Saxony, where they had been further developed and improved. Of particular interest are the Merinos that were bred at Rambouillet, France since 1786. Some of these were brought to the United States in 1840, but the large importations from both France and Germany were made in the 1890's. These Rambouillets and Merinos have been the foundation of the western sheep industry, and they and their crossbred descendants today constitute a large majority of this country's ewe flocks—especially those of the western range states and Texas. In many states of the Middle West and East also, the Western ewe is the chief producer of lamb meat and wool for market.

The early importations of sheep from England were not of the quality of modern English breeds (Connor, 1918), because they were made before the breeding improvements, which began in England in the eighteenth century. Until this time, the English sheep were relatively coarse, leggy, late-maturing animals, but with good foraging qualities. Long-wool types from the marsh regions of Kent, Leicestershire, and Lincolnshire, as well as the intermediate types such as the Ryeland, Dorset, and Southdown were evident in the early American sheep. The increase of industrial development in the United States, with the influx of foreign labor accustomed to eating mutton and lamb, resulted in large imports of mutton sheep. Continued importation of other medium-wool mutton breeds such as the Hampshire, Suffolk, Oxford,

Shropshire, and Cheviot now account for most of the meat type breeds that are most common in the United States.

Wentworth, in *America's Sheep Trails* (1948), presents a comprehensive review of the history of sheep production from the early development in the eastern colonies, and the movements into the Middle West and Southwest in the early 1800's, to the completion of the westward movement at the close of the Civil War. In the following years sheep numbers fluctuated from 35 million head to a peak of more than 50 million at the beginning of World War II. In 1950, numbers declined to under 30 million head. Since then, numbers have averaged around 31 million head, with a peak of more than 33 million head in 1960, followed by a decline to about 28 million head in 1964.

9-2. CLASSIFICATION OF BREEDS AND TYPES OF SHEEP

Sheep breeds are classified by origin and use and, particularly, according to fineness and length of wool. The fine-wool breeds, such as the Rambouillet and Merino, produce wool used largely in apparel manufacture. These breeds are characterized by their heavy, dense fleeces with staple lengths ranging from 2 to 4 inches. Fine-wool breeds are angular in form, have relatively slow growth rates, and reach maturity slowly.

Meat breeds such as the Suffolk, Hampshire, and Southdown produce wool that is medium in fineness and length, but often with fairly lightweight fleeces. These breds are noted for their thick, blocky, low-set conformation, which is usually associated with meatiness and ability to fatten readily.

Sheep breeds are generally classified as wool, mutton, or other type according to special use, but most breeds in the United States are dual-purpose, being valued for both meat and wool. Meat production generally accounts for 60 to 80% of the income from sheep, although in the Southwest wool production may almost equal meat production in economic returns to the producer. This has led to emphasis on both meat and wool in all domestic breeds, although more western ewes are wool types. On the other hand, less attention is paid to wool in the meat breeds that provide sires for the bulk of slaughter lambs. The crossing of these types provides a convenient and efficient means for maximum production of meat and wool without the difficulty of attaining proficiency for both in the same breed.

Sheep breeds specialized for products other than meat and wool are uncommon in the United States, although milking of sheep is common in many countries of the world. These breeds are triple-purpose, being also kept for meat and wool. The production of fur pelts and skins from sheep are important in many parts of the world, particularly in the U.S.S.R. and the Middle East. Specialized breeds for fur pelts such as the Karakul are found in the United States but have little economic importance here.

9-3. SHEEP BREEDS OF THE WORLD

Hundreds of breeds of sheep exist over the world today and include a wide variety of sizes, shapes, types, and colors. Mason (1951) lists most of these with information on their origin, characteristics, and uses. Wool-type breeds have become most important in the southern hemisphere, particularly in Australia, New Zealand, South Africa, and Argentina and Uruguay in South America, although both fine- and long-wool breeds are found all over the world. Fat-tailed and fat-rumped sheep are common in the desert and semidesert areas of the Near East, Asia, and Africa. They usually produce carpet wool, although wool yields are generally low and the sheep are valued for many purposes. Dairy breeds of sheep are more common in Central and Southern Europe. Northern short-tailed varieties, noted for their prolificness, are found in the Scandinavian countries. Meat breeds originating from Britain, which are common there as well as in the United States and Canada, are widely distributed throughout the world.

9-4. SHEEP BREEDS IN THE UNITED STATES

Information on breeds of sheep in the United States is tabulated in Tables 9-1 and 9-2. Quantitative traits vary widely, of course, according to environmental conditions, and are therefore given in general terms. There is considerable overlapping, and wide variations exist within most breeds. Briggs (1958) has discussed in detail the history, characteristics, distribution, and importance of sheep and goat breeds.

Fine-wool breeds, of which the Rambouillet is the most common, are of predominantly Merino origin and are noted for their high yields of uniform wool. Rambouillets are large but lack the conformation of the meat breeds and also grow and mature somewhat more slowly. Their wool ranges in average fineness from 64's to 80's or finer (see footnote to Table 9-1), with staple lengths from 2 to $3\frac{1}{2}$ inches. Skin folds, which were once common, are now almost nonexistent. Classification into A, B, and C types according to size and number of folds is no longer necessary, since present-day sheep are all of the smooth or C type. Unfortunately, wool covering on the faces is still common although it has been amply demonstrated by Terrill (1949) and many others that openfaced sheep, free from wool blindness, produce more lambs and only slightly less wool. Polled Rambouillets are becoming more numerous and have definite economic advantages, particularly where damage from screwworms and fleece maggots is common (Terrill 1953). Unfortunately the polled gene is sometimes linked with that for cryptorchidism, a condition in which one or both testes fail to descend normally. Rambouillets have provided the foundation stock for most western range sheep, and relatively pure Rambouillets are common in Texas and the Southwest.

Merinos of the American and Delaine types are somewhat smaller and

TABLE 9-1. | *Information on breeds of sheep in the United States.*

Classification	Breed	Country of Origin	Approximate Date of Origin	Approximate Date of First Importation	Fleece Fineness*	Fleece Length†	Fleece Weight**
Fine wool	Rambouillet	France	1786	1840	Fine	Medium	Heavy
	Merino	Spain	Early	1801	Fine	Medium	Heavy
	Debouillet	United States	1920	—	Fine	Medium	Heavy
Meat, medium wool	Suffolk	England	Early 19th cent.	1888	Medium	Short	Light
	Hampshire	England	Early 19th cent.	1881	Medium	Medium	Medium
	Shropshire	England	1860	1860	Medium	Medium	Medium
	Southdown	England	Late 18th cent.	1803	Medium	Short	Light
	Dorset	England	About 1815	1887	Medium coarse	Medium	Medium
	Cheviot	England	End of 18th cent.	1838	Medium	Medium	Medium
	Oxford	England	1836	1846	Medium coarse	Medium	Medium heavy
	Tunis	North Africa	Ancient	1799	Medium	Medium	Medium
Crossbred, dual-purpose	Corriedale	New Zealand	1880–1910	1914	Medium	Medium long	Heavy
	Columbia	United States	1912	—	Medium	Medium long	Heavy
	Panama	United States	1912	—	Medium	Medium long	Heavy
	Targhee	United States	1927	—	Medium fine	Medium	Heavy
	Romeldale	United States	1915	—	Medium	Medium	Heavy
	Montadale	United States	1933 on	—	Medium	Medium	Medium
Long-wool	Romney	England	Early	1904	Coarse	Long	Heavy
	Lincoln	England	Early	Late 16th century	Coarse	Long	Heavy
	Leicester	England	1755–1790	Late 18th century	Coarse	Long	Heavy
	Cotswold	England	Early	1832	Coarse	Long	Heavy
Carpet wool, mixed wool, fur	Scottish blackface	Scotland	Early	1861	Coarse	Long	Heavy
	Navajo	United States	Early	1540, 1598	Coarse	Long	Medium
	Karakul	Asia	Ancient	1909	Coarse	Long	Medium

* In general, fine wool averages 64's and finer, medium wool from 50's to 62's, and coarse wool 48's and coarser in U.S. Numerical Grades.

† In approximate terms of one year's growth, long wool would be 6 in. and longer, medium wool from 2 to 6 in., and short wool under 2 in.

** Heavy fleeces generally average over 12 lb., medium fleeces from 8 to 12 lb., and light fleeces under 8 lb., for one year's growth from mature ewes producing lambs.

TABLE 9–2. | *Information on breeds of sheep in the United States.*

Classification	Breed	Body Size*	Body Type	Color Traits	Horns	Number of Registrations in 1963†
Fine wool	Rambouillet	Large	Angular-blocky	White	Rams only	8,164
	Merino	Medium	Angular	White	Rams only	1,500‡
	Debouillet	Medium	Angular	White	Rams only	1,164††
Meat, medium wool	Suffolk	Large	Blocky	Black face and legs	Polled	31,539**
	Hampshire	Large	Blocky	Black face and legs	Polled	25,701
	Shropshire	Medium	Blocky	Black face and legs	Polled	6,928
	Southdown	Small	Blocky, lowset	Brown face and legs	Polled	10,067
	Dorset	Medium	Blocky	White face and legs	Rams and ewes: polled and horned	5,002
	Cheviot	Medium-small	Blocky	White face and legs	Polled	4,308‡‡
	Oxford	Large	Blocky	Brown face and legs	Polled	1,698‡‡
	Tunis	Medium	Angular, blocky	Red or tan face	Polled	192
Crossbred, dual-purpose	Corriedale	Medium-large	Blocky	White	Polled	12,236
	Columbia	Large	Blocky	White	Polled	7,914
	Panama	Large	Blocky	White	Polled	—
	Targhee	Medium-large	Blocky	White	Polled	512
	Romeldale	Medium-large	Blocky	White	Polled	—
	Montadale	Medium-large	Blocky	White	Polled	2,691
Long wool	Romney	Medium-large	Blocky	White	Polled	1,045
	Lincoln	Large	Blocky	White	Polled	381
	Leicester	Large	Blocky	White	Polled	—
	Cotswold	Large	Blocky	White	Polled	—
Carpet wool, mixed wool, fur	Scottish Black face	Medium-large	Blocky	Blackface and legs	Rams and ewes	—
	Navajo	Medium	Angular	Variable	Rams and ewes: polled and horned	—
	Karakul	Large	Angular	Black, brown	Rams only	547

Note: Breeds were not listed here if no information on registrations was available. Karakul registrations were listed for only one association In 1963.

* Large mature ewes in good condition would generally average over 150 lb., medium ewes from 110 to 150 lb., and small ewes under 110 lb.

† Registration numbers were tabulated by the Breeder's Gazette, March, 1964.

** Registrations are totaled for three associations.

‡ Registrations for 1961, the last available figures.

†† Registrations for the Suffolks are the total from two associations, although some dual registration occurs.

‡‡ Registrations for 1962, the last available figures.

Fig. 9-1. Fat-rumped sheep from U.S.S.R.

have more angular forms than Rambouillets. They produce extremely fine wool with relatively high grease content. They were once quite common in Ohio and Texas but are tending to be replaced by the larger Rambouillet in Texas and by crossbred breeds in Ohio and other areas. The Debouillet, a relatively new breed, has resulted from crossing the Rambouillet and Delaine Merino and is most common near its area of origin in New Mexico.

Medium-wool meat breeds are kept in purebred or high-grade flocks throughout the farm states in the Middle West and East. The larger black-faced Suffolk and Hampshire breeds are used all over the country as sires of slaughter meat lambs, particularly when mated to western white-faced ranged ewes. These western ewes or their crossbred ewe offspring are transported in large numbers to the Middle West, East, and Southeast for use as mothers of slaughter lambs. The fleeces of these black-faced breeds and their crosses often contain black fibers that restrict their use to darker fabrics. Shropshire and Southdown rams are used as sires of crossbred slaughter lambs to a much lesser extent. Southdowns are valued for showing, particularly by youth clubs. Dorsets have a reputation for less restricted breeding seasons than the other meat breeds. The polled strain, which originated in North Carolina, is increasing in numbers. The Cheviots, which are

Fig. 9-2. Rambouillet ram. [Courtesy USDA.]

Fig. 9-3. American Merino ewe. [Courtesy USDA.]

of fairly small body size, and the Oxfords, with faces that are generally covered, are somewhat less numerous than the other meat breeds. The Tunis breed is quite rare although there is an active breed registry organization.

Crossbred dual-purpose breeds have increased in popularity in all parts of the United States in recent years. Their heavy fleeces of medium wool, along with rapid growth rates, good meat conformation, open faces, and high lamb production, have made them both profitable and popular. These breeds all originated from crosses of long-wool and fine-wool breeds.

The Corriedale, the oldest of these breeds, resulted from crossing Leicester or Lincoln rams on Merino ewes and was imported from New Zealand in 1914. It was more common in the western range states during the early years, but in recent years has become much more numerous in the farm states of the Middle West.

The Columbia was developed by the United States Department of Agriculture by crossing Lincoln rams on Rambouillet ewes (Marshall, 1949). The crossbred offspring were then interbred. This, the first breed to be developed in the United States, is noted for its large size and heavy fleeces. The breed Association formed in 1941 has pioneered with an inspection system of registration that requires that animals meet a standard of excellence as

Fig. 9-4. Suffolk ram. [Courtesy USDA.]

Fig. 9-5.   Hampshire ram. [Courtesy USDA.]

determined by an Inspector of the Association, as well as being offspring of registered parents. A similar breed, the Panama, formed by crossing Rambouillet rams on Lincoln ewes, is most common near its area of origin in Idaho.

The Targhee was developed by the United States Department of Agriculture by crossing Rambouillet rams on Lincoln-Rambouillet, Corriedale-Lincoln-Rambouillet, Corriedale, and Columbia ewes (Terrill, 1947). Some other breeding was used, but the breed is made up of about three-quarters fine-wool and one-quarter long-wool breeding. This breed also has an inspection system of registration and permits flock as well as individual registration. The Targhee is well established in the western states. Crossbred types or strains of sheep similar to the Corriedale, Columbia, Panama, and Targhee are common throughout the western ranges, particularly in the Intermountain and Northwest areas.

The Romeldale, developed by A. T. Spencer of California by crossing Romney rams on Rambouillet ewes, is not common outside its area of origin.

The Montadale resulted from crossing the Cheviot and Columbia (Mattingly, 1945), and is found largely in the farm states.

Fig. 9-6.   Shropshire ewe. [Courtesy USDA.]

Fig. 9-7. Southdown ram. [Courtesy USDA.]

A tailless or "No Tail" sheep has been bred by the South Dakota Agricultural Experiment Station, using fat-rumped sheep as foundation animals. This breed has not been released to private breeders, and is being discontinued.

The long-wool breeds, although extremely important in the formation of crossbred types or strains, are bred pure in only small numbers. Breeds such as the Leicester and Cotswold have largely disappeared. The Romney is most numerous in the warm damp climate of the Willamette Valley in Oregon. This breed has not been as successful in crossing to produce range sheep as the Lincoln. The Lincoln breed, although small in number, is often in demand for crossing with Rambouillets or fine-wool range sheep to produce first-cross replacement ewes. These first-cross ewes are highly productive of both lambs and wool.

Carpet-wool, mixed-wool, and fur breeds of sheep are of little importance in the United States, because importation of duty-free carpet wools has resulted in low prices for domestic wools of these types. Scottish Blackface sheep have largely disappeared, and only a few Karakul flocks remain. Navajo sheep of New Mexico (Blunn, 1940, 1943) have been used to produce wool for weaving the famous Navajo blankets. However, Navajo sheep have been top-crossed with Rambouillet rams and only a small flock of

Fig. 9-8. Dorset ewe. [Courtesy USDA.]

relatively pure but improved Navajo sheep still exists at the Southwestern Range and Sheep Breeding Laboratory at Fort Wingate, New Mexico. Improved, coarse-weaving-wool sheep have been developed by this laboratory.

Purebred sheep are far more important to production than their numbers indicate because they provide sires for a large part of the sheep population. In 1920 there were almost 500,000 registered purebred sheep in the United States, a number representing slightly more than 1% of the total sheep and lambs (Spencer et al., 1924). The proportion seems to be similar today, although data are unavailable.

Some of the prominent breeds of today were well established in the United States 40 years ago. In 1920 the Shropshire, Rambouillet, Merino, and Hampshire made up 86% of the registered purebred sheep (Spencer et al., 1924). In 1963 these breeds made up 36% of the registrations; additional breeds, either absent or of minor importance in 1920, the Suffolk, Corriedale, and Columbia, now make up 45% of the registrations. The leading breeds, based on number of registrations in 1963, are the Suffolk, Hampshire, Corriedale, Southdown, Rambouillet, Columbia, Shropshire, and Dorset breeds, which account for 93% of the sheep registered (*Breeder's Gazette,* March, 1964).

Experimental breed comparisons of sheep have been uncommon. Cooper and Stoehr (1934) under range conditions found Columbias to be heaviest in body weight, followed by Rambouillets and Corriedales. Columbias and Rambouillets produced heavier fleeces than Corriedales, and Columbias and Corriedales produced greater numbers and pounds of lamb per ewe than did Rambouillets. Losses from dead and missing sheep were lowest for Rambouillets and highest for Columbias. Terrill and Stoehr (1942), studying the same breeds under the same conditions, found that Corriedales produced more pounds of lamb per pound of ewe than did Columbias or Rambouillets, and Rambouillets produced more pounds of wool per pound of body weight than did Corriedales or Columbias. Bennett et al. (1963) found Targhees, Rambouillets, and Columbias ranked in that order in lambs weaned per

Fig. 9-9. Cheviot ewe. [Courtesy USDA.]

Fig. 9-10. Corriedale ram. [Courtesy USDA.]

ewe bred. The order for average weaning weight and pounds of clean wool was Columbia, Targhee, and Rambouillet.

Branaman (1940) compared Hampshires and Southdowns under farm conditions. Hampshires produced more lambs that reached market weight earlier. The estimated total digestible nutrients required for each hundred pounds of final feed-lot weight were practically the same for Southdown and Hampshire single lambs, but the yield of lean meat was slightly higher for the Southdowns. Sidwell et al. (1962, 1964) found that Hampshires, Merinos, Shropshires, and Southdowns ranked in that order in lambs weaned per ewe bred, and in the same order, except that Shropshires preceded Merinos, in average weaning weight (lambs weaned per 100 ewes: 100.3, 98.0, 80.3, and 78.6, respectively).

These studies are among the few in which controlled experiments have been used to compare breeds of sheep. Progress in sheep production may have been retarded by the paucity of such comparisons, from which producers could more adequately discriminate among breeds.

9-5. USEFULNESS OF PUREBRED SHEEP OF THE VARIOUS BREEDS

Purebred sheep are useful in providing relatively uniform types adapted to special uses or to unique climatic conditions, and they can be expected to breed true in their particular characteristics. Fine-wool breeds, such as the Rambouillet, are adapted to warm, dry climates, while coarse long-wool types, such as the Romney, are adapted to damp, humid climates. It seems especially important that the type of wool be suited to the climatic conditions. The crossbred intermediate type like the Columbia and Targhee thrive in cool and temperate climates under a wide variety of conditions. In general, breeds of medium to small size, such as the Merino, are better suited to sparse feed conditions while larger breeds, such as the Columbia or Lincoln, do best under lush feed conditions. White-faced wool breeds, such as

Fig. 9-11. Columbia ram. [Courtesy USDA.]

the Rambouillet, Columbia, and Targhee, are suited to western range conditions, probably because of their Merino blood. The meat breeds are generally kept pure for commercial production only in the farm states. The western purebred meat breed flocks are kept under more ideal nutritive conditions to produce sires of crossbred slaughter lambs.

The breeder can choose among about 15 to 20 common breeds of sheep in the United States to find one which best fits his particular situation. A safe guide for the beginner is to choose the breed that is already most common in the area. Market demand for breeding stock is an important consideration. Commercial flocks of high-grade or nearly purebred sheep can be readily established by continually topcrossing with a particular breed of sire. Three or four top crosses will generally produce sheep very similar to purebreds while five or six top crosses can be expected to produce those almost identical to purebreds.

The gain in production or hybrid vigor from crossing breeds is an important advantage made possible by pure breeds. Sheep breeders have long recognized that crossbreeding results in progeny that are more fertile and have higher growth rates than the average of purebred parents. Increased wool production and milk production have also resulted from crossbreeding.

Fig. 9-12. Targhee ram. [Courtesy USDA.]

Fig. 9-13. Lincoln ram. [Courtesy USDA.]

The crossing of a third breed on two-breed cross ewes or a fourth breed on three-breed cross ewes has given added advantages (Sidwell et al., 1962, 1964). In addition, improved meat type of offspring of wool-type ewes, such as Rambouillet or Merino, is obtained by crossing with the meat breeds, thereby increasing the efficiency of production of both lambs and wool. The many comparisons that have been made of sire breeds for production of lambs for meat from a variety of native ewes have been reviewed by Rae (1952). In general these comparisons show Hampshire or Suffolk sires to be superior in the United States

Interbreeding of crossbred animals has not resulted in the marked increase in variability which is often cited as a disadvantage of crossbreeding (Rae, 1956), although variability of wool fineness in the flock may increase, particularly where widely different wool types are crossed. In general, interbreeding of crossbred animals does not provide the predictability of characteristics in the offspring that is found with pure breeding. On the other hand two- and three-breed crosses may be quite predictable and the offspring may be as uniform as purebreds. Crossbreeding, particularly to produce the wool-type mother of the market lamb, has not been as fully exploited in the United States as has the use of the meat breed sire on the wool-type ewe. Experimentation is needed to develop and test patterns of crossbreeding so that there may be maximum efficiency in production of high-quality lambs and wool.

Crossing of existing breeds to form still more new breeds continues. Present efforts seem to be aimed at the development of white-faced meat breeds and meat breeds that have less restricted breeding seasons than the British mutton breeds. Outcrossing to improve existing breeds, which is so common in the Soviet Union, is generally not practiced in this country. An exception is the use of a number of long-wool and medium-wool breeds to improve the Navajo sheep of New Mexico.

Constant improvement of breeds continues with selection, aided by performance and progeny testing (Terrill, 1958). Those breeds that show the greatest improvement in efficiency of production are apt to grow in popularity while those whose breeders continue to emphasize fancy points and uneconomic traits, such as wool covering on the face, small scurs, or color on the legs, can be expected to decline. Breeds are in competition with each other and survival in the long run will probably depend more on their relative productivity and efficiency of production than on promotion efforts or on show-ring performance. Many breeds in other parts of the world, such as the Dormer, German Whitehead Mutton, Ile-de-France, Precoce, Texel, East Friesian, and Finnish Landrace, could probably be used to advantage in the United States, either as pure breeds in crosses with present breeds to produce hybrid vigor or to produce additional new breeds. This has generally been prevented by restrictions on importation of sheep from most of the world in order to avoid the introduction of exotic diseases.

9-6. BREEDS OF GOATS

Goats are similar to sheep and some are difficult to distinguish from sheep, although the tail of the goat is said to turn upward always, and that of the sheep downward. It is not certain that crosses of sheep and goats have produced viable offspring, but fertilization and early embryonic growth do occur (Gray, 1954). Goats, as well as sheep, produce a wide variety of products; milk, mohair, meat, and skins are most important.

9-7. MILK GOATS

Milk goats are most frequent in the suburbs of large cities and in small towns. A good milk goat will supply sufficient milk for a large family and can be kept where it would be impossible to keep a cow. Goat dairies are also common near large cities. Goat's milk differs from cow's milk, particularly in the small size of the fat globules and the soft curd. These probably contribute to its ease of digestibility. It is often used for infants, invalids, and those allergic to cow's milk. It is estimated that approximately one million goats are kept for milk production in the United States. Much of the information on milk goats presented here is taken from Potts and Simmons (1955). Information on characteristics of breeds of milk goats are summarized in Table 9-3.

The Saanen and Toggenburg breeds are most prominent in the United States, followed by the Nubian and various Alpine breeds. The Murcian breed, introduced from Spain and adapted to hot temperatures, is of fairly minor importance. Another minor breed is the Norska, which was brought from Norway many years ago. Common American and Spanish goats are kept in many parts of the United States. These are sometimes used for milk, although they are primarily meat animals, and are valuable for the clearing of brush land. These common breeds can be graded up for milk goats by topcrossing with dairy breed bucks.

TABLE 9-3. | *Information on breeds of milk goats in the United States.*

Breed	Country of Origin	Approximate Date of First Importation	Color Markings	Horns	Size	High Record of Production for Approximately 10 Months* Milk (lb.)	Fat (lb.)
Saanen	Switzerland	1904	Cream to white	Hornless, but some exceptions	Large	4,905	191.2
Toggenburg	Switzerland	1893	Brown with white stripes on each side of face; white legs	Hornless, but exceptions common	Medium	5,750	202.5
Nubian	Nubia (Upper Egypt, Ethiopia)	1896	Black, dark brown, or tan; with or without white markings	Hornless, but some horned males	Large	4,362	193.9
French Alpine	France	1922	White to black; often with spotting	Hornless, but exceptions common	Large	4,655	171.5
Swiss Alpine	Switzerland	ca. 1920	Brown with black markings	Usually hornless	—	2,016	128.0
Rock Alpine	United States	—	White to black; often with spotting	—	—	2,793	95.3
American La Mancha	United States	—	Black, brown; with or without white markings	—	—	3,295	118.5

Source: Information taken from Potts and Simmons (1955), except for production records.
* From American Milk Goat Record Association Handbook (1964).

Fig. 9-14. Saanen doe. [Courtesy USDA.]

The Saanen is one of the leading breeds of dairy goats. During the years when the United States Department of Agriculture maintained a herd at Beltsville, Maryland, the grade and purebred Saanen does milked from 8 to 10 months after kidding and produced an average of 5.6 lb. of milk per day. The average butterfat content of the milk was approximately 3.5%. Mature Saanen does will average 120 lb. and bucks 185 lb. in body weight.

The Toggenburg breed, including many high-grade animals, is probably the most numerous of the goat breeds in numbers. Wattles or appendages from the neck are generally characteristic of this breed. Records from the United States Department of Agriculture herd at Beltsville, Maryland, show that grade and purebred Toggenburg does have milked from 7 to 10 months after kidding and produced an average of 4.1 lb. of milk per day. The milk-fat content of the milk has averaged about 3%. The average weight of mature does over a three-year period was 96 lb.

The Nubian breed is characterized by large drooping ears and a peculiarly shaped head. Although a large breed, it appears to be less hardy than the breeds of European origin. The distinct odor so prevalent in the male goat, particularly during the breeding season, is somewhat less pro-

Fig. 9-15. Toggenburg doe. [Courtesy USDA.]

nounced in the Nubian. The milk is especially high in milkfat, yielding 5% in some cases.

The French Alpine breed is large and hardy, with heavy bone, and shows a good capacity for milk production. It appears to be increasing in popularity. The Rock Alpine has resulted from a combination of the Swiss and French Alpine. A few British Alpine goats have been imported. These have resulted from general crossbreeding, but with the Saanen and Toggenburg predominating.

The American LaMancha breed was developed in the United States from a few LaMancha goats that came from Spain to Mexico prior to 1925. Crosses were made with high-producing individuals from other breeds in California during the 1930's. A herdbook was opened by the American Milk Goat Record Association in 1954. Standards for the breed were approved in 1957 and a group of basic American LaManchas were selected.

9-8. ANGORA GOATS

Angora goats appear to have originated before Biblical times, as a number of references to the use of goats' hair are found in the Bible (Thompson, 1901). The area of origin surrounding Ankara (Angora) in Asiatic Turkey is a high plateau of about 3,000 feet elevation, with mountains and deep valleys. The climate, which is dry with extremes in temperatures, was probably favorable for the development of the long, lustrous mohair.

Angora goats were first imported into the United States in 1849. They spread westward after the Civil War, particularly to Texas and California. Texas has been the leading state ever since. In recent years, Texas, with more than 4 million head, has had about 95% of the Angora goats in the United States. Other leading states are Arizona, New Mexico, Missouri, Oregon, California, and Utah.

The Angora goat with its long locks of mohair presents a distinctive appearance. The locks are of different types, such as ringlet, flat, or web, as

Fig. 9-16. Angora buck. [Courtesy Sheep and Goat Raiser.]

described by Gray (1959). The mohair locks grow at the rate of 6 to 12 inches per year, and the goats are normally clipped twice each year. Mohair becomes coarser with age, and the fine kid mohair is most valuable. Quality, covering, weight of fleece, and freedom from kemp is emphasized in selection. The average mohair clip per goat has increased from about 5 lb. in the late 1940's to about 6.5 lb. in the early 1960's. Both sexes have horns. Open faces are associated with higher kid production (Shelton, 1960). Size and weight for age are also emphasized in selection. A yearling buck should weigh at least 80 lb., mature bucks from 125 to 175 lb. Yearling does should weigh at least 60 lb. and mature does from 80 to 90 lb.

9-9. SUMMARY

A brief account has been given of the development of breeds of sheep and goats, especially in the United States. A variety of available breeds have been described from which the producer may choose the one that best fits his particular needs, whether for pure breeding, for grading up to high-grade flocks, or for taking advantage of hybrid vigor from crosses. Production traits are emphasized because breeds of sheep or goats are useful only as they contribute to efficient production of meat, milk, wool, and mohair.

REFERENCES AND SELECTED READINGS

References marked with an asterisk are of general interest.

*AMGRA Handbook, 1964. American Milk Goat Record Association.

Breeder's Gazette, 1964. Annual Purebred Review. March.

Bennett, J. A., D. H. Matthews, and M. A. Madsen, 1963. *Range Sheep Breeding Studies in Southern Utah.* Utah Agr. Exp. Sta. Bull. 442.

Blunn, Cecil T., 1940. Improvement of Navajo sheep. *J. Heredity,* 31(3):99–112.

———, 1943. Characteristics and production of old type Navajo sheep. *J. Heredity,* 34(5):141–152.

Branaman, G. A., 1940. Some factors in lamb production associated with size and type in mutton sheep. *J. Agr. Research,* 60(7):473–486.

*Briggs, Hilton M., 1958. *Modern Breeds of Livestock.* Revised ed., Macmillan, New York.

*Carmen, Ezra A., H. A. Heath, and J. Minto, 1892. *Special Report on the History and Present Condition of the Sheep Industry of the United States.* USDA Bureau of Animal Ind., Washington, D.C.

*Connor, L. J., 1918. *A Brief History of the Sheep Industry in the United States.* Annual Report of the American Historical Association.

Cooper, J. M., and J. A. Stoehr, 1934. *Comparison of Rambouillet, Corriedale, and Columbia Sheep Under Intermountain Range Conditions.* USDA Circular 308.

Gray, A. P., 1954. *Mammalian Hybrids.* Technical Communication No. 10 of the Commonwealth Bureau of Animal Breeding and Genetics, Edinburgh.

Gray, J. A., 1959. *Texas Angora Goat Production.* Texas Agricultural Extension Service B-926.

Marshall, F. R., 1949. The Making of Columbia Sheep. *National Wool Grower,* 39(4):9, 32.

*Mason, I. L., 1951. *A World Dictionary of Breeds, Types and Varieties of Livestock.* Technical Communication No. 8 of the Commonwealth Bureau of Animal Breeding and Genetics, Edinburgh.

Mattingly, E. H., 1945. The Montadale. A new breed. *Sheep Breeder,* 65(3):4.

*Potts, C. G., and V. L. Simmons, 1955. *Milk Goats.* USDA Farmers' Bulletin 920.

Rae, A. L., 1952. Crossbreeding of sheep. *Animal Breeding Abstracts,* 20(3):197–207 and (4):287–299.

———, 1956. The genetics of the sheep. *Advances in Genetics,* 8:189–265.

*Reed, Charles A., 1959. Animal domestication in the prehistoric Near East. *Science,* 130(3389):1629–1639.

Shelton, Maurice, 1960. The relation of face covering to fleece weight, body weight and kid production of Angora does. *J. Animal Sci.,* 19(1):302–308.

Sidwell, G. M., D. O. Everson, and C. E. Terrill, 1962. Fertility, Prolificacy and Lamb Livability of Some Pure Breeds and their Crosses. *J. Animal Sci.,* 21(4):875–879.

Sidwell, G. M., D. O. Everson, and C. E. Terrill, 1964. Lamb Weights in Some Pure Breeds and Crosses. *J. Animal Sci.,* 23(1):105–110.

*Spencer, D. A., M. C. Hall, C. D. Marsh, J. S. Cotton, C. F. Gibbons, O. C. Stine, O. E. Baker, V. N. Valgren, R. D. Jennings, G. K. Holmes, W. B. Bell, and Will C. Barnes, 1924. The sheep industry. In *United States Department of Agriculture Yearbook, 1923,* pp. 229–310.

Terrill, C. E., 1947. Breed crosses used in the development of Targhee sheep. *J. Animal Sci.,* 6:83–92.

———, 1949. The relation of face covering to lamb and wool production in range Rambouillet ewes. *J. Animal Sci.,* 8(3):353–361.

———, 1953. The relation between sale price and merit in Columbia, Targhee and Rambouillet rams. *J. Animal Sci.,* 12(3):419–430.

*———, 1958. Fifty years of progress in sheep breeding. *J. Animal Sci.,* 17(4):944–959.

———, and J. A. Stoehr, 1942. The importance of body weight in selection of range ewes. *J. Animal Sci.,* 1(3):221–228.

Thompson, G. F., 1901. *Information Concerning the Angora Goat.* USDA Bull. 27.

*Wentworth, E. N., 1948. *America's Sheep Trails.* Iowa State College Press, Ames, Iowa.

Chapter 10

Breeds of Horses

See from the first yon high-bred colt a field,
His lofty step, his limbs' elastic tread:
Dauntless he leads the herd, still first to try
The threatening flood, or brave the unknown bridges
But no vain noise affrighted; lofty-necked,
With clean-cut head, short belly, and stout back;
His sprightly breast exuberant with brawn.

VIRGIL, *Georgics, III*

10-1. HISTORICAL ASPECTS AND SOME CHARACTERISTICS OF FORMATION OF BREEDS OF HORSES

Early ancestral forms of the horse are known to have existed in both western and eastern hemispheres, but apparently evolution in the eastern hemisphere ended in early extinction. Thus the evolution of the modern horse from primitive forms took place largely in the Americas, and from there the most recent in line of descent migrated to Asia and Europe during the early Pliocene time by way of land connections through the Bering Sea. Later, during the Ice Age, the horse became extinct in the Americas but survived in the eastern hemisphere (Riggs, 1932).

Primitive Forms and Their Relative Importance in Formation of Modern Breeds. The nature of the immediate predecessors of the modern horse has not been determined, but a widely accepted view is that three distinct primitive stocks have contributed to present-day breeds (Ridgeway, 1905).

The Celtic horse, which was found in the Hebrides and a part of Ireland, is thought to have been the progenitor of pony breeds. The ancient horse of Europe, the Near East, and Britain probably derived from another wild stock that was widely distributed in Europe and Asia. A third type is thought to have originated in the deserts of North Africa and to have given rise to the horses of the Barbary States, Egypt, Arabia, and Persia. There is some historical evidence that the horses of North Africa were introduced into Europe by invaders from North Africa as early as the third century B.C. and were probably imported often thereafter, mainly by way of Spain. Thus there was probably some admixture of the stocks in Europe, as well as in Asia Minor, from very early times. Usually, however, the North African type is given credit for having played the more important role in the formation of modern breeds of light horses; the European type is credited with having been the ancestor of draft breeds, as well as having played a lesser role in development of light-horse breeds. In general, draft breeds are thought to have been developed directly from native horses in western Europe and Britain, only the Percheron breed having been improved by the use of Arabian breeding.

Improved Stocks and Their Use in Breed Formation and Improvement

BARB, ARAB, AND RELATED STRAINS. All of the important light horse breeds throughout the world have descended from or have been largely improved by descendants of the North African horse. These horses, the Barb of North Africa and the Arab and related strains, the so-called Eastern or Oriental horses, were greatly improved during several centuries by intelligent selection and breeding practices. The most refined of these, the Arab, is not a large horse but is characterized by a beautiful head and neck, strong shortly coupled back, good chest capacity, and excellent feet and legs. Though not unusually fast, the Arab has always been noted for great refinement and unusual stamina. The earliest introduction of these horses to Spain, and the subsequent importations during the years of the conquest and occupation by the Moorish invaders, resulted in the great improvement of native stocks so that horses of Spain became recognized as being of excellent quality. These were the horses, brought to America by Spanish conquerors and colonists, that provided the seed stock for the western mustang and the horses that eventually spread throughout the eastern part of the United States.

THOROUGHBRED. At a somewhat later time the Arab and related horses were used in a more methodical program of improvement of horses native to Europe and Britain and in formation of new breeds including the English Thoroughbred. The Thoroughbred horse resulted from the breeding of Barb, Arab, and Turk stallions to mares indigenous to the British Isles at that time (1660–1750) as well as to imported mares of breeding similar to their own. The primary impetus to develop this breed appears to have been the desire to breed a horse with great speed and stamina for use on the race-course. By 1750 this aim had been realized to such a degree that the importation of more Eastern sires failed to improve the stock greatly. The Thoroughbred has played a large and important role in the development of all modern breeds of light horses found in America except the Appaloosa and, of course, the Arabian.

Development of New Breeds

STIMULI FOR BREED FORMATION. Most light horse breeds have developed in response to fairly specific stimuli—for example, the Thoroughbred in answer to the requirement for speed and stamina on the racecourse. Later the development of the Morgan breed occurred in response to the need for an all-round useful horse and road horse in America, at about the time of the American Revolution. As roads improved the demand for speed in harness—both for transportation and in racing—increased, and the larger, faster American Standardbred was developed, drawing heavily upon the Thor-

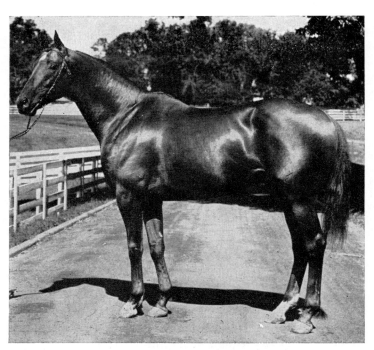

Fig. 10-1. Swaps, a fine example of the Thoroughbred horse, Kentucky Derby winner with earnings of over $800,000. Now at stud, he has sired numerous stakes winners including Chateugay, winner of the 1963 Kentucky Derby. [Courtesy Mr. John W. Galbreath.]

Fig. 10-2. Go Man Go, a top money-winning racing Quarter Horse and now a leading sire. He was sired by the Thoroughbred horse Top Deck. Note heavy muscling of rear quarters, a desirable conformation trait for short distance racing. [Courtesy the American Quarter Horse Association.]

oughbred for size, speed, and stamina. Similarly the American Saddle Horse breed was formed to meet the need for a good, smooth, and easy traveling horse under saddle. Impetus for the greatest development of the American Quarter Horse as a breed arose from the need for a horse suitable for working cattle. The characteristics desired were at the same time compatible with tremendous speed for short distances. Often the requirements that were largely responsible for the formation of a breed have disappeared or have become relatively unimportant; for example, the need for horses solely for transportation, either on horseback or in horse-drawn vehicles, no longer exists. For this reason the Standardbred owes its continued existence to its use in harness racing and in the horse-show roadster classes. Principal uses of American Saddle Horses are now in the horse shows and as pleasure

Fig. 10-3. Cuco Britches, a working Quarter Horse in action, is a good representative of the stock horse type with innate affinity for working with cattle.

horses. In more recent years uses of Quarter Horses have tended to become categorized so that most are bred either for racing short distances or for stock-horse work, including showing at halter and in western performance classes, rather than for both.

Some of the more recently formed breeds and registries have been established not on the basis of functional requirements, but on the basis of characteristic color or color patterns. Some, such as the palomino, are in fact color registries only, and horses registered in any of several other breed registries may also be listed in them. Thus they include an extremely heterogeneous genetic group. The Appaloosa registry, which also falls in this general category, lists three acceptable basic color patterns, but these cover a wide range. An extreme range in conformational characters is also permitted. A factor of importance in selection of foundation stock for formation of breeds for special purposes has been the natural tendency of certain individual horses or breeds to perform specific natural gaits well, or to learn acquired gaits more readily.

SELECTION OF SUITABLE FOUNDATION STOCK. Whenever the need for a horse to perform a job has been a force for breed formation, the foundation stock has been selected from any suitable source available. For this reason all major breeds of horses developed in America have a number of other breeds and usually a considerable number of individuals of unknown breeding in their foundation stock. Usually the Thoroughbred or breeds tracing strongly to the Thoroughbred have provided the main basis for refinement, intelligence, speed, and stamina.

FORMATION OF BREED ORGANIZATIONS AND REGISTRIES. Most of the breed organizations and registries were formed late in the developmental stages of the breeds, and selection and admission of so-called foundation stocks took place "after the fact." These organizations have pursued various policies in order to develop the type of horse considered desirable for the breed.

In the early stages of the development of a breed, provisions are frequently made for allowing the registration of animals whose parents have not been registered. The Quarter Horse Association, for example, maintains an open registry. Prior to 1962 eligibility for registration of outside animals was based upon a complicated system which included the evaluation of performance and conformation (Briggs, 1958).

Since 1962 no *outside* animal may be entered in the registry except those which have as one parent a registered Quarter Horse and the other a registered Thoroughbred. Such animals must achieve a specified level of performance and pass inspection for suitable conformation prior to permanent registration in the American Quarter Horse Stud Book and Registry.

Examples of other breeds that retain procedures whereby desirable individuals may be admitted to registry are the Standardbred and the American Saddle Horse. In practice relatively few horses are admitted to the Standardbred and American Saddle Horse Registries; outside blood, now restricted to Thoroughbred, is used freely in Quarter Horses. In 1964 about 20% of racing Quarter Horses that qualified for register of merit were sired by registered Thoroughbreds (*Quarter Horse Journal,* 1965). The reciprocal cross is also popular, and many racing Quarter Horses have one half or more Thoroughbred breeding from crosses in the second and later generations. At the other extreme of the Quarter Horse registry, many halter and performance register-of-merit qualifiers have had no outcrosses for at least several generations. Thus the degrees of homogeneity of type and genetic constitution vary greatly from breed to breed. Generally it may be said that progress toward homogeneity is greatest in breeds that are selected on the basis of performance, especially when objective measures of performance are available.

SUCCESS OF SOME BREED ORGANIZATIONS IN THE ATTAINMENT OF DESIRED TYPE AND HOMOGENEITY OF GENETIC BASE. The Thoroughbred, by virtue of its long-closed stud book and selection for racing, probably has the most homogeneous genetic base of any breed. The American Standardbred, American Saddle Horse, and Tennessee Walking Horse have been rigidly selected for specific purposes, and reasonable constancy of type has resulted, at least as far as the specialty of the breed is concerned. The conformation of Standardbred or the Tennessee Walking Horse is relatively less important to breeders than is performance; as a result there are wide variations in size and conformation in these breeds. Because it has long been, first and fore-

Fig. 10-4. Plainview's Julia, winner of the $10,000 five-gaited stake at the Kentucky State Fair in 1960, shows the extreme refinement and elegance of the American Saddle Horse. The long, nicely slopping shoulders and pasterns, so important to a smooth, comfortable ride, are well demonstrated. [Courtesy the American Saddle Horse Breeder's Association.]

most, a show horse, conformation and quality of performance of varied gaits have been important in selection of breeding stock in the American Saddle Horse breed. Today this breed has no equal in the number and style of gaits it performs, nor does any breed equal it in elegance and refinement of conformation.

The American Quarter Horse breed must be considered to be still in its formative stage. While breeders of stock-horse types have relied to a great extent upon animals already in the registry for breeding stock for several generations, racing Quarter Horses are commonly not more than a generation or two removed from the Thoroughbred. No doubt if the stud books were closed to Thoroughbreds, dilution of racing Quarter Horse strains with stock-horse strains would result in a slower breed, and recently set track records might stand for a long time. The problem of fixing the breed type appears to be a serious one because of the divergent ideas of members of the breed organization on just what the uses of the Quarter Horse should be.

10-2. TYPES OF HORSES AND THEIR TRAITS

The primary uses of the horse as a source of power for the accomplishment of work or as a means of transportation—either for pleasure, sport, or facilitating performance of the rider's task—require that horses be of good substance and properly conformed so that they function efficiently and without defects in their manner of travel or performance. Often faults of conformation either directly cause or are highly correlated with defective ways of going. Forefeet that toe out or in will cause the feet to travel in an inward or outward arc, respectively, and to be predisposed to interference. Thickness at the shoulder causes a rolling action and resultant lost motion. Short upright pasterns are associated with rough, choppy gaits, and a steep croup often causes a horse to overstep and thus to interfere. Equally important is the fact that all of these faults of conformation strongly predispose a horse to unsoundnesses.

Because the tasks they perform are so varied, horses must possess special traits that suit them to particular jobs.

Draft Horses. The draft horse is almost unique because the primary requirement for it is the development of a great deal of power. Some of the most important factors in fulfilling this requirement are relatively heavy weight, heavily muscled back, loin, and quarters, and—as in all horses—sound feet and legs. The chief means of developing the power required is the forward displacement of the center of gravity. This is achieved by the support of much of the weight of the forequarters by the muscles of the back, loin, and rear quarters, and by the extension of the rear legs by muscles of the rear quarters. It may be seen that for efficient function the requirement for conformation of head, neck, and forequarters is such that a rela-

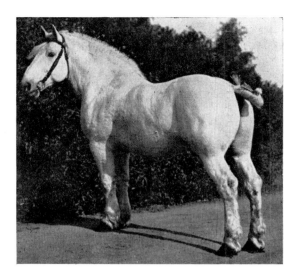

Fig. 10-5. Fernand, an imported Percheron stallion that stood at the University of California, Davis, from 1916 to 1918, when draft horses were the primary source of farm power.

tively larger proportion of the animal's total weight must rest upon the fore-limbs than in light horses. For the same reason a slightly greater length of back may be tolerated in draft horses, provided there is ample muscling in that region.

Since great power for relatively long periods is required, stamina is important. Most horsemen look for indications of this almost intangible quality in a deep heart girth and roomy middle. This seems reasonable because such configuration is assumed to provide ample respiratory capacity, space

Fig. 10-6. Of historic interest are the large hitches of draft horses and mules such as these; they were a common part of the agricultural scene in the western United States prior to their displacement by power machinery during the early part of the twentieth century.

TABLE 10-1. *Characteristics and uses of the more important modern breeds of horses.*

Breed	Approximate Period of Breed Formation or Most Important Development	Colors	Height (hands)	Weight (lb.)	Most Important Fields of Use
		LIGHT HORSES			
Arabian	100–600 A.D.	Bay, gray, chestnut, brown	14–2″ to 15–2″	850–1,000	General pleasure
Thoroughbred	1660–1800	Bay, chestnut, brown, black, gray, roan	15–2″ to 17	1,000–1,300	Racing, hunting, polo, general pleasure, stock horse
Morgan	1790–1850	Bay, chestnut, black, brown	14–2″ to 15–2″	950–1,150	General pleasure
Standardbred	1800–1875	Bay, chestnut, brown, black, gray, roan	14–2″ to 16–2″	850–1,200	Harness racing, horse-show roadsters
American Saddle Horse	1840–1890	Chestnut, bay, brown, black, other	15 to 15–3″	1,000–1,150	Gaited saddle and fine harness, horse shows; pleasure riding
Tennessee Walking Horse	1890–1935	Bay, chestnut, black, brown, roan, gray, sorrel, white, yellow	15 to 16	1,000–1,200	Horse-show walking horse, pleasure riding
Hackney	1760–1885	Bay, chestnut, black, brown	15 to 16		Horse shows, heavy harness
American Quarter Horse	1850 to date	Bay, chestnut, brown, black, roan, gray, dun, palomino	14–2″ to 15–2″	1,000–1,250	Stock horses, rodeo, short racing, pleasure riding
Appaloosa*	1938–1949	Various patterns of spotting, speckling, mottling on various colors of background	14–2″ to 16	900–1,200	General pleasure, parade, stock horse

DRAFT HORSES

Percheron	1800–1885	Black, gray, brown, chestnut, bay	15–2″ to 17	1,600–2,200	Draft
Clydesdale	1720–1880	Bay, brown, black, other	15–2″ to 17	1,700–2,000	Draft
Shire	1780–1880	Bay, brown, black, other	16–2″ to 17	1,800–2,200	Draft
Belgian	1850–1900	Chestnut, roan, bay, brown, gray	15–2″ to 17	1,900–2,400	Draft
Suffolk	1770–1880	Chestnut	15–2″ to 16–2″	1,500–1,900	Draft

PONIES

Shetland	1870–1871	Black, brown, bay, chestnut, mouse, spotted	9–2″ to 10	300–400	Child's mount, horse shows
Welsh	Unknown	Chestnut, bay, gray, black, roan	11 to 13	350–500	Child's mount, horse shows
Hackney	1760–1885	Bay, Chestnut, black, brown	11–2″ to 14–2″	450–850	Heavy harness pony in horse shows

* Strictly speaking, Appaloosa is not a breed but a color registry.

for vital organs, and the ability to handle adequate amounts of nutrients.

Although appearance is secondary to utility, most of today's few purebred draft horses in this country are used for show, and balance of conformation, refinement, and breed type are emphasized. Since strict utility, usually in teamwork, has always been demanded of draft types, a calm, tractable disposition and intelligence are essential.

Stock Horses. The stock horse of the kind found on western cattle ranches and in rodeo competition and horse shows is required to perform a great variety of tasks: he must be able to start and stop quickly; he must be able to show a great deal of speed for short distances; he must be extremely agile and capable of changing direction rapidly; he must have weight and strength enough to hold a steer on the end of a rope. Since he is usually expected to put in a full day at these activities, he must have great stamina and must necessarily be of calm, even, and tractable disposition, yet alert and ready for action.

Probably most important for rapid maneuvering, starting, stopping, turning, early speed, and so on, are powerfully muscled hindquarters and short, heavily muscled back and loins, since muscles of the rear quarters provide propulsive power, and shortness and strength of the back maximize lightness and handiness of the forequarters. Good feet and legs are important. For maximum speed there should be considerable length of limb. However, it appears that most often sheer power makes up for a lack in this respect in horses used solely for stock-horse purposes.

The ideal appearance of stock horses results when the conformation requisite to desired characteristics is achieved—that is, a relatively short, compact, powerfully muscled but well-balanced horse with well-defined withers to hold a saddle, with well-muscled, well-sloped shoulders for maximum support and absorption of concussion resulting from violent maneuvers, and with head carried moderately low to provide visibility and space for action of the rider and, incidentally, to enhance balance and speed.

Obviously the complexity of the tasks performed by stock horses makes a high degree of intelligence and learning ability a necessity. The great majority of stock horses in use on ranches and in sports and shows are of American Quarter Horse breeding with a few Thoroughbreds, Arabians, Morgans, and Appaloosas constituting the small remainder.

Race Horses. The demands made by racing are undoubtedly among the most strenuous of any made on the horse, especially on his feet and legs. Speed, great stamina, intelligence, and certainly feet and legs of the soundest kind are prerequisite. In general appearance there is considerable variation in conformation of horses used for racing; for instance, compare the

Thoroughbred distance horse with the racing Quarter Horse or harness racing Standardbred. Nonetheless, certain characteristics are common to each and distinguish them from other types of light horses. As a rule height is somewhat greater in proportion to length in racing horses than in other types, because of the requirement for speed. A good slope of shoulder is necessary for maximum length of stride as well as for absorption of concussion at racing gaits. Flat, smooth muscling without bulkiness of the shoulder is also essential for clean, unhampered action of the forelimb, and to reduce burdensome weight. Racing horses tend to have somewhat longer necks than other types but the heads are carried moderately low for best balance. Characteristically, muscling should be long, flat, and smooth in horses that are raced for longer distances of a mile or more. This tends to be true for Standardbred horses and for Thoroughbreds racing these distances. Muscling often tends to look heavier, rounder, and thicker in Thoroughbreds used for sprinting the shorter distances, and this type of muscular development reaches a maximum in racing Quarter Horses, where extreme speed for distances no greater than a quarter of a mile is most important. Thoroughbreds, particularly sires, that have shown great speed as sprinters have been used extensively in the development of the racing Quarter Horse and thus the two breeds tend to have many characteristics in common.

Pleasure Horses. The category of pleasure horse today covers an extremely wide range of types, and many pleasure types are synonymous with types produced only for show. The American Saddle Horse and the Tennessee Walking Horse, for instance, are probably used more in horse shows than in pleasure riding.

Because the requirements for pleasure horses vary so markedly from those for horses serving other purposes, it may be helpful to discuss briefly some of the desired traits.

GAITS. The natural tendency noted earlier of certain individuals and/or breeds to perform specific natural gaits well, or to learn acquired gaits readily, is a factor of special significance in the developing of some breeds of pleasure horses. The natural gaits are these:

1) *The walk*—a slow 4-beat (each foot strikes the ground separately) gait.
2) *The trot*—a rapid 2-beat gait in which the diagonal front and hind legs move in unison.
3) *The gallop*—a rapid 3-beat gait.
4) *The canter*—a restrained gallop. The American Saddle Horse performs this gait in a rocking-chair manner.
5) *The pace*—a rapid 2-beat gait in which the laterals move in unison. This is a natural gait for only a few individual horses, especially Standardbreds.

Fig. 10-7. A trotter in action. The two-year-old Standardbred, "Ayres," considered to have one of trottings "picture" strides. [Courtesy U.S. Trotting Association.]

Some acquired gaits are the following:
1) *The rack*—a fast and animated 4-beat gait. It is a striking gait, characterized by extreme knee action. Much stress is placed on the performance of this gait in judging 5-gaited horses.
2) *The slow gait.* There are actually several slow gaits that may constitute the fifth gait of the 5-gaited horse. They include the running walk, fox trot, and stepping pace. The running walk is a slow 4-beat gait in which the horse appears to be breaking out of the walk into the run. The fox trot is a slow, slightly broken, rhythmic trot. The stepping pace is the most common. There is less swaying than in the true pace, and the cadence of the laterals is slightly broken.

CONFORMATION. To the extent that conformation is related to performance, it is important in all breeds of pleasure horses, or, for that matter, in all breeds and classes of horses. In some breeds, such as the Tennessee Walking Horse, the main emphasis is on performance, and this breed thus lacks the refinement (especially in the head, neck, and topline) of the American Saddle Horse.

TEMPERAMENT. In pleasure horses it is desirable to have animation combined with gentleness and tractability.

COAT COLOR. Owners of pleasure horses take special pride in the appearance of their animals. Stylish action, animation, and spectacular coat color are prized especially in parade horses. Pinto, Palomino, and Appaloosa are distinctively colored horses.

Ponies. With the exception of the Hackney Pony, whose sole use is in horse shows as a heavy harness pony, the primary uses of ponies are for children's mounts and in horse shows. Characteristics desirable in a child's pony are about the same as those in general pleasure horses, with distinct emphasis on gentleness, tractability, and reliability. Since many of these ponies are shown in competition, they are also required to have an attractive appearance and to show breed characteristics. There has been a rather large increase of interest in ponies recently, with a resulting increase in numbers registered and, generally, improvement in quality.

Fig. 10-8. A pacer in action. The world record holder at one mile, Adios Butler. His record for the mile on a one mile track is 1:54¾. Compare action with that in Fig. 10-7. [Courtesy U.S. Trotting Association.]

REFERENCES AND SELECTED READINGS

References marked with an asterisk are of general interest.

* Briggs, H. M., 1958. *Modern Breeds of Livestock*. Macmillan, Revised Ed., p. 668.

1964 Register of Merit Qualities. *The Quarter Horse Journal,* 1965. 17(7):152.

*Ridgeway, W., 1905. *The Origin and In-*fluence of the Thoroughbred Horse*. Cambridge University Press. 538 pp.

*Riggs, E. S., 1932. *The Geological History and Evolution of the Horse*. Leaflet No. 13, Field Museum of Natural History, Chicago.

Chapter 11

Types, Breeds, and Crosses of Poultry

The cock, astrologer in his own way,
Began to beat his breast and then to crow,
And Lucifer, the messenger of day,
Began to rise and forth her beams to throw
And eastward rose, as you perhaps may know.
                    CHAUCER, *Troilus and Criseyde*

11-1. CHICKENS

Origin and Domestication. Domestication of fowl in India dates back at least to 1000 B.C., and records indicate that chickens were raised by the Chinese as far back as 1400 B.C.

Cockfighting has had as much to do with the domestication of the fowl as has its use as a source of food. The sport of cockfighting is many centuries old and is still popular in parts of Asia, in the islands of Sumatra, Java, and the Philippines, as well as in Mexico and Puerto Rico. Cockfighting existed in England for many years and was especially popular during the reigns of James I and Charles II. By act of parliament, cockfighting in England was declared unlawful in 1849.

Darwin believed that the wild jungle fowl of India (*Gallus gallus*) was the primary ancestor of all domesticated chickens. Because the Leghorn and other Mediterranean breeds closely resemble *Gallus gallus,* Darwin's theory would seem plausible. However, because the Cochin and other Asiatic breeds differ so markedly from the Mediterranean breeds, some authors believe that they had ancestors other than *Gallus gallus,* that resembled the Aseel breed.

Breed Formation in America. Hutt defines a breed as "a group of fowls related by descent and breeding true to certain characteristics which the breeders agree to recognize as the ones distinguishing the breed." In general, body type or shape determines the breed, although there are some exceptions. Within each of the breeds there are different varieties, distinguished by different plumage colors, patterns, and comb shape.

The origin of recognized breeds and varieties is rather obscure except for those developed in England and America. However, enough information is available so that the influence of imported stocks on the poultry in the United States can be fairly well traced.

Poultry shows have played a very important part in the development of breeds in America. The first poultry exhibit in the United States was held in Boston in 1849 and this city has held a show each year since that time. Subsequently, other poultry shows have flourished in essentially all parts of the United States and Canada. Competition among exhibitors has been an important stimulus to the perfection of recognized breeds and to the creation of new ones.

The American Standard of Perfection. The American Poultry Association was organized in 1873 by representatives of different sections of the United States and of Canada; its primary objective was to standardize breeds and varieties of domestic fowl shown in exhibition. This organization sponsors the publication of the *American Standard of Perfection,* first printed in 1874 and since revised many times. The *Standard* is principally a guide for judging the type and color characteristics of the breeds and varieties recognized by the Association. Breeds of fowl having common geographic origin are grouped together into *classes.* Thus we have breeds belonging to the Mediterranean, Asiatic, American, and English classes. However, some class designations, such as the class of Game Bantams, are only for convenience.

In the 1958 edition of the *American Standard of Perfection* a total of 230 breeds and varieties of chickens, turkeys, ducks, geese, and guinea fowl are described. This includes 125 varieties of chickens belonging to 44 breeds. The problem of describing all these varieties is not as great as it might seem because there are relatively few standard color patterns. For example, the silver-penciled plumage pattern is common to the Dark Brahma and to the Plymouth Rocks and Wyandottes of the silver-penciled variety. Likewise, the columbian pattern is common to the Light Brahma and Columbian Rock. Figure 11-1 illustrates some of the common plumage patterns.

In the poultry industry there are no flock registry associations like those for purebred large farm animals. In a sense, the *American Standard of Per-*

Fig. 11-1. Various forms of feather markings as found in the plumage of the domestic fowl. Numbers 1 to 7—laced feather patterns; 8–10—crescentric form of penciling; 11–12—mottled; 13–14—spangled; 15, 17 to 21 —barring or horizontal penciling; 16—diagonal spangling. [Courtesy American Poultry Journal, Chicago, Ill.]

fection takes the place of such registry associations by recognizing as "pure-bred" only those individuals that show characteristics conforming to those given in the *Standard of Perfection*. However, no attempt is made to guarantee the blood purity of breeds of poultry.

Important Breeds and Varieties. Of the 125 varieties of chickens described in the *American Standard of Perfection,* only 6 or 7 are of real economic importance. Table 11-1 gives the regional distribution in the United States of the breeds and crosses participating in the National Poultry Improvement Plan in 1959. Commercially, pure breeds—or more correctly speaking, pure strains of pure breeds—are markedly declining in number. At the same time, the number of strain crosses, crossbreds, and incrossbreds has greatly increased. Almost all our commercial broilers come from cross-mated flocks of different breeds. Furthermore, Leghorn strain crosses, although qualifying as "pure breeds," are generally classified as arising from cross-mated flocks.

LEGHORN. The Single Comb White Leghorn is by far the most important breed kept for egg production in America, as well as in most countries of Europe. The Leghorn is characterized by an active and flighty disposition, early sexual maturity, excellent laying ability, and a relatively small body size. It is well adapted to the extremes in the climate of North America, lays white eggs, and is nonbroody; the chicks are early feathering and grow rapidly.

A greater total effort has gone into breeding high-production strains of Leghorns than into the development of any other breed. Leghorns were first imported into America from Italy in about 1835, and since then many outstanding egg-laying strains have been developed in this country.

The relatively small size of the Leghorn is an economic advantage to the producer, as are its early maturity and rate of egg production. Leghorns commence laying considerably earlier than larger breeds, such as the Rhode Island Red or Plymouth Rock. Small birds tend to be more economically efficient egg producers because less feed is required for body maintenance.

TABLE 11-1. | *Breed distribution in the United States (percent of each breed or cross) of chickens participating in the National Poultry Improvement Plan in 1959.*

Region	New Hampshire	White Leghorn	White Plymouth Rock	Barred Plymouth Rock	Rhode Island Red	Cross Mated	Incross Mated	Other
North Atlantic	0.9	20.3	12.2	1.8	1.8	57.5	3.7	1.8
E. North Central	3.3	24.1	10.6	0.7	1.0	44.9	13.0	2.4
W. North Central	1.4	26.3	3.6	0.3	1.0	47.1	18.7	1.6
South Atlantic	1.2	5.9	4.5	0.03	0.9	86.2	1.1	0.22
South Central	1.0	6.2	8.1	0.3	0.4	80.1	3.6	0.3
Western	5.4	30.5	5.2	0.1	0.9	50.7	6.7	0.5

Source: ARS 44-2 leaflet (1960) of the USDA.

The Leghorn, however, is inferior as a meat bird and most of the hens, after completing their production year, are utilized in the manufacture of chicken soup and other prepared foods.

BARRED PLYMOUTH ROCK. The Barred Plymouth Rock was developed as one of the first dual-purpose breeds in America, mainly from crosses of the Black Java and the Black Cochin with the American Dominique. Plymouth Rocks have single combs and lay brown-shelled eggs. Body type and plumage color was fixed by inbreeding selected families from the Essex strain, traced as far back as 1878. Since the Barred Rock carries the sex-linked barring gene, this variety is used to produce chicks with sexes distinguishable at hatching.

Much of the early development of the poultry industry in America was centered on the Barred Rock. As a dual-purpose breed it has been used extensively for both egg and meat production. Some strains figured largely in the development of the broiler industry, while others were selected primarily for egg production. Despite the importance of the Barred Plymouth Rock in the development of the poultry industry, it has declined in number recently.

Barred Rocks were introduced into England about 1879 and won widespread popularity in that country. They are also a major breed in the poultry industry of Japan.

RHODE ISLAND RED. The states of Rhode Island and Massachusetts were most prominent in the development of the Rhode Island Red. Importations of Cochins and Malays from the Orient were crossed with the native fowl in the formation of this breed. The Rhode Island Red was originally developed as a dual-purpose fowl, as were all the other American breeds. It has yellow skin and red ear lobes, lays a medium-brown egg, and has a quiet disposition. Through selection, broodiness has been largely eliminated and some strains with high egg production have been developed. They reach sexual maturity relatively early, although later than most strains of White Leghorns. The breed was admitted to the *American Standard of Perfection* in 1905.

The Rhode Island Red has been a favorite dual-purpose fowl as well as a fancier's breed. It has been important for crossbreeding and in the formation of commercial inbred lines used in hybridization. During the past 20 or 25 years the Rhode Island Red has declined in numbers, as have all dual-purpose breeds.

THE NEW HAMPSHIRE. The New Hampshire is one of the newest breeds of the American class, having been admitted to the *American Standard of Perfection* in 1935. This breed was derived solely from the Rhode Island Red: there is no record of any other blood ever having been introduced into it. In color the New Hampshire is similar to the Rhode Island Red but of a much lighter shade. The foundation stock from which the New Hampshire was developed was especially selected for early maturity, rapid feathering, and fast rate of growth. The most popular strains have been closely allied to the development of the broiler industry. During recent years, however, the New Hampshire has lost much of its popularity to the White Plymouth Rock.

THE WHITE PLYMOUTH ROCK. The White Plymouth Rock was originally developed as a dual-purpose breed, probably by outcrossing Barred Rocks to white birds, followed by successive backcrosses to the barred variety. Recently this breed has attained first importance in the broiler industry, although it has been almost completely abandoned as a dual-purpose breed.

The modern White Plymouth Rock has been selected very intensively for rapid growth, early feathering, and good feed conversion. Since mature body weight is correlated with early growth, modern broiler strains have large bodies. Hens frequently weigh as much as 8 lb. and cockerels as much as 12 lb. Such strains are relatively poor egg producers.

The White Plymouth Rock is literally the mother of today's enormous broiler industry, having largely replaced the New Hampshire as the female parent for the production of commercial broilers. The main reason for this change is the preference for white feathers. White birds dress out into a more desirable appearing carcass, without the black pin feathers found in the New Hampshire and the Barred Plymouth Rock.

AUSTRALORP. The Australorp, developed in Australia, was derived from the dual-purpose English breed, the Black Orpington. Australian breeders placed primary emphasis on high egg production. The Australorp is not of importance in this country as a pure breed but is of some importance in the production of crossbreds.

CORNISH. The Cornish fowl, generally known abroad as the Indian Game, originated in Cornwall, England, and appears to have been developed from several different breeds, including the Aseel, the Old English Game, and the Malay. Both sexes of the Cornish are similar in conformation, being closely feathered, very compact, distinctive in shape, and extremely plump breasted. The three varieties are the Dark, the White, and the White Laced Red.

The Cornish has become important in this country in the development of male lines used for crossbreeding in the broiler industry. Most such male lines probably contain 50% or more Cornish blood. These strains closely resemble the original Cornish type but have white plumage.

NONSTANDARD BREEDS. Not all of our commercially important breeds of chickens have been officially recognized in the *American Standard of Perfection*. The California Grey is one such breed. This is an egg-laying strain widely used by hatcherymen for crossing. Developed by the Dryden Poultry Farm of Modesto, California, from initial crosses of the White Leghorn and the Barred Plymouth Rock, the breed retains the barred plumage of the Plymouth Rock and the white-shelled egg of the Leghorn. The California Grey crosses successfully with many White Leghorn strains and frequently has been used in such crosses by hatcherymen in preference to a Rhode Island Red or a Plymouth Rock. This is because the Grey-Leghorn cross progeny lay white eggs, while the other crosses produce tinted or cream-colored eggs, to which some markets object.

Undoubtedly the California Grey would be eligible for recognition by the American Poultry Association if the developers of the strain desired this type of recognition. Because the breeding of male and female parent lines for the production of both egg-laying flocks and broilers is under the complete control of private breeding concerns, there has been little interest in obtaining recognition of such strains in the *American Standard of Perfection.*

In the formation of an inbred line, the blood of one or more breeds may be used, but since such inbreds are used only in the production of commercial hybrid crosses there would be little point in recognizing any of the lines as new breeds. In fact, so-called Standard breeds are coming to be of less and less significance in the American poultry industry of today.

Commercial Types of Chickens. Practically all chickens produced in the United States today are either bred exclusively for egg production or bred exclusively for broiler meat production. In either case, the final product may be a pure strain, a strain cross, a breed cross, or an inbred hybrid combination.

PURE STRAINS. A strain of chickens generally takes the name of the breeder who developed it. A pure strain is not necessarily any more "pure" genetically than a pure breed; however, if a strain has remained closed to outside blood for several years, then, in this sense, such a strain might be called pure. Pure strain breeding has been important for the production of both meat and eggs in the past. However, the past 15 to 20 years has witnessed marked changes in systems of producing commercial types of chickens. Pure strains are now used almost exclusively as parents of commercial crosses. In fact, today it is virtually impossible to buy breeding stock of any of the well-known pure strains of chickens developed in the United States.

STRAIN CROSSES. The progeny of the cross of two different strains of the same breed is called a strain cross, although such chickens may still be classed as purebreds. If the strains used in crossing are somewhat inbred or if they differ in recent origin, the cross progeny may show some favorable hybrid vigor in egg production or other economically valuable characteristics. Consequently, this method of breeding has an advantage over pure-line breeding for the production of commercial stock. The majority of chickens produced today for commercial egg production are Leghorn strain crosses. Strain-cross Leghorns are popular not only because of increased production resulting from hybridization but also because they retain the uniformity of white-shelled eggs, plumage color, and other characteristics ordinarily associated with a purebred. Commercial strain crosses usually involve only two strains but three or even more may be used.

BREED CROSSES. Progeny from crosses of strains representing different breeds

a

b

c

d

e

Fig. 11-2. Some breeds of chickens: (*a*) Single-combed White Leghorn hen; (*b*) Barred Plymouth Rock pullet; (*c*) Dark Cornish hen; (*d*) Single-combed Rhode Island Red hen; (*e*) Columbian Wyandotte cock. [Courtesy Robert F. Delancey, Poultry Press, York, Pa.]

are called crossbreds. Such crossbreds ordinarily show hybrid vigor or heterosis for egg production, rate of growth, and certain other characters. Some of the more important commercial types of crossbreds are listed here.

Sex-linked cross. When a Rhode Island Red or a New Hampshire male is mated to a Barred Plymouth Rock female, the male progeny are barred like their mothers but the females are nonbarred like their fathers. This cross has been used extensively over the past years, especially in New England, the cockerels being raised for meat production and the pullets kept as layers. The crossbred pullets are usually black and carry a variable amount of red or gold in the terminal feathers of the neck hackle.

The reciprocal mating of this cross—that is, a Barred Rock male mated to a Rhode Island Red or New Hampshire female—produces all barred crossbred progeny. This was a popular cross for the production of broilers in the early days of the broiler industry.

Leghorn-Red cross. Crosses from a Leghorn male and a Rhode Island Red female have been used extensively in the Midwest as medium-weight layers. This cross has proved to be an exceptionally good layer, and when the hens are marketed for meat they usually command a better price than Leghorns. However, the eggs are intermediate in color between the white of the Leghorn and the brown of the Rhode Island Red. Because certain markets discriminate against a cream-colored or tinted egg, the use of this cross has not been widespread.

It is interesting that the performance characteristics of the cross of a Leghorn male and a Rhode Island Red female are quite different from those of the reciprocal mating. The former crossbreds show a tendency toward early sexual maturity and nonbroodiness, but frequently have rather high adult mortality. In contrast, the crossbreds from a Rhode Island Red male and a Leghorn female mature more slowly and are much more inclined to broodiness, but, surprisingly, have a lower adult mortality. Table 11-2 illustrates this point with data from an Iowa 3-year experiment comparing

TABLE 11-2. | *Performance of reciprocal crosses of White Leghorns and Rhode Island Reds.*

	White Leghorn Male × RI Red Female	RI Red Male × White Leghorn Female
Number of birds	244	242
Age at first egg (days)	178	188
Number broody hens	21	3
Percent broody hens	8.6	1.2
Number of deaths	95	43
Percent laying house mortality	38.9	17.8

Source: Iowa Agr. Expt. Sta., 1957.

Fig. 11-3. (*a*) White leghorn ♂ × Australorp ♀ to produce White Austra shown in (*b*); (*c*) and (*d*) Special broiler-type cross used to produce broiler progeny (*e*). (*opposite*).

e

f

g

Neither male nor female parent lines can be identified as belonging to a particular breed, yet each has been especially bred for broiler qualities. (f) Leghorn-type hybrid. (g) Strain cross. Most commercial egg layers today are classified as either straincrosses or hybrids. [*Acknowledgments:* (a) and (b), courtesy Richardson's Poultry Farm, Redlands, Calif.; (c), courtesy Peterson Breeding Farm, Decatur, Ark.; (d) and (e), courtesy Ames In-Cross Inc., Des Moines, Iowa; (f) courtesy DeKalb Agricultural Association, DeKalb, Ill.; (g), courtesy Welp's Breeding Farm, Bancroft, Iowa.]

reciprocal crosses of the two breeds. Whether the difference in adult mortality between the crosses is due to sex linkage or to a maternal effect transmitted through the female Rhode Island Red parent has not been definitely established.

Austra-white. This cross from an Australorp male and a White Leghorn female has proved to be extremely hardy and able to withstand the rugged farm conditions frequently encountered in the Midwest. The Austra-white is a good layer but tends toward excessive broodiness. Dark shank color of the pullets and tinted eggs are disadvantages of the cross in regions where there is market discrimination against these.

Because it has greater resistance to respiratory infection than a strain-cross Leghorn, this cross has attained some popularity, especially in southern California.

The White-Austra is a cross of a Leghorn male on an Australorp female. It does not exhibit the high broodiness of its counterpart, the Austra-white, yet is apparently quite resistant to respiratory diseases (see Fig. 11-3).

INBRED HYBRIDS. An inbred hybrid is defined as the progeny produced from a mating of two inbred lines or from a mating of first-generation inbred line crosses. However, because the term "hybrid" is used in describing crosses between breeds and varieties, the United States Federal Trade Commission in its trade-practice rules for the poultry industry recommends that the word hybrid should be qualified by stating, "in immediate conjunction therewith the type of cross used in the production of the industry product, such as 'inbred line-cross hybrid' or 'inbred hybrid,' 'cross-bred hybrid,' 'strain-cross hybrid,' 'line-cross hybrid,' etc." The Federal Trade Commission also defines an inbred line as: "A group of inbred chicks resulting from breeding closely related poultry and in which the individuals in question have an average coefficient of inbreeding of 37.5% (equivalent to two generations of brother-sister matings)."

Some of the outstanding commercial chickens in the country have been produced by methods similar to those used in the development of commercial hybrid seed corn. The use of inbreeding as a tool for developing egg-laying chickens attained considerable importance shortly after World War II, particularly in the Midwest. Actual breeding details of commercially inbred hybrids are trade secrets of the companies producing them. Generally, hybrid varieties are given a numerical designation, and are described in terms of plumage color, color of the eggs, and other characteristics that distinguish the hybrid.

Inbred hybridization gained momentum through the late 1940's but some of this was lost after random-sample egg-laying tests became popular about 1952. Results show that the best performing entries in random-sample tests are not necessarily inbred hybrids. Strain-cross entries frequently prove superior to some of the inbred-hybrid entries. Thus, whether intensive inbreeding, as used by hybrid seed-corn producers, is justified with chickens is still an unsettled issue; it is being challenged by breeding systems using lesser amounts of inbreeding. Commercial breeders today are more concerned with the performance of their product than with the system of breeding.

11-2. TURKEYS

Origin. The modern domesticated turkey is thought to be descended from two differing wild subspecies, one found in Mexico and Central America and the other found in the United States. The southern species is smaller than the species native to the United States, which also has a characteristic bronze plumage.

Darwin held to the view that the turkey was domesticated by the original inhabitants of America. He suggested that the wild forms found in Mexico and Central America were early ancestors of domesticated varieties. However, since the modern strains of Bronze turkeys correspond closely to the wild subspecies found in the United States, it is probable that the latter was used largely in the development of American varieties.

Varieties of Turkeys. Only one breed of turkeys is recognized by the *American Standard of Perfection* and hence we speak only of different varieties. The most important of these are the Bronze, White Holland, and the Beltsville Small White. Other varieties recognized by the *American Standard of Perfection* are the Narragansett, the Black, the Slate, and the Bourbon Red.

THE BRONZE. The Bronze is by far the most popular variety of turkeys in this country. It is also the largest. In recent years a subvariety, called the Broad-Breasted Bronze, has almost completely replaced the original Bronze variety. The Broad-Breasted Bronze has been especially selected for rapid growth, high feed conversion, and body conformation but is deficient in fertility, hatchability, and egg production. The adult plumage of the Bronze is basically black with the surface showing an iridescent sheen of red, green, bronze. There are distinct bands of copper bronze on the tail and back feathers; the wing feathers are black-and-white barred. The females may show white edging on the breast feathers.

WHITE HOLLAND. Just as there is a preference for broiler chickens with white plumage there is a trend to white plumage in market turkeys. The original White Holland was a smaller bird than the Bronze, but the variety has been made commercially more competitive by selecting for large size and broader breast development. At the same time breeders have tried to maintain better egg production than in the Broad-Breasted Bronze. Actually the improvement of the White Holland has mainly been accomplished by repeated backcrossing to the Broad-Breasted Bronze. The new type, known as the Broad-Breasted White is but slightly less deficient in hatchability and fertility than the Broad-Breasted Bronze. Essentially, the only difference between the Bronze and the White Broad-Breasted varieties is in color, although the former tends to be somewhat heavier, and the latter shows slightly better fertility.

BELTSVILLE SMALL WHITE. This is a separate variety recognized by the *American Standard of Perfection* and developed by the United States Department of Agriculture at Beltsville, Maryland. The originators of this variety set out to develop a breed that would satisfy the demand for lightweight birds. The Beltsville has been used to produce turkey fryer-roasters of 6- to 8-lb. market weights dressed for the New York market. Because Beltsville Whites are rather small when mature, they do not attain heavy market weight as quickly or as economically as do the larger varieties, and thus have not become popular. On the other hand, they are good egg producers and are superior in hatchability to the broad-breasted varieties.

11-3. WATERFOWL

Ducks. The major part of the duck-raising industry in the United States is concentrated on Long Island, New York, where most of the young roasting ducks for the New York market are produced. Ducks may also be kept principally for egg production, but duck-egg production has never attained the importance in this country that it has in other parts of the world.

According to Jull, the wild mallard duck (*Anas boschas*) is the ancestor of all domestic breeds. Mallards must have been domesticated at a very early time, since the Romans referred to them more than 2,000 years ago. It is believed that commercial duck raising has been practiced longer in China than in any other country. Ducks are raised extensively in Belgium and the Netherlands, where farms are located along the canals. The most important breeds are the Pekin and the Khaki Campbell. Other well-known breeds are the Indian Runner, the Rouen, the Aylesbury, the Cayuga, and the Muscovy.

PEKIN. This breed, used exclusively in the Long Island duck industry, originated in China and was first imported into this country about 1873. The Pekin is noted for very rapid growth, pure white plumage, and yellow skin. It produces an attractive carcass. Ducks averaging 6 lb. at 8 weeks of age are not unusual.

THE KHAKI CAMPBELL. This is the most important egg-laying breed of ducks in Europe. Khaki Campbells are especially popular in the Netherlands and in England. They were developed from crosses involving Indian Runners and are similar to them. The Jansen strain of Khaki Campbells developed in Holland has shown such remarkable egg-production records that the question has been raised whether the chicken is really the best choice for the production of man's egg supply. Reports show that the egg production of the Jansen strain of Khaki Campbell ducks far exceeds the best production records of chickens. These ducks have averaged 320 eggs per bird per year;

moreover, their eggs are larger than those of chickens. Adult mortality averages only about one-third that of chickens.

MUSCOVY. The Muscovy (*Cairina mouschata*), which belongs to a different species than the other commercial breeds of ducks, is thought to have originated in Brazil. The females rarely quack and the drake does not have the curled tail feathers characteristic of drakes of all other breeds. The drake is about one-third larger than its mate.

Geese. With the possible exception of the chicken, the goose was the first bird to be domesticated and was regarded as a sacred bird in Egypt 4000 years ago. The Romans learned to value goose liver as a delicacy and large numbers of geese were placed in pens and fattened to increase the size of the liver; they also learned to use the feathers for filling mattresses and

a

b

c

Fig. 11-4. Some breeds of poultry: (*a*) Toulose gander; (*b*) White Emden gander; (*c*) Broad Breasted Bronze Turkey tom. [Courtesy Robert F. De Lancey, Poultry Press, York, Pa.]

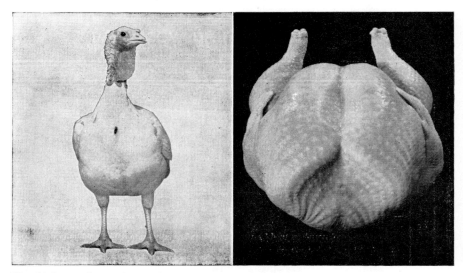

Fig. 11-5. Body conformation in a modern commercial strain of turkeys: (*a*) live turkey; (*b*) dressed turkey. [Courtesy Nicholas Turkey Breeding Farm, Sonoma, Calif.]

cushions. The domestic goose became well distributed over all of Europe early in the Christian era. Goose raising is still an important enterprise in many European countries, including France and Hungary.

The production of geese has never been concentrated in specific areas of the United States as has that of other classes of poultry. They are raised in relatively small numbers on farms widely scattered throughout the United States and Canada, and the total number in the United States has consistently declined since about 1930. According to the 1950 census, there were only slightly more than 1 million geese raised in 1949; geese were present on only 1.8% of the farms in the United States. Goose raising has limited opportunities because the demand for goose meat is largely restricted to urban areas having a substantial foreign-born population.

Breeds most commonly found in this country are the Toulouse, Embden, and Chinese. Other breeds are the African, Roman, Pilgrim, and Sebastopol.

TOULOUSE. This breed originated in the south of France and was named after the French city of that name. It has grayish plumage, darker on the back and lighter in the fluff (see Fig. 11-4). The adult gander averages 26 lb. and the adult goose 20 lb. The Toulouse is the most common breed in the United States. They are not as good layers as the Embden, generally producing from 20 to 35 eggs per season. They are a hardy and vigorous breed but are somewhat slow in growth and thus not adapted for marketing at an early age.

EMBDEN. This is a white-plumaged breed originating in Germany. Standard weights are less than those of the Toulouse. Egg production averages from 30 to 50 eggs per season. The white feathers of the Embden are more valuable than those of colored breeds.

11-4. GUINEA FOWL

The guinea fowl, native to Africa, was brought to Europe by the Portuguese toward the end of the Middle Ages.

There are three domestic varieties, the Pearl, the White, and the Lavender. The most common is the Pearl, which appears to be the original variety developed from the wild African species. Guinea fowl are bred in very limited numbers in this country. They are valued mostly for the delicate and somewhat gamy flavor of their meat.

REFERENCES AND SELECTED READINGS

American Poultry Association, 1958. *The American Standard of Perfection.* Box 337, Great Falls, Montana.

Bateson, W., and R. C. Punnett, 1906. Experimental studies on the physiology of heredity. *Poultry Repts. Evol. Comm. Roy. Soc.,* 3:11–30.

Brown, E., 1906. *Races of Domestic Poultry.* Arnold, London.

Jull, M. A., 1927. The races of domestic fowl. *National Geographic Magazine,* 51: 379–452.

Inheritance of Livestock Traits and Livestock Selection

Chapter 12

Introduction to Modes
of Inheritance

"Mendelian theory, it seems clear, was resisted from the time of its announcement, in 1865, until the end of the century, because Mendel's conception of the separate inheritance of characteristics ran counter to the predominant conception of joint and total inheritance of biological characteristics."

BARBER, B., *Science, 134:596, 1961*

12-1. INTRODUCTION

One of the most interesting characteristics of living things is their ability to reproduce themselves with reasonable accuracy. There is little difficulty in distinguishing between species—for example, a calf and a lamb. One has little difficulty distinguishing between most breeds—Hereford and Angus calves, or Leghorn and Cornish chicks, for instance. Each has inherited easily recognized color patterns from its parents. However, when we examine in detail the characteristics of a group of calves or pullets of similar breed and age, we are struck by the variation in certain traits, and the uniformity of others among individuals in the group. In a group of Angus calves, there is little variation in color, except for a rare red calf; but weaning weights in a group of uniform age may vary 100 pounds or more. A group of Leghorn pullets may show little, if any, variation in color, but the age at the laying of the first egg may vary several months among the group.

Certain facts emerge when we study the reproduction of animals. One is the striking ability of species and breeds to reduplicate distinctive traits such as color. The other is that a great deal of variation in many character-istics occurs in groups that appear similar in general make-up. Genetics is the biological science concerned with the study of inheritance of traits. In seeking to understand how these likenesses and variations are brought about, geneticists must study the traits of offspring of controlled matings. They must take into account differences in feeding, weather, and other outside factors that affect the development of offspring, and they must study the variation between more and less related animals.

The sources of likeness and variation may be broadly divided into two categories: heredity and environment. The most difficult but still the most important job of the geneticist or animal breeder is to learn to distinguish accurately between the effects of these two factors on the development of animal characteristics so that he may interpret and properly use his breeding results. Seldom indeed do we have a situation in which one factor is exerting an influence without the other. One such exception consists of clones of plants, which are propagated vegetatively by cuttings or graftings. All mem-bers of the clone of Bellflower apples, for instance, are identical in their heredity since in reality they are all parts of the same plant. Thus all of the differences we see among them in tree size, shape, and so forth, must be due to differences in environmental factors such as soil, climate, and cultural methods. Another exception—in animals—is identical twins, both of which develop from the same fertilized egg cell. All differences between them are due to environmental factors. In almost all of the cases in which we are ob-serving and trying to use variation for improving animals, however, both heredity and environment are involved and are difficult to separate. Environ-ment ordinarily has little effect on the coat color of cattle, and most of the differences we see are hereditary, but extreme environmental differences may influence such variations as fading of hair pigment by bright sunlight or by certain nutritional conditions.

Perhaps a working definition of the two sources of variation is needed at this point. *Heredity* will be considered to comprise all inborn tendencies and capabilities that are derived from the parents of an individual. *Environment* will include all those external conditions, including nutrition, that surround an animal and have an effect on its growth, development, and production.

12-2. THE BRIDGE BETWEEN GENERATIONS

Each animal grows entirely from one cell: a fertilized egg, or zygote. This cell must divide and duplicate itself many, many times—until the daughter cells number many millions—in order to develop into an adult individual (see Fig. 12-1). During this time, no new living material is added to the organism. Nutrients are absorbed into the cells and converted to living matter, but no new living material is added from outside sources. Thus, the single cell from which the new individual grows is the only bridge between

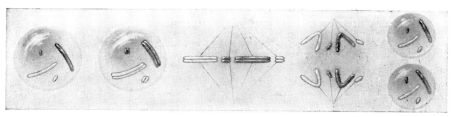

Fig. 12-1. Behavior of chromosomes in an ordinary growth division (mitotic division) of a body cell. (Only two pairs of chromosomes shown). Each and every chromosome duplicates itself so that each daughter cell has the same number as the parent cell. [From Srb and Owen, *General Genetics,* Freeman, San Francisco, 1952. Redrawn after Sharp, *Fundamentals of Cytology,* McGraw-Hill, New York, 1943.]

offspring and their parents. All inherited characteristics, tendencies, and abilities must be contained in the fertilized egg. The controlling mechanism for every trait of the offspring, including its ability to grow into a complicated set of organs and tissues, resembling its parents more or less closely, must be contained in some way in the single-celled zygote.

Let us examine this fertilized egg a little more closely. It is derived from the union of an egg from the female parent and a sperm from the male. The egg is a relatively large cell, and most of its mass is composed of cytoplasm. Embedded in the cytoplasm is the nucleus, which contains a set of rod-shaped structures called chromosomes. The sperm, on the other hand, is small, with very little cytoplasm, and the part which enters the egg at fertilization is composed mainly of chromosomes. Detailed microscopic studies of these cells have shown that the contribution of the sperm and egg to the fertilized egg or zygote, while very unequal in cytoplasm, is almost identical in chromosome content. With the exception of some species, such as the chicken, in which half of the ova contain no sex chromosome, an equal and approximately identical set of chromosomes is contributed to the zygote by each parent cell. Painstaking work by many scientists in the past several decades has shown conclusively that these chromosomes contain the real hereditary bridge between the generations. All of the factors necessary to ensure the growth of an individual animal of the proper kind from the egg are carried from generation to generation in the material of the chromosomes. Like modern-day computers that have complicated directions for mathematical operations "memorized" or built in, the chromosomes carry myriad directions for growth, for reactions to environmental forces, and for the development of specific characters. In this way all the inherited characteristics are passed from parent to offspring, from generation to generation down through the ages. The bridging of generations by chromosomes is illustrated in Figure 12-2.

Chromosome bridges between generations—one pair only shown

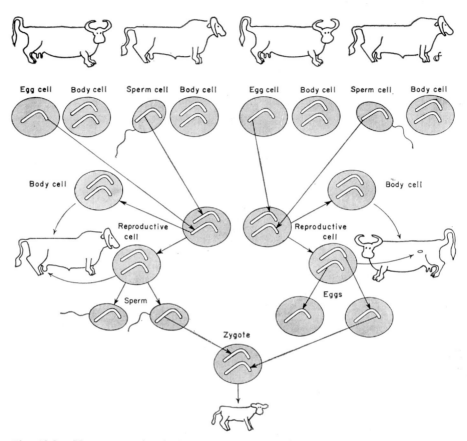

Fig. 12-2. How one pair of chromosomes acts as bridge of heredity between generations. One of each pair of chromosomes is in each sperm and egg. When the two come together to form the zygote which will grow into a new individual, they re-form each pair of chromosomes. Thus a sample of all the heredity of each pair of parents is transmitted to each offspring.

12-3. CHROMOSOMES IN INHERITANCE

One set of chromosomes comes to the fertilized egg from each of the two parents. In our domestic livestock, except domestic birds, the set from the female parent contains exactly the same number of chromosomes as that from the male parent. Each chromosome in a set is a little different in its shape, size, or composition from each other one. In other words, the chromosomes in the egg of a cow are 30 individual units, all slightly different from each other. For each individual chromosome in the egg, there is one like it in the sperm of the bull. Thus, when the two sets come together in the zygote, they make up 30 pairs of chromosomes. This pairing is very important, since it accounts for the behavior of inherited characteristics in crosses.

In short, chromosomes are paired, one member of each pair coming from each of the parents of the individual. When the cells in the reproductive organs of the individual divide to produce sperm or eggs, the members of each pair of chromosomes separate and go to different sperm or egg cells. This results in sperm and eggs having only half the number of chromosomes that body cells contain. The type of cell division that occurs in sperm and egg formation is illustrated in Fig. 12-3. Geneticists call this half number

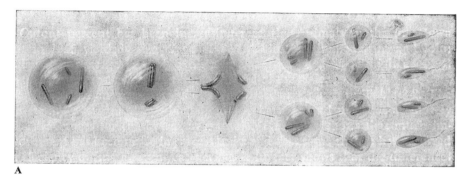

A

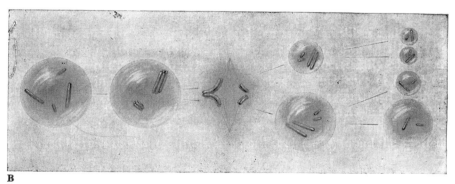

B

Fig. 12-3. Reduction division (meiosis) in the male (*A*) and female (*B*), showing behavior of the chromosomes. (Only two pairs of chromosomes shown in each case). Note that in both sexes, chromosomes undergo pairing and separation of the pairs to the daughter cells as well as a duplication, so that division of one mother cell results in four daughter cells, each with half the number of chromosomes of the mother cell. In the male, all four develop into sperm, whereas in the female three fail to develop and the result is only one functional egg cell. [From Srb and Owen, *General Genetics*, Freeman, San Francisco, 1952. Redrawn after Sharp, *Fundamentals of Cytology*, McGraw-Hill, New York, 1952.]

the "haploid" number. When the sperm and egg unite to form the zygote, pairs are made up again and the normal "diploid" number of chromosomes is restored. In all cell divisions during growth of the individual (see Fig. 12-1), each and every chromosome divides so that all body cells have diploid chromosome numbers and only the cells that are becoming sperm or eggs in the reproductive organs have haploid numbers. By this process a constant number of chromosomes is maintained in the species and hereditary material from the two parents is brought together so that both contribute equally to the hereditary traits of the new individual.

12-4. VARIATION IN CHROMOSOME SETS

It is often said that no two people are exactly alike. It is unlikely also that any two individuals of any species of domestic animals have exactly the same hereditary makeup (except identical twins). One reason for this may be seen when we consider the variety of combinations of chromosome sets that may be contributed by one animal to its offspring. Let us consider a female fruit fly, for instance, whose normal body cells have four pairs of chromosomes. The fruit fly is used as an example because of the simplicity of its chromosome makeup. In the formation of an egg cell in the reproductive organs, which member of each pair will go to this cell is determined entirely by chance, and each member of a pair may contain slightly different hereditary material. The list in Table 12-1 shows that there are 16 possibilities for different combinations of chromosomes in the set that may be in a given egg or sperm.

In swine, with 20 chromosome pairs, there would be 1,048,576 combinations possible. In cattle, whose cells have 30 pairs of chromosomes, there are proportionally more combinations of chromosomes possible—1,073,741,824 different ones, to be exact. This great number of possible combinations of

TABLE 12-1.	*Possible chromosome combinations which may appear in ova of the fruit fly (Drosophila melanogaster) according to the law of independent assortment of chromosomes. The somatic or body cells in this species have four pairs of chromosomes: A_2A_1, B_2B_1, C_2C_1, D_2D_1.*

$A_1B_1C_1D_1$	$A_1B_2C_1D_1$	$A_2B_1C_1D_1$	$A_2B_2C_1D_1$
$A_1B_1C_1D_2$	$A_1B_2C_1D_2$	$A_2B_1C_1D_2$	$A_2B_2C_1D_2$
$A_1B_1C_2D_1$	$A_1B_2C_2D_1$	$A_2B_1C_2D_1$	$A_2B_2C_2D_1$
$A_1B_1C_2D_2$	$A_1B_2C_2D_2$	$A_2B_1C_2D_2$	$A_2B_2C_2D_2$

Note: Each letter stands for a chromosome; A_1 and A_2 are the two members of one chromosome pair. Since the two members of a pair of chromosomes almost always carry slightly different hereditary material, each combination shown would make a slightly different contribution to the inheritance of the offspring. The same number of possible combinations may appear in the male.

hereditary material gives us an idea how much hereditary variation there can be in the characteristics of animals. It also explains why there is so little chance of two individuals being exactly alike.

12-5. CROSSING OVER BETWEEN CHROMOSOMES

If we consider an individual pair of chromosomes, one member of the pair has come from the sire and one from the dam. We might think, then, that an individual chromosome is passed on intact from generation to generation. The process is not that simple, however. Because of a phenomenon called *crossing over,* material may be exchanged between the members of the chromosome pair in the formation of sperm or egg (Fig. 12-4). Thus, although a chromosome in an individual comes from its sire, it may or may not have come intact from the grandsire or granddam. Sometimes material is exchanged between chromosomes of a pair during formation of the egg or the sperm, so that part of the chromosome comes from one grandparent and part from the other.

Considering an individual's whole complement of chromosomal heredi-

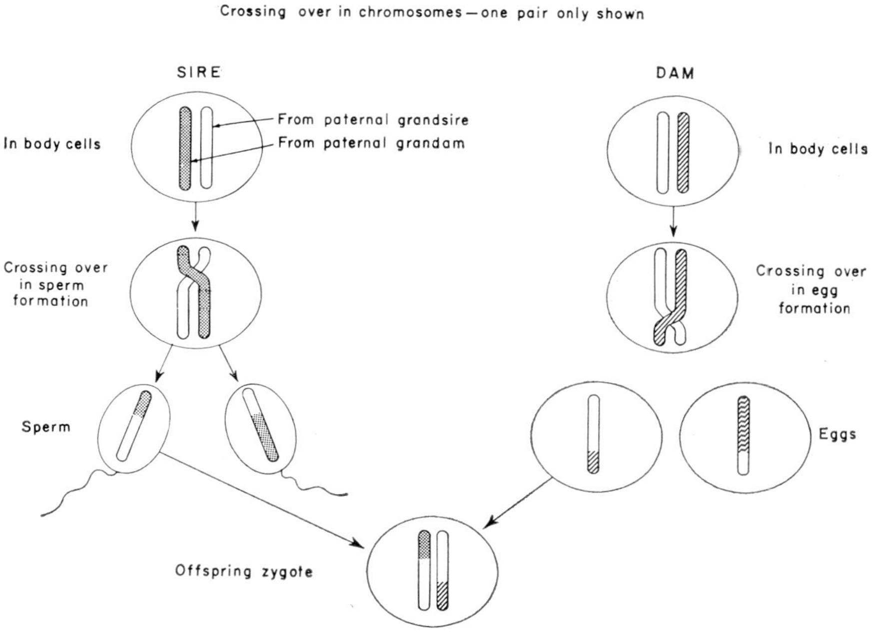

Crossing over in chromosomes — one pair only shown

SIRE   DAM

In body cells

From paternal grandsire
From paternal granddam

In body cells

Crossing over in sperm formation

Crossing over in egg formation

Sperm

Eggs

Offspring zygote

Fig. 12-4. Diagram shows schematically how material is exchanged between the members of a pair of chromosomes during the development of sperm or eggs from cells in the reproductive organs of the parents. Thus it is possible that some hereditary material from both of the grandparents may be passed on by a parent to its offspring, in a single chromosome.

tary material, exactly half comes from the dam and half from the sire, because one member of each pair comes from each parent. Were it not for crossing over, a given chromosome in a given sperm (or egg) would have come entirely from either the grandsire or granddam. As Fig. 12-4 shows, however, a portion of a chromosome may be derived from each of the grandparents. It is thus extremely unlikely that a sire may transmit to any offspring only material from its sire or only material from its dam. This is important, when we are considering the possible influence of an outstanding animal in the pedigree of one we are thinking about for purchase or selection. If the outstanding ancestor is more than one or two generations back, we can see that its influence on the heredity of the individual is very much diluted.

12-6. NUMBER OF CHROMOSOMES

Different species of animals differ in the number of pairs of chromosomes in their normal body cells. How these differences arose in the evolution of species is still not known. There are related groups of species of plants in which the chromosome numbers are multiples of each other; this indicates that the number may somehow become double through some accident in development. Indeed, plant scientists have learned how to double chromosome numbers by treatment with chemicals such as the alkaloid colchicine, thus producing new plant types with slightly different characteristics. So far this manipulation of hereditary material has not been achieved in animals. Chromosome numbers of our domestic species are listed in Table 12-2.

TABLE 12-2.	*Chromosome numbers in certain animals, including man. Considerable variation in the number of chromosomes is seen between different species. Owing to the difficulty of preparing cells in which chromosomes can be counted accurately in mammals and birds, different scientists have found slightly different numbers of chromosomes.*

Species	No. of Chromosomes in Body Cells	No. of Chromosomes in Sperm and Eggs
Man	46	23
Cattle	60	30
Sheep	54	27
Swine	40	20
Horse	64	33
Goat	60	30
Chickens		
male	78	39
female	77	38 or 39

Source: Most of the figures cited in this table were determined by Mathey, quoted by Makino, S., 1951, in *Chromosome Numbers in Animals*, Iowa State College Press, Ames, Iowa.

12-7. SEX

In domestic animals a difference in the composition of one pair of chromosomes is associated with sex determination. In all other pairs of chromosomes the two members of the pair are indistinguishable in their appearance, even though they differ in their content of hereditary material. In the sex chromosomes, however, there is often a difference in appearance in one sex. In mammals the two pair members are identical in the female but differ in the male. In birds, such as poultry, a different situation exists. The female has only one sex chromosome, but the male has two, which are identical. The sex chromosomes are usually designated as X and Y (see Fig. 12-5). The following chromosome makeups then appear:

	Male	*Female*
Mammals	XY	XX
Birds	XX	XO*

* Most workers have found no Y chromosome in the hen, thus leaving her with one less than the male.

In cattle, then, there can be only one kind of chromosome in the unfertilized egg—the X type chromosome. Males, on the other hand, may produce sperm containing either an X chromosome or a Y chromosome. Thus the male parent determines the sex of the offspring according to which type of sperm fertilizes the egg. The unfertilized egg of the female bird, however, may contain either an X chromosome or no chromosome at all; the sperm can contain only the X chromosome. Thus, the hen determines the sex of the

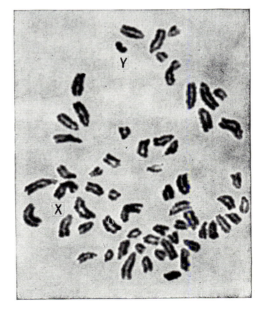

Fig. 12-5. Photomicrograph of chromosomes of a bull of the Swedish Lowland Breed. These 60 chromosomes are each dividing in an ordinary growth division of a body cell. The X and Y chromosomes are indicated. [Photo courtesy Dr. Yngve Melander, University of Lund, Sweden.]

offspring. Attempts by research workers to control the sex of the offspring of mammals have centered on trying to find some way to separate X- and Y-bearing sperm. The most promising method to date involves the migration of the two types of sperm to different poles of an electrical field. Sex control, however, has not reached the stage of practical application.

12-8. GENES

We call the basic hereditary unit in the chromosome a gene. Intensive study by scientists using such new tools as the electron microscope and microchemical analysis is coming progressively closer to learning exactly what it is. The nucleic acid DNA (deoxyribonucleic acid), present in the chromosomes, is the material of which the genes are composed (Fig. 12-6). It is a substance whose molecules consist of long coiled chains, capable of duplicating themselves and of sending to the cytoplasm specific directions for the formation of protein molecules through which gene effects are produced. Although we do not know as yet how much DNA constitutes a specific gene, we have learned a great deal about the location and behavior of genes.

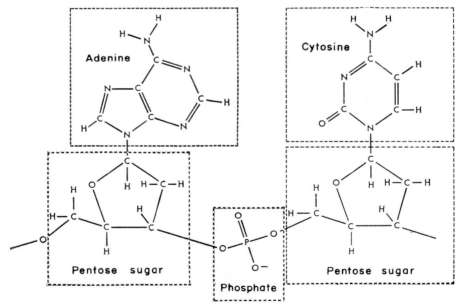

Fig. 12-6. Small portion of chain of deoxyribonucleic acid (DNA) which shows the manner of linking the three main constituents: bases such as adenine or cytosine, a pentose sugar (deoxyribose) and phosphate. [From "The Structure of the Hereditary Material" by F. H. C. Crick. Copyright © Oct. 1954 by Scientific American, Inc. All rights reserved.]

A specific gene is located at a particular spot, which we call a locus, on a certain chromosome. Chromosomes are paired and, since there is a corresponding spot on each of the members of the chromosome pair, genes are paired also. Just as chromosomes duplicate themselves when cells divide, so genes also duplicate themselves. The members of each pair of genes, working together, influence a specific characteristic of the animal.

All the animals of a species considered, there may be one or several types of genes available for a specific locus on a particular chromosome pair. If there is only one type of gene available, then all the animals of the species will be alike in the trait which this gene pair controls. Many such cases probably exist, but unfortunately we can only recognize the existence of a gene when there are several genes at a specific locus. All the genes at a specific locus are called *alleles*. They all have some effect on the same characteristic, but their effects differ. In any individual they appear as a pair, and the particular pair of genes that happen to be together will affect the resulting characteristic; that is, different alleles interact with each other. Their effects may have different strengths so that one will obscure the other (dominance), or the effects of two alleles may be simultaneously expressed (no dominance). Examples of these interactions follow.

Dominance. Probably the most familiar type of allelic gene interaction is that in which one dominates the other. A well-known example is eye color in humans: the allele for brown is dominant over the allele for blue. Conventionally, the dominant gene is indicated by a capital letter and the recessive gene by a small letter. Let us call the blue eye gene "b." and the brown eye gene "B." Since both alleles are available in various individuals within the population, the following genetic types of individuals can occur:

BB—Brown eyes
bb—Blue eyes
Bb—Brown eyes

Individuals carrying two genes for brown, BB, or two genes for blue, bb, are said to be "homozygous" for eye color. Gene interaction is evident in the third type, Bb. The B gene is dominant over b and covers its effect more or less completely in individuals where both genes are present. The b gene is said to be recessive and blue is said to be the recessive eye color. A person carrying genes for both brown and blue eyes, Bb, is described as "heterozygous" for eye color. Such a brown-eyed, heterozygous individual can be distinguished from a brown-eyed homozygous individual, BB, only if he or she has a blue-eyed child.

Of course, even though there might be many alleles or genes affecting eye color, the child has only those genes that are contained in the chromosomes of its particular parents. It must receive one of these genes on the chromosome that it receives from its father and the other on the chromo-

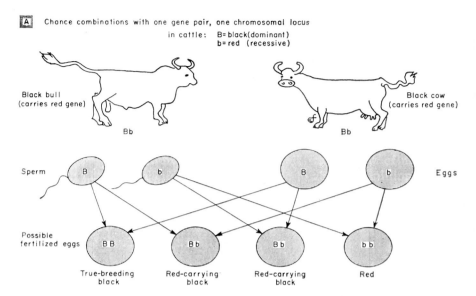

A Chance combinations with one gene pair, one chromosomal locus

in cattle: B= black(dominant)
b= red (recessive)

Black bull
(carries red gene)
Bb

Black cow
(carries red gene)
Bb

Sperm   B   b   B   b   Eggs

Possible
fertilized eggs   BB   Bb   Bb   bb

True-breeding   Red-carrying   Red-carrying   Red
black   black   black

B   Other possible matings considering only this locus and same genes.

Male parent		Female parent		Offspring	
1.	BB	×	BB	All BB	Black, non-carriers
2.	BB	×	Bb ⎫	1/2 Bb	Black, red carriers
	Bb	×	BB ⎭	1/2 BB	Black, non-carriers
3.	BB	×	bb ⎫	All Bb	Black, red carriers
	bb	×	BB ⎭		
4.	Bb	×	bb ⎫	1/2 Bb	Black, red carriers
	bb	×	Bb ⎭	1/2 bb	Red
5.	bb	×	bb	All bb	Red

Fig. 12-7. Part A shows possible chance gene combinations which may be formed in the mating of two black animals each carrying the recessive red gene. Each possibility is equally likely to occur. Part B gives the possible matings which can occur when these two genes are present in the population for this locus on the chromosome and the possible ratios of offspring which might occur if large numbers were produced.

some that it receives from its mother. Fig. 12-7 illustrates a case of dominance in color inheritance in cattle.

Two genetic terms may be introduced at this point—genotype and phenotype. Genotype refers to the genes that an individual carries for a specific trait or traits. For eye color in humans, the genotypes are BB, Bb, and bb. The phenotype refers to the visible trait actually expressed in an individual. Thus, the genotype of an individual heterozygous for eye color is Bb, and the phenotype is "brown-eyed."

No Dominance. Sometimes there is no dominance between allelic genes, or dominance may be incomplete. An example is the roan color in Shorthorn cattle. Considering that there are genes N for white and n for nonwhite (or solid color) in the population, there are three possible gene makeups.

> NN—Animals' color will be white
> Nn—These have white hairs mixed in with the colored
>   ones—roan color
> nn—These animals will be solid colored. The usual color
>   with Shorthorns would be red

In some crossbreds from black breeds, these genes could operate to produce "blue-roans," where black hairs instead of red are mixed with white in the Nn type. In no dominance, neither gene covers up the effect of the other, but in the animal possessing both genes the effects are mixed.

More than Two Alleles. An example in which three genes may be available for a given locus in the population is the coat color in rabbits. Genes for full color, C; white with black points, Himalayan, c^h; and albino, c, are known. The C gene is dominant over both the others, and c^h is dominant over c. Genotypes and phenotypes that may be found are as follows:

> CC —Full colored
> Cc^h—Full colored
> Cc —Full colored
> c^hc^h—White with black points (Himalayan pattern)
> c^hc —White with black points (Himalayan pattern)
> cc —Albino

Crossing colored, CC, to Himalayan, c^hc^h, results in all colored offspring. Breeding these offspring together results in 3 colored and 1 Himalayan. Further, all offspring from Himalayan, c^hc^h, crossed with albino are Himalayan in color. If these offspring are bred together, the results are a ratio of 3 Himalayan to 1 albino offspring. Large numbers of offspring are necessary to obtain the mathematical ratios. These genes are termed an allelic series. A number of examples involving even larger numbers of genes are known.

Interaction Between Different Pairs of Genes. The gene makeup of an individual at one chromosome locus may have an effect on the expression of the effect of genes at another locus. For instance, in sheep there is a color locus that has an overpowering effect on other color inheritance. An animal that is WW or Ww has a white fleece. However, if the gene makeup at this locus is ww, then other color genes at other loci or on other chromosomes may have an effect: the animal may be black or brown, depending on its gene makeup at another locus. However, when W is present, it suppresses the action of these other genes. Other types of interaction between different pairs of genes are known. Genes at several chromosome locations may affect the same characteristic. Many pairs of genes at different chromosome locations affect milk production in dairy cattle. Since milk production is a complex process, it is influenced by genes affecting a number of different characteristics and body processes, such as body size, appetite, the secretion of hormones affecting udder development, lactation, and general metabolic level of the animal. For each locus involved the genes available may be considered as affecting production in a plus or minus way. The genetic potential for milk production, then, is the result of the net sum of all the plus and minus effects of all the genes present. This is called "additive gene action." There are probably additional gene pairs involved in this important character which involve some type of allelic gene interaction such as dominance.

The effect produced by genes is influenced by the environment. Genes for high milk production or egg production are not able to make their effects appear in animals that are starved or subjected to other extremes of environment. Hereford calves, which are genetically white-faced and red-bodied, may become very yellow if they feed on grass or hay with a high content of the element molybdenum. Other less drastic environmental effects are also of great importance in the expression of the genetic potential of individuals.

Mutation. If genes are capable of reduplicating themselves accurately millions of times when cells divide, how then do different types get into a population? In other words, where did this variety of genes come from? Once in a great while, in the process of duplication of a gene, some accident occurs that changes the character of the gene so that it will affect the trait it conditions in a new way. This accidental change is called a mutation. Mutations are the source of all the genetic variation that exists in a population. Estimates of the frequency with which they occur in certain genes have been made. For most loci, these accidents occur only once in millions of gene duplications. This low frequency of change in the genetic material is necessary to the stability of the genetic system of a species. If mutations occurred often, a species would not be able to maintain its form and characteristics. The whole system of cell division and reproduction is geared to maintaining stability in a species. That mutations do not occur easily can be seen from the drastic treatments that are required to cause them artificially. Intensive X-ray treatment will produce an increased number of mutations. Chemical treatments with strong poisons, such as some of the war gases, will also increase mutations.

Probably most of the mutations that do occur in a given population are lost because their effect makes the animals less adapted to survival. Once in a while, however, one will occur that does not decrease adaptation or may even make animals possessing it better fit to survive. This new gene may then become one of the available genes in the population of the species.

By this process, new genetic variation may gradually come into the population.

Chance in Inheritance. The formation of the particular complex of inherited material of any new offspring from a mating is a matter of chance. The two parents between them have twice as many genes for any particular chromosome locus as the offspring will have. Except for mutation, the offspring can have only genes possessed by the parents. But which of the parental genes appears in the offspring is a matter of chance.

To illustrate the chance distribution of genes to the offspring, let us consider the inheritance of red color in Aberdeen Angus cattle. A single autosomal recessive gene, b, is involved. Black, B, is dominant to red, b. Figure 12-7 shows the type of gametes furnished by heterozygous parents, as well as the result of the mating. In the bull two types of sperm are formed: one carries the gene for black, B, and the other carries the gene for red, b. The two types of sperm are formed in equal numbers and it is equally likely that either type will fertilize the egg in a particular mating. The distribution of the B and b genes in the ova is the same as in the sperm. Thus there are four combinations of sperms and eggs and each is as likely to occur as the other. One of the four possible combinations will result in a red animal; in other words, the chance of obtaining a red Angus by this mating is one chance in four. The second part of the diagram (*B*) shows the other possible matings in a herd in which parents with all possible genotypes for this characteristic are present. For instance, the chance of obtaining a red animal from mating No. 4 is one chance in two.

More than One Locus. Consider Fig. 12-8, where the possibilities of a mating involving genes at two different chromosome loci are diagramed. If the two gene pairs are located on different chromosomes, then the distribution of one pair of genes to the offspring has no effect on the distribution of the other pair. They are completely independent of each other. Since this is true, each of the parents produce four different types of sperm or eggs, when we consider genes for both pairs. There are sixteen ways in which the two sets of four types of sperm and four types of eggs may combine to form a particular offspring, and each of these combinations is just as likely to occur as any other. Thus the chance of any particular combination occurring is the number of times it appears in the sixteen possible combinations. The combinations are summarized at the bottom of the diagram. If we consider gene makeup, there is one chance in 16 that any particular offspring will be WWFF, but 4 chances in 16 that it will be WwFf. We thus see how chance enters into the formation of the gene makeups of offspring. Note that the chance for a given phenotype is not the same as for a given gene makeup. Where dominance is involved, as with the genes chosen for this diagram, several gene makeups produce the same appearance in the off-

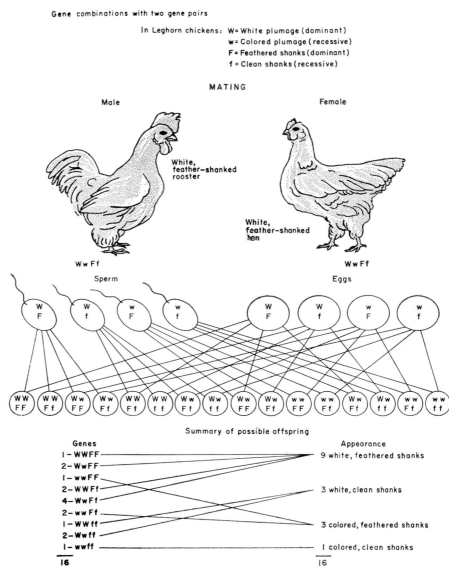

Gene combinations with two gene pairs

In Leghorn chickens: W= White plumage (dominant)
w= Colored plumage (recessive)
F= Feathered shanks (dominant)
f = Clean shanks (recessive)

MATING

Male   Female

White,
feather–shanked
rooster

White,
feather–shanked
hen

Ww Ff   Ww Ff

Sperm   Eggs

Summary of possible offspring

Genes		Appearance
1 – WWFF		9 white, feathered shanks
2 – WwFF		
1 – wwFF		
2 – WWFf		3 white, clean shanks
4 – WwFf		
2 – wwFf		
1 – WWff		3 colored, feathered shanks
2 – Wwff		
1 – wwff		1 colored, clean shanks
16		16

Fig. 12-8. Possible combinations of offspring when considering two pairs of genes which are independent of each other, and the ratio of different types of offspring which would be most likely to be produced if a large number were born to the mating shown.

spring. Thus there are 9 chances in 16 that an offspring of this mating will be white with feathered shanks. Four of the different gene combinations produce this same appearance.

Of course there are many ways in which genes can interact to produce offspring that differ in appearance. Many ratios in the occurrence of a trait result from different matings. However, which gene goes from a parent to an offspring is purely a matter of chance, and this is the basis on which the probability of any offspring combination occurring can be calculated.

Linkage. When two pairs of genes are located on the same chromosome, they cannot be distributed independently, like those in Figure 12-8. By being located on the same chromosome, one member of one gene pair is physically linked to one member of other gene pairs on this chromosome. If the loci of the two gene pairs are close together on the chromosome they are closely linked. If they are far apart on the chromosome, then there may be exchanges of material between the members of the chromosome pair during the formation of sperm or eggs. This is called "crossing over." However, the chances of producing particular offspring combinations are different than if they were independent—that is, located on different chromosomes. This is important in animal breeding. If a desirable gene is closely linked to one that has an undesirable effect, the two may be very difficult to separate in selective breeding.

Mendelian Genetics. Up to now we have been concerned primarily with the laws and principles of inheritance that is controlled by only one or a few pairs of genes, each having an effect that may be individually recognized. This aspect of the science of genetics is called classical or Mendelian genetics, after its founder. Gregor Mendel, a monk in what is now Brno, Czechoslovakia, made experimental matings of plants in the monastery gardens. He observed and reproduced the genetic ratios that have been discussed above, and summarized his work in a paper published in 1865. However, he received no recognition until his results were rediscovered in 1900. This was the beginning of the science of genetics—one of the younger scientific disciplines.

After the rediscovery of Mendel's work, many investigators expanded the knowledge of heredity. It was soon seen, however, that many complex characters, such as growth or milk production, could not be explained in simple ratios produced by a few genes. To deal with these problems, the branch of genetics known as quantitative or population genetics has developed.

Multiple Factor Inheritance. Most of the economic characters that we are interested in changing in selective breeding programs—milk production, egg production, growth, and feed efficiency—are apparently governed by the action of a large number of gene pairs, each having an effect too small to isolate from those of the others. In a herd of dairy cattle, for example, the amount of milk produced may vary widely, with no separable classes based upon milk production of cattle within the herd. These quantitative traits are more susceptible to influence by environmental forces than is, for instance, coat color. This susceptibility makes it more difficult to identify individual gene effects.

We have seen that, as we go from one pair of genes to two, the number of possible offspring combinations to consider goes from 4 to 16. If three gene pairs were involved, there would be 64 combinations. With the large numbers of genes that are almost certainly involved in determining traits such as milk production, there are probably millions of possible combinations and it becomes impractical to raise enough calves to identify all combinations. How, then, can we study and make practical use of the hereditary differences in these characteristics that are inherited in such a complicated way?

Though the individual genes involved cannot be identified, they no doubt follow the same rules as those we can identify. Thus we know that the offspring receives one gene of each pair in its entire complement from each of its parents. That is, each parent is equally important in the inheritance of an offspring. This is true even for characteristics like milk production, where one of the parents does not exhibit the character: bulls have just as much influence on the milk production of their daughters as do cows. Half of the genes that a sire passes to an offspring have come from his dam and half from his sire, although, because of the way genes interact and recombine, the traits coming from either parent or from any grandparent may be impossible to distinguish.

Generally, however, since we cannot work with the individual gene makeups and effects for these characters of economic importance, we must use a statistical approach to breeding for them. We must take into account herd and population averages, and the amount of variation that comes from genetic and environmental sources in different groups in order to formulate good breeding programs.

12-9. THE ROLE OF ENVIRONMENT

Any characteristic observed in an animal is determined by that animal's genetic makeup and by the environment to which it is subjected. We can be safe in saying that if any animal had been raised in a different environment—hotter, colder, or with different feed or treatment—that many of its characteristics would have been different, even though it had exactly the same genetic makeup. In a breeding program, we wish to see through the masking effect of environmental differences, and we are constantly attempting to measure the true genetic worth of an animal.

Even such things as coat color may be changed by environmental extremes. If we remove some hair from the backs of Himalayan rabbits and place them in a cold room, the hair that grows back will be black, but if the rabbits are kept in a warm environment, the hair that grows in will usually be white. Black sheep that are maintained on a copper-deficient diet or on

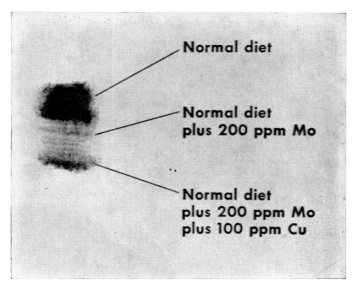

Fig. 12-9. Effect of molybdenum and copper on wool color. [Courtesy Dr. Harold Goss, University of California at Davis.]

one high in molybdenum will grow new wool of a steely gray color rather than black. When returned to a normal diet, they grow their usual black wool (see Fig. 12-9). If cattle are put on the same type of diet, that is, one with an excess of molybdenum or deficiency of copper, the hair becomes yellowish if it is normally red, and grayish if it is normally black (see Fig. 12-10). These are examples of extremes in environment. However, the more common differences we see in breeding programs are of more economic concern to us.

As an example, let us consider the selection of ewe lambs for flock replacements on the basis of their inherited ability to grow fast. One might think that it would be simple to select those that are largest at a given age. However, twin lambs are smaller at an early age than single lambs (Table 12-3). Similarly, those whose mothers are young will be smaller than those from mature dams. In both instances the smaller size has no relation to gene makeup, but is due to a smaller supply of milk being available to the lamb while it is growing. This is an environmental effect and must be taken into account in trying to select the genetically best replacements.

Pullets that have been raised on high energy diets may begin to lay at a different age and at a different rate than those raised on the range. Different light-control plans may cause pullets of exactly the same breeding to lay earlier or later, and to produce a smaller or larger number of eggs. These differences are due to environment and if we attempt to pick the ones that are genetically best for breeders, allowances must be made for these

Fig. 12-10. Even with relatively stable inherited characters such as coat color, extremes of environment may cause modification. Both of these animals are genetically black and appeared like the one nearest the fence at weaning time. A diet high in the element molybdenum caused the color change you see in the nearer animal after a period of several months feeding. [Photo courtesy University of Nevada.]

environmental differences in order to distinguish genetic differences. The animal breeder must constantly seek to remove, or evaluate, the effects of differences in environment in order to determine the true breeding value of his stock.

12-10. PROBLEMS AND OPPORTUNITIES

In this discussion, we have presented a glimpse of the basic knowledge of heredity that has been accumulated since the founding of the science of genetics during the early part of the twentieth century.

Certain traits have been shown to be inherited in a simple understandable fashion, and the ratios of the types of offspring resulting from specific matings can be estimated. It is known, for example, that red coat color in the Angus breed of beef cattle depends upon a single recessive gene. Therefore, if black Angus males and females carrying this recessive gene are mated, one-

TABLE 12-3.	*Average weights at 120 days of age of Suffolk lambs raised under farm conditions at the University of California, Davis, 1952–1954.*		
		Average Weight (lb.)	*Difference Between Groups (lb.)*
Average of all lambs in flock		92.4	
Average of all single lambs		98.7	
Average of all twin lambs		86.1	12.6
Average of lambs from 2-year-old ewes		83.8	
Average of lambs from 4-year-old ewes		94.5	10.7
Average of all male lambs		99.7*	
Average of all female lambs		85.1*	14.6

* Sex is determined by heredity and thus this is not an environmental factor.

fourth of the offspring will be red; if a black Angus bull, which is heterozygous for this gene, is mated to red Angus cows, one-half of the progeny will be red. Furthermore, if the genotype for this trait is unknown in a bull, it can be determined by mating him to a definite number of females that carry the red gene or to a fewer number of red Angus females. Thus, a means of purging the red gene from a herd is available. Other traits such as the horned condition in cattle can be handled in essentially the same manner.

Many of the traits of economic importance, such as growth and milk production, depend upon many genes as well as upon environmental factors. Before deciding upon the best means of approach to improve the trait, it is important to know the extent to which the trait is determined by heredity; that is, a "heritability estimate" is made. After this knowledge is available, a program for improvement of the trait can be planned. This is the field of population genetics, which will be introduced in the following chapter.

REFERENCES AND SELECTED READINGS

Bogart, R., 1959. *Improvement of Livestock.* Macmillan, New York.

Castle, W. E., 1940. *Mammalian Genetics.* Harvard University Press, Cambridge.

Iltis, H., 1932. *Life of Mendel.* Translated by E. and C. Paul. Norton, New York.

Rice, V. A., F. N. Andrews, E. J. Warwick, and J. E. Legates, 1957. *Breeding and Im-* *provement of Livestock.* McGraw-Hill, New York.

Scheinfeld, A., 1950. *The New You and Heredity.* Lippincott, Philadelphia.

Sinnott, E. W., L. C. Dunn, and T. Dobzhansky, 1950. *Principles of Genetics.* McGraw-Hill, New York.

Srb, A. M., R. D. Owen, and R. S. Edgar, 1965. *General Genetics.* 2nd ed. W. H. Freeman and Company, San Francisco.

Chapter 13

Selection and Mating Systems

"Understanding the origin, nature, and potentialities of germ plasm is the most important activity of man on this earth, infinitely more important than the probing of space with rockets."

HENRY A. WALLACE

13-1. INTRODUCTION

Powerful means for developing more efficient farm animals lie in the hands of the breeder. Just as the designer and engineer may blueprint and develop more efficient tools and machines, the breeder may outline programs for developing improved animal tools—better equipped to produce food and fiber of higher quality and at less cost. Genetic "engineering" is a field of development barely tapped in the improvement of most farm animals.

Among the pioneer scientists who have developed theories and experimental testing of procedures that are the most useful in applying the results of Mendelian genetics to animal improvement are Sewall Wright of the University of Wisconsin, R. A. Fisher of Cambridge University, and J. L. Lush of Iowa State University. They have developed and explained many of the concepts concerning selection and mating systems that apply to the genetics of herds and populations. George Harrison Shull and Edward M. East caused a genetic revolution in the breeding of plants, which has affected animal breeding as well, by their discoveries of the power of inbred lines when hybridized to produce increases in corn yields of as much as 20% or more (Kiesselbach, 1951).

Population genetics is the term used for this special area of genetics so important to animal improvement. While its biological soundness is rooted in Mendelian genetics, it differs from the classical studies of the effects of single genes in F_1 and F_2 generations. Population genetics is concerned with genetic changes in the total herd—the effect on the average of all animals, and the variations that occur. The population approach was found to be necessary because early experiments indicated ratios of phenotypes such as 3:1 and 9:3:3:1 did not occur for most of the economically important traits. Rather, a continuous distribution of phenotypes over a wide range was observed, which were best explained by the actions of many pairs of genes. For example, variations in milk production indicate involvement of 7 to 200 pairs of genes. With total estimates of 10,000 to 20,000 pairs of genes occurring in animals, it becomes obvious that the breeder is dealing with populations of great numbers of genes as well as populations of animals. Appraising individual actions of genes in these animals seems an impossible task. The most useful avenue, to date, has been to study the gross effects of the actions of many genes under effects of selection and mating systems.

13-2. HERITABILITY ESTIMATES

Further complexities in applying Mendelism to the improvement of economic traits result from the variation caused in these traits by environment as well as by inheritance. Even identical twins, animals that have the same genotypes, differ. These differences illustrate conclusively the effect of environmental variation on the expression of traits and contribute to the popularly debated subject: which is more important—heredity or environment? "Which is more important" is generally interpreted to mean which contributes more to the individual's own deviation from the general average.

Population geneticists and animal breeders (Lush, 1948) have found mathematical and statistical tools that have helped them appraise the relative amount of variation attributable to genes and to environment. The relative importance of heredity, they have found, varies with the trait. In traits such as variations in the amount of white on Holsteins, 90 to 95% of the differences have been found to be due to the heritable differences between animals. In traits such as litter size of pigs, heritability is only 10 to 15%. Heritability is the term coined to describe the degree of phenotypic resemblance among relatives, as between offspring and parents. It is a term now widely used and accepted in animal breeding. Heritability indicates the percentage of response or improvement expected by the breeder from exerting a given amount of selection pressure on a trait in his herd. For example, if a breeder's unselected sheep flock averages 8 lb. of wool and rams averaging 11 lb. are selected from the flock to mate with selected ewes averaging 9 lb., the average selection differential between that selected group and the general average is $([11 + 9]/2) - 8 = 2$ lb. If heritability were 100%, the progeny of the selected group would average 2 lb. above the mean of the progeny of an unselected flock. However, the average of the offspring is usually nearer the mean of the herd than is that of their parents. That is, heritability is less than 100% and the progeny of selected parents do not, under the same conditions, on an average produce as well as the selected parents. In observations of many experimental flocks, about 40% of the selection differential or "reach" in wool production has been reflected in the progeny of the selected group. The heritability then is 40%. One method of computing heritability is by noting the regression of each offspring's record on the average of its parents. The selection differential must be multiplied by the heritability to estimate the progeny average. Since .40 × 2 = .8, .8 lb. improvement in wool yield in the next generation is expected, rather than 2 lb.

Much of this regression toward the mean is attributed to the selection of animals that have had, by chance, a better environment than the average within the herd. The environmental contribution to their superior phenotype is not inherited.

Since, for a given amount of selection, more is gained by selecting for traits with high heritabilities than for those with low, the breeder might simply conclude that these are the traits to emphasize. This is true only if they have approximately the same economic worth, for the breeder is an applied economist as well as an applied geneticist and must combine these considerations in a comprehensive selection program. In Hereford cattle, the shade of red is highly heritable, about 70%, but variations in the trait have yet to be shown to be important in the production of beef. On the other hand, weaning weight is much less heritable but is one of the more economically important traits in beef cattle. The breeder might well decide to emphasize in selection the trait with only moderate or even low heritability.

TABLE 13-1. | *Some average heritability estimates of farm animals.*

BEEF CATTLE		CHICKENS	
Weaning weight	0.21	Mortality	0.10
Feedlot gain	0.68	Egg production	0.22
DAIRY CATTLE		SHEEP	
Milkfat production	0.24	Fleece weight	0.33
Mastitis resistance	0.26	Type score	0.10
Services per conception	0.05	Face covering	0.53
	SWINE		
	Litter size, 56 days	0.17	
	Feed economy	0.27	

Source: Averages of all estimates from Table 88, *Handbook of Biological Data,* W. B. Saunders, 1956. These estimates vary with the material from which the data are gathered. For example, earlier in this chapter, the heritability estimate for fleece weight (from another source) was given as 0.40 instead of 0.33. In Chapter 6, the heritability of feedlot gain is given as 0.45.

TABLE 13-2.	*Estimated ranking of importance of single traits to the breeder based on economic worth and heritability.*

Traits	Relative Economic Worth (r^2)*	Heritability (g^2)	Index of Importance $(r^2 \times g^2)$
Weaning weight	0.64	0.30	0.19
Size of dam	0.10	0.70	0.07
Daily gain	0.14	0.45	0.06
Days to finish	0.21	0.25	0.05
Percent calf crop	0.64	0.07	0.04 (estimated)
Feed per pound of gain	0.04	0.39	0.01
Carcass cut-out value	0.08 (estimated)	0.25–0.50	0.02–0.04 (estimated)
Slaughter grade	0.21	0.00	0.00

Source: Lindholm, H. B., and H. H. Stonaker, 1957. Economic importance of traits and selection indexes for beef cattle. *J. Animal Sci.*, 16:998–1006.
* Squared correlation coefficient between each trait and net income.

In one study in which steers were individually fed until they had reached low choice grade, variations in several traits were correlated with the net income per 100 lb. live weight. Squaring these correlations gives an estimate of the variation in net income attributable to a given trait. This is one procedure by which relative economic importance of the traits can be estimated. The breeder must, in addition, consider the likely response to selection as estimated by the heritabilities of these traits; thus overall ranking of importance of single traits to the breeder depends upon both economic importance and heritability.

In addition, there are genetic situations that help to explain some of the regression of offspring toward the mean. Offspring of hybrids are known to show some of this regression. Second-generation seed from hybrid corn yields only about 85% as much as first-generation seed (Kiesselbach, 1951). Superior appearance or performance due to specific interactions between genes are not usually reproduced in the offspring at the average level of the selected parents.

Heritability values thus are not abstract or theoretically derived. They are based on actual observations from selection experiments or resemblances of relatives in herds.

The ranking in Table 13-2 was obtained in a single experiment and may not be applicable in other situations. The example shows, however, that one trait may deserve 5 to 10 times more emphasis than another. Although single-trait selection is not advocated as the most efficient procedure for general improvement, emphasis on a trait or traits must take the factors of economic worth and heritability into consideration when they are to be incorporated into a total score or selection index for an animal.

Selecting for one trait probably does not often leave other traits unaffected.

This is explained by the fact that genes may influence more than one characteristic. For example, selecting for rate of gain in swine will cause a genetic improvement in efficiency of feed conversion, but will also slightly increase fatness. In dairy cattle high production is negatively correlated genetically with percentage of fat in milk. In beef cattle, selection for daily gain increases efficiency and mature size, but selection for mature size does not increase efficiency very much.

These associations between traits have a bearing on overall results from selection as well as on heritabilities and economic considerations. A milestone in animal improvement was reached when L. N. Hazel in 1943 invented a selection-index procedure that would maximize selection opportunities by appropriate weighting of traits according to heritability, economic worth, and the correlations between traits. An example of a selection index is provided in Chapter 16.

While it is difficult experimentally to appraise the increase in efficiency of selection by this method, estimates in sheep indicate at least a 20% to 50% increase. In some breeds or herds, undoubtedly even greater improvements have been or could be brought about by the use of a selection index.

13-3. SELECTION SYSTEMS

Selection on Basis of Progeny Tests or of Phenotype. Hidden traits such as carcass characteristics or sex-limited traits such as milk production cannot be accurately predicted from the animal's appearance. Evaluation of a breeding animal for these traits must be delayed until the progeny have been tested. On the other hand, several important economic traits in sheep, such as growth rate, fleece character, and face covering can be observed. Thus the selection of breeding rams on the basis of phenotypes will result in a shorter generation interval and thereby hasten improvement. This point is discussed in Chapter 17.

The Danish system of progeny testing for carcass quality in swine (Clausen, 1953) offers a classic example of meat animal improvement by adherence to a consistent procedure of progeny-testing swine litters for many generations. Systematic testing for back fat, belly thickness, length of side, and efficiency of feed over almost 60 years illustrates changes made in the direction desired. Estimations of changes effected in the Danish Landrace during a 25-year period are shown in Table 13-3.

In dairy cattle, progeny testing of sires based on the production of their daughters is essential. Rapid advances in procedures have been made in recent years. The long-followed sire index—based on the comparison of daughters' records with their dams' records—has practically been discredited because of the confounding effects and important influences that changes in the herd environment have had upon the daughters' records. More promising and accurate appraisal of sires' progenies are being made through the

TABLE 13-3. | *The improvement in carcass quality of the Danish Landrace pig.*

	Carcass Qualities	1926–27	1951–52
	Loss at slaughter (%)	27.2	26.4
	Export bacon (Wiltshire sides) %	59.5	61.3
	Length of body (cm)	88.9	93.4
	Thickness of back fat (cm)	4.05	3.42
	Thickness of belly (cm)	3.06	3.30
Judged by points. Maximum score, 15 pts.	Firmness of back fat	12.7	13.6
	Shoulders	12.2	12.7
	Distribution of back fat	—	12.8
	Thickness & quality of belly	12.0	13.1
	Hams	12.3	12.5
	Fineness (heads, bone, skin)	12.5	13.1
	Amount of lean meat	12.4	12.9
	General bacon type	12.2	12.6
Percentage of pigs	Too lean	0	3
	Very good	50	83
	Too fat	28	12
	Much too fat	22	2
		1929–30	1951–52
Feed units per pound live wt. gain		3.39	3.06

Source: Clausen, H., 1953. *The Improvement of Pigs.* Marjory Boyd, Belfast, Ireland.

use of artificial insemination, for it is possible to breed to several different bulls within a given herd. Thus different progenies having nearly the same environment may be compared. This new type of sire evaluation holds promise of a new era in dairy cattle improvement.

Progeny testing of sires based on sufficient numbers of offspring from selected mates and raised concurrently under similar environmental conditions can theoretically reveal precisely the additive genotype of the parent. No other system can do this. The number of progeny required to reveal this additive genotype varies with the heritability of the trait and relative sampling errors. Generally speaking, greater genetic gains can be expected by subordinating the sire's evaluation to opportunities to select among more sires. If 100 cows are available for the progeny testing of bulls, it is usually better to mate a few cows to each of a relatively large number of bulls than many cows to few bulls.

These relationships are illustrated in Table 13-4. In the example, 100 cows are available for progeny testing and 2 progeny-tested bulls are ultimately needed. The 100 cows could be mated to only 2 bulls. This would give a very accurate appraisal of the 2 bulls, but since 2 bulls are needed

TABLE 13-4. | *An example of genetic progress as a result of using different numbers of sires in progeny tests with 100 females.*

Number of Bulls to Progeny Test	Number of Progeny per Male	Percent of Males to be Retained	Selection Differential on Sires* (A)	Correlation Between Sire's Genotype and Daughters' Av. Performance† (B)	Relative Expected Progress (A × B)
2	50	100	0	1	0
4	25	50	0.8	0.82	0.66
8	12	25	1.3	0.73	0.95
16	6	12	1.6	0.60	0.96
32	3	6	2.2	0.47	1.05
100	1	2	2.4	0.30	0.73

* Lush, J. L., 1945. *Animal Breeding Plans.* Iowa State College Press.
† Lush, J. L., 1931. The number of daughters necessary to prove a sire. *J. Dairy Sci.*, 14:209–220.

both must be used, and no selection on progeny test is possible. Thus, a zero selection differential is obtained as shown in Column 4. The greater selection differentials are obtained by progeny testing more bulls, even though each individual progeny test is a less accurate appraisal of that particular bull's genotype.

Column 5 gives an example of the expected correlations between a sire's genotype and the average of varying numbers of offspring. This shows that for a trait of a given heritability the reliability of a progeny test increases with increased numbers of progeny. The overall relative rate of progress from progeny testing is shown in Column 6 as the product of the selection differential and the correlation between sire's genotype and offspring average. As can be seen, a peak in rate of genetic improvement was achieved by using the test herd of cows to produce about 3 progeny per sire. There was not much difference in achievement between 3 and 12 progeny per sire, but rate of improvement was decreased if the number of sires tested was cut down in order to produce as many as 25 progeny per sire. The correlations are influenced by the heritability of the trait: they are low for traits with low heritabilities and high for traits with high heritabilities.

Selection on Basis of Pedigree. Pedigrees are useful aids in selection but are limited in predictive value because of the sampling nature of inheritance, the influence of environmental factors on traits, and, to a degree, the reliability of the pedigree itself.

Pedigrees are most useful for traits that are sex limited, low in heritability, or greatly influenced by inbreeding and hybridizing. In recent years they have been widely used for lowering the incidence of such undesirable recessives as dwarfism in cattle. The probability of an animal being a heterozy-

TABLE 13-5. | *The probability of an animal being a dwarf carrier based on its relationship to a recessive dwarf.*

Relationship to a Recessive Dwarf	Probability of Being a Carrier (Percent)
Parent	100
Full brother or sister	67
Half brother or sister	50+
Son or daughter of a half sib*	30–40
Average normal appearing animal in major beef breeds	15–25

Source: J. L. Lush and L. N. Hazel, Dwarfism in beef cattle, Iowa State University [Mimeo.]
* Above 25% because of probability of obtaining recessive gene from other parent.

gote can be indicated from pedigree information as shown in Table 13-5.

This table indicates that a breeder could better his chances of avoiding dwarfism, a trait that has not been uncommon in occurrence, if the trait had not been found "close up" in the pedigree. Pedigrees thought to be clear of dwarf production have been worth a considerable premium.

Stressing ancestors many generations removed usually approaches faddism; that is, in promoting the sale of an animal, a breeder may emphasize the animal's relationship to some famous ancestor even though, because of the halving of relationships each generation the relationship is too remote to be of any significance. Exceptions occur in linebreeding, where relationships even to rather remote ancestors are highly regarded in breeding systems devised for that purpose; examples of such exceptions are given in Section 13-4, where linebreeding is discussed in more detail.

One cannot determine precisely the genotype of an offspring from a given mating regardless of how much is known about the animals in the pedigree. The genotype of that offspring is only a sample of those of the parents, and this is the limiting biological factor in prediction.

13-4. MATING SYSTEMS

There are traits which give little or no response to selection. This might seem to be very discouraging to the breeder, but the fact that a trait has low heritability does not mean that it offers little chance for genetic manipulation or control. In fact the greatest commercialization of genetics in farm animals, thus far, has been with low heritability traits—egg production in chickens and litter size in swine. This is because characteristics influencing fertility generally have a low heritability but a considerable amount of heterosis, or hybrid vigor, which results from crossing of breeds or inbred lines. *Heterosis is that extra performance obtained in the cross above the average of parents raised under a comparable environment.*

The study of mating systems has to do with an aspect of genetics and animal improvement that is quite apart from the selection process. There are many ways in which breeding animals may be paired for matings. They may be paired by relationship in order to produce progeny that are more inbred than average, for example brother to sister, sire to daughter, son to dam. They may be so paired as to be different genetically and to produce a species hybrid or outbred, as by mating a Shorthorn bull to a Brahma cow. They may be paired in order to be dissimilar in appearance, such as a racing-type Quarter Horse to a "bulldog" type: this tends to produce intermediate types of offspring, and is sometimes called corrective mating. Phenotypically similar animals may be paired; such pairs cause more extreme types in the total population than found otherwise. If tall men were to select tall wives, and short men short wives, we would have an example of assortive mating, or the mating of likes. Such mating would tend to create greater extremes among the offspring within the population than if mating were random with regard to the trait. Different mating systems are used to accomplish different ends. In most programs selection is used in conjunction with a mating system.

Inbreeding and outbreeding or hybridizing are the best known mating systems. They are the opposite of one another in pairing procedure and in their genetic effects: inbreeding makes animals more homozygous, outbreeding makes them more heterozygous. A strong incentive to investigate inbreeding vs. hybridizing effects in farm animals has resulted from the independent findings of Shull and East. Shull (1909) reported the phenomenon of hybridizing from using but 2 inbred strains of corn for crossing. Later, in noting results of crosses of 8 inbred strains, he found some crosses that produced markedly more than outbred open-pollinated corn. This hybrid vigor in corn has led to widespread experimentation in hybridization of most of the farm animals. Successful commercial application has been achieved with chickens and swine. One West Coast chicken breeder, long noted for the excellence of his outbred strain of birds, which were selected without inbreeding, now produces nothing but hybrid birds obtained by crossing inbred lines.

Why should the hybrid be able to produce more than such rigidly selected outbred populations? This question has not been completely answered, but there are two theories that are important in explaining heterosis.

In the study of lower forms of life, such as the molds and bacteria, clues are developing that indicate that the life processes may be enhanced by the biochemical action of different alleles, each contributing something to an increased efficiency of the organism's development. An example in man is the greater resistance to malaria of individuals heterozygous for the sickle-cell gene, which causes hemoglobin cells to have a distinctive sickle shape. In malarial ridden sections of Africa, these heterozygotes have a higher survival rate than individuals without the sickle-cell gene. Individuals that are homozygous for the sickle-cell gene, however, have a high rate of death from anemia. Thus the heterozygote is more fit than either homozygote. Geneticists call this overdominance, or the extra performance of the hybrid due to heterozygosity at a given locus.

Another type of hybrid vigor results because favorable genes often have a degree of dominance and unfavorable genes are likely to be recessive. Hybrids thus have more dominant genes at the many loci involved than do their more inbred parents. Thus, they have a greater dosage of dominant genes at many loci than the parents possessed.

The tremendous commercial importance of hybrids in poultry and corn is a result of the great economic importance of certain low-heritability traits that influence fertility or reproductive rate. Heritabilities of egg production and corn yields are known to be low and yet these are the most important commercial traits. Should the economic importance of reproductive rate in swine, sheep, and cattle be found to be as great as in poultry and corn, there will be need to study applicability of similar methods. In beef cattle, reproductive rate and weaning weight already have been indicated by some to be the most important traits economically, and these are traits that show a great amount of heterosis. The general rule is that traits high in heritability do not show as much heterosis as do traits of lower heritability. Nor do traits that show little heterosis show much inbreeding regression. We may then accept the general proposition that inbreeding and heterosis are opposite and exclusive of one another in their effects. They actually may be considered as gradation of the effects from much homozygosity to little homozygosity.

Traits that require considerable heterosis for their fullest expression show an increase in the amount of vigor with a decrease in the relationship between animals in the cross. Crosses between breeds give greater hybrid vigor than crosses between families within a breed. Crosses between inbred strains from different breeds give even greater heterosis. But if crosses are made between extremely unrelated animals—as between different species or different genera—there may be high rates of embryonic loss and other incompatibilities that limit the ability to cross. For example, in crossing bison bulls on domestic cows, hydramnios, the development of excessive amniotic fluid, frequently occurs. Also, F_1 males from the cross are often sterile.

Various types of matings listed are shown in order of probable increasing heterozygosity produced in the progeny.

Inbred: The progeny resulting from the mating of closely related animals. Linebreds are inbred, but with a high relationship to a particularly admired animal.

Outcross: The mating of relatively unrelated animals within the same breed or variety.

Topcross: The mating of a male of a specified family to females of another family of the same breed.

Topincross: The progeny resulting from the mating of inbred sires with noninbred dams of the same breed.

Incross: The progeny resulting from the crossing of individuals of inbred lines within the same breed.

Crossbred: The progeny resulting from the mating of different breeds.

Topcrossbred: The progeny resulting from the mating of inbred sires with noninbred dams of different breeds.

Incrossbred: The progeny resulting from the crossing of individuals from inbred lines of different breeds.

The maintenance of a crossing system requires breeders of parent seed stock to maintain noninterbred stocks from which crosses can be made. This is how the purebred breeder and the registry society fulfill needs of the commercial producer. They serve to maintain a closed population that is not permitted to cross with other populations. Although the degree of homozygosity obtained is not high—probably not more than 8 to 12% in many breeds—it is high enough to maintain a source of material that makes crossbreeding feasible in some parts of the livestock industry.

Many breeders obtain further control over the inheritance of their herds by linebreeding to the most admired animals within their herds. The idea here is not to inbreed purposely, but to maintain a close relationship to the best animals. Homozygosity increases more than it would in the usual purebreeding methods, but the prepotency of the breeding stock from the selected group of animals should be enhanced in the process. In the history of many herds, linebred families have become famous for their breeding performance and individual excellence as well.

Robert Kleberg purposely linebred the King Ranch Quarter Horses to Old Sorrel, a son of a Thoroughbred mare and Old Hickory. The pedigree of a Quarter Horse colt bred by King Ranch illustrates the results of this long-continued linebreeding program. A high degree of relationship to Old

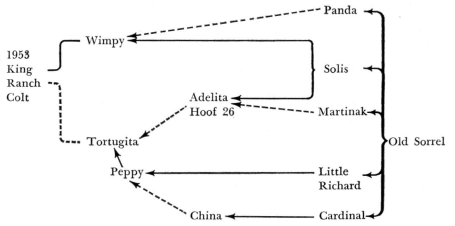

Fig. 13-1. A pedigree of a Quarter Horse. This illustrates a system of line breeding with a minimum of inbreeding.

Sorrel (.45) has been maintained, and yet the inbreeding of the colt (F = .09) has not been greatly increased.

The development of inbred families within the breed serves to increase or maintain the genetic uniformity within the strain and to increase the breeding predictability of that strain within itself and in crosses with other strains (Fig. 13-2), but experiments indicate that the conformation and producing ability within inbred strains is hurt by inbreeding. In chickens, and to a lesser degree in swine, there has been considerable development of inbred lines to be used in crossing to produce useful hybrids. Brother-sister, sire-daughter, and son-dam matings are the closest that can be made in livestock. These cause a decrease in heterozygosity of 25% per generation in contrast to the 50% reduction that can be made by selfing, or self fertilization in corn. Relationships from matings of relatives are as follows:

	Mating	Relationship Between Mates	Inbreeding of Progeny
	Uncle-Niece	.125	.062
	Half Brother-Sister	.250	.125
	Full Brother-Sister	.500	.250
2 generations	Full Brother-Sister	.600	.375
3 generations	Full Brother-Sister	.727	.50

To compare the systems of mating and selection, we rely today largely on experimental evidence from laboratory animals. It has been found that "elite" hybrids in Drosophila have consistently outproduced strains carefully selected without inbreeding for egg production, a trait low in heritability. Elite hybrids are superior individuals obtained by crossing specific inbred lines; the best inbred lines to use to produce elite hybrids are determined by trial and error. For traits higher in heritability, such as egg size, selection has been highly effective. Where both high and low heritability traits with both high and low heterosis are involved, a combination of selection and crossing seems indicated. In Table 13-6, the single crosses are crosses of two

TABLE 13-6. | *Relative performance in various selected traits after 16 generations of selection under 4 different methods.*

Methods of Selection	Number	Mean Daily Fecundity	Mean Egg Size	Performance Index
Closed population	582	90.8	38.93	2,309.2
Reciprocal cross	573	97.7	38.42	2,314.5
Recurrent cross	544	102.1	39.16	2,366.9
Single cross I	161	104.3	38.58	2,346.0
Single cross II	164	101.3	38.38	2,324.1

Source: Bell, A. E., C. H. Moore, and D. C. Warren, 1955. The evaluation of new methods for the improvement of quantitative characteristics. Cold Spring Harbor Symposia on Quantitative Biology XX. Biol. Lab. Cold Spring Harbor, N.Y.

Fig. 13-2. Inbred herd bulls from two distinct inbred families are the products of many years of inbreeding coupled with selection for heavy weaning weights and fast feedlot gains. (Top) Linecross sons of this bull were the fastest-gaining cattle in a feed test involving 21 progeny groups. (Bottom) This bull, used through artificial insemination, increased weaning weights in a commercial herd by 40 lb. and sired high-gaining progeny in several feeding tests. For commercial beef production, linecross daughters of one strain should be bred to bulls in the other strain. This results in hybrid vigor within a breed. [Courtesy Colorado State University.]

inbred lines. Reciprocal crosses are between two large noninbred populations that have been selected specifically to combine well with each other. Recurrent crosses are crosses between two strains in which one strain is selected to cross well with a specific inbred tester strain.

13-5. SUMMARY

It is obvious that breeding techniques that are useful under one set of circumstances cannot necessarily be considered appropriate for another objective. In general, selection for desired traits can be applied by all breeders, but in order to maintain reasonably high selection differentials, they must restrict the number of traits selected. They should select animals raised in an environment similar to that in which they expect their commercial progeny to perform. Owing to the differences between herds in environmental conditions, the animals generally should be ranked on a "within herd basis." Usually the breeder of commercial animals should follow a mating system different from that of the purebred breeder, in order to maintain a greater degree of hybrid vigor in his market animals. The commercial breeder generally should outcross, whereas the purebred breeder should linebreed to hold high relationships to his best animals; some purebred breeders actually form closely inbred families in order to supply the commercial breeders with the inbred lines needed to provide the greatest opportunities for exploiting hybrid vigor through crossing.

REFERENCES AND SELECTED READINGS

References marked with an asterisk are of general interest.

* Clausen, H., 1953. *The Improvement of Pigs.* Marjory Boyd, Belfast.

* Kiesselbach, T. A., 1951. A half-century of corn research. *Am. Scientist,* 39:629–655.

Lindholm, H. B., and H. H. Stonaker, 1957. Economic importance of traits and selection indexes for beef cattle. *J. Animal Sci.,* 16:998–1006.

Lush, J. L., 1931. The number of daughters necessary to prove a sire. *J. Dairy Sci.,* 14:209–220,

*————, 1945. *Animal Breeding Plans.* Iowa State Univ. Press.

————, 1948. Genetics of populations. Iowa State Univ. Mimeograph, pp. 271–284.

Shull, G. H., 1909. A pure line method in corn breeding. *J. Am. Breeders' Assn.,* 5:51–59.

Spector, W. S., 1956. *Handbook of Biological Data.* Saunders, Philadelphia, pp. 111–113.

Genetic Improvement in Beef Cattle

14-1. PURPOSES OF SELECTION IN BEEF CATTLE

Beef cattle are raised primarily for the purpose of converting roughages—hay, silage, pasture, crop residues such as cornstalks and cottonseed hulls—and various by-products of crop processing and milling, all of which are inedible by man, into palatable and nutritious human food. Grains are also used in the final stages of the beef production process to improve the finish and flavor of beef, but beef cattle are less efficient in converting grains into meat than are chickens or swine. Thus, the fact that 80% or more of the nutrients required to produce a pound of beef come from roughage gives the cattle industry its important place in our economy. According to a recent estimate, about 23% of the gross income of American farmers and ranchers comes from cattle and calves. An estimated four-fifths of this comes from the beef industry and the remainder from the dairy industry.

Selection in beef cattle should be made directly for those traits or characteristics that will result in either (1) improved production efficiency, that is, a greater output of meat per unit of feed consumed, or (2) improved quality of product. Every industry must produce a product of the quality desired by consumers, and at a price they can afford, if it is to prosper.

To date, the beef industry has done well in these respects. There is reason to believe that the adoption of modern selection procedures and the abandonment of some practices that have hindered progress in the past will result in further improvement of the genetic potential of beef cattle.

The basic problem of selection in beef cattle, as in other species, is one of estimating as accurately as possible the genotypes of the available potential breeding animals. The animals with probable superior transmitting ability for those characters considered important must then be intermated in combinations calculated to result in maximum improvement in the next generation. In commercial herds, interest centers on (1) production of superior market animals and (2) the production of replacement heifers that will further improve performance in the next generation. In purebred or seedstock herds the sole objective is to produce animals of improved breeding value. However, since selection ideals in the two kinds of herds are quite similar, we will discuss them in general, making special reference to the kind of herd only in cases where practices should differ.

Family and pedigree selection means selecting an animal for breeding purposes wholly or partly on the basis of performance of relatives—either ancestors or collateral relatives, including full and half brothers and sisters, uncles, aunts, and cousins (Lush, 1947). Historically, as discussed in Chapter 6, Breeds of Beef Cattle, pedigree selection has sometimes been used unwisely in beef cattle breeding, with undue emphasis on particular pedigrees or family names. Pedigree information, to be useful, must be based on the performance of close relatives. Animals with superior sires, dams, or sibs have a higher probability of having superior genotypes than animals of equal individual merit from inferior parents or with inferior sibs. With traits of medium to high heritability it is unlikely that an inferior animal from superior parents will have a superior genotype. Thus if family selection in beef cattle breeding is used properly, it gives a little extra consideration to animals from superior families and penalizes somewhat the superior individuals from below-average families. The pedigree is a supplement rather than a substitute for individual selection. Usefulness of pedigree or family selection is likely to be limited by incomplete information on relatives.

The indicators of genotype most useful in beef cattle selection are the individual's own characteristics, in both appearance and performance. Fortunately, both sexes of beef cattle express in their phenotypes many of the characters on which selection is based, and heritability of several important traits is medium to high. An individual, no matter what its pedigree, should therefore ordinarily be average or above in observable characteristics to warrant trial as a breeder.

The third basis for selection is progeny performance. In theory this is the best method of selection, since animals with superior progeny have demonstrated their transmitting ability. In practice with beef cattle there are, however, limitations on the usefulness of progeny testing. First, only a limited number of sires can be progeny tested and these must be selected on the basis of individuality or pedigree so that much of the possible selection pressure has already been applied before progeny-test information is available. Second, cows will be well advanced in years before they can have produced enough offspring for an accurate appraisal of transmitting ability. Third, generation interval is lengthened by progeny-test procedures. For rate of gain, a reasonably highly hereditary trait, about one of every three or four bulls selected for high gains failed to produce above-average calves (Kincaid and Carter, 1958). Thus every breeder should use progeny performance to weed out breeding animals whose offspring fail to live up to expectations. Progeny testing becomes of increasing usefulness in artificial insemination programs in which extensive use is made of those sires proved superior on the basis of progeny (Warwick, 1960).

14-2. SELECTION FOR CHARACTERS RELATED TO ECONOMICAL PRODUCTION OF BEEF

Fertility and Longevity. Specific, simply inherited defects leading to sterility or lowered fertility (Eriksson, 1943) and hereditary tendencies toward the development of cystic ovaries and other conditions that produce at least temporary sterility (Erb et al., 1959) have been found in dairy cattle. Presumably, hereditary low fertility in certain strains of Bates-bred Shorthorn cattle contributed to their loss of popularity in the 1870's. For a time low fertility was considered a desirable trait, however, since it kept numbers low in certain lines of popular pedigrees and thus kept the prices per head high!

In spite of these examples, however, the studies made to date (Table 14-1) in beef cattle, as well as more numerous studies in dairy cattle, have usually found the apparent influence of heredity on reproductive efficiency to be small. There is, and presumably has always been, automatic selection for fertility. Infertile animals leave fewer offspring to propagate the next generation. Conceivably, under normal management conditions a hereditary equilibrium has been reached in which further intentional selection will be ineffective or nearly so. Under these circumstances, much emphasis (beyond what is automatic) on reproductive records of sires and dams in selecting animals for breeding purposes will probably slow total progress, since it will not improve reproductive rates and will decrease selection intensity for traits that respond to selection.

The foregoing conclusions may require modification for those breeds or strains that have recently been taken from unfavorable environments to

TABLE 14-1. | *Heritability estimates for beef cattle characters.*

Character	Approximate Heritability General Level	Percent
Calving interval	Low	0 to 15
Birth weight	Medium	35 to 40
Weaning weight	Medium	25 to 30
Weaning conformation score	Medium	25 to 30
Maternal ability of cows	Medium	20 to 40
Summer pasture gain of yearling cattle	Medium	25 to 30
18 month weight of pastured cattle	High	45 to 55
Mature cow weight	High	50 to 70
Cancer eye susceptibility	Medium	20 to 40
Steers or Bulls Fed in Drylot from Weaning to Final Age of 12 to 15 Months		
Feedlot gain	High	40 to 60
Efficiency of feedlot gain	High	40 to 50
Final weight off feed	High	50 to 60
Slaughter grade	Medium to high	35 to 45
Carcass grade	Medium to high	35 to 45
Area ribeye per lb. carcass weight	Medium to high	30 to 50
Fat thickness over rib per lb. carcass weight	Medium to high	25 to 45
Tenderness of lean	Medium to high	30 to 70

Source: Summarized from many published sources. Wider averages indicate characters for which fewer studies have been made so that probable average heritability is less well established.

improved conditions, and possibly for breeds based on crosses of those breeds that have been removed to more favorable environments. If feed supplies are inadequate, low reproductive rates may have survival value. Perhaps the cow that calves annually will be unable to survive for very long under such conditions while the every-other-year calver will be able to build up body reserves between calves and survive longer. Cattle with this type of background, when brought into an improved environment, may require several generations to reach a new equilibrium and, during this period, selection for fertility might well be effective. This possibility has not been experimentally tested.

No studies are available on the effects of heredity on length of life in beef cattle. Studies in dairy cattle have produced conflicting results, but an analysis by Parker et al. (1960) does not indicate heredity to be an important factor. Here again, natural selection is a determining factor because animals with a long productive life leave more offspring and hence contribute more genes to the next generation. Whether breeders can afford to put additional selection pressure on longevity is presently uncertain.

Preweaning Growth and Factors Affecting It. The beef cow has two functions to perform: (1) to calve regularly, preferably each year, and (2) to raise her calf to weaning age. The overhead of maintaining a cow—both fixed costs and feed—are relatively little affected by whether or not she calves or by the size of the calf weaned. The equivalent of about 3 to 3½ tons of dry roughage is necessary to maintain the average beef cow a year. She must raise a heavy calf each year if efficient production is to be achieved.

Weaning weight is affected by both inherent growing ability of the calf and the maternal qualities of its dam. Weaning weights are correlated with milk production of dams (Gifford, 1953) to a considerable degree, but probably other less obvious characters of cows also affect growth of their calves.

Heritability of calf weaning weight, based on individual calf performance, and the maternal qualities of cows, as evaluated by weights of their calves, are both moderately high (Table 14-1) and selection should be effective in improvement.

Calf weights are rather highly repeatable. Cows ranked in the lower one-tenth to one-fourth of a herd on the basis of one or two offspring can be safely culled. Only rarely would they rank in the upper half of the herd with subsequent calves (Botkin and Whatley, 1953).

Evaluation of weaning weight presents problems, since it is affected by several nonhereditary factors. Bull calves gain faster than heifers. Mature cows in the age range of five to nine years raise heavier calves than those either younger or older. Thus adjustments of records are necessary if selection of calves or culling of cows is to be based on true genetic merit. Unfortunately, the amount of adjustment necessary apparently varies from area to area and even from herd to herd in the same general region. The adjustment factors given in Table 14-2 are based on a large body of data from Virginia (Marlowe et al., 1958). They are illustrative of general trends but should not be applied directly without collateral information indicating that they are applicable to the herd in question.

Growth to weaning age can be expressed either as average daily gain from birth to weaning or as weight at a standard age, such as 210 days. Either is satisfactory and personal preference and convenience should determine the choice.

Postweaning Gaining Ability. Carcasses most in demand in the United States are those of cattle in the 900 to 1200 lb. range in live weight. This means that the average slaughter animal makes half or more of its total weight gain after weaning. Rapid gains during this period are desirable because the length of time cattle must be grazed or fed is reduced, with resulting reductions in labor and overhead. In year-round feedlot opera-

TABLE 14-2. *Examples of factors used to adjust preweaning daily gains of calves. Factors given are for spring-born, noncreep-fed calves. To adjust calf gains to basis of steer calves from mature dams (6 to 10 years of age), multiply actual daily gain of each calf by the factor given.*

	Sex of Calf		
Age of Dam	Bull	Heifer	Steer
2 years	1.15	1.29	1.20
3 years	1.06	1.19	1.10
4 years	1.02	1.14	1.05
5 years	0.99	1.11	1.03
6 to 10 years	0.96	1.08	1.00
11 to 13 years	1.01	1.13	1.05
14 years & over	1.10	1.24	1.15

Source: From Marlowe *et al.*, 1958.

tions, rapid gains permit feeding a greater number of cattle in the course of a year. A more important reason for desiring rapid gains is the apparent relation between rate and efficiency of gain; this will be discussed in the next section.

Extensive studies have shown rate of gain in the feed lot to be one of the more highly hereditary traits in beef cattle (Table 14-1). Direct selection for gaining ability—either high or low—has been shown to be effective (Kincaid and Carter, 1958; Shelton et al., 1957). Heritability of gain on pasture is somewhat lower but is still high enough for selection to be effective. Apparently heritability tends to be higher if nutritional levels are high, allowing full expression of inherent differences.

Evaluation of postweaning gaining ability in beef cattle is complicated by the questions of when to test, length of test period, and influence of pretest environment on gains. Much research has been done on these questions but final answers are not available; indeed, there may not be single answers to fit all conditions.

Various types of feed tests have been devised: one is carried out during fixed age periods (age-constant), one during certain fixed time periods regardless of age (time-constant), one during fixed weight periods (weight-constant), and one until a certain final weight or a certain final degree of finish has been reached.

If environmental conditions are uniform, gains vary greatly between animals, but individual beef cattle of conventional types will gain at a reasonably constant rate to at least 1000 to 1200 lb., after which gains will decrease as maturity is approached. Under constant environmental conditions, then, any of the above tests will give a good evaluation of inherent gaining ability. Variations in weather make it impossible to maintain uniform environmental conditions for long periods. In time-constant tests, which begin on a given date and end on a given date, all animals on test are exposed to the same environment. Time-constant tests are also the simplest to conduct because all animals go on and off test on the same dates. In all other tests animals go on and off test individually, as they reach given ages, weights, or degrees of finish. For these reasons time-constant tests have been the most popular when evaluation of gaining ability has been the only objective. They are satisfactory if the length of test period is adequate, if the animals tested have only a relatively small initial variation in age and weight, and if they are conducted while the animals are in the age and weight ranges in which gains of individual animals tend to be rapid.

Research workers have not reached agreement on the preferable length of time-constant postweaning gain tests. Knapp and Clark (1947) found that heritability of gain increased during consecutive 84-day periods of a 252-day feeding test. On the other hand, in the direct selection experiments referred to above, tests of 140 and 196 days proved satisfactory.

Of greater importance than type of test or exact length of test is the problem of effect of pretest environment on gains. The phenomenon of "compensatory gains" has long been recognized. Cattle held for a period of time on restricted nutritional levels will gain at rates above their long-time inherent potential when later put on adequate rations. Thus, if a period of undernutrition precedes a gain test, an erroneously inflated appraisal of an animal's gaining ability will be obtained.

This problem of variation in pretest environment can never be completely eliminated. Prior to weaning some cows will have given enough milk for their calves to have gained at a nearly maximum rate, while other calves will have been receiving less than an optimum amount of food. Even if calves are put on gain evaluation tests immediately after weaning, compensatory gain may result in less accurate estimates of inherited gaining ability than desired. The problem is greatly magnified, however, if gain tests are delayed and a period of undernutrition occurs between weaning and the test. This is not particularly serious if all animals in the test groups are treated alike and test gains are compared only within the group, but results may not be too meaningful if these two conditions are not met.

The use of lifetime gains, from birth to the end of postweaning gain tests, is a means of reducing error and is now being widely used. Limited evidence indicates that heritability of this final weight may be higher than either weaning weight or postweaning gain alone.

The foregoing discussion applies primarily to young bulls being tested to aid in determining which will be used for breeding purposes, and to steers being fed out for slaughter to progeny test their sires.

Although by no means conclusive, some research results (Chambers et al., 1960) suggest that heavy feeding of young heifers damages their future milking and perhaps their reproductive abilities. This possibility, together with the cost of feedlot testing, which puts on more condition than needed for normal growth, strongly suggests that gaining ability in heifers should be evaluated by weight at a standard age of 14 to 18 months under normal herd management rather than in feedlot tests.

Efficiency of Gain. Accurate direct evaluation of efficiency of gain, defined here as amount of feed consumed per pound of gain, is difficult in beef cattle. Individual feeding must be practiced and this may influence feed consumption. It is a different feeding procedure than is used in feedlot conditions in industry and it may possibly rank cattle differently (Fig. 14-3).

Feed required for maintenance is thought to increase according to about the three-fourths power of live weight. Thus, as animals get heavier a higher proportion of the feed consumed is required for maintenance and apparent efficiency is reduced. More feed energy is required to produce fat than other body tissue. As an animal puts on fat, gain in relation to feed intake is reduced.

On the assumption that individual feeding is a satisfactory evaluation method, the ideal method of evaluating efficiency would be to feed beef animals of comparable body composition individually through a fixed weight

Fig. 14-1. A productive 3-year-old Angus cow and her first calf raised without creep feeding. This calf graded high choice and on a mature-dam basis had a daily gain of 2.09 lb. Productive cows such as this are money-makers. [Photo courtesy Tennessee Agriculture Experiment Station.]

range under constant environmental conditions. Unfortunately this is a physical impossibility, since some animals finish at lighter weights than others. Further, since animals would go on and off test at various times, it would be impossible for them all to be exposed to exactly the same environment.

In view of the foregoing, every system of direct evaluation of efficiency yet devised involves some compromises. Animals must be individually fed under all proposed systems. One of the earlier proposals (Black and Knapp, 1936) was to feed from 500 to 900 lb., ignoring differences in finish. Guilbert and Gregory (1944) recommended feeding to a constant degree of finish. This proposal suffers from two drawbacks: the difficulty of evaluating finish accurately in live animals, and the possibility of some animals being either lighter or heavier than markets prefer when the specified degree of finish is reached.

Others have recommended feeding through a time-constant or age-constant period and adjusting feed consumption figures statistically for differences in live weight, ignoring degree of final finish.

None of these procedures are ideal. It is uncertain whether any of them are superior to the more generally used system of basing selection on rate of gain of group-fed animals. There is a rather high correlation between rate and efficiency of gain (Koch et al., 1963), so that selecting for rate of gain results in considerable indirect selection for economy of gain. In one study there was a saving of about 8% in feed per unit of gain for each $\frac{1}{4}$ lb. increase in average daily gain.

There is, however, considerable disagreement among research workers on this point. At present, few breeders can afford to feed individually, but they should select for efficiency indirectly through selection for rate of gain. If future research develops better methods of estimating efficiency, this general statement will be subject to modification.

Hereditary Defects. Any hereditary defect will reduce efficiency of beef production. Some of these are inherited simply (Gilmore, 1950), usually as recessives. Dwarfism (Pahnish et al., 1955) is the trait that has been of greatest concern to the beef industry in recent years. In general, the frequency of such defects can be reduced in a commercial herd by eliminating all animals that have produced defective calves. In purebred herds more drastic measures—including elimination of animals related to those producing defective calves and using cows that have previously produced defective calves to progeny test bulls introduced to a herd—are necessary, since the purebred breeder, in addition to stopping the birth of defective calves in his own herd, has the obligation to eliminate the gene so that he will not be selling carrier animals to his future customers. When suspected hereditary abnormalities occur, a breeder should get the best professional advice available on its mode of inheritance and apply known genetic principles to its elimination.

Fig. 14-2. Hereditary differences are economically important. When bred to comparable groups of cows, the bull on the left sired calves worth an average of $10.49 more at weaning (210 days of age) and his steer calves had a $10.08 greater return in the feedlot when full fed for six months as compared to progeny of the bull on the right. [Photo courtesy Oklahoma Agricultural Experiment Station and the Animal Husbandry Research Division, USDA.]

Other traits, such as cancer eye (Anderson, 1960), are partly hereditary and partly dependent upon environment. Selection against them in much the same manner used for other quantitative traits should reduce their frequency.

Cattlemen believe that some types of feet and legs, which can be observed in young animals, are likely to become unsound later in life. There is little or no scientific evidence on this point.

14-3. SELECTION FOR QUALITY BEEF

Kinds of Beef in Greatest Demand by Consumers. Surveys have almost always shown that American consumers want beef with a high percentage of lean, a minimum of waste fat, sufficient marbling, and a high degree of tenderness. Since fat may be associated to some degree with tenderness and flavor, it is difficult to meet all these requirements. Within the carcass, some cuts—the rib, loin, and round—are preferred by consumers, and it would be desirable to produce cattle with as high a percentage of these as possible.

It is only in very recent years that serious attention has been given to the possibility of using scientific approaches in attempts to breed for improved carcass qualities in beef cattle. Inevitably, research has exposed fallacies in traditional standards without developing evidence that will permit clear-cut recommendations on improved procedures. It is, however, a fast-developing field and the alert student will give careful consideration to new developments.

Conformation in Beef Cattle Selection

HISTORICAL ASPECTS. Until very recently most beef cattle improvement was based upon visual appraisal of animals, and those that were considered superior were selected for breeding purposes. History tells us that Robert Bakewell and other pioneer beef cattle improvers were conscious of carcass quality problems nearly 200 years ago. Bakewell is said to have pickled the joints of certain animals in order to have them for future reference. Today we can only guess what influence the semiobjective estimates of beef quality may have had on their selections. Since objective methods of evaluating beef quality were not then available, it seems probable that ideas of what constituted desirable beef types developed more from observation than from objective evidence. Further, there is every reason to assume that ideas then developed at least in part on the basis of what seemed logical rather than on exact measures.

As is probably true of every procedure that develops as an art rather than a science, certain things about conformation in beef cattle long ago were suggested as probabilities and after being repeated many times came to be accepted as fact when they were actually untrue or only partly true. For example, it was generally believed that animals of what was considered superior beef type—wide, deep, compact, and low-set—were more efficient in use of feed than other kinds. Actually, it was shown at the Iowa Station as long ago as 1893 that dairy steers made as efficient gains as beef steers (Wilson and Curtiss, 1893); other experiments since then have shown little or no relationship between conformation and efficiency of gains. Judging lore is full of ideas about the relationship of such characteristics as head shape, thickness of hide, fineness of hair, and so forth, to gaining ability and efficiency. Without exception, controlled experiments have shown little or no relationship.

It is thus a mistake to expect that conformation evaluation can serve as a substitute for objective measures of production abilities. It is an equally serious mistake to jump to the conclusion, as some have done, that because it has limitations, conformation evaluation is of no value or consequence in beef production. Carcasses of beef-type and dairy-type steers fed similarly show smaller differences in many characteristics than might have been expected from differences in live appearance. However, beef-type carcasses have usually had thicker muscles, a higher proportion of lean to bone, and lower proportions of kidney, caul, and intestinal fat. These differences have presumably resulted from selection for conformation.

RELATION OF CONFORMATION TO YIELD. The term "yield" in beef cattle usually refers to dressing percent or the weight of the dressed carcass in relation to live weight. We shall also use it in relation to the percent of the most valuable cuts—rib, loin, and round—in relation to carcass weight.

Dressing percent is greatly influenced by finish, with fatter animals usually having higher yields. Its relationship to conformation is less clear, al-

Fig. 14-3. Calves being individually fed to determine both rate of gain and efficiency of gain. Either tie-in feeders of this general type or stalls for each calf are necessary to determine individual feed consumption and efficiency. [Photo courtesy Nebraska Agricultural Experiment Station and Animal Husbandry Research Division, USDA.]

though beef animals with width and thickness and minimal development of the middle are usually thought to dress higher.

Conformation was long thought to be related to the percentage of preferred cuts in the carcass, and it was generally believed that the more thickness, depth, and shortness of body and low-setness an animal exhibited, the better its carcass would be. During the late 1930's, extra short, low-set types (variously known as Comprests or Compacts) appeared in at least two breeds of beef cattle and for a time were thought by many to represent the ultimate in beef type. Research showed, however, that when fed to the same degree of finish, cattle of this kind had no higher percentage of preferred wholesale cuts (Stonaker et al., 1952). Since they also had several production disadvantages, they soon disappeared from the scene.

Other work has indicated that conformation that accords with previously accepted standards is not a good indication of yield of preferred wholesale cuts (Casida et al., 1957). There are indications, however, that animals that are wide at shoulders, loin, and hooks, and round and deep in the twist, tend to be superior in yield of preferred cuts (Green, 1954). Thus there appears to be a definite possibility of visually selecting for improvement in conformation if proper standards are developed.

RELATION OF CONFORMATION TO EATING QUALITY OF BEEF. Flavor, tenderness, and juiciness are the factors usually considered to be important in determining palatability of beef to consumers. Apparently, tenderness is what a majority of consumers want above all else. Tenderness appears to be rather highly heritable (Table 14-1) but has little or no relation to conforma-

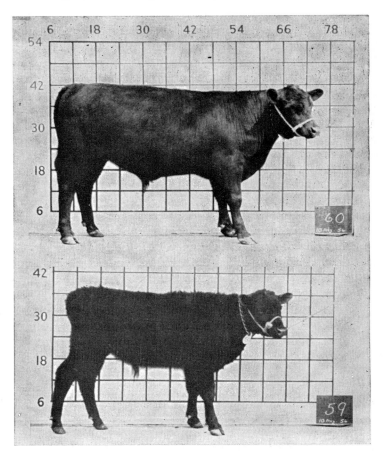

Fig. 14-4. A pair of identical (monozygotic) twins illustrates how differences in fat influence apparent conformation. Twin above had been on liberal ration for six months, while his twin below had been on a submaintenance diet. Had they been fed the same ration, appearance would have been nearly identical. [Photo courtesy Animal Husbandry Research Division, USDA.]

tion. Recent experiments have shown that the lean tissues of even such extremes as beef and dairy steers fed for equal lengths of time do not differ appreciably in eating quality. Therefore in selecting for eating quality we must use criteria other than conformation.

EVALUATION OF CONFORMATION AND ITS IMPORTANCE TO CATTLE BREEDERS. From the foregoing it is apparent that selection of cattle on the basis of standards of conformation used in the past leaves much to be desired. However, conformation is a very important factor in marketing purebred, feeder, and slaughter cattle today. Because changes in preferences tend to evolve slowly, cattlemen cannot depart far from accepted standards without running the risk of severe financial loss. Today it definitely appears that beef cattle can have greater length of legs and body than previously preferred and still produce superior carcasses, and that the extreme depth of chest and middle sometimes favored in the past are not necessary for economical production and probably result in increases of brisket and other cheap cuts of meat. Width of body, particularly near the top line and in the region of the rump and round, is desirable and should receive emphasis. Indications are that these facts are being applied in industry and that the types favored in the 1940's would not be popular today.

The old Scottish saying that "fat is a good color" is a succinct way of posing the greatest problem in evaluating conformation of beef cattle. Fat cattle always look better and it is very difficult from visual appraisal to differentiate those cattle with an abundance of natural lean tissue from those which are merely well covered with fat. The problem can be minimized by raising all animals to be compared under the same environmental conditions and systems of management. Fatter animals will then tend to be those that are superior in other respects.

Systematic conformation evaluation of animals in a herd is desirable when they are weaned and again at the time they approach market weights at some standard age of 12 to 18 months. Several scoring systems are in use, differing primarily in the numbers or letters used to designate the various grades. (See Albaugh et al., 1956, and Marlowe et al., 1958, for two widely used grading systems and for discussion of other aspects of performance testing.)

Methods of Selecting for Improved Carcass Merit. The reader may get the impression that improvement of carcass quality is an impossibility. It is true that carcass quality is unrelated, or nearly so, to items important in economical production of beef and that visual appraisal of conformation is of only limited usefulness for effecting further improvement. But this is not to imply that selection for conformation was not effective in years past, when average merit was lower.

Admittedly, carcass improvement will not be easy, and very few new techniques have been developed. It now appears that the best approach is to make initial selections of breeding animals for traits important in economical production on the basis of their own performance and whatever family information may be available. Attention can also be paid to conformation with emphasis on those factors having at least some relation to carcass quality. The most promising sires can be progeny tested on randomly selected groups of cows, and a sample of at least six steer progeny fed out to normal market weights, slaughtered, and their carcasses evaluated. The sires whose progeny had the best combination of productive traits and carcass characters could be used extensively. In carcass evaluation, the animals with the best ratings will be those yielding the highest percentages of lean meat of satisfactory eating quality, the least waste fat, and cuts of satisfactory shape.

Several agricultural extension and breed association beef improvement programs have incorporated progeny testing for carcass characteristics, using the general procedures outlined above.

It is estimated that progress per year in changing carcass characteristics can be only one-fifth to one-fourth as rapid as would be feasible if it were possible to estimate accurately the carcass characteristics of living animals and to use the superior ones for breeding without having to wait for progeny-test information. Research is active in this area (see Chapter 2). Techniques include the use of ultrahigh-frequency sound devices to estimate thickness of fat and lean in living animals, use of small biopsy samples of tissue on which chemical determinations of potential tenderness can be made, and refinements of methods of taking live animal measurements and conformation scores in hopes that some can be found that will more accurately predict carcass composition. None of these are ready for use in industry as yet, and breeders can only hope that some of them, or other new techniques yet untried, will be useful in the future. Until this happens breeders should use the progeny-test technique to make as much progress as possible.

REFERENCES AND SELECTED READINGS

References marked with an asterisk are of general interest.

Albaugh, R., H. T. Strong, and F. D. Carroll, 1956. A Guide for Beef Cattle Improvement Programs. Calif. Agr. Ext. Serv. Cir. 451.

Anderson, David E., 1960. Studies on bovine ocular squamous carcinoma ("cancer eye"). V. Genetic aspects. *J. Heredity* 51(2):51–58.

Black, W. H., and B. Knapp, Jr., 1936. A comparison of several methods of measuring performance in beef cattle. *Proc. Amer. Society Animal Prod.*, pp. 103–107.

Botkin, M. P., and J. A. Whatley, Jr., 1953. Repeatability of production in range beef cows. *J. Animal Sci.*, 12:552–560.

Casida, L. E., O. B. Butler, D. E. Brady, and J. H. Knox, 1957. Symposium: The meat type steer. *J. Animal Sci.*, 16:224–248.

Chambers, D., J. Armstrong, and D. F. Stephens, 1960. Development of replacement heifers for expression of maternal traits. In *Rpt. Okla. Agr. Expt. Sta. 34th Annual Feeders Day*, pp. 89–94.

* Clark, R. T., James S. Brinks, Ralph Bogart, Lewis A. Holland, Carl B. Roubicek, O. F. Pahnish, James E. Bennett, and Ross E. Christian. *Beef Cattle Breeding Research in the Western Region.* Western Regional Research Publication; Oreg. Agr. Expt. Sta. Tech. Bul. 73, 1963.

Erb, R. E., P. M. Hinze, and E. M. Gildow, 1959. *Factors Influencing Prolificacy of Cattle. II. Some evidence that certain reproductive traits are additively inherited.* Wash. Agr. Expt. Sta. Tech. Bull. 30.

* Eriksson, Karl, 1943. *Hereditary Forms of Sterility in Cattle.* Hakan Ohlossons Boktyckeri, Lund, Sweden.

Gifford, W., 1953. *Records-of-Performance Tests for Beef Cattle in Breeding Herds. Milk production of dams and growth of calves.* Ark. Agr. Expt. Sta. Bull. 531.

* Gilmore, L. O., 1950. Inherited non-lethal anatomical characters in cattle—a review. *J. Dairy Sci.,* 33:147–165.

Green, W. W., 1954. Relationships of measurements of live animals to weights of grouped significant wholesale cuts and dressing percent of beef steers. *J. Animal Sci.,* 13:61–73.

Guilbert, H. R., and P. W. Gregory, 1944. Feed utilization tests with cattle. *J. Animal Sci.,* 3:143–153.

Kincaid, C. M., and R. C. Carter, 1958. Estimates of genetic and phenotypic parameters in beef cattle. I. Heritability of growth rate estimated from response to sire selection. *J. Animal Sci.,* 17:675–683.

Koch, Robert M., L. A. Swiger, Doyle Chambers, and K. E. Gregory, 1963. Efficiency of feed use in beef cattle. *J. Animal Sci.,* 22(2):486–494.

Knapp, B., Jr., and R. T. Clark, 1947. Genetic and environmental correlations between growth rates of beef cattle at different ages. *J. Animal Sci.,* 6:174–181.

Lush, Jay L., 1947. Family merit and individual merit as bases for selection. *Amer. Natur.,* 81:241–261, 362–379.

Marlowe, T. J., C. M. Kincaid, and G. W. Litton, 1958. *Virginia Beef Cattle Performance Testing Program.* Va. Agr. Expt. Sta. Bull. 489.

Pahnish, O. F., E. B. Stanley, C. E. Safley, and C. B. Roubicek, 1955. *Dwarfism in Beef Cattle.* Ariz. Agr. Expt. Sta. Bull. 268.

Parker, J. B., N. D. Bayley, M. H. Fohrman, and R. D. Plowman, 1960. Factors influencing dairy cattle longevity. *J. Dairy Sci.,* 43:401–409.

Shelton, M., T. C. Cartwright, and W. T. Hardy, 1957. *Relationships Between Performance Tested Bulls and the Performance of their Offspring.* Tex. Agr. Expt. Sta. Prog. Report No. 1958.

Stonaker, H. H., M. H. Hazaleus, and S. S. Wheeler, 1952. Feedlot and carcass characteristics of individually fed comprest and conventional type Hereford steers. *J. Animal Sci.,* 11:17–25.

*Warwick, E. J., 1960. *Genetic Aspects of Production Efficiency in Beef Cattle. Proc. of Conf. "Beef for Tomorrow."* In Natl. Acad. Sci., Natl. Research Council Pub. 751, pp. 82–92.

Wilson, J., and C. F. Curtiss, 1893. *Steer feeding.* Iowa Agr. Expt. Sta. Bull. 20.

Chapter 15

Genetic Improvement in Dairy Cattle

The present discoveries in science are such as lie immediately beneath the surface of common notions.

FRANCIS BACON, *Advancement of Learning* (1605)

15-1. INTRODUCTION

The economic justification for specialized dairy cattle is their great proficiency in transforming feeds, mostly inedible by man, into one of man's most nutritious foods. The dairy cow vies closely with the hog in leading other classes of domestic animals in economy of production of human food. From an equivalent amount of feed consumed, the dairy cow returns nearly four times as much edible food as does the steer.

Constructive breeding should have for one of its goals the development of an animal that manufactures products for human consumption more economically during a longer life span. Even though the genetic makeups responsible for a large flow of milk or for varying the chemical constituents of milk are not completely understood, remarkable progress has been made in improving the economy of production.

If one wishes to improve type and production simultaneously, the rate of improvement in both characteristics is impeded. Most dairy cattle breeders want their animals to conform to the approved dairy type in addition to being high and profitable milk producers; this complicates selection since type and production are not highly correlated. When a breeder is selecting for two or more characteristics, his progress depends primarily on the heritabilities of these characteristics, the genetic correlation between them in the same individual, and the actual intensity of selection.

Most of our economically important characteristics are of the difficult or complex type of inheritance called multifactor inheritance. They are controlled by many genes, and the individual effects of a single gene are seldom, if ever, recognized. Milk production is a good example of multifactor inheritance. Many genes and many kinds of gene interactions are involved. In addition to the genetic influences, environment also plays an important role in milk production. Feeding and management practices as well as herd health can cause great variation in total milk yield. In spite of the difficulties, enormous progress has been made in increasing production and improving the conformation of dairy animals. The cows that existed before cattle were used for dairy products of any kind produced only enough milk to nourish their calves, whereas today many thousands of cows average more than 10,000 lb. of milk in a 305-day lactation. The production record of one outstanding cow is almost 43,000 lb. of milk in 365 days.

The dairy cattle breeder striving for genetic improvement in his animals has two basic tools at his disposal—selection and mating systems. Selection is choosing the animals that will constitute the next generation. The basic problem of selection, which is the same for all farm animals, is estimating as accurately as possible the genotype of the animals. After the animals that are to be the parents of the next generation are selected, the breeder must decide which mating system (outcrossing, linebreeding, inbreeding, or crossbreeding) will be most apt to result in maximum improvement in the next generation.

Dairy cattle breeders are usually trying to improve several qualities in their cattle—high production, persistency, regularity of breeding, longevity, disease resistance, good conformation, easy milking, and good disposition. Even though individual qualities are considered in the selection of breeding animals, each animal must be rejected or retained on the basis of the sum of all its qualities. Animals that have the best combination of desired qualities must be selected, since only rarely is an individual best in more than one or two traits. The relative emphasis to be given to each trait is dependent on the amount of improvement that is likely to result from selection. Some traits may have such a low heritability that attempts at improvement by selection would be ineffective. Another point to be considered is the increase in profit to be expected from improvement of a trait. In herds where the sale of milk is the main source of income the dairyman should concentrate on production, with emphasis on conformation only as it will influence the productivity and usefulness of his cattle. But if the breeder is selling breeding stock or exhibiting cattle he may pay more attention to conformation and color markings than does the commercial dairyman.

15-2. SELECTION FOR GENETIC IMPROVEMENT

Traits to Be Used in a Selection Program

MILK PRODUCTION. Since dairy cattle are maintained primarily for their production of milk, this character should be given major emphasis in breeding programs. If their production merits are to be accurately evaluated, dairy cattle must be production tested. The tests can range from checking periodic milk weights, if only milk production is to be considered, to monthly tests for milk, milk fat, solids-not-fat, and proteins, if the constituents of milk as well as milk production are to be considered (see Chapter 7).

Numerous studies have shown that the heritability values for production traits in dairy cattle are not high. Most heritability estimates for milk production range from 0.20 to 0.30. These values for heritability indicate that selection for milk production should bring about a gradual increase in milk yield.

If production records of individual animals are to be accurately compared, the records must be standardized to common conditions. Most records are converted to conform with those of a mature cow milked twice daily for 305 days. Conversion tables are available (Rice et al., 1957) by which records can be standardized. On the average a cow milked three times daily will produce 17 to 20% more milk than if she is milked twice daily. Since most records are now made on twice-daily milking, conversion tables for times milked will not be listed. Conversion tables are also available (Rice et al., 1957) to adjust records that are less than 305 days or more than 305 days to a 305-day basis. However, the merit of adjusting short records to a 305-day basis is questionable and such adjustment may greatly distort estimates. Since it is not possible to compare a large number of cows at any given age, age conversion factors are used to adjust production records to a mature basis. The age conversion factors given in Table 15-1 were developed by Kendrick for use in the DHIA Sire-proving Program.

The production record of a cow results from genetic and environmental effects and their interaction. In order to evaluate more accurately a cow's genetic potential for production, a dairyman must evaluate environmental factors. Some of the more important of these are age of cow at calving, length of gestation, length of lactation period, number of times milked daily, season of calving, length of preceding dry period, and health. Many of these environmental effects can be minimized by the use of conversion factors. By the use of more than one record (lifetime averages) some of the biasing effects of environment can be overcome.

MILK COMPOSITION. Heritability estimates of from 0.20 to 0.30 for milk fat have been reported by many research workers. For solids-not-fat and total milk solids. estimates of 0.34 and 0.36 have been reported (Johnson, 1957). These values indicate that selection for these individual production traits should bring about a gradual improvement in them. However, a negative

TABLE 15-1. *Age conversion factors for 305-day production records.*

	Factor				
Age	Ayrshire	Brown Swiss	Guernsey	Holstein	Jersey
2–0	1.30	1.45	1.24	1.31	1.27
3–0	1.18	1.23	1.12	1.10	1.15
4–0	1.10	1.10	1.06	1.08	1.06
5–0	1.03	1.04	1.02	1.02	1.02
6–0, 7–0	1.00	1.00	1.00	1.00	1.00
8–0	1.00	1.00	1.01	1.00	1.01
9–0	1.02	1.01	1.02	1.02	1.02
10–0	1.03	1.02	1.04	1.04	1.04
11–0	1.04	1.04	1.06	1.06	1.06
12–0	1.06	1.06	1.08	1.09	1.08
13–0	1.07	1.08	1.10	1.12	1.10
14–0	1.09	1.10	1.12	1.15	1.12

Source: Condensed from table developed by Kendrick for use in the D.H.I.A. Sire Program.

genetic correlation exists between milk yield and percent of milk fat, which makes it difficult to improve both of these traits simultaneously (Johnson, 1957). Genetic correlations mean the correlations between the sets of genes that affect two characteristics in the same animal. Genetic correlations should not be confused with phenotypic correlations, since phenotypic correlations between traits result from a combination of genetic and environmental correlations. The relationship between milk yield and solids-not-fat appears to be independent or slightly positive. That between milk fat and solids-not-fat is positive but not linear; that is, an increase in milk fat of 1% does not necessarily imply that solids-not-fat will increase approximately the same amount (Johnson et al., 1961).

CONFORMATION. The type or conformation of an animal is given a great deal of emphasis in the show ring and in the various herd classification programs. How much selection pressure the breeder should devote to improving the conformation of his cattle will depend on his major goal. If he wishes to show cattle and sell breeding stock, type may receive considerable attention. On the other hand, if the primary goal is maximum milk production, only such components of type related to usefulness, such as conformation of feet and legs, body size, or udder size and attachment, would usually be considered. The components of type generally considered in dairy cattle classification are listed in Table 15-2, with their estimated heritability values and their genetic correlation to butterfat production (Johnson and Fourt, 1960).

The breeder can improve the conformation of his animals by careful selection. However, conformation is composed of many single parts, and the

TABLE | *Heritabilities of type and the components of type of 3161 daughter-dam pairs of*
15-2. | *Brown Swiss cattle.*

Characteristic	Heritability Estimates	Genetic Correlation to Milk fat Production*
Type	0.35	0.242
General appearance	0.33	0.450
Dairy character	0.30	0.410
Body capacity	0.23	0.377
Rump	0.36	0.210
Feet and legs	0.19	0.237
Mammary system	0.23	0.410
Fore udder	0.28	0.461
Rear udder	0.35	0.480
Milk fat	0.28	

Source: Johnson and Fourt, 1960.
* Relationship between the sets of genes that affect these components of type and milk fat in the same animal.

more of these single factors the breeder tries to improve simultaneously the less effective is his selection for any one factor. Then, too, most dairy cattle are of little value unless they are high milk producers, no matter how outstanding their conformation is. The dairy cattle breeder usually aims to increase or at least maintain the present level of milk production in his herd while improving conformation of the cattle.

Estimates of genetic correlation between type and production have been quite variable: -0.52 (Freeman and Dunbar, 1955), 0.18 (Harvey and Lush, 1952), and 0.24 (Johnson and Fourt, 1960). The high negative correlation of -0.52 means it would be impossible to improve type and production simultaneously. If the 0.24 correlation were correct, selection for type alone would require approximately four generations to obtain the genetic improvement in production that could be expected in one generation of selection for production alone.

DISPOSITION AND EASY MILKING. The dairyman desires many traits in his cattle that are not directly related to the amount of milk a cow will produce. He wants cows that are easy to milk, having teats that are well placed and that are neither too large nor too small. He prefers cows that are easy to handle and not too high strung and nervous. With the increased emphasis on mechanization and efficiency of labor in the handling and milking of dairy cattle, more emphasis will undoubtedly be put on improving their disposition and rate of milking. More experimental work is needed in this area before definite conclusions can be reached about the effectiveness of selection for these traits.

HEREDITARY DEFECTS. Hereditary defects can be listed as two types, lethal and nonlethal. Lethal characters are those inherited traits that result in premature death. There are a great many known lethals in dairy cattle (Gilmore, 1952). The economic significance of lethals can be great. The average cow produces only 1.3 females that reach milking age in a lifetime. Lethals that prevent the perpetuation of a particular line of breeding containing genes of desired characters could result in great economic loss. The total loss caused by inherited lethals is not known; but recognition of the importance of the problem of inherited lethals in dairy cattle has increased as a result of the rapid expansion (or growth) of artificial breeding. Because the use of frozen semen now makes it possible for a male to sire thousands of offspring in his lifetime, it is very important to ascertain that he is not a carrier of lethal genes.

Nonlethal hereditary defects are those that do not cause death but do reduce the efficiency of an animal. Some defects are inherited as a simple recessive (Gilmore, 1950). Dairy cattle do not have any one defect that is of greatest concern, like dwarfism in beef cattle; however, there are many traits that are of concern: sickled hocks, flexed pasterns, winged shoulders, sloping rumps, poor udder shape or attachment. Selection against these traits in much the same manner used for other quantitative traits should reduce their frequency.

Traits for Which Selection Is Impractical

SELECTION FOR FERTILITY AND LONGEVITY. Heritability studies of breeding efficiency have not produced consistent results; estimates have ranged from almost zero to 0.32. No significant differences were found in breeding efficiencies among 19 cow families for services per conception, days from calving to first breeding, or days from first breeding to conception (Tabler et al., 1951).

Since fertility or, more accurately, lowered fertility, has many causes, it is difficult to determine which cause is responsible for infertility in specific instances. It is not easy to obtain heritability estimates. Two of the most common ways to measure reproductive efficiency are by number of services required for each pregnancy and by the length of time in months between calvings. Both of these methods are seriously affected by such individual factors as purposely not breeding a female for a given length of time in order to change season of calving, breeding to older males of known low fertility, health of the animals, and feeding and management practices.

Several inherited characters cause sterility or reduce reproductive efficiency (Gilmore, 1952). Hereditary tendencies toward the development of cystic ovaries and other conditions that produce at least temporary sterility have been found in dairy cattle (Erb et al., 1959).

It is very doubtful that the breeder will gain much by trying to select for reproductive efficiency since the heritability at best is low. Rather, he should use the selection pressure for traits that respond better to selection. Some automatic selection for fertility and longevity will naturally occur since sterile animals do not reproduce and hence are culled; also, animals of low fertility leave fewer offspring in the herd to propagate the next generation. Studies on the effect of heredity on longevity in dairy cattle have produced varied results, but a recent analysis (Parker et al., 1960) did not indicate heredity to be an important factor.

The low heritability values point out that most of the variation found in the measuring of reproductive efficiency is nongenetic. At present it appears that application of genetic principles in selecting for improved breeding efficiency has little or no effect.

DISEASE RESISTANCE. Little attempt has been made in practical dairy husbandry to develop resistant lines. The use of preventive and therapeutic measures has been relied upon to control most of our cattle diseases. Mastitis in dairy cattle is a serious disease problem that costs the dairymen of the United States more than all the rest of the diseases combined. It is a difficult disease to control because so many different bacteria can produce the disease. Special instances have indicated that the susceptibility to mastitis manifested by related animals may have a genetic cause. Other reports and common observations indicate that pendulous udders predispose the cow to mastitis by permitting the udder and teats to be subjected to infection and injury. Udder size, the nature of the teat sphincter, and the strength of udder attachment also appear to be genetically influenced and related to mastitis susceptibility.

The role of inheritance in most of the diseases of dairy cattle has not been given much consideration in the past. Increasing the frequency of genes for resistance to diseases, however, would be of great practical importance. Future studies may justify more emphasis on disease resistance in dairy cattle. It should be kept in mind, however, that improvement in milk production may be slowed if selection for disease resistance is made simultaneously.

Methods of Selection. Because dairy cattle breeders are usually trying to improve several qualities in their cattle, the intensity of selection possible for each is lowered. Selection may be done in at least three general ways. One method is to cull simultaneously but independently for each of the characteristics. For each characteristic culling levels are established, below which all individuals are culled, no matter how good they are in other characteristics.

A second method is to select for one characteristic at a time until that is improved, then for a second characteristic, later a third, and so forth, until finally each has been improved to the desired level. This is known as the tandem method of selection.

The tandem method is by far the least efficient of the three, even when the characteristics are not affected by any of the same genes and it can be assumed that the improvement made in the first one will not be lost later in selection for improvement in the others. Selection for one characteristic at a time will improve that one faster than will any other method of selection, but while that is being done other traits may deteriorate.

The third method is to establish a selection index or total score to measure net value. This is done by adding the animal's score for its merits in each characteristic. Those with the lowest total scores are culled.

The selection index method is more effective than the method of independent culling levels, because it permits unusually high merit in one characteristic to make up for slight deficiencies in the other. The method of independent culling levels has a practical advantage, which may be important under some circumstances: culling on each characteristic may be done whenever that characteristic develops and without waiting to measure or score the later characteristics.

A combination of these methods is usually most desirable. A selection index can be established according to the goal of each breeder. The value assigned to the various characteristics can be changed as the need arises. When a selection index is used, some of the advantages of culling on independent levels can be made use of by culling the very worst in each characteristic as that develops, but leaving the doubtful cases to be decided later by the selection index.

Evaluation of Genotype. The success of any breeding program depends primarily on the ability of the individual breeder to evaluate properly the animals that are to be selected as parents of the next generation. Three commonly used methods of evaluating animals are (1) individual merit, or phenotype, (2) pedigree, and (3) progeny test.

PHENOTYPE. Selection based on individuality, or phenotype, is practiced with dairy cattle when cows are culled on the basis of their record of production or type. The success of selection by individual merit depends upon (a) the ability of the breeder to choose the right animal, (b) the number of individual animals to choose from, and (c) the narrowness and definiteness of the standard adhered to.

The phenotypic characteristic upon which selection is based often varies in expression from time to time. The amounts of milk and fat that a cow produces in different lactations and the changes in certain components of type in different stages of lactation are examples. Most of the variations from one time to another are due to changes in the environment that prevail at the time, or have prevailed just previously to the time the observations are made.

The use of an average of many repeated observations as a basis for selection is one of the most effective ways of overcoming mistakes and confusion that would otherwise result from the effects of temporary environmental conditions (Lush, 1945). This method requires that records be kept, and culling must be postponed until two or more observations have been made.

Lifetime averages are most useful in helping to overcome temporary environmental effects, but they do not keep the breeder from being deceived by permanent effects of environment.

PEDIGREE. A second method of evaluating an animal is by pedigree. A pedigree is a genealogy chart that presents the names of ancestors in an order that shows the number of generations each is removed from the animal being evaluated. The real value in pedigrees lies in associating with the ancestors some measure of their economic value such as milk production or type. When pedigree selection is used the intensity of individual selection is reduced. The average individual merit of those selected is lowered when superior animals are rejected or inferior animals kept for breeding on the basis of their pedigree.

Because most of our economically important characteristics are controlled by many genes and these genes are heterozygous in most animals, the sampling nature of inheritance limits the accuracy of pedigree selection. It is generally agreed that pedigree selection should be used as an accessory to individual selection. It can best be used in making decisions on animals of similar individual merit. Pedigree selection is particularly useful when selecting young animals for traits that are sex-limited or that are exhibited only after sexual maturity. Udder shape and attachment is an example of such a trait; milk production is another.

PROGENY TESTS. A third method of evaluating an animal is by progeny tests, which means estimating the individual's heredity by studying its offspring. The progeny tests are valuable for selection for such quantitative characters as milk production or milk constituents, because (1) heritability is low and individual selection is thus not very accurate, (2) the character is expressed in only one sex, (3) the time between generations is long, and (4) the female has few offspring. Because of the relatively large number of genes involved, very little is known about the mode of inheritance of milk and milk constituent production. When properly used, progeny tests make selection more accurate; they prevent the breeder from being deceived by the effects of environment, and of dominance and epistasis as much as he might otherwise be. In practice, even for these characteristics, progeny testing should be used only as a supplement to other types of selection rather than as the sole basis. Lush (1935) carefully investigated the theoretical accuracy of progeny testing as compared to individual selection. He concluded that when the likenesses between the performance of offspring (correlation between phenotypes) was due to only the genes they received in common from their parents, and the heritability of the tract being considered was low, then at least five tested progeny were necessary to make the selection of these parents more accurate on a progeny test basis than selection on the basis of their own individuality. If there is a correlation as large as 0.25 between progeny due to nongenetic factors, progeny testing cannot possibly be as accurate as individual selection.

Selection of Females. Improving dairy cattle depends, first, on the breeder's ability to recognize animals that are genetically superior and, second, on the effectiveness of permitting these superior animals to reproduce.

Records of production offer the most reliable means of selecting profitable cows. If the animals in a herd are handled as a unit, so that certain individuals are not given special attention, the average production for the herd is a reliable basis upon which to compare individual cows in the herd. When a cow has several records, more dependence should be placed on her average as an indicator of her real producing ability. Ranking cows in the herd on the basis of their producing abilities is an effective aid to selection. A selection standard can be determined for each herd and will be relative rather than rigid.

Cows that have milked extremely poorly under apparently normal conditions should be culled. Under some conditions it may be desirable to allow a second chance to the borderline cases. However, by the time the next lactation is three or four months along, it should be fairly evident whether or not these cows should be culled.

Pedigree selection is the fastest method of selection; but most of the time it is less dependable than selection based on the individual's own characteristics or on the characteristics of its progeny. Under certain conditions an appraisal of a cow's relatives, in addition to her own records, may be used to determine her average breeding value. For traits having heritability values as high as 0.20, the individual's own performance should still receive most emphasis in predicting the animal's breeding value.

When the major source of income is from the sale of milk, selection is simplified. The cows can be ranked from high to low on the basis of milk production, and the least profitable cows sold. There is no reason to keep a cow that is not profitable or potentially profitable.

A breeder selling breeding stock or classifying his animals for type must pay more attention to conformation. A selection index, in which each characteristic to be selected for is given a value, is recommended. The problem of selection becomes more difficult when a breeder is trying to improve two or more characteristics simultaneously. The progress made depends primarily on the heritabilities of the characteristics, the genetic correlations between them in the same individual, and the actual intensity of selection. The use of pedigrees and progeny testing as accessories to individual selection are very beneficial, but major emphasis should still be on individual selection.

Selection of Sires. Genetically the sire and dam contribute equally to the genetic makeup of their offspring. However, since a sire can have many more offspring in his lifetime than a cow can in hers, it is through the bull that the most genetic improvement in a herd is made. Every dairyman should be concerned with the quality of the breeding of the bulls he uses. The problem of selecting a herd sire that will improve a herd has not been completely solved, but new and modern techniques have increased the chances of selecting or using a herd-improving sire.

In general, the dairyman has three sources from which to obtain sires. He can use artificial insemination services, he can purchase a herd sire, or he can raise one. Since all dairymen should set a goal that is a little higher than their present herd average, they must face this very fundamental question in selecting future herd sires: bulls that will transmit higher levels of production than the present herd average must be selected. This is essential for gradual improvement in a dairy herd.

PROVED SIRES. Perhaps no subject in the field of dairy-cattle breeding has received more written and verbal comment than that of proved sires. Most sires are proved by the Division of Dairy Herd Improvement Investigations of the United States Department of Agriculture, which uses lactation records sent in from Dairy Herd Improvement Associations from all over the country.

Official proofs are issued whenever a bull used in artificial breeding has 25 or more daughter-dam comparisons or a bull used in natural service has 10 or more daughter-dam comparisons.

A proved sire is one on which there is sufficient unselected information to indicate his transmitting ability. The greater the amount of information, the more reliable the proof. It is often assumed that a proven sire has increased production in the herds in which he has been used. His genotype cannot be estimated, however, until the conditions under which the proof was made are analyzed. The level of production of the herds in which the bull was used is an important key to evaluating the proof. A desirable proof on a bull is presumed to be an indication that he increased production significantly. A sire that had been used in a high producing herd, for which records showed very little, if any, increase, could be called a desirable sire because it is difficult even to maintain production in high producing herds. On the other hand, when a bull's daughters show an increase in a low producing herd, it is difficult to know how good the bull really is.

A desirable proof is one in which the average production of the bull's daughters is well above the average for the breed.

There are several methods of evaluating proved sires:

Daughter Average. The average production of the daughters of a particular sire are a valuable indication of that sire's breeding value, provided the daughters are unselected and there are enough of them. There should be 100 daughters for an A.I. (artificial insemination) proved sire and 15 to 20 for a naturally proved sire.

Daughter-Dam Comparison. This was one of the first methods of evaluating sires and is still in use today. This program has some disadvantages when compared to other programs such as daughter averages or herdmate comparisons. Some disadvantages are these: (*a*) The dams and daughters usually are not milking at the same time, which means that they make their records under different environmental conditions. (*b*) Comparisons are usually made in one or a few herds under a limited range of environmental conditions. Many dams have no records; therefore, some of the daughters with records cannot be used, since there are no records on the dam for comparison.

Herdmate-Daughter Comparisons. This is the newest approach to the evaluation of a sire. Herdmates of a sire's daughters are the other cows in the herd that calved in the same year and season as the sire's daughters did. This has a decided advantage over the daughter-dam comparison because in the herdmate-daughter comparison the records being compared have been made at the same time and under the same management conditions, whereas in the daughter-dam comparison the records of the daughters and the records of the dams may have been made several years apart, and in the interval the management conditions may have changed drastically.

YOUNG SIRE. Although it is advantageous for every dairyman to use a good proved bull, there are times when he may wish to choose or produce a young sire through selective mating.

Pedigree information is the major factor used in selecting a young sire. Research has shown that the first and most important thing that dairymen should look for in the pedigree of a young bull calf is the proof of his sire. One should not settle for anything less than a son of a good proved sire. This sire should have as many producing daughters as possible. The daughters should have produced considerably more than their dams if the bull was bred to low-producing or mediocre cows. If he was bred to high-producing animals, then the daughters should show production well above the present herd average of the dairyman making the selection.

The young bull should be evaluated for type in order to avoid physical defects and poor type. Personal inspection of the close relatives of the calf should be made. There should be attempted some estimate of the environment of the herd in which the bull calf's relatives performed and of how this environment compares to the herd in which he is to be used. A large proportion of the differences between herd averages is in feeding and management, which cannot possibly affect the genetic make-up of the young bull. After the young bull has been chosen he should be used enough to get a representative sample of daughters. His service should then be restricted until he is proven, that is, until his daughters have production records.

15-3. MATING SYSTEMS

Inbreeding. Inbreeding is defined as the mating of animals more closely related than the average of the population from which they came. The primary purpose of inbreeding is to increase the probability that the offspring will inherit the same outstanding characteristics from sire and dam. This is accomplished by the increase in the percentage of gene pairs that are homozygous and a decrease in the percentage that are heterozygous. The speed with which this occurs is determined by the closeness of relationship between the individuals mated.

The general effects of inbreeding are a decline in vigor as shown by lowered yields of milk and fat, more disease, less growth, higher mortality during calfhood, and a greater percentage of calves born dead. These effects are the results of undesirable genes becoming homozygous. A decrease of 210 lb. of milk and 4.9 lb. of milk fat per 1% of inbreeding was reported by Laben et al. (1955) as the results of an inbreeding experiment that was conducted in California for a long period of time.

Although the general effects of continued inbreeding are usually undesirable, inbreeding has an important place in dairy cattle breeding. Some reasons for practicing inbreeding are these: (1) it forms uniform and distinct families, so that interfamily selection may be possible in a more effective way; (2) it is necessary if relationship to a desirable ancestor is to be kept high; (3) it increases prepotency. Prepotency is the power of an animal to stamp its own characters on the offspring to the exclusion of those of the other parent, and since it depends upon homozygosity of dominant genes, inbreeding is the only known method of increasing it. Inbreeding is also necessary to hold together desirable gene combinations and to propagate them through future generations. A good rule to remember when trying to determine if inbreeding should be practiced is this: since inbreeding intensifies what is present, never start an inbreeding program with poor or mediocre cattle.

Linebreeding. Linebreeding is very popular and is in common use among dairy cattle breeders. Linebreeding is a means of maintaining relationship in a herd to some ancestor regarded as unusually desirable. Since the only way to accomplish this is by intermating among that ancestor's descendants, some degree of inbreeding is inevitable. The degree of inbreeding is kept down by using for parents animals that are both closely related to the admired ancestor but are little, if at all, related to each other through any other ancestor.

Linebreeding, more than any other breeding system, combines selection with breeding. A great deal of its success will depend on the breeder's ability to select the animals that have been outstanding in his herd. The better the animals, the better the results from linebreeding.

Linebreeding tends to separate the herd into distinct families, between which effective selection can be practiced. It will also build up homozygosity and prepotency but usually not as fast as close inbreeding. Linebreeding should be used only in herds that are well above average for type and production. The herd should be large enough to permit rigid culling and to prevent inbreeding from occurring too rapidly.

Outcrossing. This system of breeding has been used more extensively than any of the others in dairy cattle breeding. Outcrossing means mating of unrelated or distantly related animals. The usefulness of outcrossing depends almost wholly on the effectiveness of selection. More variation is expected between individual offspring because the mates are more heterozygous than in linebreeding or inbreeding. Since it is unlikely that unrelated animals will carry the same genes, the chief advantage of this system of breeding is that it tends to cover up undesirable recessives. It is the mating system recommended for the average or below-average purebred herd. In herds of this type the owner has the problem of maintaining individual merit rather than of making undesirable genes homozygous.

The disadvantage of a continuous outcrossing program is that it is not likely to lead to any significant improvement through the fixation of desirable genes.

Crossbreeding. Crossbreeding is the mating of animals belonging to different breeds. Like any other form of outbreeding, it tends to lower the prepotency of the individual (its ability to transmit its characteristics to its offspring), thus making selection among the crossbred individuals less effective. It does, however, promote individual merit, because of the general dominance of genes favorable to size, fertility, vigor, and so forth.

There has been less experimental crossbreeding of dairy cattle than of most other classes of farm animals. No clear-cut cases of heterosis or superior crossbred production are apparent from records of trials to date, although experimental work has provided conclusive evidence that three or four breeds of dairy cattle can be combined in crosses without harmful and possibly with beneficial results.

This evidence has led to the general recommendation that for commercial dairymen who have no special interest in a particular breed, but are primarily concerned with increasing the efficiency of their cows and establishing a high level of milk and butterfat production, the best practice to follow is to breed each cow to the best sire available. Artificial insemination, with semen available from bulls of several breeds, makes it possible to follow any systematic crisscross breeding or rotational crossbreeding system that is desired without excessive investments in sires. More experimental work is needed before other more definite conclusions can be reached.

Remarkable progress has been made through breeding, feeding, and management in developing the high-producing dairy cow of today from an ancestor that produced only enough milk to nourish its young. In spite of such progress, much still remains to be done. Many of the problems that remain to be solved are difficult ones—which is why they still remain unsolved. Some of the goals of the dairy cattle breeder and research worker are to increase resistance to disease, especially mastitis; to prolong the useful life of the dairy cow; to increase the percentage of certain constituents in milk, especially protein; to increase fertility and breeding efficiency; to increase the economy of production through better feed utilization; to develop more accurate methods of evaluating sires. These are but a few of the many problems challenging the dairy cattle breeder and research worker.

REFERENCES AND SELECTED READINGS

Erb, R E., P. M. Hinze, and E. M. Gildow, 1959. *Factors Influencing Prolificacy of Cattle. II. Some Evidence that Certain Reproductive Traits are Additively Inherited.* Wash. Agr. Expt. Sta. Tech. Bull. 30.

Freeman, A. E., and R. S. Dunbar, 1955. Genetic analysis of the components of type, conformation, and production in Ayrshire cows. *J. Dairy Sci.,* 38:428.

Gilmore, L. C., 1950. Inherited non-lethal anatomical characters in cattle—A review. *J. Dairy Sci.,* 33:147.

———, 1952. "Reproductive efficiency," in his *Dairy Cattle Breeding.* Lippincott, Chicago. Chapter 7.

———, 1952. "Lethals," in his *Dairy Cattle Breeding.* Lippincott, Chicago. Chapter 9

Harvey, W. R,, and J. L. Lush, 1952. Genetic correlations between type and production in Jersey cattle. *J. Dairy Sci.,* 35:199.

Johnson, K. R., 1957. Heritability, genetic and phenotypic correlations of certain constituents of cow's milk. *J. Dairy Sci.,* 40:723.

———, and D. L. Fourt, 1960. Heritability, genetic and phenotypic correlations of type, certain components of type, and production of Brown Swiss cattle. *J. Dairy Sci.,* 43:975.

———, R. A. Hibbs, and R. H. Ross, 1961. Effect of some environmental factors on the milk fat and solids-not-fat content of cow's milk. *J. Dairy Sci.,* 44:658.

Laben, R. C., P. T. Cupps, S. W. Mead, and W. M. Regan, 1955. Some effects of inbreeding and evidence of heterosis through outcrossing in a Holstein-Friesian herd. *J. Dairy Sci.,* 38:525.

Lush, J. L., 1935. Progeny test and individual performance as indicators of an animal's breeding value. *J. Dairy Sci.,* 18:1.

———, 1945. "Aids to selection—The use of lifetime averages," in his *Animal Breeding Plans.* Iowa State College Press, Ames. Chap. 13.

Parker, J. B., N. D. Bayley, M. H. Fohrman, and R. D. Plowman, 1960. Factors influencing dairy cattle longevity *J. Dairy Sci.,* 43:401.

Rice, V. A., F. N. Andrews, E. J. Warwick. and J. E. Legates, 1957. "Selecting dairy cattle," in their *Breeding and Improvement of Farm Animals.* McGraw-Hill, New York. Chap. 19.

Tabler, K. A., W. J. Tyler, and G. Hyatt, Jr., 1951. Type, body size and breeding efficiency of Ayrshire cow families. *J. Dairy Sci.,* 34:95.

Chapter 16

Genetic Improvement in Swine

*Here Lies
All That was Eatable
Of a Prize Pig.*

*He was Born
On February 1, 1845:*

*He was Fed
On Milk, Potatoes, and
Barley-Meal:*

*He was slaughtered
On December 24, 1846,
Weighing 20st. 9lbs.*

Stop, Travellor!

*And Reflect How Small a Portion
Of This Vast Pig
Was Pork Suitable
For Human Food.*

YOUATT, *Punch* (1855)

16-1. INTRODUCTION

Swine differ from the other farm animals in several respects that must be considered in devising methods of genetic improvement. Let us look at some of these differences:

1) Swine are polyovulatory (litter-bearing) animals; that is, several eggs are released at each estrus and several offspring are produced at each parturition, rather than one as in cattle and horses and one or two as in sheep.

2) Meat is their sole product. They do not perform work or produce fiber, milk, or eggs for human use.

3) Their suckling period is much shorter and less important in total weight gain than is that of cattle and sheep.

4) Swine are concentrate eaters. Their digestive tract is not adapted to utilize large amounts of roughage.

5) They have a greater tendency to produce overfat carcasses than do other meat animals. Fat in excess of the amount desired is becoming more of a problem in the breeding of all meat animals, because of changes in eating habits of people in many countries, particularly in the United States. Nevertheless, for beef cattle and sheep a finishing period on a high-energy ration is commonly required in order that they may have enough finish for desired meat quality, whereas with swine it may be necessary not only to select for "meatiness" but also to modify feeding during the finishing period to reduce the amount of fat in the carcass.

These unique qualities suggest that plans for improvement through breeding will differ in some respects from those for other animals.

16-2. DEFINITION OF TRAITS

Swine traits may be classified in a general way into three groups: (1) productive, (2) reproductive, and (3) structural.

The productive traits include rate of gain, efficiency of feed utilization, and milk production. Research experience has indicated that these productive traits in swine have intermediate heritability. Thus, they can be improved by selection, but they will also show some response in crosses.

The reproductive group includes number farrowed per litter and number weaned per litter. Number farrowed is influenced by the number of eggs shed and the survival rate of the embryos during development. Reproductive traits have low heritability. Therefore, they show the greatest response of any of the traits to crossing. The response in crossing is particularly strong where the sow is herself a crossbred.

The structural traits include carcass characteristics such as fatness and leanness, muscle size, and mature body size. These structural traits are highly heritable and can be improved by selection. Their response to crossbreeding is less than for either reproductive or productive traits.

Table 16-1 contains heritability estimates from many studies.

16-3. SELECTION

Selection is simply keeping some animals and culling others, and is effective in improving traits in proportion to their heritability. One can expect fairly rapid improvement to result from selection of animals superior in conformation and carcass traits (provided they are accurately measured), moderate improvement from selection for rate and efficiency of gain, and very slow improvement from selection for larger litters.

TABLE 16-1. | *Estimates of heritability of swine traits, based on reports of investigations from many breeds and several countries.*

Traits	Heritability Percentage	
	Range	Approximate Average
Productive Traits		
Weight of pig at 5–6 months	3–66	30
Growth rate (weaning to 180–200 lb.)	14–58	29
Economy of gain	8–72	31
Reproductive Traits		
Number of pigs farrowed	0–24	15*
Number of pigs weaned	0–32	12*
Conformation and Carcass Traits		
Length of legs	51–75	65
Number of vertebrae		74
Conformation scores	10–35	29
Type, within herds of similar type		38
Type, between herds of small, intermediate, and large type (Poland China)		92
Length of carcass	40–81	59
Loin eye area	16–79	48
Thickness of backfat	12–80	49
Thickness of belly	39–72	52
Percent of carcass weight		
Loin	51–65	58
Shoulder	38–56	47
Fat cuts	52–69	63
Lean cuts	14–76	31†

Source: From Craft, W. A., 1958. Fifty years of progress in swine breeding. *J. Animal Sci.*, 17:960.
* Probably high.
† Probably low.

There are three systems of selection that a breeder may use, either separately or in combination. These three (previously discussed in Chapter 15) are (1) index, (2) independent culling levels, and (3) tandem. Some practical illustrations of their use in swine improvement follow.

At the Iowa Swine Testing Station an index was developed that provided for one-half of the selection emphasis on backfat thickness, one-fourth on rate of gain, and one-fourth on efficiency. The greatest emphasis was placed on the trait with the highest heritability—backfat thickness—but some emphasis was also placed on the other two traits. The Iowa Index, as originally used, was:

$$260 + (35 \times \text{gain}) - (75 \times \text{BF}) - (40 \times \text{EFF}) = \text{Index}.$$

The 260 is a constant, chosen to give an average index of about 100. The other figures used as multipliers are weighting figures for the items included in the index. Rapid growth rate, low backfat, and efficiency (a small number of pounds of feed used per pound gained) will contribute to a high index. To illustrate the use of the index, suppose a choice is to be made between two boars with the following records:

No. 1—2.0 lb. gain/day, 1.2 in. backfat, 3.2 lb. feed/lb. gain.

No. 2—1.8 lb. gain/day, 1.0 in. backfat, 3.0 lb. feed/lb. gain.

No. 1 has an index of 112 and No. 2 an index of 128, indicating that the leanness and efficiency of No. 2 more than compensate for his lower rate of gain. Thus the index removes the guesswork from combining information on different traits and makes selection decisions easier. Of course, some common sense must be used in the application of any index, and the presence of a serious defect in conformation, or some unsoundness, will eliminate a few animals, regardless of their index scores. Generally, such animals will be eliminated at an early stage, before the indexes are computed.

In the application of the Iowa Index, it was felt that some pigs were passed that were too fat or too slow in growth rate, even though they met the standard of 100. A system of independent culling levels, in addition to the index, was therefore established. All pigs with less than 1.6 lb. daily gain, more than 1.45 inches backfat, or more than 3.25 lb. feed per lb. of gain, would fail the test. Thus the index gave a general description of the pig, weighted according to the importance and heritability of the traits concerned, but the independent culling levels eliminated some high-indexing pigs that were badly deficient in one of the three traits.

The principal advantage of the index in selection is that it allows a particularly good score in one trait to compensate for a deficiency in another. The use of culling levels curtails this advantage, and is a less efficient system than the index method. It is true, of course, that certain traits may be of more importance to one breeder than to another, because of the level of performance of the herd for that trait, and there is therefore justification in some cases for considering individual traits as well as the index. For example, a breeder whose herd has an excellent average for gain and efficiency but in which the hogs are much fatter than desired will wish to select boars that are especially lean.

A summary of the first eight years' tests at the Iowa Swine Testing Station is given in Table 16-2. The general testing procedure has been to test a set of February and March farrowed pigs in the spring and summer and a set of August and September farrowed pigs in the winter.

Inspection of the table points up two pertinent facts:

1) A marked change has occurred in the meatiness of the tested pigs in the eight years of testing, and efficiency has been improved somewhat.

TABLE 16-2. | *Season comparisons of rate and efficiency of gain and carcass traits of pigs for successive years. (S, spring; F, fall.)*

	Pen Average		Boar Average		Barrow Cutout			
	Gain	Eff.	Probe	Index	Carcass Length (inches)	Backfat (inches)	Ham & Loin (percent)	Loin Eye (sq. inches)
56S	1.89	292	1.46	101	29.1	1.64	32.3	3.22
56F	1.95	303	1.31	109	28.9	1.61	33.2	3.50
57S	1.80	294	1.24	113	29.2	1.60	33.5	3.40
57F	1.79	319	1.11	112	29.1	1.51	34.5	3.80
58S	1.80	285	1.22	118	29.4	1.51	34.3	3.62
58F	1.76	323	1.19	106	29.1	1.50	34.9	3.81
59S	1.80	296	1.25	119	29.5	1.50	35.3	3.63
59F	1.87	297	1.17	125	29.1	1.48	36.2	3.94
60S	1.77	277*	1.19	128	29.3	1.48	35.6	3.77
60F	1.94	286*	1.20	130	29.3	1.61	35.8	3.94
61S	1.80	284*	1.17	125	29.3	1.48	35.6	3.87
61F	1.87	280*	1.06	140	29.2	1.41	37.3	4.08
62S	1.89	262*	1.08	149	29.7	1.35	36.8	3.96
62F	1.99	274*	1.04	154	29.6	1.39	38.8	4.22
63S	1.96	269*	1.08	156	29.9	1.36	39.4	4.13
63F	1.97	277*	.92	157	29.7	1.35	39.8	4.25

Source: Iowa Swine Testing Station, Ames, Iowa.
* Adjusted to "boar equivalent" basis by subtracting 5 for each barrow from the actual feed efficiency of the pen.

2) Spring pigs have generally been somewhat meatier and less efficient than fall pigs farrowed in the same year.

Thus the index, designed to cause improvement in meatiness, has apparently been effective. As meatiness has improved, the index has been changed to give less emphasis to backfat thickness and more to the other traits. Standards have also been raised, and adjusted to accommodate the poorer winter efficiency.

The present index is as follows:

1) *Boars with littermate barrows*
   Index = 117 + 50 (gain) − 50 (feed efficiency) − 40 (probe) + 3(H&L%)
2) *Boars with half brother barrows*
   Index = 240 + 50 (gain) − 50 (feed efficiency) − 50 (probe)

The original index was weighted to give most emphasis to backfat probe. If all the emphasis had been placed on this trait until the desired level of performance was reached and then selection attention directed toward a second trait, such as gain, this would have been tandem selection. This method of selection has the advantage that more rapid improvement in a specific trait occurs, but it is the least efficient method of improving total merit and is not recommended. It has the further disadvantage that if any desired performance traits are genetically negatively correlated with the trait under selection, they will decline while the selected trait is being improved.

16-4. MATING SYSTEMS

Under mating systems we consider the degree of relationship between the sows and boars mated. This can range from close inbreeding, where the sows and boars are close relatives such as brother and sister, to the mating of un-related individuals within a breed, and to crossbreeding, in which boars of one breed are bred to sows of another breed or cross.

The effects of inbreeding and crossbreeding have been investigated in many experiments, and although it would be desirable to have more infor-mation on how specific breeds combine in crosses, in general the effects of these different mating systems in swine are well known. Much of the avail-able information in this area, particularly with respect to the effects of inbreeding and of crossing inbred lines, has come from the Regional Swine Breeding Laboratory, a cooperative research project between the USDA and several state experiment stations, which was set up in 1937 to investigate swine-breeding methods.

In general, inbreeding decreases performance traits, with the greatest de-crease coming in litter size and survival, followed in order by rate of gain, feed efficiency, and carcass characteristics, the last being scarcely if at all affected. It may be noted that this is in inverse relationship to the her-itability of the traits, those of high heritability showing little inbreeding effect and those of low heritability being seriously affected. It has been found that inbreeding the equivalent of one generation of full brother-sister mating decreases litter size raised by about 1.1 pigs per litter (Dickerson et al., 1954; Bradford et al., 1958). The effect seems to be greater in first-litter gilts than in mature sows. Inbreeding does lead to greater genetic uniformity than outbreeding systems, but because of its effect on performance is not generally recommended.

Crossing inbred lines leads to a recovery of the vigor lost through inbreed-ing and possibly some additional gain, but because of the cost of developing inbred lines and their poor performance, the use of inbred lines to produce hybrid hogs, according to the system used to produce hybrid corn, has not become widespread. Swine producers can and do utilize hybrid vigor, how-ever, by crossing animals of different breeds; in fact, most commercially raised hogs in the United States are crossbreds of one kind or another.

Crossbreds may be two-breed crosses, produced by mating purebred boars of one breed to purebred sows of another breed, three-breed crosses, pro-duced by breeding purebred boars of one breed to crossbred sows represent-ing two other breeds, or more complex crosses. Because the advantages of crossbreeding are greatest in reproductive traits, the crossing program should use crossbred females, as in the three-breed cross. Two other crossbreeding systems that permit the use of crossbred females are (1) crisscrossing, in which boars of two breeds are used alternately, and (2) rotational crossbreeding, in which boars of three or more breeds are used in rotation; in each system the females from the previous cross constitute the sow herd.

Examples of specific crossbreeding plans for swine, with comments on these plans, are given in Table 16-3. The contributions of the breeds involved (expressed as percentages) to the different kinds of crosses are summarized in Table 16-4.

Rate of gain exhibits some hybrid vigor, but feed efficiency is only very slightly affected. Carcass characteristics are usually intermediate between those of the parents; this emphasizes the importance of careful selection of the breeding animals used for crossing.

In spite of the many experiments on crossbreeding, the actual effect of crossbreeding, in terms of percentage increase of crossbred over purebred performance, is difficult to arrive at accurately. This is because the experiments have differed in such respects as breeds used, experimental conditions,

TABLE 16-3. | *Examples of crossbreeding plans for commercial hog production.*

Plan	Example	Comments
Two-breed repeat crossing	Breed Hampshire sows to Yorkshire boars. Market all crossbred pigs. Buy Hampshire replacement gilts and Yorkshire boars.	Two-breed crossing is the starting point for all crossing schemes. It is not recommended as a terminal program because it does not use crossbred females
Crisscrossing	Breed Duroc sows to Landrace boars; breed Landrace X Duroc sows to Duroc boars; breed three-quarters Duroc, one-quarter Landrace sows to Landrace boars; continue alternate use of boars of these two breeds.	Uses crossbred females, therefore is preferred to two-breed repeat crossing. Recommended where only two suitable breeds are available.
Three-breed rotational crossing	Same as three-breed crossing, except that three-breed cross gilts are kept and bred to boars of one of the original breeds.	About the same amount of hybrid vigor as three-breed crossing except that crossbred sows are always used and preferred, because purchase of females is not required after first cross. Recommended where three suitable breeds are available.

TABLE 16-4.	*Outline of various methods of crossbreeding, with the percentage contribution of the parent breeds. Original sows are purebred. All boars used are purebred.*

Percentage Contribution of Parent Breeds	Two-Breed Repeat Crossing	Crisscrossing	Three-Breed Rotational
Breeding of sows after first generation	Purebred (probably purchased)	Crossbred (raised)	Crossbred (raised)
Percent contributions parents—first generation	50:50	50:50	50:50
Percent contributions of parents—second generation	50:50	75:25	25:25:50
Percent contribution of parents—third generation	50:50	37.5:62.5	62.5:12.5:25
Percent contribution of parents—fourth and later generations	50:50	67:33 (Approximately)	57:29:14 (Approximately)

characteristics measured, and whether purebreds of both parent breds were available for comparison. A summary by Carroll and Roberts, as presented by Dickerson (1952), is shown in Table 16-5.

Note that the increase in total litter weight weaned is larger than that for any of the other traits listed. Both components of litter weight—number of pigs in the litter and individual pig weight—are increased. Litter weight at 5 to 6 months of age would show an even greater effect, since postweaning gain increased more than preweaning gain.

The data in Table 16-5 do not show the effect of crossbreeding the sow or her litter production. In a study of data from commercial herds, Bradford et al. (1953) found that 3005 straightbred gilts raised an average of 6.53 pigs per litter, while 836 crossbred gilts raised an average of 7.20 pigs per litter. It would take many generations of selection to increase litter size by this amount; this illustrates the importance of crossbreeding in improving productivity.

A sound crossbreeding program involves selection of the best breeds and individual representatives of those breeds available as foundation stock. Since breeds differ in different traits and no one breed is outstanding in all traits, breeds should be selected so that the strong points of one parent breed compensate for the weaknesses of the other parent breed or cross. In the initial cross it is wise to select the more prolific breed as the female parent.

TABLE 16-5. *Results of crossbreeding experiments with swine, as summarized by Carroll and Roberts (1942).*

Factors of Production	No. of Experiments	Mean of Two Pure Breeds	Mean of Crossbreds	Relative Performance of Crossbreds*
No. pigs per liter	12	9.74	9.48	97.3
Birth weight of pigs (lb.)	6	2.77	2.79	100.6
Survival (%)	15	76.3	80.2	105.1
Weaning wt of pigs (lb.)	15	32.5	33.12	101.8
Weaning wt of litters (lb.)†	13	235.6	254.1	107.9
Av. daily gain (lb.)	9	1.38	1.44	104.0
Feed for 100 lb. gain (lb.)	6	374.1	368.6	101.5
	Danish pig-testing stations			
Av. daily gain	32	1.3	1.38	101.5
Feed per 100 lb. gain (lb.)	32	354.4	344.3	99.7

* Performance of purebreds = 100.
† From the original publications of these experiments.

There is a belief, once very common among stockmen and still strongly held by some, that crossbreeding leads to increased variability and eventual degeneration of the stock. While indiscriminate crossing of inferior stocks will lead to poor results (just as will the breeding of inferior purebreds), a systematic crossbreeding program based on carefully selected purebred boars produces pigs that are superior in total production potential to purebreds, and no more variable in performance traits than straightbred commercial hogs. Some variability in such characteristics as color pattern and shape and size of ear do occur, but there is no evidence of increased variability in rate of gain, litter size, or carcass traits. The general acceptance of cross-breeding by swine producers is undoubtedly an advantage to this industry.

To summarize the discussion on selection and mating systems: selection is effective in improving carcass traits, but relatively ineffective for reproductive traits; selection is moderately effective in improving rate of gain and efficiency of feed utilization; crossbreeding is effective in improving litter size, and moderately effective in improving rate of gain, but has little effect on carcass traits or feed conversion.

To illustrate, the gains that might be expected from one generation of selection, and from crossing unselected representatives of different breeds are presented for three traits in Table 16-6. The table shows that cross-breeding might be expected to result in about the same increase in 5-month pig weight (or average daily gain) as one generation of selection, and that using crossbred sows will result in an increase in litter size that would take several generations to achieve by selection.

Crossbreeding combined with selection—that is, the crossing of stocks se-

TABLE 16-6. | *Gains to be expected in three traits from one generation of selection, and from crossing unselected representatives of different breeds.*

		Increase over Straightbreds of:	
Trait	Expected Gain from Selection*	Crossbred from Purebred Dam	Crossbred from Crossbred Dam
No. pigs raised per litter	0.06 pig	0.035 pig†	0.67 pig**
Five-month pig weight	6.9 lb.	6.6 lb.†	Probably similar to value for two-breed cross
Backfat probe	0.14 inches	‡	‡

* Assuming: (a) Top 20% of gilts and 2% of boars raised are selected. (b) Standard deviation of 2.0 pigs, 24.0 lb. and 0.3 in. for litter size, 5-month pig weight and backfat probe, respectively. (c) Heritability values as given in Table 16-1. (d) Selection is based on 4 traits, that is, these three and one other, such as conformation score.

† Based on an increase of 5% in survival and 4% in postweaning rate of gain, as given in Table 16-5.

** From comparison of litters from crossbred and straightbred gilts, as reported by Bradford, G. E., A. B. Chapman, and R. H. Grummer, 1953. *J. Animal Sci.*, 12:582.

‡ No data available. Little or no difference expected between crossbreds and straightbreds in this trait.

lected as carefully as assumed here for the calculation of expected gains from selection alone—would be expected to result in improvement equal to the sum of the expected improvements shown for selection and for cross-breeding.

Selection as intense as that shown would be possible only in a large closed herd where all boars and gilts born were available for selection. In actual practice in the industry, where most boar pigs are castrated and the boars used are produced in a relatively small number of herds, selection would, on the average, be less intense than has been assumed for this example. Thus these estimates of progress from selection are probably optimistic.

16-5. IMPROVEMENT PROGRAMS

Systematic swine improvement programs designed to measure and select for performance traits are a relatively recent innovation in the United States. One of the first—certainly the first in which a selection index was used in the selection of commercially raised livestock—was the Wisconsin Swine Selection Cooperative, started in 1947 under the direction of Dr. A. B. Chapman of the University of Wisconsin. In this program, replacement gilts are selected on the basis of an index combining number of pigs farrowed, number raised, and litter and pig weight at five months. Backfat probe information has recently been added to the index. This program is used for both commercial and purebred herds, the majority being commercial herds.

Several other countries have had improvement programs for many years. Denmark initiated a progeny-testing plan in 1907 that has been in continuous operation since, except for the years of the two world wars. In this program, pigs submitted by breeders throughout the country are raised at central test stations under uniform conditions. The pigs are evaluated according to rate of gain, feed conversion, and carcass quality, and breeding animals are selected on the basis of performance of their offspring or litter mates at the test station. Changes in mean performance for all pigs tested in Denmark from 1923 to 1935, as reported by Lush (1936), were as follows:

Item	Approx. 1923 mean	Approx. 1935 mean
Average daily gain	1.2 lb.	1.4 lb.
Feed/lb. gain	3.6 lb.	3.3 lb.
Body length	89.5 cm.	92.0 cm.
Backfat thickness	4.2 cm.	3.6 cm.

Since feeding and management were quite uniform during this period, all or nearly all of this improvement must have been genetic improvement resulting from the selection practiced. Improvement has continued in most of these traits, so that levels of performance today are well ahead of those for 1935.

A program patterned closely on the Danish system has also been in operation in Canada for more than two decades.

The Danish and Canadian systems are based on the progeny test which, while accurate, is slow, since the breeder must wait until a boar or sow has several progeny evaluated before deciding whether that animal is good enough to add to the breeding herd. The development of the backfat probe by Dr. L. N. Hazel of Iowa State College, in 1952, has provided an important impetus to performance programs in the United States. It was known from earlier research on carcasses that thickness of backfat is a good indication of lean cut yield of the carcass. The probe provides a simple, accurate measurement of backfat in the live animal and thus a good estimate of potential carcass quality of an animal, which can then be used for breeding.

The first organized modern swine testing station in the United States was established in 1954 by a group of breeders at Forrest, Illinois. The station, which was operated by the breeders, tested boar and barrow litter mate pairs; the boar was probed and the barrow slaughtered.

The Ohio Agricultural Experiment Station established a Swine Evaluation Station in 1954, in which all the test animals were slaughtered. In this station all data were coded, and no information identified with breed or breeder was released to the public. The Iowa Boar Testing Station was established in 1955 and tested its first pigs in 1956. Here the test unit was, at first, two boars and one barrow per pen. Later it was changed to three boars and one barrow per pen. This was the first station to use an index as a system for classifying the tested animals, and also the first to castrate animals not meeting the standard set by the board of directors. All animals meeting the standard were sold at auction, and all information was published.

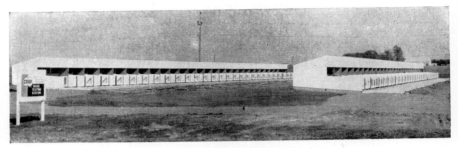

Fig. 16-1.   View of Eagle Grove, Iowa Swine Testing Station.

Many stations have been established since 1956, among which are two operated by a private concern in Iowa. Figure 16-1 shows a picture of the station at Eagle Grove, Iowa.

The testing stations have served two functions: education and direct improvement. Education has probably been the more important function, as many hog raisers have developed a working knowledge of the principles of swine breeding through their work with the stations. Many breeders have set up their own facilities for on-the-farm testing, patterned on those of the large testing stations.

Concurrent with the development of central stations was the instigation in several states of such on-the-farm tests. In these tests the breeder either weighs and probes his pigs himself, or joins with his neighbor to hire someone to do the work. Many commercial producers use this on-the-farm program to aid in correct selection of sow herds.

Another program that has received considerable attention in recent years is the Swine Certification Program. Sponsored by the purebred swine breed associations, it was implemented largely through the efforts of the late Rollie Pemberton, formerly secretary of the Hampshire Swine Record Association.

For pigs to qualify for certification, they must meet certain standards for number farrowed and weaned, and weaning weight of litter. They also must meet certain growth rate standards and, upon slaughter, meet certain carcass standards. The first hurdle is in number farrowed, number weaned, and weaning weight. The last item varies between breeds, depending on whether birth weight, 21-day weight, 35-day weight, or 56-day weight is used.

Growth and carcass standards are fairly uniform for all breeds:

> Weight—200 lb. at not more than 180 days.
> The animal must be slaughtered between 180 and 220 lb.
> Carcass specifications: Backfat—1.6 in. maximum
> Loin eye—4.0 sq. in. minimum
> Length—29 in. minimum

More stringent standards are used for pacesetter litters.

The program has been widely publicized and is extensively used by breeders in advertising. It has several advantages: (1) It is relatively simple for the breeder to keep the necessary records. (2) It has the wholehearted support of the breed secretaries. (3) It gives some unity to testing programs. Its disadvantages are these: (1) The standards are too low for rate of gain; thus, a breeder could restrict food intake to make his pigs lean enough to meet the carcass standards. (2) The best pigs in a litter may be slaughtered, as a basis for certifying litter mates that may not be as good. (3) The emphasis on sow productivity traits is probably not warranted; the low heritability of these traits is likely to prevent much improvement in them, and emphasizing them may result in reducing emphasis on other more highly heritable ones. The more traits considered, the less the attention which can be given to each. Those that are capable of being improved by selection should be emphasized.

Some selection for litter size will occur whether or not the breeder makes an effort to select pigs from large litters, because there are more pigs available for selection from large litters than from smaller ones. This may be regarded as natural selection, and its occurrence throughout the history of

Fig. 16 2. Weighing time at the Ida-Grove, Iowa Swine Testing Station. [Photo courtesy Bob Casey, Station Manager.]

the species is the most probable explanation for the low heritability of this trait. Those genes that, acting in an additive fashion, contribute to large litter size in each breed have been fixed by this natural selection, leaving only those genes not readily fixed as a cause of genetic variation in the trait. On the other hand, some attention to litter size in a selection program may be justified on the basis that litter size and weight—at least weigh at early ages —are negatively associated. That is, pigs from larger litters tend to be smaller because of less uterine space and/or nutrients and less milk per pig during the suckling period. Selection of the heaviest pigs without regard for the size of the litter from which they came could lead to selection of animals from smaller than average litters, which should probably be avoided.

Largely as a result of one or all of these improvement programs, thousands of swine breeders and commercial swine raisers today have some understanding of swine breeding principles. A perusal of a modern swine association journal indicates that the emphasis is now being placed on records, in contrast to the late 1940's when the emphasis was almost altogether on show ring winnings. Swine breeders have adopted some objective methods of improving their stock. Much improvement has been made, and more will be made in the quality of pork available to the consumer and in the economy with which it is produced.

REFERENCES AND SELECTED READINGS

References marked with an asterisk are of general interest.

Bradford, G. E., A. B. Chapman, and R. H. Grummer, 1953. Performance of hogs of different breeds from straightbred and crossbred dams on Wisconsin farms. *J. Animal Sci.*, 12:582.

———, 1958. Effect of inbreeding, selection, linecrossing and topcrossing. 1. Inbreeding and selection. *J. Animal Sci.*, 17:426.

Dickerson, G. E., 1951. Effectiveness of selection for economic characters in swine. *J. Animal Sci.*, 10:12.

———, J. L. Lush, and C. C. Culbertson, 1946. Hybrid vigor in single crosses between inbred lines of Poland China swine. *J. Animal Sci.*, 5:16.

*———, 1952. "Inbred lines for heterosis tests," in *Heterosis*. Ed. J. W. Gowen. Iowa State College Press. Chap. 21.

*———, C. T. Blunn, A. B. Chapman, R. M. Kottman, J. L. Krider, E. J. Warwick, and J. A. Whatley, 1954. *Evaluation of Selection in Developing Inbred Lines of Swine*. Mo. Research Bull. 551. (North Central Regional Publ. No. 38.)

Durham, R. M., A. B. Chapman, and R. H. Grummer, 1952. Inbred versus non-inbred boars used in two sire herds on Wisconsin farms. *J. Animal Sci.*, 11:134.

* Lush, J. L., 1936. *Genetic Aspects of the Danish System of Progeny-Testing Swine.* Iowa Agr. Expt. Sta. Res. Bull. No. 204.

Chapter 17

Genetic Improvement in Sheep and Goats

While the pride of man is humbled by the reflection, that the most profound works of art are but feeble imitations of nature, he will derive some consolation from the consideration, that God has condescended in some sort to render him his agent, and to give him extensive powers over the animal and vegetable creation; not only in subjecting them to his control, but even in enabling him, within certain limits, to change and alter their natures, so as better to adapt them to his own use, without subjecting them too far to his whims.
ROBERT R. LIVINGSTON, *"Essay on Sheep"* (1813)

17-1. INTRODUCTION

Sheep and goats are undoubtedly much more useful to man than they were when our ancestors first tamed them some seven to eight thousand years ago. Tremendous improvement has even been made since they were first brought to this continent less than 500 years ago. In fact, in 1810 the average wool clip was reported as 2 lb. per head (Connor, 1918). USDA statistics show that this had increased to 7 lb. near the beginning of this century, and now approaches 9 lb. Gains in fleece weight have certainly been accompanied by improvement in staple length and clean yield for respective grades. Similar gains in mohair production by Angora goats are indicated by increases from 4 lb. clip per goat in the 1930's to 4.5 to 5 lb. in the 1940's, 5 to 6 lb. in the 1950's, and an average of 6.5 lb in the early 1960's. The reduction of cryptorchidism by selection in an Angora goat flock has been shown by Warwick (1961).

Lamb production has shown an upward trend from 83–89 lambs saved per 100 ewes, one year of age and older, in the 1920's to 88–97 lambs saved per 100 ewes in the 1950's and early 1960's. Slaughter weights of lambs have increased also. Market lambs now may weigh as much as 10 to 20 lb. more at a given age than they did some 20 years ago. Improvements in production of milk goats are shown by higher records of production in more recent years; see the review of the American Milk Goat Record Association, 1962.

Average values for Rambouillet rams in the Sonora, Texas ram tests show definite gains during a 15-year period in average daily gain, clean wool weight, staple length, and skin folds (Menzies, 1963). Only face covering showed no evidence of improvement.

Improvements in efficiency of production have also resulted from increased production per animal, as many costs are on a per head basis. These gains in productivity and in efficiency of production can be attributed to better feeding and management practices, including better control of parasites and diseases, as well as to gains from genetic improvement. Notwithstanding these gains in efficiency of production, further improvements can and must be made if the sheep and goat industries in the United States are to survive and grow.

17-2. SYSTEMS OF BREEDING

Pure breeding is practiced by most professional animal breeders. Pure breeds of sheep and goats provide the foundation and continuity through which most genetic improvements from selection are made. The presence of these purebred stocks also permits further gains from crossbreeding. Pure breeds of sheep and goats are almost entirely closed to outside breeding and therefore some breeding of relatives or inbreeding occurs. In fact, most purebreeders linebreed in order to concentrate the blood of outstanding sires. Yet they usually avoid close or continued inbreeding. Studies of inbreeding in the Rambouillet and Hampshire breeds have shown that a slow increase of 0.4% to 0.5% in inbreeding has occurred in these breeds per generation (Dickson and Lush, 1933; Carter, 1962). A slightly higher increase was found for Toggenburg goats in Britain (Mason, 1954). It appears from these studies that a little separation into families within breeds has been taking place, although this has probably been due more to a geographic separation than to a deliberate attempt to form distinct families.

Inbreeding is usually avoided by practical breeders because of its detrimental effects on most productive traits. A review by Rae (1956) shows a decline in merit in most traits, the decline being sharpest for body weight, type, and condition. Fleece traits were less affected than body traits. Inbreeding had little effect on staple length, face covering, and skin folds. Inbreeding generally results in a reduction in vigor and viability, because the more inbred animals grow at a slower rate and are more susceptible to death from conception on. The inbreeding effects remain fairly constant throughout life (Terrill et al., 1948). The lost vigor is generally regained by crosses among inbred lines or by top crosses on noninbred stock. Recessive traits such as black color, overshot jaws, and cryptorchidism are more likely to be revealed with inbreeding.

Sheep breeders have for a long time been concerned with the possibility of improving sheep through developing and crossing inbred lines. Such concern has been stimulated by successes with hybrid corn. During and since the 1930's relatively large numbers of inbred lines of sheep have been developed by state and federal agencies. Many of these were started at the U.S. Sheep Experiment Station, Dubois, Idaho. Testing of these lines through linecrossing and topcrossing to unrelated stock is still far from complete. It seems possible that crosses among the better crossing lines or topcrosses from some lines may excel selected outbred groups. Gains from crossing lines are most important for traits such as lamb production, growth rate, and viability. However, it is not yet clear how inbred lines of sheep can best be utilized in practical breeding procedures.

Outcrossing within pure breeds is a much more common breeding practice than inbreeding. This probably tends to counteract the effects of any inbreeding that does occur and also to maintain heterosis at a relatively high level. Heterosis cannot be measured directly and therefore one can only speculate as to its importance.

The beneficial effects of heterosis can be obtained by crossing breeds as well as by crossing inbred lines, and crossbreeding is widely used in commercial sheep production (Rae, 1952). Crossbreeding of sheep generally leads to increases in fertility, milk production, growth rate of the lambs, body weight, and wool production (Terrill, 1958b).

17-3. INHERITANCE OF TRAITS IN SHEEP AND GOATS

Many traits that are inherited in simple Mendelian fashion have been identified in sheep. These, which have been reviewed by Rae (1956) and Terrill (1958b), include color and pattern, horns, ear length, multinipples, lethals or sublethals, various abnormalities, chalk face, entropion, jaw defects, birth coat and fleece structure, skin folds, and blood traits. The knowledge of the mode of inheritance of these traits aids the breeder in eliminating those that are undesirable, such as black color and various other defects, and in fixing those that are desirable, such as the polled trait. Of course a variety of colors and color patterns have been studied. Recessive black is of interest in all important breeds of sheep in the United States that are white. It appears that at least two pairs of recessive genes for black exist in domestic breeds. A dominant black gene is in evidence in the black marker sheep used on western ranges. This gene may have been derived from the Karakul breed. Furthermore, there is evidence that several dominant genes can produce black.

The inheritance of horns in the Rambouillet and Dorset breeds is of interest, although horns, particularly scurs, may also appear in polled breeds. This is especially true of breeds with Rambouillet blood, such as the Columbia and Targhee. Horns are apparently recessive to the polled trait in both Rambouillets and Dorsets although Rambouillet ewes homozygous for the horned gene have horn knobs and those carrying the polled gene have depressions in the skull in place of the knobs. The inheritance of scurs has not been well defined, but scur or horn growth appears to be greater in rams heterozygous for horns than in homozygous polled rams.

Lethal and semilethal traits, which usually result in early death, are generally inherited as simple recessives and include muscle contracture, earlessness, cleft palate, paralysis, rigid fetlocks, dwarfism, nervous incoordination, congenital photosensitivity, and blindness.

Cryptorchidism, or failure of descent of one or both testicles, appears to be due to recessive genes, although it is sometimes linked with the polled trait. Entropion, or turned-in eyelids, is inherited, although not as a simple recessive. Wattles, which appear in Navajo sheep and in Toggenburg goats, appear to be inherited as a dominant. Jaw inequalities, particularly a short lower jaw, appear to be due to several pairs of genes. Hermaphroditism in milk goats appears to be inherited as a simple recessive character (Eaton and Simmons, 1939).

Blood antigens and other blood traits in sheep have received increased attention in recent years. These studies have laid a foundation for identifying parentage and for checking on changes in homozygosity for those traits that are not evident to practical breeders. Only slight, if any, relationships have been shown between blood types and productivity of sheep.

Chromosome number seems established as 54 for the diploid number for sheep (Borland, 1964) and 60 for goats (Makino, 1951).

Early genetic studies of sheep attempted to work out the mode of inheritance of many traits of economic importance in a simple Mendelian manner. The F_1 and F_2 generations following breed crosses were studied. Particular interest was shown in twinning or in number of lambs born from single and twin mothers. Gradually the efforts to explain the inheritance of all traits in a simple manner were abandoned upon the realization that many production traits were dependent on many genes and that special statistical methods aided the study of these traits. The results of such studies have found useful application in selection for genetic improvement.

17-4. IMPROVEMENT THROUGH SELECTION OF SHEEP

Selection is the most important way in which the sheep breeder can improve his products and efficiency of production (Terrill, 1958a). It has been repeatedly demonstrated that effective selection methods will lead to permanent gains, not only in quantity but also in quality of lamb and wool produced. The amount of selection that can be practiced for any one trait is limited. Threfore it is important that traits for which the greatest progress can be made and that are most valuable are emphasized in selection.

Estimates of heritability, which estimate the proportion of gain made in selection of parents that is passed on to the offspring, are useful in determining the relative progress that can be made in selection to improve various traits. Thus, emphasis can be given to the traits with which the most progress can be made. In sheep, estimates of heritability have usually been obtained from relationships among relatives. Estimates are available for a large number of traits but many are based on relatively small numbers under varying conditions, and therefore show considerable variability. Nevertheless, rough groupings of the various traits can be made according to their relative heritability (Terrill, 1958b). These are presented in Table 17-1. In general, heritability estimates over 40% have been classified as high, those from 20 to 40% as moderate, and those under 20% as low. It is noted that most traits of economic importance are moderately or highly heritable except multiple births and lambs weaned, conformation, and fatness.

Traits important to income must be emphasized if selection is to be effective in producing more profitable sheep. Lamb production is most important. Selection gains for more open faces will increase number and pounds of lamb per ewe with only slight loss in fleece weight (Terrill, 1949b). Larger

TABLE 17-1. | *Relative estimates of heritability of traits in sheep.*

High	Moderate to High	Moderate	Low to Moderate	Low
Face covering	Yearling weight	Average daily gain	Weaning weight	Birth weight
Staple length	Grease fleece weight	Carcass traits	Milk production	Type or conformation
Birth coat	Clean fleece weight	Fur traits	Number of nipples	Condition or fatness
Crimps per inch	Clean wool yield	Index of over-all merit		Multiple births
	Fiber diameter	Skin folds		Lambs weaned
	Early lambing	Color on legs		
	Cannon bone length	Resistance to parasites		

Fig. 17-1. Face covering of weanling Rambouillet lambs. A and B, open face with score of 3 for ram and ewe lamb; C and D, partially covered face with score of 4 for ram and ewe lamb; E and F, covered face with score of 5 for ram and ewe lamb. [Courtesy USDA.]

ewes will produce more and heavier lambs. In fact, for each pound increase in the weight of a yearling ewe, she may produce about one-half pound more of lamb per year (Terrill and Stoehr, 1942). Market weights of lambs have generally increased through the years and will probably increase still more. Further gains should also come from increasing the number of lambs marketed per ewe. Thus it will pay to increase production of twins. Ewes having twins can be expected to wean an average of about 40 lb. of lamb per ewe-year more than ewes of the same age having singles (Sidwell, 1956; Terrill, 1949a, 1957). Twinning has low heritability but small improvements may be worthwhile. To select for twin production, one should favor young ewes that have had twins and rams born as twins from young mothers. Twinning at first lambing is less frequent than in older animals; thus, mothers having twins at an early age are more likely to transmit this trait.

Quality and quantity of lamb meat deserve consideration in selection, but unfortunately we know little about how to measure these, particularly in the live animal where selection can be applied. Thickness through the hind legs may offer promise. More facts are needed on the relationship of various conformation traits to carcass value.

Clean fleece weight is the most valuable wool trait to consider in selection. Neale's device (Neale et al., 1958) for estimating clean fleece weight provides a quick easy method for identifying high-producing sheep (Fig. 17-2). Improvements in staple length, fleece density, and uniformity of length and fineness will also probably lead to economic gains. Improvement in fleece quality will no doubt be given more emphasis by producers when prices paid for wool are more commensurate with quality.

Any selection program should include the culling of unsound and unthrifty as well as low-producing animals. Since unthrifty animals probably would not leave many offspring, their elimination may not greatly increase genetic progress. However, the remaining animals will give higher average production and there will be less risk of transmitting defects.

17-5. SYSTEMS OF SELECTION

After it has been decided which traits are to be emphasized in selection, there are various systems which may be followed in determining which animals are best suited to transmit these traits. The use of records on relatives

Fig. 17-2. Neale's device for estimating clean fleece weight by measuring volume of fleece under constant pressure. [Courtesy USDA.]

Fig. 17-3. Measuring staple length on mature Rambouillet ram. [Courtesy USDA.]

are often useful, particularly if selection must be practiced before the trait can be measured in the offspring. When pertinent records become available on the lambs these should receive more attention than any information from their pedigree. Consideration may be given to full sisters or brothers when traits such as milk production, lambing rate, or semen production cannot be measured in both sexes. Progress from selection is generally slower if selection must be based on the records of relatives than it is if the trait can be measured directly in the animals to be selected.

One of the most commonly used methods of selection on records of relatives is that based on records of the progeny or progeny testing. This method is usually more accurate than selection on the animal's own phenotype because a measure of the animal's breeding worth is obtained before selection is practiced. However, it is slower because decisions cannot be made until after progeny are produced, and a part of the flock must be preserved for producing these progeny (Dickerson and Hazel, 1944). Breeders, selecting within their own flocks, can use progeny tests on outside sires to compare them with sires of their own breeding before using them extensively.

Selection of superior animals usually involves the use of an index of some kind because one must combine the values of various important traits simultaneously or balance the strong points against the weak points in order to rank the animals from best to poorest. A calculated index combining objective measures will, generally, be more accurate and effective than any mental combining of values by a breeder ranking animals under observation.

The use of an index, as shown by Hazel and Lush (1942), is a more efficient way of selecting for several traits at the same time than the use of independent culling levels or tandem selection for one trait at a time. An index permits a constant and objective degree of emphasis on each trait considered in selection. Without an index, ideals are more likely to shift from year to year. Also, one is likely to overemphasize more obvious traits like body type or color and to underemphasize traits of greater economic importance, such as average daily gain or milk production.

Lush (1945) has presented an objective basis for determining emphasis on each of several traits in an index by giving weight to these characters in proportion to their heritability times economic importance (see Chapter 16).

The use of an index on Rambouillet lambs at Dubois, Idaho, was responsible for increasing over-all selection pressure by about 28% for weanling traits. The primary change brought about by index selection was to shift emphasis from neck folds to face covering and weaning weight. The use of an index on Wisconsin farm sheep based on weaning weight and staple length has been accompanied by improvements in these traits (Felts et al., 1957; Chapman, 1958).

Fig. 17-4. Weighing weanling lamb on Wisconsin farm. [Courtesy V. L. Felts, University of Wisconsin.]

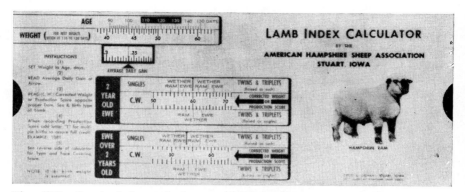

Fig. 17-5. Lamb index calculator or slide rule. [Courtesy American Hampshire Sheep Association.]

17-6. FACTORS INFLUENCING RATE OF PROGRESS

Progress from selection for improvement of sheep and goats may not be as rapid as it is for improvement of animals that reproduce at a faster rate, such as poultry or swine. However, improvements can be made, especially if factors that influence the rate of progress are optimum. Factors that may influence progress are an optimal environment that allows a high proportion of a generation to reach breeding age and genetic factors such as intensity of culling for a specific trait or the heritability of the trait. Selection ideals must remain constant for a period of years for selection to be effective. The exhibition of sheep at fairs and shows has not only stimulated interest in sheep breeding but also serves to set breed ideals and standards. Unfortunately, the show ring has sometimes led to emphasis on traits that are not economically important such as head shape, scurs, or complete face covering or on traits of low heritability, such as conformation and fatness. Greater emphasis on production traits that cannot be assessed in the show ring would be desirable.

Progress in improving individual traits may depend in part on the accuracy with which the trait can be measured, although the rapidity and economy of the measure are also important. Attention is being given to the development of better measures for surface area, body form, carcass traits, wool fiber length and fineness, and clean wool yield. The device (Fig. 17-2) for estimating clean fleece weight developed by Neale et al. (1958) has been a significant gain in this area. Also important has been the development of a fertility index for rams by Hulet and Ercanbrack (1962).

17-7. PERFORMANCE TESTING

Performance testing offers another means of increasing the accuracy of records on which selections are made, especially for stud sires. Ram tests.

such as those conducted in Texas and Utah, permit the breeder to compare his rams with those of other breeders under standard conditions. Of course, such tests of the breeder's own rams may be conducted by on-the-farm tests, as is done in Wisconsin, Ohio, and other states. Ideally, performance tests should include, under standard conditions, information on the ram's own performance, such as daily gain, efficiency of feed use, clean fleece weight, staple length, and grade of wool, as well as similar information on his progeny. This could be done by mating the rams when they are lambs to sample groups of ewes at the start of the test and then evaluating both the ram and his offspring about 9 or 10 months later. Such a test could include carcass information on the offspring. Such tests would be costly, but they might be worthwhile, particularly on important stud rams.

If adjustments are not made for age, type of birth, and age of dam, these factors tend to confuse nongenetic with hereditary effects. Single lambs generally weigh 8 to 10 lb. more at weaning than twins, and lambs from mature dams excel those from young dams by 6 to 8 lb. When lambs are gaining over 0.5 lb. per day at weaning, a few days difference in age can make an appreciable difference in weight. If single lambs are favored because they are bigger it is probable that no progress will be made in improving weaning weight because the twins, even though smaller, might still be better genetically than the single lambs.

Age has important effects on body weight, fleece weight, staple length, fineness, and pounds of lamb raised per ewe. Thus, when comparing dams of different ages, one should take into account the fact that body weights, fleece weights, and lamb production increase until the animal is 3 to 5 years of age or older. Generally, staple length decreases and fleeces become coarser with age. The mohair of Angora goats markedly increases in coarseness with age. Effects of age on phenotype are fully as important for sires as dams. Lack of adjustment for age will favor 3- to 5-year-old rams over those 1 and 2 years of age. Younger rams of the next generation will generally be genetically superior if rapid improvement is being made.

If these nongenetic effects are not taken into account, the selected sheep may owe their advantages to favorable environmental factors and thus no genetic improvement will result. Years may make quite a difference. There is very little genetic difference from year to year in sheep because only a part of the parents can be changed each year, but there may be large environmental yearly changes. Clean fleece weight can vary as much as $1\frac{1}{2}$ to 2 lb. in different years simply because feed or other environmental conditions are better in one year than another.

One way to minimize these environmental effects is to select within groups of the same age or within singles or twins by saving the same proportion of each. Statistical adjustments are more precise and can easily be applied to an index.

Generations should be turned rapidly for greatest gain, as genetic progress is made only from one generation to the next. Length of generation is defined as the average age of the parents when the offspring are born. This average is about 4 years in sheep. Ram generation length can be reduced to 2 years by using only the best yearling rams each year. Reducing generation length by one half would double the genetic gain per year. If improvement is being made and a reasonably large number of offspring are produced, the best son should be better than his sire. The more quickly the change is made the more rapidly one may take advantage of this gain. Thus, in many circumstances more rapid progress can be made by selecting solely on the basis of phenotype than by the slower progeny-test procedure. The breeder also often tends to make repeated use of an outstanding sire when a quicker change to his best son might give greater improvement.

The sheep or goat breeder must often choose between the frequent selection of outstanding young rams on their own merit and the slower procedure of selecting sires on the performance of their progeny. Progeny testing is

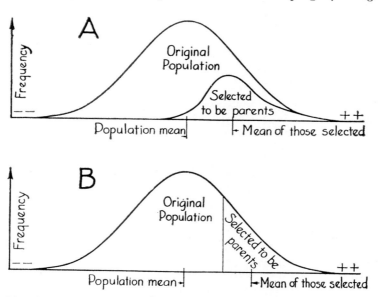

Fig. 17-6. Two ways in which the merits of those chosen to be parents by rather intense selection might be distributed with respect to the merit of the original population from which they were taken. The better individuals are to the right, the poorer to the left. *A* indicates the usual kind of selection, in which at least a few mistakes are made and in which some attention must be paid to characteristics other than the one for which merit is indicated here. *B* is the most extreme form of selection conceivable. No mistakes are made, and selection is entirely for the characteristic for which degrees of merit are indicated along the horizontal scale. [From Lush, 1945.]

most effective in selecting for traits that are low in heritability, such as conformation or fattening ability. It is usually unnecessary in selecting for highly heritable traits, such as face covering or staple length, where the offspring simply confirm what has already been observed in the parents. Progeny testing is usually desirable when selecting among rams whose records were made in different herds or farms. Adjustment for environmental herd differences in production records is difficult. Selection for crossing ability between lines or breeds necessitates the use of progeny tests. Similarly, progeny testing is important when a trait cannot be observed in both sexes, such as milk production, or where it can only be observed in a group of offspring, such as lamb mortality. Progeny testing may be most important in selecting for carcass traits, which cannot be adequately measured in the live animal. Artificial insemination is safer with progeny-tested sires, and the greater accuracy in appraisal of individual sires is more justified where sires are to be used so widely. The selection of inferior rams, which sire large numbers of lambs (as from artificial insemination), can be very costly. On the other hand, when there is selection among relatively large numbers of rams for natural mating within a closed flock for moderately or highly heritable traits, more rapid progress can usually be made by mating only the best yearling rams each year. Only a few, if any, rams with the very best progeny can be expected to excel the best yearling rams available in the next year.

17-8. FACTORS AFFECTING SELECTION DIFFERENTIALS

The aspect of selection that can be most influenced by the breeder is the selection differential. Selection differential represents the difference in a trait or group of traits between the herd average and the average of the animals selected for producing the next generation. Any progress made is always a fraction of the selection differential. Much of the effort in selection is aimed at obtaining as large a selection differential as possible for the important traits. One measure of the effectiveness of a selection index is the size of the selection differentials for the respective traits resulting from its use, as compared with alternative methods of selection. In comparing selection differentials in terms of pounds of wool or inches of staple length it may be desirable to convert the selection differential to some standard unit of measure such as a percentage of the mean or a multiple of the standard deviation.

The amount of selection that can be practiced is limited by the size of the flock. For example, at least 3 rams usually need to be used within a closed flock to avoid close inbreeding. As each ram can be mated to 50 to 75 ewes the total flock size must be in the range of 150 to 225 ewes, or a lower selection differential for rams will result. Of course, selection takes advantage of genetic variation and the larger the breeding group, the larger will be the total range of variation. Furthermore, in large flocks the best rams can be mated to the best ewes to produce rams for the entire flock. However, this superflock should be large enough that considerable selection can be practiced among the rams produced.

The higher the reproductive rate is, the greater are the selection differentials that can be obtained, because the number of replacement animals needed remains constant. Feeding and management practices that increase the lambing rate are therefore usually an aid to greater gains from selection.

It pays to select for as few traits as possible. Opportunity for selection is definitely limited and attention to unimportant traits such as color on the legs or shape of head can reduce the attention to important traits such as pounds of lamb or wool produced. Progress for any one trait decreases as the number of traits selected for increases. The decrease is equal to one divided by the square root of the number of traits involved (Lush, 1945). For example, if a breeder selects for four traits he makes only one-half as much progress in each one as if he selected for only one trait; selection for nine traits gives only one-third as much progress in any one trait as if only one was considered. It is not profitable to select for fancy points or for things for which the lamb or wool producer is not paid.

It is also important to select under the feed and environmental conditions under which the sheep are going to produce. Unfortunately, conditions under which purebred rams are generally raised are usually better than the range or farm conditions under which their offspring will be raised. Thus they may be selected for characteristics that are not the most desirable in the ordinary or poorer environments in which most market lambs and wool are produced.

The use of accurate and objective records will increase the effectiveness of selection. Some of the most important records are those of lambing, weaning weight, staple length, and fleece weight. Neale's fleece-squeezing machine gives an accurate estimate of clean fleece weight very quickly. With this machine it is feasible to obtain clean fleece weight estimates on every sheep shorn. Certainly it would pay to obtain this measure—or at least grease fleece weight—on every ram fleece.

Demanding accuracy in records can be carried too far. As accuracy is increased, generally the cost becomes higher also. A cheap, objective, approximate measure may be far better than an expensive, highly accurate one. This is especially true if the cheaper measure is used more widely and consistently.

To be most effective, maximum selection pressure should be applied to both rams and ewes. However, selection of rams is of greater importance. About 80 to 90% of the gains that can be made in a flock closed to outside breeding comes from the selection of rams (Terrill, 1951). One ram can be mated to as many as 50 to 75 ewes in a limited breeding period, thereby requiring the use of only 3 to 4% of the rams produced to sire the next generation. More than half of the ewes produced must be retained to become mothers of the next generation. Thus, much larger selection differentials can be obtained for rams than for ewes, and hereditary or permanent gains from selection come largely from the choice of rams to become sires.

The selection differentials for sires can be increased still further by the use of artificial insemination. In this way several thousand offspring may be obtained from each sire in one season. A record of 17,000 ewes artificially inseminated by semen from one ram during a 115-day period has been reported from the Soviet Union. Artificial insemination of sheep is advantageous chiefly because it permits more intense selection of outstanding sires (Terrill, 1960). Justification is difficult on any other basis. Any advantage of the breeding value of a sire can be extended to a much greater number of offspring by artificial insemination. There also may be disadvantages to the wide use of individual sires through artificial insemination. Rams that have not been thoroughly tested may spread undesirable traits—especially recessive defects, which may not be revealed under ordinary use—far more widely under this method.

17-9. RATE OF PROGRESS FROM SELECTION

Rate of progress expected from selection on an annual basis can be estimated by multiplying heritability times the selection differential and then dividing by the generation length. Because the last two generally differ for rams and ewes, it is convenient to add the selection differentials together and then multiply by heritability. The result must be divided by the sum of the generation lengths for sires and dams to obtain the expected progress per year. These estimates cannot be very precise if a number of related traits are considered in selection, because exact estimates of heritabilities and phenotypic and genetic correlations among traits are not generally available for sheep. Even generation length may be difficult to calculate accurately. However, such estimates may still be very useful as guides to the best selection procedure to follow in any particular situation.

Actual progress from selection is usually determined by trends of values for each trait over a period of years. This is often unsatisfactory, because genetic trends are often confounded with environmental trends (Terrill, 1951). The use of unselected control groups or random repeat matings are thought to be essential for selection experiments (Terrill and Ercanbrack, 1960) to obtain estimates of annual environmental changes. Even such control groups may be subject to bias by natural selection and by genetic drift.

17-10. IMPROVEMENT THROUGH SELECTION OF DAIRY GOATS

The principles that have been emphasized for improvement of sheep through selection also apply to dairy goats. Milk production should be emphasized. It would be expected to be moderately heritable as it is in sheep and dairy cattle. Not only is high milk production per day desirable but persistency of production for 8 to 10 months or longer is also important. The udder should be large and the teats should be large enough to make milking easy.

Bucks should be selected from high-producing does and those who sire superior milking offspring should be retained for extensive use. Hornlessness is preferred by most breeders, but this trait appears to be linked with the factor leading to intersexes (Eaton, 1945). Large size, high feed capacity, vigor, long life, and fertility are desirable in both sexes. Good dairy conformation is generally preferred, but direct selection for high milk production should receive the major emphasis.

17-11. IMPROVEMENT THROUGH SELECTION OF ANGORA GOATS

Again, the same principles that apply to improvement of sheep through selection also apply to Angora goats. Gray (1959) has outlined many important considerations in selecting Angora goats for mohair and kid production. Most of the income from Angora goats comes from the sale of mohair and therefore weight of fleece, which is moderately heritable, is the most important trait to emphasize in selection. Individual fleece weights—at least those for the bucks—should be recorded by breeders. Length of staple contributes to fleece weight and thus is important. Mohair becomes coarser with age, but the fine quality of the kid mohair is preferred and can be improved by selection. Goats with short staple and with fleeces that are light or kempy, or that contain any colored fibers should be discriminated against. The body and belly should be well covered and the covering should be dense. The web-lock or intermediate type of fleece holds up well in production and is often preferred by commercial producers. Angora goats have more of a tendency to shed than do sheep, and animals with this tendency should be culled.

Size, vigor, and fertility are important to efficient production and should be emphasized in both sexes. Open faces should be stressed in selection, because they are related to the production of both kids and mohair. Does with open faces are larger, produce heavier fleeces, and drop a much higher proportion of kids than does with covered faces (Shelton, 1960). Meat conformation should receive some attention in selection, since many Angora goats, particularly the wethers, are sold for meat.

17-12. FUTURE NEEDS

Although much is known about genetic improvement of sheep and goats, further advancements are urgently needed if these animals are to compete effectively with alternative sources of food and fiber. Not only does the rate of reproduction need to be increased but more consistent methods of controlling it are necessary. The artificial control or alteration of the sex ratio would be advantageous. Many advances in breeding and selection techniques would be facilitated by practical methods for storage, culture, and transfer of germ cells and gonadal tissues from both the sheep and the goat. The inability to store ram semen effectively either in the liquid or frozen state is a deterrent to more widespread use of artificial insemination as a selection aid.

Inbred lines of sheep show some promise, but practical methods for their use in improvement programs need to be developed. Selection studies show promise of continued genetic gains, but more rapid and effective methods of selection are required. Tremendous possibilities exist, not only to improve present breeds of sheep and goats, but to develop new and more useful breeds. Comprehensive comparisons of the productive ability of breeds and their crosses are particularly needed. Such studies may lead to much more effective and extensive use of crossbreeding for more efficient commercial production of meat, fiber, and milk from sheep and goats.

17-13. SUMMARY

Although there is convincing evidence that large gains have been made in improving the productivity and efficiency of production of sheep and goats in recent years, much more improvement is possible and desirable. The pure breeds and pure breeding provide the basis for continued gains. The production and crossing of inbred lines offer hope for the production of superior stock, but tests are not yet conclusive. Crossbreeding offers a means for improving commercial production. A knowledge of the inheritance of many traits that depend on only one or two pairs of genes can aid the breeder in eliminating undesirable traits and in fixing those that are desirable.

Selection procedures are aimed mainly at production traits, which depend on many genes. The amount of selection that can be practiced is limited, and therefore highly heritable traits and those that are most important to income should be emphasized in selection. These include lamb production, face covering, body weight, carcass quality, and clean fleece weight. Progress from selection can be maximized by the use of accurate records of production, by adequately accounting for environmental effects, by turning generations rapidly, by selecting consistently for only the most important traits, and by concentrating on the selection of males. Selection is at present the most important way in which the sheep and goat breeder can improve his products and the efficiency of production.

REFERENCES AND SELECTED READINGS

References marked with an asterisk are of general interest.

AMGRA Handbook, 1962. American Milk Goat Record Association.

Borland, R., 1964. The chromosomes of domestic sheep. *J. Heredity,* 55:61–64.

* Carter, R. C., 1962. Breed structure and genetic history of Hampshire sheep. *J. Heredity,* 53:209–214.

* Connor, J. J., 1918. *A Brief History of the Sheep Industry in the United States.* Annual Report of the American Historical Association.

Dickerson, G. E., and L. N. Hazel, 1944. Effectiveness of selection on progeny performance as a supplement to earlier culling in livestock. *J. Agr. Research,* 69:459–476.

Dickson, W. F., and J. L. Lush, 1933. Inbreeding and the genetic history of the Rambouillet sheep in America. *J. Heredity,* 24:19–33.

Eaton, O. N., 1945. The relation between polled and hermaphroditic characters in dairy goats. *Genetics,* 30:51–61.

———, and V. L. Simmons, 1939. Hermaphrodism in milk goats. *J. Heredity,* 30(6): 261–266.

Felts, V. L., A. B. Chapman, and A. L. Pope, 1957. Estimates of genetic and phenotypic parameters for use in a farm flock ewe selection index. *J. Animal Sci.,* 16(4):1048 (Abstract).

* Fraser, A. S., and B. F. Short, 1960. *The Biology of the Fleece.* Animal Research Laboratories Technical Paper No. 3. Commonwealth Scientific and Industrial Research Organization, Melbourne, Australia.

Gray, James A., 1959. *Selecting Angora Goats for Increased Mohair and Kid Production.* Texas Agr. Expt. Sta. MP-385.

Hazel, L. N., 1943. The genetic basis for constructing selection indexes. *Genetics,* 28:476–490.

———, and Jay L. Lush, 1942. The effi-

ciency of three methods of selection. *J. Heredity,* 33(11):393–399.

Hulet, C. V., and S. K. Ercanbrack, 1962. A fertility index for rams. *J. Animal Sci.,* 21(3):489–493.

*Lush, Jay L., 1945. *Animal Breeding Plans.* Iowa State University Press, Ames, Iowa.

* Makino, S., 1951. *An Atlas of the Chromosome Numbers in Animals.* 2nd Ed. Iowa State College Press, Ames, Iowa, 290 pp.

Mason, I. L., 1954. A genetic analysis of the British Toggenburg breed of goats. *J. Heredity,* 45(3):129–133.

Menzies, J. W., 1963. *Improvement of Sheep Through the Selection of Performance-tested and Progeny-tested Breeding Animals, 1962–63.* Texas Agr. Expt. Sta. MP-645.

Neale, P. E., G. M. Sidwell, and J. L. Ruttle, 1958. *A Mechanical Method for Estimating Clean Fleece Weight.* New Mexico Agr. Expt. Sta. Bull. 417.

Rae, A. L., 1952. Crossbreeding of sheep. *Animal Breeding Abstracts,* 20:197–207, 287–299.

*———, 1956. The genetics of the sheep. *Advances in Genetics,* 8:189–265.

Shelton, M., 1960. The relation of face covering to fleece weight, body weight and kid production of Angora does. *J. Animal Sci.,* 19(1):302–308.

Sidwell, G. M., 1956. Some aspects of twin versus single lambs of Navajo and Navajo crossbred ewes. *J. Animal Sci.,* 15(1):202–210.

Terrill, C. E., 1949a. *Selecting Rambouillet Ewes for High Lamb Production.* Proc. Western Section, Amer. Soc. of Animal Prod., Moscow, Idaho.

———, 1949b. The relation of face covering to lamb and wool production in range Rambouillet ewes. *J. Animal Sci.,* 8(3): 353–361.

———, 1951. Effectiveness of selection for

economically important traits of sheep [abstract]. *J. Animal Sci.*, 10(1):17–18.

———, 1957. Sheep improvement through selection. *National Wool Grower*, 47(5): 14–17.

———, 1958a. Recent advances in sheep breeding. *California Livestock News*, October 14.

———, 1958b. Fifty years of progress in sheep breeding. *J. Animal Sci.*, 17(4):944–959.

* ———, 1960. *The Artificial Insemination of Farm Animals*. Ed. E. J. Perry. Rutgers, New Brunswick. Third Revised Edition.

———, and S. K. Ercanbrack, 1960. *Operation and Usefulness of a Random Bred Control Population of Rambouillet Sheep.* Presented at the Joint Meeting of the Beef Cattle Breeding Technical Committees, Stillwater, Oklahoma, July 25, 1960.

———, G. M. Sidwell, and L. N. Hazel, 1948. Effects of some environmental factors on traits of yearling and mature Rambouillet rams. *J. Animal Sci.*, 7(3):311–319.

———, and J. A. Stoehr, 1942. The importance of body weight in selection of range ewes. *J. Animal Sci.*, 1(3):221–228.

* Turner, H. N., 1963. Influence of breeding on productive efficiency of sheep. *Proc. World Conf. Animal Production*, 2:175–188. European Association for Animal Production, Corso Triest 67, Rome, Italy.

Warwick, B. L., 1961. Selection against cryptorchidism in Angora goats. *J. Animal Sci.*, 20(1):10–14.

Chapter 18

Genetic Improvement in Horses

18-1. INTRODUCTION

Until the end of World War I the draft horse was one of the dominating forces in United States agriculture. Millions of acres of land were used to produce feed for 23 million horses. With the advent of the tractor, truck, and automobile, the draft and carriage horse industry collapsed and this in turn resulted in a decline of interest in research on horses. This explains why we know less about the nutrition, physiology and genetics of horses than of any other large farm animals. But now the horse industry is again growing, because of an intense interest in light horses and ponies for recreation and sport. This revived interest in the horse has called attention to the appalling paucity of knowledge about its biology. Genetic research should be undertaken that will lead to the improvement of the horse in such traits as speed in racehorses, performance of gaits in saddle horses, coat color, anatomical and physiological soundness, reproductive fitness, disease resistance, and behavior patterns.

18-2. COAT COLOR INHERITANCE IN HORSES

Over a twenty-year period, Castle and colleagues published a series of papers dealing with color inheritance in horses. Under Castle's guiding genius a three-pronged approach was developed. First, the ample information on color in the stud book records of various breeds was studied. This approach—the usual one—to the genetic explanation of horse colors had a disadvantage in that it ignored the much more substantial and exact knowledge of color inheritance that had been derived from experimental studies made on other mammals, rodents in particular. Second, Castle, following previous attempts by Wright (1917) and Odriozola (1951), succeeded in presenting a genetic framework for explaining horse colors that was in accord with that derived from the experimental study of other mammals. Third, as the theory developed, various breeders collaborating with Castle (Castle and King, 1951) enabled him to verify some of his hypotheses with test matings. He concluded that the wild type coat color of most mammals, including horses, results from the activity of four basic color genes A, B, C and E.

The following account of how specific genes interact to produce coat color in mammals generally and in horses particularly is largely abstracted from Castle (1954) and Castle and Singleton (1961).

Gene C, the Color Gene. The colors of mammals are derived in general from the presence in the coat of two different groups of pigments, black-brown and red-yellow. Both are end products of a process of oxidation of a substance capable of producing color when so oxidized. The initial step in this process is controlled by a dominant color gene C, which may, by mutation, become inoperative and result in complete or partial albinism. In a complete albino, genotype cc, no pigment whatever is formed, so that the hair is white and the skin and eyes (iris) are pink because there is no pigment to conceal the red hue of blood under the skin. Familiar examples of complete albinism are white rabbits, rats, and mice. An animal whose genotype includes cc will be albino regardless of what color genes it may have. In the various rodent species studied, the C gene has mutated to several different forms with effects that are intermediate between C– and cc. As mentioned in Chapter 12, for example, an allele of C is c^h, which when homozygous produces the type of albino exemplified by the Himalayan rabbit. Its body is white and its eyes pink, but the fur on the extremities (ears, nose, tail, and feet) is slightly pigmented. Castle believed that all horses are CC [1] (that is, none are albinos) despite the fact that horse fanciers classify several types of horses as albinos. The inheritance of the so-called albino will be discussed later in this section.

Gene B, the Black Gene, and Gene A, the Agouti or Wild Pattern Gene. The coats of most mammals in the wild state are predominantly black pigmented. In the production of black pigment, which is found in the eyes and in the skin of the extremities as well as in the coat, gene B acts in conjunction with gene C. In the absence of genes that regulate the distribution of black pigment, the entire coat will be black. Actually very few wild mammals are uniformly black in color. This is because of the action of other genes

---

[1] Since we are assuming all horses to have CC in their genotypes there is no need to specify this when writing genotypes.

that restrict the distribution of black to particular areas of the coat. When, as a result of such gene action, black ceases to be produced in a particular area, it is automatically replaced in that area by red-yellow, a different end product of the oxidation process that occurs in cooperation with gene C. The coat thus becomes a mosaic of black and yellow pigments, which has concealing value in the struggle for existence, allowing a predator to steal upon its victims unobserved, or a prospective victim to escape notice of the predator.

A concealing coat of this sort is the agouti pattern of rabbits and other rodents, which has counterparts in the coats of wolves and many other groups of mammals. Its production is controlled by gene A acting in conjunction with C and B. We may regard this as a gene for wild-coat pattern. It limits the production of black pigment to (1) particular parts of the coat, chiefly dorsal and peripheral, and (2) particular parts of the individual hairs. The result is a generally gray coat on the dorsal area, with individual hairs having an intensely black tip, a subapical band of yellow, and a dull black base. The ventral areas of the coat are lighter; they often lack the black hair tips, resulting in a whitish-yellow belly color. This is a very effective concealing pattern in a natural environment. It results entirely from the action of the dominant gene A in cooperation with the genes C and B.

Loss of the wild pattern of the coat may result from a recessive mutation of A to a, leaving the coat a uniform black in the genotype aaBC,[2] which is known as nonagouti black or recessive black.

In rabbits and many other rodents, as well as in dogs, the wild pattern gene A has undergone a mutation, designated a^t, which results in the black-and-tan coat pattern. In this mutant, when it is homozygous $a^t a^t$, the individual hairs of the dorsal areas lack the yellow band and the animals have a black back and a light belly, with banded hairs along the sides between back and belly. This mutation is dominant to uniform black, aaBC, but recessive to ordinary agouti, A.

Castle has adopted Odriozola's (1951) hypothesis that in horses the gene A is represented by four alleles, A^+, A, a^t, and a. (1) A^+ is the allele found in the wild ancestral stock, exemplified by the surviving Prejvalski horse of Mongolia. It has the coat pattern of the domestic bay together with markings commonly lacking in the bay, such as a dark spinal stripe, a dark vertical bar on the shoulders, and transverse zebra bars on the front legs. These as a group may be referred to as zebra markings; one or more of them are of frequent occurrence in certain domestic breeds and in feral horses in Argentina. (2) A is the allele that is responsible for the coat pattern of the domestic bay horse, AB; zebra markings are commonly lacking in such horses. (3) The allele, a^t, is responsible for the coat pattern of the seal brown

---

[2] When a dominant gene occurs in the genotype its symbol will be written only once if we are solely interested in the corresponding phenotype. For example, in this case, aaBBCC has the same phenotype as aaBbCc.

horse, a^{t}B. This genotype has zebra markings more often than the bay. (4) The allele a, when homozygous, aa, causes the bay coat pattern and zebra markings to be entirely absent. The genotype aaB is that of the recessive uniformly black horse.

In rodents and many other mammals including horses, there has occurred in gene B a recessive mutation to an allele b, which when homozygous, bb, replaces black pigment with brown throughout the body, in the eyes, skin, and coat. When this happens the black agouti pattern becomes brown agouti known as cinnamon—chestnut in horses—genotype Abb. And the non-agouti or uniform black phenotype, genotype aaB, becomes chocolate— uniform brown or "liver" in horses—genotype aabb.

Gene E, the Extender of Dark Pigment Gene. Gene E, acting in conjunction with genes B and C, governs the extension (or restriction) of dark pigment in the coat of mammals. It has three known alleles, E, e, and E^D. (1) E governs the full extension of black (or brown) pigment throughout the coat, as is found in horses of the ancestral wild type, and in domestic bay, black, and chestnut varieties. (2) The recessive allele of E is e. When it is homozygous, ee, it causes restriction of dark pigment to peripheral parts of the coat, leaving the central body region yellow-red. In horses, it causes the difference between mealy bay, ABE, and red bay, ABee. (3) E^D, dominant extension, is a mutation of gene E; it was first clearly demonstrated in the rabbit, but has since been found to be of rather wide occurrence among mammals. It results in such a marked increase in the production of dark pigment that when homozygous, E^{D}E^D, it may completely hide the presence of an A (agouti) coat pattern gene. In the heterozygote AE^{D}E, or AE^{D}e of the rabbit, the agouti pattern shows but slightly; this phenotype is designated by Punnett as agouti-black.

In horses the E^D allele may be either homozygous, E^{D}E^D, or heterozygous, EE^D or E^{D}e, in the dominant black genotype which, unlike the recessive black genotype, is born with a jet black coat that will not fade.

Mutations Resulting in White Hairs or White Areas in the Coat. The following mutations result in white hairs or white areas in the coat: (1) Dominant white, gene W, coat entirely white, eyes colored; probably lethal when homozygous. (2) Dominant dilution, gene D, peculiar to horses; less effective in heterozygotes than in homozygotes. Palomino and buckskin horses are heterozygotes (Dd); the corresponding homozygotes (DD) are so-called albinos of type A and B respectively. (3) Dominant dilution, gene S (for silver), found in Shetland ponies; first observed as a mutation about 1886 in a mare described as fawn color with white mane and tail. This gene is stronger in its diluting action than gene D. Instead of being incompletely dominant, as gene D is, it is almost completely dominant, the homozygous SS being only slightly more dilute than the heterozygous Ss. (4) Roan, gene R, a type of congenital, non-progressive silvering; probably lethal when homozygous like gene W. (5) Gray, gene **G**, a type of dominant non-congenital but progressive silvering, usually dappled because of an independent modifier. (6) Recessive white markings (white feet, blaze), genes undetermined. (7) Dominant white spotting, gene **P** (piebald, pinto), white areas extensive;

TABLE 18-1. | *Color classes resulting from different combinations of genes A, B, and E, with or without gene D.*

Combinations of A, B, and E		Combinations of A, B, E, and D	
Genotype	*Phenotype*	*Genotype*	*Phenotype*
A⁺BE*	Ancestral type, bay with zebra markings		
ABE	Dark or mealy bay	ABEDd**	Dun, body light sooty yellow, mane and tail black, dorsal stripe
ABee	Red bay	ABeeDd	Buckskin, body clear light yellow, mane and tail black, dorsal stripe
aᵗBE	Seal brown, light areas inconspicuous	aᵗBEDd	Light seal brown
aᵗBee	Seal brown, light areas inconspicuous	aᵗBeeDd	Light seal brown
aaBE	Recessive black, uniform	aaBEDd	Mouse, uniform dilute black
aaBee	Recessive black, mane and tail darker than body	aaBeeDd	Mouse, "Grullo," mane and tail darker than body
AbbE	Chestnut	AbbEDd	Palomino, mane and tail white, body sooty yellow-red
Abbee	Sorrel, light mane and tail	AbbeeDd	Palomino, body clear cream to golden yellow, mane and tail white
aᵗbbE	Chestnut "brown"	aᵗbbEDd ⎫	
aᵗbbee	Sorrel "brown"	aᵗbbeeDd ⎭ Claybank	
aabbE	Chestnut uniform, "liver"	aabbEDd	Light chestnut uniform
aabbee	Sorrel uniform	aabbeeDd	Light sorrel uniform
ABEᴰ	Dominant black, intense and uniform, "jet" black	ABEᴰDd	Dominant black
		ABEDD ⎫	Perlino or
		ABeeDD ⎭	Albino type B
		AbbEDD ⎫	Cremello or
		AbbeeDD ⎭	Albino type A

* When a dominant gene occurs in the genotype its symbol will be written only once if we are solely interested in the corresponding phenotype. For example AAbbee has the same phenotype as Aabbee.

** D is incompletely dominant.

homozygotes probably whiter than heterozygotes as in similar varieties of white spotted rabbits and dogs.

In Table 18-1, color classes resulting from different combinations of genes A, B, and E with and without D are presented.

18-3. HEREDITARY ANATOMICAL AND PHYSIOLOGICAL DEFECTS AND RESISTANCE TO DISEASE

Table 18-2 presents cases on record, for the horse, of hereditary anatomical and physiological defects and resistance to disease. Other defects and diseases have been reported over the years, many of them in *Animal Breeding Abstracts,* but are not included here because of the tenuous nature of the evidence.

Hemolytic disease of the newborn foal, hereditary in nature, is associated with an incompatibility of blood groups. It is discussed in the following section.

18-4. THE INHERITANCE OF EQUINE BLOOD GROUPS AND HEMOLYTIC DISEASE OF THE NEWBORN FOAL

Scientists at several laboratories are currently studying the genetics of equine blood groups (Stormont and Suzuki, 1964). The results developing are similar to those already found for man, cattle, chickens, and sheep: namely, that blood groups are associated with specific genes represented by a series of multiple alleles.

Although study of equine blood groups began shortly after the discovery of the four ABO groups in man in 1901, it is only recently that interest has shifted to their genetics. This shift has been stimulated by recognition of the potential use of blood groups as aids in solving problems of questionable parentage; and by the discovery in 1948 that hemolytic disease of the newborn foal is due to isoimmunization of mares during pregnancy resulting from a genetic incompatibility similar to the Rh incompatibility in humans that causes hemolytic anemia of the newborn infant (*erythroblastosis fetalis*).

The genetic basis of the incompatibility is illustrated in Fig. 18-1. This depiction oversimplifies the real situation. There is evidence that more than one gene is involved (Stormont, Suzuki and Rhode, 1964), although according to Franks (1962) 82% of the 40 cases of incompatibility he studied appeared to be due to one gene. His work also showed that in 58 mares "in general . . . although a first hemolytic foal may be produced at any time in a mare's breeding life except during the first pregnancy, the large majority (78%) of first hemolytic foals are produced in the fourth to seventh pregnancies."

TABLE 18-2. | *Hereditary anatomical and physiological defects and resistance to disease.*

Trait	Mode of Inheritance	Reference
Aniridia with secondary cataract Malformation of the iris accompanied by opacity of the lens (cataract) causes more or less reduced vision in affected foal. With advancing age, most affected horses become totally blind.	Simple dominant	*Nord. Vet.-Med.* 7:773
Atresia coli A complete lack of development of the intestine in the region of the ascending colon. Foals die within a few days.	Simple recessive	*Bibliographia Genetica* 6:123
Big head (osteoporosis) Gradual development of enlarged, softened and porous bones in the head.	Simple dominant	*J. Heredity* 19:159
Bleeding Bursting of blood vessels (due to thinness of walls), especially in the nasal mucosa. Found in English Thoroughbred race horses. Studies of 185 horses with the condition showed that 17 died of it.	Simple recessive	*Bibliographia Genetica* 6:123
Encephalomyelitis An epizootic, mosquito transmitted, virus disease marked by high fever and central nervous symptoms.	Statistically significant breed difference in resistance (of horses managed alike) to encephalomyelitis suggested a genetic basis for resistance.	*J. Heredity* 30:349
Mallenders A disease affecting the skin in the flexure of the carpus or knee—resembles psoriasis to some extent. Only found in draft horses as a rule.	On the basis of family data, believed to be recessive in nature.	*Tierzucht* 9:195
Roaring (laryngeal hemiplegia) A functional disease characterized by inspiratory dyspnea which is intensified with exercise. Usually not apparent prior to appearance of permanent teeth—increases in severity with age. Unless corrected by surgery of the larynx it can eventually stop the career of a racer.	Simple recessive	*Bibliographia Genetica* 6:123
Wobbles Irregularity in gait due to bilateral incoordination of hind legs. More frequent in males than in females.	On the basis of family data, believed to be recessive in nature.	*J. Heredity* 41:319

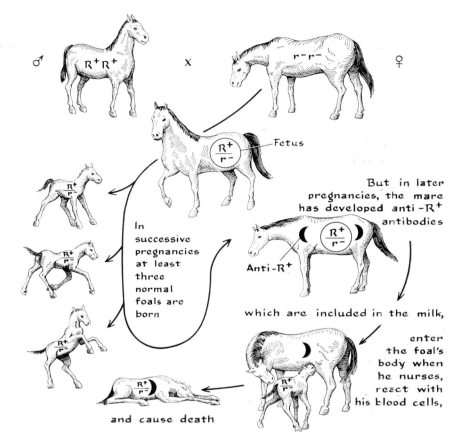

Fig. 18-1. A foal sometimes dies because antibodies that react with his blood cells are contained in his mother's milk. [From *General Genetics* by Adrian M. Srb and Ray D. Owen. San Francisco. W. H. Freeman and Company. 1952.]

18-5. THE GENETICS OF GESTATION LENGTH

The genetics of the gestation length of horses has been extensively studied. It has been found that the genotype of the foetus affects the length of time it is carried in utero. This implies that the stallion exerts a genetic influence on gestation length.

Two important nongenetic factors affecting gestation length are season of breeding and level of nutrition of the mare. Howell and Rollins (1951), summarizing the results of four studies in widely located areas of the northern hemisphere, reported that gestation lengths resulting from spring breedings (March through May) are on the average two weeks longer than those result-

Source of Variance	Variance Components*	Percent of Total Variance
Season of breeding	53.91	43.4
Level of nutrition of mare	6.53	5.2
Level of nutrition × season of breeding	0.00	0.0
Genotype of foal		
Additive	22.74	18.3
Dominance deviations	8.84	7.1
Permanent maternal traits	15.14	12.2
Residual (unexplained)	17.14	13.8
Total	124.30	100.0

*The variance, a measure of variation, can be subdivided into parts, each of which is proportional in size to the importance of the respective cause or source of variation.

ing from summer and fall breedings (June through November). This seasonal effect is independent of level of nutrition of the mare. Well-fed mares have shorter gestations than those on a maintenance ration.

Rollins and Howell (1951) estimated the relative importance of sources of variation in gestation lengths of horses (Table 18-3).

18-6. PERFORMANCE TESTS: THE INHERITANCE OF SPEED

Over the years, performance tests have been used extensively in the formation and improvement of many breeds and classes of horses, ponies, and mules. Among traits for which performance tests are currently being made or have been made in the past are endurance in cavalry horses, speed in racehorses, performance of gaits in saddle horses and roadsters, and draft ability in work horses and mules.

However, very few formal genetic studies of performance tests for horses have been reported. This may be partly due to the drastic decline in importance of horses during the second quarter of this century—at the time when adequate techniques in quantitative genetics and statistics capable of coping with analyses of such tests were first making their appearance. The few studies that have been reported are concerned with racing.

Eight heritability estimates of racing ability, reported in four studies, average 34% and range from 19% to 63%. These results suggest that the racing performance of a stallion or mare (due allowance having been made for nongenetic factors) is a useful criterion in selection. One of the studies

(Gilmore, 1947), in addition, produced evidence[3] that "nicking" (a mating of a particular stallion to a particular mare that results in especially outstanding progeny) may also be of importance.

However, this wide range of heritability estimates, 19% to 63% suggests inaccuracies of estimation, which are due to difficulties in comparing the speed (and earning power) of horses under equitable conditions. According to Steele (1962), ". . . running races are conducted irrespective of the weather, except in emergencies. In a few well established events, horses of the same age race under the same weight and may be further restricted to the same sex, but conditions of most races are less uniform.

"Since horses are so different in their racing capacities and may also change over a period of time, those better on the basis of past performance are usually handicapped by the assignment of extra weight whereby all are supposed to have the same chance of winning. There are numerous other weight allowances in particular races: filly competing with colts, age difference, earnings less than a specified amount, foaled or owned within a particular state, apprentice jockey, and so on. Distances vary widely and the same is true of purses. No other form of animal performance has such complications to limit genetic analysis."

18-7. THE HORSE, ASS, AND MULE KARYOTYPES AND THEIR RELATION TO THE STERILITY OF THE MULE

Among the common domesticated animals, the mule has always attracted especial attention. Its stamina and vigor are well known, yet it is notoriously sterile.

The mule's vigor is compatible with much of the genetic evidence from plants and animals that the more genetically diverse the cross, the greater the resulting hybrid vigor. It has long been suspected that its sterility is related to an incompatibility in chromosome pairing at meiosis due to the genetic diversity of the horse-ass cross. Recent studies, to be discussed below, seem to support this explanation of the mule's sterility.

For meiosis to proceed smoothly, the chromosomes the individual receives from each of its two parents must be similar in number and morphology. The number of chromosomes typical for a species of plants or animals along with a description of the morphology of the chromosomes is spoken of as the karyotype of the species. Karyotypes of the horse, ass and mule are presented in Table 18-4.

Metacentric and acrocentric refer to the location of the spindle fiber attachment to the chromosome at metaphase (in both mitosis and meiosis). In

[3] The evidence involves a comparison of regressions of offspring's performance on sire's performance and offspring's performance on mid-parent's performance respectively.

TABLE   2n *chromosome number* (*body cells*) *and distribution of metacentric and acrocentric*
18-4.   *autosomes.*

		Autosomes	
Animal	2n Chromosome Number	Metacentric	Acrocentric
Horse	64	26	36
Ass	62	38	22
Mule	63	32	29

Source: Based on Makino *et al.*, 1963, and Benirschke *et al.*, 1964.

the metacentric chromosomes, attachment is towards the middle of the chromosome, and in the acrocentric chromosomes, it is at the end. In Fig. 18-2 the metacentric chromosomes look like X's, the acrocentric chromosomes like V's.

As illustrated in Fig. 18-2c, the mule receives a haploid autosomal set of 19 M (metacentric) and 11 A (acrocentric) chromosomes from its ass parent and 13 M and 18 A from its horse parent. This means that at meiosis in the mule, completely harmonious pairing of the chromosomes of the two sets is obstructed by (1) the different location of the spindle fiber attachment within some of the pairs and (2) the fact that there is one more chromosome in the horse set than in the ass set.

According to Makino et al. (1963), a study of the hybrid between the domestic fowl and the common pheasant showed that the degree of dissimilarity between parental chromosomes seemed to be related to the sterility of the resultant hybrids. The degree of difference in the chromosome complement between the horse and the ass was observed to be considerably greater. He further states, "it was shown by histological studies of testes of an adult mule male that the degeneration of male germ cells which directly leads to sterility took place during early stages of the meiotic prophase . . . thus the meiotic division was entirely absent in the mule testes and no spermatozoa were produced."

Prior to the 1950's it was impossible to obtain good cytological preparations of the chromosome contents of mammalian cells; hence accurate knowledge of mammalian karyotypes was sorely lacking.

During the last decade, according to Hsu et al. (1963), "several methodological improvements have opened a new horizon for mammalian karyology: (1) the application of tissue culture techniques to cytology, providing the investigators with ample cells in mitosis, (2) the use of hypotonic solution pretreatment to spread the chromosomes, (3) the employment of colchicine to contract the chromosomes, and (4) the application of squash methods or methods of air drying to force the chromosomes into one plane of focus."

It is possible that during the next decade some animal breeding problems in disease and reproduction, which are related to metabolic disturb-

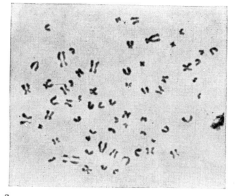

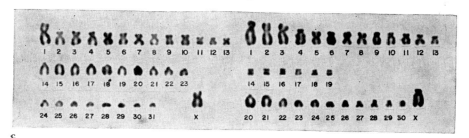

Fig. 18-2. (*a*) Metaphase plate of male hinny, (Orcein, x 2000). (*b*) Karyotype of male hinny prepared from cell shown in (*a*). (*c*) Karyotype of female mule; 63 chromosomes, consisting of two haploid sets, horse set (left) and ass set (right). [Photographs (*a*) and (*b*) courtesy of Dr. Kurt Benirschke, Department of Pathology, Dartmouth Medical School, Hanover, New Hampshire. Photograph (*c*) courtesy of Dr. Sajiro Makino, Zoological Institute, Faculty of Science, Hokkaido University, Sapporo, Japan.]

ances, may be solved by the presently expanding study of mammalian chromosomes.

Using the newly found techniques, Benirschke et al. (1964) made a chromosome study of an alleged fertile mare mule and concluded that she was not a mule, but a jenny.

REFERENCES AND SELECTED READINGS

Benirschke, Kurt, Richard J. Low, Margaret M. Sullivan, and Ross M. Carter, 1964. Chromosome study of an alleged fertile mare mule. *J. Heredity*, 55:31–38.

Castle, W. E., 1954. Coat color inheritance in horses and in other mammals. *Genetics*, 39:35–44.

Castle, W. E., and F. L. King, 1951. New evidence on the genetics of the Palomino Horse. *J. Heredity*, 42:60–64.

Castle, W. E., and W. R. Singleton, 1961. The Palomino horse. *Genetics,* 46:1143–1150.

Franks, D., 1962. Horse blood groups and hemolytic disease of the newborn foal. *Ann. N. Y. Acad. Sci.,* 97:235.

Gilmore, Robert O., 1947. Statistics on sires and dams. *The Thoroughbred of California,* Vol. 6, No. 8, 24–25, 66; Vol. 6, No. 9, 22–23, 58.

Howell, C. E., and W. C. Rollins, 1951. Environmental sources of variation in the gestation length of the horse. *J. Animal Sci.,* 10:789–796.

Hsu, T. C., Helen H. Rearden, and George F. Luquette, 1963. Karyological studies of nine species of Felidae. *The American Naturalist,* Vol. XCVII, No. 895, 225–234.

Makino, S., T. Sofuni, and M. S. Sasaki, 1963. A revised study of the chromosomes in the horse, the ass and the mule. *Pro-*

ceedings of the Japanese Academy, 39: 176–181.

Odriozola, M., 1951. *A los Colores del Caballo.* 435 pp., Madrid.

Rollins, W. C., and C. E. Howell, 1951. Genetic sources of variation in the gestation length of the horse. *J. Animal Sci.,* 10:797–806.

Steele, Dewey G., 1962. Genetic improvement in horses. In *Introduction to Livestock Production.* (1st Ed., 1962). Ed. H. H. Cole, W. H. Freeman and Company, San Francisco. Chap. 18.

Stormont, Clyde, and Yoshiko Suzuki, 1964. Genetic systems of blood groups in horses. *Genetics,* 50:915–929.

Stormont, Clyde, Yoshiko Suzuki, and E. A. Rhode, 1964. Serology of horse blood groups. *Cornell Vet.,* 54:439–452.

Wright, S., 1917. Color inheritance in mammals. VII. Horse. *J. Heredity,* 8:561–564.

Chapter 19

Genetic Improvement in Poultry

19-1. GOALS OF THE POULTRY BREEDER

Broad goals of the commercial breeder are to increase product output per animal, to increase the efficiency of producing a product, and to improve the quality of an existing product. Improvement in fertility, hatchability, growth rate, body conformation, egg yield, meat yield, feed conversion, egg quality, meat quality, and viability are all facets of these three broad goals.

Interest in breeding for type, for breed characteristics, and for freedom from defects and standard disqualifications is secondary, because the modern breeder realizes that undue emphasis on characteristics of questionable economic value slows down the possible rate of genetic improvement in important economic traits.

19-2. ECONOMIC ASPECTS OF MODERN POULTRY BREEDING

The poultry industry in the United States consists of three major segments—concerned separately with the production of eggs, broilers, and turkeys—and for each of these there is a special kind of breeding enterprise. The poultry industry has exploited the sciences of biology for the development of high-performing genetic stocks, which has been much greater than that for any other class of livestock. It has been quick to apply modern breeding schemes and concepts, including inbreeding and hybridization, reciprocal recurrent selection, and many aspects of the theory of quantitative inheritance and even blood-typing techniques, to the possible improvement of performance. As a result, a moderate demand has arisen for the employment of college-trained geneticists for commercial research and development.

The Foundation Breeder and the Franchise Hatcheryman. The breeding segment of the poultry industry includes the foundation breeder, concerned with genetic improvement, and the hatcheryman, who multiplies the stock supplied by the foundation breeder. The producer, or farmer, buys his chicks or poults from the hatcheryman, which he then grows out as commercial egg layers, broilers, or market turkeys.

Not more than fifteen years ago the foundation breeder and the hatcheryman were independent operators. Breeders produced typically "pure strains" or "pure lines." The hatcheryman was free to purchase pure-line stock from any breeder he chose. It was his problem to decide whether the final product he would put together for the farmer customer would be a pure strain, a strain cross, or crossbred.

Today, the foundation breeder and the hatcheryman are typically associated in a franchise arrangement. A contract, signed by the two parties, specifies that the breeder will provide the hatcheryman with the breeding stock for his hatchery supply flocks. Such supply flocks are commonly referred to as parent flocks because the hatcheryman uses the chickens as parents of the commercial chicks he sells. The males going into the parent-flock matings are usually of different breeding than the females, so that the commercial chicks produced are strain-crosses, crossbreds, or inbred hybrids.

Franchises are relatively new in the poultry breeding and hatchery business. This arrangement gives the breeder essentially complete control of the stock his associate hatcheries sell. Control is dependent on the use of a strain cross or a hybrid combination. The hatcheryman finds the franchise arrangement desirable because this allows him to sidestep the technical problems of breeding and to concentrate his efforts on chick multiplication and selling. On the other hand, since he must sign a contract with the foundation breeding farm, the hatcheryman sacrifices some of his business independence.

The foundation breeder recognizes the importance of a trade name. Most large commercial breeding farms today have their trade names copyrighted so that competitors will not infringe upon them.

The franchising system of selling breeding stock to associate hatchery outlets has proved highly successful. In Iowa, for example, of the 260 hatcheries that are members of the Iowa Hatchery Association, 215 (or 83%) operate under a franchise arrangement.

The key to a successful franchise arrangement hinges on hybridization in one form or another. There are two kinds of breeders. One is the strain-cross breeder, who believes that intensive inbreeding is not prerequisite to hybrid vigor. The other is the inbred hybridizer, who believes that intensive in-

Fig. 19-1. A foundation White Leghorn breeder establishment: (*a*) aerial view showing layout of buildings—all birds are confined; (*b*) a windowless trap-nest house; (*c*) inside a windowless trap-nest house; (*d*) birds in trap-nest. [Courtesy Heisdorf Nelson Farms, Inc., Kirkland, Wash.]

breeding of the lines to be crossed is necessary to produce a commercial chicken with maximum hybrid vigor.

The modern foundation poultry breeding farm has many of the characteristics of a large business. The majority of such breeding farms are incorporated and, as is typical of a large corporation, they are departmentalized. Specially trained personnel have charge of separate departments of sales, advertising, office management, purchases, and so forth. Most of the large poultry breeders have sales representatives in each important state or region of the country. Many breeding farms have foreign sales outlets with production and sales staffs stationed in different foreign countries.

The business of genetic improvement is channeled through a department of breeding and development. Such a department may be directed by a college-trained geneticist, with key personnel consisting of physiologists, pathologists, and veterinarians, as well as additional geneticists. Important tools of the breeding department include data-processing equipment to handle the thousands of individual records of new breeding combinations tested for performance and to compare the new combinations with their current commercial product and perhaps samples of competitors' stocks.

Random Sample Performance Tests. Random sample performance tests are commonly used in this country today. In 1960 there were 15 egg-laying tests, 10 meat-production tests for broilers, and 8 turkey performance tests. Random sample performance tests are largely a development of the 1950–1960 decade, the first having been started in California in 1947.

The purpose of the tests is to provide information on the performance of commercial chicks and poults under uniform testing conditions. Test information serves as a guide for the relative performance of commercial stocks offered for sale. The random sample aspect of the test provides assurance that biases are eliminated in comparing different commercial varieties. Random sample tests have largely replaced the old "standard" egg-laying test for which the breeder hatched, selected, and reared his stock under conditions different from those of his competitors.

The typical random sample egg-laying test consists of 50 chicks or poults per entry, which are reared to maturity and then kept separate in a laying pen until about 500 days of age, at which time the entries are ranked in over-all net income based on the entire test period.

A new kind of random sampling is the Multiple Unit Poultry Test, in which testing is carried out at more than one farm or location. Such tests are conducted in Iowa, Kansas, and New Hampshire; in Iowa each entry of 100 or more pullets is placed on four typical farms. Because they are tested on several farms, differences in performance indicate more reliably the per-

Fig. 19-2. A chicken house of the California Random Sample Poultry Test. [Courtesy California Poultry Improvement Commission. Modesto, Calif.]

formance that poultry raisers can expect from a strain under average farm conditions.

There seems little doubt that the random sample performance test, particularly with reference to the egg-laying phases of the poultry industry, has acted as a powerful stimulus to breeders in developing high genetic merit of commercially sold poultry. The outstanding performance of the average chicken today, compared with the chicken of 15 or even 10 years ago, can be partly credited to the interest of breeders and hatcherymen in official random sample performance tests.

Not everything about the random sample test is advantageous. Winners of random sample tests have naturally expanded and become larger, but many breeders, some large and some small, have been forced out of business because of mediocre or poor performance of their flocks in random sample tests. Thus, this type of official comparison of commercial products has led to some monopolism; some breeders are becoming larger, but the total number of breeders is fewer. Just where this will ultimately lead remains to be seen.

19-3. CHARACTERS SHOWING SIMPLE MENDELIAN BEHAVIOR IN THE FOWL

Some of the classic Mendelian principles were first applied to poultry after the rediscovery of Mendel's laws, about 1900. Simple monohybrid and dihybrid ratios illustrating Mendel's laws were reported by Bateson and Punnett. The classic examples of sex-linked inheritance and complementary gene effects were also illustrated very early with examples from the fowl.

Some evidence indicates that the fowl has 6 pairs of chromosomes, including the sex chromosome. This agrees with the fact that only 6 linkage groups have so far been discovered in the fowl. In all avian species the female has a single sex chromosome (the heterogametic sex) and the male has two sex chromosomes. The turkey is thought to have 9 pairs of chromosomes. However, existing evidence has not yet confirmed the true number in either the chicken or the turkey.

Inheritance of Plumage Color. Only those color characteristics that have some economic significance are discussed in the following paragraphs.

WHITE. Genetically, there are two kinds of pure white plumage—a dominant white and a recessive white. In addition, there is a Columbian plumage, which is mostly white with some black feathers. White plumage has become the favored color in all segments of the poultry industry. The White Leghorn or crosses involving the White Leghorn are preferred for egg production; the White Rock and specially developed "dominant white" meat-type male lines are preferred for broiler production. Even for turkeys and ducks the preferred color is white. All ducks of the Long Island duck industry belong to one breed, the White Pekin. The trend in the turkey industry is to replace the bronze plumage with white.

Dominant White. This plumage color is represented typically by the White Leghorn and is determined by the gene I. When a homozygous White Leghorn, II, is crossed to a Black Australorp, ii, the resulting progeny, Ii, are predominantly white with varying amounts of black flecking. This shows that I is incompletely dominant in the F_1 genetic background of half-Leghorn and half-Australorp. However, if the F_1 is backcrossed to the Leghorn in successive generations, leading to a genetic background of Leghorn blood, then the heterozygous Ii individuals are indistinguishable from the II individuals. On the other hand, if the F_1 is successively back-crossed to the Australorp, the heterozygotes, Ii, show a large amount of nonwhite, proving that in the genetic background of the Australorp, *I* no longer behaves as a dominant (see Fig. 19-3).

Recessive White. The White Wyandottes and the original White Plymouth Rock have a recessive white plumage determined by a pair of genes, cc. Crosses of these two breeds with colored varieties yield colored progeny, as expected.

Columbian plumage. This plumage color is represented by the Columbian Plymouth Rock, the Light Brahma, the Delaware, and other breeds. The plumage of these breeds is white with some black in the hackle feathers, wing primaries, and tail.

Two pairs of genes are responsible for the Columbian pattern, a restrictor for extension of black, e, and a gene for white, called silver, S. The restrictor is the allele of black plumage, E, and limits black to the areas indicated. Silver, S, is sex-linked and dominant to its recessive allele gold, s.

Since the female is the heterogametic sex, a Columbian hen would have the genotype eeS– and the cock would be eeSS or eeSs.

BLACK. Black plumage is thought to be due to a single autosomal gene, E (extension of black), which is dominant to restriction of black, e. In addition, black plumage requires the gene for color, C, which is the allele of the recessive white cc gene. Examples of breeds with black plumage are the Australorp and the Black Minorca.

BARRING. The common form of barred plumage is due to a sex-linked dominant gene, B. This gene acts as a restrictor for black, causing black-and-white barred pattern in the feathers. The Barred Plymouth Rock is the best known variety carrying this gene. The genotype is represented as BBSSEE (male) and B–S–EE (female). The silver gene, S, determines the white bars of the pattern. If gold, s, replaces silver, the plumage is black-gold barring.

The barring gene may be used to produce sex-linked crosses distinguishable at hatching time. The cockerel progeny of a cross between a New Hampshire male and a Barred Plymouth Rock female are barred like the female parent, and the pullet progeny are nonbarred like the male parent

Fig. 19-3. Dominance in white plumage as influenced by genetic background. In White Leghorns (right column) I is completely dominant, but in the Spanish breed (left column) I is incompletely dominant. [From Poultry Department, Iowa State University, Ames, Iowa.]

Inheritance of Skin and Shank Color. There are two skin pigments. The carotenoids produce the yellow color of the skin and melanin the dark color. Melanin is formed in special cells known as melanophores.

Presence or absence of carotenoids is determined by an autosomal pair of genes, W,w. The dominant gene, W, inhibits carotenoid deposition, causing white skin and shanks, while the double recessive, ww, permits yellow skin and shanks to occur. Most breeds common in America have yellow skin and shanks—for example the White Leghorn, White Plymouth Rock, Barred Plymouth Rock, and the Rhode Island Red. The White Minorca, having the genotype WW, is a breed with white skin and shanks.

Melanin is considered to be controlled by a sex-linked partially dominant gene pair, D,d. D inhibits melanin in the dermis of the skin and shanks. The Black Australorp, therefore, has the genotype d– (female) and dd (male).

In some breeds and crosses green shanks are found. The green is due to the presence of yellow pigment in the epidermis and melanin in the dermis. The genotype for males would therefore be wwdd.

Rate of Feather Development. Early feathering is due to a sex-linked recessive gene, k. Chicks showing early feathering can be easily identified at hatching by the length of the covert feathers of the wing in proportion to the length of the primary feathers. At 10 days of age, late-feathering chicks show a short tail; early-feathering chicks, a longer one (see Fig. 19-4).

All modern broiler strains now carry the sex-linked early feathering gene. Early feathering is essential for minimizing pinfeathers on the dressed carcasses.

Since the K,k genes are sex-linked, they may be used in determining sex at hatching. Early-feathering males, kk, mated to late-feathering females, K–, produce slow-feathering male progeny and early-feathering pullets. This mating scheme is useful for the breeding of egg-laying chickens since the cost of vent sexing can be avoided: the late-feathering cockerels would ordinarily be discarded at hatching.

19-4. INHERITANCE OF ECONOMICALLY IMPORTANT CHARACTERS

Characters of economic importance are egg yield, egg size, exterior and interior egg quality, hatchability, fertility, viability, and rate of growth. All these factors enter into the economics of poultry production and all have a hereditary basis. Hence the poultry breeder is concerned with them.

Studies from random sample egg-production tests show that about 90% of the variation among entries in income over feed and chick costs can be accounted for on the basis of variation in four traits. Listed in order of

Fig. 19-4. Influence of sex-linked gene, K, k, on feathering. Cockerel (left) is homozygous for rapid feathering, kk. Cockerel (right) is heterozygous, Kk. [Courtesy Poultry Department, Kansas State University, Manhattan, Kansas.]

their importance, they are (1) egg production, (2) viability, (3) body weight, and (4) egg size. Egg quality, fertility, hatchability, and other traits would, under average circumstances, be less important. Of course any economic trait may assume major importance to a breeder if his strain or cross is markedly deficient in some respect.

All economic characters are quantitative in nature; that is, they are not classifiable into distinct categories. In the study of their inheritance the statistical approach (quantitative genetics) is used, since geneticists have not been successful in identifying single-gene effects. For this reason the genetic basis of quantitative traits is assumed to be "polygenic"—that is, these traits are influenced by many genes, each with small effects. Furthermore, differences between individuals of identical genotypes are obscured by variable factors of environment. An important concept in studying the inheritance of quantitative traits is heritability (see Chapter 13). The degree of heritability brings into proper perspective the relative importance of heredity and environment as they influence the observed variation in a particular trait

Egg Production. For egg-type strains, rate of production is naturally the most important character. The fowl, through hundreds of years of domestication, has developed into the efficient egg-laying machine that it is today. An important factor of domestication affecting rate of egg production is that of removing the eggs regularly from the nest. Even a wild bird such as the sparrow has the capacity to lay several times its ordinary egg output in a season if the eggs are removed from the nest. It is important to recognize that environmental influences such as proper care and feeding are responsible along with breeding for the high production records we find in the modern strains of chickens today.

Egg yield is the product of two factors: intensity or rate of laying and the time interval in the laying period. The interval between the first egg that a hen lays and the last one she lays before she goes into a molt is known as the "biological year." Factors tending to increase rate of production or length of the biological year result in greater total egg production.

COMPONENTS OF EGG PRODUCTION. During the past 40 or 50 years, many attempts have been made to determine the mode of inheritance of egg production. In 1919 Dr. Goodale of Massachusetts, working with Rhode Island Reds at the state experiment station, developed the theory that the inheritance of egg production can best be studied in terms of five different components of egg production: (1) age at sexual maturity, (2) intensity of production, (3) persistency of production, (4) winter pause, and (5) broodiness.

Goodale assumed that each of these five different traits was independently inherited, and that genetic improvement in total egg production could be accomplished most effectively by separate consideration of the inheritance of each of these components.

There is some question whether winter pause is really a genetic entity. Also, breeders find difficulty in measuring persistency of production because many hens may continue laying through their fall molt periods. For these reasons, most commercial breeders today focus attention on the other three.

Age of sexual maturity. A pullet is said to be sexually mature when she lays her first egg. The earlier the pullet commences laying, the longer will be the biological year and the more eggs it will be possible for her to produce. On the average, Leghorns become sexually mature between 170 and 185 days. Dual-purpose breeds, such as the Rhode Island Red and the Barred Plymouth Rock, reach sexual maturity two weeks or more later. There are important differences between strains in age at sexual maturity.

Evidence from reciprocal crosses between early- and late-maturing strains indicates that some genes for maturity are sex-linked. For this reason, whenever a cross is planned between two strains differing in this trait it is wise to mate the males of the earlier-maturing strain to females of the late-maturing strain. The female progeny from such a cross would then carry the chromosome having early sexual maturity genes from the sire. Age at sexual maturity has been reported to be about 30% heritable (see Table 19-2). As one might expect, environmental factors greatly influence this trait. For example, early-hatched chicks tend to mature earlier than those hatched late. This is explained by the difference in the amount of daylight during the growing period of early- and late-hatched chicks.

Intensity or rate of production. This factor is measured by the number of eggs laid by a hen during a standard time interval or by the percentage of eggs produced during a variable time interval. Rate of egg production is one of the most important factors determining the profitability of an egg-laying strain, but since this character is highly subject to environmental influence, the heritability of this trait is low—perhaps no higher than 10% —although some studies indicate the heritability to be greater.

Broodiness. The Mediterranean breeds of fowl such as the Leghorn are characteristicaly nonbroody, but Asiatic breeds show a strong tendency toward broodiness. Strains of American breeds, having originated from crosses of Mediterranean and Asiatics, show variable amounts of broodiness. Since the hen takes time out from laying when she goes broody, this shortens the biological laying year and hence the total potential annual egg yield. Also, broody hens can be very troublesome because of their demands on the nests in a laying house.

Evidence suggests that broodiness is determined by complementary effects of genes. Strains of Rhode Island Reds developed in Massachusetts during the period from 1920 to 1950 have been selected for freedom from broodiness. However, crossing these strains to Leghorns produces progeny showing more broodiness than the parents, which suggests that gene interactions are involved. Furthermore, broodiness has a sex-linked basis. Thus, the crossbred progeny of Rhode Island Red males and Leghorn females show higher broodiness than the reciprocal cross (see Table 19-1). Again, the high broodiness of the Austra-White (Australorp male × Leghorn female) is due to sex linkage.

TABLE 19-1. | *Broodiness in reciprocal crosses of White Leghorns and Rhode Island Reds.*

	White Leghorn Male × RI Red Female		RI Red Male × White Leghorn Female	
	No. Birds	% Broody	No. Birds	% Broody
Test 1	37	0	44	18
Test 2	45	0	42	10

Source: Iowa Agr. Expt. Sta., 1956.

TABLE | *Heritability of economic traits in chickens.*
19-2.

Heritability	Trait	Percent
High	Adult body weight	60
	Average egg weight per hen	50
Intermediate	Growth rate to broiler age	35
	Sexual maturity	30
	Annual egg production	20
Low	Hatchability	12
	Total mortality	8
	Leucosis mortality	5
	Fertility	5

Source: Estimates based in part on data from the *Handbook of Biological Data*, 1956. Saunders, Philadelphia, p. 111.

MEASUREMENTS OF EGG PRODUCTION. Egg production may be measured on a flock or on an individual hen basis. In order to obtain individual hen records, trap-nesting is necessary. A few years ago almost all poultry breeders trap-nested seven days a week, each week in the year, in order to obtain a complete record for every hen. However, studies have shown that a satisfactory measure of the egg-producing ability of a hen is possible by trap-nesting less than seven days a week. Most breeders today trap-nest only two or three days per week. This reduces labor costs.

Viability. Although great progress has been made in the control of diseases of baby chicks and growing birds, mortality—especially of the laying flock—is still one of the most important problems of the commercial poultry geneticist. It is not at all uncommon for flocks to show 30, 40, and even 50% mortality for the laying year without diagnosis of an epidemic disease such as cholera or Newcastle. Instead, the diagnostic reports may show a high incidence of leucosis and reproductive disorders. Since these do not respond to usual methods of disease control, they are of special concern to the commercial breeder. The most practical method of control seems to be in the development of genetically resistant strains and crosses.

Leucosis is a cancerous disease of chickens that occurs in several different forms. The most important are "fowl paralysis" or neurolymphomatosis, which affects the nerves of growing pullets and laying birds, and visceral lymphomatosis. Birds affected by visceral lymphomatosis characteristically show an enlargement of the liver; hence the common name, "big liver disease."

Prolapsis of the oviduct is a common reproductive disorder. Poultrymen refer to afflicted birds as "pickouts," since the everted oviduct protruding

TABLE 19-3. | *Breed differences in adult mortality due to prolapse of the oviduct.*

Year	Breed	Total Birds	Total Mortality	Number Dead From Prolapse	Percent Prolapse of Total Mortality
1957	Leghorn	1380	269	1	0.4
	Fayoumi	629	64	16	25.0
1958	Leghorn	1731	208	0	0
	Fayoumi	1037	119	24	20.2
1959	Leghorn	1913	207	15	7.2
	Fayoumi	1467	173	29	16.8
All years	Leghorn	4024	684	16	2.3
	Fayoumi	3133	356	69	19.4

Source: Data from the Iowa Agr. Expt. Sta.

through the vent is generally the focal point of cannibalism of pen mates. A heritable basis for this ailment is suggested by the breed differences shown in Table 19-3. The Fayoumi breed, native to Egypt, has consistently shown a higher incidence of prolapsis than Leghorns at the Iowa Station. However, the Fayoumi has a higher natural resistance to leucosis.

Studies indicate that general laying-flock viability has a low heritability, perhaps around 5%. This indicates that most effective genetic improvement requires an evaluation of average performance of sire progeny families. Many breeders deliberately expose chicks to mature birds to increase disease contamination in order to identify genetically resistant families.

The possibility of developing genetically resistant strains has been well demonstrated at Cornell University. Figure 19-5 shows the difference between strains selected for resistance and for susceptibility to leucosis. In 1955 the resistant strains showed only 2 to 3% mortality from leucosis, compared with more than 25% for the susceptible strain.

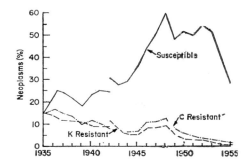

Fig. 19-5. Mortality from leucosis in White Leghorn lines selected for resistance and susceptibility. Breaks in curves at 1942 represent two levels of exposure. [From Hutt's *Genetic Resistance to Disease in Domestic Animals,* Cornell University Press, Ithaca, N.Y., 1958.]

A rather remarkable fact is that reciprocal crosses from the Leghorn breed and the dual-purpose breeds (Plymouth Rocks, Rhode Island Reds, Australorps, and others) show a striking difference in laying-house mortality. Table 19-4 gives the results of 2-year experiments conducted in Iowa, which show that crossbred progeny produced in 1956 from Leghorn males mated to dual-purpose females had 24% mortality over a 9-month laying period, compared with 13% for the reciprocal crosses. Comparative figures were 39% and 18%, respectively, in 1957. Other workers have noted that similar differential response is due either to a maternal influence of the dual-purpose female, perhaps in the form of disease transmission, or to a genetic effect associated with the sex chromosome of the Leghorn male parent.

Body Size, Conformation, and Growth Rate. Body size and conformation are highly hereditary traits. Consider the size and shape differences between breeds, and contrast, for example, a Bantam, a Leghorn, and a Cochin.

BODY SIZE. Large body size is of primary importance to broiler and turkey breeders, mainly because mature body size is correlated with rate of growth. Also, broilers and turkeys produced from strains having large adult body size show the most efficient feed utilization.

On the other hand, in egg-laying strains of chickens, small or intermediate body size is preferred. Such birds have lower body-maintenance cost, but sufficient mature size is still necessary to insure the production of market eggs of satisfactory size.

Between individuals of the same breed and strain, differences in adult body size have been shown to be highly heritable. Most estimates indicate the heritability to be between 40 and 70% (see Table 19-2). This means that the breeder may effectively change body size within a strain by simple mass selection.

CONFORMATION. Conformation is especially important in turkeys. The broad-breasted turkey common on the market today—but unknown twenty years ago—bears witness to the accomplishment of modern turkey breeders (see Fig. 11-5). However, whether this great emphasis on broad breasts is altogether advantageous is open to question, and many breeders now

TABLE 19-4. | *Laying-house mortality in reciprocal crosses of Leghorns and heavy breeds.* *

Test Year	No. Birds	Leghorn Male × Heavy Breed Female	No. Birds	Heavy Breed Male × Leghorn Female
1956	246	24%	178	13%
1957	244	39%	242	18%

Source: Iowa Agr. Expt. Sta.
* Heavy breeds: New Hampshire, R.I. Red, B.P. Rock, W.P. Rock.

wonder whether it is responsible for the poor reproductive performance found in many strains of turkeys today.

Body conformation in broiler chickens is of secondary importance owing to the practice of marketing broiler meat in the cut-up and packaged form. In egg-laying flocks, body conformation is of no direct importance. Most of the egg-laying fowl when marketed are purchased by the canning industry and are used in the manufacture of chicken soup and other prepared foods.

GROWTH RATE. Rate of growth is of major importance in the breeding of meat chickens as well as turkeys. Rapid growth means a saving in time, labor, feed consumption, and overhead in the production of meat. Growth rate is apparently about 30% heritable in most strains of broilers and recent work suggests the same degree of heritability in turkeys. This suggests that an important part of the high performance in growth rate and feed efficiency, attained by broilers and turkeys for the past two decades, can be credited to genetic improvement. However, the important contributions of modern rations to high performance should not be underestimated.

For all animals, rate of growth is characterized by a period of acceleration followed by a period of deceleration. This accounts for the typical growth curve, which is illustrated in Fig. 19-6 for broilers and turkeys.

In broilers the periods of acceleration and deceleration are roughly divided at 12 weeks of age, and in turkeys the periods divide at about 16 weeks of age. The most economic growth takes place during the accelerated growth period. For this reason, broilers marketed at 12 weeks or less, or turkey fryers grown to 16 weeks of age, can be produced on less feed per pound of gain than, say, more mature roasting chickens. Evidence that has accumulated during the past 20 years suggests that efficiency of feed utilization is hereditary. Breed differences, strain differences, and differences between sire progeny have been observed. For example, Jull (1952) found

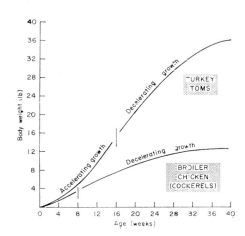

Fig. 19-6. Comparative growth of turkeys and broiler chickens.

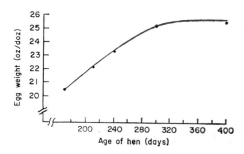

Fig. 19-7. Egg size increase in a laying flock. [Data from the Iowa Multiple Unit Poultry Test.]

as much as 24% difference in feed consumed between two groups of progeny of different Barred Plymouth Rock sires. In general, feed efficiency is highly correlated with growth rate.

Egg Size. The graph in Fig. 19-7 shows how egg size increases to the maximum from the time that the first egg is laid. Weight of the first egg laid by the pullet is about 75% of the maximum reached when the pullet is mature. The speed at which a pullet reaches mature egg size is influenced by hatch date and by age at maturity. Birds hatched in March and April reach maximum egg size in a shorter period of time than chickens hatched in October or November. Size of the first egg is closely associated with body growth, which in turn is influenced by time of hatching. The time of hatching is an important factor because of its relation to the amount of normal daylight during the growth of a bird.

A number of management factors may also influence egg size. For example, eggs from hens in cages may frequently be larger than those from hens under floor management; results reported by the California Random Sample Test show this to be true. Another example, from the Iowa Multiple Unit Poultry Test, is given in Table 19-5.

19-5. CORRELATED RESPONSES TO SELECTION IN POULTRY

The problem of the breeder would be much simpler if each economic trait was inherited independently. Then, if a strain is deficient in, say, egg size, a correction could be made by selection for larger eggs without worrying about the consequences to other traits of importance. Recent evidence indicates that all traits are genetically correlated—either in a positive or negative sense. This seems especially true of populations that have been subjected to selection over many generations—for example, strains of chickens selected for egg production. In such populations most of the remaining genetic covariation would be caused by genes having pleiotropic effects. Over long periods of selection, genes favorable to each of two traits would have been fixed by selection. Then only those pleiotropic genes favorable

TABLE 19-5. | *Egg size in floor vs. cage management (ounces per dozen) from 3 pairs of farms.*

Test	Floor	Cage
1	24.8	26.8
2	25.7	25.8
3	25.5	26.6
Av.	25.3	26.4

Source: First Iowa Multiple Unit Poultry Test, 1958.

to one trait but unfavorable to another would be unfixed. This leads to negative genetic correlations.

From a practical standpoint, this means that the breeder wanting to select for improvement in one trait must see that no serious regression has occurred in important correlated traits. It appears that breeders have already encountered some difficulties. The relatively poor egg production of broiler strains appears to be a consequence of "all-out" selection for growth and feed efficiency. The poor fertility in broad-breasted strains of turkeys is probably a consequence of overemphasis on conformation.

Examples of correlated responses to selection are given in Figs. 19-8 and 19-9. White Leghorns from a single foundation stock were selected for egg

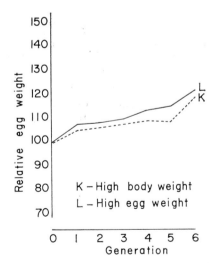

Fig. 19-8. Direct and correlated effects of selection for egg weight and body weight.

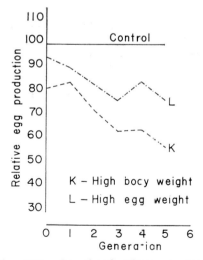

Fig. 19-9. Correlated effects on egg production from selection for egg weight and body weight.

weight or body weight: Figure 19-8 shows that the birds selected for high egg weight (Line L) have responded as expected because egg weight is highly heritable. At the same time those birds selected for high body weight also had an increased egg weight (Line K). This is a correlated response to selection for body weight. Especially significant is the fact that both lines have declined in egg production (Fig. 19-9). This is a further correlated response and illustrates what frequently happens to a reproductive trait when strong selection is made for a morphological trait.

It is clear, then, that such a simple trait as body weight cannot be changed without some alteration in other traits unless they are under selection. Commercial geneticists must be cautious in making genetic changes in their breeding stock—especially for highly heritable traits, since these are most likely to cause marked correlated responses.

As long as the genetic correlation between two traits is positive, simultaneous improvement in both traits is possible. However, the opportunity to advance genetically on both traits is decreased in proportion to the degree of negative correlation.

REFERENCES AND SELECTED READINGS

References marked with an asterisk are of general interest.

*Hagedoorn, A. L., and G. Sykes, 1953. *Poultry Breeding.* Crosby Lockwood and Son Ltd., London.

*Hutt, F. B., 1949. *Genetics of the Fowl.* McGraw-Hill, New York.

*————, 1958. *Genetic Resistance to Disease in Domestic Animals.* Comstock Publishing Associates, Ithaca.

*Jull, M. A., 1952. *Poultry Breeding.* 3rd ed. Wiley, New York.

Lerner, I. M., 1951. *Principles of Commercial Poultry Breeding.* Agricultural Experiment Station and Extension Service Manual I. Univ. of California, Berkeley.

*Warren, D. C., 1953. *Practical Poultry Breeding.* Macmillan, New York.

————, 1958. A half century of advances in the genetics and breeding improvement of poultry. *Poultry Sci.,* 37:3–20.

Physiological Mechanisms and Livestock Production

Chapter 20

Nervous and Hormonal Controlling Mechanisms

The most important attributes of animals, whether common to all or peculiar to some, are, manifestly, attributes of soul and body, in conjunction, e.g. sensation, memory, passion, appetite and desire in general, and in addition, pleasure and pain.

ARISTOTLE, *On Sense and the Sensible*

20-1. INTRODUCTION

The nervous and endocrine systems are concerned with the adjustment of the animal body to external and internal changes. Certain changes in the external environment may be of little importance to the animal body, but others, such as changes in food availability, temperature, light intensity, or impending physical danger, are of extreme importance—they are changes to which the animal must adjust in order to survive. A change in internal environment may be observed after violent exercise. The actively contracting muscles markedly increase the use of oxygen and nutrients and the production of carbon dioxide and other waste products. Since the greater mass of the active body is made up of contracting muscles, there very soon occur small but significant changes in blood composition in the body. Such changes in internal environment set into motion nervous and hormonal mechanisms, which increase the heart rate and the blood flow to the lungs and active muscles, thereby renewing and tending to maintain a constant blood composition.

Rapid body adjustments to changes in the external or internal environment are initiated for the most part by the nervous system. The nervous system also provides in the cerebrum a center for storing and associating sensations, making possible the mental processes of memory, learning, and thought. The endocrine system is designed to regulate certain body functions, such as body growth, for which time is not a major consideration. The secretions of the endocrine organs, the hormones, although not directly concerned with higher mental processes, can modify them considerably; note the severe psychological disturbances that may be associated with the cessation of gonadal function during the menopause or climacteric in the human, conditions that are relieved by the administration of estrogens or androgens, the hormones normally secreted by the gonads. But the adjustment of body function by the nervous and endocrine systems is dependent upon the mechanisms that detect changes in the internal and external environment. Therefore, let us first consider the receptor organs of the nervous system.

20-2. THE NERVOUS SYSTEM

The Receptor Organs. Changes in the environment are of many types and must be sorted out or recognized; that is, the animal body must have receptor mechanisms capable of detecting the kinds of change and the degrees of change in the environment. The most complex receptors of the animal body are the eyes and ears, receptors sensitive to changes in the environment in the intensity and quality of light and sound. There are, however, a multitude of receptors sensitive to changes in the external environment that are far simpler in structure and less evident. Receptors sensitive to chemical change (taste and smell), temperature, and pressure are located in discrete or diffuse areas of the body surface.

The internal environment has many receptors, in the heart, arteries, muscles, hollow organs (digestive tract, reproductive tract, and bladder), the middle ear, and the brain, which are sensitive to changes in pressure and in concentration of oxygen, carbon dioxide, and other chemicals in the blood and to the degree of stretching or contraction of muscles. Receptors, therefore, are strategically placed to detect changes in both the external and internal environments, changes that constitute stimuli capable of exciting specific receptors. Excitations set up in the receptors by stimuli excite, in turn, nerve fibers attached to or in contact with the receptor. The excitation, propagated along the nerve fiber to the central nervous system at a speed up to 120 m/sec, is known as a nerve impulse.

Details concerning the nature of the nerve impulse are beyond the scope of this text. Suffice it to say that (1) the speed of transmission is generally faster in the neurons with myelinated axons (Fig. 20-2) and will vary from 10 to 120 m/sec; (2) a change of electrical potential (action potential) is associated with the passage of the nerve impulse, and provides a means of detecting the passage of a nerve impulse along a nerve fiber; (3) the change in electrical potential depends upon a shift of potassium ions from the axon (Fig. 20-1); and (4) these changes in electric potential result in a localized electrical current, which in turn causes ionic shifts in adjoining portions of the axon. Thus the excitation, or impulse, is propagated along the axon in much the same manner that a flame proceeds along a fuse, the conspicuous difference being that the changes in the fuse are irreversible whereas the nerve is restored to its resting condition in fractions of a second.

The Neuron. Most of us are familiar with the nerves in the arms and legs. More precisely, they are really nerve trunks containing hundreds of extensions or processes of highly specialized body cells called neurons. Neurons are the units of the nervous system and are basically alike, whether concerned with highly involved functions of the nervous system such as memory, learning, or thought, or the simpler functions of conduction to and from the central nervous system.

All the cells of the body possess the ability to react to stimuli and to conduct the excitation initiated by stimuli the length of their cells. The neuron exhibits this property of conduction to the highest degree. The extensive ramifications of neuron processes throughout the body and their multiple connections within the brain and spinal cord provide an excellent communication system between widely separated portions of the body as well as between the body and the external environment.

The neuron has a cell body and, typically, two types of cell projections —dendrites and axons (Fig. 20-2). Neurons conducting impulses from the body surface and from the various receptors to the spinal cord and brain are called afferent or sensory neurons. Many afferent neurons have elongated dendrites, frequently referred to as axons, some of which extend from a receptor in the skin, run the length of a limb, and terminate at the point of the cell body. Thus the neurons are unique cells in that their processes may be several feet in length.

Efferent or motoneurons, which conduct impulses to muscles and other "effectors," originate in the spinal cord or brain, where the cell body of the neuron is located. The efferent neurons have short dendrites, and the axons emerging from the cord or brain vary in length according to the distance of the muscles and glands from the central nervous system.

A third type of neuron, the association or interneuron, makes connections between afferent and efferent neurons within the spinal cord and brain. Neurons are similar in that they have a cell body and cell processes but differ from each other in the number of dendrites, the length of their axons, and whether or not the processes are covered with a myelin sheath. Myelin is a lipoprotein material covering the axon and imparting a glistening white appearance to the nerve fiber. In peripheral neurons, as illustrated in Fig. 20-2 (*a* and *d*), the myelin sheath is interrupted at regular intervals by evenly spaced constrictions known as the nodes of Ranvier. The gray and white

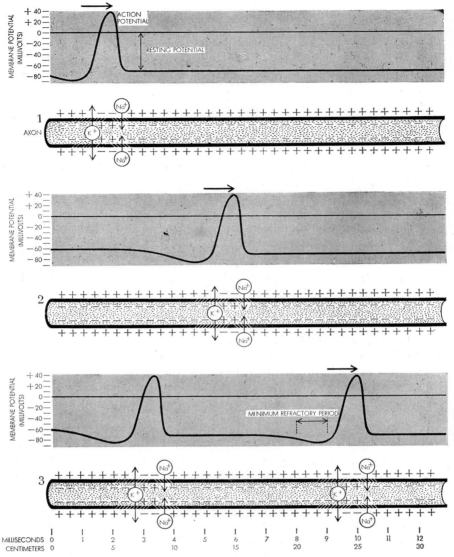

Fig. 20-1. Propagation of nerve impulse coincides with changes in the permeability of the axon membrane. Normally the axon interior is rich in potassium ions and poor in sodium ions; the fluid outside has a reverse composition. When a nerve impulse arises, having been triggered in some fashion, a "gate" opens and lets sodium ions pour into the axon in advance of the impulse, making the axon interior locally positive. In the wake of the impulse the sodium gate closes and a potassium gate opens, allowing potassium ions to flow out, restoring the normal negative potential. As the nerve impulse moves along the axon (1 and 2) it leaves the axon in a refractory state briefly, after which a second impulse can follow (3). The impulse propagation speed is that of a squid axon. [From "How Cells Communicate" by B. Katz. Copyright © Sept. 1961 by Scientific American, Inc. All rights reserved.]

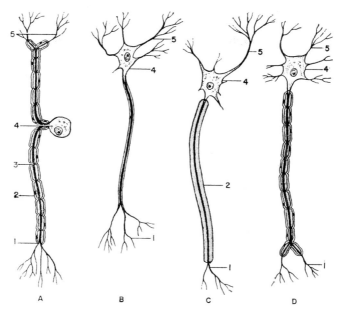

Fig. 20-2. Types of neurons: (*A*) Myelinated afferent neuron of peripheral nerve; (*B*) nonmyelinated interneuron of the gray matter of the spinal cord; (*C*) multipolar myelinated motor neuron of the brain; (*D*) myelinated motor neuron of peripheral nerve. Parts of the neuron: (1) axon; (2) myelin sheath; (3) node of Ranvier; (4) neuron cell body; (5) dendrites.

matter of the nervous system is dependent upon the distribution of nonmyelinated neuron cell bodies, dendrites, and axons (gray matter) and myelinated axons (white matter).

There are varied and extremely complicated connections of afferent, interneuron, and efferent neurons within the nervous system. The basic functional unit of the nervous system, however, the reflex arc, demonstrates the functional relationship between afferent neuron, integration center, and efferent neuron (Fig. 20-3).

The Reflex Arc and the Basic Spinal Reflex. Many responses to changes in the external environment are protective mechanisms. This is exemplified by the abrupt withdrawal of a limb after contact with a sharp object. The body's reaction to this type of stimulus is very rapid and is usually accomplished before the individual is aware of the pain. The rapidity of response indicates that the nervous pathway from the stimulated receptor to the integrating center and back to the contracting muscles must be rather direct and uncomplicated. This type of response is a basic spinal reflex.

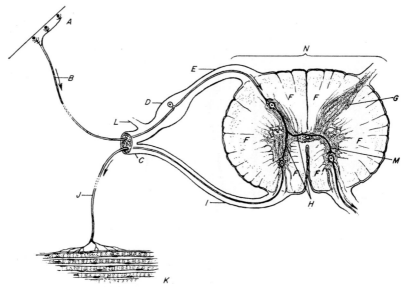

Fig. 20-3. The spinal cord, the spinal nerve, and the reflex arc. (*A*) Pressure receptor in the foot; (*B*) sensory neuron from the body surface; (*C*) spinal nerve; (*D*) afferent nerve cell bodies in dorsal nerve; (*E*) dorsal root of spinal nerve; (*F*) white matter of spinal cord; (*G*) gray matter of spinal cord; (*H*) interneuron in gray matter of spinal cord; (*I*) ventral root of spinal nerve; (*J*) motor neuron; (*K*) muscle of leg; (*L*) autonomic pathway to and from the spinal cord; (*M*) interneuron connections with opposite side of spinal cord; (*N*) spinal cord.

A basic spinal reflex is dependent upon the integrity of neurons present in a reflex arc. Figure 20-3 illustrates the simplest components of the reflex arc—receptors, afferent neurons, interneurons, efferent neurons, and effectors. Pressure receptors in the sole of the foot are stimulated by changes in pressure occurring during walking or by any localized pressure such as a sharp object. Sufficient excitation of the pressure receptors induces an excitatory state in a sensory nerve fiber, an excitation that is passed the entire length of the neuron as a nerve impulse. Fine endings of the axon of the sensory neuron terminate in close contact with the dendrites of an interneuron within the gray matter of the spinal cord, forming a synapse, or conjunction. The excitation is propagated across the synapse and is conducted in the interneuron to its synaptic connection with the cell body of an efferent neuron in the ventral horn of the gray matter of the spinal cord. The efferent neuron, which leaves the spinal cord by the ventral root, terminates in skeletal muscle and stimulates it to contract, thus removing the foot from the sharp object.

In view of the nervous conduction rate of 50 to 70 m/sec and the limited number of neurons involved, it is not surprising that the basic spinal reflex is almost instantaneous. The example given is oversimplified; many receptors would be activated, several afferent neurons, interneurons, and efferent neurons would be involved, and numerous skeletal muscles would contract in the removal of the foot. Nevertheless, the basic component of the reflex arc remains the same in all reflex actions of the body, although interneurons leading to other centers, as well as efferent neurons leading to other tissue, may be activated by the initial stimultaion of afferent neurons. Fig. 20-4, though still oversimplified, gives a better understanding of the complexity of pathways within the nervous system.

Parts of the Nervous System. Embryologically the central nervous system is a continuous tube extending from the anterior to the posterior portion of the body. Dilations, outgrowths, and considerable foldings of the anterior portion of the tube in the developing individual give rise to a prominent anterior portion, the brain, and a more posterior portion, the spinal cord (Fig. 20-5).

The spinal nerves leave the spinal cord at regular intervals—at each vertebra from the tail bone to the neck region—and supply afferent and efferent neurons to the tissues of the body. The reflex centers are located in the gray matter of the brain and spinal cord and communicate freely with each other by ascending and descending neurons in the white matter of the spinal cord (myelinated nerve fibers traveling in the outermost portion of the spinal cord). This intercommunication of the reflex centers makes possible the coordinated activities of skeletal muscle innervated by several spinal or cranial nerves. Furthermore, it is a means by which higher integrating centers in the brain are informed of changes controlled by lower spinal centers. This is why strong pressure applied to the sole of the foot would result in a contraction of several muscles in the leg to bring about withdrawal of the foot, a shifting of the body weight by the contraction of muscles in the hips, waist, and body trunk, and contraction of muscles in the opposite leg in order to maintain balance as the center of gravity of the body is shifted. Also, mental awareness of the stimuli results from afferent impulses in ascending neurons to higher brain centers.

The areas of the body supplied by neurons of each spinal segment can be mapped out, with the control of skeletal muscles in each area assigned to definite spinal nerves. Quite often, reflex testing by a practitioner can pinpoint damage to a particular level of the spinal cord by testing the functional integrity of the various reflex arcs of the spinal cord.

The spinal reflex centers control the primitive movements of the animal body, including the contractions of skeletal muscles involved in protective responses. Spinal centers, with a minimal number of neurons involved in the reflex arc, are very well suited for rapid responses to external

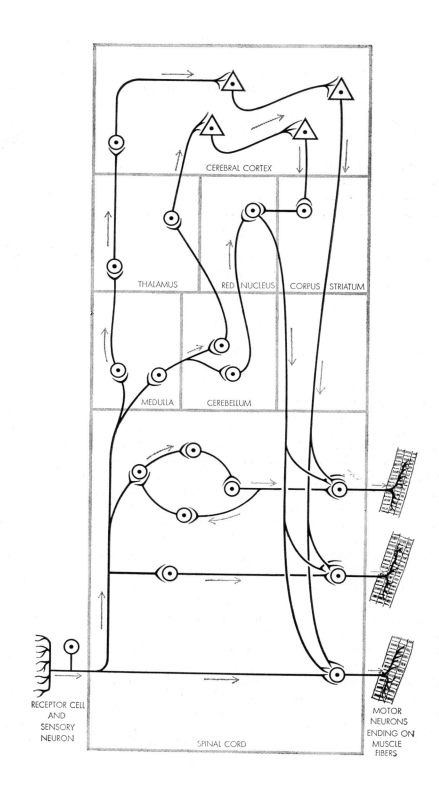

CEREBRAL CORTEX

THALAMUS

RED NUCLEUS   CORPUS STRIATUM

MEDULLA   CEREBELLUM

RECEPTOR CELL
AND
SENSORY
NEURON

MOTOR
NEURONS
ENDING ON
MUSCLE
FIBERS

SPINAL CORD

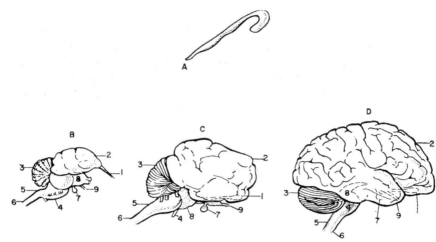

Fig. 20-5. Cerebral dominance in higher species. (*A*) Undifferentiated neural tube in all embryos; (*B*) pigeon brain; (*C*) pig brain; (*D*) human brain; (1) olfactory process; (2) cerebrum; (3) cerebellum; (4) pons; (5) medulla oblongata; (6) spinal cord; (7) pituitary gland; (8) mid-brain. (9) optic nerves.

stimuli. Finer adjustments of reflexes are made by higher integrating centers, aware of the total sum of environmental changes impinging upon the body and influenced by the processes of experience, learning, and willful intent. Reflex centers in the brain rather than the spinal reflex centers integrate the activities associated with several vital body functions such as respiration, circulation of blood, and regulation of heart beat.

The brain and the spinal cord cannot be considered as separate systems. Many of the functions of the brain are mediated through neurons descending the spinal cord and synapsing with spinal neurons, and all the functions of spinal reflex centers are influenced by brain centers. One of the greatest differences between the spinal cord and brain is in the amount and type of physiological activity that each controls. The spinal cord can be considered

Fig. 20-4. Simplified flow diagram of nervous system barely hints at the many possible pathways open to an impulse entering the spinal cord from a receptor cell and its sensory fiber. Rarely does the incoming signal directly activate a motor neuron leading to a muscle fiber. Typically it travels upward through the spinal cord and through several relay centers before arriving at the cerebral cortex. There (if not elsewhere) a "command" may be given (or withheld) that sends nerve impulses back down the spinal cord to fire a motor neuron. [From "How Cells Communicate" by B. Katz. Copyright © Sept. 1961 by Scientific American, Inc. All rights reserved.]

as a series of reflex centers concerned with skeletal muscle contraction. A spinal center on the right side of the spinal cord is duplicated by another on the left side; that is, there is a spinal center for each set of duplicate muscles. A portion of the brain is also made up of duplicating reflex centers controlling muscles in the head region. Other centers, however, control specific body functions and are not duplicated; there are, for example, several centers in the brain concerned with the flow of blood through the body. Each center has a definite role, controlling heart rate, the diameter of the arteries, or the shunting of blood flow from nonactive to active tissues, but the actions are complementary or antagonistic.

The principle of the reflex arc applies to all reflex centers in the brain; that is, afferent neurons carry impulses to brain centers and efferent neurons carry impulses away from the brain centers. The main difference between brain and spinal centers is in complexity; extraordinarily complex connections, by means of association neurons, are evident between brain centers. A greater degree of complexity in neuron interaction is apparent from the lowest to the highest portion of the brain.

The afferent and efferent neurons of the brain do not necessarily travel together in one nerve as do many of the spinal neurons. The brain has 12 pairs of cranial nerves. Some cranial nerves are purely sensory (olfactory, optic, or acoustic nerves), others, almost entirely motoneurons (oculomotor nerve—to muscles of the eye), and others, mixed nerves (facial nerve—sensory from the taste buds in the mouth and motor to the muscles of the scalp and salivary glands). The cell bodies of the neurons occupy the center of the spinal cord (the gray matter), but the reverse is true in brain tissue; that is, the outermost portion of the brain is made up of gray matter and the interior portions are made up of ascending, descending, and association myelinated nerve tracts. This is one reason why it is not too difficult to study the functions of some areas of the brain by placing electrodes directly on the skull and noting the responses obtained by electrical stimulation of specific neuron cell centers.

The medulla oblongata, the portion of the brain connecting with the spinal cord, contains reflex centers that affect the regulation of breathing, the strength and rate of heartbeat, the flow of blood through body vessels, and the secretions of several glands, and centers controlling coughing, sneezing, vomiting, and the adjustment of posture.

The cerebellum, lying just anterior to the medulla oblongata, contains reflex centers involved in the maintenance of skeletal muscle tone. A state of partial contraction of skeletal muscles is necessary for reflex or voluntary contraction to avoid the jerkiness that would exist if the muscles contracted from a completely relaxed state; the contraction must be slightly opposed by antagonistic muscles. The cerebellum is also involved in the fine adjustment of skeletal muscle activity as seen in locomotion as well as in voluntary muscle activity requiring extreme dexterity. It is the coordinating center for many spinal reflexes in the body.

The midbrain has reflex centers controlling specific body functions and coordinating centers exhibiting considerable control over lower brain centers in the cerebellum and medulla oblongata. The anterior portion of the midbrain, the thalamus, is a relay center for the majority of sensory impulses to the cerebrum. It sorts the various sensory impulses to appropriate conscious areas of the brain and also initiates various body responses as a result of intense or extreme afferent stimulation. A large aggregation of neuron cell bodies make up the hypothalamus, which regulates extremely complicated body functions, such as body temperature, sleep, blood flow, heartbeat, and hunger, and controls the production or release of hormones from the anterior pituitary gland. Some of this regulation is indirect and is brought about by stimulation or inhibition of centers in the medulla oblongata.

The functions of the involuntary muscles and glands are controlled by the autonomic nervous system; it includes afferent and efferent neurons and reflex centers in the central nervous system (as does the somatic nervous system, which controls the voluntary, or skeletal, muscles). Most of the involuntary muscles and some glands, in contrast to voluntary muscles, are under the control of two sets of motoneurons, the sympathetic and parasympathetic neurons. The parasympathetic neurons leave the central nervous system in certain of the cranial nerves and sacral spinal nerves in the posterior part of the body. The sympathetic motoneurons leave the central nervous system through spinal nerves in the thoracic and lumbar regions of the body. The parasympathetic system is mainly concerned with body functions of an animal in a resting state and regulates functions such as the secretion of salivary and digestive juices and the normal movements of the intestinal tract. The sympathetic system, on the other hand, is involved in what may be considered as emergency reactions; all the functions activated by stimulation of the sympathetic system are those that allow the animal to exhibit its greatest functional capacities. Sympathetic fibers also control the release of epinephrine from the adrenal gland. Epinephrine acts similarly to the tissue activators released by sympathetic neuron endings. The result of epinephrine release, therefore, is an intensification and reinforcement of all of the independent actions of the sympathetic system and a coordinated "alarm reaction." Body activities such as increased heart action, diversion of blood from nonactive to active tissues, mobilization of body energy sources, the secretion of sweat, maximal accommodation of the eyes for near vision, and the erection of hair or feathers are enhanced upon stimulation of the sympathetic system. Pain, fright, loud noise, or any intense afferent stimulation can result in the "alarm reaction," with stimulation of the sympathetic system and release of epinephrine.

For the most part, the sympathetic and parasympathetic systems act in antagonism to each other in the control of vital body functions, but together they complement general body welfare: vital body functions under their control are not allowed to come to a complete halt nor to overfunction seriously. A good example of this is seen in the control of the heartbeat. Both of the systems have afferent and efferent neurons to heart musculature, but the parasympathetic efferent neurons are inhibitory and sympathetic neurons stimulatory to heartbeat. In mammals, as contrasted to birds, there is considerable parasympathetic tone to the heart at all times; that is, the heart rate is always being depressed when an animal is in the resting state. One obvious result of this tone is that whereas the bird has an extremely high heart rate (very close to its maximal rate) even at rest, the mammal, when active, can triple the resting heart rate. Under conditions of extreme afferent stimulation of the midbrain—as in trauma, fright, loud noise, or willful physical exertion—the parasympathetic centers are depressed and heartbeat increases. Initially this alarm reaction is of short duration and the parasympathetic system renews the inhibitory role unless the emergency persists. If physical exertion is continued the parasympathetic system is inhibited by other brain centers aware of increased needs for a high blood pressure and increased blood flow.

The cerebrum, that most higly developed part of the brain in birds and mammals, constitutes the major part of the brain; in fact, it represents about three-quarters of the total mass of nervous tissue in the body. The high degree of intelligence found in man can doubtlessly be attributed to its relatively large size. The lower brain and spinal cord are mostly reflex centers that react predictably to impulses reaching them by sensory and motoneurons (from other brain centers). Functions of the cerebrum, however, which include learning, memory, thought, and purposeful action, cannot easily be explained on the basis of the reflex arc and reflex center, since the activities of the cerebrum are for the most part unpredictable and are based upon previous experiences of the animal. The cerebrum exerts considerable influence over the lower brain and spinal centers, not only in willful movement, but also in the perfect adjustment of lower brain and spinal cord reflex action. The cerebrum also makes possible modifications of basic reflex action.

Observations of animal activities demonstrate that there are considerable variations among animals in their responses to the same stimuli. This does not imply that basic reflex arcs are lacking or abolished in different animals and other responses substituted; we can assume, unless damage or developmental abnormalities are apparent, that all cattle, for example, are born with almost the same number and location of neurons. It does imply, however, that there are modifications of reflex responses in animals that are dependent upon experience. The conditioned reflex is a basic learning phenomenon. Once an animal has been exposed repeatedly to certain indifferent stimuli, this exposure will influence the response to the same situation occurring at a later date. Conditioned reflexes can therefore be considered as all those reflexes that are learned by the animal in the course of its existence, and they are modifications that contribute in large part to the personality or special activities of the animal.

The establishment of a conditioned reflex was early demonstrated by Pavlov (1910) in experiments on dogs. Certain chemical receptors in the tongue (taste buds) are stimulated by the presence of food in the mouth and, through a simple reflex, this results in the release of saliva. The normal stimulus for salivation, the placing of food in the mouth, is termed an unconditioned stimulus. If, when a dog is fed, a bell is rung, and if this association is made repeatedly (the bell rung each time the dog is fed), eventually the ringing of the bell alone (the conditioned stimulus) will cause a secretion of saliva. The same response can be elicited if the conditioned stimulus is a particular sound, a light, or a tactile stimulus. In a conditioned response the efferent or motor side of the reflex arc remains the same; that is, in the dog experiment, the end result is the stimulation of the release of saliva, regardless of the conditioned stimulus. It is therefore obvious that the afferent side of the reflex arc can vary, and simply becomes integrated with the efferent side of the reflex arc already established. This is an uncomplicated example of a conditioned reflex, but much more complex relations dealing with the inhibition, reinforcement, and establishment of conditioned reflexes have been demonstrated.

Another conditioned reflex seen in domestic animals is the letting down of milk as the result of some repetitive conditioned stimulus associated with milking, such as the clanging of milk pails, the movement toward the milking shed, or any external stimuli (the conditioned stimuli) associated with the unconditioned stimulus, which is the manipulation of the teats at milking time. This reflex is considered in more detail in Chapter 23.

According to Pavlov, the formation and maintenance of conditioned reflexes are dependent upon an intact cerebral cortex. Theoretically there could be as many conditioned reflexes in an animal as there are separate reflex arcs. This would indicate, if conditioned reflexes can be established over the efferent limb of the many body reflexes, that neurons from the cerebral cortex must have direct or indirect synaptic connection with all of the lower brain and spinal cord reflex centers. This, in turn, would indicate a tremendous amount of involuntary cerebral control of body function.

20-3. THE ENDOCRINE SYSTEM

Endocrine Glands and Their Hormones. There are available to the body cells many substances that modify their activities. Nervous excitation or inhibition of muscles or glands are brought about by the release of compounds from nerve endings close to the tissues. Various products of metabolism, carbon dioxide, for example, can inhibit metabolic activity in the majority of body cells but can stimulate activity in other cells, notably, the chemical receptors. However, the majority of the body's constituents supplied to the tissue fluids—minerals, amino acids, proteins, fats, and vitamins—although of immense importance to the normal function of the body cells, do not act as regulators of cell function.

One group of compounds in the animal body that regulate or direct the physiological activities of tissues are the hormones. They are organic compounds produced by a tissue or tissues of the body, passed into the blood stream, exciting or inhibiting the physiological activity of some tissue and not as far as is known directly involved in metabolic processes. The term hormone means, "I excite," but it is known that a given hormone may stimulate one physiological activity and inhibit another. Hormones are believed to act either by increasing or decreasing the concentration of enzymes or the availability of substances necessary for normal cellular function within the stimulated or inhibited tissue.

The glands or tissues producing hormones, the endocrine glands, are not of any one type of tissue. The hormones also differ markedly in chemical structure. Figure 20-6 shows diagrammatically the location of the various endocrine organs in a sow. Notice that the hypothalamus and the pituitary glands, glands exhibiting a marked control over the functions of the other endocrine glands, are located within the skull; that is, on the ventral surface of the midbrain. The hypothalamus is chiefly a nerve center composed of many neuron cell bodies, whereas the pituitary gland is a more typical endocrine organ, storing or producing the "tropic hormones," which control a large number of the other endocrine organs in the body. The hypothalamus, in addition to many other functions, controls the release of several of the hormones from the pituitary gland.

The endocrine tissues, the hormones they produce, and the type of tissue and responses they stimulate are listed in Table 20-1. The complementary and antagonistic actions of the various hormones are too numerous to consider here. It should be pointed out, however, that the resultant physiological effect of endocrine control in the intact animal is due to the interaction of several endocrine glands and their hormones.

Hormones may influence a few specific tissues, or all the tissues of the body; the tissues they influence are called their "target tissues." Growth hormone, the adrenal cortical hormones, thyroxin, and insulin influence most if not all the tissues of the body, whereas the gonadotropins, ACTH, TSH, estrogens, androgens, and progesterone, restrict their actions to a limited number of body tissues. These are, of course, generalizations, since it has been shown that androgens may influence protein metabolism in all cells of the body and that estrogens, at least in birds, strongly influence the mobilization and distribution of lipid during normal growth and egg production. As our knowledge of tissue functions expands, we will, no doubt, find

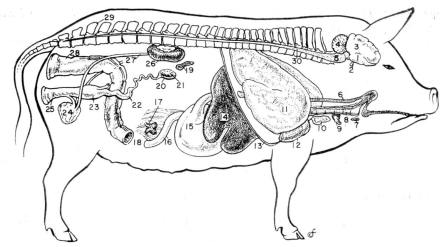

Fig. 20-6. The endocrine glands (marked with an asterisk) and other organs in a sow. (1) Hypothalamus*; (2) pituitary gland*; (3) cerebral cortex; (4) cerebellum; (5) medulla oblongata; (6) esophagus; (7) parathyroid gland*; (8) trachea; (9) lobe of thyroid gland*; (10) thymus gland; (11) lung; (12) heart; (13) diaphragm; (14) liver; (15) stomach; (16) small intestine; (17) islets of Langerhans in pancreas*; (18) approximately 55 ft. of small and 12 ft. of large intestine removed; (19) adrenal gland*; (20) ovulation bursa of oviduct; (21) ovary*; (22) right uterine horn; (23) cervix; (24) urinary bladder; (25) vagina; (26) kidney; (27) ureter; (28) rectum; (29) vertebral column; (30) spinal cord. Hormones are secreted by the gastric and intestinal mucosa—thus these tissues also are endocrine glands. [Drawn by C. A. Dick.]

that all hormones are much less limited in function than was at first supposed.

Means of Controlling Hormone Secretion. The secretion of certain hormones is believed to be a continuous process, and the activities they support are of longer duration than the activities stimulated or inhibited by the nervous system. The exceptions to this generalization are apparent when one considers hormones involved in the development of the reproductive organs and mammary glands, tissues continually changing in patterns influenced by age, season, environmental conditions, and conception. The secretions of hormones are controlled by three means: (1) the nervous system, (2) other hormones reaching the glands through the blood, and (3) changes in the chemical composition of the blood supplying an endocrine organ.

NERVOUS CONTROL. An example of this type of control is the stimulation of ovulation in rabbits by the act of mating. The afferent neuron excitation

TABLE 20-1. | *The endocrine tissues in mammals, the hormones, and the tissues of the body directly influenced.*

Endocrine Tissue	Hormones Secreted	Tissues Influenced	Major Physiological Actions
Pituitary gland			
Anterior portion	Growth hormone (GH)	All tissues	Stimulates general body growth.
	Adrenocorticotropin (ACTH)	Adrenal cortex	Maintenance of functional adrenal cortex.
	Follicle stimulating (FSH)	Ovaries	Germinative function (oogenesis).
		Testes	Germinative function (spermatogenesis).
	Luteinizing (ICSH or LH)	Ovaries	Maturation and ovulation of follicles (estrogen secretion) and establishment of corpora lutea.
		Testes	Secretion of androgens from Leydig cells.
	Lactogenic or luteo-tropin (LTH)	Ovary (corpus luteum)	Maintenance and progesterone secretion.
		Mammary gland	Formation of milk in alveoli.
		Crop gland (pigeon)	Formation of "crop milk."
	Thyrotropin (TSH)	Thyroid gland	Stimulates secretion of thyroid hormone.
Posterior portion*	Oxytocin	Mammary gland	Initiates let-down of milk into ducts and cisterns.
		Uterus	Stimulates contraction of uterus.
	Vasopressin or anti-diuretic	Peripheral blood vessels	Constricts; increases blood pressure.
		Kidney tubules	Promotes water resorption.
Thyroid glands†	Thyroxin and some triiodothyronine	All tissues	Increases rate of cellular metabolism.
Parathyroid glands†	Parathyroid hormone	Intestine, bone, kidney	Increases blood calcium levels.
Adrenal glands			
Adrenal cortex	Glucocorticoids	All tissues	Antistress action hormones, anti-inflammatory; mobilizes energy sources; increases blood glucose.
	Electrocorticoids	Kidney, all tissue indirectly	Salt and water balance.
Adrenal medulla	Epinephrine	Muscles of cardiovascular system	Heart vessels dilate, heart muscle increases rate and strength of contraction.
		Skeletal muscles	Mobilizes energy sources; increases strength of contraction.
		Liver	Glycogen mobilized for energy.
	Norepinephrine	Smooth muscle and glands	Stimulates smooth muscle and glands.
		Peripheral blood vessels	Functions in maintenance of blood pressure.
Ovaries	Estrogens	Mammary glands	Development of duct system.
		Genital tract	Growth of accessory reproductive organs.
		Skin	Female fat distribution.
	Progesterone	Mammary glands	Development of alveolar system.
		Genital tract	Preparation of tract for fertilized ova, nutrition of zygotes, and maintenance of pregnancy.
	Relaxin	Cartilage and ligaments of pelvic girdle	Dissolution and relaxation of cartilage and ligaments to facilitate parturition
Testes	Androgens	Accessory sex organs	Maturation, storage, and nutrition of sperm.
		Body conformation	Male muscle and fat distribution.
	Estrogens	Accessory sex organs	Action not obvious in normal male.
Placenta	Gonadotropins, adreno-corticotropin, estro-gens, progesterone	Ovaries and uterus and adrenals	Complements action of anterior pituitary and ovaries.
Gastrointestinal tract	Secretin	Pancreas	Alkaline secretions for digestion.
	Pancreozymin	Pancreas	Enzyme secretions for digestion.
	Cholecystokinin	Gallbladder	Evacuates bile into intestine.
	Enterogastrone	Stomach	Inhibits motility and acid secretions.
	Gastrin	Stomach	Stimulates acid secretion.
Islets of Langerhans	Insulin	Muscle and adipose tissue	Lowers blood glucose
	Glucagon	Liver	Raises blood sugar

* It now appears that oxytocin and vasopressin are produced in the hypothalamus and stored in the posterior pituitary.

† Substances that lower blood calcium (calcitonin and thyrocalcitonin) have been extracted from both the parathyroid and thyroid.

** The intimate attachment of fetal and maternal membranes for purposes of nutritive and excretory exchange. The hormones produced can therefore be of fetal or maternal origin.

brings about a release of gonadotropins from the anterior pituitary, and this in turn causes final maturation of ova and ovulation about twelve hours after mating. Other examples of nervous control are considered in Chapter 23.

HORMONE CONTROL. The control of hormone secretion from an endocrine gland by hormonal means is exemplified by the anterior pituitary thyrotropic, adrenocorticotropic, and gonadotropic hormones, which control the production and release of thyroxin, adrenal steroids, and estrogens and androgens from their respective glands. Thyrotropin (TSH), from the anterior pituitary, stimulates the release of thyroxin from the thyroid glands, a hormone that increases the general metabolic rate of the cells of the body. The amount of thyroxin circulating in the blood in turn determines the amount of TSH secretion from the anterior pituitary; any drop in thyroxin stimulates further release of TSH, and any rise in thyroxin concentration inhibits further production of TSH. This feedback mechanism to the gland producing the tropic hormone is referred to as "reciprocal control" and is an excellent mechanism for regulating the amounts of hormone in the blood stream. The feedback mechanism is illustrated as follows: if a normal animal is given increased amounts of thyroxin, the supply of thyroxin to the blood will slowly shut off the release of TSH from the anterior pituitary gland. When the animal is fed an amount equal to or exceeding the amount of thyroxin it is normally producing, the secretion of TSH from the anterior pituitary gland is almost completely shut off. If this thyroxin feeding is continued with a resulting deficit of TSH in the blood, the thyroid gland will atrophy and stay in a nonfunctional state, but the metabolic rate of the body will be maintained by the thyroxin. Sudden discontinuation of the thyroxin feeding will result in an animal completely unable to produce its normal complement of thyroxin and a hypothyroid state will result—that is, an abnormally low metabolic rate in all tissues of the body.

CONTROL BY CHANGING COMPOSITION OF THE BLOOD. Examples of endocrine glands regulated by "chemical control" are the pancreas, controlled by the level of glucose in the blood, and the parathyroid glands, controlled by the level of calcium in the blood. Let us consider the control of insulin secretion by the pancreas. Increased amounts of sugar in the blood supplying the pancreas cause a release of insulin, which stimulates the conversion of blood sugar to glycogen and to fat. This release tends to lower blood sugar to normal levels. We can say, then, that the normal control of the release of insulin from the pancreas is dependent upon the level of sugar in the blood supplying this organ, since any increase will stimulate the secretion of insulin.

There are several other endocrine organs in the body—the adrenal gland, the anterior pituitary, and the pancreas itself—that secrete hormones tending to increase or spare the level of blood glucose, by the mobilization of glycogen to blood glucose in the liver and or by utilization of body fat and protein for energy purposes. There seems to be delicate balance in adjustment of proper blood glucose that is not readily apparent unless one of the antagonistic components becomes faulty. Certain conditions of pancreatic insufficiency result in diabetes mellitus. In this particular disease the functioning of the pancreas is faulty and it does not respond to high levels of blood glucose by the release of insulin. The antagonistic hormones that tend to increase blood glucose are therefore not opposed, and blood glucose reaches levels two to three times that found in the normal individual. The amount of glucose in the blood reaching the kidney is far above the amount that the kidney can resorb. This leads to the pronounced symptoms of diabetes mellitus—a tremendous increase in the volume of urine and a high amount of sugar in the urine. That the major cause of this disease lies with the failure of the pancreas is indicated by the complete reversal of blood glucose levels and the disappearance of glucose from the urine following the injection of insulin. The regulation of the diet, plus adequate insulin therapy in the diabetic, results in normal metabolism and in the conversion of glucose to its storage compounds, glycogen and fat. With this method of therapy, however, extreme difficulties can be encountered with improper use of the insulin. Hypoglycemic shock can result in a diabetic who is either on normal levels of insulin with increased muscular activity or on too high levels of insulin alone. In either case the level of blood glucose is reduced below that level required for normal nervous activity, the end result being unconsciousness and death if glucose is not administered.

Hormone Action and Interaction. Although the exact mechanism of hormone action is not known, there are in general two types of body changes that follow endocrine gland activity or treatment with hormones. One change is the obvious increase or decrease in a specific type of metabolism within the body. Examples of such responses are the general increase in nitrogen retention in body cells, together with a decreased nitrogen loss in the urine, following treatment with growth hormone; the rapid conversion of body fat and protein to energy sources, expressed as an increase in nonprotein nitrogen and glucose in the blood, following treatment with adrenal hormones; the active resorption of calcium from the bones after treatment with parathyroid hormone. These changes are accomplished without any obvious increase in the functional sizes of the target tissues stimulated. The other type of change stimulated by hormone secretion or treatment is the increase in the number and size of the cells of the target tissues, a compensation that enables them to increase functional capacity. The action of the tropic hormones of the anterior pituitary (ACTH, TSH, FSH, LH, and LTH) stimulates this type of response. Adrenocorticotropin injection results in an increase in size of the adrenal gland and consequent increased production of adrenal hormones. The size and functional capacity of this endocrine organ are dependent upon ACTH, just as the thyroid gland is dependent upon TSH. Removal of the anterior pituitary gland (the source of the tropic

NEURAL-HORMONAL CONTROL OF LACTATION

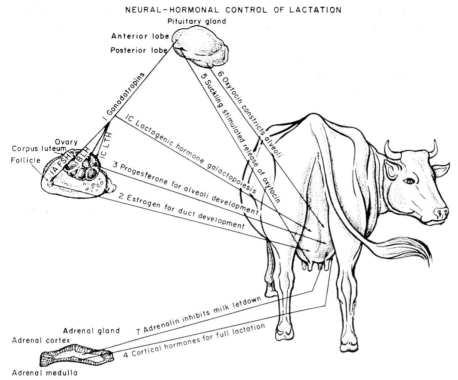

Fig. 20-7. Neural-hormonal control of lactation. (1) Pituitary gonado-tropins: (*A*) FSH—develops follicles; (*B*) LH—final maturation and ovula-tion of follicles and establishment of the corpus luteum; (*C*) LTH—luteo-tropin activity maintains corpus luteum and lactogenic activity stimulates milk formation in developed mammary gland. (2) Ovarian estrogens de-velops mammary duct system—slightly during normal cycle and markedly during pregnancy. (3) Progesterone develops mammary alveolar system, the milk secreting portion, by midpregnancy. (4) Adrenal glucocorticoid hormones necessary for normal lactation. (5) Suckling act or manipulation of teats, by nervous means, stimulates release of oxytocin from posterior pituitary. (6) Oxytocin, reaching the alveoli through the blood, constricts the smooth muscles surrounding them and lets down the milk. (7) Milk let-down can be inhibited by fright, by rough or poor management fac-tors, or by release of epinephrine.

hormones) is therefore followed by a regression in size and functional ca-pacity of the adrenal and thyroid glands.

It is also possible in normal endocrine control to have one or more hor-mones stimulating the development of a tissue and other hormones maintaining the development and activity of the same tissue. A response of this type is seen in the mammary gland: estrogens and progesterone from the ovaries develop the duct and alveolar systems of the gland; lactogenic hormone, from the anterior pituitary, can then induce the functional activity —that is, the formation of milk within the gland—and yet another hormone, oxytocin, is necessary for the squeezing out or the let-down of milk (Fig. 20-7).

REFERENCES AND SELECTED READINGS

Pace, D. M., and B. W. McCashland, 1960. *College Physiology*, Part 3, Nervous Co-ordination; Part 4, Receptors. Crowell, New York.

Pavlov, I., 1910. *The Work of the Digestive Glands*. Second English Translation by W. H. Thompson. C. Griffin and Co., London.

Scheer, B. T., 1953. Humoral Integration. In *General Physiology*. Wiley, New York, Chap. 19.

Chapter 21

Reproduction

21-1. INTRODUCTION

The ultimate value of any farm animal depends upon its ability to reproduce. Our knowledge of the factors affecting reproductive efficiency is increasing through the accumulation of information on endocrine relationships and the interaction of nutritional and genetic factors on the reproductive process. The entire process is regulated by interlocking hormone systems that synchronize the function of the sex mechanism. The details differ in various species, but the problems in the ewe and in the cow or chicken are basically the same. Eggs must be produced and shed by the female, and a mechanism must be present to ensure the presence of viable sperm. The rapid development of artificial insemination in domestic animals has stimulated much research on reproduction in recent years. Some knowledge of anatomy and physiology as well as endocrinology is important, if we are to take full advantage of advances in this area of husbandry.

21-2. ANATOMY OF THE MALE REPRODUCTIVE ORGANS

In the male the primary organs of reproduction are the paired testes, where the sperm cells are produced. These organs are located in an outpocketing of the abdominal wall known as the scrotum. Other portions of the male reproductive system are the paired accessory glands, the duct system, and the penis (Fig. 21-1).

Testes. The testes develop from the sexually undifferentiated gonads and have a dual function: (1) production of spermatozoa in the seminiferous tubules, and (2) secretion of testosterone, the male sex hormone, by the

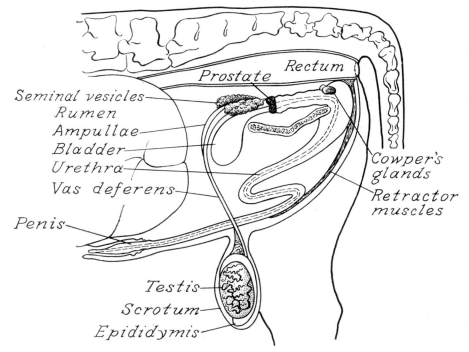

Seminal vesicles
Rumen
Ampullae
Bladder
Urethra
Vas deferens

Penis

Prostate
Rectum

Cowper's glands
Retractor muscles

Testis
Scrotum
Epididymis

Fig. 21-1. The reproductive organs of the bull. [From Salisbury and VanDemark, 1961.]

interstitial cells that lie between the tubules. In adult domestic mammals the testes are located in the scrotum; however, they develop in the dorsal region of the abdominal cavity and migrate into the scrotum near the time of birth. In birds the testes do not descend, but remain in the body cavity near the kidneys. The scrotum is a thermoregulatory organ, and its principal function is to maintain the temperature of the testes below the body temperature. In the ram normal body temperature is about 102°F, but that of the testes averages 94°F. It has been known for a long time that cryptorchids, animals whose testes remain inside the body, are sterile. This condition occurs in all mammalian species and may be caused by hormonal deficiencies or by anatomical obstructions in the inguinal canal that prevent the testes from descending into the scrotum.

Sperm cells are produced in the seminiferous tubules, which make up most of the testicular mass (Figs. 21-2 and 21-3). A mature sperm cell consists of the head, midpiece, and tail. Among mammalian cells, spermatozoa are unique because of their mobility provided by the flagellate tail. The total length of a bovine sperm cell is about 70 microns (.07 mm), with the head being approximately 9 microns in length. The head of the bull, boar, ram,

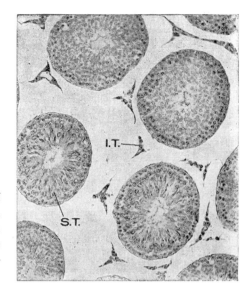

Fig. 21 2. Cross section of testes of rat at 48 days of age just before sexual maturity. Some sperm are evident in the seminiferous tubules (S.T.). The interstitial tissue (I.T.) is still undeveloped. [Courtesy H. H. Cole.]

and stallion sperm cell is oblong and flattened. The head of cock spermatozoa is cylindrical in shape.

Male Accessory Glands. The male accessory sex glands include the paired seminal vesicles, the prostate gland, and the paired bulbo-urethral or Cowper's glands. There is considerable variation in the relative sizes of the glands in different species. In the boar the seminal vesicles are greatly enlarged and

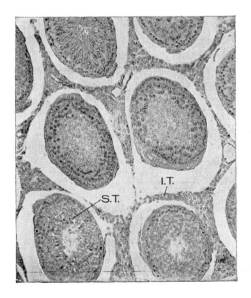

Fig. 21-3. Cross section of testes of rat at 48 days of age that has been treated with pregnant mare serum gonadotropin for about 10 days. Note the enormous development of the interstitial tissue (I.T.) as compared to that seen in Fig. 21-2. This hormone has very little effect, however, on the seminiferous tubules (S.T.). [Courtesy H. H. Cole.]

contribute to the large volume of semen in this species. The male accessory organs contribute the seminal plasma of the ejaculate, which has two main functions: it serves as a suspending and activating medium and supplies an energy source. Semen volume and sperm cell concentration vary according to the relative activity of the accessory organs in various species.

Removal of the testes, castration, effects significant changes in the male: (1) there is permanent sterility, (2) sexual desire or libido is greatly reduced or lacking, and (3) the secondary sexual characteristics or masculinity of body form are less prominent. Sterilization of a male without subsequent loss of libido and secondary sexual characteristics can be effected by removing a section from each vas deferens. In a vasectomized animal the cells that produce testosterone continue to function.

21-3. ANATOMY OF THE FEMALE REPRODUCTIVE SYSTEM

The female reproductive system consists of the ovaries and a duct system (Fig. 21-4). The ovaries are homologous (having similar origin) to the testes in the male, and also have a dual function—that is, production of (1) eggs or ova, and (2) hormones.

The Ovaries. The ovaries are lobulated in structure in the sow and in birds. They are relatively smooth and almond- or bean-shaped in the cow, ewe, and mare. Prominent structures in the ovary are the follicles, which contain the ova and corpora lutea, the solid bodies of tissue originating from the walls of ruptured follicles after ovulation (Figs. 21-5 to 21-8). The presence of a corpus luteum is necessary for initiation of pregnancy in all placental mammals and for successful maintenance of pregnancy in many of them. The diameter of a mature follicle in the cow is 1.8 cm; in the ewe and sow, 1.0 cm; and in the mare, 3.0 cm. The diameter of a fully developed corpus luteum or yellow body is similar to that of the mature follicle for

TABLE 21-1. | *Average semen production and site of deposit in female.*

Male	Volume per Ejaculate (ml)	Sperm Cell Concentration (millions/ml)	Site of Semen Deposit in Female
Boar	250	300	Uterus
Bull	5	800	Anterior vagina
Ram	1	2000	Anterior vagina
Stallion	75	200	Forced into uterus
Cock	1.5	4000	Everted vagina

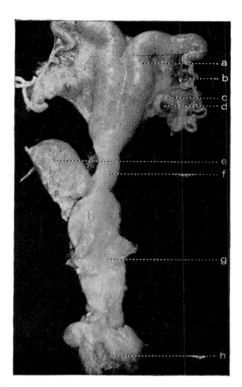

Fig. 21-4. Reproductive tract of the ewe: (*a*) uterine horn, (*b*) oviduct, (*c*) ovary, (*d*) corpus luteum, (*e*) bladder, (*f*) cervix, (*g*) vagina, and (*h*) vulva. (¾ ×).

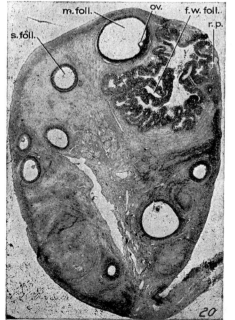

Fig. 21-5. Ovary of ewe on the first day of metestrus. The point of rupture (r.p.) of the definitive follicle is still visible. The wall of the ruptured follicle is folded (f.w.foll.) and most of the follicular fluid has escaped. s.foll.—small follicle; m.foll.—medium sized follicle; ov.—ovum attached to the wall of the medium sized follicle by the discus proligerus. [From Cole, H. H., and R. F. Miller. *Am. J. Anat.* 57: 39–97, 1935.]

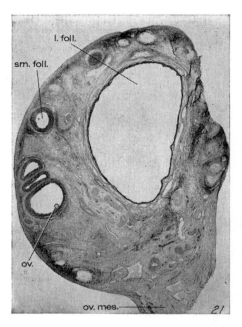

Fig. 21-6. Ovary during anestrus. Note that the ovary contains large follicles (l.foll.) and small follicles (sm. foll.) but corpora lutea are lacking. The ovary is attached to the body wall by means of the ovarian mesentery (ov.mes.). [From Cole, H. H., and R. F. Miller. *Am. J. Anat.* 57: 39–97, 1935.]

the different species. Corpora lutea are not formed in the bird. The left ovary is functional; the right remains rudimentary.

There is very little difference in the size of ova produced by the various placental mammals; the rabbit and the cow are nearly equal in this respect. The diameter of the cow ovum is about 150 microns. Distinguishing features of the ovum are the cytoplasm enclosed by the vitelline membrane, which constitutes the ovum proper, and the surrounding thick, highly re-

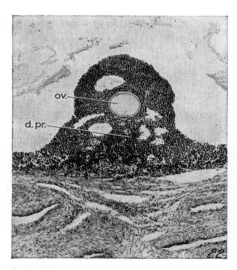

Fig. 21-7. Ovum (ov.) of ewe sacrificed during proestrus. The ovum is firmly attached to the follicle wall by the discus proligerus (d.pr.). The nucleus of the ovum is clearly visible. [From Cole, H. H., and R. F. Miller. *Am. J. Anat.* 57: 39–97, 1935.]

Fig. 21-8. Ovum (ov.) of ewe within ovary, near end of estrus. The fine granular inner portion of the ovum is the cytoplasm, which is surrounded by a band known as the zona pellucida. Outside the zona pellucida, the granulosa cells of the discus proligerus (d. pr.) radiate out. These cells are known as the zona radiata and accompany the ovum into the oviduct when the follicle ruptures. The cells of the discus proligerus are undergoing cytolysis and therefore this ovum is nearly free from the follicle wall.

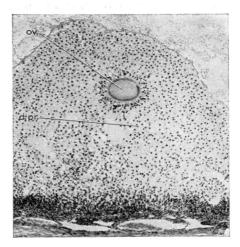

tractile membrane known as the zona pellucida. The presence of sperm cells in the zona pellucida after mating is suggestive that fertilization will occur, and normal cleavage of the ovum proper is strong evidence that fertilization has occurred. To obtain this information, however, ova must be obtained from the reproductive tract, following slaughter, or by operative procedure, and examined microscopically. (See Fig. 21-9.)

The Duct System. The infundibulum is the funnel-shaped end of the oviduct near the ovaries. It picks up the ova and the oviduct transports them

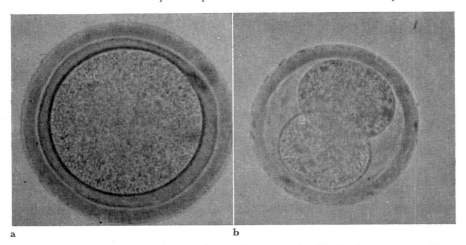

a   b

Fig. 21-9. Photomicrographs of (*a*) an unfertilized sheep ovum, (*b*) a two-celled sheep ovum. Sperm heads can be seen in the zona pellucida of the fertilized ovum (430 ×).

to the uterus. The latter is lined with a thick mucosal layer or endometrium, which contains many glands. The inner surface of the uterus in the cow, ewe, and other ruminants contains approximately 100 slightly projecting circular areas known as caruncles. During pregnancy these provide for the attachment of fetal membranes by interlocking with corresponding areas, known as cotyledons, on the surface of the chorion. Uterine glands are present only in the noncaruncular area. The sow uterus contains no caruncular areas. The uterine lining of the pregnant mare contains raised structures known as endometrial cups. These take no physical part in fetal membrane attachment, but secrete a hormone, equine gonadotropin. The uterus of any mammal also contains a well-developed outer longitudinal layer of muscle and an inner circular layer of muscle, which assist in expelling the young at the termination of pregnancy. The cervix may be considered the neck of the uterus; it effectually closes the external opening and protects the uterine contents (Fig. 21-4). The vagina is the posterior part of the reproductive tract and serves as the female organ of copulation.

The female duct system receives the sperm cells and conveys them to the site of fertilization. The external genitalia are composed of the labia majora or vulva and the clitoris, which is a rudimentary organ located at the ventral junction of the vulva and is homologous to the glans penis of the male. The clitoris is capable of limited erection and the labia, because of an increased flow of blood, becomes turgid during estrus. The female genital tract is suspended from the dorso-lateral wall of the pelvic canal by the broad ligament.

21-4. SEX DIFFERENTIATION AND DEVELOPMENT

Each individual is potentially bisexual; that is, the necessary primordia or rudimentary structures are present for the complete development of either the male or female reproductive system. The gonads and two sets of ducts are present. One set, the Mullerian ducts, will predominate and de-

TABLE 21-2. | *Reproductive traits in farm animals.*

Female	Length Cycle (days)	Duration of Estrus	Time of Ovulation	Length of Gestation (days)	Age at Puberty (months)
Cow	21	14 hr	14 hr after end of estrus	280	6–8
Sow	21	2–3 days	Usually before end of estrus	114	5–7
Ewe	16.5	1–2 days	Near end of estrus	146	Usually first autumn
Mare	22	5–7 days	1–2 days before end of estrus	336	12

Fig. 21-10. A freemartin. The heifer was born twin to a bull. The vulva is modified into a sheathlike structure. Usually the external genitalia of the freemartin appear normal. The gonads of this heifer were located in the inguinal region. [Courtesy Dr. L. E. Casida, University of Wisconsin.]

velop if the embryo is to become a female; another set, the Wolffian or mesonephric ducts, will develop if the embryo is to become a male. Genes for maleness or femaleness influence the differentiation of the gonads and the genital ducts into that of the male or female sex. If a gonad is to become a testis, chordlike masses of cells are formed and eventually become the seminiferous tubules. If the gonad is to become an ovary, these chordlike masses form the medulla of the ovary, and additional primordial germ cells grow into the gonad, where they become differentiated into ova. Accidental interference with this embryological process can result in intersexuality.

The external genitalia and duct system, like the gonads, develop from a stage at which all of the rudimentary anatomic features of both sexes are present. Occasionally, however, developmental accidents happen, and a system that should be rudimentary, differentiates and gains considerable prominence. This may lead to aberrations in the duct system that may be either slight and harmless, or great enough to cause complete sterility, as in the "freemartin" in cattle (Fig. 21-10), described by Lillie (1917).

When cattle produce twins of opposite sex, approximately 11 out of 12 genetic females have abnormal reproductive organs and are sterile: they are called freemartins. Placental membranes of cattle twins often join, resulting in a fusion of the blood vessels and in an intermixing of blood of the two fetuses. If they are of opposite sex, the male hormone has an arresting effect on the ovaries and the female genital tract and causes various portions of the male system to develop in the female.

Owen (1945) discovered, and subsequent studies show, that about 90%

of twins of opposite sex in cattle have identical blood types. This knowledge and the recent finding that varying proportions of the freemartin's white blood cells contain male, XY, chromosomes instead of female, XX, chromosomes may eventually lead to better understanding of sex abnormalities. Typing the blood of twins of opposite sex in cattle at birth can provide useful information in predicting the possible future fertility of a female whose anatomical deviations are not apparent. A mosaic of blood cells is evidence that the blood of the twin fetuses was intermixed.

21-5. SEX RATIOS

In sexual reproduction the mechanism of reduction or meiotic division during formation of the reproductive cells should lead to the formation of equal numbers of male- and female-producing sperm cells. In mammals, the male sex is heterogametic, producing two types of sperm cells. Genetic sex of the young is thus determined by the type of sperm that unites with the egg. In chickens the opposite is true, since the female produces two types of ova. Sex ratios at birth based on large numbers of animals often show slight deviations from equality. A summary of data shows that in horses, sheep, and chickens, there is a slight deficiency of males, while cattle and swine show an excess of males. If the X- and Y-bearing sperm cells are produced in equal numbers, then lack of random fertilization or differential embryo mortality must be responsible for deviations from equality in sexes.

Many unsuccessful attempts have been made to control sex, including electrophoretic, mechanical, and chemical separation of the two types of sperm cells. Wide deviations from a fifty-fifty sex ratio can occur by chance alone; hence, we should view with considerable skepticism any method or theory that purports to control sex determination. Obviously, a solution of the problem of controlling the sex of offspring would have tremendous significance in the field of animal production.

21-6. HORMONES IN REPRODUCTION

Reproduction and lactation in domestic animals are regulated by hormones. The blood carries these chemical messengers to areas of the body—"target" organs—where they produce their effects.

Of the endocrine glands, the pituitary is the most important. In the cow it is about the size of an acorn and is located at the base of the brain. It has two main parts, the anterior and the posterior pituitary. If the pituitary gland is removed from young animals, the ovaries and testes fail to develop to normal size and do not produce eggs or sperm. If the gland is removed from mature animals that are producing eggs or sperm, these functions cease and the ovaries and testes degenerate. The growth of the ovaries and testes and their functions after sexual maturity or puberty are determined by the secretion of the gonadotropic hormones from the pituitary gland and by interaction with hormones produced by the ovary and the testis. Successful pregnancy is also largely a result of hormone activity.

Types of Hormones. Two types of hormones are concerned with reproduction: (1) gonadotropic hormones, which are complex proteins and affect the gonads, and (2) gonadal hormones, which are steroids, chemically, and are secreted by the gonads. Both types of hormones produce marked effects on the reproductive organs and on the behavior of the animal. Gonadotropins stimulate the accessory reproductive organs indirectly by stimulating the secretion of gonadal hormones.

Other hormones may modify the animal's metabolism so that normal reproductive function may be impaired. Of these hormones, thyroxin, secreted by the thyroid gland, has been most extensively studied. Hormones of the adrenal cortex are of vital importance to survival of the animal; however, their specific effect upon reproduction must await further study.

Conclusive evidence is lacking on the nature of the hormonal imbalance involved in the development of cystic ovaries (Fig. 21-11). Deficiency of luteinizing hormone has been suggested as a cause, because injection of pituitary extracts rich in this hormone has a beneficial effect in many instances.

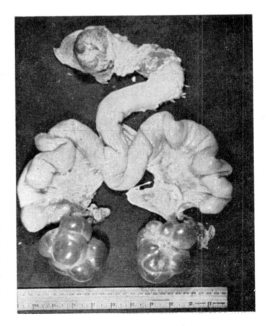

Fig. 21-11. Reproductive tract of sow, showing cystic ovaries. The follicles have failed to ovulate and release their ova. Many of the follicles are enlarged to more than 1 inch in diameter.

TABLE | *Glands and hormones affecting reproductive processes.*
21-3. |

Gland Secreting Hormone	Name of Hormone	Function of Hormone
	Follicle stimulating hormone (FSH)	Initiates follicular growth and influences spermatogenesis.
Anterior pituitary	Luteinizing hormone (LH)	Stimulates growth of interstitial tissue of gonads in both sexes. Synergizes with FSH in follicular growth. Causes ovulation and corpus luteum formation.
	Lactogenic hormone (prolactin)	Maintains functional capacity of the corpus luteum in some species. Stimulates secretion of milk. Causes growth of crop gland in pigeon.
Posterior pituitary	Oxytocin (pitocin)	Stimulates milk let-down and contraction of uterine muscle.
	Estrogen	Induces estrus. Responsible for development of accessory reproductive organs of female.
Ovary (and during pregnancy the placenta)	Progesterone	Inhibits release of gonadotropin(s). Causes growth of uterine glands and alveoli of mammary glands, maintains pregnancy.
	Relaxin	Relaxes public ligaments and dilates cervix. Many of its actions influenced by estrogen and progesterone.
Testes	Testosterone	Stimulates sexual libido. Responsible for development of accessory reproductive organs and of male sex characteristics.

21-7. THE SEXUAL SEASON

In the majority of wild mammals and birds, both sexes have an alternating sexual and nonsexual season. At the time when reproductive function is quiescent, the animal is said to be in anestrus. The females of most breeds of sheep have a distinct sexual season, which in the northern hemisphere occurs in the fall and early winter. The mare also shows a tendency toward seasonal breeding, with short periods of anestrus and irregular cyclic patterns during fall and early winter. Species that exhibit this trait are said to be seasonally polyestrous in their reproductive behavior. Estrous cycles are repeated regularly throughout the year in the sow and in the cow, and they are known as polyestrous or continuous breeders.

The causes of seasonal reproduction are not fully known. There can be no doubt that light plays an important role in determining reproductive periodicity of animals. The pituitary itself has been shown to respond to varying lengths of daylight; in some species breeding activity is initiated by increasing the length of day, and in others, by decreasing it. Nerve impulses resulting from the stimulation of the retina in the eye by light somehow influence the pituitary gland. Nalbandov (1964) has put forth the hypothesis that alteration in the kinds of hormones secreted, rather than secretory activity of the pituitary gland, may be responsible for this ostensibly inconsistent pattern. Changing environmental temperature has also been shown to alter the time of sexual activity in some breeds of sheep. It is important to point out that no single factor can explain the phenomenon of season breeding activity in all species. Light, temperature, food supply, neural stimuli, and other factors may also be responsible.

21-8. THE ESTROUS CYCLE

The estrous cycle consists of growth and maturation of the follicle, ovulation, development, and subsequent regression of the corpus luteum. The cycle may be divided into four phases: (1) proestrus—the interval between cessation of functional activity of the corpus luteum and the onset of estrus, (2) estrus—the period during which the female will accept the male, (3) metestrus—the interval between end of estrus and the time when the corpus luteum becomes secretory, and (4) diestrus—the longest phase of the cycle and the period during which the corpus luteum is functional.

One tentative explanation of the control of the estrous cycle is as follows. When estrogen is produced by the ovary under the stimulus of the follicle-stimulating hormone, FSH, its concentration in the blood increases until it causes the animal to exhibit heat, and, theoretically, it inhibits the secretion of FSH by the pituitary gland. The pituitary then produces the luteinizing hormone, LH, which causes ovulation and growth of the corpus luteum. When the corpus luteum is present and producing progesterone, this hormone affects the pituitary gland and prevents it from secreting LH, thus inhibiting ovulation. Upon regression or cessation of secretory activity of the corpus luteum, the follicles again begin to enlarge, under the synergistic action of FSH and LH, and produce estrogen in sufficient amounts to cause onset of the estrous period. Thus, the cyclic nature of reproductive phenomena depends on a delicate balance in the amounts of hormones that are secreted at various times.

21-9. OVULATION AND FERTILIZATION

After the ovum is discharged from the ovarian follicle, it enters the infundibulum and begins to travel down the oviduct toward the uterus. If sperm cells are present, fertilization will occur in the upper portion of the oviduct. Spermatozoa deposited in the female are carried up the reproductive tract by means of uterine movement, as well as by means of their own motility. Billions of sperm cells may be deposited in the female at the time of breeding, and many may penetrate the outer shell or zona pellucida of the ovum. However, only one spermatozoon unites with the egg nucleus in fertilization.

After ovulation, the ovum is capable of being fertilized for a period of five to ten hours. On the other hand, sperm cells are capable of retaining fertilizability for a day or two in the female tract. Fertilization has been reported in the hen up to 32 days after mating, but the sperm cells of mammals, with the possible exception of bats, do not remain viable in the female tract more than a day or two. The relationship of time of breeding to ovulation is important in view of the short life of the ovulated ovum and sperm cells. It has recently been demonstrated that spermatozoa placed in the cervix of a cow reach the upper end of the oviduct a few minutes later. Trimberger and Davis (1943) reported that cows bred artificially at intervals from onset of estrus to 48 hours after end of estrus were most fertile when bred during mid-estrus. Fertility in cows bred 6 hours or more after the end of estrus was decidedly lower, and none conceived when bred later than 36 hours after the end of estrus.

21-10. PREGNANCY

As the fertilized egg passes down the oviduct, it divides with a resulting increase in cells but no increase in size. By the fourth day the fertilized egg or zygote has entered the uterine horns or the uterus proper. The stage of development during which the zygote is enclosed in the zona pellucida is approximately 10 days in length. After the tenth day the zona pellucida disappears and the membranes of the developing embryo enlarge rapidly. Attachment of the placental membranes to the endometrium of the uterus is a gradual process in all farm animals. Amoroso (1952) has described fusion between the embryonic and maternal tissues as being almost complete by 22 days in the sheep, 24 days in the pig, 28 days in the cow, and only partially complete by 63 days in the mare.

During pregnancy much larger amounts of certain hormones may be produced. These additional amounts of hormones are not formed in the pituitary gland or in the ovaries, but in structures associated with the developing fetus. The blood of mares at 45 to 140 days of pregnancy contains large quantities of a gonadotropic hormone called equine gonadotropin or pregnant mare serum, PMS. The presence of equine gonadotropin in the blood of the mare is the basis for a biological test for pregnancy between the 45th and 140th days after breeding. From the 140th day to term, pregnancy can be diagnosed in this species by the presence of large amounts of estrogen in the urine. Unfortunately no reliable biological tests for pregnancy are available for other domestic animals.

During pregnancy the uterus is closed by a cervical plug, which protects its contents against infection from external sources.

21-11. FETAL MEMBRANES AND PLACENTA

The young zygote exists for a short time by absorbing nutrients contained in the fluids of the uterus, but as the embryo increases in size it is imperative that it establish a more adequate source of nutrition. Extraembryonic membranes develop for this purpose, and they also serve as surrounding membranes that protect the embryo. Fetal membranes are made up of the amnion, allantois, and the chorion. The umbilical cord joins the embryo to the fetal membranes and is composed mainly of the umbilical veins and arteries.

The amnion is the inner, fluid-filled sac that surrounds the embryo, forming a protective cushion against external shocks; the fluid also prevents adhesions between the surface of the embryo and surrounding tissues. At parturition the amnion commonly known as the "water bag," acts as a wedge in opening the cervix. The allantois is an outgrowth of the hindgut. In mammals its principal function is to carry the fetal circulatory system into the chorion to form the fetal placenta. The outer tissue covering the fetus is the chorion or the allantochorionic membrane. Through this membrane gaseous and nutrient exchange between the fetal and the maternal blood system occurs.

Attachment of the fetal placenta to the uterus varies according to species. The mare and sow have a diffused type of placenta, in which the entire chorionic surface contains ridges that fit into corresponding folds in the endometrium of the uterus. The cow and the ewe have a cotyledonary type of placenta. This type is attached to the uterine wall only at localized regions on the surface of the chorion known as cotyledons. The attachment of cotyledonary type of placenta is considered more intimate than the diffused type, since fewer layers of tissues separate the maternal and fetal blood systems (Fig. 21-12).

21-12. PARTURITION

Maintenance of pregnancy is dependent upon the presence of functional corpora lutea—except in the mare, where the corpus regresses at about 150 days of pregnancy. Analysis of the exact mechanisms concerned with parturition shows that they involve complex events that are incompletely understood. However, a change in hormone balance is clearly associated with the onset of parturition. In the ewe, cow, and sow the corpus luteum recedes near the end of gestation, and progesterone secretion is reduced. In the

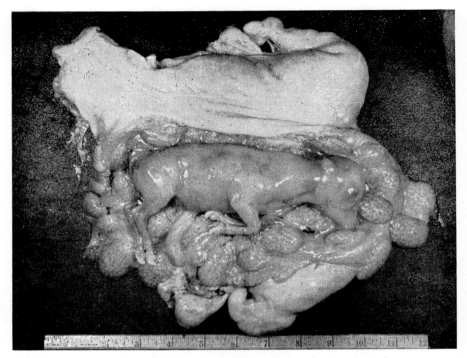

Fig. 21-12. A 100-day calf fetus. The placental membranes have been cut open to show the fetus and the umbilical cord. The areas resembling peanut shells are the cotyledons.

mare multiple corpora lutea regress at about the 250th day of gestation, but the placenta may secrete progesterone in this species. Actually, pregnancy can be prolonged and parturition delayed by injecting progesterone. The hormone relaxin, secreted in some species by the uterus and in others apparently by the ovary, causes relaxation of the cervix and softening of the connective tissues of the pelvic region, which allows expansion to create a passageway for the fetus. With the decline in progesterone the active estrogen in the body increases and, at some point in this sequence, the uterine muscles become sensitive to the action of pituitrin or oxytocin. A series of uterine muscular contractions aided by abdominal muscles finally result in expulsion of the fetus.

A diffuse placenta may pull away from the uterus more rapidly than the cotyledonary type, resulting in cessation of gaseous exchange. For this reason, once the birth act has begun it should be fairly rapid in the mare and in the sow, while it may take somewhat longer in the ewe and in the cow without causing the death of the newborn by suffocation. In the mare the amniotic and chorionic membranes do not fuse, and as a result of this the foal may be born enclosed in the amniotic sac. The presence of an attendant to free the newborn from this membrane so that it does not suffocate is advisable. In the ewe and cow the amnion fuses with the chorion later in pregnancy, and the young in these species are usually born free of any membrane covering.

21-13. FACTORS AFFECTING REPRODUCTIVE EFFICIENCY

Fertility of farm livestock is one of the major problems facing the breeder. Reproductive rate is determined by (1) litter size, and (2) length of the interval from parturition to the next successful conception. The number of young born depends upon (1) number of eggs ovulated, (2) number of eggs fertilized, and (3) number of fertile eggs that implant and survive through gestation to be born as living young.

Fertility cannot be comprehensively considered with respect to one sex without considering the other. In the male, anorchidism (absence of testes) or cryptorchidism may prevent the formation of sperm. Anatomical sterility in the male may also result from deformities or complete absence of part of the reproductive tract. High environmental temperatures result in low fertility or temporary sterility in rams. Exposure to high temperature also has a detrimental effect on fertility and early embryo survival in ewes.

In the female various conditions involving the reproductive tract may interfere with or prevent conception. Cystic follicles or persistent corpora lutea may prevent follicles from maturing and cause temporary sterility. Nutritional deficiencies may delay onset of puberty and retard growth in general and affect ovulation rate in some species. A review of research by Reid 1960) shows that mildly restricted feeding does not have an adverse effect on reproduction in cattle and swine but that feeding animals until they become overly fat is definitely detrimental to their reproductive performance. Wallace (1948) has shown that certain fetal tissues, notably nervous tissue and the skeleton, are less affected than other tissues by undernutrition in the dam. Hammond (1944) has put forth the theory that the fetus can exercise a demand for nutrients on the maternal organism and is not entirely at the mercy of the unfavorable environment to which the dam may be exposed. This whole concept is extremely important in relation to maternal nutritional level and reproductive efficiency.

Pathological disorders may also interfere with fertility through inflammation of various parts of the reproductive system, resulting in blockage of tubes. Disease organisms such as *Brucella abortis, Vibrio fetus, Trichomonas fetus,* and *Leptospira pomona* interfere with gestation and result in abortion of the fetus. Initiation of the estrous cycle is often an indication of early termination of pregnancy, even though an aborted fetus is not evident.

Occasionally cows will ovulate without showing estrous behavior. The cause of these "silent heats" is not clearly understood, but evidence points to incomplete regression of luteal tissue in the ovaries as a possible factor. This condition results in a lengthening of the interval between breedings.

Recently much attention has been given to the hard-to-settle or "repeat-breeder" animal. These individuals require two, three, or more services before they conceive. Normal estrous cycles occur, and a large percentage of such females are apparently free of any abnormality of the tract that would preclude fertilization. The reasons for their inability to conceive are unknown; however, failure of fertilization and high embryonic mortality have been observed in studies involving repeat-breeding sows and cows (Casida, 1953).

For maximum fertility, it is recommended that cattle not be bred earlier than 60 days following parturition, even though they may exhibit heat earlier. A disturbing problem with beef cattle is the excessively long period before the first postpartum estrus, since it becomes difficult to breed for a yearly calf crop if the interval is longer than 90 days. Mares usually have a heat period 5 to 10 days following foaling. Even though some mares will conceive when bred during "foal heat," many breeders feel that it is advantageous to delay breeding until the second heat after foaling. Except for the postpartum heat, which occurs about 3 to 7 days after farrowing, sows fail to exhibit estrus while nursing a litter. Breeding sows at the postpartum or 3-day heat is not successful, because ovulation usually fails to take place.

21-14. ARTIFICIAL INSEMINATION

Artificial insemination is the deposit of male reproductive cells in the female reproductive tract by mechanical means rather than by natural mating. The first research in artificial insemination of domestic animals was conducted by an Italian physiologist in 1780. The primary advantage of artificial insemination is that of speeding up the rate of livestock improvement. This can be accomplished because the number of sperm cells required for insemination is less than that provided in natural mating. Bull semen can be routinely diluted 1:100 without any significant drop in fertility. With an accepted volume of 1 ml of diluted semen per insemination, it is possible to breed 500 cows from one ejaculate. Other advantages that have resulted are continued use of superior sires that are no longer able to serve naturally, long-distance breeding, and the development of techniques for freezing semen so that a sire can be used long after his death.

Techniques of artificial insemination are well established for cattle, but information on problems of spermatozoa storage, dilution rate, and number of sperm necessary for insemination is limited for other species. Limited sexual response among ewes and sows in the absence of a male poses a problem in the use of artificial insemination in these species. The same problem exists, to a lesser degree, among beef cattle, since close observation by a

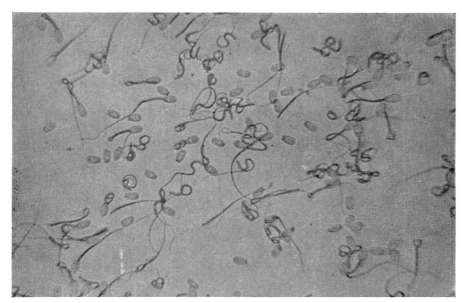

Fig. 21-13. Sperm cells of a ram that show damaging effects of high summer temperature—tailless heads and bent and coiled tails. This ram was temporarily infertile. (430 ×)

herdsman is often not practical and estrus may not be detected. Artificial insemination, combined with techniques of spermatozoa preservation by freezing, may overcome some problems of infertility associated with environment (Fig. 21-13).

21-15. HORMONAL CONTROL OF THE REPRODUCTIVE PROCESS

The phenomenal success of artificial insemination for exploiting the reproductive powers of the bull in the dairy industry has stimulated interest in the possibility of similar exploitation of the female in order to obtain more offspring. Superovulation, the act of ovulating more than the normal number of ova during one heat period, and ova transplantation or inovulation have been suggested as means to this end (Willett, 1953). Since the report on alteration of the estrous cycle in sheep by Dutt and Casida (1948), methods for synchronizing estrous cycles of farm animals are being evolved so that they may be bred artificially in groups on a single day. One such method of synchronization of estrous periods that has been investigated is the injection of appropriate amounts of progesterone to facilitate ova transplantation or mass insemination. Recent development of orally effective progestational compounds may simplify this problem. Injection of gonadotropic hormones has been employed to increase twinning rate in sheep and in those dairy cattle that produce veal or feeder calves. An alteration in light-dark ratio will cause a reversal in the breeding season in sheep. Hormones have also been employed to induce conception in anestrous ewes, but results have not been uniformly successful. Possibly the greatest benefit derived from these studies would be the use of these techniques as a tool for research in the physiology of reproduction and in genetic studies of domestic animals.

REFERENCES AND SELECTED READINGS

References marked with an asterisk are of general interest.

*Amoroso, E. C., 1952. *Marshall's Physiology of Reproduction*, Vol. 2. Ed. A. S. Parkes. Longmans, Green and Co., London.

*Asdell, S. A., 1946. *Patterns of Mammalian Reproduction*. Comstock Pub. Co., Ithaca.

Casida, L. E., 1953. Prenatal death as a factor in the fertility of farm animals. *Iowa State College J. Sci.*, 28:119–126.

*Cole, H. H., and P. T. Cupp (Eds.), 1959. *Reproduction in Domestic Animals*. Vols. I and II. Academic, New York.

*Corner, G. W., 1946. *The Hormones in Human Reproduction*. Princeton University Press.

Dutt, R. H., and L. E. Casida, 1948. Alteration of the estrual cycle in sheep by use of progesterone and its effect upon subsequent ovulation and fertility. *Endocrin.*, 43:208–217.

Hammond, J., 1944. Physiological factors affecting birth weight. *Proc. Nutrition Soc.*, 2:8–14.

*——— (Editor), 1957. *Progress in the Physiology of Farm Animals*. Vols. 2 and 3. Butterworth's Scientific Publications, London.

Lillie, F. R., 1917. The free-martin; a study of the action of sex hormones in the foetal life of cattle. *J. Exp. Zoology*, 23:371–452.

*Marshall, F. H. A., 1952. *Physiology of Reproduction*. 3rd Ed., Vol. 2. Ed. A. S. Parkes. Longmans, Green and Co., London.

*Nalbandov, A. V., 1964. *Reproductive Physiology*. 2nd ed. Freeman, San Francisco.

Owen, R. D., 1945. Immunogenetic consequences of vascular anastomoses between bovine twins. *Science*, 102:400–401.

Reid, J. T., 1960. Effect of energy intake upon reproduction in farm animals. *J. Dairy Sci.* (Suppl.), 43:103–122.

*Robson, J. M., 1947. *Recent Advances in Sex Reproduction Physiology*. Blakiston, Philadelphia.

Salisbury, G. W., and N. L. VanDemark, 1961. *Physiology of Reproduction and Artificial Insemination of Cattle*. W. H. Freeman and Company, San Francisco.

*Sisson, S., and J. D. Grossman, 1938. *The Anatomy of Domestic Animals*. 3rd Ed. Saunders, Philadelphia.

*Snook, Robert, and H. H. Cole, 1964. Endogenous gonadotropic activity in mare serum subsequent to chromic treatment with gonadotropin. *Endocrin.*, 74:52–63.

Trimberger, G. W., and H. P. Davis, 1943. *Conception Rate in Dairy Cattle by Artificial Insemination at Various Stages of Estrus*. Nebr. Agr. Expt. Sta. Research Bull. 129.

Wallace, L. R., 1948. The growth of lambs before and after birth in relation to the level of nutrition. *J. Agr. Sci.*, 38:93, 242, 367.

Willett, E. L., 1953. Egg transfer and superovulation in farm animals. *Iowa State College J. Sci.*, 28:83–100.

Chapter 22

Egg Laying

We, however, commence with the history of the hen's egg . . . for as eggs cost little, and are always to be had, we have an opportunity from them of observing the first clear and unquestionable commencements of generation, how nature proceeds in the process, and with what admirable foresight she governs every part of the work.
WILLIAM HARVEY, *On Animal Generation*

22-1. INTRODUCTION

Success or failure of a species is measured by its ability to preserve itself and undergo some numerical expansion. The success of birds as a class is largely due to the fact that they have evolved physiological mechanisms that cause them to lay eggs at a time of season when such factors as weather and food supply are optimal and when maximal survival of the young can be expected. The numbers of eggs laid and incubated are commensurate with the physical capabilities of the hen for brooding the eggs and caring for the chicks; the species that are most successful are those that tax these capabilities to the utmost and allow for the highest reproductive rate. There are mechanisms that cause the hen to *want* to brood eggs and care for chicks; finally, provisions are made for the return of the ability of hens to lay eggs after a cycle of mothering chores has been completed. Obviously such a complex situation requires a system of signaling mechanisms that tell a hen what time of the year it is, when it has laid enough eggs to start brooding them, and when the chicks no longer need parental protection so that she can again start the cycle of laying eggs. Such a signaling system cannot be simple, and neuroendocrine feedback mechanisms have evolved in both mammals and birds. The nervous system maintains communications among parts of the animal body, and the endocrine system, using hormones as messengers and as local organ representatives, sees to it that this or that part is stimulated to the right degree at the right time. The neuroendocrine control systems have the tasks of synchronizing events within the body and relating the internal events to the external environment in which the population lives. Unfortunately, it will not be possible to treat in detail the many elegant methods that have evolved to permit optimal functioning of animal organisms, but a brief synopsis should encourage the interested student to go deeper into the fascinating problems of reproductive physiology.

The eyes of birds serve as receptors of light intensity. Birds can thus tell the difference between seasons: increasing intensity and duration of light foretell spring, and decreasing light signals the approach of fall and winter. In either event, the excitations resulting are transmitted via the optic nerve to the hypothalamus, one of the major control centers of the neuroendocrine system (Fig. 22-1). From the hypothalamus, chemical substances are released and eventually reach the anterior lobe of the pituitary gland. If the amounts of light are increasing, the anterior lobe responds by increased secretion of the hormones responsible for gonadal growth.

Many interrelated events occur in the reproductive cycle of hens, and there are many different messages received by and transmitted from the hypothalamus. For instance, in the hen's wild ancestor, which laid about twelve eggs in a nest, the pressure of the eggs against the breast initiates a message to the effect that about as many eggs have been laid as a hen can hatch—that is, cover with her body. These messages, when properly decoded and relayed, cause the pituitary gland to switch from secretion of gonad-stimulating hormones to the secretion of the hormone prolactin, which is responsible for the manifestation of the maternal instinct. The flow of prolactin is at first maintained by the continued tactile messages caused

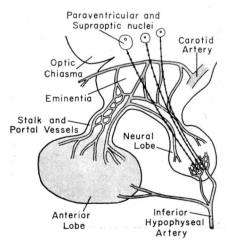

Fig. 22-1. The hypothalamo-hypophyseal system in birds. Note that the neural lobe is separate from the anterior lobe, that both lobes are connected to the hypothalamic nuclei by blood vessels, but that only the posterior lobe is innervated.

by the pressure of eggs against the breast, and later by the presence of the hatched chicks. As the chicks become more independent, the chirping and crowding around the mother for shelter diminishes, and the signal for continued secretion of prolactin weakens. The hen's pituitary gland may now go back to secreting gonad-stimulating hormones and the laying of another set of eggs, *provided* messages from the outside still tell her that enough time remains to raise a second brood; that is, provided there has been no marked decrease in amount of light. If the brood becomes self-sufficient at a time when the amount of light that impinges on the eye is decreasing, the pituitary gland is not sufficiently stimulated to cause secretion of adequate amounts of gonadotropic hormone; the ovary remains regressed until the following spring, when the whole cycle begins again.

The only connection between the hypothalamus and the anterior lobe of the pituitary gland is humoral, and the connection between the hypothalamus and the posterior lobe of the pituitary gland is neural (Fig. 22-1). Thus, the nerves connecting the shell gland (or uterus) to the hypothalamus may carry the message that the shell gland contains a hard-shelled egg that is ready to be expelled. This information is transmitted directly via the connecting nervous system from the hypothalamus to the posterior lobe, which instantly responds by releasing appropriate amounts of the hormone oxytocin. This hormone causes contraction of the muscles in the shell gland, which causes expulsion of the egg.

In a very general and sketchy way, this short discussion should introduce the kinds of problems we will examine in the hope of understanding the physiology of egg production. We must, of course, keep in mind that the domesticated hen is a development from its ancestor, the Indian Jungle Fowl, which was modified extensively by man before it achieved its present commercial usefulness. In the concluding section of this chapter we shall return to the problem of how this hen—whose ancestors laid at most 50 eggs a year—became an organism capable of producing as many as 366 eggs in 365 days.

22-2. THE OVARY AND THE LAYING CYCLE

Components of the Egg. The prime aim in life of mammals and birds alike is procreation, but the reproductive problems faced by birds are completely different from the problems faced by mammals. In the mammal, one or more young not only begin their existence in the uterus, but must continue to develop and be nourished in it. Thus the mammalian egg can be small, since the developing fetus can obtain all of its nourishment from the maternal organism, to which it becomes attached by means of a placenta. The avian embryo develops completely outside of the maternal organism, and for this reason the bird egg must be sufficiently large to contain all the nutrients that will be needed by the growing embryo throughout its development. Avian eggs also must be enclosed by a shell rigid enough to withstand the vicissitudes of the incubation period. The nutrients contained in the egg are the lipoproteins of the yolk, the protein of the egg white (albumen), and the calcium carbonate of the egg shell. All these substances come from the maternal organism, and birds have mechanisms by means of which the different building materials are made available to the proper organ at the proper time (Table 22-1).

The chemical components of the yolk, the albumen, and the shell are mobilized either directly from the gut or from maternal body reserves such as the skeleton, the liver, or the fat depots. These components must be transformed into chemical compounds that can readily pass through cell membranes and that can be transported through the blood stream. After the precursors of yolk and albumen appear in the blood stream, how are the lipoprotein globules suspended in the blood stream consistently directed to the growing follicles where they will be transformed into yolk and deposited into the ovum, and why are the precursors of egg albumen selectively removed from the blood stream in the oviduct and nowhere else? Similarly, how is the calcium carbonate filtered out by the shell gland and deposited around the finished egg? These questions remain, for the time being, without definite answers. Endocrinologists call the ovary, the oviduct, and all other hormone-supported structures, "genetically conditioned end organs," but this term is meaningless since it explains neither the physiology nor the biochemistry of these processes. At best, it simply calls attention to the fact that the ovary has the built-in ability to distinguish between yolk and calcium carbonate, to accept and to use the yolk and to reject the calcium carbonate. In the same mysterious way the ovary can differentiate among the multitude of hormones that are carried by the blood stream. It responds only to certain hormones, and remains completely unaffected by all the others that bathe it.

TABLE 22-1. *Comparison of average length of parts of the chicken oviduct, their contribution to the egg, and the time spent by egg in each section.*

| Part of Oviduct | Average Length of Part (cm) | Contribution | | | Time Spent by Egg in Each Part (hours) |
		Kind	Total Amount (gm)	Percent Solids	
Infundibulum	11.0	Chalaza	32.9	12.2	$\frac{1}{4}$
Magnum	33.6	Albumen			3
Isthmus	10.6	Shell membrane	0.3	80.0	$1\frac{1}{4}$
Shell gland	10.1	Calciferous shell	6.1	98.4	18–22
Vagina	6.9	Mucus	0.1		$\frac{1}{60}$

Source: Nalbandov, A. V., 1958. In *Comparative Endocrinology*, Wiley, New York.

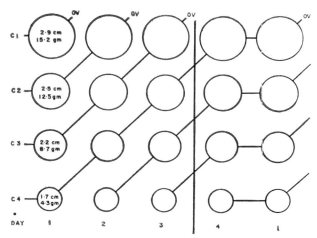

Fig. 22-2. Diagrammatic presentation of a hen's follicular hierarchy and clutch sequence. As the largest follicle leaves the hierarchy by ovulation, lower members move up one size notch. At the vertical line the clutch ends and the C4 follicle of the last clutch becomes the C1 follicle of the next one. [From Nalbandov, in *Comparative Endocrinology*, A. Gorbman, ed., Wiley, New York, 1959.]

Processes of Egg Formation. Near the time of ovulation, the ovary of a mammal contains one or two ripe follicles if the female is monotocous (producing a single offspring at one time), or several follicles of ovulatory size if the female is polytocous. In contrast, the ovary of the laying hen consists of a series of follicles (called the follicular size hierarchy), ranging from microscopic size to the size of the follicle that is destined to ovulate next. This follicle weighs about 15 g, the next member of the size hierarchy about 14 g, and the third about 10 g. Below this size there may be several follicles of the same weights, the different weight classes ranging down to follicles weighing less than 1 mg, and eventually down to a multitude of follicles of microscopic size (Fig. 22-2).

The interesting and significant fact about this size hierarchy is that only one follicle reaches ovulatory size each day. Only after ovulation of the largest follicle do the smaller follicles move up one notch in size and reestablish the hierarchy as it existed just before ovulation (Fig. 22-2). The interval between the ovulation of two successive eggs is usually 24 to 28 hours. As a rule hens lay one egg a day on several successive days before there is a day on which no egg is laid. The uninterrupted series of successive eggs laid is called a clutch. The tendency of some birds to lay short clutches (1 to 3 eggs) and others to lay longer clutches (6 and up to 100 or more) is a heritable characteristic, and each hen tends to repeat her typical clutch length throughout most of her productive life. The hens with the longest clutches will of course be able to produce the largest number of eggs each laying year, because they have the fewest number of nonproductive days; the hens with 1- or 2-egg clutches cannot produce more than 180 or 240 eggs a year, even if there is no time off for molt or winter pause. In order for the hen to be able to produce 300 or more eggs annually, the clutch length must be 5 or more eggs.

The laying cycle of the hen is, to some degree at least, related to light. Under normal day-night conditions, ovulations always occur early in the morning. The yolk then begins its trip down the oviduct, where it acquires the egg white, the soft shell membranes, and eventually the calciferous shell. This whole process from ovulation to oviposition (laying the egg) takes an average of 26 hours. Because in the majority of hens ovulations are held in abeyance until the previous egg has been laid, each subsequent ovulation and oviposition occurs a little later in the day than did the previous one, so that eventually the next scheduled ovulation would have to occur late in the afternoon. For unknown reasons, however, this ovulation does not take place and there is a break in the clutch of 24 to 36 hours. Hens with a very short interval between oviposition and ovulation lay longer clutches, while hens in which the interval between oviposition and ovulation is long —2 or more hours—tend to have short clutches and a lower total annual egg production.

These facts raise two very important and interesting problems. One is the problem of the mechanisms involved in establishing and maintaining the follicular size hierarchy for prolonged periods of time. The hen must be able to distribute the hypophyseal hormones circulating in the blood stream in such a way that some follicles get a larger quantity of hormones, so that they grow faster and attain larger size, and other follicles receive a smaller quantity. The net result of this rationing system is that there is established and maintained a follicular size hierarchy in which the position of the individual follicle is determined by the amount of hormone stimulating it. The second problem concerns the possible mechanisms involved in the timing of the intervals between ovulations.

The following information is necessary before we can try to provide answers to these problems.

22-3. ENDOCRINE CONTROL OF FOLLICULAR GROWTH AND OF OVULATION

Hormones Involved in Follicle Growth. The endocrine control of follicular growth and of ovulation is by no means simple and is not yet completely understood. Much of what will be said in this section can be documented by data, but such proofs are beyond the scope of this discussion and the interested student should consult recent symposia and textbooks if he wishes to separate scientific facts from educated guesses.

The neuroendocrine mechanisms governing reproduction in birds are very similar to those of mammals. In birds, the anterior and posterior lobes of the pituitary gland are two anatomically distinct structures that are separated by a septum. Each of these lobes is connected by a separate stalk to the hypothalamus. Through the stalk of the posterior or neural lobe run nerve fibers connecting this lobe to the hypothalamic nuclei. The anterior lobe is not innervated and its only connection to the hypothalamic nuclei is through the vascular system, in which the blood flows only in one direction, from the hypothalamus to the anterior lobe of the pituitary gland. The posterior lobe serves as a storage reservoir for the posterior pituitary hormones (vasotocin and oxytocin), which are formed in the area of the hypothalamus and pass into the neural lobe along the nerve fibers connecting it to the hypothalamus.

The hormones of the anterior lobe are probably formed directly in the anterior lobe and released from it into the peripheral circulation. Although the anterior lobe secretes five or six different hormones—the somatotropic, adrenotropic, thyrotropic, gonadotropic, and lactogenic hormones—we will be concerned here mainly with the gonadotropic complex. Frequently this is subdivided into the follicle-stimulating hormone, FSH, associated predominantly with follicular growth, and the luteinizing hormone, LH, mainly responsible for causing ovulation. For the sake of simplicity we shall use the term "gonadotropic complex," GTC, throughout this discussion.

Proofs of Hormonal Control of Follicular Growth. The gonadotropic complex from the pituitary gland is responsible for the growth and maturation of the ovarian follicle. That the ovary and its follicles depend upon the gonadotropic hormone is demonstrated by the fact that removal of the pituitary gland (hypophysectomy) leads to a rapid and complete degeneration of the ovarian follicles. This process of follicular degeneration is called atresia and is illustrated in Fig. 22-3. Conversely, one can prevent follicular atresia in hypophysectomized hens by the injection of gonadotropic hormones (Fig. 22-4). Experiments of this type allow us to conclude that the rates at which hens lay eggs depend on the amount of the gonadotropic complex secreted by the pituitary gland.

The vascular networks covering the larger follicles (Fig. 22-5) are much more extensive and intricate than the networks supplying the smaller follicles. Ovulation of the largest follicle of the hierarchy results in an abrupt shutting down of the very extensive vascular network supplying that follicle. Thus, following ovulation, the amount of blood flowing through the vascular networks of smaller follicles of the hierarchy increases and more hormone becomes available for the lesser members of the hierarchy. How some

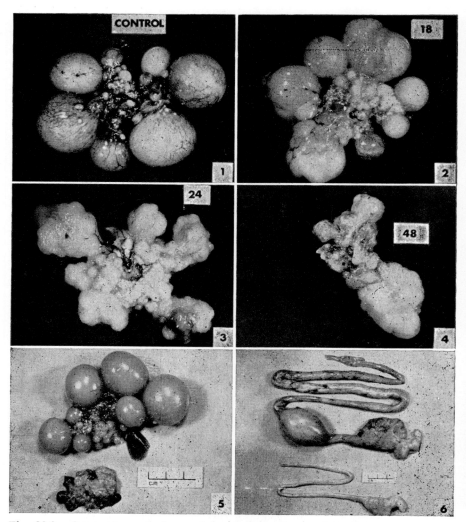

Fig. 22-3. Comparison of sizes and conditions of normal ovary (1 and 5) under the effects of hypophysectomy. Note the rapidity with which the ovary degenerates 18, 24, and 48 hours after hypophysectomy. In (5) and (6) the ovaries and oviducts of normal and hypophysectomized chickens are shown 6 days after hypophysectomy. [From Opel and Nalbandov, *Proc. Soc. Exptl. Biol. and Med.* **107**:233, 1961.]

follicles gain more than their mates is not quite clear, but it is reasonable to think that this is a simple matter of chance and may depend on the initial proximity of some of the follicles to a larger blood vessel, allowing them to grow faster than some of their neighbors.

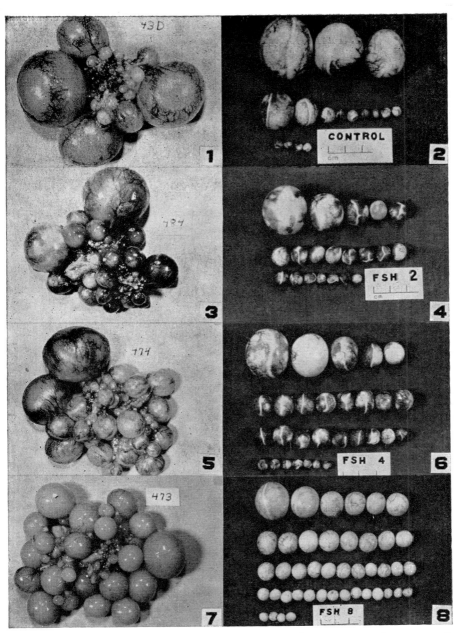

Fig. 22-4. Follicular atresia caused by hypophysectomy can be prevented. (1) Shows the ovary of a normal hen. In (2), the size gradation of follicles (hierarchy) is illustrated. Attempts to substitute for the hen's own pituitary by injection of mammalian GTH (3 to 8) are only partly successful; atresia is prevented, total number of follicles is increased, but the hierarchical distribution of follicular sizes is different (compare 1 with 3, 5, and 7, and 2 with 4, 6, and 8). [From Opel and Nalbandov, *Endocrinol.* **69**:1016, 1961.]

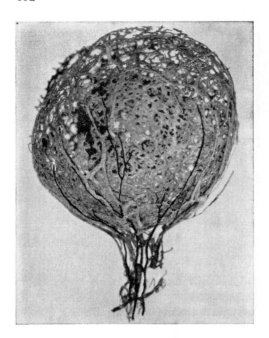

Fig. 22-5. Resin cast of a follicle of near ovulatory size, showing the intricate venous system. The arterial system is shown in black. [From Nalbandov, 1964.]

Hormonal Control of Ovulation. Next we can examine the question of how ovulation is triggered and how the intervals between ovulation are kept at about 24 to 28 hours. We already know that the normal laying hen ovulates only one follicle each day, and that without hormonal support the follicles degenerate. We can study the mechanisms involved in follicular maturation and in ovulation either by injecting gonadotropic hormones, GTH, into normally laying hens, or by hypophysectomizing hens and then replacing the missing hormones by injection in an attempt to initiate the normal sequences of events involved in follicular maturation and in ovulation. If GTH is injected subcutaneously for several days into normally laying hens, the normal follicular size hierarchy is abolished; four or even ten or twelve follicles can be caused to reach ovulatory size simultaneously. (This fact is an excellent argument in favor of the assumption that the total amount of hormone normally available for the whole ovary as well as for the individual follicle is limited. When the ovary and the individual follicles are flooded with exogenous hormone—introduced artificially from the outside—one can increase the number of follicles reaching ovulatory size in accordance with the amount of hormone injected.) If hens pretreated with GTH are given an intravenous injection of this hormone, there will be ovulation of as many follicles as reached ovulatory size under the pretreatment. In laying hens with a normal follicular hierarchy, an intravenous injection of GTH usually hastens the ovulation of the largest follicle, which normally would have ruptured a few hours later; but it is never possible to cause the ovulation of the second- or third-largest follicles of the hierarchy. This demonstrates that follicles that have been nurtured to ovulatory size by proper endogenous or exogenous GTH can be made to ovulate if the organism is flooded by GTH at the proper time. This can be done experimentally by the intravenous injection of GTH. This suggests that in the normal laying hen the anterior pituitary releases increased quantities of GTH at certain times during the maturation of the largest follicle, and that this causes ovulation of this follicle.

It has been established that the interval between release of endogenous GTH and ovulation (or the injection of exogenous GTH and ovulation) is between 8 and 12 hours. During this interval the hormone causes certain changes in the wall of the largest follicle, permitting it to break open and to release the ovum. What these changes are is not known, but a possible answer may be provided by the data presented in Table 22-2. These data were obtained from laying hens, which were hypophysectomized and given a single injection of GTH at different intervals after the operation. It was found that when the GTH was injected fewer than 3 hours after hypophysectomy, the hens ovulated only 1 ovum; but when the GTH injection occurred 3 or more hours after the operation, an increasing number of hens ovulated 2 or more ova. In other words, the longer the ovary remains without support of GTH from its own pituitary gland, which was removed, the easier it is to cause multiple ovulations by exogenous GTH. By comparing (Table 22-2) the weights of the yolks, whose ovulation was induced by hormone injection (I_1 and I_2), with the weights of ova released prior to hypophysectomy (C_1 and C_2), it is found that, in the absence of the hen's own pituitary gland, not only the largest follicles, but also immature members of the follicular hierarchy, can be caused to ovulate. This is in distinct contrast to normal intact hens, in which injection of GTH can only hasten the ovulation of the largest follicle but can never cause the rupture of the second or third members of the follicular hierarchy. (On the basis of this in-

TABLE 22-2. | *Effect of increased interval between hypophysectomy and LH injection on ovulability of follicles.*

Interval (hours)	No. of Hens	No. of Single Ovulations	No. of Double Ovulations	Average Weight of Ova (gm)			
				C_1	C_2	I_1	I_2
0	8	8	0	17.8	17.8	16.9	
2	8	8	0	18.4	17.7	16.3	
3	8	6	2	17.0	16.3	14.9	7.7
4	8	6	2	18.4	17.4	15.0	11.1
6	8	1	7	17.6	17.0	15.4	13.7
12	8	2	5	17.5	17.2	14.9	10.2

Source: Opel and Nalbandov, Endocrinology 69:1029 (1961).

formation, try to formulate a theory concerning the mechanism of ovulation before reading the explanation given below.)

These data suggest that the changes preceding follicular rupture occur more rapidly in the absence of the pituitary gland; that is, in the absence of follicular stimulation by GTH. Thus, ovulation may be normally an aging process, which is initiated by the absence of GTH of hypophyseal origin. This would mean that as long as a follicle is supported by some GTH it is incapable of ovulating. Therefore, in the normal hen, so long as the blood stream carries adequate amounts of GTH, the smaller follicles are stimulated to grow actively and are incapable of ovulating. The largest follicle approaching ovulatory size has the most extensive vascular system and receives the largest amount of hormone. As this follicle reaches its maximum size, the pressure of the accumulating yolk inside the follicular membranes partially squeezes shut some of the blood vessels supplying the follicle walls. This leads to diminishing blood flow, hence to diminishing hormone stimulation, hence to a breaking down of parts of the follicle wall, especially the stigma. All of these events finally culminate in ovulation. The smaller follicles of the hierarchy do not ovulate because their vascular systems carry enough GTH not only to maintain but to stimulate the follicle wall to further growth. So long as they receive such stimuli, the follicle wall cannot break down (or "age" or become ischemic) and ovulation cannot occur.

There remains the question of the mechanism involved in the timing of intervals between ovulation. We can present a plausible theory for which there is good evidence, but much additional work remains to be done before we can be certain of all the factors involved in the timing mechanism.

In the great majority of normal ovulations, the largest follicle ruptures within a few minutes, or at most a few hours, *after* the previous egg has been laid. We have learned earlier that the interval between release of ovulatory doses of hypophyseal GTH and ovulation is about 8 to 12 hours. The time sequences in Tables 22-1 and 22-2 show that this GTH release occurs during the time when the egg is in the process of acquiring the hard shell in the uterus.

There is experimental evidence to show that the oviducal nervous system participates in signaling instructions to the pituitary gland, and that instructions *not* to release amounts of GTH adequate for ovulation may originate in the oviduct. We already know that such signaling systems are frequently involved in phenomena controlled by hormones and that they are part of the neuroendocrine feedback mechanism. In the hen it seems to operate, in brief, as follows. If a yolk is present in the albumen-secreting portion of the oviduct, a nerve-conducted signal goes from the oviduct to the hypothalamus, and from there to the pituitary gland, informing the latter that there is already one ovum in the oviduct and that no new ovulations should be permitted (that is, no ovulatory doses of GTH should be released) until the oviduct is ready for the next egg. Soon after the egg leaves the oviduct and enters the uterus or shell gland, these signals stop and the pituitary gland now releases the amounts of GTH needed to accomplish ovulation some 8 to 12 hours later; that is, after the hard-shelled egg is laid. As soon as the next ovum enters the oviduct, the ovulation-blocking signal again becomes effective and holds further ovulations in abeyance.

22-4. FUNCTION OF THE OVIDUCT

The oviduct consists of five anatomically distinct portions, four of which are histologically different. The vital statistics of these areas, as well as their function, are summarized in Table 22-1. It is amazing that the magnum is able to contribute 32.9 g of a proteinaceous substance to the egg in only 3 hours. Even though the albumen is mostly water, the task of filtering this amount of albumen from the blood stream into the glands of the magnum, and from there into its lumen through which the yolk is passing, seems truly phenomenal. It is interesting that the magnum is incapable of distinguishing between a yolk and any other foreign body. Thus, it will deposit albumen around a cork or a wadded-up piece of paper, or even around a cockroach, as occurred in one experiment when one somehow found its way into the magnum. The size of the laid egg is largely determined by the size of the ovum passing down the oviduct. Relatively less albumen will be deposited around a small yolk (as in the case of pullet eggs), and the resulting finished egg will weigh considerably less than the egg laid by a mature hen, which tends to ovulate larger yolks. There is a slight tendency for successive eggs of a given clutch sequence to become smaller, because the yolks ovulated late in the clutch sequence are likely to be smaller than those ovulated earlier. This relationship holds statistically true for populations of birds, but individual hens may lay progressively larger eggs in a clutch.

The physiological function of the oviduct is controlled by the ovarian steroid hormones. The ovary normally secretes estrogen, androgen, and perhaps progesterone. Estrogen alone is able to cause the morphological enlargement of the oviduct, which is tremendously sensitive to estrogen. Even small doses of exogenous estrogenic hormone are capable of enlarging the oviduct of immature chicks several hundred percent. This hormone alone, however, is incapable of stimulating the development of the oviducal glandular apparatus. To develop the glands, and to cause them to secrete albumen, the shell membranes, or the calcareous shell, requires the interaction of at least two of the ovarian hormones. One of these is certainly estrogen, but it remains unclear whether the second cooperating hormone is androgen or progesterone. The ovary is known to secrete the male sex hormone, androgen, which is responsible, among other things, for comb size and its bright red color in laying hens. Because both androgen and estrogen are normally secreted by avian ovaries, it appears plausible that the interaction of these two hormones is normally responsible for the size of the oviduct, its glandular development, and its ability to secrete the various components of the completed egg. Under experimental conditions, it is possible to substitute progesterone for androgen and to obtain oviduct growth and albumen secretion even in sexually very immature female chicks, from either the estrogen-androgen or the estrogen-progesterone combination.

Having acquired the layers of albumen in the magnum, the egg is enclosed in the soft-shell membranes in the isthmus, and the calcareous shell is deposited around it in the shell gland or uterus. It is possible to cause the premature expulsion of the partially calcified egg from the uterus by the injection of the posterior pituitary hormone, oxytocin, which causes contractions of the musculature of the shell gland. It seems probable that this hormone is normally responsible for the expulsion of the finished egg and that uterine nerves are involved in signaling the posterior lobe of the pituitary that the egg is fully formed and can be laid. It remains unknown whether it is the thickness of the hard shell or the time spent by the egg in the uterus that is responsible for the initiation of this neural signal. There is a striking similarity between the process of oviposition in birds and parturition in mammals: essentially the same hormones and the same mechanisms (uterine musculature) are involved in the expulsion of the fetus. In neither case is it known exactly how the signal for release of oxytocin is timed.

22-5. CONCLUDING REMARKS

The ancestor of the domestic hen laid from 25 to 50 eggs a season, hatched them into chicks, and took care of them as long as they needed protection. The enormous plasticity of the species, which could be converted by domestication from such primitive beginnings into an efficient egg-producing machine capable of producing up to 366 eggs a year, is astonishing. The physiologic principles underlying this transformation have been discussed earlier and need only be summarized. (It is suggested that the student attempt to make such a summary before reading what is to follow.)

When the chicken was taken into the household of early man, it was noticed that by removing the eggs from the nest of the hen, the onset of broodiness could be postponed and more prolonged periods of egg laying could be obtained. For a long time, this was probably the only way in which an increase in productivity was obtained, but it is unlikely that the productivity of the individual hen could have been pushed much beyond 100 or, at most, 150 eggs per year. It is only within the present century, and especially since the advent of incubators, that the maternal instinct has become an unnecessary and even undesirable attribute of hens, because hens do not lay during the time of incubating eggs and caring for chicks. It was noted that those hens that were better mothers, and that spent more time on being broody, consequently laid fewer eggs. By giving the hens with a poorly developed mothering instinct a greater opportunity for reproduction, it became possible to increase productivity of the population by gradually eliminating the broody instinct. We know now that by eliminating the broody individuals, the early breeder was actually eliminating those hens and strains of hens that had genes capable of causing the pituitary gland to secrete copious quantities of prolactin.

The ability of hens to lay the largest number of eggs is due to the presence or absence of at least four genetically controlled characteristics. High-producing strains must be free from broodiness, they must have a short winter molt, they must be persistent (that is, able to lay throughout most of the year), and finally, they must have the ability to lay long clutches. The intensity of the expression of these characteristics is determined by the rates of production of certain hormones, which are more or less prominently concerned in determining the degree to which these characteristics are manifested. The early breeders, who were interested in improving domestic chickens, unknowingly selected individuals whose pituitary glands secreted less prolactin, more gonadotropic hormone, and reduced amounts of thyroxine, known to control rate and intensity of molt. Like most other quantitative characteristics, such as milk production or growth rate, high egg production requires the complex interaction of a great number of genes, which determine not only the rates at which glands secrete their hormones but also the sensitivity with which end organs respond to the hormonal or neural stimuli acting on them.

REFERENCES AND SELECTED READINGS

Breneman, W. R., 1955. Reproduction in birds: the female. In *Comparative Physiology of Reproduction and the Effects of Sex Hormones in Vertebrates.* Cambridge University Press.

Fraps, R. M., 1955. Egg production and fertility in poultry. In *Progress in the Physiology of Farm Animals.* Vol. 11. Butterworth, London.

Nalbandov, A. V., 1964. *Reproductive Physiology.* 2nd ed., W. H. Freeman and Company, San Francisco.

Chapter 23

Lactation

Nor shall the brood-kine, as of yore, for thee
Brim high the snowy milking pail, but spend
Their udders' fullness on their own sweet young.
                                  VIRGIL, *Georgics,* III

23-1. INTRODUCTION

Lactation, or milk secretion, is the sole physiological function of the mammary gland. From these glands, the young of these vertebrate animals we call mammals derive their first nutrients for the maintenance of life. In order to understand better the process of milk secretion, this chapter will consider first a brief discussion on the anatomy of the mammary gland, including the suspensory system of the udder, the milk-collecting system, teat structure, and the vascular, lymphatic, and nerve supply. The embryological and postnatal development of the mammary gland will be considered. The subject of milk secretion—the neural and hormonal relationships of the initiation and maintenance of lactation, and milk expulsion (milk let-down) from the secretory and storage areas of the mammary gland—will be discussed. Since cows do not secrete milk that is always of the same composition, or that is up to the level of their inherited capacity, the influence of various physiological and environmental factors on milk production will be considered.

The mammary gland is a *skin* gland because it originates from specialized cells (primordia) on the outside layer of the body surface during embryological development. Other skin glands are the sweat glands and the sebaceous glands, the latter being responsible for the oily secretions observed on the body surface. All three types of skin glands are similar in their development.

The skin glands are exocrine in nature; that is, they pour their secretory products to the exterior, in contrast to the endocrine glands (see Chapter 20), which release their products into the blood stream.

23-2. ANATOMY OF THE MAMMARY GLAND

The number of mammary glands and their position on the external body surface is peculiar to the species concerned. In cattle there are 4 glands located in the inguinal region, each having 1 teat or passageway to the body surface. The mare also has 4 glands in the inguinal area, but only 2 teats; each teat contains 2 ducts, one from each gland served by the teat Inguinal mammary glands are also present in sheep and goats. In these, only 2 glands are present, each supplied with its own teat. The mammary glands of humans and elephants are located in the pectoral region. In the bitch, sow, rat, and others, the mammary glands number from 10 or more and are disposed on the ventral body surface, on either side of the midline from the pectoral region to the inguinal region. It is interesting to note that each mammary gland of humans and bitches has several ducts opening on the body surface rather than the single opening in the teats of cows and goats.

Gross Anatomy. It is beyond the scope of this chapter to pursue the individual features of the mammary glands of all the species of mammals. Since the production of milk by dairy cows is a major agricultural enterprise, the mammary system of the cow will serve as the model to be discussed.

The 4 mammary glands of the cow are grouped together as a unit and are referred to collectively as the udder. The udder is divided into halves with 2 glands or quarters present in each half. In high-producing cows, an udder just before milking may weigh as much as 150 lb. If the skin alone was responsible for suspending such a high-producing udder, it would hang far below the abdominal surface of the body. However, usually this does not happen because of the lateral suspensory ligaments and the median suspensory ligament (Fig. 23-1).

SUSPENSORY SYSTEM. When viewed from behind, the halves of the udder usually can be clearly distinguished because of the presence of a slight or marked groove between them, the intermammary groove (Fig. 23-1). The lateral suspensory ligaments, primarily composed of tough, strong, fibrous tissue, extend from the pelvic bones down the right and left sides and rear of the udder, and continue underneath the udder and up into the intermammary groove. The median suspensory ligament, which is firmly attached to the abdominal floor, is found between the two halves and is well supplied with elastic as well as fibrous tissue.

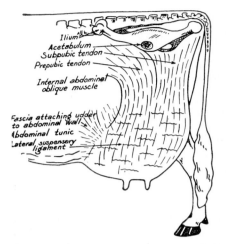
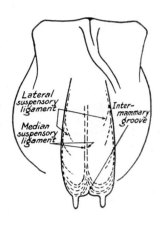

Fig. 23-1.   Suspensory apparatus of the udder. [From V. R. Smith, 1959.]

GLAND STRUCTURE. The four glands of the bovine udder are distinct, separate units. In a cross section of the udder, the boundary between the two
halves is clearly marked by the median suspensory ligament. However, between the glands of each half no clear line of demarcation is present. For
gross examination it is necessary to inject one quarter of each half with a
dye in order to show that the glands are not functionally connected.

Each mammary gland is divided into lobes and each lobe into lobules.
A mammary gland might be compared to a tree, where the entire tree would
be equivalent to a complete mammary gland, the large branches comparable
to the lobes, and the small branches to the lobules.

DUCT SYSTEM. The duct system not only conveys the milk from the secretory units of the udder to the teat cistern, from which it can be removed,
but it also stores part of the milk secreted during the milking interval.

Each secretory unit of the udder is connected to the remainder of its
lobule by small intralobular ducts. The lobules in turn drain their contents
into the large ducts present in each lobe, the "intralobar" ducts. Connecting
all the lobes in each gland are the interlobar ducts, which open into the
large milk-collecting ducts leading to the gland and teat cisterns. The complexity of the milk collecting system can be seen in Fig. 23-2.

THE TEAT. In the cow, each mammary gland is provided with a teat. It
not only provides the means whereby the milk can be withdrawn from each
gland, but it also performs the unique functions of keeping the milk in and
foreign material out. These functions are accomplished by means of a strong
sphincter muscle surrounding the teat opening (external meatus). Also aiding to keep the milk in are folds of tissue (Fürstenberg's rosette) at the top
of the streak canal. These folds are so arranged that whenever pressure is

applied to the teat the folds cover the streak canal like a cap. Figure 23-3 illustrates the main features of the bovine teat.

Vascular System of the Udder

ARTERIES. The synthesis of milk requires that a constant supply of nutrients be supplied to the udder. Most of the recent information concerning the volume of blood necessary to produce a volume of milk has been obtained from experiments conducted on the udders of goats. The noteworthy research work of Linzell (1960b) has demonstrated that, on the average, 650

Fig. 23-2. A plastic cast on an udder showing the milk-collecting system. The quarters of the left half have been injected with vinyl acetate plastic of the same color, whereas the quarters of the right have been injected with different colors. The plastic was permitted to solidify and the mammary tissue dissolved away with concentrated hydrochloric acid. It can be seen clearly that each quarter is a distinct secretary unit. [From Yapp and Nevens, *Dairy Cattle, Selection, Feeding and Management,* Wiley, New York, 1955.]

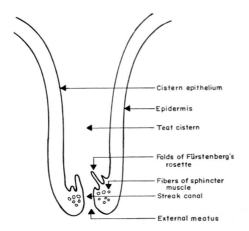

Cistern epithelium

Epidermis

Teat cistern

Folds of Fürstenberg's rosette

Fibers of sphincter muscle

Streak canal

External meatus

Fig. 23-3. A sagittal section of a teat, showing the important anatomical structures.

pounds of blood must pass through the udder for each pound of milk secreted. Early in the lactation period the ratio may be 400 : 1 or less, and when milk secretion declines at the end of the lactation period a ratio of 1000 : 1 or greater can be expected. Efficient udders will have a low ratio of blood to milk, and in inefficient udders the ratio will be high. This tremendous volume of blood reaches the udder primarily through the inguinal ring (an opening through the abdominal wall, just dorsal to the udder, permitting communication between the udder and abdominal cavity) through the external pudic arteries (right and left). A secondary and almost incidental supply reaches the udder through the perineal arteries (right and left). All of these arteries are branches of large arteries leading from the abdominal aorta, the large vessel carrying oxygenated, nutrient-rich, arterial blood from the heart (Fig. 23-4).

VEINS. Actively metabolizing tissue produces a large volume of by-products which, in general, is waste. It is the function of the venous portion of the vascular system to carry these wastes to areas of the body from which they can be discarded. The venous drainage of the udder depends on several factors of which the following are the most important: (1) the position of the animal, that is, whether she is standing or lying; (2) age; (3) the number of offspring. The principle veins carrying blood from the mammary glands are the external pudic veins and the subcutaneous abdominal or milk veins (Fig. 23-5). These large veins contain valves periodically spaced along their length that direct the flow of blood. In the subcutaneous abdominal veins, from a point just posterior to the umbilicus, the valves are disposed so that all blood must flow toward the external pudic veins, which begin at the base of the udder. From a point just anterior to the umbilicus, the valves in the subcutaneous abdominal veins are placed so that the flow of blood is toward the heart in a cephalad (toward the head) direction. In young females, or in virgin females of all ages, these valves are intact and effectively control the flow of blood as described above. When lactation begins, following the birth of the first offspring, the increased activity of the mammary glands creates such a demand for blood that the valves in the subcutaneous abdominal veins between the umbilicus and the udder are overwhelmed and become nonfunctional. Whenever this occurs, the flow of blood in the veins draining the udder is determined by the position of the animal. While it is standing, the blood flows from the udder to the heart, almost completely by way of the subcutaneous abdominal veins; in other words, the pull of gravity overcomes the ability of the valves to direct the blood toward the external pudic veins. On the other hand, when it is in the lying position, blood flow is impeded in one or both of the subcutaneous abdominal veins because of compression caused by the weight of the animal on them; blood then returns to the heart almost exclusively via the external pudic veins (Linzell, 1960a).

LYMPH VESSELS. Lymph is body fluid that has passed from the capillaries out into the tissues. Lymph finds its way back to the heart through lymph vessels but, because it is almost colorless, the lymphatics are difficult to trace. Lymphatic vessels are periodically supplied with nodes that serve as filters for removing foreign materials; they also produce large quantities of lymphocytes, one of the types of white blood corpuscles. Lymph enters the blood stream through a duct in the vena cava in the thoracic region. The chief lymph nodes or glands in the udder, the supramammary lymph nodes, are located at the base of the udder next to the abdominal floor and just posterior to the inguinal canal. There are usually one or more such nodes per mammary gland. Vessels carrying lymph from the supramammary lymph nodes associated with each mammary gland usually converge into one lymph vessel per gland and pass through the inguinal canal, as do the external pudic vessels and certain nerves, to lymph nodes deep in the abdominal cavity.

The rate of mammary gland lymph flow has been measured in sheep, goats, and cattle, but the data are difficult to compare because the stages of lactation and udder weights have not always been given. It is clear, however, that the rate of lymph flow varies with the rate of milk secretion and the mean mammary gland blood flow (Linzell, 1960c). This is to be expected, since lymph comes from blood and during increased mammary gland activity the requirement for blood is elevated. In cows early in the lactation period, the rate of lymph flow from the udder may be as high as 50 liters per day (Lascelles et al., 1964).

Innervation of the Udder. A general outline of the somatic and autonomic nervous systems has been considered in Chapter 20.

MOTOR NERVES. Like all skin glands, the mammary glands receive only motor nerve fibers of sympathetic nervous system origin. They reach the udder primarily by way of the inguinal nerves and the perineal nerves; however, lateral cutaneous nerves, which reach their designated areas after fol-

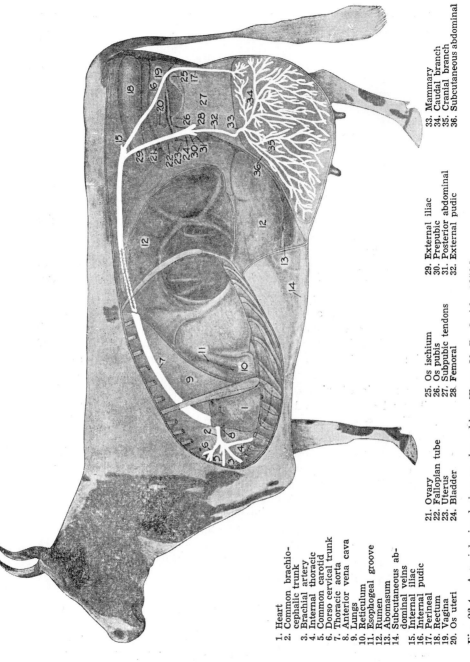

Fig. 23-4. Arterial circulation to the udder. [From V. R. Smith, 1959.]

1. Heart
2. Common brachio-
   cephalic trunk
3. Brachial artery
4. Internal thoracic
5. Common carotid
6. Dorso cervical trunk
7. Thoracic aorta
8. Anterior vena cava
9. Lungs
10. Reticulum
11. Esophogeal groove
12. Rumen
13. Abomasum
14. Subcutaneous ab-
    dominal veins
15. Internal iliac
16. Internal pudic
17. Perineal
18. Rectum
19. Vagina
20. Os uteri

21. Ovary
22. Fallopian tube
23. Uterus
24. Bladder
25. Os ischium
26. Os pubis
27. Subpubic tendons
28. Femoral
29. External iliac
30. Prepubic
31. Posterior abdominal
32. External pudic

33. Mammary
34. Caudal branch
35. Cranial branch
36. Subcutaneous abdominal

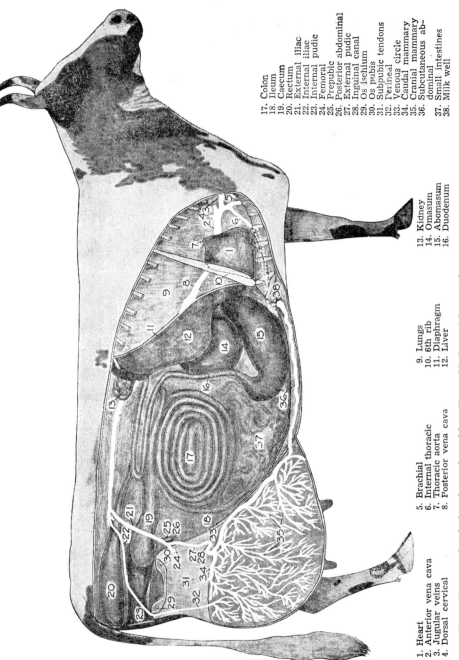

Fig. 23-5. Venous circulation from the udder. [From V. R. Smith, 1959.]

1. Heart
2. Anterior vena cava
3. Jugular veins
4. Dorsal cervical

5. Brachial
6. Internal thoracic
7. Thoracic aorta
8. Posterior vena cava

9. Lungs
10. 6th rib
11. Diaphragm
12. Liver

13. Kidney
14. Omasum
15. Abomasum
16. Duodenum

17. Colon
18. Ileum
19. Caecum
20. Rectum
21. External iliac
22. Internal iliac
23. Internal pudic
24. Femoral
25. Prepubic
26. Posterior abdominal
27. External pudic
28. Inguinal canal
29. Os ischium
30. Os pubis
31. Subpubic tendons
32. Perineal
33. Venous circle
34. Caudal mammary
35. Cranial mammary
36. Subcutaneous abdominal
37. Small intestines
38. Milk well

lowing down the abdominal wall, may be a secondary source of motor innervation.

The lumbar spinal nerves L-3 and L-4 contain the sympathetic nerve fibers or the inguinal nerves. The sacral spinal nerves S-3 and S-4 contain the sympathetic nerve fibers for the perineal nerves. The sympathetic nerve fibers, which are in the lateral cutaneous nerves, originate primarily from spinal nerves L-2. Fig. 23-6 illustrates the disposition of these nerves.

There is no evidence that indicates that the secretory tissue of the udder is dependent upon nervous stimulation for normal activity.

SENSORY NERVES. It is the function of the somatic sensory nerves to carry impulses from their respective receptors to areas of the nervous system responsible for orienting the entire organism to changes in its environment. For example, if noxious stimuli are applied to the udder of a cow, she will attempt to move away from the source of the stimuli. On the other hand, if the udder is massaged gently by hand or nuzzled by a calf, the response will be the expulsion of milk. Without sensory nerves, neither of these reflex acts could be accomplished.

The sensory nerves of the udder, like the motor nerves, are in the inguinal, perineal, and lateral cutaneous nerves (Fig. 23-6). The sensory components of each inguinal nerve serve all the tissues of half an udder with little or no overlap to the other half. Each perineal nerve sensory component serves the posterior part of half an udder, but also overlaps to some extent to the other

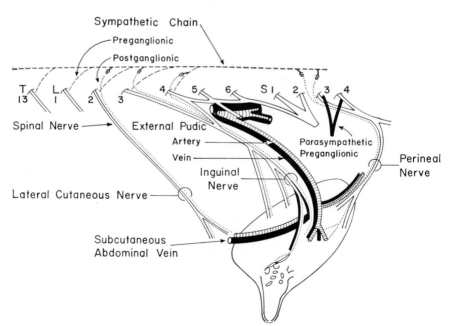

Fig. 23-6. Diagram of the innervation of the udder. [From Linzell, 1959.]

half. Some of the lateral cutaneous sensory nerve fibers may reach anterior areas of the udder (Linzell, 1959). The disposition of the nerve supply to the udder has been shown by Linzell (1959) and St. Clair (1942).

Structure of Milk Secretory Units (Alveoli). Among the few structures of the bovine udder that cannot be seen without the assistance of a microscope are the milk-secreting alveoli and their tiny tubules, which lead to the intralobular milk-collecting ducts. In shape an alveolus is approximately spherical, and it is lined with the actual milk-secreting cells. Each alveolus is well endowed with capillaries, which furnish the raw materials for the various milk components.

Surrounding each alveolus is a basketlike network of strands of tissue called myoepithelium. The contraction of these fibers is believed to be responsible for the expulsion of milk from the alveoli at the time of milking.

A diagram showing the relation of the myoepithelium and capillaries to an alveolus is shown in Fig. 23-7, which points out two salient features: (1) the myoepithelium is in intimate contact with the basement membrane of the alveolus; (2) the capillaries (only the arterial side is shown in Fig. 23-7) overlay the myoepithelium. Thus, the myoepithelium can contract without influencing the delicate capillary bed.

A photograph illustrating the actual relationship of the myoepithelium to an alveolus is shown in Fig. 23-8.

23-3. MAMMARY GLAND DEVELOPMENT

Embryological Development. For an evaluation of the embryonic and fetal development of the bovine mammary gland, the student is referred to the excellent contributions of Turner (1930, 1931). In general, the work of Turner shows that by the time the heifer fetus is six months old, the gland cisterns, teat cisterns, and the beginnings of the duct system can be clearly differentiated. At birth all the nonsecretory structures of the udder are very near their mature form.

Postnatal Development

BIRTH TO PUBERTY. Until recently, it had been generally believed that mammary glands undergo only slow change during this period of growth. Much of the increase in udder size up to the time of puberty has been attributed to the deposition of fat; however, the duct system appears to increase in magnitude, unquestionably contributing to the general size increase (Fig. 23-9).

Near the onset of puberty there appears to be a sudden increase in metabolic activity of the mammary gland, as shown by Cowie (1949) in his experiments with rats. He found that the growth of the mammary glands at this stage of development was three times that of general body growth. It

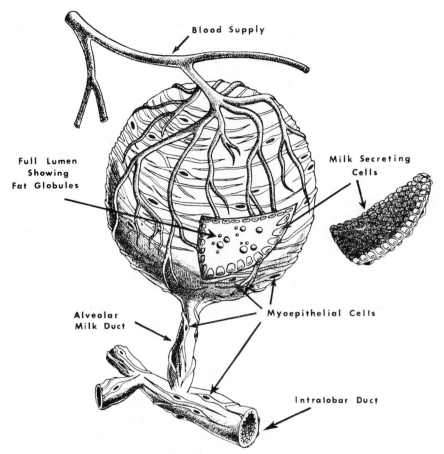

Fig. 23-7. A mammary gland alveolus. Only the arterial side of the capillary bed has been shown. [From C. W. Turner, 1954–55.]

was suggested that the increased activity of the anterior pituitary gland might be responsible for the increased rate of mammary growth. Whether the same situation exists in cattle remains to be shown.

EFFECT OF RECURRING ESTROUS CYCLES. The onset of puberty in cattle is associated with the periodic appearance of "heat" (estrous) symptoms. During this time, until the cow becomes pregnant, udder growth is confined to the extension of the duct or milk-collecting system (Fig. 23-10). It is believed that estrogen and possibly progesterone secreted by the ovary are responsible for this stage of udder growth.

EFFECT OF PREGNANCY. The state of pregnancy in the cow is associated with the presence of a functional corpus luteum in the ovarian tissue. During the first pregnancy, the duct system is completed in the first half of the

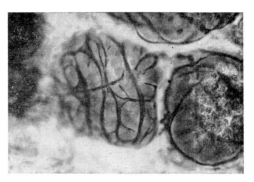

Fig. 23-8. Photograph of part of the surface of a small contracted alveolus (goat), showing a myoepithelial cell with nucleus and branching processes. [From K. C. Richardson, 1949.]

gestation period; the milk secreting units, the alveoli, develop during the last half (Fig. 23-11). Experimental evidence indicates that the final stage of the development of the mammary glands of cattle during the first pregnancy is dependent upon the relative amounts of the two hormones estrogen and progesterone. Shortly before parturition, the alveoli and ducts become distended with a fluid that only slightly resembles milk. The alveolar secretory product is called "milk" only after it has certain physicochemical properties.

In succeeding pregnancies the cow is lactating for most of the gestation period, and the mammary glands are nonfunctional for only a few weeks before parturition. The period of nonfunctioning of the mammary gland is commonly referred to as the "dry period."

EFFECT OF PARTURITION. Following parturition, the udder is stimulated into intensive secretory activity, and there is a gradual increase in milk production from the first to approximately the sixth week. After this time there is a gradual decline in milk production, until the cow is permitted to "dry up" for her ensuing parturition and next lactation period (Fig. 23-12).

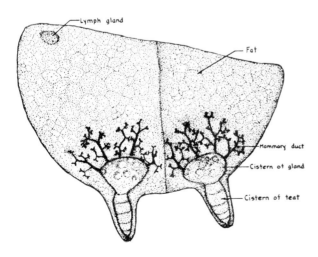

Fig. 23-9. A cross section of the udder of a heifer that has not reached sexual maturity. At this stage the glands consist of a small teat and milk cistern. The ducts leading out from the cistern are small and short, with few branches. [From C. W. Turner, 1934.]

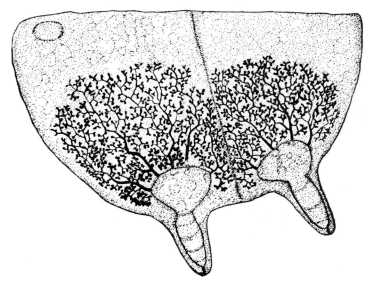

Fig. 23-10. A cross section of the udder of a heifer after many "heat periods." The duct system shows extensive development but the lobule-alveolar system is not stimulated to growth. [From C. W. Turner, 1934.]

INVOLUTION. The return of the udder to the nonfunctional state is called involution. When the udder is involuted, the milk-collecting system remains intact; however, it is impossible to differentiate the milk-secreting alveoli.

23-4. MILK SECRETION

It has been pointed out that a mammary gland is an exocrine skin gland; it is also an apocrine gland. In apocrine glands the larger secretory products accumulate at one end of the cells of which the gland is composed. The smaller, less complex molecules are secreted into the alveolar lumen. At some time, and for some reason unknown at present, the tips of the cells containing the secretory products rupture, and the products—as well as part of the cellular cytoplasm—escape into the alveolar lumen. Following this phase, the nucleus and cytoplasm begin again their activity of forming secretory products.

Physiological Considerations of the Secretory Process. The mechanisms whereby the constituents of milk are formed and passed into the lumen of the mammary alveoli have been the subject of much controversy. It appears that the formation of milk is due to filtration, selective absorption, synthesis, and secretion.

It first must be understood that before milk is formed there must be a source of raw materials for its formation. These materials can come only from the blood present in the capillaries adjacent to the alveoli.

Very uncomplicated molecules, such as water and mineral ions, can pass from the blood through the alveolar cells and into the alveolar lumen by simple filtration. More complex milk precursors are selectively taken from the blood by the alveolar cells and synthesized into substances more complex than the precursors. For example, the monosaccharide glucose is absorbed from the blood and formed into the disaccharide lactose.

After the characteristic milk constituents have been synthesized, they are secreted into the lumen of the alveoli by the method previously described for apocrine glands. The secretory activity continues until the pressure in the alveoli, resulting from the increased volume of fluids, exceeds the secretory pressure exerted by the alveolar cells.

The Initiation of Lactation. As we have noted earlier, estrogen and progesterone are responsible for the growth and development of the mammary

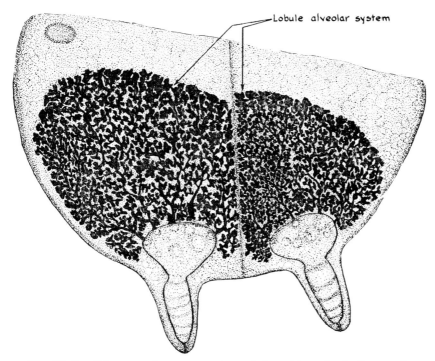

Fig. 23-11. Diagram of a cross section of the udder of a heifer at about the fifth month or the middle of pregnancy. The lobule-alveolar system has begun to develop. [From C. W. Turner, 1934.]

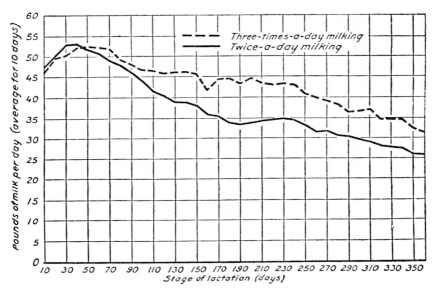

Fig. 23-12. Lactation curves illustrating the effect of frequency of milking on lactation persistency. [From Woodward, 1931.]

gland. Shortly before parturition, the secretion of milk begins, as a result of the action of another hormone, prolactin, secreted by the anterior pituitary. The cause for prolactin secretion is the subject of considerable debate. Suffice it to say that, according to most investigators in this field, changes in the levels of either (or both) estrogen and progesterone in the blood somehow influence prolactin secretion.

The importance of the anterior pituitary in the initiation of lactation cannot be overrated. For example, laboratory animals that have had their pituitary glands removed in late pregnancy deliver their young but do not lactate. Pituitary hormones other than prolactin that are important in lactation are somatotropin or growth hormone and adrenocorticotropic hormone, ACTH.

Maintenance of Lactation

ANTERIOR PITUITARY GLAND. The integrity of the anterior pituitary is essential for the maintenance of lactation. Experimentally, it is possible to cause the termination of lactation by surgical removal of the anterior pituitary. In goats that have undergone such surgery, it is possible to maintain lactation by replacement therapy. One might expect that the administration of prolactin alone would cause normal secretory activity of the mammary gland, but this has not been found to be so. In fact, crude extracts of the anterior pituitary gland have been found to be more efficient than purified prolactin for maintaining milk secretion.

From the experiments of Cowie, Knaggs, and Tindal (1964) with hypophysectomized (pituitary gland removed) goats and purified materials, it is now known, at least in this species, that the following hormones are required to maintain a normal rate of milk secretion: the pituitary hormones prolactin and somatotropin or growth hormone, STH; an adrenal glucocorticoid; thyroxine from the thyroid gland; insulin, a hormone from the islet cells of the pancreas.

In the animal whose pituitary gland is intact, adrenal glucocorticoid release is controlled by the pituitary hormone adrenocorticotropin, ACTH, and thyroxine by thyrotropin, TSH, also a pituitary hormone. Thus the reason the pituitary gland occupies a central position in the maintenance of lactation is that it is responsible for the formation of prolactin, STH, ACTH, and TSH.

ADRENAL GLAND. Removal of the adrenal glands in experimental animals usually results in an immediate reduction of lactation, which can be returned almost to normal by the administration of minerocorticoids, such as aldosterone, and glucocorticoids, such as cortisone. Minerocorticoid release is not disturbed to any great extent following hypophysectomy, as is that of the glucocorticoids, because it is controlled by an area of the brain called the hypothalamus. The mechanism by which these hormones act in maintaining lactation is not known at this time. However, it is known that the adrenal hormones are important in gluconeogenesis (the formation of glucose from noncarbohydrate sources), lipogenesis (fat synthesis), and sodium and potassium balance.

THYROID GLAND. The thyroid gland has the important function of assisting in the regulation of general body metabolism. It is doubly important to glands that are already functioning at a rapid rate. Since actively secreting mammary glands fall into this category, it is not surprising to discover that milk production may be depressed as much as 75% when the thyroid gland is removed. Milk secretion in thyroidectomized cows can be returned to normal by feeding or injecting thyroxine or other compounds having thyroxine-like activity.

Thyroxine is a galactopoetic material in that it is capable of causing increases in milk production that are above normal level when it is administered to cows whose thyroid gland is intact. Its use is not considered to be advantageous as a routine practice; for a more thorough discussion of this subject, the student is referred to an excellent review by Blaxter et al. (1949).

THE NEURAL ASPECT OF THE MAINTENANCE OF LACTATION. Up to this point, consideration has only been given to the hormonal influence on the maintenance of lactation, but certain neurohumoral mechanisms are also important.

It was proposed by Selye (1934) that prolactin secretion is periodically stimulated by manipulation of the teats, and that, at the time of suckling, receptors in the region of the teat are stimulated into activity. Impulses resulting from this stimulation are then carried centrally to areas of the brain responsible for the release of prolactin from the anterior pituitary gland. He submitted the following evidence: unsuckled rat mammary glands could be maintained in secretory activity for only as long as other mammary glands were routinely suckled; furthermore, if all the teats of lactating mammary glands were tied off in order to prevent the removal of milk, the glands could be maintained in an active state by periodic suckling.

There is some evidence—which supports Selye's theory—that the hormone oxytocin may have a function in the maintenance of lactation. It has been suggested that since oxytocin is released from the posterior pituitary immediately before each milking as a result of teat manipulation, this hormone exerts a periodic stimulating effect on the cells of the anterior pituitary responsible for the formation of prolactin (McCann, Mack, and Gale, 1959). The attractive feature of this theory is its simplicity. Although they recognize that oxytocin may be involved with the maintenance of lactation, the opponents of this theory suggest that oxytocin exerts its effect as a result of direct action on the mammary gland itself (Meites and Hopkins, 1961).

Milk Expulsion (Milk Ejection, Milk Let-Down). Milk expulsion involves all the processes whereby milk contained in the mammary gland at the time of suckling or milking is made available for withdrawal. It is a complex reflex and involves neuroendocrine mechanisms.

Though Gaines pioneered in the field, the modern concept of the reflex nature of milk expulsion was first proposed by Ely and Petersen (1941). In their experiments they denervated half an udder of a cow and found that no milk could be expressed from the denervated half until the intact half had been stimulated by suckling or manipulation of the teats. They theorized that manipulation of the teats stimulated sensory receptors in this region, which were in turn responsible for stimulating the release of oxytocin from the posterior pituitary. Oxytocin then traveled by means of the blood to the tissue of the udder responsible for expressing milk.

Since 1941 considerable effort has been expended in attempts to more fully understand milk expulsion. The general theory of Ely and Petersen (1941) has been confirmed many times. We now know that sensory impulses are sent to the brain by the receptors in the vicinity of the teat. The hypothalamic area of the brain is then responsible for causing the release of oxytocin from the posterior pituitary into the blood. When it reaches the udder, oxytocin stimulates the myoepithelial cells surrounding the alveoli to contract, with the result that the milk is squeezed into the ducts, from which it can easily be removed by proper milking procedure (Fig. 23-13).

Milk let-down can be easily developed into a conditioned reflex: the cow begins to associate the sights, sounds, and odors that occur just before milking with the procedure itself. In other words, the stimuli arriving to

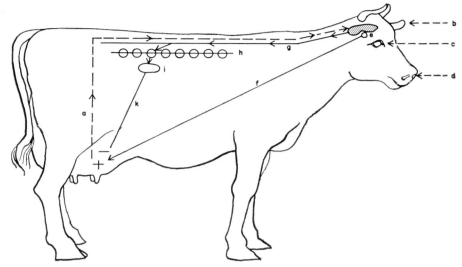

Fig. 23-13. A diagrammatic sketch illustrating the important factors that influence milk expulsion. Impulses resulting from tactile stimulation of the teats and udder (*a*), as well as impulses arising from auditory (*b*), visual (*c*), and olfactory (*d*), stimuli, cause the release of oxytocin from the posterior pituitary gland (*e*). Oxytocin is carried in the blood (*f*) to the udder, where it effects the expulsion of milk. Pain and other disagreeable sensations (fright, for example) result in impulses being sent to the adrenal gland (*i*), via the spinal cord (*g*) and sympathetic ganglia (*h*). Epinephrine from the adrenal medulla is carried to the udder in the blood (*k*), where it antagonizes the action of oxytocin. [After H. Ziegler and W. Mosimann, 1960.]

the brain from the receptors in the eyes, ears, and nose at the time of milking are, after a period of training, just as efficient in eliciting the release of oxytocin from the posterior pituitary as are impulses that arrive as the result of teat manipulation (Fig. 23-13). As a matter of fact, milk frequently drips from the teats of cows as they enter the milking area.

A stimulus that is capable of causing pain or fear will result in the inhibition of milk expulsion, as will the injection of adrenaline. It is known that noxious stimuli (resulting in pain sensation) cause the release of adrenaline from the adrenal medulla in order that the animal can take action necessary to protect itself from the unfavorable circumstances. On the basis of this evidence it is believed that adrenaline is the substance in the blood responsible for the inhibition of milk let-down whenever noxious stimuli are present (Fig. 23-13).

23-5. MILK

Physiological Factors Affecting Quantity and Composition of Milk. It has been observed, in general, that whenever there is a decrease in the percentage of fat in the milk, there is a concomitant decrease in the solids-not-fat. This has not yet been fully explained. The discussion of qualitative aspects will be concerned primarily with the changes in the percentage of fat in the milk, with the assumption that the percent solids-not-fat also changes.

STAGE OF LACTATION. The greatest variation in the composition of milk takes place following parturition. The secretory product found in the udder at calving is referred to as colostrum; it is not milk as we know it. It is richer than milk in the following products: globulins, vitamins A and D, iron, calcium, magnesium, chlorine, and phosphorus. Conversely, it contains less lactose and potassium than milk does. Approximately five days after parturition, the mammary secretory product is called "milk."

Total production generally increases for the first 30 days of lactation and then declines slowly. A typical graph of the rise and fall of milk production over an entire lactation period is shown in Fig. 23-12.

During the lactation period, the fat percent of the milk is usually inversely related to the amount produced. Whenever production is high, fat percent is low, and vice versa.

PERSISTENCY. This term refers to the degree at which the rate of milk secretion is maintained as a lactation period progresses. It is an inherited characteristic that can be strongly affected by environment, such as plane of nutrition. After the peak of production has been reached, each month's production should be approximately 90% of the preceding month if persistency is satisfactory.

EFFECT OF PREGNANCY. Carrying a calf in a normal pregnancy does not appear to influence the composition of milk. In respect to total production, it has been estimated that the energy requirements of the fetus and the cost of pregnancy to the animal are equivalent to 400 to 600 lb. of milk. Expressed in other terms by another observer, the energy cost of pregnancy is approximately 3% of the total production.

Despite the strong stimulus for prolactin secretion by periodic removal of milk from the mammary glands of the cow, the hormones of pregnancy exert an inhibitory effect on milk production. On the other hand, most unbred lactating cows can be expected to lactate almost indefinitely, although at a much lower rate than that observed during earlier phases of a lactation period.

FIRST- AND LAST-DRAWN MILK. The percentage of fat is higher in the milk removed from the udder last. The reason for this has not been ascertained; however, it has been proposed that the fat and other large milk particles adhere closer to, or are actually in, the tips of the alveolar cells just before milk expulsion. Whenever the myoepithelial cells are stimulated by oxytocin, the water-soluble portion of the milk is readily forced into the large ducts. As this portion of the milk is removed from the udder, it is possible for the larger particles of milk to enter the lumen of the alveoli either as the result of reduced intra-alveolar pressure or because the tips of the cells are ruptured by the sudden contraction of the myoepithelium surrounding the alveoli. Perhaps both of these mechanisms are involved.

AGE. With advancing age there is a gradual increase in milk production until maturity is reached (6 to 8 years old). At this time production begins to decline but the rate is slower than the increase up to maturity. Conversion tables for the various breeds of dairy cattle are available for adjusting the records of immature cows to their expected mature production or mature equivalent (Kendrick, 1953). It is estimated that, as a general rule, at 2 years a cow produces approximately 70% of her mature production, at 3 years, 80%; 4 years, 90%; 5 years, 95%; and at 6 years, her mature record.

After maturity there is a slight decrease in the percentage of fat and other major constituents of milk. Sodium chloride, albumin, and nonprotein nitrogen, on the other hand, gradually increase in concentration in the milk as the cow grows older. Presumably this increase reflects an increase in permeability of the alveolar cells for materials from the tissue fluid or a gradual reduction in the activity of the cell responsible for rejecting these products.

SIZE. Large cows within a breed usually produce more milk than small cows, but the increase in production is not in direct proportion to body weight. Brody (1945) found that for each 100-lb. increase in body weight, production only increased 70% of the proportional increase in body size. Of course, this increase applies only to animals that have the same efficiency of production.

BREED. There are genetic factors, as evidenced by breed differences, that modify unknown physiological mechanisms influencing the quantity and quality of milk. The effect of breed on the average lactation curve is illustrated in Fig. 23-14. The influence of breed on average milk production and composition is discussed in Chapter 7, "Breeds of Dairy Cattle."

ESTRUS. The effect of estrus on milk production cannot be predicted for any individual cow; however, in the majority of cows, milk production drops slightly. Accompanying the drop in production is an increase in fat percentage.

Environmental Factors Affecting the Quantity and Composition of Milk.
It is only under the most favorable circumstances that a cow attains its inherited capacity for production. A purchased cow may produce more or less milk than she produced in her previous environment. Environmental conditions include, for example, milking techniques, nutrition, length of dry period, number of times milked daily, and disease.

LENGTH OF DRY PERIOD. The length of time that a cow is dry between

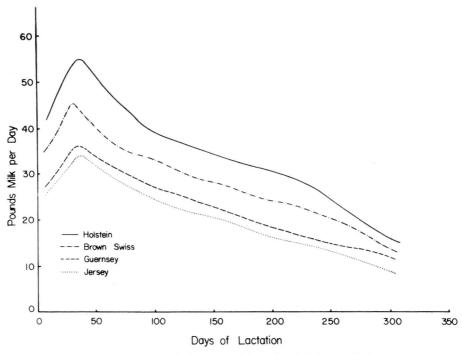

Fig. 23-14. Lactation curves of the various breeds of dairy cattle kept at the University of Arkansas Agricultural Experiment Station. [From Rakes et al., 1963.]

lactation periods affects production in the ensuing lactation. A short dry period does not allow her to build up body reserves for the tremendous drain of the next lactation period. As a result, production usually suffers. Conversely, excessively long dry periods will lower her lifetime production record. A dry period of approximately 60 days is optimal.

NUTRITIONAL CONDITION AT THE TIME OF CALVING. It is only reasonable to expect cows in a poor nutritional condition at the time of calving to produce less milk than cows in good condition. On the other hand, excessive conditioning prior to calving has not regularly resulted in increased production.

PREPARTUM MILKING. The removal of mammary secretory products late in the dry period before parturition is called prepartum milking. It has been argued that prepartum milking is an aid in the relief of udder congestion; however, experimentally it has not been found to be a good general practice. When prepartum milking is performed, the composition of the mammary secretory product at the time of parturition is found to be similar to milk, especially if the cow is milked a week or so previous to calving. Prepartum milking can rob the newborn calf of the colostrum so important to its early well-being; therefore, if prepartum milking is found to be necessary, all of the secretory product should be saved and fed to the calf after it is born. The incidence or severity of milk fever has not been found to be reduced by prepartum milking.

THE INTERVAL BETWEEN MILKINGS. The recent reviews of Elliott (1959a, 1959b) serve to indicate the controversial nature of this subject. In order to determine the true effect of "interval between milkings" one must rule out differences caused by the length of experimental periods, experimental design, management, age of cows, inherited level of production, and other unknown factors. Kendrick (1953) considered many of these variables when he reported: ". . . on the average, at two years of age a cow will produce approximately 20 per cent more milk if she is milked three times a day than if she is milked only twice a day; at three years of age she will produce approximately 17 per cent more; and at four years of age and over, she will produce approximately 15 per cent more. Similarly, at two years of age a cow will produce approximately 35 per cent more milk if she is milked four times daily than if she is milked only twice a day; at three years of age she will produce approximately 30 per cent more; at four years of age and over, she will produce approximately 20 per cent more." The increased production resulting from more frequent milking has been attributed to the more frequent reduction in intramammary pressure, which has been shown to inhibit milk secretion. Therefore, by relieving the intramammary pressure more often, secretion continues for a greater proportion of time. Milking once a day reduces milk production approximately 50%.

ENVIRONMENTAL TEMPERATURE. Cows of breeds popular in this country produce most efficiently at an environmental temperature of about 50°F. When the temperature exceeds 80°F, most cows drop off in production; accompanying this is an increase in fat percentage. On the other hand, Holsteins can endure temperatures of 8°F with little or no effect on production. Cows of the Jersey breed exhibit a drop in production whenever the environmental temperature goes below 40°F. Again, fat percentage increases as the total production declines.

The decrease in production observed at temperatures exceeding 80°F is probably the result of decreased metabolic activity influenced possibly by a decreased activity of the thyroid gland; feed consumption is also decreased. Both of these mechanisms result in a net decrease in heat production When the temperature is low, more energy is required to maintain body temperature and thus less is made available for milk production. Feed consumption is stimulated by low temperatures.

SEASON. It is difficult to separate such factors as management, nutrition, temperature, humidity, and exercise from a purely "seasonal effect" on milk production. The overall effect is, presumably, the result of a combination of all these factors. For example, cows usually drop off in milk production during the hot periods of late summer. This decline could be caused by the increased temperature or a decrease in nutrients from pastures, which are generally in poor condition this time of the year.

It is clear that more research work is needed to evaluate the influence of season on milk and milk fat production.

DRUGS. Many types of compounds that exert a pharmacological effect have been used on cows in attempts to increase milk and milk fat production. Most have no effect. However, certain hormones such as thyroxine and oxytocin, can temporarily increase yields of both milk and fat. Unfortunately, oxytocin must be administered just after each milking in order to get the residual milk; its administration procedure is both an expensive and a time consuming procedure. Tests have shown that even though thyroxine can temporarily increase production, increased nutrients are required to prevent the drain of milk precursors from the cow's body. Whenever thyroxine administration is ended, there is an immediate drop in production, which counterbalances any temporary increase.

Recently research has been conducted on the use of stilbesterol and tranquilizers for milk production, but the results are not encouraging.

FEEDING. The adherence to the principles of good feeding practices is particularly important for the maximum production of milk of uniform composition. It will therefore be considered more completely in Chapter 32, "Dairy Cattle Management."

DISEASE. It is difficult to assess the effect of "disease" in general on milk secretion without considering the many manifestations of each individual disorder. In general, digestive disturbances and diseases that affect the entire cow reduce the yield of milk and elevate the fat percentage. Diseases of the udder, such as mastitis, not only reduce the yield of milk but drastically alter its composition.

REFERENCES AND SELECTED READINGS

References marked with an asterisk are of general interest.

Blaxter, K. L., E. P. Reineke, E. W. Crampton, and W. E. Petersen, 1949. The role of thyroidal materials and of synthetic goitrogens in animal production and an appraisal of their practical use. *J. Animal Sci.*, 8:307–352.

*Brody, S., 1945. *Bioenergetics and Growth.* Reinhold, New York.

Cowie, A. T., 1949. The relative growth of the mammary gland in normal, gonadectomized and adrenalectomized rats. *J. Endocrin.*, 6:145–147.

Cowie, A. T., G. S. Knaggs, and J. S. Tindal, 1964. Complete restoration of lactation in the goat after hypophysectomy. *J. Endocrin.*, 28:267–279.

Elliott, G. M., 1959a. The direct effect of milk accumulation in the udder of the dairy cow upon milk secretion rate. *Dairy Sci. Abstracts*, 21:435–439.

————, 1959b. The effect on milk yield of the length of milking intervals used in twice a day milking, twice and three times a day milking and incomplete milking. *Dairy Sci. Abstracts*, 21:481–490.

Ely, F., and W. E. Petersen, 1941. Factors involved in the ejection of milk. *J. Dairy Sci.*, 24:211–223.

*Folly, S. J., 1956. *The Physiology and Biochemistry of Lactation*. Oliver and Boyd, London.

Kendrick, J. F., 1953. *Standardizing Dairy-Herd-Improvement-Association Records on Proving Sires*. USDA B.D.I. Inf. 162.

Lascelles, A. K., A. T. Cowie, P. E. Hartmann, and M. J. Edwards, 1964. The flow and composition of lymph from the mammary gland of lactating and dry cows. *Res. Vet. Sci.*, 5:190–201.

Linzell, J. L., 1959. The innervation of the mammary glands in the sheep and goat with some observations on the lumbosacral autonomic nerves. *Quart. J. Exp. Physiol.*, 44:160–176.

Linzell, J. L., 1960a. Valvular incompetence in the venous drainage of the udder. *J. Physiol.*, 153:481–491.

Linzell, J. L., 1960b. Mammary-gland blood flow and oxygen, glucose and volatile fatty acid uptake in the conscious goat. *J. Physiol.*, 153:492–509.

Linzell, J. L., 1960c. The flow and composition of mammary gland lymph. *J. Physiol.*, 153:510–521.

McCann, S. M., R. Mack, and C. Gale, 1959. The possible role of oxytocin in stimulating the release of prolactin. *Endocrin.*, 64:870–889.

Meites, J., and T. F. Hopkins, 1961. Mechanism of action of oxytocin in retarding mammary involution: Study in hypophysectomized rats. *J. Endocrin.*, 22:207–213.

*Nevens, W. B., 1951. *Principles of Milk Production*. McGraw-Hill, New York.

*Petersen, W. E., 1950. *Dairy Science*. 2nd Ed. Lippincott, Philadelphia.

Rakes, J. M., O. T. Stallcup, and W. Gifford, 1963. *Persistency and the Lactation Curve of Dairy Cows*. Univ. of Ark. Agr. Expt. Sta. Bull. 678.

Richardson, K. C., 1949. Contractile tissues in the mammary gland, with special reference to myoepithelium in the goat. *Proc. Roy. Soc.* (London), B 156:30–45.

Selye, H., 1934. On the nervous control of lactation. *Am. J. Physiol.*, 107:535–538.

*Smith, V. R., 1959. *Physiology of Lactation*. 5th Ed. Iowa State Univ. Press, Ames, Iowa.

St. Clair, L. E., 1942. The nerve supply to the bovine udder. *Am. J. Vet. Research*, 3:10–16.

*Turner, C. W., 1930. *The Anatomy of the Mammary Gland of Cattle. I. Embryonic Development*. Mo. Agr. Expt. Sta. Research Bull. 140.

————, 1931. *The Anatomy of the Mammary Gland of Cattle. II. Fetal development*. Mo. Agr. Expt. Sta. Research Bull. 160.

————, 1934. *The Causes of the Growth and Function of the Udder of Cattle*. Mo. Agr. Expt. Sta. Bull. 339.

*————, 1939. *The Comparative Anatomy of the Mammary Glands, with Special Reference to the Udder of Cattle*. Univ. Coop. Book Store, Columbia, Mo.

————, 1962. *Harvesting Your Milk Crop*. Babson Bros. Dairy Research and Educational Service, Chicago, Ill.

*Ziegler, H., and W. Mosimann, 1960. *Anatomie und Physiologie der Rindermilchdrüse*. Paul Parey, Berlin.

Chapter 24

Growth

"... *new ideas need the more time for gaining general assent the more really original they are.*"

VON HELMHOLTZ

24-1. INTRODUCTION

Growth is a physiological activity of great practical importance in all classes of livestock but of special significance in meat-producing animals such as swine, beef cattle, and sheep. A fast rate of gain is the key to success. It is essential, therefore, that students in livestock production be introduced to the complexities of the phenomenon of growth.

Any definition of growth is almost certain to be inadequate, since to be correct it must include all the intricate and complex processes and changes involved in this phenomenon. One definition by Brody (1945) describes growth as "relatively *irreversible* time change in magnitude of the measured dimension or function." To state it more simply, growth is an increase in size during a specific period of a part or the whole of whatever is being measured. This concept is useful for purposes of quantitative analysis since it conveys the idea of an increase in population, an increase in cell numbers or size, an increase in linearity, or an increase in weight—all as a function of time. It excludes those irregularities occasioned by a changing food supply or different physiological states such as gestation or lactation.

This definition, however, is far too general since it is limited only to a measure with respect to some reference system, and tells us nothing about the types of changes or the processes that cause them.

"True" growth can be described as a composite of many diverse physiological biochemical processes, and represents, in a restricted sense, a net accumulation of body substance in which protoplasmic reproduction exceeds destruction. Since the term "protoplasm" refers only to the "living" portion of the organism, this definition would rule out not only such things as food or water in the digestive tract and waste products not as yet excreted, but also such substances as glycogen stored in the liver, fat in the fat depots, and calcium in the bones. Thus, it is implied that "true" growth represents some permanent increase in the protoplasmic mass.

24-2. MEASURES OF ORGANISMIC GROWTH

Measurement of growth must depend upon the selection of a unit that best describes the type of physiological change being evaluated. When dealing with the organism as a whole, almost any unit chosen will represent the algebraic sum of many changes—reflecting loss as well as increase of body substance, and not necessarily indicating an accumulation of protoplasm per se. For example, much of the weight increase in mature beef cattle is in the form of stored fat rather than in that of protein or skeletal gain.

In spite of obvious discrepancies, the growth curve, showing rate of weight or linear increase, is a useful expression and deserves some analysis. The student, however, should continue to bear in mind the limitations imposed by the unit of measure selected.

The Growth Curve. If one plots the cumulative increase of the measured unit with time, generally a sigmoid curve, S-shaped (Fig. 24-1), is obtained regardless of whether the units are numbers of individuals, weight of an individual, or height of an individual.

The fact that these curves are similar in appearance might be anticipated. In a population, the unit is represented by an individual organism, and the shape of the curve reflects the changes in the numbers of these individuals or in the size of the population with time. Similarly, the change in size of a single individual or organism can be reflected by the number of individual cells existing at any time, although the unit chosen may be a weight or linear function. Accordingly, the rate of increase is essentially proportional to the number of units capable of duplicating themselves at any given time. Since this increase does not proceed indefinitely, the curves eventually show a rate decrease resulting from some inhibiting force. The student should realize, however, that the shape of the curve depends upon changes in growth processes throughout the entire lifetime of an individual or population. When the age of an individual or the size of a population is artificially limited, as is the case in domestic animals, the curve no longer has a characteristic S shape.

Growth can be represented and mathematically expressed in several ways: (1) as an average rate of increase in the measured dimension per unit of

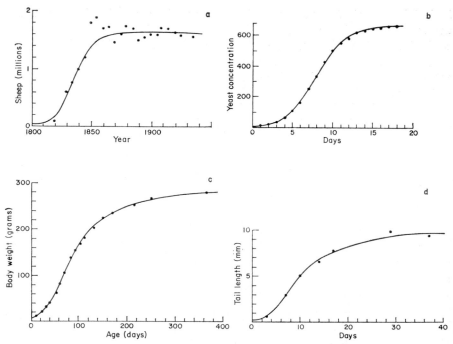

Fig. 24-1. Growth curves illustrating the similar S-shaped characteristic irrespective of the independent variable selected. (*a*) Sheep population of Tasmania. [Data from Davidson, *J. Tr. Roy. Soc. S. Australia,* **62**:342 (1938).] (*b*) Yeast growth. [Data from Carlson, *Biochem, Ztschr.,* **57**:313 (1913).] (*c*) Rate of body weight increase in the male white rat. [Data from Pearl, *The Biology of Population Growth,* Knopf, New York, 1925.] (*d*) Regeneration of tadpole tail. [Data from Durbin, *J. Exptl. Zool.,* **7**:397 (1909.]

time, (2) as a percentage increase of an initial dimension per unit of time, and (3) as a cumulative increase throughout a prescribed period of time.

The average rate of increase is the expression most commonly used by the animal husbandman, who speaks of the average daily gain made by an individual or group of animals. To obtain this value, one need only subtract an initial weight (W_1) from a final weight (W_2) and divide this difference by the number of days ($t_2 - t_1$) between these two measurements:

$$\text{average rate of gain} = \frac{W_2 - W_1}{t_2 - t_1}.$$

The percentage rate of increase can be expressed as

$$\frac{W_2 - W_1}{W_2} = \text{percentage of growth rate.}$$

Since the values this expression represents may be exaggerated or minimized—depending on the time periods selected for the initial and final measurement—it is seldom used. Before it can be utilized, it is essential to define the time interval in the life of the individual to which the expression applies, in order that the period of early, rapid growth can be compared with that of mature, slow growth.

The cumulative growth curves (Fig. 24-1) show by their characteristic S shapes that growth is rapid during the early phase of life but, with advancing age, size increase becomes slower and slower until finally growth appears to cease entirely. For purposes of comparison, one usually divides the curve into three parts; a self-accelerating phase, a point of inflection or reversal, and a self-inhibiting phase.

During the self-accelerating phase, size increases by some power of the size at the time of measurement. The reason for this exponential increase can be illustrated rather simply. By assuming that an organism begins as a single cell capable of dividing itself into two cells and that each of these, in turn, divides into two more, and so on, one can write a mathematical equation expressing this relationship: $S = 2^n$, where $S = $ size and $n = $ number of times division has occurred. But all cells do not multiply at the same rate, and rate of weight gain or linear increase is not strictly a function of the total number of cells at any given time. So this formula is a great oversimplification of the growth phenomena.

Mathematical expressions to describe the cumulative growth curve have been formulated. From these equations growth constants can be derived and are useful to predict, under a given set of conditions, growth rate as well as mature size for an individual animal as determined by its genetic makeup.

A more appropriate expression for the self-accelerating phase can be given by the equation $dw/dt = kW$, which means that the instantaneous rate of increase is proportional to the number of reproducing units—represented by weight for the growth of an animal.

The point at which growth acceleration stops and growth deceleration begins is known as the point of inflection; in higher animals this point coincides with the time of puberty. The reason for this break is not clearly established but its association with a time of changes of major endocrine events strongly suggests that the production of the hormones active in reproductive processes exerts an important influence.

The self-inhibiting phase, like the self-accelerating phase of the growth curve, can also be mathematically defined. Since the growth rate at this time is proportional to the growth yet to be made, it can be expressed as a function of the difference between the mature size, A, and the size at the time of estimation, W. The general equation denoting this relationship is as follows:

$$\frac{dW}{dt} = -k(A - W).$$

These expressions can be solved by the calculus. By integrating the expression $dW/dt = kW$, one obtains

$$\int_A^W \frac{dW}{W} = k \int_0 dt; \qquad \ln W = \ln A + kt; \qquad W = Ae^{kt},$$

Where ln W = natural log of weight at time t, ln A = natural log of weight at time 0, t = time, and k = instantaneous growth rate. Similarly,

$$\frac{dW}{dt} = -k(A - W)$$

$$\int \frac{dW}{A - W} = -k \int dt$$

$$\ln (A - W) = -kt + \ln B$$

$$A - W = Be^{-kt}$$

$$W = A - Be^{-kt}.$$

Because of the complexities and vagaries of biological processes, attempts to express changes in terms of relatively fixed, often complicated mathematical relationships may be difficult to interpret. For this reason, the use of simple mathematical approximations to express growth rates or age equivalents can be as useful as complex equations. An example of this type of treatment is the calculation of physiological age. Since animals of different breeds or species do not mature at the same rate, it is sometimes advantageous to make comparisons on the basis of physiological rather than chronological age.

Although physiological age can be compared by complicated mathematical devices, these methods do not greatly improve upon the use of a simple ratio of the mature ages of different types of animals. For example, if a Holstein steer reaches mature size at 24 months and a Hereford steer at 18 months, then the ratio 18/24 or .75 can be used as a constant to equate approximately the ages of the two animals. Thus, a 12-month-old Holstein steer has the same physiological age as a 9-month-old Hereford steer.

Linear Increase: Changes in Form. Organismic growth may also be measured in terms of linear increase. This measure, in the domestic mammal, is represented principally by changes in skeletal size; it is useful as a means of quantitating animal conformation in terms of correct body proportions.

In judging an animal an individual makes visual estimates of size and structure of different body parts and forms an opinion of its value. Many attempts have been made to find measures that would express excellence of body conformation quantitatively but thus far subjective visual estimates have not been improved upon. In meat animals our greatest limitation lies in our inability to estimate carcass excellence in the living animal. Some new approaches that hold considerable promise are mentioned in Chapter 2. Size and shape of the loin eye and the nature of fat deposition, extremely important factors in determining carcass value, are especially difficult to estimate in the living animal.

A marked advance in our concepts of growth was made by Hammond (1932), who formulated a theory of differential or heterogonic growth based on measurements of different organs and tissues of sheep sacrificed and dissected at different times of development. His observations confirmed the results of earlier studies which showed that different parts of the body grow at different rates, and provided fundamental information on the nature of the developmental changes of these different parts. For example, maximum growth rate of nervous tissue occurs at an early age, with bone, muscle, and fat following in that order. From this type of study a quantitative basis to account for changes in conformation can now be established.

Measurements of these differences have further practical considerations. They have been suggested as being useful to predict the age of maturity the growth potential, and the functional capacity as related to structural changes in the animal.

Measures of Composition of Gain. Although the animal producer is vitally interested in rate of gain as determined by daily weight increase, he is also concerned with the composition of this gain. For a meat animal to be acceptable to the consumer it must meet certain requirements, for example, beef must have sufficient fat deposition together with a high ratio of lean meat to bone.

Current practices utilize subjective methods to evaluate these properties. Recently, however, some methods depending upon objective measures have been investigated. Most of these techniques utilize the fact that the percent of water in the fat-free body is constant. Since the amount of water in fat tissue is very small in comparison to the amount of water in the total body, any variation in the percent of total body water must be largely due to the amount of fat present.

Methods for measuring body water in the live animal involve dilution techniques. For example, an animal is injected with a known amount of a chemical tracer substance such as antipyrine or deuterium, and after a short interval the blood is sampled to determine the distribution of this tracer. From the ratio of the original concentration to the final concentration of this tracer material, it is a simple matter to calculate the total water. Unfortunately, dilution techniques have not been consistently successful in large ruminant animals, because of the difficulty of eliminating rumen water content.

Another method that utilizes carcass specific gravity to predict carcass and body composition has proved valuable in many instances, but since it is useful only after slaughter, one cannot follow changes at intervals throughout the growing or fattening phases in the individual animal. Nevertheless, as a simple method for estimating carcass fat, it is important and deserves some discussion.

The specific gravity of fat, S_f, and the specific gravity of the fat-free body, S_{ff}, are relatively constant for a particular species. The specific gravity of the carcass, S_c, can be determined easily by calculating the ratio of the difference between its weight in air and weight in water to the weight in air. From these three factors one can calculate the percent of fat by the following formula: percent fat $= (S_{ff} - S_c)/(S_{ff} - S_f)$.

Lofgreen and Garrett (1954) have determined the specific gravity of steer fat (0.894) and fat-free body (1.155). The percent fat can be determined from these values. For example, if the specific gravity of the carcass is found to be 1.059; then percent fat $= (1.155 - 1.059)/0.261 = 36.8\%$.

Although the specific gravity method is convenient when estimating total fat, measurements of carcass composition are time-consuming. The techniques involve the separation of representative parts of the carcass or the whole carcass, either by physical or chemical means, into its constituents of fat, lean, and bone.

In future studies, refinements and variations of these techniques may provide useful tools for evaluating changes in chemical growth in meat animals. For example, knowledge of the effects of altering the nutritional state, the influence of various hormone treatments, and genetic effects on differential changes in the major body components would provide important information to the animal producer. One review (Reid et al., 1955) and a recent article (Meyer et al., 1960) present excellent and detailed discussions of the application of these techniques. The interested student will find them to be valuable references.

To treat in any detail these various concepts requires far more space than that allotted. The student, however, should become aware of these occasionally abstract ideas of growth and seek fuller explanations from such detailed sources as Brody (1945), Pomeroy (1955), and Pálsson (1955).

24-3. FACTORS INFLUENCING ORGANISMIC GROWTH

With this background it is now pertinent to consider how growth rate and body composition may be influenced by nutritional and hormonal factors. It is common knowledge that one of the effects of disease is a reduced growth rate; but since this result is due in the final analysis to a physiological upset, which usually is associated with a reduction of appetite, the effect of pathological factors on growth need not be discussed specifically.

The Influence of Nutrition. Of the factors influencing growth, plane of nutrition plays the most important and obvious part. Since, however, a consideration of the nutrition of livestock is included in Section V, no attempt will be made to discuss specifically the influence of the various nutrients in this chapter. Only the general aspects of nutrition on growth and body composition will be covered.

Effect on Fetal Growth. The effect of undernutrition of the dam on fetal growth is usually a stunting of the fetus. This effect has been studied most extensively in sheep. Wallace (1946, 1948) demonstrated in a series of experiments that undernourishment, particularly during the last one-third of pregnancy, resulted not only in a reduction of the weight of the mother but also of the lamb. On the other hand, overfeeding the mother during pregnancy does not result in fetal growth above normal.

Influence of Plane of Nutrition on Body Composition. As pointed out earlier, different parts and tissues of the animal body develop at different rates, leading to changes in the proportions and composition of the body as the animal grows. By altering the nutritional state of the animal it is possible to control, to some degree, the rate at which these changes occur. For example, McMeekan (1940) showed that when pigs were maintained on a high plane of nutrition during the first 16 weeks, a time when bone and muscle were rapidly developing, rate of growth of these tissues was stimulated; if maintained on a low plane, growth rate was depressed. At a later age a high nutritional plane stimulated and a low nutritional plane depressed fat formation. If pigs were started on a high plane and changed to a low plane, they produced carcasses with much lean and little fat; if the nutritional planes were reversed, fat carcasses resulted. When lambs are treated in a similar fashion comparable results are obtained, and the body proportions of cattle can be altered by changing the nutritional plane at different stages of development.

Clearly, therefore, controlling the amount of incoming nutrients makes it possible to affect the development of parts and tissues of the body, as determined by the particular metabolic activity of the part or tissue during the time of control (Fig. 24-2).

Maternal Influence on Fetal Size. Because the early formative stages of growth occur in the uterus of the mammal it is important to consider the effects of the uterine environment upon fetal size.

In the rabbit and in other species bearing multiple young, the size of young at birth is determined by the number in the litter; the larger the litter, the smaller the individual fetus. This effect on fetal size appears to be a result of a limited blood supply due to the greater number of placentae.

In animals bearing single young, the problem becomes more complicated. In the horse, however, there is good evidence that the size of the dam exerts an important effect upon the size of the fetus. Walton and Hammond have clearly demonstrated that reciprocal crosses between the large Shire horse

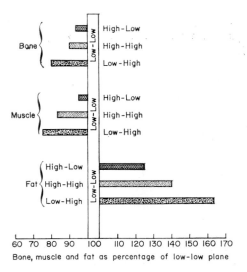

60 70 80 90   100   110 120 130 140 150 160 170
Bone, muscle and fat as percentage of low-low plane

Fig. 24-2. Effect of plane of nutrition on relative composition of pig carcasses. [Reprinted by permission of the Cambridge University Press from McMeekan, *J. Ag. Sci.*, **30**: part IV, p. 519 (1940).]

and small Shetland pony produced cross-foals from the Shire mare which were three times the size of cross-foals from the Shetland mares. Whether this maternal regulation can be accounted for by the availability of the supply of nutrient material to the fetus or whether it is dependent upon cytoplasmic inheritance has not been adequately answered (Fig. 24-3).

Although permanent differences between the size of these two crosses persisted, the crossbred foals from the Shire mares grew less rapidly than purebred Shire foals and the crossbred foals from the Shetland mares grew more rapidly than purebred Shetland foals. Stated differently, ultimate body size depends upon the genetic composition of the fetus as well as its environment *in utero*.

24-4. THE ENDOCRINE CONTROL OF GROWTH

Of the physiological processes influencing body growth, the endocrine system plays one of the most important parts. All the endocrine glands secrete hormones that influence metabolic activity, and most of them exert, with the possible exception of the posterior pituitary, either direct or indirect effects upon growth processes. It is difficult to discuss the physiological action of a single hormone upon growth as an independent effect, since many if not most endocrine secretions influence a response by modifying the action of, or interacting with, other endocrine glands. Thus, it is for convenience only that each endocrine gland is considered separately in the discussion to follow; for lack of space, only those hormones exerting the most obvious effects on general body growth will be included. For this reason the student should extend his knowledge by referring to recent physiological or endocrinological textbooks.

Growth Regulation by the Anterior Pituitary. All hormones of the anterior pituitary gland are growth-promoting, but the action of several are directed more or less specifically toward particular organs or tissues. For example, adrenocorticotropin, ACTH, causes an increase in weight and functional activity of the adrenal cortex; thyrotropin, TSH, exerts a similar effect upon the thyroid; and the gonadotropins—follicle-stimulating hormone, FSH, luteinizing hormone, LH, and luteotrophin, LTH—stimulate the gonads. Growth hormone, STH, on the other hand, has a more general action on total body growth. In addition to having many diverse metabolic effects, it stimulates both muscle and bone development. Because it thus contributes to a general increase in body mass, growth hormone may appear to be the most important for animal production. This view, however, is not necessarily correct, for the other tropic substances stimulate secretion of target-gland hormones, which in themselves may be equally important in influencing rates of bone, fat, or protein formation. Nevertheless, because of the well-recognized somatotropic effect of growth hormone, a somewhat detailed consideration will be given to it here.

Hypophysectomy, or surgical removal of the pituitary gland, results in a cessation of growth, reflected in cessation of further increase in bone or

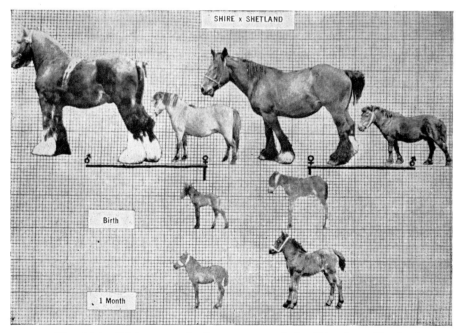

Fig. 24-3. The maternal effects on growth in Shire horse-Shetland pony crosses. [Reprinted by permission of the Royal Society of London from Walton and Hammond, *Proc. Roy. Soc. London,* Ser. B, No. 840.]

muscle mass. If, on the other hand, either hypophysectomized or normal animals are given pituitary extracts containing growth hormone, weight gain is accelerated as a result of increased protein synthesis and bone growth. Recent studies have clearly demonstrated a partial species specificity for growth-hormone action. For example, primates will not respond to growth hormone of bovine or ovine origin but will show changes in nitrogen retention when extracts of human glands are given.

Excessive secretion or additional administration of growth hormone produces multiple effects. It stimulates skeletal growth: during early life, it causes an abnormal increase in the length of long bones; after adulthood, a disproportionate overgrowth of the skeleton. It has a protein anabolic action, which is reflected in an increased nitrogen retention. It may also result in a slight increase of appetite. Chronic administration of large doses will produce in some species, such as the dog, rat, and cat, a diabetogenic effect manifested by hyperglycemia (elevated blood sugar), glycosuria (sugar in the urine), and ketonuria (ketone bodies in the urine). In addition to these well-known effects, it potentiates the action of the other pituitary tropic hormones on their specific end organs.

Although instances of inherited pathological changes in stature, such as dwarfism in mice and humans, have been demonstrated to be a result of a deficiency of growth hormone, clear-cut evidence has yet to be produced that genetic dwarfism in cattle is associated with a similar deficiency.

Other studies on the content of pituitary growth hormone in domestic species have suggested an important relationship between this substance and growth processes. For example, the pituitary glands of swine selected for rapid growth contained larger amounts of growth hormone per unit of body weight than did those selected for slow rates of gain (Baird et al., 1952). Similarly, differences in rates of growth of Holstein heifers were correlated with differences in pituitary content of both growth and thyrotropic hormones (Armstrong and Hansel, 1956).

Although these were studies of pituitary levels, they suggest that one may predict the growth rate of animals by determining the level of growth hormone. To predict growth rate successfully it will be necessary to determine levels of growth hormone in the blood, and recent studies (Read and Bryan, 1960), indicate that it may soon be possible to quantitate such levels. Conceivably, such a procedure might greatly accelerate progeny testing of our meat animals.

The Thyroid. Two biologically active compounds are secreted by the thyroid gland: (1) thyroxine, which is the principal circulating hormone, and (2) tri-iodothyronine, which is present in the blood in relatively small amounts. The principal action of these two substances is to increase energy production and oxygen consumption of most body tissues. Because of the general nature of these effects, it is obvious that any over- or underproduction of hormones by the thyroid gland will result in a multiplicity of symptoms, among which not the least important is an effect upon growth. A deficient thyroid secretion will prevent the attainment of normal adult size. Excessive fattening is frequently associated with reduced metabolic activity.

Following the development of an inexpensive source of a thyroid-active substance, iodinated casein, considerable interest concerning its use as a growth stimulant in domestic animals was generated. On the basis of the best available evidence, however, the addition of this compound to the rations of beef cattle, sheep, or swine appears not to be justified.

Antithyroid drugs, or goitrogens, are substances capable of suppressing thyroid activity. This action has suggested a possible effect of these compounds on fattening processes. Although a few studies have indicated some increase in the economy of gain following the feeding of the antithyroid drug thiouracil to swine, most authorities report unfavorable results, such as reductions in appetite and growth rate. In ruminants, antithyroid compounds have been of little benefit.

As growth stimulants or as means of altering body composition, neither thyroid-active nor thyroid-suppressing compounds would appear to offer much opportunity for consistently useful applications to livestock production.

The Adrenal Cortex. The adrenal glands are each composed of two parts, the cortex and the medulla. The medulla, or center, secretes adrenaline, a hormone that controls blood pressure. The cortex, which forms the outer layers of each gland, plays an essential role in the metabolic activities of the body. In its absence, life cannot continue unless supportive therapy is maintained. More than forty steroids have eben isolated from it. These include glucocorticoids, mineralocorticoids, androgens, estrogens, progestogens, and others whose biological activity is not yet understood.

These steroids—excluding the androgens, estrogens, and progestogens—can be divided into two categories: (1) those regulating electrolyte and water metabolism (mineralocorticoids), and (2) those influencing carbohydrate, protein, and fat metabolism (glucocorticoids).

In cases of adrenal cortical insufficiency, a number of characteristic symptoms occur: extreme muscular weakness, hypoglycemia, hemoconcentration, gastrointestinal disturbances, reduced blood pressure and body temperature, and kidney failure. Adult animals lose body weight and young animals cease to grow. The administration of cortical extracts will correct these conditions. Since glucocorticoids promote glucose formation from tissue protein, they may be considered protein catabolic, but in addition they may increase fat deposition. Obesity is a frequent occurrence in humans who suffer from excessive adrenal activity, and recent investigations have demonstrated that the administration of cortisone, a glucocorticoid, to cattle and sheep results in a significant increase in body fat. These are but two of many possible examples of how the adrenal cortex can influence growth and body composition; because of its important role in the metabolic activities of the whole body, it is apparent that almost any alteration will result in multiple responses.

The Testes. The chief hormones produced by the testes are androgens, so called because of their masculinizing effects. Testicular hormones exert a stimulatory action upon the growth of male accessory sexual tissue, such as the prostate and the seminal vesicles. They bring about, in addition, pronounced changes in general body metabolism by their influence upon protein synthesis. The administration of androgens such as testosterone or methyl testosterone decreases urinary nitrogen loss, suggesting nitrogen storage in the form of tissue protein. This effect of androgens is of obvious importance and may explain the difference in size between males and females.

A number of experiments have indicated that the administration of various androgenic steroids can stimulate gains in ruminants but not in swine: the carcasses of treated swine contain less fat and more protein. The use of androgens as growth stimulants in ruminants has not been popular, because the same effect can be accomplished more economically with diethylstilbestrol.

Effects of Castration. The student may wonder why the practice of castration of animals produced for meat is almost universal, since with the removal of the testes an important source of endogenous androgens is lost. It is well established that rams, bulls, and boars grow at a faster rate than wethers, steers, and barrows, respectively. This difference in rate of size increase can be accounted for principally by greater muscle development. Advantages to be gained by orchidectomy may be more important than growth increase. Male aggressive behavior is inhibited; the meat is usually tenderer; undesirable odor, such as that encountered in boar meat, is prevented; quality and palatability may be improved. For some purposes, however, the noncastrate animal may be desired. For example, bull meat is an essential ingredient of salami or bologna, and if animals are slaughtered at a very early age, the objectionable features of the meat of a mature, intact male are not particularly apparent.

Ovarian Hormones. The principal steroid hormones produced by the ovaries are the estrogens and progestogens. Both of these substances not only exert pronounced effects upon the growth of female reproductive tissue but, in addition, may also produce a variety of effects that are manifested as changes in body weight or skeletal growth.

For example, the effects of estrogens in a bird are striking. Following the subcutaneous administration of either natural (estradiol, estriol, estrone, and other) or synthetic (diethylstilbestrol, dienestrol, hexestrol, and other) estrogens, a rapid elevation in blood lipid level occurs. This response is accompanied by an increased fat deposition, although the parallelism between the two effects is not necessarily complete. In addition, estrogen causes hyperossification of endosteal bone, but, strangely enough, large doses increase bone fragility. The latter response may be due to increased bone resorption preceding hyperossification, but this is not definitely established.

The extra amount of subcutaneous fat, more tender skin, and improved finish as a result of estrogen treatment in chickens results in higher-grading carcasses, an effect of commercial importance. However, the use of these substances to treat birds is no longer allowed by the Food and Drug Administration because it is alleged that estrogenic residues in the meat might possibly produce cancer in those who eat it.

The response obtained when estrogens are administered to beef cattle or lambs is of great commercial importance. In contrast to chickens, ruminants show no increase in fat deposition as a result of estrogen treatment. Instead, a marked increase in protein formation and a reduction in carcass fat content are the general responses. The effects of estrogen in the ruminant are in many respects similar to those of growth hormone and androgens in other species. The subcutaneous implantation of from 24 to 30 mg of diethylstilbestrol or the feeding of 10 mg per head per day to beef steers brings about an increase of 10 to 20% in rate of gain, causes a greater nitrogen retention, increases appetite, and improves apparent gross feed efficiency.

The use of stilbestrol in increasing growth in cattle will be discussed in greater detail in Chapter 31.

The variety of effects of estrogen upon growth is difficult to explain. In some species, such as swine and the rat, high doses result in a reduction of body weight gain; in the chicken the response is principally stimulation of fat deposition; in the ruminant, nitrogen retention and protein synthesis is clearly the result (Fig. 24-4). This last effect explains the stimulation of rate of gain with no appreciable increase in feed consumption and repre-

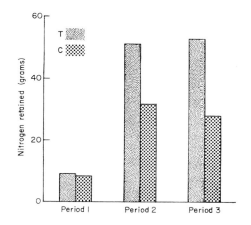

Fig. 24-4. The effect of 60 mg implants of stilbestrol on nitrogen retention in beef steers. The values for treated animals, T, are compared with untreated controls, C, and represent three different collection periods. Period 1, before treatment; period 2, 14 days after treatment; period 3, 31 days after treatment. [Data from Clegg and Cole, *J. Animal Sci.*, **13**:108 (1954).]

sents one of the most dramatic advances in the efficiency of meat production
of the twentieth century.

The Effect of Pregnancy or Ovariectomy on Growth. In the rat, pregnancy
has a stimulating effect upon both skeletel and tissue growth in spite of the
extra burden of the fetuses. Thus, growth increase surpassing that of non-
bred littermate controls continues at a fairly constant rate for the first six
pregnancies and is due chiefly to an increased appetite, which occurs as
early as 48 hours after copulation (Cole and Hart, 1938). Feed consumption
and body weights of pregnant and of lactating rats are compared with those
of nonpregnant controls in Figs. 24-5 and 24-6. Several striking features of
these results may be noted: (1) the increase in appetite during pregnancy
occurs before there is an extra demand for food; (2) feed consumption
throughout pregnancy exceeds demands, thus causing increased growth of
the mother; (3) near the end of lactation the feed consumption of *ad libitum*
fed rats is nearly three times that of their nonlactating littermate controls.
The factors that control these changes in feed consumption are still unex-
plained.

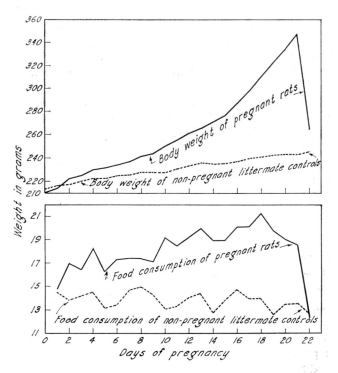

Fig. 24-5. Feed consumption and body weights of pregnant and nonpregnant rats fed *ad libitum*. [From Cole and Hart, 1938.]

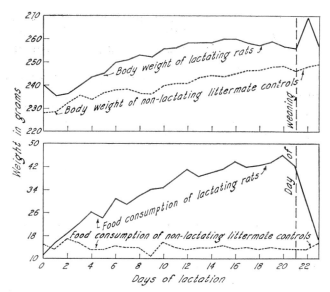

Fig. 24-6. Feed consumption and body weights of lactating and nonlactating controls fed *ad libitum*. [Courtesy H. H. Cole.]

On the other hand, pregnancy in cattle (Hart et al., 1940) or in swine (Heitman, unpublished) does not cause an increase in appetite or food consumption, and the slight increase in weight during pregnancy over that of unbred animals can be accounted for entirely by the increase in weight of the gravid uterus. The practice of breeding heifers and gilts to be slaughtered should therefore be discouraged.

Another important difference between the rat and cow is the effect of ovariectomy or spaying upon body weight gains. Spayed female rats grow more rapidly than intact controls. Although the stimulation of weight increase is largely associated with increased fat deposition, some stimulation of skeletal growth also occurs. When heifers are spayed no changes in their growth rate can be demonstrated and no benefit occurs as a result of this operation in feed-lot animals.

With the development of basic information on the role of the endocrine system on growth, important applications of the use of hormones in domestic animals have and will be found. A recent review (Casida et al., 1959) summarizes much of the current knowledge and should serve the student as a useful reference.

24-5. CONCLUSION

Growth is difficult to define in a specific fashion since it includes many diverse metabolic and physiological processes. Since "true" growth usually involves an increase in cell numbers, changes in mass may not describe properly the characteristic change taking place. For example, increase in weight may include protein synthesis, fat storage, or food and water in the digestive tract.

An obvious requirement for the optimal development of an organism is an adequate and proper nutritional intake. When this requirement is met, the endocrine system plays a major role in the physiological regulation of growth by its influence upon mineral, carbohydrate, fat, and protein metabolism.

REFERENCES AND SELECTED READINGS

References marked with an asterisk are of general interest.

Armstrong, D. T., and W. Hansel, 1956. The effect of age and plane of nutrition on growth hormone and thyrotropic hormone content of pituitary glands of Holstein heifers. *J. Animal Sci.,* 15:640.

Baird, D. M., A .V. Nalbandov, and H. W. Norton, 1952. Some physiological causes of genetically different rates of growth in swine. *J. Animal Sci.,* 11:292.

*Brody, S., 1945. *Bioenergetics and Growth.* Reinhold, New York.

*Casida, L. E., F. N. Andrews, R. Bogart, M. T. Clegg, and A. V. Nalbandov, 1959. *Hormonal Relationships and Applications in the Production of Meats, Milk and Eggs.* Publ. No. 714, Nat. Acad. Sci., Wash., D.C., 53 pp.

Clegg, M. T., and H. H. Cole, 1954. The action of stilbestrol on the growth response in ruminants. *J. Animal Sci.,* 13:108.

Cole, H. H., and G. H. Hart, 1938. The effect of pregnancy and lactation on growth in the rat. *Am. J. Physiol.,* 123:589.

*Hammond, J., 1932. *Growth and Development of Mutton Qualities in the Sheep.* Oliver and Boyd, Edinburgh.

Hart, G. H., H. R. Guilbert, and H. H. Cole, 1940. *The Relative Efficiency of Spayed, Open and Bred Heifers in the Feedlot.* Calif. Agr. Expt. Sta. Bull. 645.

Lofgreen, G. P., and W. N. Garrett, 1954. Creatinine excretion and specific gravity as related to the composition of the 9, 10, 11th rib cut of Hereford steers. *J. Animal Sci.,* 13:496.

McMeekan, C. P., 1940. Growth and development in the pig, with special reference to carcass quality characters. *J. Agr. Sci.,* 30:276.

Meyer, J. H., G. P. Lofgreen, and W. N. Garrett, 1960. A proposed method for removing sources of error in steer experiments. *J. Animal Sci.,* 19:1123.

*Pálsson, H., 1955. Conformation and body composition. In *Progress in the Physiology of Farm Animals.* Vol. 2. Ed. J. Hammond. Butterworth's Scientific Publications, London.

*Pomeroy, R. W., 1955. Live weight growth. In *Progress in the Physiology of Farm Animals.* Vol. 2. Ed. J. Hammond. Butterworth's Scientific Publications, London.

Read, C. H., and G. T. Bryan, 1960. The immunological assay of human growth hormone. In *Recent progress in hormone research,* 15:187–218. Ed. by G. Pincus. Academic, New York.

Reid, J. T., G. H. Wellington, and H. O. Dunn, 1955. Some relationships among the major chemical components of the bovine body and their applications to nutritional investigations. *J. Dairy Sci.,* 38:1344.

Wallace, L. R., 1946. The effect of diet on fetal development. *J. Physiol.,* 100:34.

———, 1948. The growth of lambs before and after birth in relation to level of nutrition. *J. Agr. Sci.,* 38:93, 243, 267.

Walton, A., and J. Hammond, 1938. The maternal effects on growth and confirmation in Shire horse—Shetland pony crosses. *Proc. Roy. Soc. of London,* Series B, No. 840, 125:311.

Chapter 25

Physiology of the Digestive System

In health the muscular activity of the bowel seems often to be soothing to the brain.

ALVAREZ, W. C., *An Introduction to Gastroenterology*

25-1. INTRODUCTION

The process of digestion takes place in the digestive tract, or alimentary canal. It requires little imagination or knowledge of anatomy to see that the animal body is built around this canal, or hollow tube. The alimentary canal of mammals includes the mouth, esophagus, stomach, small intestine, caecum, large intestine, and rectum in all domestic animals. Ruminants and birds have additional modifications, which will be discussed later. The overall functions of the tract are (1) to store food for a short time, (2) prepare the food for absorption, (3) assimilate the useful products, and (4) reject the undigestible portion. The statement of these functions appears to be simple enough, but knowledge of the physiology of digestion had to await the development of modern chemistry, the discovery of enzymes and study of their properties, the discovery of hormones and their biological activities, development of knowledge concerning nervous control of the digestive tract, and, finally, the development of the study of cellular physiology.

Some of the early attempts to study digestion make very interesting reading (Fulton, 1930). Reaumur, a distinguished French scientist, experimented in the year 1750 with a pet kite, a bird of prey which he had trained to swallow small sponges. After a short time the bird would eject the sponges. When Reaumur squeezed out the fluid from the sponges, he was rewarded to find that the juice recovered had the power to liquefy meat. Here was preliminary evidence that there were substances—later discovered and named enzymes—present in the digestive tract that would break down foodstuffs to simpler materials.

A few years later an Italian investigator, Spallanzani, studied digestion on himself. He sewed up food in small linen bags, which he swallowed. These bags were recovered after they had passed through the body and it was found that food such as meat disappeared from the bags without mastication and therefore must have been liquefied in passing through the tract. Next followed the discovery of free hydrochloric acid in the stomach by Prout in 1824 and the isolation in concentrated form of pepsin by Schwann in 1835.

One of the most fascinating experimental studies of digestion of all time was that of an American Army doctor, William Beaumont, in 1824. He had under his care a young man who, as a result of an accidental discharge of a shotgun, was left with a permanent opening (fistula) into his stomach. Beaumont carried out many physiological and chemical studies of digestion on this subject under one of the most fantastic legal contracts ever written.

The patient bound himself in Dr. Beaumont's employ for a period of one year to "serve, abide and continue with the said William Beaumont, wherever he shall go, . . . and to submit to such philosophical or medical experiments as the said William shall direct or cause to be made on or in the stomach of him, the said Alexis . . ." For this, he was to receive one hundred and fifty dollars a year and board and lodging.

With the subsequent discovery of additional enzymes, the chemical basis of digestion was accepted and attributed to the action of a number of digestive enzymes secreted in various parts of the tract.

Many chemical reactions need a minute amount of a starter or a key to activate them. Such a substance is called a catalyst. As an example, the hardening of vegetable oils to make a solid substance (hydrogenated oil) by the addition of hydrogen requires a catalyst, such as finely divided palladium, to initiate the reaction.

25-2. ENZYMES

Enzymes are defined as organic catalysts, produced by living cells but independent of living cells in their action. A list of digestive enzymes is given in Table 25-1. So far, all the enzymes that have been isolated in crystalline or chemically pure form are found to be proteins. They act as keys in opening specific chemical bonds or in activating certain reactions without being used up in the process. Thus, a given weight of a digestive enzyme will split many times its own weight of food but, being somewhat unstable organic compounds, they eventually lose activity. Their action is completely de-

TABLE | *The digestive enzymes.*
25-1. |

Food Substance (Substrate)	Name of Enzyme	Source	Products Produced
Carbohydrates			
Starches	Salivary amylase	Saliva	Maltose
	Pancreatic amylase	Pancreatic juice	Maltose + some glucose
	Amylase	Succus entericus	Maltose
Disaccharides			
Maltose	Maltase	Succus entericus	Glucose
Lactose	Lactase	Succus entericus	Glucose + galactose
Sucrose	Invertase (sucrase)	Succus entericus	Glucose + fructose
Fats and oils	Pancreatic lipase	Pancreatic juice	Glycerol + fatty acids
Proteins			
Native proteins	Pepsin	Gastric juice	Polypeptides, primary protein derivatives
	Rennin	Gastric juice	Polypeptides
Partially degradated	Trypsin	Pancreatic juice	Polypeptides, proteoses
proteins from	Chymotrypsin	Pancreatic juice	Polypeptides, etc.
gastric digestion			
Polypeptides	Carboxypeptidase	Pancreatic juice	Peptides + some amino acids
	Amino peptidases	Succus entericus	Dipeptides + amino acids
Dipeptides	Dipeptidase	Succus entericus	Amino acids
Residues from nucleo-proteins			
Nucleic acids	Polynucleotidases	Succus entericus	Nucleotides
Nucleotides	Nucleotidases	Succus entericus	Nucleosides + phosphoric acid
Nucleosides	Nucleosidases	Succus entericus	Purines + pyrimidines and phosphoric acid

stroyed by heat and by many chemical substances, such as salts of lead, copper, or mercury.

Digestive enzymes were among some of the first enzymes studied and isolated. They were given common names—ptyalin, pepsin, and trypsin— long before their true chemical nature was established. Later, enzymes were named according to the action they proomted, or were named from the compound on which they act, with the ending "ase." For example sucrase acts upon sucrose only, maltase is specific for maltose; pepsin and trypsin are two of the digestive enzymes that act on proteins and are classed as proteinases.

25-3. SECRETION OF DIGESTIVE FLUIDS

The classical experiments of Pavlov in the early part of this century revealed the process of physiological control of gastric secretion. Cannon developed his well-known method of showing movement of the stomach by fluoroscopic studies, a method widely used today in hospitals and clinics to diagnose gastric ulcers. Bayliss and Starling discovered the hormone secretin, which stimulates the pancreas to secrete its digestive fluid into the intestine. These three major contributions did much to extend our knowledge of digestion. There remained the problem, among others, of where and how the products of digestion were absorbed into the body from the digestive tract. While great progress has been made in answering these questions with the use of modern methods of research in enzyme chemistry and isotopic-labeled compounds, there still remain many unanswered questions.

25-4. CHEMISTRY OF DIGESTION

Before the discovery of enzymes, digestion was considered to be the result of combined mechanical action and some unknown chemical action, often referred to as "fermentation," a word borrowed from the ancient art of wine making. Now it is known that the enzymes found in gastric juice—in pancreatic fluid and in the secretions of the small intestine—cause the splitting of certain chemical bonds in foodstuffs, a reaction in which water is used up and compounds of smaller molecular size are produced. This process is called hydrolysis. An example of hydrolysis is the digestion of the sugar maltose with the aid of the enzyme maltase to form two molecules of glucose:

$$C_{12}H_{22}O_{11} + H_2O \xrightarrow{\text{(maltase)}} 2C_6H_{12}O_6$$

$$\text{maltose} + \text{water} \longrightarrow \text{glucose}$$

Maltose is not absorbed as such but the glucose formed is readily absorbed.

In a similar manner, starches are hydrolyzed to maltose by enzymes called amylases; fats are hydrolyzed to glycerol and fatty acids by lipases. Large protein molecules are broken down by steps to polypeptides, peptides, and finally to the amino acids, the so-called building stones of all proteins.

As mentioned above, a number of enzymes have been isolated in pure crystalline form and in this pure state they have all so far been found to be proteins. This explains some of the properties of enzymes and the characteristics of enzyme action. Their activity is slowed down as the temperature is lowered and finally ceases at temperatures below freezing. At body temperature, enzyme action is rapid, but at boiling temperature action is stopped permanently. Heavy metals, such as copper, lead, and mercury, precipitate proteins and therefore stop enzyme action. Acid and alkali have a pronounced effect on rate of digestion with enzymes. Thus, pepsin, which acts on proteins while they remain in the stomach, is very effective in the presence of hydrochloric acid produced in gastric juice, but this action is stopped when the stomach contents are neutralized in the small intestine. Here other enzymes begin their work in the more neutral medium.

Enzymes are powerful catalysts. It has been shown, for example, that one part of a crystalline enzyme such as pancreatic amylase will split up to 4,000,000 times its weight of starch.

Digestion in Various Species. Although the chemical principles involved in digestion in different species are primarily the same—that is, the hydroly-sis of food substances to smaller units which can be absorbed into the system —the anatomical features that bring about these chemical reactions in various species vary considerably. It will be necessary first to describe the digestive processes of a single-stomached animal, such as the pig, dog, or man, and then to consider the anatomical differences in the digestive tract of ruminants and birds.

Digestion in the Mouth. The digestion of insoluble particles is known to take place on their surfaces. Hence the advantage of mastication is to reduce the solid particles in size and thus increase the surfaces exposed to enzymes. Chemical action in the mouth is not very important since only a very few species produce ptyalin, a starch-splitting enzyme, in their saliva. Ptyalin is not present in the saliva of ruminants. Even in animals that do produce ptyalin, the food is in the mouth such a short time that the action is small and is stopped by the hydrochloric acid of the stomach.

Gastric Digestion. Gastric digestion in this discussion refers to digestion (1) in the stomach of those animals with a simple stomach—man, the pig, horse, dog, and others (see Fig. 20-6), (2) in the fourth or true stomach of ruminants and (3) in the glandular stomach or proventriculus (Fig. 25-1) of birds. The peculiarities of digestion in ruminants and poultry will be taken up further on.

Gastric juice is secreted by cells of the gastric glands embedded in the inner wall of the stomach, the gastric mucosa. The constituents of gastric juice are (1) hydrochloric acid (HCl), which lowers the pH of the stomach contents for optimal action of the enzyme, pepsin, (2) pepsin, a proteolytic enzyme, which breaks down proteins to smaller molecules, polypeptides, (3) rennin, another proteolytic enzyme, resopnsible for the clotting of milk pro-teins and no doubt of special significance in the newborn, (4) mucus, a pro-tective secretion coating the internal lining of the stomach and presumably preventing the destruction of the stomach wall by HCl. If this lubricant is not secreted in sufficient quantities or if HCl is secreted in abnormal amounts, erosions of the stomach lining, gastric ulcers, may result.

Although some fat splitting in the stomach has been attributed to the

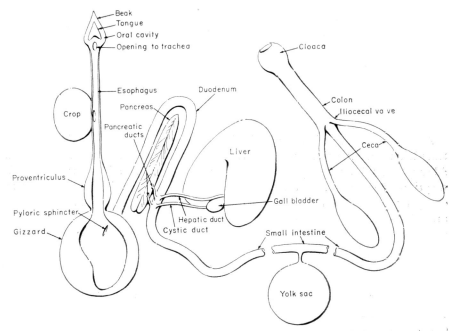

Fig. 25-1. Digestive tract of a chick. The yolk sac is in process of being resorbed.

presence of a gastric lipase, this action is minor. Gastric juice does not appear to have any action on carbohydrates.

Digestion in the Small Intestine. The major part of digestion takes place in the long tube, the small intestine. Here digestion is the result of secretions containing enzymes that are (1) produced in the pancreas and reach the small intestine by way of the pancreatic duct; (2) formed in the liver and secreted as bile into the intestine, sometimes by a common duct with the pancreas and in other species by a separate duct; and (3) produced by intestinal glands lining the small intestine. The excess of acid in the partly digested material leaving the stomach is neutralized by pancreatic juice, bile, and the secretions of the small intestine. In this nearly neutral medium, enzymes now present in the small intestine carry on the remainder of the digestion processes. Starches are split to maltose and this is further split to glucose. Sucrose, if present, is split to invert sugar and any lactose to glucose and galactose. Fats are digested to glycerine and various fatty acids, and proteins are finally broken down to simple units, the amino acids. There are no enzymes secreted in the digestive tracts of animals that have been found to digest cellulose, the principal constituent of the fibrous part of foodstuffs. However, a small amount of cellulose may be broken down by bacterial action in the digestive tract of simple-stomached animals, especially in the caecum and colon, but in ruminants a considerable amount of cellulose is digested. The action is attributed to enormous numbers of microorganisms that inhabit the rumen and not to any cellulose-splitting enzyme produced in the digestive juices.

25-5. ABSORPTION OF DIGESTED PRODUCTS

Although the chemistry and physiology of absorption from the digestive tract has been intensively studied, there are many questions yet unanswered. Absorption takes place principally from the small intestine. Very little absorption of digestion products occurs in the mouth, gullet, or stomach; an exception is the absorption of fatty acids across the rumen wall. Some drugs are directly absorbed from these areas but the principal mechanism for absorption of the end products of digestion is found in the microscopic fingerlike projections lining the small intestine, called villi (Fig. 25-2). These projections into the interior of the canal are lined with columnar epithelial

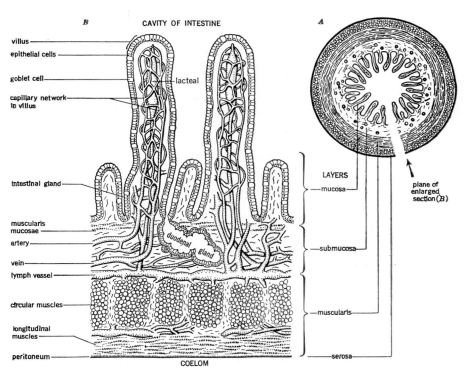

Fig. 25-2. A villus in the small intestine, lined by a single layer of epithelial cells. [From Storer and Usinger, *General Zoology*, McGraw-Hill, New York, 1957.]

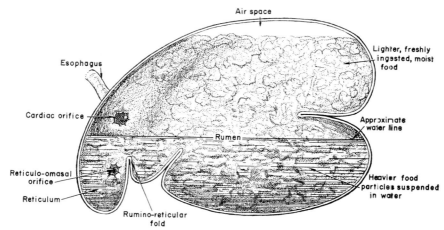

Fig. 25-3. Relationship of the rumen and reticulum. Note that the lighter roughage fills the dorsal portion of the rumen except for the small air space. The rumen and reticulum are only partially separated by the rumino-reticular fold. A syringe-like action of the reticulum forces freshly swallowed food posteriorally into the dorsal rumen. [From Cole, Mead, and Kleiber, "Bloat in cattle," *Calif. Agr. Exp. Sta. Bull.* 662 (1942).]

cells through which the food substances must pass. Inside the villi are small capillary blood vessels and lymph ducts, which collect the absorbed material. The study of the physiology of the epithelial cells of the villi is a fascinating subject. Absorption through these cells is found to be a highly specific process. Digestion is not simply a process producing water-soluble materials. Sucrose and lactose, for example, are water-soluble but are not absorbed until they are split to hexose sugars. Glucose and galactose are molecules of identical size but the rates of absorption are quite different. Most of the fatty acids produced by fat digestion are water-insoluble but are readily absorbed from the intestine. We see then that absorption is a selective process dependent upon properties of living cells and undoubtedly due in some manner to enzymes produced in the process. Absorption, therefore, is not a matter of simple diffusion through a membrane. The mechanism involved represents one of the unanswered problems to which newer methods of study—using compounds containing isotopic labels—are giving valuable clues.

25-6. PECULIARITIES OF RUMINANT DIGESTION

In herbivorous animals such as cattle and sheep, the compound stomach appears to be an evolutionary modification of the simple stomach for providing a compartment in the alimentary canal where fibrous foods may be held to undergo a soaking and "fermentation" before passing on through the canal. The ruminant stomach is actually divided into four compartments. The rumen, or first compartment, is very large in the adult animal and may hold up to 50 or 60 gallons of soft food material (Fig. 25-3). The reticulum is the second compartment, much smaller in size. The third compartment is known as the omasum, somewhat larger than the reticulum. The fourth compartment, called the abomasum, or true stomach, is similar in function to the stomach of monogastric animals. The total capacity of the four compartments is rumen, 80%, reticulum, 5%, omasum, 7%, and abomasum, 8%. The rumen, reticulum, and omasum are nonglandular and thus do not produce acid or digestive juices. Because proteolytic enzymes and hydrochloric acid are absent, they do, however, provide excellent compartments for the growth of many types of microorganisms—both bacteria and protozoa—that are taken in with the food. Through the action of these microorganisms, coarse materials from roughages, containing complex polysaccharides, cellulose, and lignocelluloses, are broken down. Starches and simpler carbohydrates are also attacked by the microorganisms. Recent studies by Otagaki et al. (1955) have shown that proteins, too, may be acted upon by rumen microorganisms. Since no digestive enzymes are produced by the rumen, the action taking place resembles a fermentation involving quite a different series of chemical reactions than those resulting from the action of digestive enzymes secreted by the digestive glands. End products are many in number, but principally short chain fatty acids—acetic, propionic, and butyric—are produced. It is well known that rumen microorganisms can also use simple sources of nitrogen such as ammonium salts and urea, building these up to complex proteins. Thus the ruminant is provided with a variety of proteins derived from the bodies of microorganisms. On passing into the true stomach and into the intestines, these organisms—which have multiplied in the rumen, reticulum, and omasum—are digested, and their bodies serve as a source of food protein. Several of the B vitamins are also synthesized in the rumen.

Fermentation in the rumen results in the liberation of considerable heat. This phenomenon is of special significance in the adaptation of ruminants to the environment. If ruminants are well fed, the resulting heat allows them to withstand very cold climates. In hot climates, however, heat production increases the stress on the animal. The heat liberated following roughage feeding is much greater than that following ingestion of concentrates. Thus, in hot climates the feeding of rations low in roughage permits better adaptation to the environment.

25-7. DIGESTION IN BIRDS

Digestion in the chicken has been studied more thoroughly than that in any other bird. The digestive organs of chickens are quite different in several respects from those of mammals (Fig. 25-1). However, the same types of chemical reactions take place as have been described for other types of livestock above. In the chicken, teeth are not present; hence mechanical reduction of the food must be accomplished in other ways. The esophagus has a pouchlike enlargement or crop where solid food is stored temporarily and somewhat softened. Passing on down the gullet, the food enters the stomach, which is actually made up of two separate parts, quite unlike that of mammals. The first part is the proventriculus (or glandular stomach). It has a straight narrow passageway not much larger than the esophagus, with a thick wall containing glands that secrete gastric juice and hydrochloric acid. Food does not accumulate in the glandular stomach as it does in the stomach of mammals but, after mixing with gastric juice, passes on into the ventriculus or gizzard. This is a large, thick-walled muscular organ, which grinds the solid food with the aid of pieces of grit that the bird has swallowed. The gizzard does not secrete any digestive fluids but simply grinds and mixes the solid food with gastric juice already supplied by the proventriculus. By contractions of the gizzard, which occur two or three times per minute, food is passed on into the first part of the small intestine, the duodenum. Here gastric digestion, started in the proventriculus and gizzard, continues in an acid medium; neutralization does not occur as it does in mammals. Pancreatic fluid and bile enter the duodenum near the second loop. The pancreatic fluid is only weakly alkaline and bird bile is acid in reaction, so very little neutralization of the acid-mixed food takes place. It has been suggested that trypsin, produced by the chick pancreas, may be different in its optimum reaction to that produced in mammals. In any event, the intestinal contents remain somewhat acid throughout the whole tract, in contrast to that of mammals.

The digestive tract of birds, compared with that of mammals, is short and compact, a characteristic of their body, which is streamlined for flight. Digestion and absorption must therefore take place in a shorter time than in mammals, and the undigested residues, which would be dead weight, are ejected with very little delay. Digestion in the small intestine is further aided, as in mammals, by secretion of intestinal digestive fluids, but proceeds, as mentioned above, in an acid medium.

In the chicken, a pair of caeca, or blind sacks, about six to eight inches long, open from the small intestine at the junction with the colon. These thin-walled pouches are lined in their narrow portion with villi, which indicate that the caeca may be concerned with absorption. Little is known, however, concerning their function. The colon in birds is very short, and because it, too, is lined with villi, absorption must take place here.

25-8. CONTROL OF THE SECRETION OF DIGESTIVE JUICES

Both the nervous and endocrine systems are involved in controlling the secretion of digestive juices. Space will not permit a detailed discussion of these controlling mechanisms, but examples will be discussed briefly. In most instances both the nervous and endocrine systems are involved.

Control of Salivary Secretion. Let us first consider the control of salivary secretion. Though one frequently encounters the statement that the flow of saliva is regulated solely by the nervous system, this is not strictly true. As early as 1867, Eckhard reported that one of the three sets of salivary glands in sheep, the parotid glands, continues to secrete after the nerve supply has been severed. According to more recent studies, this so-called paralytic secretion depends at least to some extent upon the stimulation by the hormone, adrenaline, secreted by the adrenal medulla. If this view is correct, hormones do play some role in controlling the flow of saliva.

The increased flow of saliva associated with eating, however, is dependent upon the nervous system. This can be simply demonstrated by showing that the chewing of palatable food does not evoke the usual increase in the flow of saliva if the nervous connections to the salivary glands have been severed. The rate of flow of saliva, therefore, depends upon the stimulation of the taste buds in the mouth by chemicals of ingested food. Some foods such as fresh meat cause the secretion of saliva rich in mucin, whereas other foods cause a more profuse watery flow. The "conditioned" flow of saliva is described in Chapter 20.

Control of Gastric Juice Secretion. In gastric juice secretion, we have the best example of the complementary action of the nervous and endocrine systems. The flow of gastric juice is initiated by nervous and maintained by hormonal means.

The first phase, often referred to as the cephalic phase, is under nervous control. Proof of this was given by the great Russian physiologist, Pavlov. He used a dog with an esophageal fistula and with a small gastric pouch, known widely as the Pavlov pouch. When this dog was fed, the food was swallowed and was passed to the outside; even though food did not enter the stomach, gastric juice was secreted. Denervating the stomach abolished secretion produced in this manner. From this one may conclude that food in the mouth and pharynx stimulates nerve fibers, resulting in a reflex flow of gastric juice.

An experimental animal with a denervated stomach and a Pavlov pouch can be used to prove the existence of the second means of stimulating gastric juice secretion, the gastric phase. If certain foods are placed in the Pavlov pouch of this animal, gastric juice is secreted. The hormone gastrin, produced by the internal lining (gastric mucosa) of the stomach, is responsible for stimulating the gastric glands in this instance.

As the partially digested food reaches the small intestine, more gastrin is produced by the intestinal mucosa and transported to the gastric glands via the blood. Thus, gastrin produced by the stomach and intestinal wall maintains the secretion of gastric juice after eating has ceased.

25-9. MOVEMENTS OF THE DIGESTIVE ORGANS

Throughout the entire length of the alimentary canal, the walls of the organs have muscular layers. (The role of voluntary muscular activity associated with mastication and swallowing of food is well known.) Muscular contractions serve two main purposes: (1) mixing of the ingested food (ingesta) with the digestive juices, and (2) movement of the food along the digestive tract. In ruminants contraction of the omasum may bring about further breaking down of food particles, and the contraction of the gizzard of birds plays a similar role. Additional specialized contractions, especially significant in ruminants, facilitate regurgitation of food for further mastication and the eructation of gas accumulating in the rumen.

Mixing of the chyme (ingested food plus the digestive secretions) is performed by contractions of segments of the digestive organs. Segmental contractions are especially prominent in the small intestine. Pumping-like movements of the villi of the small intestine also help in mixing the chyme, as no doubt do their pendular (swaying) movements. All of these movements can occur without nervous activation. In other words, they are dependent upon intrinsic properties of the musculature itself; thus, we refer to them as being automatic or, more specifically, myogenic.

Peristaltic movements occur throughout the digestive tract. In 1899, two prominent English physiologists, Bayliss and Starling, described peristalsis as a wave of dilation preceding a wave of contraction. Alvarez (1940), using improved techniques, questions whether the wave of inhibition is a normally occurring phenomenon, and most recent investigators conclude that the more important feature of peristalsis is a wave of contraction that sweeps the chyme caudally.

The control of peristaltic movements is independent of the central nervous system, CNS; that is, all nervous connections can be severed without abolishing them. Specialized nerve networks, Auerbach's plexuses, located in the walls of the digestive organs and with no apparent connections with the CNS, synchronize the motility so that an orderly wave of movement occurs.

One might conclude from the above discussion that muscular activity of the digestive organs is not influenced by the CNS. This is not true. Though the contractions may occur independently of the CNS, their magnitude and rate are influenced by nerve impulses over the autonomic nervous system—sympathetic fibers inhibit their activity and parasympathetic fibers carry impulses that increase muscular activity.

25-10. SPECIAL FEATURES OF RUMINAL, RETICULAR, AND OMASAL ACTIVITY

The movements of the rumen, reticulum, and omasum, unlike those of the true stomach and intestines, are not governed locally but rather are dependent upon the CNS. These pouchlike organs are considered to be out-croppings of the esophagus. Thus, since the esophagus is at least partially under the control of the CNS, this type of control is not surprising.

Motility of the rumen mixes the ingesta, moves it from one compartment of the rumen to another, and moves food to the second stomach, the reticulum. Motility of the reticulum is largely concerned with the movement of food from the cardia to the posterior rumen and, conversely, from the reticulum to the omasum. Much of the early information on motility of these organs was obtained by studying cows with rumen fistulas (Fig. 25-4).

Contractions of both the rumen and reticulum are concerned in effecting the reflex acts of regurgitation and eructation. Since these acts are of special significance in the ruminant animal, a brief description will be given.

Rumination. The stomachs of the ruminant are designed to store large amounts of feed. Further, inasmuch as the food may be regurgitated and ruminated, the animal may consume the feed rapidly with a minimum of chewing before swallowing.

Regurgitation of food by the ruminant is a tranquil process when compared to the rather violent process of regurgitation or vomiting in man and other monogastric animals. By making use of animals with rumen fistulas, Schalk and Amadon (1928) showed that the stimulus for this reflex was coarse material in the rumen (Fig. 25-4). The proof consisted of inducing regurgitation and rumination by the simple procedure of inserting the arm into the rumen through a rumen fistula and rubbing a wisp of hay over the inner surface of the rumen. More recently, Cole et al. (1942) showed that the number of hours a cow spends ruminating depends upon the coarseness of the roughage; animals on green alfalfa tops with soft stems and leaves ruminate very little, if at all, whereas cows on Sudan grass hay with scabrous (barblike) leaves ruminate much longer. Animals on a hay-and-grain ration ruminate for about 15 minutes at each ruminating period and spend 6–8 hours daily ruminating (Schalk and Amadon, 1928).

How is regurgitation brought about? An extra strong contraction of the reticulum floods the cardiac orifice with the soupy ingesta. Concurrently the cardia dilates, an inspiratory movement lowers the pressure in the esophagus, and finally an antiperistaltic wave sweeps the soupy ingesta to the mouth. The cow squeezes out the excess water, which she swallows, chews the cud (usually about 35 champs of the jaws per bolus), and then swallows this remasticated food. Immediately upon swallowing, the act of regurgitation is repeated and new material is brought up for rumination.

Fig. 25-4. A cow with a permanent opening (fistula) into the rumen. To prevent loss of food and to prevent drying of the rumen mucosa, the opening is closed with a pneumatic plug. This cow delivered several calves while fistulated. [From Cole, Mead, and Kleiber, "Bloat in cattle," *Calif. Agr. Exp. Sta. Bull.* 662 (1942).]

One may logically inquire as to whether the process of rumination can be controlled voluntarily. Apparently there is little if any voluntary control. A cow on very soft feed, on which there is a minimum of rumination, will frequently extend the muzzle forward and open her mouth as though attempting to regurgitate. The fact that the cow fails to accomplish the act, together with other information mentioned, provides strong presumptive evidence that the act is largely involuntary and dependent upon the extent to which coarse feed stimulates nerve fibers terminating in the rumen mucosa.

Eructation. As about one cubic foot of gas may be produced in the rumen hourly after a full feed, it is mandatory that a means is provided for expelling this gas. Though a small amount of rumen gas may pass across the rumen wall, enter the bloodstream, and then be exhaled with expired air, the major portion of it must be expelled by belching (eructation). Eructation is a reflex act but the normal stimulus is not known. It has been demonstrated that eructation becomes more frequent when the rumen pressure is elevated, but pressure could scarcely be the normal stimulus inasmuch as belching continues when the rumen is opened to the outside air pressure by means of a rumen fistula. Cole et al. (1942) suggested that scabrous roughage in the rumen was the normal stimulus, but convincing evidence is still lacking. The proposal was based upon the fact that bloat does not occur when sufficient coarse roughage is present in the diet. Thus it seems logical that both regurgitation and eructation would have a common stimulus.

One should not infer from this that the two acts have other characteristics in common. During regurgitation, as previously mentioned, the reticulum is contracted, whereas at the moment of eructation the reticulum is dilated and the rumen is contracted. Dilation of the reticulum lowers the level of ingesta around the cardia while contraction of the rumen forces the gas to the cardia from the air pocket in the dorsal rumen.

Bloat. Inflation of the rumen with accumulated rumen gas is known as bloat (Fig. 25-5). Any anatomical or physiological defect that interferes with normal eructation results in a condition of chronic bloat. On the other hand, acute bloat occurs only under specific dietary regimes as, for example, on green legumes. Some animals bloat readily, others rarely. Thus acute bloat depends both upon "animal" and "plant" factors. Weiss (1953) was the first to propose that lack of sufficient salivary flow was a predisposing factor in bloat and Mendel (1961) has submitted data to confirm this. Presumably saliva tends to inhibit foaming of ingesta. Ingesta from legumes is particu-

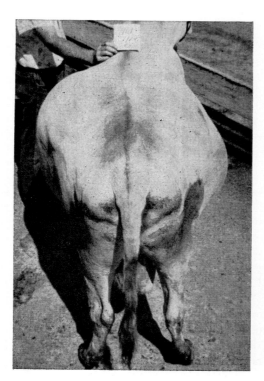

Fig. 25-5. This cow bloated after consumption of green alfalfa tops. Note the marked distension of the left body wall. Foaming of ruminal ingesta following consumption of green legumes makes eructation of gas difficult.

larly prone to foam. All agree that foaming of the ingesta with consequent blocking of the cardiac orifice is a major cause of bloat. Some bloat-prevention procedures, such as the use of antibiotics, may be effective by virtue of their inhibition of microorganism activity and a consequent reduction in gas production; other procedures, such as use of vegetable and animal oils, act by preventing foaming; the effectiveness of grass hays may be due to the increased reflex flow of saliva, as suggested by Weiss, the saliva in turn reducing foam formation.

25-11. CONTROL OF FOOD INTAKE

In our discussion of digestion, we have assumed the presence of food in the alimentary tract, but we have not considered the factors that determine food intake. The efficiency of our meat- and milk-producing animals depends in large measure upon the amount of food ingested. Let us consider the factors that determine the amount of food consumed when animals are offered a free choice.

There are many false concepts about what determines the ability of an animal to consume large amounts of food. In judging of cattle, for example, the consideration of "body capacity" is based on the inference that visual inspection will enable one to determine the amount of food the animal will consume. Unfortunately, there is no one chemical or physical test that will provide us with reliable measures of the capacity of an animal to consume food.

Most writers on the subject state that food consumption depends upon "appetite" or "hunger." There is no uniformity on the usage of these terms. We believe it preferable to use the term "appetite" to describe the sensation of the pleasant anticipation of food. "Depraved appetite" would refer to an unnatural desire for a specific food as the craving of sugar by the diabetic. "Hunger," on the other hand, would refer to an unpleasant sensation resulting from prolonged starvation and involving hunger contractions of the stomach.

Some of the factors controlling appetites are as follows: (1) Two centers in the hypothalamus have been postulated—a feeding center and a satiety center. Damage to the first reduces food intake and damage to the latter in rats results in overeating and extreme obesity. Though many theories have been suggested, little is known about the mechanisms controlling the activity of these centers. (2) In the rat, pregnancy increases appetite long before there are any increased demands for nutrients (Chapter 24). Thus it seems unlikely that the increased appetite in this instance depends upon changes in the composition of the blood. One unconfirmed report indicated that the injection of blood from pregnant into nonpregnant rats increased the food consumption of the latter. This indicates that a hormonal mechanism might be the cause of increased appetite during pregnancy. (3) Food consumption is increased during lactation of the rat (Chapter 24). Unfortunately reliable data are not available to show the difference in food consumption between lactating and nonlactating cattle. However, a 1,000 pound lactating cow will consume nearly twice as much as a 1,000 pound fat steer. This fact in itself shows the fallacy of the concept that food intake can be predicted by visual appraisal of "body capacity." Possibly, the taking up of nutrients by the mammary gland of the lactating cow may change the composition of the blood, which in turn will have an influence on the food centers in the hypothalamus. (4) Some have felt that stretching of the stomach walls by food stimulates nerve fibers that have an effect upon the hypothalamic centers. Campling and Balch (1961) obtained some information that could be interpreted as favoring this postulate. Hay was removed through a rumen fistula in cows as rapidly as it was swallowed. The cows consumed 177% of the normal intake. (5) High environmental temperature and many types of sickness affect appetite, but little is known of the mechanisms involved.

In conclusion, we may say that our information is meager concerning mechanisms that control appetite. It is a field of great importance in livestock production and is deserving of more intensive investigation.

REFERENCES AND SELECTED READINGS

References marked with an asterisk are of general interest.

*Alvarez. W. C., 1940. *An Introduction to Gastroenterology, being the Third Edition of the Mechanics of the Digestive Tract.* Hoeber, New York.

Balch, C. C., and R. C. Campling, 1962. Regulation of voluntary food intake in ruminants. *Nutr. Abst. Rev.,* 32:669.

Campling, R. C., and C. C. Balch, 1961. Preliminary observations on the effect on the voluntary intake of hay, of changes in the amount of reticulo-ruminal contents. *Brit. J. Nutr.,* 15:523.

Cole, H. H., S. W. Mead, and W. M. Regan, 1942. Production and prevention of bloat in ruminants. *J. Animal Sci.,* 2:285–294.

Eckhard, C., 1944. Cited by B. P. Babkin, in *Secretory Mechanism of the Digestive Glands.* Hoeber, New York, p. 178.

*Fulton, J. F., 1930. *Selected Readings in the History of Physiology.* Charles C. Thomas, Springfield, Chap. 5.

Mendel, V. E., and J. M. Boda, 1961. Physiological studies of the rumen with emphasis on the animal factors associated with bloat. *J. Dairy Sci.,* 44:1881–1898.

Otagaki, K. K., A. L. Black, H. Goss, and M. Kleiber, 1955. In vitro studies with rumen microorganisms using carbon-14-labeled casein, glutamic acid, leucine and carbonate. *Agr. and Food Chem.,* 3:948–951.

*Pavlov, I. P., 1910. *The Work of the Digestive Glands.* Second English translation, by W. H. Thompson, C. Griffin and Co., London.

*Schalk, A. F., and R. S. Amadon, 1928. *Physiology of the Ruminant Stomach (Bovine). Study of the Dynamic Factors.* N. Dakota Agr. Expt. Sta. Bull. 216. 64 pp.

Weiss, K. E., 1953. The significance of reflex salivation in relation to froth formation and acute bloat in ruminants. *Onderstepoort J. Vet. Research,* 26:241–250.

Chapter 26

Adaptation to
the Environment

Physioclimatology, bioclimatology, or environmental physiology, a rather neglected subject, is worthy of cultivation in its own right to furnish an intellectual basis for understanding and controlling the world.

SAMUEL BRODY

Of all the known forces which have directed the evolution of man and his ever-changing civilization, none have had more effect over a long period of time than the factors that constitute the climatic environment. Day-to-day changes in the environment—in temperature, light, moisture, air movement, available nutrients, a variety of radiations from the sun and outer space, and the long-range weather patterns that characterize a climate—have a profound effect on plants, animals, and man. The entire known evolution of the earth and its inhabitants is the history of soil and plant development, animal origins, evolution, and distribution, based largely on changes brought about by interactions with the environment.

The climatic environment influences nearly every economic aspect of plant and animal agriculture, crop yields and composition, animal growth, reproduction, milk production, egg production, and the efficiency of conversion of foodstuffs to economic units.

Primitive man learned to modify his environment by the use of fire as a source of heat and light, but it is only in the last century that man has utilized central heating, the electric light, mechanical refrigeration, and, more recently, large-scale cooling and dehumidification. During the nineteenth and early twentieth centuries man subscribed to the philosophy that the control of his own private climate was possible, but that animals should be selected for specific climatic conditions. As a result, there are literally hundreds of breeds of cattle, sheep, swine, and poultry scattered over the globe. Most of them have been selected for survival under local environmental conditions. Unfortunately, the great majority of these livestock breeds are of extremely low productivity and are unknown outside their local areas.

Our present domesticated animals are the result of thousands of years of natural selection. Those best suited to a particular environment survived, and those that were poorly adapted either moved to a more favorable environment or perished. During the past two centuries man has made considerable progress in the selection and propagation of animals for a particular environment, and during the past several decades has learned to modify the environment for the mutual benefit of himself and animals. The role of livestock breeding in adaptation to the environment has been discussed in the chapters dealing with animal selection. Methods of modifying the environment will be presented in the chapters involving livestock management.

This chapter will emphasize the physiological mechanisms of adaptation, such as heat production, the maintenance of a uniform body temperature, and the effects of the environment on growth, reproduction, and performance. First, however, we will discuss the various components of the environment.

26-1. THE NATURE OF THE ENVIRONMENT

It is becoming increasingly difficult to define the physical environment, because scientists continue to discover important new environmental factors. Primitive man obviously recognized that the sun and fire provided both heat and light, that body heat could be conserved by draping the body in animal skins, and that trees and caves provided protection from the sun, wind, rain, or snow.

Hippocrates, the Greek physician who is regarded as the father of medical science (460–377 B.C.), was one of the first to describe the nature and effects of the climatic environment in his essay *Airs, Waters, and Places*. He recognized many of the environmental factors or influences that affect plants, animals, and man.

Thermal Environment. All of the higher animals are homoiotherms— they attempt to maintain a constant body temperature. This requires a delicate balance between the heat produced within the animal, the heat gained from the environment, and the heat lost by the animal to the environment.

In certain areas of the earth, high environmental temperatures create problems in adaptation. The digestion and assimilation of food is accompanied by a marked increase in heat production, called the heat increment of feeding. In ruminants the process of fermentation is accompanied by considerable heat production, and in all animals cellular activity of the various body systems, especially muscular activity and work, are accompanied by heat production. All productive functions, whether milk or egg production or rapid growth or fattening, are also accompanied by heat production. In addition, especially during the summer months or in southern or tropical environments, solar radiation is a major factor in affecting livestock productivity because it complicates the problem of heat loss.

The first million years of man and beast, at least in the productive temperate zones, were devoted to the conservation of heat and to staying warm. In other words, the problem was one of adaptation to cold. A cow that calved in the spring and dried up a few months later, or a hen that laid a nest of a dozen eggs and then hatched the chicks, was under little climatic stress during the summer. It is correct to generalize that the animals native to North America, the British Isles, or Northern Europe are cold-weather animals. They are admirably adapted for the production of heat and surviving, when without shelter, in very cold climates.

Solar Environment. The sun performs some essential and some harmful functions for animals. It is a source of heat, the available amount depending on the distance of the animal from the sun; as reviewed by Wright (1954), the mean annual temperature decreases approximately 1°F for each degree of latitude north or south of the equator. The nature and color of the animal covering affects solar heat absorption and thus affects body temperature. Dark pigments are most absorptive of solar energy, and light-colored skin, hair, or hair tips are most reflective. A black animal would thus be at a disadvantage in the tropics. However, solar radiation may produce severe burning of the skin and may be a major causal factor to cancer of the skin (Blum, 1954). Thickened, pigmented skin is more resistant to the sun's rays than white or nonpigmented skin.

The sun may perform a useful function in animal nutrition. The exposure of animals to the ultraviolet rays of the sun activates the precursors of vitamin D in the skin to form vitamin D, necessary for the prevention of rickets. Until livestock producers learned to supply vitamin D in the ration, rickets was common in areas where sunlight is reduced during the winter.

Seasonal changes in length of day have a profound effect on some domestic animals. This is especially true with respect to the reproductive processes of chickens, turkeys, and other avian species, and to those of sheep.

Fig. 26-1. Average age of first egg (sexual maturity) for groups of pullets hatched at weekly intervals throughout 21 months and reared with either natural daylight, O, or with supplementary artificial light from the age of 16 weeks onward, L. Sexual maturity was advanced by providing extra light beginning at the 16th week, except for birds hatched at a season when daylight is increasing—for example, for birds hatched in January. [Morris and Fox, "Light and sexual maturity in the domestic fowl," *Nature*, 181:1453 (1958).]

Birds usually lay and hatch their eggs as day length increases. Prior to 1920 the entire poultry industry was essentially a seasonal business. Eggs were abundant and low-priced in the spring and scarce and high-priced in the winter; poultry meat was a holiday luxury. One of the first applications of environmental control was the use of electric lights in poultry houses for the purposes of stimulating fall and winter egg production (Fig. 26-1).

Sheep, unlike poultry, are stimulated to reproduce as day length decreases (Fig. 26-2). Most breeds of North American sheep have a fall breeding season and the lambs are born in the spring. Changes in day length have little known effect on the reproduction of cattle and swine.

Precipitation and Humidity. Under natural conditions animals were entirely dependent on natural feed supplies, and flourished when foodstuffs were abundant. In the United States the corn belt became an important producing and fattening area for livestock because of a favorable climate for both plants and animals. The great range areas of the West and Southwest were limited by seasonal fluctuations in forage and a lack of feed grains

Altitude. It is difficult to assess the effects of the several climatic factors associated with changes in altitude. With increasing altitude there is a decrease in barometric pressure, a decrease in temperature, and a change in vegetation. Because of the reduction in barometric pressure, animals have difficulty in meeting oxygen requirements. This problem can be solved by increasing the oxygen content of the air or by increasing the atmospheric pressure. Depending upon location with respect to the equator, there may be a marked increase in solar and cosmic radiation at increasing altitudes, and the danger of severe sunburn and even skin cancer may increase.

The effects of altitude upon man have been widely studied. As reviewed by Monge (1954), people who live in the Andes Mountains of South America

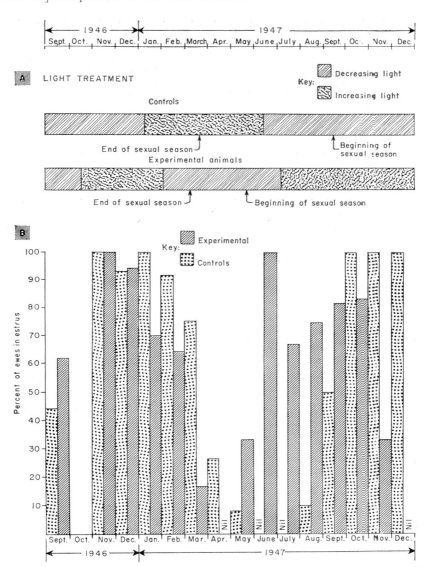

Fig. 26-2. The influence of length of daylight on the sexual season of the ewe. *A,* Control ewes were subjected to normal light conditions; the number of daylight hours in the experimental group was modified by placing animals in a darkened room during part of the day or by placing them in a lighted room to extend the daylight hours. Note that restricting light of the experimental ewes resulted in the sexual season beginning in May as contrasted to September for the control ewes. *B,* Number of heat periods expressed as a percentage of number theoretically possible for each month. [Adapted from Yeates, *J. Agric. Sci.* **39:**1 (1949).]

at altitudes up to 17,000 ft have developed special adaptation mechanisms. Andean man is characterized by a large chest and lung capacity, greatly increased blood hemoglobin, and a highly efficient vascular system. Men going to high altitudes must become acclimated or risk serious illness or death, and, similarly, mountain men must be gradually adapted to conditions at sea level.

As described by Phillips (1949), the yak, llama, alpaca, and vicuña flourish at altitudes above 6,000 ft. The llama is the chief work animal of Andean man and is so well adapted to high altitudes that it is seldom seen below 12,000 ft.

Cosmic Radiation. Knowledge of the occurrence and nature of cosmic rays and their effects upon living organisms has been gained by scientists during the past 50 years. It is clear that the entire surface of the earth is constantly bombarded by extremely high-energy cosmic rays and secondary radiations from cosmic rays. It is equally clear that such radiations may have harmful effects on living material and that at altitudes above 60,000 ft might be dangerous to space travelers.

In 1927 Muller showed that the exposure of organisms to X-ray irradiation affected the rate of gene mutations. He was later awarded a Nobel prize for this discovery, and it led to a series of other researches, which indicate that cosmic radiation has and does play a part in gene mutations and is one of the mechanisms involved in natural selection of plants and animals.

Macro- and Microclimates. With the exception of climatologists, most people think of the world as being made up of a few large climatic areas. We recognize the general boundaries of the tropical zones north and south of the equator, and most of us would classify the southernmost parts of the United States in the subtropical zone. In general, however, we regard most of the continental United States as being in the temperate zone and do not think of Massachusetts as being greatly different from New York, Illinois, or parts of the Pacific Coast area. If we are to understand the role of the climatic environment in livestock production, we must learn to understand the nature of the immediate environment in which a particular animal exists.

The State of California might illustrate the point. In this state the temperature may vary from $-36°F$ to $134°F$; altitudes range from 276 ft below sea level in Death Valley to 14,495 ft at the top of Mount Whitney. The average yearly precipitation varies from less than 2 inches to 109 inches. In some mountain areas as much as 449 inches of snow have been recorded. It could be said that livestock in California might be unable to exist under the natural conditions of certain areas and might be ideally adapted to other regions.

A microclimate is the climate of the immediate area in which a plant or animal lives. The area may be small or large, and many microclimates may exist on a single farm. Most grazing animals seek shade during the summer daylight hours. Trees or wooded areas provide effective shade and cooling because they reduce the warming effects of solar radiation; they are microclimates. Swine spend much of the time in a recumbent position. A warm or cool floor, a wallow, or a mist-type spray are examples of special microclimates for swine. A southern slope that is warmer than a northern slope in winter, any type of shade, building, heat lamp, chick brooder, or hole in the ground—all are examples of microclimates. Fattening cattle in the Imperial Valley of California gain more rapidly when effective shades are used; the shade (a microclimate) modifies the macroclimate, thus improving the performance of cattle exposed to the high daytime temperatures of the desert.

26-2. PHYSIOLOGICAL AND PHYSICAL MECHANISMS OF ADAPTATION

Homoiothermic Animals. Most higher animals, including all the farm animals, maintain a reasonably constant body temperature. The temperature is usually lower in the early morning hours (4 to 6 A.M.) than in the evening (6 to 8 P.M.). This is called diurnal variation. There are differences between mammals; in man the normal temperature is $37\,^\circ C \pm 1\,^\circ C$, and in swine approximately $39\,^\circ C \pm 1\,^\circ C$. In birds body temperature is somewhat higher than in mammals; it is about $41\,^\circ C \pm 1\,^\circ C$ in the domestic fowl.

Physiological Means of Regulating Heat Production. Any living organism might be considered as a system of continuous irreversible reactions, each of which is accompanied by the production of heat. When the animal is at complete muscular rest and in a state when absorption of food from the gut is not going on, we consider the heat being produced as "basal" heat and use the term basal metabolic rate to describe the condition or state. Muscular activity is accompanied by varying degrees of heat production. This heat is called activity heat. The body temperature may rise because of increased tissue oxidation, which is in reality increased heat production, or because of a decrease in heat loss from the body. Males usually have a higher metabolic rate than females or castrate males. The newborn and the senescent animal frequently produce less heat than the growing, highly productive animal. The high-producing dairy cow may produce twice as much heat as a dry cow of the same size. The endocrine and the neuroendocrine systems are vitally concerned in total heat production. An increase in the secretion of the thyroid, anterior pituitary, or adrenal glands, or the male or female gonads may increase metabolic rate. Food is, first of all, a source of animal energy. Following the ingestion of food, especially protein, there is an almost immediate increase in heat production. This rapid increase is called the dynamic effect, or the heat increment of feeding.

It is well known that all nutrients do not have the same gross heat-producing properties. A gram of digestible fat produces about $2\frac{1}{4}$ times more heat than a gram of carbohydrate and 1.6 times more than protein.

The complex process of fermentation within the rumen also produces considerable heat. In man the term basal metabolism is quite appropriate, since the postabsorptive state is reached in the early morning hours. In the ruminant it is a practical impossibility to wait until fermentation stops in the rumen, and the term resting metabolism is more appropriate. When body temperature increases beyond the normal range we say that a fever has been produced. Extreme muscular activity, exposure to high environmental temperatures, or interference with heat loss from the body may cause a temporary increase in body temperature. Shivering is characterized by rapid muscular contractions and is a means of increasing heat production.

One of the most classical procedures of diagnosis in clinical medicine is the measurement of body temperature. Many pathogenic organisms produce an increase in the temperature of the infected host; this symptom is called a fever. In some diseases there may be a pronounced and constant temperature elevation, and in others, such as brucellosis, the temperature may rise and fall (undulant fever).

Physical Means of Regulating Heat Loss. We have discussed the problem of maintaining a reasonably constant rate of heat production in the homoiothermic animal. In the maintenance of a constant body temperature it is extremely important that heat loss be in balance with heat production. Since the farm animals most common in North America evolved in a temperate or cool climate it is not surprising that they are more efficient in heat production than in heat loss. The practical problem of keeping these animals comfortable and highly productive during the summer is one that is receiving considerable attention by research workers.

Heat is lost or removed from living or nonliving objects according to certain physical laws. The exchange of heat from a steam radiator to the surrounding environment is not greatly different than the process of heat transfer from an animal to its environment. In animals heat is lost by radiation, convection, conduction, and evaporation. The most obvious difference between a machine and an animal is the presence in the animal of the vascular and respiratory systems and their regulation by the nervous system.

RADIATION. Heat radiates from the animal surface to the surrounding environment (Fig. 26-3). In man the radiations are in the infrared range between 5 and 20 microns. The amount of heat lost from an animal depends upon the surface area of the animal, the temperature of the skin, the nature of the coat (hair, wool, feathers), and the temperature of the environment.

RADIANT HEAT SOURCES ON ANIMALS
IN SUN & SHADE

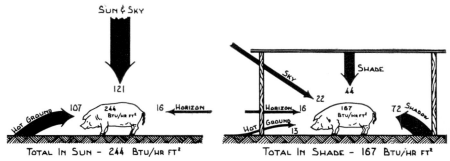

Fig. 26-3. Radiation received by an animal in the sun and in the shade on a typical August day in Imperial Valley in California. [From Bond, Kelly, and Ittner, "Radiation studies of painted shade materials," *Agric. Eng.*, **35**:389 (1954).]

The larger the animal the smaller the surface area with respect to the mass of the animal. A fat, compact beef animal has less body surface per unit of weight than a thin, rangy bovine, and therefore has more of a problem of heat loss by radiation in warm weather. A single animal pulls himself into a "humped" posture in cold weather in order to reduce radiant heat loss. A group of animals may huddle together as a means of conserving heat by reducing radiant and convective heat loss.

CONVECTION. The air immediately adjacent to the skin or coat of the animal is generally warmer than the air in the surrounding environment. The replacement of the layer of warm air by cooler air removes heat from the animal body by convection (Fig. 26-4). As in radiant heat loss, the surface area of the animal and the surface and air temperatures are very important. In addition, the velocity of the air moving over the animal surface will affect the warm air layer close to the animal and will thus regulate convective heat loss.

CONDUCTION. Heat flows from a warm to a cool medium. An animal loses heat by physical contact with the environment. Still air is a very poor conductor of heat. If the body is protected by a thick coat of hair, wool, feathers, or fat, the loss of heat by conduction to the air is small. Cold water is a highly efficient cooling medium. An animal, unless protected by fat as is a whale, will become severely chilled if placed in ice water. A cold wet concrete floor may be an effective cooling device for swine during the summer because it speeds up conductive and evaporative heat loss. During cool weather a cold floor may result in chilling in young animals such as pigs and chickens.

Fig. 26-4. A fan is used to increase air movement for cooling dairy cattle by convection. The warm air next to the skin is replaced by cooler air in this manner. [Photo courtesy T. E. Bond.]

EVAPORATION. The loss of heat from the animal body by the evaporation of moisture from the body surface or from the respiratory system is extremely important (Fig. 26-5). The evaporation of water from the body surface at 91 to 92°F removes 580 gram calories of heat for each gram of moisture evaporated. Evaporative heat loss is a complicated phenomenon. In man, a species with highly developed sweat glands, large quantities of water are lost by visible perspiration. The farm animals, with the exception of the horse, are classified as nonsweating. However, moisture is lost from the body surface of nonsweating species and is called insensible perspiration. Considerable quantities of water are evaporated from the lungs in the expired air. Under most environmental conditions inspired air is cooler and less saturated with moisture than the air in the lungs. Some moisture and heat is always lost from the respiratory system and assists in body heat loss.

The amount of heat lost by evaporation will depend on the moisture in or on the skin, the nature of the animal cover, the temperature, saturation, and velocity of the surrounding air, the respiratory rate and volume, and the temperature and humidity of the inspired and expired air.

Physiological Means of Regulating Heat Loss. The temperature of a well-designed and well-constructed mechanical device is usually regulated by a sensing element called a thermostat. In a modern air-conditioning system the thermostat activates the appropriate equipment for supplying cool or warm air on demand. The correct operative temperature of an automobile motor is controlled by thermostats in the cooling system. In an animal the temperature-sensing mechanism is located in the hypothalamic area of the brain.

The blood is the medium by which heat is distributed rather uniformly throughout the body. As the temperature of the blood flowing through the hypothalamus incrases above the normal range, a series of nerve impulses are dispatched to the body systems concerned with heat loss. If the blood temperature drops below the normal range, the hypothalamus activates those body systems concerned with heat conservation. In addition, there are sensory receptors at the surface of the body that are also concerned with

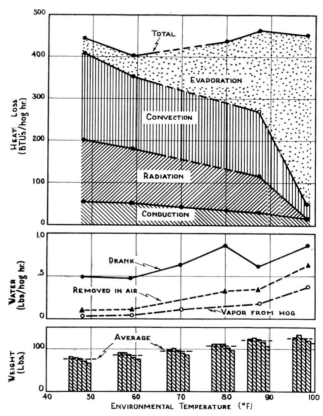

Fig. 26-5. Heat is lost from the body by conduction, radiation, convection and evaporation. Note that as the environmental temperature increases, evaporation accounts for the major portion of heat loss. [From Bond, Kelly, and Heitman, "Heat and moisture loss from swine," *Agr. Eng.*, **33**:148 (1952).]

the flow of blood in the skin, the activity of the sweat glands, and the erection of the hair or feathers.

Let us consider the adaptations an animal would make under several different environmental conditions. At ambient temperatures of 60 to 65°F there is little problem of heat loss. The blood vessels in the skin are constricted; the great mass of blood is confined to the muscular and visceral organs; the sweat glands, if present, are inactive; respiration is normal; and heat production and heat loss are in approximate balance. This temperature range would be in the thermoneutrality zone (comfort zone) for mature farm animals in North America. At lower temperatures, especially between 0 and 50°F, food intake and activity increase in order to increase heat production. The blood vessels in the skin become increasingly constricted and circulation in the extremities may be greatly reduced. It is well known that the fingers, toes, nose, and ears of man are particularly subject to freezing at subzero temperatures. As the body surface becomes chilled, shivering is initiated and body hair or feathers are erected or fluffed in an attempt to trap the layer of warm air next to the skin. This reduces heat loss by convection.

At environmental temperatures between 65° and 85°F, farm animals bring various cooling systems into play. With rising ambient temperature there is increasing dilation of the blood vessels in the skin. Since the blood is an effective medium of heat transfer, bringing the blood to the body surface increases heat loss by radiation, convection, and conduction. In the nonsweating farm animals there is an increase in insensible perspiration or moisture loss from the body surface. Since the principal loss of moisture comes through evaporation from the lungs, the respiration rate increases in order to speed up evaporation and, consequently, heat loss. A high-producing dairy cow requires large amounts of feed for the maintenance of lactation. Feed is a source of energy and high milk production is accompanied by high heat production. Unless effective means of cooling are available, the cow must attempt to reduce heat production as well as to increase heat loss. It is inevitable that, at high temperatures, milk production must decline.

At temperatures above 85°F the body must make drastic adjustments if a state of well-being is to be maintained. Man is fortunate in having well-developed sweat glands. Sweating increases as ambient temperature rises. If the humidity or moisture saturation of the air is low, or if evaporation is increased by air movement, the evaporation of moisture from the body surface has a considerable cooling effect. If evaporation does not occur, heat loss may be insufficient for the maintenance of normal body temperature and an artificial fever is produced. Nonsweating animals must rely on other means of cooling. In hot weather the rate of respiration of cattle, sheep, swine, and poultry is increased beyond the normal range and panting eventually occurs. Well before the panting stage, animals make whatever other adjustments are possible. They seek shade and avoid solar radiation, they reduce feed intake and lie quietly for long periods of time, and if water is available they will wade or lie in it. This increases heat loss both by evaporation and conduction.

The covering of the body surface and the tissues beneath the skin may undergo a variety of modifications. The skin may be pigmented to reduce the damage following solar radiation; this is important at the higher altitudes near the equator. Animals that evolved near the equator may have a sparse hair coat, in contrast to northern mammals. Tropical breeds of sheep have a hairlike wool rather than the dense fleece of temperate-zone sheep. Arctic birds, especially those that are aquatic, have an especially water-repellent and dense feather cover. Animals may deposit fat in thick layers beneath the skin. The fat depots may be a source of energy during periods of food shortage, but they also have a role as insulation for the prevention of heat loss. Aquatic polar animals, especially the whale, are protected from the cold by subcutaneous fat.

Critical Temperature. It has been emphasized that the homoiothermic animal attempts to maintain a constant body temperature. It has also been emphasized that the maintenance of body temperature is an extremely complicated process. The concept of critical temperature cannot be defined or described in a few words. The nutritionist usually defines the critical temperature as that environmental temperature at which the body must increase the rate of cellular oxidation by chemical means for the maintenance of normal rectal temperature. We are aware that there are wide species differences in rectal temperature and in the adaptability of the various species to fluctuating environmental temperatures. Within a species temperature may fluctuate with time of day, age, sex, and individuality. The temperature of the deep viscera is significantly greater than that of the skin, limbs, tail, comb, scrotum, and other specialized areas. Species differences in surface covering, sweat glands, surface area, genetic makeup, and adaptability, and differences in environmental conditions, such as solar radiation, humidity, air movement, and air temperature, have a marked effect on how and when the physical and physiological mechanisms of temperature regulation become activated. The concept that there is a specific and constant critical temperature at which the rate of chemical heat production is increased is unrealistic.

However, there is a temperature range within which a man or animal is most productive or most comfortable (Fig. 26-6). In man this is often called the comfort zone. A naked man at rest would probably select a temperature of 85°F as being the most satisfactory; a worker in a steel mill might prefer a temperature of 50°F or less. The animal physiologist uses the term thermoneutrality zone or physiologically effective temperature. This is the environmental temperature range at which a specific animal is in thermal bal-

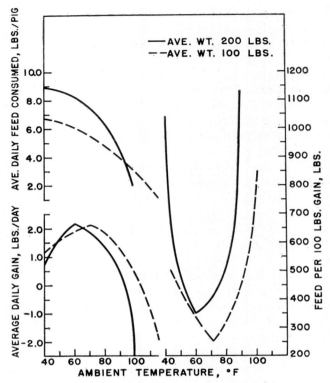

Fig. 26-6. Ambient temperature and its effect on feed consumption, average daily gain and efficiency of food utilization. [From Heitman, Bond, and Kelly, "Effect of temperature on swine," *Calif. Agric.,* **8:**8 (1954).]

ance and is capable of maintaining a normal body temperature without utilizing chemical means of thermoregulation. A highly productive dairy cow would have a much lower optimum temperature than a young calf.

Mechanisms for Adaptation to High Altitudes. Relatively little is known about the specific physiological mechanisms by which animals such as the yak and llama have become adapted to high altitudes. Studies on Andean man by Monge (1954) and current investigations in aviation physiology and space biology have given us a good basic understanding of the problem. With increasing altitude, barometric pressure and oxygen pressure drop. Man, in flight, may compensate for the reduced oxygen pressure by breathing pure oxygen, by increasing the alveolar pressure of oxygen in the lungs, or by pressurizing the surrounding environment. Andean man is forced to compensate for the reduced oxygen pressure in the atmosphere in other ways.

Men living above 10,000 ft undergo an increase in the anteroposterior diameter of the thorax, the length of the sternum, and the thoracic volume. This makes possible a larger exchange of air in the lungs. The capillaries in the lungs increase and expose the air to a greater capillary bed. The numbers of red blood cells per unit volume of blood are also increased, and there is an accompanying increase in total blood volume as well. As blood cell numbers increase, total blood hemoglobin increases and the tissue needs for oxygen are thus provided for. We cannot discuss the role of the blood as a source of tissue nutrients, or its role in the body buffer and excretory systems. It is of interest to note, however, that the adaptation of a man to altitudes above 10,000 ft requires an additional 2 liters of blood, 676 g of hemoglobin, a 100% increase in blood bilirubin, and an 80% increase in blood pyruvic acid (Monge, 1954).

The nonadapted man, flying from sea level to high altitudes, may experience only mild discomfort, or he may become severely ill. There is an immediate increase in respiration rate in an attempt to overcome the reduced oxygen pressure. Even moderate activity may be accompanied by muscular weakness because of oxygen shortage. Rapid, deep respiration may produce hyperventilation and disturb the normal acid-base balance, and chronic mountain sickness may follow.

26-3. SPECIAL ADAPTATIONS OF ANIMALS TO THEIR ENVIRONMENT

Size and Shape. The large animal, having a relatively small surface area with respect to body mass, is best adapted to cold climates. The European breeds of cattle and those native to the colder regions of Asia are, in general, compact and smooth skinned. Zebu cattle and other types that have become adapted to the tropics are more rangy in type and have a larger surface area (Fig. 26-7). Zebu cattle have, in addition to a hump, an extremely loose, pendulous skin with large folds along the ventral surface. The types of sheep that have developed in the cold or temperate zone are not necessarily large, but they are compact and have short necks, ears, and tails.

Type of External Coat. Sheep and goats exhibit marked variation in body covering. Tropical and desert breeds of sheep are covered by short hair rather than wool and some breeds have very little cover on the belly wall. In addition, sheep and goats in warm areas tend to have long legs, necks, ears, and tails. These modifications increase heat loss by radiation and convection. The breeds of sheep that developed in the temperate zone have dense coats of fine wool, and goats from this zone have heavy coats of long hair.

a   b

Fig. 26-7. Brahman cattle (*a*) are better adapted to hot climates than are the more compact breeds such as Herefords (*b*) Crosses of Brahmans with Herefords (*c*) result in animals with better conformation than the Brahmans and better heat resistance than Herefords. Animals of this cross are referred to as "Brafords." [From Ittner, Guilbert, and Carroll. *Calif. Agr. Exp. Sta. Bull.* 745, 1954.]

c

Skin Thickness and Pigmentation. Cattle breeders in temperate areas have tended to stress the desirability of a thin, pliable skin, whereas in parts of South America and Africa a thick hide has been preferred. In tropical and subtropical regions losses due to insects, brush injuries, and solar radiation are sometimes serious. Although conclusive evidence is not available, it does appear that skin type is an important factor in adaptation. A white coat or skin has superior properties with respect to reflection of solar energy, but nonpigmented skin is particularly sensi·.·ᴇ to both burning and skin cancer. Skin pigmentation is desirable in tropical and desert areas.

Adaptation to Feed Shortage. In only a few areas of the world is feed produced at a uniform rate throughout the year. Animals have always adapted themselves to feed surpluses and feed shortages. The farm animals common to North America tend to become fat during the summer, and draw upon the fat depots, when necessary, if food is scarce. The fat-tailed and fat-rumped sheep of Asia and Africa, zebu cattle, and the camel have unusual modifications for fat storage. The fat-tailed sheep develops a massive fat depot when feed is abundant, but after a prolonged feed shortage the tail becomes thin and devoid of fat. The Arabian camel has long been well known for its adaptability to the desert. Its long neck, rangy conformation, and short coat adapt it for heat loss; its broad feet fit it for desert travel, and its fatty hump and efficient moisture conservation fit it for survival with a minimum of feed and water.

26-4. THE CHANGING GEOGRAPHY OF LIVESTOCK PRODUCTION

Under conditions of natural selection, species of plants and animals tend to concentrate where they are best suited. Until the end of the nineteenth century the pattern of crop and livestock distribution was reasonably stable. Dairy cattle and poultry were produced near the centers of population, even though they were not grain-producing areas. The commercial corn belt was reasonably well defined in the central states, and because the supply of corn was localized, the production of swine and the fattening of cattle became centered in the Corn Belt. Beef cattle and sheep were propagated in the range states because there was no alternative use for the land. The South was not regarded as suitable for livestock because of disease and parasite problems, but the primary reason for the absence of livestock was the existence of a cotton and tobacco economy and a lack of feed grains for livestock.

The situation has changed remarkably in the last quarter century. The development of hybrid corn has extended the Corn Belt to southern, western, and tropical areas. The more recent development of grain sorghums, the establishment of the soybean, and improved methods of storing and utilizing forage have completely revolutionized our pattern of livestock production.

The broiler industry has established itself in the Deep South; Florida has become a leading cattle state; cattle feeding has spread to the range and Pacific Coast states; and swine production appears to be moving west and south.

Some of the new environments are more favorable and some less favorable than the old. The heating requirements of broiler houses in Georgia are less rigid than in New Hampshire. However, summer farrowing problems in Georgia are entirely different from those in northern Indiana or Illinois. Several points stand out. Our present strains of farm animals, including poultry, are better adapted to cold than to hot climates. Increased temperatures reduce feed intake, accompanied by reduced growth rate, milk or egg production, and, to some degree, fertility. However, the increasing supplies of feed in the warmer areas, year-round grazing, lower building costs, and the development of new markets are advantages that must be considered.

None of the environmental problems are insurmountable. Many have been overcome and others will be as soon as we understand them. We have developed new crops for these new areas by selection. We can expect to make additional progress by animal selection. The problems we cannot solve by selection we should attempt to solve by improved feeding and management. We should be able to determine the optimum environment for every stage of the life cycle for each species. We should investigate the nutritional requirements under a variety of environmental conditions, and we should develop practical means of environmental control wherever and whenever they are economically justified.

REFERENCES AND SELECTED READINGS

References marked with an asterisk are of general interest.

Blum, H. F., 1954. Effects of sunlight on man. *Meteorological Monographs*, 2:43–49. Amer. Meteorological Soc.

Brody, S., 1956. Climatic physiology of cattle. *J. Dairy Sci.*, 39:715–725.

Findlay, J. D., and W. B. Beakley, 1954. Environmental physiology of farm animals. In *Progress in the Physiology of Farm Animals*. Vol. I. Ed. J. Hammond. Butterworth's Scientific Publications, London. Chap. 6, pp. 252–298.

Hutchinson, J. C. D., 1954. Heat regulation in birds. In *Progress in the Physiology of Farm Animals*. Vol. I. Ed. J. Hammond. Butterworth's Scientific Publications, London. Chap. 7, pp. 299–362.

Ittner, N. R., T. E. Bond, and C. F. Kelly, 1958. *Methods of Increasing Beef Production in Hot Climates*. Calif. Agr. Expt. Sta. Bull. 761.

Johnston, J. E., 1958. The effects of high temperatures on milk production. *J. Heredity*, 49:65–68.

Marshall, F. H. A., 1937. On the change over in the oestrous cycle in animals after transference across the equator, with further observations on the incidence of the breeding seasons and the factors controlling sexual periodicity. *Proc. Roy. Soc. London*, 122:413–418.

McDowell, R. E., 1958. Physiological approaches to animal climatology. *J. Heredity*, 49:52–61.

Means, T. M., F. N. Andrews, and W. E. Fontaine, 1959. Environmental factors in the induction of estrus in sheep. *J. Animal Sci.*, 18:1388–1396.

Monge, M. C., 1954. Man, climate and changes of altitude. *Meteorological Monographs*, 2:50–60. Amer. Meteorological Soc.

*Phillips, R. W., 1949. *Breeding Livestock Adapted to Unfavorable Environments*. FAO Agr. Agricultural Studies No. 1, Food and Agriculture Organization of the United Nations.

*Rhoad, A. O., 1955. *Breeding Beef Cattle for Unfavorable Environments*. University of Texas Press, Austin.

Wright, N. C., 1954. The ecology of domestic animals. In *Progress in the Physiology of Farm Animals*. Vol. I. Ed. J. Hammond. Butterworth's Scientific Publications, London. Chap. 5, pp. 191–251.

Yeates, N. T. M., 1954. Daylight Changes. In *Progress in the Physiology of Farm Animals*. Vol. I. Ed. J. Hammond. Butterworth's Scientific Publications, London. Chap. 8, pp. 363–392.

Section V

Nutrition and Livestock Production

Chapter 27

General Nutritional Considerations

The life of animals, then, may be divided into two acts—procreation and feeding; for on these two acts all their interests and life concentrate. Their food depends chiefly on the substance of which they are severally constituted; for the source of their growth in all cases will be this substance.

ARISTOTLE, *History of Animals, Book VII*

27-1. HISTORICAL

Good nutrition is one of the fundamentals of livestock production; it involves the wise use of available feeds in formulating a palatable, economical, and nutritionally balanced ration for livestock and poultry.

Civilized man has long recognized the essentiality of food for survival for himself and his animals; to absorb nutrients is a basic attribute of living matter. Availability of food and animal feedstuffs has historically influenced man's ability to survive and has limited his attempts to extend the boundaries of his domain for habitation. However, with the domestication of animals and the selective improvement of species based on optimum utility or beauty, the efficient utilization of feeds in animal production became an increasingly critical challenge to husbandmen.

Scientific principles of nutrition have evolved through time, but experimentation is a product of relatively modern times, that has developed largely during the last 150 years. Husbandmen have long been keen observers and have provided a legacy of beliefs and practices for scientific agriculture to test, to quantify, and to elucidate in terms of physiological and chemical significance.

It is worthwhile here to cite, even in a limited way, a few of the historical milestones that have contributed to an understanding of the modern nutritional concepts that presently constitute the foundation for animal nutrition.

Nutrition as a science draws heavily upon the basic sciences of chemistry, physiology, and biochemistry. These sciences began to flourish in the early part of the nineteenth century, and it is not surprising that scientists using these new tools began a detailed, systematic study of the composition of natural products, such as feedstuffs, soils, plants, and animals. The concept that matter was composed of discrete elements early suggested that food might benefit the animal and human by providing specific food elements—proteins, fats, carbohydrates, and minerals—needed for energy or to build its own protoplasm.

By 1850 the German scientist, Justus von Liebig (frequently referred to as the father of agricultural chemistry) began to advance theories on respiration and the production of heat from foods and to speculate on the source of carbon dioxide in exhaled air. He suggested that fats and carbohydrates were used for energy, and he visualized body tissue as a dynamic substance capable of breaking down food proteins and building from them. He and his students published food composition tables based on the concept of nutritive value and equivalent food value that was accepted at that time.

McCollum, in his *History of Nutrition,* credits the brilliant Frenchman, J. B. Boussingault (1802–1887), with being the "founder of scientific agriculture." He fed cows protein-deficient rations—single feeds, such as potatoes or sugar beets—and showed that these animals could not use atmospheric nitrogen, that they needed nitrogenous materials—proteins—for growth and maintenance of body tissue. Also, he determined the amount of different foods needed by an animal per unit of time. The latter observation was a method for quantifying the quality factor of foods and feeds. Similar pioneering studies in scientific agriculture were conducted in England at the now famed Rothamsted Agricultural Experiment Station, between 1840 and 1890, by its founders, J. B. Lawes and J. H. Gilbert. These men, working with cattle, sheep, and hogs, carried out extensive feeding experiments designed to measure the efficiency of conversion of feed to live animal weight. Lawes and Gilbert analyzed the animal body and estimated the production of different body constituents from various feeds.

Another milepost in the history of nutrition was the establishment during the period between 1870 and 1910 of the concept that animals have unknown food requirements not measurable by the chemical techniques then in use and not included in the major nutrient grouping of minerals, carbohydrates, fats, and proteins.

Lunin in the 1880's fed mice a diet of recombined milk constituents with fatal results, while mice fed whole milk survived. This led him to conclude correctly that animals require more than casein, fat, lactose and mineral salts.

Studies in England by Hopkins, and in the United States by Babcock, Hart, Steenbock, Osborne and Mendel, McCollum, and McCay, clearly established that natural feeds contained special nutritional entities—probably in small amounts—essential for good animal nutrition.

In 1912 Funk suggested that the protective principles in foods now known to prevent human maladies, such as scurvy, beriberi, and pellagra, should be called "vitamines" from "vita" and "amine." It took more than 40 years to isolate and to characterize all of the known vitamins present in natural feedstuffs. Nutritional investigations of vitamins helped to establish conclusively that animals require an assortment of nutrients in definable quantities.

27-2. NUTRIENTS DEFINED

Crude proteins, fats, carbohydrates (including crude fiber), minerals, and vitamins are the major components of feeds. These constituents can be determined by using standardized chemical procedures, which insure uniformity in methods of analysis and in reporting of results. The animal consumes the feeds and, following the digestive processes, utilizes specific nutrients (chemical compounds or elements) supplied from these broad categories of feed constituents: for example, amino acids from crude proteins, monosaccharides such as glucose from starches of the carbohydrate group, fatty acids from the fats, and calcium and phosphorus from minerals.

An understanding of the term nutrient is important. A nutrient can be defined as a specific chemical element or compound supplied by, or derived from, the diet and absorbed into the blood from the digestive tract to be used by the body tissues to support physiological processes. Care must be exercised in defining nutrient, because, in common usage, crude protein is referred to as a nutrient in feeds, but it is not utilized in that form by the animal. The large protein molecules, as ingested, are not absorbed in appreciable quantities across the intestinal wall; it is the end products of protein digestion, amino acids, that are absorbed. Thus, amino acids, derived from the breakdown of proteins or formed from nonprotein nitrogenous compounds by the action of rumen microflora, are the nutrients.

27-3. ESSENTIAL AND NONESSENTIAL NUTRIENTS

Nutrients are classified as essential and nonessential. An essential nutrient cannot be synthesized by the body and hence must be supplied in the ration. Essential nutrients are those chemical elements or compounds that are required by a given animal species for its proper nutrition and that are not synthesized in the body; or, if they are synthesized, the rate of synthesis is not adequate to meet the daily requirements. Certain of the amino acids (see Table 27-1), minerals, fatty acids, and vitamins are essential nutrients. A source of energy is essential for animal survival; however, carbohydrates, fats, and under certain conditions, amino acids, can be utilized as nonspecific energy sources in meeting the basic need.

Nonessential nutrients are chemical elements or compounds that are needed by the animal for growth and maintenance, but that can be synthesized by body tissues or are supplied as by-products of the microbial reactions affected by the microflora indigenous to the digestive tract. These chemicals need not be supplied in the ration although many of them are present in natural feedstuffs. Certain amino acids are also in this category (Table 27-1). Similarly, fatty acids, specific carbohydrates (disaccharides, sucrose, starches, and cellulosic feeds for ruminant animals), and certain vitamins serve nonspecific functions or can be synthesized in adequate amounts to meet daily nutritional requirements.

It must be stressed, though, that both groups of nutrients serve specific physiological functions in the body; hence the dietary classification refers to the nutritional and not to the metabolic significance of these biologically important chemicals. Again the amino acids provide an example: nonessential amino acids, such as glutamic acid and alanine, are a part of every protein molecule and are therefore needed by the animal to synthesize body proteins.

The essentiality of nutrients is not the same for all animal species, nor does the requirement remain the same quantitatively through the animal's life-span. Quantitatively, the requirement is influenced by physiological need; for example, it is higher during the rapid growth phase of the immature animal, and in the mature animal the requirement is usually increased by the stress of work, reproduction, or lactation. There are also marked differences between species—especially between ruminants and nonruminant animals.

The basic difference between ruminants and simple-stomached animals is the rate and extent of microbial synthesis of certain nutrients by the microflora of the gastro-intestinal tract. In addition, microflora in the rumen can digest cellulose, a carbohydrate found in forages, but very little cellulose is digested by monogastric animals. As an illustration of the importance of microbial synthesis, ruminants do not require the essential amino acids listed in Table 27-1 because these acids can be synthesized in the rumen from nonprotein nitrogen compounds such as urea. Even amino acids from proteins undergo changes in the rumen. Much remains to be learned about ruminal synthesis, but it is established that the anatomical differences in digestive systems markedly influence nutritional requirements. Again it must be stressed that both the ruminant and nonruminant need a similar assortment of nutrients for growth, maintenance, and reproduction; but methods of supplying the daily nutrient allowances differ. These differences are reflected in the nutritional requirements of the animal.

These basic concepts of nutrition will be further amplified as the physiological role of the various nutrients are discussed.

27-4. ROLE OF NUTRIENTS IN THE BODY

Amino Acids. Nutritionists and producers recognize the importance of adequate protein intake on growth, reproduction, and efficiency of production for all classes of livestock. (This aspect of nutrition will be discussed in greater detail in Chapter 30.) The protein portion of all feedstuffs serves as a source of the amino acids, the fundamental chemical units composing all proteins. Schematically, one can represent an amino acid as

$$R-\underset{\underset{NH_2}{|}}{\overset{\overset{H}{|}}{C}}-\overset{O}{\overset{/\!/}{C}}-OH$$

where **R** represents one or more carbon atoms or combinations of carbon atoms, additional amino groups, hydroxyl groups, or sulfur. There are at least 20 known amino acids that can be linked together in all possible sequences and chain lengths to form the many kinds of proteins found in the plant and animal kingdoms.

The characteristic chemical structure of proteins suggests some of the essential functions of amino acids in the body. The synthesis of tissue protein, for example, is accomplished by enzymatic processes whereby amino acids are combined in specific sequences into chains (polypeptides) characteristic of the species or specific tissue: for example, the proteins of glands, hair, muscles, and blood. Genetic factors predetermine the amino acid sequences of proteins.

Amino acids under certain conditions are metabolized to yield energy for general body use in the same way as carbohydrates and fats are used (Chapter 29).

NEW TISSUE PROTEIN. Growth, reproduction, maintenance of viable cells and components, and tissue repair each involves the formation of new cells (tissues) to which the protein component helps give structure and function. Since amino acids are necessary for protein synthesis, it is pertinent to note the nutritional classification of amino acids on the basis of their essentiality in the ration or diet for monogastric animals. In Table 27-1 the current state of knowledge is summarized under three headings: essential, semi-indispensable, and nonessential or dispensable according to the scheme used by Block et al (1956) and Frear (1950).

TABLE 27-1. | *Nutritive classification of the amino acids.*

Essential (Indispensable)	Semi-indispensable	Nonessential (Dispensable)
Histidine	Arginine*	Glutamic acid
Lysine	Tyrosine†	Aspartic acid
Tryptophan†	Cysine†	Alanine
Phenylalanine	Glycine*	Proline
Methionine	Serine*	Hydroxyproline
Threonine		
Leucine		
Isoleucine		
Valine		

* Arginine and glycine are essential for chicks and turkeys. Serine will spare or replace glycine.
† Tyrosine will spare but not completely replace phenylalanine. Cystine will spare but not completely replace methionine. Nicotinic acid will spare but not completely replace tryptophan.

INTERSTITIAL BODY FLUIDS, BLOOD, AND CELLULAR COMPONENTS. Amino acids are also needed to build the protein in essential regulatory functions of the body and its cells. Proteins in interstitial fluids (between cells) help maintain body fluid balance. The proteins in a colloidal state exert an influence on the osmotic pressure of solutions, intra- and extra-cellular, thereby affecting the movement of water across semi-permeable cell wall membranes.

The two major components of the blood, plasma and red blood cells, have important and physiologically significant characteristics that result from their protein constituents. Hemoglobin in the red blood cells and myoglobin of the muscle are examples of proteins that contain minerals in the molecule (conjugated proteins). Hemoglobin contains iron; it has the capacity to combine loosely with oxygen, and it functions as a mechanism for transporting oxygen from the lungs to the various tissues of the body.

All cell components, such as the nuclei, chromosomes, and cell wall membranes, contain protein.

PROTEINS AS ENZYMES. Enzymes are biological catalysts and regulators essential in chemical reactions in the body. Since they are proteins, they can be formed only if certain amino acids are available. As mentioned in Chapter 2, a part of the protein of muscle consists of enzymes.

PROTEINS AS HORMONES. Hormones regulate certain physiological activities of the body (Chapter 20). Some of the hormones are proteins.

Carbohydrates. Carbohydrates are the most important source of energy in livestock rations. Chemically, carbohydrates occur in foods and feeds, such as monosaccharides, disaccharides, and polysaccharides. Glucose and fructose are examples of a monosaccharide; sucrose one molecule each of glucose and fructose) and lactose (one molecule each of glucose and galactose) are common disaccharides and are the predominant carbohydrates in cane and beet sugar and in milk sugar, respectively; the two most important polysaccharides are starch and cellulose. Whether carbohydrates are absolutely essential components of the diet is still subject to question. The body has the ability to convert most amino acids and the glycerol component of fats to glucose. The question becomes somewhat academic, however, because such carbohydrates as starch and cellulose are usually cheaper sources of energy than are either protein or fats. The manner in which carbohydrates supply energy will be discussed in Chapter 29.

The animal tissues derive energy from the metabolism of fats, deaminated amino acids, and carbohydrates. Energy, regardless of its source, serves the animal's requirements for energy used in growth, muscular activity, the maintenance of body temperature, digestion and metabolism, and reproduction and lactation. Energy-yielding compounds can have a sparing effect on the requirement for other nutrients. To illustrate, the level of carbohydrate in the ration can influence the efficiency of utilization of proteins. Reference has already been made to the derivation of energy from the metabolism of the amino acids. If the energy level of a ration is below the requirements for a given level of production or activity, the use of amino acids (proteins) for energy competes with other physiological needs for amino acids, such as tissue formation, to the possible detriment of the animal. In immature animals this use of amino acids for energy might induce less efficiency by substitution of more costly protein for carbohydrates. In the adult (mature) animal, the need for energy and protein is somewhat less and hence less critical.

Fats (Lipids). The fat or lipid component of feedstuffs provides the animal with two basically different types of nutrients—essential fatty acids and a nonspecific source of energy. A few brief comments about the chemical makeup of fats will help in understanding their role in nutrition. Fats are made up of glycerides, which, chemically, are esters composed of fatty acids and glycerol. When these glycerides are split enzymatically by lipolysis during digestion, they yield the fatty acids and glycerol, both of which can be metabolized to yield energy (calories) to the body pool for use as discussed in the preceding section. The metabolizable energy value of fatty acids is about 2.25 times more per unit of weight than that of carbohydrates (1 g of fat contains 9 kcal; 1 g of carbohydrate, 4 kcal; 1 g of protein, 4 kcal).

Besides being a nonspecific energy source, dietary fat is a source of fatty acids that appears to function as an essential nutrient under certain conditions. Animals (rats have been used primarily in these studies) that have been fed fat-free diets have developed deficiency symptoms. These conditions respond favorably to supplementation with the unsaturated fatty acids, linoleic (two double bonds) and linolenic (three double bonds). One may deduce that other fatty acids required in physiological processes are synthesized by animal tissues. It is known also that the vitamin, pyridoxine, favorably influences the fatty acid requirement and presumably the efficiency of utilization or synthesis of linoleic and/or linolenic acid. The specific fatty acid requirement is the most critical in the young, growing animal; the needs of the adult animal appear to be satisfied more easily.

Another function of fat is that it aids in the absorption of vitamins A, D, and E, presumably because these vitamins are fat soluble.

Minerals. The subject of mineral requirements of livestock is an intriguing aspect of animal nutrition. Minerals (ash) constitute approximately 5% of the animal body weight. The mineral elements can be classified as macro- or micro-elements, depending upon the quantitative requirement for each. Trace amounts are expressed frequently in parts per million (ppm) or in milligrams per pound of ration, whereas requirements of macro-elements— such as calcium, phosphorus, and sodium—are expressed as a percent of the ration, for example, 0.25 to 1.0%. Usually classified with the minerals are the halogens (iodine, chlorine, and fluorine), which are important nutritionally.

MACRO-ELEMENTS. The macro- or major elements that have been demonstrated as being required by the animal body are calcium, sodium, potassium, phosphorus, chlorine, sulfur, and magnesium.

MICRO-ELEMENTS (trace minerals). This classification includes iodine, zinc, manganese, cobalt, copper, iron, and molybdenum. Other micro-elements that seem to be biological requirements are selenium and fluorine. The biological requirement for barium, strontium, and bromine is not as well supported by experimental evidence.

ROLE OF MINERALS. The physiological role of minerals is diverse, and a few examples are cited here to illustrate the importance of minerals in animal nutrition. The various functions of minerals can be classified as follows.

Body structure components. Detailed analysis shows that animal tissues contain measurable amounts of 30 elements, but of these only a few serve a significant structural role in body tissues. The most obvious example is in the skeletal structure. Calcium and phosphorus are the principal minerals in bone, occurring as a calcium phosphate; Crampton and Lloyd (1959) state that 99% of the calcium and 80% of the phosphorus in the body are located in the bones and teeth. Calcium, phosphorus, and other elements are also found in soft tissues and fluids of the body, where they play important roles. Some of these roles will be touched upon in the sections to follow.

Acid-base balance. Throughout the body, fluids—blood, digestive secretions, and saliva—maintain characteristic acid-base balances measured as pH or hydrogen ion concentration. A solution of pH 7.0 is neutral or in acid-base equilibrium. Blood, for example, is maintained at pH 7.4 with only slight deviations therefrom. In the stomach, acid conditions facilitate digestion by the enzyme, pepsin; the *p*H is normally in the acid range of 2.0 to 2.5.

The mechanisms involved in acid-base balance are complicated, but the role of minerals such as sodium, potassium, calcium, and magnesium in establishing and maintaining a given acid-base balance is significant. Bicarbonates, sulfates, phosphates, organic acid salts, and complexes with proteins form strong buffering systems that neutralize gastric acids or the acidity or alkalinity of foods and liquids ingested.

Water balance in body fluids. Water movement across semi-permeable membranes from solutions of differing ionic concentration is governed by a property of solutions called "osmotic pressure." Mineral elements—especially sodium, potassium, calcium, and magnesium in conjunction with organic compounds (acids, protein colloids)—are the chief factors in establishing the osmotic pressure of fluids in biological systems. Through changing concentrations of minerals in their ionic form (Na^+, K^+, Ca^{++}), the body has a mechanism for altering ionic concentrations and thus altering osmotic pressure. It is a known chemical phenomenon that water can be forced to move through semi-permeable membranes of the cell wall from solutions of low ionic concentration to solutions of higher concentration. The net effect of these forces, governed in part by mineral ions, is to equilibrate body-fluid movement from inside the cell to the extra-cellular fluids surrounding it, and vice versa. Sodium and potassium are especially important in maintaining ionic tissue concentrations. High sodium loss—via the urine, feces, sweat —necessitates replacement; hence, the high salt requirement, in comparison to that for potassium, which is characteristically an intracellular component and therefore more static. Cellular and tissue water movement depends, too, on the physiological role of the cells involved; for example, the water movement into kidney cells will differ from that into muscle cells.

Activators in enzyme systems. Almost every mineral element functions as an activator of an enzyme system. For example, magnesium, potassium, and phosphorus are important as cofactors (activators) in the step-by-step enzymatic reactions facilitating energy metabolism; these reactions are explained in greater detail in Chapter 29. Of the essential micro-elements, copper, iron, manganese, and zinc also function extensively as activators or cofactors of enzyme systems.

Vitamin-mineral relationships. Minerals can influence vitamin requirements. Similarly, mineral utilization can be affected by either the availability or lack of vitamins. This relationship can be illustrated by citing two well-established physiological observations. Calcium and phosphorus utilization for bone development in young animals is dependent on an adequate amount of vitamin D being present. Salenium, an element that, when ingested in excess of 5 ppm, is detrimental to most animals, functions in a synergistic manner by sparing or replacing the vitamin E requirement, as exemplified by the prevention of "white muscle disease" in lambs that are fed small amounts of selenium.

The vitamin B_{12} cobalt interrelationship is an example of another vitamin-mineral relationship in which the mineral, cobalt, is a component of the vitamin molecule. Research, mostly in the United States, Australia, and New Zealand, has established that the cobalt deficiency symptoms observed in cattle and sheep are in fact caused by the lack of vitamin B_{12}. In ruminant animals, microbial vitamin B_{12} synthesis in the rumen usually is sufficient to meet the animal's daily requirement, if adequate cobalt is supplied in the ration. In monogastric animals, vitamin B_{12} is synthesized in the lower gut by the microflora but absorption is inefficient; hence, vitamin B_{12} must be supplied in the ration of these animals.

Hormones. The metabolic regulatory hormone, thyroxine, formed by the thyroid gland, contains iodine. Dietary iodine deficiency causes the thyroid to enlarge, resulting in a condition known as goiter. The establishment of iodine as a preventative for endemic goiter is one of the significant nutritional-medical research achievements of the past 75 years.

Biological oxidation-reduction mechanisms. Minerals are part of many complex biological systems of living tissues. One such system is found in hemoglobin, a vital component of blood. Iron, as a part of the hemoglobin molecule, holds oxygen on the surface of the molecule by a weak chemical bond for transport from the lungs throughout the body. In the tissue where oxygen tension (supply) is low, the iron gives up oxygen to the tissue. Iron serves in a similar manner as a part of the highly sensitive oxidation-reduction enzyme system known as the "cytochromes." A series of these enzymes, through changes in the valences of iron (ferric–ferrous), transports electrons to effect important steps in energy metabolism. Other illustrations involving other metabolic systems could be cited for copper-containing enzymes.

Vitamins. A vitamin can be defined as an organic compound required in minute amounts and functioning as an accessory nutrient in facilitating normal physiological development and maintenance of animal and plant life. Thus, vitamins are not nutrients in the same sense as amino acids, minerals, carbohydrates, and fats, but function as "accessory nutritional factors" with no structural or energy-yielding role, as such. Vitamins function at the cellular level by catalyzing enzymatic processes that are involved in energy transformation and utilization, regulation of metabolic processes, and biosynthetic reactions of animal tissues.

DEVELOPMENT OF VITAMIN CONCEPT. The development of the vitamin concept is a fascinating story. McCollum (1957), in his history of the development of nutrition research, discusses early work on the vitamins. When early investigators found that rations composed of purified nutrients prepared from milk (protein, carbohydrates, mineral salts, fat, and water) failed to support growth, they correctly concluded that milk and other feedstuffs contained, in addition to the then known constitutents, minute amounts of unidentified substances that are essential to life. By 1913 there was evidence that two factors existed: (1) a "water soluble B," and (2) a "fat soluble A." Water soluble B was found to contain groups of compounds now identified and referred to as the B complex group. More fat-soluble vitamins, including vitamins D, E, and K, were isolated and identified and are now grouped with vitamin A, the name of the original "fat soluble A." Thiamine (vitamin B_1) was the first vitamin to be synthesized, but today all the vitamins listed in Table 27-2 are available in crystalline form, and their structures have been determined. Several vitamins, such as ascorbic and nicotinic acids, were on the chemist's shelf long before it was known that they were nutrients.

FUNCTION OF SPECIFIC VITAMINS. A few illustrations of vitamin function are cited below to amplify the major roles of these compounds. The role of the vitamins in metabolism will be discussed in greater detail in Chapter 29, "Metabolic Pathways in the Utilization of Nutrients."

COFACTORS. The water-soluble (B complex) vitamins function as enzyme cofactors (adjuncts), thus indirectly affecting such processes as physiological energy utilization, protein synthesis from amino acids, and fatty acid oxidation and synthesis. Vitamin E is believed to be involved in oxidation-reduction processes, electron transfer.

BIOLOGICAL REGULATORY MECHANISMS. The body's regulatory processes are accomplished by enzymatic processes. Vitamins function as essential components of enzyme systems responsible for affecting these reactions. Vitamin D regulates calcium and phosphorus metabolism and bone growth; vitamin A governs the photoreceptor process of the eye; vitamin B_6 pyridoxine, has an important influence on the intermediary metabolism of the essential fatty acids and the amino acids; niacin functions in the synthesis and metabolism of the amino acid, tryptophan; pantothenic acid and pteroylglutamic acid regulate the transfer of carbon atoms in the synthesis of an entire gamut of

TABLE 27-2. *The vitamins.*

Water-soluble Vitamins (B Complex)	Fat-soluble Vitamins	Other Vitamins
Thiamine (vitamin B_1)	Vitamin A	Ascorbic acid
Riboflavin (vitamin B_2)	Vitamin D	Inositol
Niacin	Vitamin E	Choline
Pyridoxine (vitamin B_6)	Vitamin K	Para-aminobenzoic acid
Pantothenic acid		
Biotin		
Pteroylglutamic acid (folic acid)		
Cobalamine (vitamin B_{12})		

biological compounds through transfer of single- and two-carbon fragments; vitamin K, along with calcium, functions in the clotting of blood.

PREVENTION OF GROSS DEFICIENCY SYMPTOMS. The expression of vitamin deficiencies in animals varies with the vitamin involved, usually reflecting directly or indirectly the physiological derangement caused by the lack of adequate vitamin intake. In immature animals, a common symptom is reduced growth rate, which is usually associated with lowered efficiency of feed utilization. Examples of other vitamin deficiency symptoms in animals are these: reproductive failure, reduced vigor, greater susceptibility to disease, bone deformities, failure of blood to coagulate normally, rough and lusterless hair coat, and skin lesions.

Some of the classic symptoms of vitamin deficiencies in humans illustrate the value of observations of gross deficiency symptoms in pinpointing research on vitamin function: the lack of thiamine was found to cause beriberi; a deficiency of niacin was one of the causative factors in pellagra; pernicious anemia patients responded to vitamin B_{12} therapy; vitamin D supplementation of children's diets prevented rickets; vitamin C from citrus fruits in the diet prevented the development of a condition called scurvy, which in early times was a plague to sailors on long voyages. Today it is difficult to find uncomplicated vitamin deficiency symptoms in populations that recognize the importance of proper diet to human health and well-being.

MICROBIAL SYNTHESIS OF VITAMINS IN THE DIGESTIVE TRACT. Ruminants differ from nonruminants also in the level of intestinal synthesis of B complex vitamins. The lower digestive tract of all animals supports a microflora that synthesizes vitamins that are absorbed from the tract and utilized by the animal. The role of intestinal synthesis of vitamins has been indicated in animal studies that used bacteriostatic additives to reduce or to eliminate microbial growth in the gut. For example, a growth response to biotin in the rat is much easier to demonstrate if sulfa drugs or antibiotics are added to the ration. Coprophagy (eating of feces) must be prevented in rats and mice in order to demonstrate a response to certain B vitamins, since the feces contain a high level of B vitamins. The capacity for B-vitamin synthesis is greater in ruminants and in horses than in other mammals. It is unnecessary to add B vitamins to supplement most cattle, horse, or sheep rations, but adequate vitamins, either as a supplement or in the natural feedstuff, are essentials for good poultry and swine rations as well as for the rations of very young ruminants, whose bacterial synthesis of these vitamins is inadequate.

27-5. HOW NUTRIENTS ARE DISCOVERED

Our understanding of the function of specific substances required for good animal nutrition has evolved at about the same rate as basic scientific knowledge in the fields of chemistry, physiology, and biochemistry. Present-day poultry and livestock are the products of adaptation and selection, beginning with a more primitive form of animal capable of obtaining its nutrition on a somewhat haphazard basis under natural conditions. The availability of feed under natural conditions was, and still is for wild species, a natural selecting force. As man learned to feed domesticated animals, he was presumably guided in the selection of feed by apparent animal preferences for feeds plus the recognition of differences among species.

Throughout the eighteenth century and well into the nineteenth, scientific developments in analytical chemistry, microscopy, inorganic chemistry, and refinements in animal production led to studies on the composition of feeds —especially those that seemed to give the best results in terms of animal production. The pioneering studies of Boussingault, Lawes and Gilbert, and Leibig have been cited in Section 27-1. As scientists began to refine their methods for analyzing feeds and to develop methods for expressing nutritive value, it became apparent that the nutritional efficacy of a ration could be estimated partly on the basis of its content of known chemical substances. The following approaches are examples of those that have been used in the attempt to determine which chemical substances are nutrients.

Body Composition. The occurrence of minerals in body tissues—for example, the occurrence of calcium and phosphorus in bone—indicates that the minerals must be provided in the diet inasmuch as they cannot be synthesized within the body. When a mineral is identified, it is necessary merely to determine in what form the mineral can be provided for its efficient utilization. On the other hand, minerals that are present in minute amounts in the body (trace minerals) may go undetected. Body composition cannot be used as an indicator of the nutritive value of specific organic substances either, because these organic substances may be synthesized within the body. Therefore, this approach to the discovery of the nutritive value of specific chemicals has limitations.

Naturally Occurring Deficiencies and Their Prevention. A great many of our nutrients have been discovered as the result of "naturally occurring" deficiencies. The discovery of thiamine, the water-soluble B vitamin, is an example. Dr. Eijkman, a Dutch physician in the East Indies, reported in 1897 that the disease known as beriberi occurred in prisoners that had been fed polished rice and that the addition of rice polishings (the outer coats of the rice kernel) to the diet resulted in an improvement in the physical condition of the prisoners. The procedure from this point was to determine what chemical constituent in the rice polishings prevented the disease. A span of 39 years elapsed before Williams and Cline (1936) established the chemical structure of the substance present in rice polishings, thiamine, which prevented beriberi.

$$
\begin{array}{ccc}
 & & CH_3 \\
N{=}C{-}NH_2 \cdot Cl & & | \\
| \quad | & & C{=\!=}C{-}CH_2CH_2OH \\
H_3C{-}C \quad C{-}CH_2{-\!-\!-}N & & \diagup \\
\| \quad \| & \diagdown & | \\
N{-}CH & & CH{-}S
\end{array}
$$

THIAMINE HYDROCHLORIDE

Thiamine is a constituent of cocarboxylase, a coenzyme, which is a metabolic essential for all species. It is not a dietary essential for ruminants, however, because it is synthesized by microorganisms in the rumen.

Accidentally Occurring Deficiencies. One could cite many examples of accidentally occurring deficiencies, but let us take vitamin E as an example. In studying the nutritional requirements of rats, Evans and Bishop (1922) found that when rats were placed on a certain diet for two generations, normal growth was attained but reproduction was impaired, owing to faulty implantation. A "substance X" in wheat germ oil and in lettuce prevented this defect. This substance is now known as vitamin E. Previously, growth had been widely used as an index of vitamin deficiency, but the vitamin E studies gave warning that a vitamin deficiency might be expressed in the breakdown of other physiological activities. One should not be misled by the term "accidentally." As is true with most "accidental" discoveries, it was the knowledge and experience of the investigators that made it possible for them to recognize the significance of their observations.

Discovery of Essential Nutrients by Use of Purified Diets. To determine which amino acids must be present in the diet (essential amino acids) and which can be synthesized within the body, W. C. Rose at the University of Illinois began, in 1930, feeding rats on purified diets that were designed to provide adequate supplies of all nutrients except that a different amino acid was missing from each (Rose, 1938). In this manner he was able to determine which amino acids must be present in the diet. His work introduced a new concept: all amino acids are nutrients in the sense that they can be ingested and play a role in metabolism but only certain of the amino acids *must* be present in the diet.

27-6. CONCLUDING REMARKS

Most chemical substances in foods are broken down to simpler substances —for example, proteins to amino acids—in the digestive tract before absorption into the blood. Minerals, such as sodium chloride, or simple organic substances, such as glucose, may be absorbed from the intestine without change. In general, any substance absorbed into the blood from the digestive tract and playing a role in metabolic processes may be considered to be a nutrient; however, some chemical substances in plants, such as the "estrogenic" substance sometimes found in subterranean clover, are functionally hormone-like. Until we know more about the differences in modes of action of nutrients and hormones, it will be difficult to make a clear-cut distinction. The fact that certain of the vitamins and amino acids may be synthesized in the rumen or in the large intestine has been a complicating factor in the study of nutrients; it is this phenomenon that sometimes determines if a given nutrient is an "essential" one.

The roles of nutrients in the body are diverse: carbohydrates and fats serve as sources of energy; amino acids are precursors of the tissue proteins; vitamins and some minerals act as coenzymes; minerals have important roles as constituents of certain proteins (iron in hemoglobin) or of vitamins (cobalt in vitamin B_{12}) and, in addition, play an important role in maintaining the osmotic pressure of the body tissues and fluids.

As with drugs, many uninformed people subscribe to the view that if a little is good, why not take a lot. Excessive amounts of nutrients can be dangerous; one is less likely to encounter problems of excess with natural foods than with purified nutrients, because the amounts in foods are not usually present in concentrations sufficiently high to be dangerous.

REFERENCES AND SELECTED READINGS

References marked with an asterisk are of general interest.

Anonymous, 1955. *Methods of Analysis.* 8th ed. Assoc. of Official Agricultural Chemists, Washington, D.C.

*Block, R. J., K. W. Weiss, H. J. Almquist, D. B. Carroll, W. P. Gordon, and S. Saperstein, 1956. *Amino Acid Handbook,* C. C. Thomas, Springfield, Illinois.

*Brody, S., 1945. *Bioenergetics and Growth.* Reinhold, New York.

*Crampton, E. W., 1956. *Applied Animal Nutrition.* W. H. Freeman and Company, San Francisco.

*———, and L. E. Lloyd, 1959. *Fundamentals of Nutrition.* Freeman, San Francisco.

———, and V. G. MacKay, 1957. The caloric value of TDN. *J. Animal Sci.,* 16:541.

Evans, H. M., and K. S. Bishop. 1922. On the existence of a hitherto unrecognized dietary factor essential for reproduction. *Science,* 56:650.

*Frear, D. E. H., 1950. *Agricultural Chemistry.* Van Nostrand, New York.

Guilbert, H. R., and J. K. Loosli, 1951. Comparative nutrition of farm animals. *J. Animal Sci.,* 10:22.

*Lamb, C. A., O. G. Bentley, and J. M. Beattie, 1958. *Trace Elements.* Academic, New York.

*Maynard, L. A., and J. K. Loosli, 1962. *Animal Nutrition.* 5th ed. McGraw-Hill, New York.

*McCollum, E. V., 1957. *A History of Nutrition.* Houghton-Mifflin, Boston.

*Morrison, F. B., 1956. *Feeds and Feeding.* 22nd Ed., Morrison Publishing Co., Ithaca.

*NRC, 1959. Joint United States–Canadian Tables of Feed Composition No. 659 and Publication No. 585, Composition of

Cereal Grains and Forages, 1958. Nat. Acad. of Sci. Nat. Research Council, Washington, D.C.

Rose, W. C., 1938. The nutritive significance of the amino acids. *Physiol. Rev.* 18:109.

*Rosenberger, H. R., 1951. *Vitamins.* Interscience, New York.

Swift, R. W., 1957. The caloric value of TDN. *J. Animal Sci.,* 16:753.

*Underwood, E. J., *Trace Elements in Human and Animal Nutrition.* Academic Press, New York.

Williams, R. R. and J. K. Cline, 1936. Synthesis of vitamin B_1. *J. Am. Chem. Soc.,* 58:1504.

Chapter 28

Estimation of Nutritional Value of Feeds

28-1. INTRODUCTION

The nutritional requirements of all animals can theoretically be met by supplying purified ingredients in the minimum proportions needed for a specific metabolic function. In practice, however, the nutritional needs are supplied by one or more feeds that contain many nutrients in varying proportions. The ultimate desire of the livestock feeder is to use these feeds so that all of the nutrients required are present, when needed, to meet the animals' needs as economically as possible. Since the utilization of nutrients in feeds varies in different animal species and the economic value of feeds fluctuates rapidly, the critical livestock feeder must constantly make changes in the ingredients of his feeding program. He must accomplish these changes without upsetting the nutrient balance of his animals and still remain competitive in the livestock-feeding business. Many measures of the nutritive value of feeds have been developed to facilitate this task. These measures include chemical composition, digestibility, energy utilization, and comparative feeding trials.

28-2. THE CHEMICAL COMPOSITION OF FEEDS

Feeds are normally sold on the basis of their chemical composition, from which their potential feeding value can be estimated. The usual bag of commercial feed is identified by name and chemical composition (Fig. 28-1).

If the feed has an unusual amount of ash, the specific mineral may be listed as well. Excessive ash has no nutritional value. If the feed is a good source of certain vitamins, the vitamin content may also be listed.

FEED SUPPLIERS, INC.

DUMORE FORMULAE

DuMore Milking Ration 15%

GUARANTEED ANALYSIS

Crude Protein.........Minimum 15.0
(Includes not more than 3% protein from nonprotein nitrogen)
Crude Fat, %...........Minimum 3.8
Crude Fiber, %..........Maximum 5.5
Ash, %................Maximum 7.2

INGREDIENTS

Manufactured by

FEED SUPPLIERS, INCORPORATED
YOURTOWN, U.S.A.

Fig. 28-1   A typical feed tag.

Although labeling requirements are specified primarily for concentrates, there is a trend toward voluntary labeling of roughages, the need for which is equally important. U.S. Federal Hay Grades are based primarily on visual appraisal, which includes color, amount of foreign material, and leafiness. The correlation between the grade of hay and its feeding value is not high, primarily because stage of maturity is given only limited consideration in the grading system. Because an inverse relationship exists between the fiber content and the feeding value of alfalfa hay, chemical determination of fiber content provides the basis for a more accurate estimate of its feeding value than can be obtained by visual appraisal (Meyer and Lofgreen, 1956).

High protein feeds are usually more expensive than energy feeds. Since protein in excess of an animal's nutrient requirement is used primarily for energy, minimum quantities should be used. Large quantities of protein are not detrimental if the protein source is economically feasible.

For ruminants, protein quality is of minor importance, but for nonruminants the amount of the different essential amino acids may be critical. For example, lysine is the limiting amino acid in cereal grains and the sulfur amino acids are the limiting acids in legume seeds, such as soybeans and peanuts. A proper balance between different feed sources usually supplies the amino acids needed without the supplementation of specific amino acids. Sources of nonprotein nitrogen may be used to supply part of the protein needs for ruminants, but these have little value for monogastric animals.

Fat is a concentrated source of energy and may be supplied by using feeds that normally contain large quantities of fat or by adding either vegetable or animal fat to other ration components. The percent of fat in a feedstuff, particularly concentrates, usually reflects the energy content of the feed. The energy supplied by fat is readily available to the animal. Feeds containing approximately 6% fat are usually more valuable than feeds containing less; an even higher percentage may be feasible for poultry and swine. After the minimum requirements of the animal have been met the additional amounts that may be desirable depend upon the economic use of fat as a source of energy. Unsaturated fats are fluid at room temperature and have a detrimental effect upon carcass quality in swine because they frequently produce a soft, oily carcass.

Fiber is utilized only to a limited degree by nonruminants. Thus the higher the amount of fiber, the lower the value of the feed. Ruminants can digest an appreciable percentage of the fiber. Lignin is usually associated with the fibrous portions of the plant and is indigestible. In addition, because lignin is frequently encrusted upon the cell walls, it decreases the availability of other nutrients.

The value of specific mineral or vitamin supplements depends upon the cost per unit of the available nutrient. Both synthetic and natural sources should be considered. Abnormal levels of certain minerals, such as flourine and molybdenum, are toxic and should be avoided, regardless of other nutrients that the feed may contain. Certain vitamins are not stable, and care should be taken to assure that their level is stabilized in the feed.

28-3. THE DIGESTIBILITY OF FEEDS

The chemical composition of feeds provides much information concerning their nutritional value, but knowledge of the composition of feeds becomes even more important if this composition is indicative of the availability of nutrients to the animal. Chemical composition of feed by itself does not indicate nutrient availability. A biological measure must be employed for this purpose. Most of the digestible nutrients are available for use by the animal for productive purposes.

The digestibility of feeds is determined on the basis of individual nutrients. In other words, the digestibility of protein of alfalfa, the energy of barley, or the cellulose in range plants is determined. The digestibility is determined by measuring the intake of a nutrient and the fecal excretion of the same nutrient. The difference is assumed to be digested. In practice, this procedure becomes more involved (Table 28-1). It is not too difficult to quantitatively weigh feed for this purpose. However, a sample of feed that is similar to the feed actually consumed must be obtained for chemical analysis. Quantitative collection of fecal material is more difficult. The different methods that have been used are described below.

Stall Collections. Animals, generally male, are confined to stalls designed for quantitative fecal collection without urinary or feed contamination. Many designs have been used. What kind is employed depends upon the animal species, the sex, and the objectives of the study. A gravity separator usually suffices for male animals, but for females, a separate urine conduit is usually necessary to prevent contamination (Fig. 28-2).

TABLE | *Procedure for determining nutrient digestibility.*
28-1. |

Step	Example	How Obtained
(1) Feed intake	10 kg. alfalfa	Feed records after weighing
(2) Feed composition	15% protein	Laboratory analysis
(3) Nutrient intake	$10 \times 15\% = 1.5$ kg. protein	$(1) \times (2)$
(4) Fecal excretion	4 kg. dry matter	Experimental records after collection, drying, and weighing
(5) Fecal composition	12.5% protein	Laboratory analysis
(6) Nutrient excretion	$4 \times 12.5\% = 0.5$ kg. protein	$(4) \times (5)$
(7) Digestible nutrient	$1.5 - 0.5 = 1$ kg. protein	$(3) - (6)$
(8) Percent digestible nutrient	$\frac{1}{1.5} \times 100 = 67\%$ digestible protein	$\frac{(7)}{(3)} \times 100$

Fecal Bag Collection. Harnesses have been developed to support special fecal collection bags and to quantitatively separate feces and urine. Although animals must be subjected to the necessary experimental equipment, the use of fecal collection bags does not restrict the animal to unnatural confinement. A heifer with this equipment is shown grazing an irrigated pasture in Fig. 28-3.

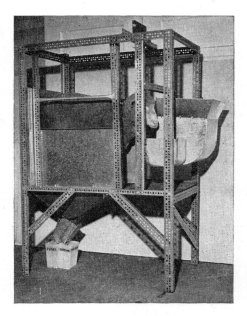

Fig. 28-2. A metabolism stall for quantitatively separating urine and feces from sheep during a digestion trial. The cage is adjustable for animals of varying size. Feces and urine may be separated in male or female sheep. [Courtesy of Dr. L. E. Harris, Utah State University.]

Fig. 28-3. A heifer grazing an improved pasture with a fecal and urinary collection bag. With the use of portable equipment, intake, digestibility, and metabolizable energy may be determined in all types of pastures. [Courtesy of University of Nevada.]

Use of Indicators. Fecal excretion can be estimated with the use of digestion indicators. Ideally, these indicators are substances normally contained in feed that are indigestible without absorption, that pass through the digestive tract at a uniform rate, and that are easily determined chemically. Although the ideal indicator does not exist, chromic oxide, lignin and chromogen(s) are indicators that have been used most extensively. If the intake of indicators is known, the fecal excretion can be calculated from the ratio of the indicator in the feces to the fecal dry matter. An appropriate formula for this calculation is as follows:

$$\text{Fecal dry matter output} = \frac{\text{amount of indicator fed}}{\%\ \text{indicator in fecal sample}}$$

For example:

$$\frac{10\ \text{g } Cr_2O_3\ \text{fed}}{0.2\%\ Cr_2O_3\ \text{in fecal dry matter}} = 5000\ \text{g fecal dry matter}$$

If the ratio of the indicator to the nutrient is determined in the feed and in the feces, the digestibility of the nutrient may be determined without quantitative measurement of either feed intake or fecal excretion. The formula for this determination is as follows:

$$\text{Digestibility} = 100 - \left(100 \times \frac{\% \text{ indicator in feed}}{\% \text{ indicator in feces}} \times \frac{\% \text{ nutrient in feces}}{\% \text{ nutrient in feed}}\right)$$

For example:
% digestible cellulose

$$= 100 - \left(100 \times \frac{5\% \text{ lignin in feed}}{10\% \text{ lignin in feces}} \times \frac{20\% \text{ cellulose in feces}}{25\% \text{ cellulose in feed}}\right) = 60\%$$

Many feeds do not constitute a complete ration when fed alone. If one of these is to be tested, a basal diet composed of a single feed, such as hay, or a mixture of feeds is fed and the digestibility determined. In a second trial, the digestibility of nutrients in the basal ration plus the feed under question is determined. It is assumed that the digestibility fo the basal ration remains constant in the two trials and the digestibility of the nutrients in the added feeds can be calculated by difference.

It is impossible to determine the digestibility of each feed ingredient, in all possible combinations with various amounts of other feed ingredients, for the different species of livestock performing greatly different functions. If the digestibility of feed nutrients is known, then the nutrient requirement can be met by a wide variety of feedstuffs. The calculation of rations for farm livestock and other animals is based on this premise.

Many factors influence the digestibility of feeds. A few of these will be mentioned and discussed, but other factors may also play a role. A nutritionally complete ration is usually more digestible than a ration deficient in one or more nutrients. Certain feeds are either more or less digestible when fed in combination with other feeds than when fed separately. As the level of protein and fat increases in the diet, their digestibility increases, but high percentages of fiber or other relatively indigestible ingredients causes the digestibility of protein and fat to decrease. The rapid movement of feed ingredients through the digestive tract, no matter what the cause—laxative ingredients, disease, poor feed preparation—decreases the digestibility. With cattle, the addition of readily digested ingredients, such as sugars or starches, decreases the digestibility of the more fibrous portions. The digestibility of rations by monogastric animals is not affected by amount of intake, but that of ruminants is variable. Amounts of feed ingested has no effect on digestibility if long or chopped roughages are fed. If ground or pelleted forages, mixed roughage and concentrate rations, or grain containing roughages like corn silage are fed, then digestibility generally decreases as feed intake increases.

Low digestibility may not always be a serious factor. Animals use nutrients first to meet their maintenance requirements. Nutrients supplied in excess of maintenance are then used for productive purposes. If low digestibility is accompanied by an increase in feed intake, then the total intake of digestible nutrients rather than the degree of digestibility of ingested nutrients becomes most critical. The use of pelleted high fiber feeds for ruminants follows this pattern. The use of a relatively inexpensive feed may decrease the percentage of digestibility of a ration, but this decrease can be compensated for by the feeding of more of this ingredient, with the net result of less expensive costs of production.

28-4. ENERGY EVALUATION

Although the availability of individual nutrients is important, it is frequently desirable to use measures that will evaluate the entire feed. Since the quantitative need for energy exceeds the requirements for all other nutrients and since many nutrients may be used for energy in addition to more specific functions, available energy is usually used as the primary measure of the value of a specific feed. The various measures of energy used for the evaluation of livestock feeds are as follows: total digestible nutrients (TDN), gross energy, digestible energy, metabolizable energy, and net energy.

Total Digestible Nutrients (TDN). The most common measure of available energy in livestock rations in the United States at this time is total digestible nutrients. The data for the calculation of TDN is obtained from digestion trials. The formula is given below:

% TDN = % digestible protein + % digestible crude fiber +
  % digestible nitrogen-free extract (NFE) + % digestible fat × 2.25

Digestible protein, crude fiber, and nitrogen-free extract have approximately equal energy value for livestock production, but digestible fat has considerably more energy value, and in the formula for the calculation of TDN, digestible fat is multiplied by the factor 2.25 to compensate for this difference.

Most roughages have a TDN value ranging from 45 to 65%. Concentrates generally range from 65 to 85%. The TDN content is inversely related to the amount of fiber in the feedstuff, and high-fiber feeds, either concentrate or roughages, usually have lower TDN values. High-fat concentrates have the highest TDN value, and in a few rare instances may actually exceed 100% because of the method of calculation. Several compilations of digestion coefficients from which TDN is calculated have been made by Morrison (1956) and Schneider (1947). The interested student is referred to these for detailed information on specific feeds. TDN values for roughages, in comparison to those for concentrates, are too high for production because of the greater amount of metabolizable energy lost as heat. On the other hand, the high TDN value of roughages is accurate for maintenance in cold climates because the extra heat is needed to maintain the body temperature.

Gross Energy. Any combustible organic material liberates energy when it burns. The total amount of heat liberated with complete combustion is referred to as the gross energy. Usually the gross energy is expressed as the number of kilocalories (kcal) per gram. A wide variety of materials, such as heating oils, heating gases, sawdust, as well as livestock feeds, have gross energy values. Of livestock feeds, the gross energy is highest in fats (average 9.4 kcal/g), intermediate in proteins (average 5.65 kcal/g) and lowest with carbohydrates (average 4.15 kcal/g).

Digestible Energy. The gross energy of a feed or other material provides no clue about the amount of energy available for livestock production. The amount of digestible energy is more useful for this purpose.

If energy considered as a single nutrient, although many nutrients yield energy, the determination of digestible energy follows the same pattern as other digestible nutrients, namely, caloric intake minus fecal calories equals the amount of calories of digestible energy.

As may be noticed in the calculation, digestible energy only accounts for losses in digestion. Although slightly different from TDN, it is similar. Few values are available for digestible energy and for routine determinations it may be estimated from the TDN content. Swift (1957) has indicated that one lb. of TDN is equivalent to 2,000 kcal digestible energy.

Metabolizable Energy. All of the digestible energy is not available for productive purposes and other energy losses occur. When energy losses in the urine and combustible gases (primarily methane, CH_4) are subtracted from the digestible energy, the remaining energy is called metabolizable energy. Generally the energy losses in the urine and combustible gases account for about 5 to 10% of the gross energy of the feed. Thus fecal energy losses are much greater. Even though metabolizable energy is a refined measure of energy utilization, under usual feeding conditions the relation between TDN, digestible energy and metabolizable energy is high. Thus as one value is known, the others can be readily determined by appropriate equations. Certain species of forage, such as range plants, may contain large quantities of absorbable essential oils that are eliminated in the urine without metabolism. For such forages metabolizable energy is far superior to digestible energy as a measure of energy availability. Sheep, and possibly some game animals, may consume such forage, but cattle generally select other species. Urine and gaseous energy losses are usually greater when cattle are fed roughages rather than concentrates. Losses are usually greater in ruminants than in nonruminants. Fecal and urine separation and evaluation can be accomplished in the same way as described for digestive studies. The amount of combustible gases can be determined through the use of a respiration chamber or calculation from digestible carbohydrates.

Net Energy. Net energy is metabolizable energy minus the heat increment. Heat increment refers to the increase in heat production following consumption of feed and includes the increased heat from fermentation in the gastro-

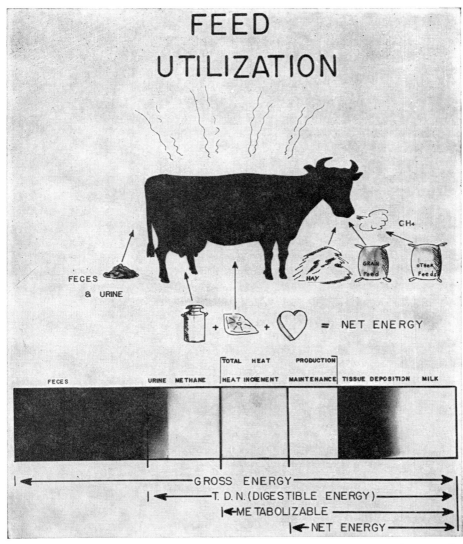

FEED UTILIZATION

CH₄

FECES & URINE

GRAIN Feed   oTHeR Feeds   HAY

+ + = NET ENERGY

FECES	URINE	METHANE	TOTAL HEAT PRODUCTION			
			HEAT INCREMENT	MAINTENANCE	TISSUE DEPOSITION	MILK

GROSS ENERGY
T. D. N. (DIGESTIBLE ENERGY)
METABOLIZABLE
NET ENERGY

Fig. 28-4. The partitioning of energy in feed stuffs by a lactating cow. It may be noted that only a small proportion of the total feed energy is converted to energy in an animal product. [Courtesy of W. P. Flatt, United States Department of Agriculture.]

intestinal tract and from nutrient metabolism (Fig. 28-4). In simplified terms, net energy is that caloric portion of feed that actually is included in some animal product or is used for maintenance of the animal. Thus, the caloric content of body gain, milk produced, wool, maintenance, or other animal products related to a specific increment of feed is the net energy of that feed.

The heat increment is determined in an animal calorimeter or indirectly through the use of a respiration chamber. These measures involve the use of much skilled labor and intricate equipment. Some portable equipment has recently been constructed for indirect measurement (see Fig. 28-5). In a respiration trial the retention of carbon and nitrogen in the animal body is measured by determining the intake and output of these two elements. From this information, the gain or loss in body fat and protein is measured, and from a knowledge of the energy equivalent of fat and protein, the energy retention can be calculated.

Another approach to the determination of the net energy of a particular ration for maintenance, growth, or fattening is by the slaughter technique. In rough outline, the procedure is as follows:

1) Start with a sufficient number of comparable animals, so that they can be split into 3 groups.

2) The first group is slaughtered at the beginning of the trial to provide a means of estimating the body composition of the remaining two groups at the start of the trial. The carcass composition can be determined by weighing the carcass in air and in water. Its specific gravity in relation to water can then be calculated. By experiment it has been found that if the

Fig. 28-5. The use of portable equipment to measure gaseous exchange and other excretory products for the determination of net energy with grazing animals. (*A*) Collection bags for urine and for feces. (*B*) Gas meter. (*C*) Tracheal cannula to collect expired air. [Courtesy of W. P. Flatt, United States Department of Agriculture.]

specific gravity is known, the percentage of fat can be determined. It also has been shown by experiment that if the composition of the carcass is known, the composition of the empty whole body can be calculated. From this composition one can calculate the caloric content of the empty whole body.

3) Feed the second group a maintenance ration (level of feeding in which there is no appreciable change in body weight). Slaughter after 150 days of feeding and determine body composition as above. This will provide the information for determining the net energy value of the feed for maintenance.

4) Feed the third group ad libitum for 150 days, slaughter, and determine body composition. By comparing the body composition of animals in the first group with those in the third group, and taking into account the amount of feed needed for maintenance, the net energy value of the feed can be calculated.

If the feed under question is not a complete feed, the net energy of the basal ration may be determined, and the test feed may be added to the basal ration, and its value determined with another group. Another method for determining net energy is to replace a feed in the ration in which the net energy is well known as a result of either calorimetric or comparative slaughter methods. The test feed is then added to the diet in quantities to give a similar response, and the two feeds are considered equal in net energy value. The values obtained can then be equated to a standard unit. The net energy of the feed is then known and can be used to compare with other feeds with similar evaluation. Somewhat similar procedures have been developed to determine the net energy of feeds for milk production.

The net energy values of livestock feeds vary with many conditions, such as the following: amount in the ration, the combination of feeds, balance of nutrients, productive function and species of animals being fed. Net energy is the only measure of available energy that is precise enough to measure these variations. The determination of net energy is the most difficult of all of the energy measures, and only limited values are available. The relation between the various energy measures is shown graphically in Fig. 28-4.

28-5. VOLUNTARY FEED INTAKE

Recent studies have suggested that factors influencing feed intake may be more critical than digestibility or chemical composition. Certainly these factors are not independent of each other. Variations in voluntary intake can modify the total available nutrients more than can the usual variation in digestibility. We know some of the factors that influence voluntary intake. For example, roughage consumption by ruminants increases with increased digestibility of the forage. The bulkiness of the ration determines to some degree the total enrgy intake. As discussed in Chapter 24, lactation increases feed consumption. Disease may also influence consumption. Feed preparation plays an important role. Animals will consistently consume more pelleted, high-fiber feeds than the same feeds in other forms. Voluntary feed intake may become a more important measure of nutritive value, as we continue to learn the many factors that influence it and how to predict its use.

28-6. COMPARATIVE FEEDING TRIALS

The nutritive value of many feeds has been determined by the use of comparative feeding trials. In these tests, the relative value of two or more feeds or two or more rations are compared by feeding them to representative animals. The response, determined by rate and composition of gain or feed per unit of energy of milk produced, or similar criteria, may be used for evaluation. If the response of the tested feed or a certain proportion of a tested feed is compared to a standard feed or ration, the results are more meaningful and may apply to a wider variety of conditions. A disadvantage of this procedure is that the results frequently depend upon the conditions under which the study was performed. The major advantage is that the results may be immediately applied to commercial livestock feeding problems.

28-7. CONCLUSION

Many measures have been proposed and used for evaluating the nutritional value of feedstuffs for livestock. Each of these has its advantages and disadvantages. In recent studies several of these measures have been used concurrently in order to have a maximum amount of information available for evaluation of feeds. This practice minimizes error and provides more complete information for evaluation by both the nutritionist and the successful livestock feeder.

REFERENCES AND SELECTED READINGS

Crampton, E. W., 1956. *Applied Animal Nutrition.* W. H. Freeman and Company, San Francisco.

Harris, Lorin, 1962. *Glossary of Energy Terms.* Natl. Acad. Sci., Natl. Res. Council Publication 1040.

Henderson, C. R., G. E. Dickerson, L. E. Casida, E. W. Crampton, I. L. Lindahl, O. G. Hawkins, A. M. Gladdis, and W. L. Sulzbacher, 1959. *Techniques and Procedures in Animal Production Research.* American Society of Animal Science.

Maynard, L. A., and J. K. Loosli, 1962. *Animal Nutrition.* 5th ed. McGraw-Hill, New York.

Meyer, J. H., and G. P. Lofgreen, 1956. The estimation of the total digestible nutrients in alfalfa from its lignin and crude fiber content. *J. Animal Sci.* 15:543–549.

Morrison, F. B., 1956. *Feeds and Feeding.* 22nd ed. Morrison Publishing Company, Ithaca, N.Y.

Schneider, B. N., 1947. *Feeds of the World, Their Digestibility and Composition.* West. Virginia Agr. Exp. Sta., Morgantown, W. Va.

Swift, R. W., 1957. The caloric content of TDN. *J. Animal Sci.* 16:753–756.

Chapter 29

Metabolic Pathways in the Utilization of Nutrients

29-1. INTRODUCTION

In one short chapter only a glimpse can be given of the complicated mechanisms involved in the utilization of nutrients. Many terms will be used that can only be completely understood after more background in chemistry, biochemistry, and physiology has been acquired. Yet, in our view, beginning students should be introduced to some general concepts concerning the roles of nutrients in the animal body and to a partial understanding of the mechanisms involved in their utilization. The student should not be overwhelmed —even experts in the field are impressed not with the depth of our knowledge but with the many areas in which our understanding is still incomplete.

The foodstuffs that animals eat and assimilate have two principal functions. They furnish the energy required to run the complex machinery of the animal cell. They provide building blocks for the processes of growth, reproduction, and lactation.

Animals need a constant supply of energy. It is sometimes difficult to envision why a nonproducing, resting organism requires energy simply for maintenance. "Resting" is a relative term; a living organism is never at complete "rest." Mammals are constantly breathing, the heart is pumping blood through the blood vessels, the kidney is removing certain products of metabolism; in fact, all tissues are maintaining a low level of activity. However, these very processes have only one main purpose—to transport nutrients to the cell and to remove the waste products formed when the nutrients are utilized.

The most basic explanation of an animal's constant requirement for energy is that life is very precarious. Without energy for maintenance, the existence of an animal, indeed of even one animal cell, with its orderly structure consisting of aggregations of large protein and fat molecules, would be virtually impossible. Thus, energy is required to maintain order within the animal cell.

29-2. THE METABOLIC EQUIPMENT OF THE ANIMAL CELL

Metabolism is the sum of the chemical changes occurring in an organism during the breakdown of food (substrate) and the synthesis of cellular material.

The metabolic equipment, which is responsible for all the chemical transformations occurring in the complex mammalian body, is not relegated to any one organ or tissue. Rather, it is distributed throughout every cell in the body. The student of elementary biology is familiar with the generalized structure of a cell, which is a unit of protoplasm consisting of a nucleus surrounded by cytoplasm, which in turn is enclosed within a cell membrane.

Metabolic processes occur principally within the cytoplasm, which, with the use of the electron microscope, has been shown to have a very intricate structure. The detail of a very small portion of the cell cytoplasm is shown in Fig. 29-1. The mitochondria, small particles that can barely be seen under the light microscope, are present in almost every living cell. A cell may contain 50 to 5,000 of these particles. In addition, most cells contain a fine network of vesicles and tubules known as the endoplasmic reticulum. The endoplasmic reticulum can be demonstrated only with the use of the electron microscope. Several mitochondria and a portion of the endoplasmic reticulum of a liver cell are shown in Fig. 29-1. The mitochondria and endoplasmic reticulum are surrounded by the nonparticulate portion of the cytoplasm or cell sap.

The substances that carry out the metabolic processes are called enzymes and are located within the mitochondria, the endoplasmic reticulum, and the cell sap. Enzymes are catalysts that are peculiar to living matter. They are produced by living cells and accelerate chemical reactions without being used up in the complete process. Enzymes are proteins, but each enzyme has its own characteristic structure, consisting of different combinations of amino acids. Over, 1,000 different enzymes have been identified in microorganisms and in plant and animal tissue. An individual cell contains a large number of enzymes—one at least for each type of chemical compound present in the cell. Dozens of enzymes participate in the breakdown of nutrients or protoplasm for fuel. Enzymes do not float around in the cell at random, but, as

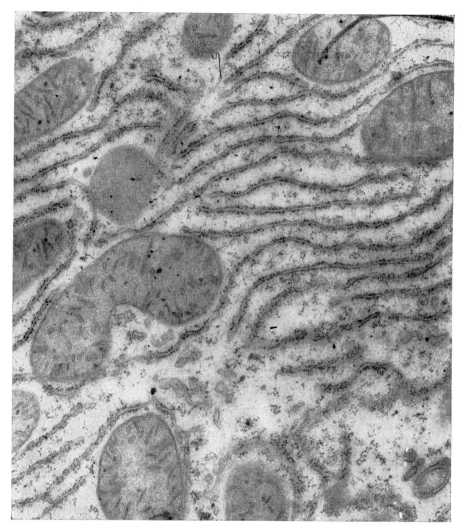

Fig. 29-1. Electron micrograph of a portion of cytoplasm of a rat liver cell. The large rounded or oval structures are mitochondria, and the filament-like strands constitute the endoplasmic reticulum. The outer surfaces of the endoplasmic reticulum are studded with small dense particles—the ribosomes—which are the sites of protein synthesis. [From Porter, K. R., The Ground Substance. In *The Cell*, v. 2, J. Brachet and A. E. Mirsky, eds., Academic Press, New York and London, 1961.]

parts in a machine, they have their own specific locations within the cell.

Many enzymes will not operate except in the presence of a nonprotein, organic (carbon-containing) substance known as a coenzyme. Often these organic compounds are derived from substances that cannot be synthesized by the animal cell, such as a B-vitamin. These vitamins must be obtained from the diet. In addition, many enzymes require the presence of metal ions or activators, which, in some enzymes, are actually part of the structure of the enzyme. These metal ions also must be included in the diet.

29-3. PATHWAYS IN THE UTILIZATION OF NUTRIENTS AS A SOURCE OF ENERGY

Phases of Foodstuff Utilization. Foodstuffs, used as a source of energy, must be broken down (degraded) to simple forms before the energy they contain can be utilized by animals. According to Krebs and Kornberg (1957), the release of energy by the degradation of fats, carbohydrates, and proteins may be said to proceed in three principal phases.

1) Large molecules[1] are hydrolyzed [2] to form much smaller compounds. Starch is broken down to glucose; fats, to fatty acids and glycerol; protein, to amino acids. This process, known as digestion, takes place in the digestive tract where foodstuffs are hydrolyzed prior to absorption, but it can also occur in the cell where protoplasm can be broken down and its energy utilized. Less than 0.6% of the energy contained in food is released during phase 1. This energy is transformed into heat and cannot be used to perform useful work. The reactions that occur in phase one are depicted in Fig. 29-2.

2) Glucose, glycerol, fatty acids, and amino acids are converted to products that can be utilized in a final common pathway (to be explained).

3) The end products of phase two are converted to carbon dioxide, CO_2, and water, H_2O, via the final common pathway.

Energy that can be utilized by the animal to perform work is trapped by the cell during phase two and phase three. Compounds formed during all three phases provide building blocks for tissue and animal product formation.

The scientific study of the biological oxidations that occur during phase three, and to some extent phase two, is said to have begun with the work of Lavoisier, an eighteenth-century French chemist. Lavoisier discovered that when a fuel, such as charcoal, is burned in a fireplace, oxygen is taken out of the air and heat, and carbon dioxide and water are given off. Furthermore, he observed in 1777 that during respiration, animals remove oxygen from the air. He stated ". . . that the production of animal heat in the body is due, at least in greater part, to the transformation of oxygen to carbon dioxide and water." In 1780 he concluded, "Respiration is therefore

---

[1] A molecule is the smallest physical unit of a substance or compound such as starch.

[2] Hydrolysis is a chemical decomposition in which each of the compounds formed when a chemical substance is split takes up the elements of water: $R—O—R' + H_2O \rightarrow R—O—H + R'—O—H$.

CARBOHYDRATE

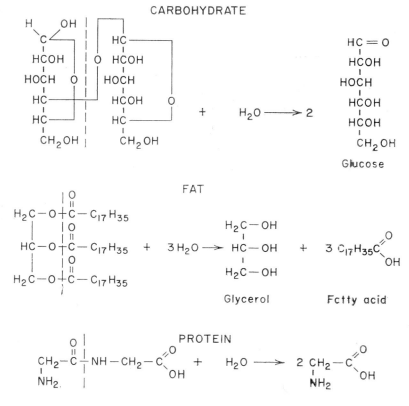

FAT

PROTEIN

Amino acid

Fig. 29-2. The first phase of foodstuff utilization. In reality, carbohydrates (such as starch) and proteins consist of many units of glucose or amino acids linked together, not just two as shown above. The letters C, H, O, and N represent carbon, hydrogen, oxygen, and nitrogen respectively. The broken line depicts the point at which splitting occurs.

a combustion, slow it is true, but otherwise perfectly similar to that of charcoal."

Consequently, the same formula can be used to show the complete combustion of glucose both inside and outside the body.

$$C_6H_{12}O_6 + 6\ O_2 \longrightarrow 6\ CO_2 + 6\ H_2O + energy$$

Almost 200 years have elapsed since Lavoisier formulated his theories. The scientific investigations conducted during the succeeding time period have demonstrated that although the endproducts of fuel combustion inside the body and within the laboratory are the same, the manner in which the animal cell carries out the combustion process is infinitely more complex.

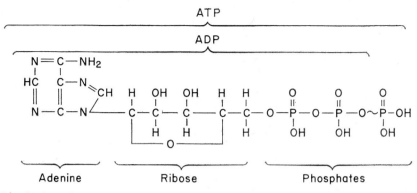

Fig. 29-3. ATP molecule. ATP has one more phosphate group than ADP, which is attached by a bond depicted above by a wavy line. The energy required to form the bond is made available again when the bond is ruptured.

How Animals Trap Energy from Food. Animals cannot use heat from the burning of fuel (food) in the performance of functions, such as muscle contraction, in the manner of heat engines. Heat engines operate under the principle that the temperature of the fuel combusted undergoes a large drop between the combustion chamber and the exhaust. An animal, on the other hand, must maintain a somewhat constant body temperature in order to avoid damaging cellular proteins and other protoplasmic constituents. Therefore, the animal has had to devise a special way to trap the energy released from glucose during combustion before it becomes heat. The animal cell very efficiently traps energy in a form that can be used in many endergonic (energy requiring) life processes, such as muscle contraction. The cell captures the chemical energy of glucose in a high energy compound called adenosine triphosphate, ATP. The complex structural formula of ATP is shown in Fig. 29-3. Theoretically, one mole[3] of ATP is formed from adenosine diphosphate, ADP, and inorganic phosphate, $HPO_4^=$, during a chemical reaction in the cell in which energy equivalent to 8,000 calories[4] is released.

$$ADP + HPO_4^= + energy \longrightarrow ATP$$

The energy that is made available when ATP is hydrolyzed to ADP and phosphate is a property of the structural changes that occur during this re-

[3] A mole is the weight of a substance in grams that is equivalent, numerically, to its molecular weight. A mole of glucose weighs 180 g and is equivalent to 6×10^{23} molecules of glucose.

[4] One calorie is that amount of heat that will raise the temperature of one gram of water from 14.5 to 15.5°C. The chemical energy of a biological compound is often defined in terms of calories, because the energy is determined by measuring the heat produced when the compound is combusted under standardized conditions in the laboratory.

action. Actually, ATP can be said to represent a charged "biological storage battery." When the energy of ATP is transferred, in the presence of a suitable enzyme system, the biological battery is in a discharged state. ADP represents the discharged battery.

According to Lehninger (1955), energy released during the hydrolysis of ATP can be used in cellular functions, which may be placed in three major categories: (1) muscular contraction, (2) biosynthesis of complex molecules such as proteins, (3) transport of substances in the body against a concentration gradient. The energy required by the nervous system can be placed in the third category. Hence, the chemical energy contained in ATP can remain as chemical energy, or it can be converted to mechanical energy, electrical energy, or heat.

The breakdown of one mole of glucose in the body results in the release of 686,000 calories of energy that can be utilized to form ATP. The instantaneous combustion of one mole of glucose in the body could result in the formation of about 85 (686,000/8000) moles of ATP. However, if the cell combusted glucose in one step—as it would be burned outside the body— energy would be released in one great burst. If such an instantaneous release of energy occurred, most of the energy would appear as heat. Not only would there be very little energy captured in the form of ATP, but also this excess heat would be detrimental to the body.

Therefore, for maximum efficiency, the cell breaks down glucose as well as other nutrients in a series of steps, some of which result in the formation of ATP. Because each step does not result in the release of exactly 8,000 calories per mole of reactant, this process itself is not completely efficient, that is, it does not result in the theoretical yield of 85 moles of ATP per mole of glucose degraded. Energy that is not trapped in a chemical form during combustion is released as heat.

Metabolic Pathways. The manner in which fats, carbohydrates, and proteins are broken down during the second and third phases of foodstuff degradation is quite complex. An entire course of study could be devoted to the details of the mechanisms involved. The pathways by which foodstuffs are degraded are illustrated below in order to show how the degradation of food can be used to provide the energy that sustains life. Rote memorization of the steps in each pathway would be impractical at this stage. It is more important to gain a general understanding of the mode of operation of each and its metabolic significance.

THE ANAEROBIC BREAKDOWN OF GLUCOSE. The complete combustion of glucose to carbon dioxide and water, with the resultant formation of ATP, occurs only in the presence of oxygen. However, glucose can be partially broken down with the release of some energy under anaerobic conditions (in the absence of oxygen). These conditions occur in muscle during heavy exercise when the energy requirement for muscle contraction exceeds the rate

ENERGY INPUT   ENERGY TRAPPED

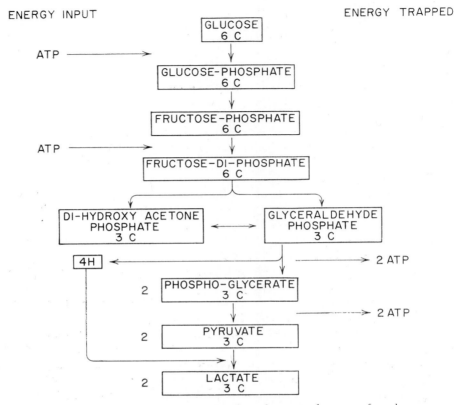

Fig. 29-4. The glycolytic pathway. During the second stage of carbohydrate utilization, glucose is broken down to two 3-carbon compounds. The number preceding the letter C in each box represents the number of carbon atoms in each compound.

of respiration. In fact, oxygen is not required during the second phase of carbohydrate degradation. This phase, which is called glycolysis, is diagrammed in Fig. 29-4. The enzymes that are responsible for carrying out this process are located in the cell sap.

Before glucose can be degraded, it must be activated. Just like fuel in a fireplace that is not combusted until it is ignited with a spark, glucose must be "ignited" in the body. The ignition is accomplished by reacting the glucose with a molecule of ATP. The activated compound that is formed, glucose phosphate, then undergoes transformation into another phosphorylated sugar, fructose phosphate. An additional ATP is required to activate fructose phosphate before it can be broken into two 3-carbon compounds, which are interconvertible. Di-hydroxyacetone phosphate must be converted to glyceraldehyde phosphate before it can be metabolized to pyruvate. The most significant accomplishments of the glycolytic (second phase) pathway are as follows: (1) Two molecules of pyruvate are formed from each molecule of glucose degraded. (2) Four molecules of ATP are formed per molecule of glucose metabolized. This results in a net gain of two molecules of ATP, since the activation reactions used up 2 molecules of ATP. (3) A total of four atoms of hydrogen are removed, two from each molecule of phospho-glyceraldehyde. The hydrogen atoms can be used to reduce[5] pyruvate to lactate.

Less than 10% of the energy of glucose is released during glycolysis. The remaining 90% of the energy of glucose is retained in the two molecules of lactate, which are formed during glycolysis. This energy is released during the third phase of carbohydrate degradation when combustion to carbon dioxide and water is completed.

Glucose phosphate can also be used to synthesize glycogen (a complex polysaccharide consisting of units of glucose linked together). Glycogen is stored, principally in the muscle or liver, and represents a very ready supply of energy that can be used during fasting or during an emergency.

AN ALTERNATE PATHWAY OF GLUCOSE METABOLISM. Enzymes for an alternate pathway of glucose metabolism also exist in the cell sap of tissues such as the liver, the adrenal gland, and the lactating mammary gland. This pathway is often called the hexose monophosphate shunt or the pentose cycle. Activated glucose, metabolized via the hexose monophosphate shunt, is oxidized in such a way that hydrogen atoms and carbon atoms are removed directly from the hexose (6-carbon sugar) molecule, itself, and a 5-carbon sugar (pentose) is formed. The pentose sugars are then the principal metabolites in a series of condensation and exchange reactions, the products of which include 3-, 4-, 6-, and 7-carbon sugars. The final product of the hexose monophosphate shunt, in addition to carbon dioxide and hydrogen, is fructose. The fructose can be converted to glucose and remetabolized via the hexose monophosphate shunt, or it can be broken down along the glycolytic pathway. The net reaction is presented below:

6 glucose phosphate + 12 NADP + 6 H_2O $\longrightarrow$

6 CO_2 + 5 fructose phosphate + 12 NADPH$_2$[6]

Estimates of the amount of liver glucose metabolized via the hexose monophosphate shunt range from 10 to 30% (White, Handler, and Smith, 1964; Fruton and Simmonds, 1958). The hexose monophosphate shunt is of greater quantitative significance in adipose tissue and in mammary gland tissue. This pathway is of particular importance because it provides pentoses

---

[5] Reduction occurs when there is acquisition of electrons; the term oxidation is defined as a loss of electrons. Each hydrogen atom contains one electron.

[6] Hydrogen atoms from glucose metabolized via the hexose monophosphate shunt are transferred to the hydrogen carrier, nicotinamide adenine dinucleotide phosphate, NADP. NADP, with hydrogens attached, is depicted as NADPH$_2$.

for the synthesis of nucleic acids (the genetic material of the cell) and hydrogen atoms ($NADPH_2$) for fat and steroid hormone synthesis.

THE TRICARBOXYLIC ACID CYCLE. The chemical energy that is not liberated when glucose is broken down anaerobically to lactate, is released during the third stage of nutrient degradation when pyruvate is oxidized in the mitochondria to carbon dioxide and water via the tricarboxylic acid, TCA, cycle. This scheme, which is diagrammed in Fig. 29-5, was originally proposed in 1937 by Krebs. The TCA cycle derives its name from four intermediate compounds—citrate, cis-aconitate, isocitrate and oxalosuccinate—which contain three carboxyl, COOH, groups as part of their structure. The student is cautioned not to be overwhelmed by the detail of Fig. 29-5. Again, it is more important to gain an understanding of the central role of the TCA cycle in metabolism than to commit to memory ever reaction that takes place.

Pyruvate does not enter the TCA cycle directly. It is decarboxylated to form carbon dioxide, CO_2, and acetyl coenzyme A, acetyl CoA. This reaction, and the structural formula of coenzyme A, are shown in Fig. 29-6. Acetyl CoA is an active form of the organic 2-carbon acid, acetic acid. The activation of acetate with coenzyme A is analogous to the activation of glucose with phosphate from ATP. All important reactions involving acetate will not proceed unless acetate is first linked to the sulfur of coenzyme A.

In its activated form and in the presence of an appropriate enzyme system, acetyl CoA condenses with oxaloacetate to form citrate, which is then metabolized via the TCA cycle, as shown in Fig. 29-5. The accomplishments of the TCA cycle may be summarized as follows: (1) regeneration of oxaloacetate, which then can condense with another molecule of acetyl CoA; thus, the TCA cycle is indeed cylic in nature; (2) formation of two molecules of carbon dioxide for every 2-carbon unit of acetate oxidized; (3) transfer of hydrogen atoms from TCA intermediary compounds to hydrogen acceptors.

The overall significance of the TCA cycle is related to its ability to act as a final common pathway by which breakdown products of glucose as well as fat and protein are completely combusted.

If, in reality, only one ATP were formed for each pyruvate oxidized via the TCA cycle, as indicated in Fig. 29-5 (during the formation of succinate from α-ketoglutarate), less than 5% of the energy released in the process would be captured. However, more than 40% of the energy released is trapped as ATP when the hydrogen atoms removed in the oxidation process are transferred to oxygen via the electron transport chain, which is also located in the mitochondria. The electron transport chain (Fig. 29-5) consists of 2 coenzymes—NAD[7] (nicotinamide adenine dinucleotide) and FAD (flavin-adenine dinucleotide)—and the cytochrome system, which is composed of a series of iron-containing proteins.

---

[7] NAD, until 1961, was designated as diphosphopyridine nucleotide, DPN+. The new nomenclature is more descriptive of the structure of the coenzyme. However, both terms —NAD and DPN+—are in current use.

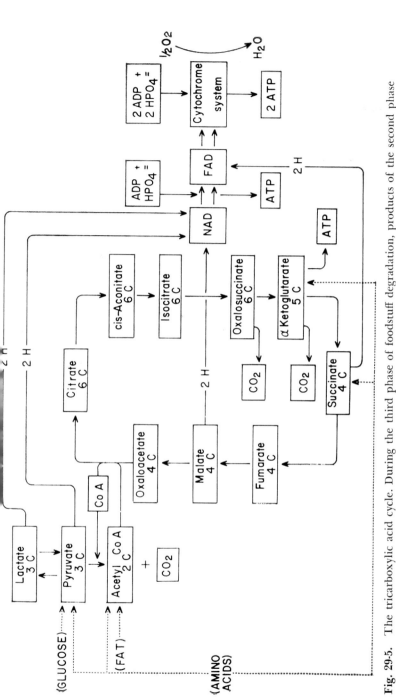

Fig. 29-5. The tricarboxylic acid cycle. During the third phase of foodstuff degradation, products of the second phase of carbohydrate, fat, and protein degradation enter the "final common pathway" (see Fig. 29-4 for second phase). Hydrogen atoms removed during this process are transported to molecular oxygen via the electron transport system. The number preceding the letter C in each box represents the number of carbon atoms in each compound. NAD and FAD denote nicotinamide adenine dinucleotide and flavin-adenine dinucleotide, respectively. [Adapted from "How Cells Transform Energy" by A. L. Lehninger. Copyright © Sept. 1961 by Scientific American, Inc. All rights reserved.]

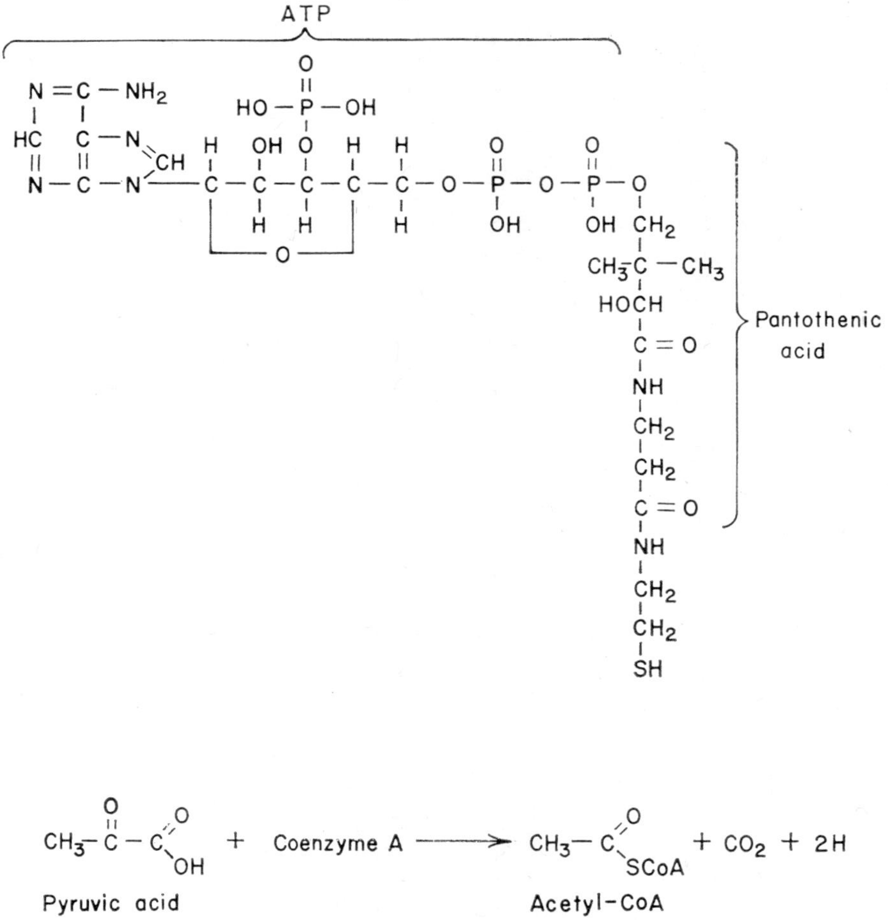

Fig. 29-6. Formation of acetyl CoA from pyruvic acid. The top part of this figure represents Coenzyme A.

Energy that is released as the hydrogen atoms are transported through the electron transport chain is used to synthesize ATP. The points at which ATP is formed during the complete oxidation of glucose to carbon dioxide and water are summarized in Table 29-1. Whenever a molecule of glucose is completely combusted in the cell, 40 molecules of ATP are formed. Since 2 molecules of ATP are required for activation during glycolysis, the net yield of ATP molecules is 38.

UTILIZATION OF FAT. Fat is a very important source of energy even though livestock rations, unlike human diets, are generally low in fat. Fat represents the principal energy store of the body. The glycogen (carbohydrate) stores of the adult human being are about 150 g under normal conditions, but about 20% of the body is composed of fat.

TABLE 29-1.	*Points of formation of ATP during glycolysis and oxidation through the TCA cycle.*

Point of Formation	No. of ATP Molecules Formed per Molecule of Glucose Combusted
Directly, during the series of reactions glucose $\longrightarrow$ lactate	4
From hydrogens removed from lactate and transferred through the electron transport system	6
From hydrogens removed from pyruvate and transferred through the electron transport system	6
Directly, during the reaction α-ketoglutarate $\longrightarrow$ succinate	2
From hydrogens removed from TCA cycle compounds and transferred through the electron transport system	22
	40

During the initial phase of fat degradation, each molecule of fat is hydrolyzed to one molecule of glycerol and three molecules of fatty acid (Fig. 29-2). During the second phase of degradation, the glycerol and the fatty acids are converted to compounds that can be oxidized via the TCA cycle, the final common pathway. Glycerol is converted in several steps to pyruvate. The fatty acids are broken down, 2-carbon units at a time, in the manner shown in Fig. 29-7. Four hydrogen atoms are transferred during the fatty acid oxidations, which precede the splitting off of each unit of acetyl CoA. One pair of hydrogen atoms is transferred to NAD; another pair, to FAD. The hydrogen atoms are eventually transported to molecular oxygen via the cytochrome system, with concomitant synthesis of five molecules of ATP.

During the third phase of fat degradation, the acetyl CoA units condense with oxaloacetate and are oxidized via the TCA cycle. The complete degradation of palmitic acid, a common 16-carbon fatty acid is summarized in Table 29-2.

Utilization of Protein. The 20 or so amino acids released from protein during the first stage of protein utilization, can, like glucose and fatty acids, be used by the body in a variety of ways. They can be used to provide energy. Unlike fat and carbohydrate, protein exceeding an animal's requirement cannot be stored in the body. Body protein increases significantly only during growth and reproduction. The increase in body protein that occurs when the level of protein in the diet is elevated never exceeds more than about 3 to 5% of the total body protein. Dietary protein, which is not converted to tissue protein or broken down and used for fuel, will be converted to body fat or carbohydrate.

Schoenheimer (1942) showed that the metabolism of dietary protein and

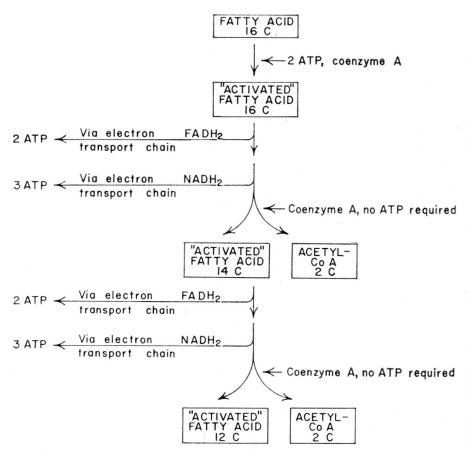

Fig. 29-7. Fatty acid oxidation. During the second phase of fat degradation, long-chain fatty acids are broken down, 2-carbon units at a time. The number preceding the letter C in each box represents the number of carbon atoms in each compound.

that of tissue protein are not separate processes. Even though an animal might daily combust an amount of protein equivalent to the protein content of its diet, part of the combusted protein might come from the tissue and a portion of the dietary protein might be used to resynthesize that tissue. According to Rittenberg 1950), ". . . even in a non-growing organism, amino acids are rapidly built into the proteins of the cell. Since the total quantity of protein remains constant this indicates that an equally rapid destruction of protein must occur to balance the synthesis. This balanced synthesis and destruction is not restricted to the proteins but is true for most of the complex constituents of the protoplasm."

TABLE 29-2. | *Degradation of palmitic acid.*

Reaction	Molecules of ATP Formed or Used Up
Palmitic acid ⟶ palmityl CoA	−2
Palmityl CoA ⟶ 8 acetyl CoA	+35
8 acetyl CoA ⟶ 16 H_2O + 16 CO_2	+96
	129

During the second stage of protein utilization, amino acids from protoplasmic or dietary proteins are also converted to products that can be metabolized via the TCA cycle. However, amino acids are not completely broken down to carbon dioxide and water in the manner of glucose and fatty acids. The animal does not have the ability to oxidize the nitrogen-containing amino group, which is removed from amino acids during the second phase of protein utilization and must therefore excrete it in the urine as urea, $(NH_2)_2CO$. Birds excrete uric acid, $C_5H_4N_4O_3$, instead of urea. Unfortunately, the excretion of the amino acid nitrogen as urea accompanies the excretion of carbon (as urea), which then cannot be oxidized, with a subsequent release of energy, to carbon dioxide. In addition, the synthesis of one molecule of urea uses up 4 molecules of ATP. Thus, the energy of protein cannot be trapped as efficiently as the energy of fat or carbohydrate.

Deamination (removal of the amino group) is the first step in the degradation of amino acids. The major pathway for deamination of most amino acids begins with the transfer of the amino group to α-ketogluterate (Harper, 1963). Glutamic acid, a nonessential amino acid, is formed. The latter can be deaminated to reform α-ketoglutarate and to form ammonia (NH_3). The ammonia can be used to synthesize urea. The carbon skeleton of the amino acid is converted to a compound, which can be oxidized via the TCA cycle.

An example of the metabolism of an amino acid (alanine) is presented below:

$$2 \, CH_3CHNH_2COOH + 6 \, O_2 + 32 \, ADP + 32 \, HPO_4^= \longrightarrow$$
$$(NH_2)_2CO + 5 \, H_2O + 5 \, CO_2 + 32 \, ATP$$

Alanine is converted to pyruvate before it enters the TCA cycle. The points at which other amino acids enter the TCA or glycolytic schemes are shown in Table 29-3.

The number of molecules of ATP yielded during combustion by the major classes of foodstuffs is compared in Table 29-4. The data presented in the table demonstrate that 100 g of fat contain more than twice the amount of energy of 100 g of carbohydrate or protein, although the efficiency

TABLE | *Point of entry of amino acids into TCA cycle or glycolytic pathway.*
29-3.

Pyruvate	*Acetyl CoA*	*α-Ketoglutarate*	*Fumarate*	*Succinate*
Alanine	Lysine	Glutamate	Tyrosine*	Methionine
Cysteine	Leucine	Histidine	Phenylalanine*	Valine
Serine	Isoleucine	Prolines		Threonine
	Tyrosine*	Arginine		
	Phenylalanine*			

Source: Information from Krebs (1964) and Harper (1963).
* The carbon skeletons of tyrosine and phenylalanine are converted to 2 molecules of acetyl CoA and one molecule of fumarate.

with which the animal traps the energy from fat and carbohydrate does not differ appreciably.

29-4. THE SYNTHETIC PATHWAYS

Certain of the products formed during each of the three phases of nutrient degradation can be used to synthesize important biological compounds. Some of these compounds and their metabolic starting materials are shown in Table 29-5.

One of the most important processes carried out by the animal cell is the manufacture of tissue protein. Protein synthesis is especially significant during growth, but it is a process that occurs continuously throughout the life span. Protein synthesis is accomplished by linking amino acids together in the manner shown in Fig. 29-2. The bond which joins two amino acids, –CONH–, is called a peptide bond.

The sites of protein synthesis are the ribosomes, which are associated with the endoplasmic reticulum (Fig. 29-1). Soon after it was shown that the protein hormone insulin contained 51 amino acid residues linked together in a specific sequence, scientists realized that amino acids are assembled on a template (a mold or pattern) during protein synthesis. The present concept of protein synthesis has been developed during the past 10 years and is still actively being studied.

TABLE | *ATP yield of the three major classes of foodstuffs.**
29-4.

Class of Nutrient	Foodstuff Degraded	Moles of ATP Formed (100 g combusted)
Carbohydrate	Starch	23.5
Fat	Tristearin	51.4
Protein	Casein	20.1

* Values from Krebs (1964).

TABLE 29-5. | *Protoplasmic constituents synthesized from metabolic intermediates.*

Starting Material (Metabolic Intermediate)	Compounds Synthesized
Essential and nonessential amino acids	All cellular proteins including enzymes, protein hormones and muscle
Acetyl CoA	Cholesterol, steroid hormones, fatty acids
"Active" succinate	Cytochromes, heme of hemoglobin
Pentose, oxaloacetate, glycine (from phosphoglycerate—see Fig. 29-1)	Nucleic acids (RNA and DNA)
Pyruvate, oxaloacetate, α-ketoglutarate	Certain nonessential amino acids

The cell constituents that make up the protein synthesizing template have not been discussed as yet. They are the nucleic acids. Nucleic acids, like proteins, are large molecules composed of smaller units linked together. Each small unit contains (1) a nitrogenous base, such as adenine shown in Fig. 29-3, (2) a pentose 5-carbon sugar, and (3) phosphoric acid, H_3PO_4. Two general types of nucleic acid are found in the cell—deoxyribonucleic acid, DNA, and ribonucleic acid, RNA. The templates are composed of RNA.

The genetic information of the cell—that information which is passed from parent to child—is coded in DNA, which is located in the nucleus. In a unique manner, still under investigation, DNA passes its coded information to RNA in the ribosome. Protein synthesis then proceeds as shown in Fig. 29-8.

The synthesis of other cell constituents, such as fat or carbohydrate, is completely carried out by enzyme systems located in the mitochondria or cell sap.

29-5. THE METABOLIC ROLE OF VITAMINS AND MINERALS

Vitamins and minerals act as coenzymes or activators for many of the enzymes participating in foodstuff degradation and protoplasm synthesis. The metabolic role of many of the B-vitamins has been well defined and is summarized in Table 29-6. For example, pantothenic acid is a constituent of coenzyme A (Fig. 29-6). As has been shown, all fatty acids must be activated with coenzyme A before they can be degraded to acetyl CoA. Acetate must be activated with coenzyme A before entering the TCA cycle.

The general functions of vitamins A, C, D, E, and K have been established. Vitamin D, for instance, increases the absorption of calcium from the intestine and the deposition of calcium in the bone. However, the role of these vitamins in the degradation of foodstuff and the synthesis of cell constituents, if such a role exists, has as yet not been well defined.

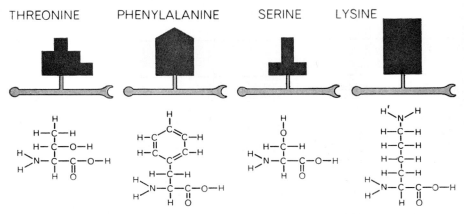

AMINO ACIDS are the building blocks of all protein molecules. Of 20-odd amino acids found in plant and animal proteins, four are symbolized here along with their chemical formulas. A typical cell can manufacture from 1,000 to 2,000 different kinds of protein.

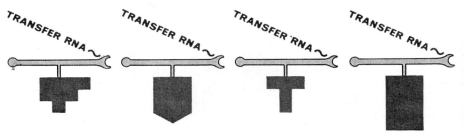

TRANSFER RNA is the name given to a soluble form of RNA that comes in some 20 varieties, one for each of the amino acids. The role of transfer RNA is to "recognize" and transport specific amino acids to the ribosomes of the cell, where they are linked up into protein molecules. Each amino is activated and joined to transfer RNA by a different enzyme.

Fig. 29-8. Protein Synthesis. [From "Messenger RNA" by J. Hurwitz and J. J. Furth. Copyright © Feb. 1962 by Scientific American, Inc. All rights reserved.]

Minerals have a variety of functions in the animal body. They are important in intermediary metabolism for two reasons: they act as activators for enzyme systems and they are constituents of many important biological compounds. Iron, for example, is an essential constituent of the cytochrome proteins. When hydrogen atoms, removed during foodstuff degradation, are transported to oxygen, each hydrogen atom temporarily gives up its electron to cytochrome iron.

The metabolic importance of magnesium, on the other hand, is not as a permanent structural component of an enzyme or coenzyme. Certain metabolic reactions, many also involving ATP, will not proceed in the absence

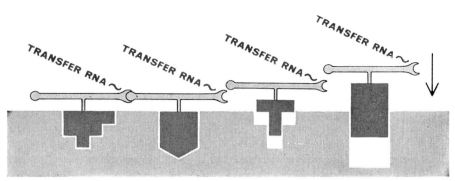

RIBOSOME

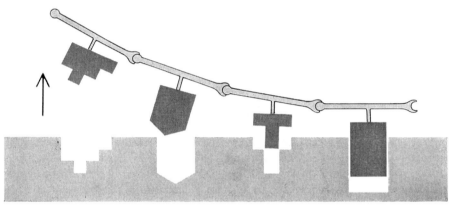

RIBOSOME

PROTEIN SYNTHESIS takes place on the surface of ribosomes, the tiny structures composed of protein and RNA. Amino acids, each carried by its own type of transfer RNA, find their proper sequence on the ribosome (*top*). The amino acids link up to form a protein chain, which then separates (*bottom*) from the ribosome.

Fig. 29-8. (continued)

of magnesium. Enzyme activators like magnesium may form a temporary complex with the enzyme or substrate.

29-6. THE ULTIMATE FATE OF FOOD

Animals convert the food they ingest to one or more of the products shown in the equation below:

Food intake = heat loss + work output +
energy stored as body tissue + animal product (milk or eggs)

All of the food that is metabolized and that is not converted to tissue or

TABLE | *The role of some B-vitamins in metabolism.*
29-6. |

Vitamin	General Function	Reaction or Pathway Requiring Participation of Vitamin
Thiamine	Coenzyme in oxidative decarboxylation	Pyruvate $\longrightarrow$ acetyl CoA α-ketoglutarate $\longrightarrow$ succinate
Riboflavin	Constituent of FAD	Electron transport
Niacin	Constituent of NAD and NADP	Electron transport
Pantothenic acid	Constituent of coenzyme A	Activation of fatty acids including acetate
Pyridoxine	Coenzyme in reactions involving amino acids	Transfer of amino groups to α-ketoglutarate
Biotin	Coenzyme in the fixation of CO_2	Fatty acid synthesis
B_{12} and folic acid	Coenzymes in one-carbon metabolism	Synthesis of nucleic acids

animal product, or used to perform work on the environment (outside the animal body) is converted to heat. Heat loss occurs when the energy released during foodstuff degradation is incompletely trapped as ATP. However, the food energy trapped as ATP will eventually be transformed to heat. For example, the energy required to synthesize tissue protein from amino acids is released as heat when the protein is hydrolyzed by intracellular enzymes during normal tissue turnover. Thus, the combustion of a given foodstuff, whether in a fireplace or in an animal, will ultimately release the same amount of heat regardless of the number of ATP's formed in the process. The only exception is protein, which is incompletely combusted in the animal.

A measurement of the heat production of the "resting" animal will indicate the amount of food or body substance that must be metabolized to keep that animal alive, that is, to maintain the animal in an improbable state.

29-7. THE UNIVERSALITY OF METABOLIC PATHWAYS

All forms of life obtain energy and/or synthetic starting materials by metabolizing biological starting material over pathways identical to or very similar to the pathways discussed in this chapter. The enzymes of the glycolytic pathway have been demonstrated in virtually all bacterial, plant, and animal cells (Korkes, 1956). The basic steps in the TCA cycle, which were identified in an experiment using pigeon-breast muscle, are practically identical with those found in mammalian tissue, insects, protozoans and many bacteria. In addition, a TCA cycle has been shown to exist in higher plants, such as potatoes, avocado fruit, and cauliflower (Krebs, 1950). The hexose

TABLE 29-7. | *The role of certain minerals in metabolism.*

Mineral	General Function	Reaction or Pathway Requiring Participation of Mineral
Iron	Constituent of cytochromes and hemoglobin	Electron transport; oxygen transport
Copper	Constituent of cytochrome oxidase	$\frac{1}{2} O_2 \xrightarrow{2H} H_2O$
Magnesium	Enzyme activator	Many reactions involving ATP, i.e. glucose $\xrightarrow{ATP}$ glucose phosphate
Zinc	Constituent of carboxypeptidase and other enzymes; activator	Peptides $\xrightarrow{\text{G.I. tract}}$ amino acids
Sulfur	Constituent of proteins (in sulfur-containing amino acids), thiamine, biotin	Reactions involving protein (enzyme reactions), thiamine, biotin
Phosphorous	Constituent of nucleic acids and ATP	Protein synthesis; reactions involving ATP
Cobalt	Constituent of vitamin B_{12}	Reactions involving B_{12}
Iodine	Constituent of thyroxine	Control of the rate of metabolism

monophosphate shunt has been found in plant, bacterial, and animal tissue.

Enzymes for the glycolytic pathway, the hexose-monophosphate shunt, and the TCA cycle are widely distributed throughout the cells of the animal body, despite the diversity in the composition of various animal organs and tissues. There are exceptions, namely, the red blood cell in which the TCA cycle has not been demonstrated.

However, the most important enzymes, such as those concerned with the breakdown of glucose, are widely distributed in all types of cells. Enzymes concerned with more specific functions occur only in certain tissues or organs. For example, enzymes that hydrolyze foodstuffs in the digestive tract are synthesized in the pancreas or intestinal mucosa, but enzymes involved with urea formation are found principally in the liver.

The cells of various animal tissues do not differ significantly, therefore, in the manner they obtain energy or in the methods by which they synthesize various cell constituents. Their diversity is a function of the following factors: (1) Chemical composition of cell constituents. For example, muscle contains 20% protein and 1% mineral, and bone consists of 30% protein and 45% mineral. Even proteins—such as proteins of the liver, blood, and skin—vary widely in composition and structure. (2) Chemical reactions of the protoplasmic constituents. For example, muscle protein is able to contract, and blood protein (hemoglobin) is able to transport oxygen. (3) The rate of metabolism, which is probably related to the ATP need of the cell or tissue.

For example, the respiratory rate of the liver and kidneys is three or four times that of the heart or diaphragm.

A study of the pathways of nutrient utilization demonstrates, therefore, that the striking diversity in structure and function observed throughout living matter has, at its basis, several very similar methods for obtaining the energy and synthetic starting material that support this diversity.

REFERENCES AND SELECTED READINGS

Fruton, J. S., and S. Simmonds, 1958. *General Biochemistry*. 2nd ed. Wiley, New York.

Greenberg, D. M., ed. 1961. *Metabolic Pathways*. Academic Press, New York.

Harper, H. A., 1963. *Review of Physiological Chemistry*. 9th ed. Lange Medical Publications, Los Altos.

Kleiber, M., 1961. *The Fire of Life*. Wiley, New York.

Korkes, S., 1956. Carbohydrate metabolism. *Ann. Rev. Biochem.* 25:685–734.

Krebs, H. A., 1950. The tricarboxylic acid cycle. *Harvey Lect.*, 44:165–199.

Krebs, H. A., and H. L. Kornberg, 1957. *Energy Transformations in Living Matter*. Springer-Verlag, Berlin.

Krebs, H. A., 1964. The metabolic fate of amino acids. In *Mammalian Protein Metabolism*. Ed. H. N. Munro and J. B. Allison. Academic Press, New York.

Lehninger, A. L., 1955. Oxidative phosphorylation. *Harvey Lect.*, 49:176–215.

Munro, H. N., 1964. General aspects of the regulation of protein metabolism by diet and by hormones. In *Mammalian Protein Metabolism,* Ed. H. N. Munro and J. B. Allison. Academic Press, New York.

Rittenberg, D., 1950. Dynamic aspects of the metabolism of amino acids. *Harvey Lect.*, 44:200–219.

Schoenheimer, R., 1942. *The Dynamic State of Body Constituents*. Harvard University Press, Cambridge, Mass.

West, E. S., and W. R. Todd, 1961. *Textbook of Biochemistry*. 3rd ed. Macmillan, New York.

White, A., P. Handler, and E. L. Smith, 1964. *Principles of Biochemistry*. 3rd ed. McGraw-Hall, New York.

Chapter 30

Meeting Nutrient Requirements for Various Physiological Activities

The human mind is often so awkward and ill-regulated in the career of invention that it is at first diffident, and then despises itself. For it appears at first incredible that any such discovery should be made, and when it has been made, it appears incredible that it should so long have escaped men's research.

FRANCIS BACON, *Novum Organum* (1620.)

30-1. INTRODUCTION

Many factors may influence the amount or kind of nutrients required by an animal. Some of these factors are inherited characteristics, such as efficiency of feed utilization, and others are environmental factors, such as climate. One of the most important factors is the function of the animal. It is obvious that the amount of nutrients required to maintain an animal, that is, to keep it from gaining or losing weight, is less than that required when the same animal is producing milk or meat. The effects of the various functions—maintenance, growth, reproduction, lactation, egg production, and work—on nutrient requirements are considered in this chapter.

30-2. MAINTENANCE

Although the primary concern of animal husbandry is the production of meat, milk, offspring, wool, or work, knowledge of requirements for maintenance is important. The maintenance requirements have many practical implications, as exemplified in the feeding of dairy cows. The total nutrient cost of lactation must also include maintenance of the animal. Therefore, a 1200 lb. cow producing the same amount and kind of milk as an 800 lb. cow needs 2.4 more lb. of TDN per day because of the difference in maintenance requirements.

Energy. The energy requirements for maintenance may be estimated by determining basal metabolic rate. The basal metabolic rate may be defined in terms of the heat produced by a resting animal in a temperature within the zone of thermal neutrality[1] and in a postabsorptive state (that is, after the digestion and absorption of the last food ingested has stopped). Equating basal metabolism to the energy requirements for maintenance is possible because energy cannot be destroyed and therefore the energy used to maintain the animal is released as heat. The conditions required for the measurement of basal metabolism are obviously not easy to control in animals. The postabsorptive state is difficult to assess in ruminants because of their long absorptive period, which may vary from 30 to 70 hours. Also, in comparison to that of humans, the activity of animals is difficult to control; some workers, however, have succeeded in training farm animals to lie quietly. Therefore, the values for animals are sometimes called standard metabolism rates or resting metabolism, but the conditions at the time of measurement must be accurately reported in order for the measurements to have any meaning.

The basal metabolic rate is an underestimation of the amount of energy required for maintenance because it does not measure efforts, such as eating and other activities, used in maintaining normal life. Therefore, an "activity factor" or "activity increment" is added to metabolic rate. The following activity increments—sheep, beef cattle, and pigs, 20 to 25%; dairy cattle, 45%; poultry, 50%; college students, 80 to 100%—have been suggested, but the values are subject to considerable variation (Mitchell, 1962).

The energy requirements have also been determined by feeding trials. Garrett et al. (1959) fed various energy levels and measured the energy retention by carcass analyses. They expressed the daily energy requirements of sheep and beef cattle for maintenance of energy equilibrium as follows:

Total digestible nutrients [pounds] $= 0.036 \times$ weight of animal in pounds$^{3/4}$

Digestible energy [kilocalories] $= 76 \times$ weight of animal in pounds$^{3/4}$

---

[1] Temperature range in which environmental temperature does not stimulate metabolism. The range varies according to species differences and the insulating quality and quantity of skin covering.

Metabolizable energy [kilocalories] $= 62 \times$ weight of animal in pounds$^{3/4}$

Net energy [kilocalories] $\quad\quad = 35 \times$ weight of animal in pounds$^{3/4}$

Protein. An animal fed a ration devoid of protein will continue to excrete nitrogen in the urine. This nitrogen is the result of tissue metabolism and is called endogenous nitrogen. The endogenous loss plus an allowance for nitrogen contained in tissues such as hair, wool, nails and feathers, which continue to grow, gives a reasonable estimate of the maintenance requirements for protein.

The protein requirements of animals may be determined either from balance trials, in which the minimum protein intake required to keep an animal in nitrogen equilibrium is a good estimate of the requirement, or from feeding trials, which should include slaughter data.

However, the physiological requirement is not for protein as such, but rather for the amino acids, which constitute the proteins, and therefore the protein must be adequately balanced in amino acid content. The ruminant is less dependent on dietary amino acids than the nonruminant animal because the rumen bacteria synthesize amino acids from dietary nitrogen.

Water. There is no one most important nutrient; the specific nutrient that is lacking is the most important one. However, water is the nutrient required in largest amounts and animals can survive longer without food than they can without water. Water is lost in the urine, feces, expired air, and from the skin. Water is obtained by drinking, from the feed and from metabolic water, that is, water formed from oxidation of nutrients. For example, carbon dioxide and water are produced when glucose is oxidized for energy.

Many factors, such as the environmental temperature, govern the amount of water an animal needs to drink. High levels of protein or minerals cause an increase in urinary water; some feeds, such as silage, contain as much as 80% water, but others, such as corn, may contain only 12%. Leitch and Thomson (1944) related the water intake to dry matter consumption and found that for sheep the ratio is 2 parts water to 1 part feed; for horses and pigs, 3:1. However, the best policy is to supply plenty of clean drinking water at all times.

Minerals. The maintenance requirement for minerals of farm animals has not been studied as extensively as those for protein and energy. There are losses of minerals in urine and feces, and these losses can be used to estimate the requirement. However, the losses of many minerals are greatly decreased by "defense" or "conservation" mechanisms during periods of deficiency. For example, during sodium depletion, hormones, such as mineral corticoids, restrict the sodium loss from the body. Other examples are iron and iodine; iron obtained from the breakdown of red blood cells is reutilized in the formation of new cells, and iodine is recycled in the metabolism of the thyroid hormones.

30-3. GROWTH

Growth is a complex biological process defined by Mitchell (1962) as "the summation of those coordinated biological and chemical processes that are initiated with the fertilization of the ovum and are terminated with the attainment of the body size and conformation and of the physiological capabilities characteristic of the species and of the hereditary background of the individual."

Only the requirements for postnatal growth will be considered in this section. Prenatal growth will be discussed in the section on nutrient requirements for reproduction. The comparative growth rates of different species

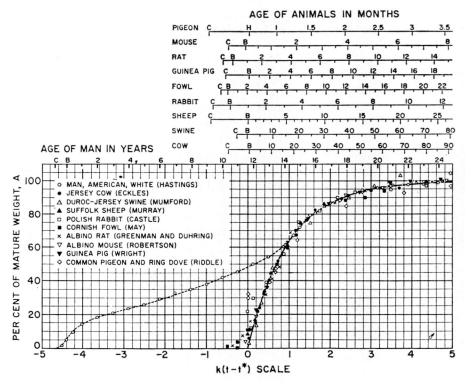

Fig. 30-1. Weight-growth equivalence of farm animals, laboratory animals, and man. [Reproduced from Brody's *Bioenergetics and Growth* (1945) through the courtesy of the Reinhold Publishing Corp., New York.]

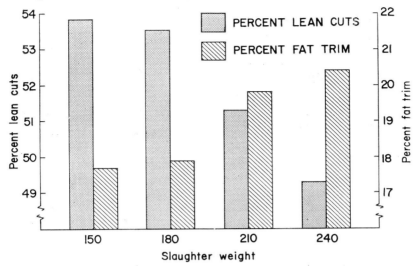

Fig. 30-2. Lean-fat relationship as influenced by slaughter weight of swine. [From Wallace, H. D., G. E. McCabe, A. Z. Palmer, M. Koger, J. W. Carpenter, and G. E. Combs, 1960. *The Influence of Slaughter Weight on Economy of Production and Carcass Value of Swine.* Fla. Agr. Exp. Sta. An. Husb. Mimeo Ser. #60-12.]

were studied by Brody (1945). He made the assumption that conception and the age at which the animal attains approximately 98% of mature weight are biologically equal in all species. He plotted the attained weights of animals against percent weight on an age equivalent basis. Some of his results are in Fig. 30-1. It is interesting that all of the curves coincide except that of man, who has a long juvenile period. The data are only approximate but are useful for comparative purposes.[2]

Energy. The energy cost of growth is mostly concerned with the energy deposited in body tissues. In the young animal, the greater part of the gain is protein, but whenever there is reasonable increase in protein mass there will be fat deposition. As an animal becomes older and increases in size the weight gain has a higher ratio of fat to protein. In the mature animal there is little increase in protein but fat deposition will continue as long as more energy is supplied than that required for maintenance. Examples of the changes in body composition as the animal increases in weight are shown in Fig. 30-2. The energetic efficiency is greater for protein formation than for fat deposition. Kielanowski (1964) estimated the energy cost of protein

---

[2] Apparently the rat data used by Brody was from virgin animals. As mentioned in Chapter 24, Cole and Hart (1938) found that both linear and weight increments occurred in rats during each of 9 successive pregnancies. None of the large domestic animals have been shown to exhibit growth during each pregnancy throughout life.

and fat deposition at 7.07 kcal/g and 14.97 kcal/g, respectively, in lambs. Therefore the energy cost of gain depends on the composition of the gain. Lofgreen (1965) fed similar rations to calves and yearlings and measured the composition of the gain. The calves weighed 410 lb. at the start of the trial and were killed at average weight of 990 lb. The yearlings weighed 644 lb. at the start and 1,258 lb. at the end. The fat content of the gain was 36% and 49% for the calves and yearlings, respectively, and the calves required only 3.7 lb. of feed above maintenance per pound of gain as compared to the yearlings' requirement of 6.1 lb.

The maintenance costs were subtracted because as the animal increases in size the maintenance requirement increases. The total energetic efficiency decreases because more food energy must be used to satisfy the maintenance requirement. In the above trial the average daily maintenance requirement per animal was 6.6 lb. for the calves and 8.3 lb. for the yearlings. The total efficiency was 6.8 lb. of feed per pound of gain for the calves and 9.8 lb. for the yearlings. Therefore, the efficiency of total energy utilization for growth of older animals is less because the type of gain costs more energy and because the maintenance costs are higher.

The species of animal may affect the total energy efficiency. Garrett et al. (1959) found the efficiency of food utilization was identical for sheep and cattle, but Brody (1945) stated that pigs have a higher energetic efficiency than cattle because they have relatively greater feed consumption capacity in proportion to maintenance cost than cattle.

Extreme environmental temperature can decrease the efficiency of utilization of food energy. The feed intake and daily gain of chicks increased as the temperature decreased from 40 to 21°C, but the greatest amount of energy was stored at 32°C (Kleiber, 1961). If the environmental temperature is very low, the energy required to maintain body temperature increases, and if the environmental temperature is too high, feed intake is decreased and maintenance requires a relatively larger percentage of food energy intake.

Protein. Like the energy requirement, the protein requirement is dependent on the composition and amount of gain. For this reason many of the protein requirements are based on feeding trials with slaughter data or nitrogen balance trials in which nitrogen retention is measured and multiplied by 6.25 to determine the amount of protein required.

The effect of age on carcass composition, as discussed earlier, is reflected in the protein requirements. For example, the protein required for a 10 to 25 lb. pig is 22% of the ration but only 14% for 75 to 125 lb. pigs.

The protein requirement also depends on the kind of protein. It has already been mentioned that the requirement is not really for protein, but for amino acids. There are 20 different amino acids commonly found in mamalian tissue, and 10 of these are called the "essential" amino acids because they cannot be synthesized in the body and must therefore be supplied in the ration. Proteins are evaluated on their ability to supply these critical amino acids; this ability is termed "protein quality." A protein containing all of the essential amino acids in reasonable proportions would be a high protein quality. The common practice is to have more than one source of protein in the ration; if one protein is low in some amino acids but high in others, a protein that is high in those amino acids deficient in the first is added. As mentioned earlier, the protein quality of the ration is much more important for nonruminant animals than for ruminants because of the synthesis of amino acids by rumen bacteria from dietary nitrogen.

There is no danger of protein toxicity resulting from supplying too much protein. The excess amino acids that are not used for protein synthesis are broken down; the nitrogen is excreted as urea and the carbon residue is used for energy. But feeding excess protein is usually not efficient, either economically or as a source of energy. Protein supplements are usually higher priced than concentrates, and they may not be as an efficient source of energy. For example, the efficiency of utilization of metabolizable energy in wheat gluten protein and starch has been estimated at 48% and 63%, respectively, for fattening of steers (Nehring and Schiemann, 1961). The decreased efficiency of metabolizable energy utilization of protein may be due to the energy needed to synthesize urea in order that the excess nitrogen can be excreted.

Minerals. The minerals required in greatest amounts during growth are calcium and phosphorus. Although most of the calcium and phosphorus are utilized in bone formation, they are widely distributed throughout the body in soft tissue and blood and have many important functions. Some of these functions are discussed in Chapter 29.

In general, most of the other minerals are supplied in adequate amounts from common feeds; however, there are exceptions, such as sodium chloride, which is needed by all animals, and zinc, which must be added to the ration of pigs. In some areas the soil or water is deficient in cobalt, copper, iodine, or magnesium, and feeds from these areas must be supplemented. Some form of iron supplementation must be given to suckling pigs to prevent anemia, because pigs are born with low stores of iron and sow milk contains low iron levels. Increasing the iron level in the ration of the sow during gestation and lactation does not increase the iron stores of the pig or the level of iron in the milk.

Vitamins. As discussed in Chapter 29, the B-vitamins act as coenzymes for reactions pertaining to energy metabolism, and therefore the requirement depends on the energy level in the ration. Ruminants are not dependent on a dietary source of B-vitamins because of rumen bacterial syntheses. The requirements for vitamin A may be met either by vitamin A or by carotene, which is converted to vitamin A in certain animal tissues. However, there are wide differences between species in the ability to convert carotene to vitamin A, and these must be considered when evaluating the vitamin A activity of a ration. There are also differences within a species in the ability to convert carotene. For example, Jersey milk has more yellow color than Holstein milk, because the Jersey is less efficient in the conversion of carotene to vitamin A. The vitamin A requirement for mammals is correlated with body weight, and therefore the required daily intake increases as the animal increases in size (Guilbert et al., 1940). Vitamin D regulates calcium-phosphorus metabolism but the requirement in the diet depends on the amount of sunlight the animal receives. The principal activity of vitamin E (tocopherol) is as a biological antioxidant to stabilize easily oxidizable components of cells. Polyunsaturated fats have a high potential peroxidizability, and therefore the vitamin E requirement is related to the degree of unsaturation of the dietary fats. Because complex interrelationships also exist between selenium, the sulfur containing amino acids, and vitamin E, dietary levels of selenium and sulfur amino acids may also influence the vitamin E requirement. Sufficient vitamin C is synthesized in the tissues of farm animals so that rations low in this vitamin do not result in scurvy as they do in man.

30-4. REPRODUCTION

Nutrient requirements for reproduction are of concern only for the female. Generally the male requires no additional nutrients beyond those to maintain normal health. The female bird requires nutrients to produce eggs; the mammal requires extra nutrients to nourish the fetus and the additional maternal tissue and to compensate for the increased heat production during gestation (presumably due to the increased level of hormonal activity).

The fetus is nourished via the placenta. Oxygen, water, monosaccharides, and some fatty acids are transferred from the blood stream of the dam to the blood stream of the fetus through the placenta. Some proteins pass through the placenta, but most are hydrolyzed to amino acids, which are transferred, and the proteins are resynthesized in the fetus. Sodium, chlorine, calcium, phosphorus, iodine, selenium, and zinc are transferred through the placenta. The placental transfer of iron is low, particularly in the sow and even high dietary levels of iron will not cause an increase. The fat soluble vitamins, A, D, E and K, also have low rates of transmission through the placenta.

Energy. A low intake of energy during early life retards the onset of puberty in cattle, has very little effect upon the age of puberty in swine, and hastens the breeding season in sheep (Reid, 1960). Breirem et al. (1961) concluded that many studies with dairy cattle show that low levels of energy (20 to 30% below current standards) during growth apparently do not impair potential fertility, but dairy cows reared on a high plane of nutrition may have a tendency to encounter breeding difficulties later in life.

The energy requirement during gestation is not constant but progresses as the fetus increases in size. The change in energy content of the uterus during pregnancy of gilts is shown in Fig. 30-3. The greatest energy needs are during the last quarter of pregnancy. Similar requirements for other nutrients during the stage of gestation have been found.

Blaxter (1957) concluded that a major nutritional cause of low reproductive performance of ruminants is insufficient food energy during the latter part of pregnancy. The increased energy requirement of ruminants during the latter part of gestation is more critical than that of swine because there is a concomitant decrease in feed intake by ruminants. According to Blaxter, in the last fifth of pregnancy voluntary food intake by cattle drops by 20 to 40%, presumably because of pressure exerted on the alimentary tract owing to the increased size of the uterus. If the dam is underfed the offspring may be weak at birth and have low survival rates.

Overfeeding is also objectionable; it is a waste of feed and may increase the embryo mortality rate. For example, Fowler and Ensminger (1951) compared the reproductive performance of gilts fed at two different levels. One group of gilts were fed *ad libitum* from weaning until they reached a weight of 150 lb.; another group was fed 70% of this amount. When the gilts reached 150 lb. both groups were put on a breeding ration and allowed to gain between 1 and 1½ lb. daily. When bred they were returned to their respective feeding levels. The *ad libitum*-fed gilts farrowed an average of

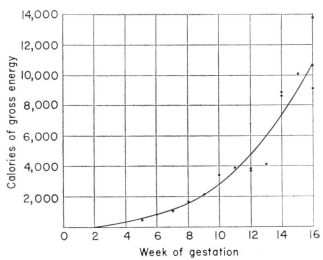

Fig. 30-3. Gross energy content of material deposited in uterus during gestation of gilts. [Mitchell *et al.* (1931).]

4.7 pigs per litter and the limited-fed gilts averaged 7.4 pigs per litter. The prenatal death rate was 62.9% for the *ad libitum*-fed gilts, and 35.3% for the limited-fed gilts. The cause for the embryonic mortality is not fully understood.

Protein. Because of the progressive rate of growth of the fetus the protein requirements are also greater in the latter part of gestation. Mitchell et al. (1931) found in the first week of gestation of gilts the daily increase of protein in the uterus was 0.54 g. In the sixteenth week of gestation the daily increase of protein was 33 g. The average daily increase for the entire period was 14 g. In cattle the daily protein retention in the fetus and membrane may be as much as 800 times greater in the ninth month than in the second month of gestation.

Minerals. Phosphorus deficiency may cause infertility. Phosphorus supplementation to cattle in phosphorus-deficient areas in South Africa increased the calf crop 30%. Phosphorus deficiency causes ovarian dysfunction, delay of puberty, and, in mature animals, irregular heat intervals and cessation of estrus.

A moderate calcium deficiency may cause depletion of the dam's bones but the young may be normal; a severe calcium deficiency, however, may cause intrauterine death. Other minerals are also important in reproduction. Iodine deficiency adults in weak young with enlarged thyroids (goiter). Selenium supplementation has increased the reproductive performance of ewes in New Zealand and Australia. Deficiencies of other minerals may cause difficulties but under most practical conditions if the ration is adequate in energy, protein, and the minerals discussed above, the ration will also contain sufficient amounts of the other minerals.

Vitamins. Vitamin A deficiency may cause irregularity of estrus, delayed breeding, sterility, abortion, or weak or dead offspring. The condition of the offspring depends on the length of the deficiency. However, no supplemental vitamin A is needed for ruminants if good quality hay is available, and the Vitamin A requirement for swine can be met by high quality alfalfa meal (Maynard and Loosli, 1962). Vitamin E deficiency causes reproductive failure in rats but vitamin E supplementation to rations containing low amounts of vitamin E has not improved the reproductive performance of cattle, sheep or goats.

The B-vitamin requirements for reproduction have little practical significance for ruminants. However, thiamine, riboflavin, choline, and pantothenic acid deficiencies in the sow will produce dead or weak pigs. Cunha (1957) stated, "Undoubtedly, a lack of vitamins in the ration [of sows] accounts for many of the small pig losses which occur each year."

30-5. LACTATION

The amount of nutrients required for lactation depends on the quantity and composition of milk produced and on the efficiency of utilization of the nutrients in the ration. The quantity and composition varies not only among species (see Chapter 3), but within species. The DHIA average for Holsteins from 1958 to 1962 was 12,471 lb. of milk containing 3.65% fat. Jerseys during the same period produced on an average of 8,066 lb. of milk with 5.16% fat. The composition and yield also change within the lactation period of the individual animal (see Chapter 23).

Energy. As energy input is increased, the milk yield increases but at a decreasing rate per unit of feed input. For example, Moe et al. (1963) proposed that a daily yield of 90 lb. of milk required 20% more lb. of TDN per lb. of milk than if the yield were only 20 lb. This decrease in return occurs because at the higher levels of feeding, less of the dietary energy is absorbed and more of the absorbed energy is convertd to body fat than at the lower levels. However, a high-producing cow is more efficient than a low producer even though the law of diminishing returns prevails, because the maintenance requirement is a smaller proportion of the total requirement for lacation as the milk yield increases.

The effect of milk composition on energy requirements is also important. Moe et al. (1963) suggested that the energy allowance for a cow producing 60 lb. of milk a day would be 0.31 lb. of TDN per lb. of milk, if the milk contained 3% fat, but 0.41 lb. of TDN if the fat content was 5%.

Protein. The protein requirement is determined by the amount of protein secreted in the milk plus any protein catabolized in the secretory process, and can be calculated from the protein content by assuming that the efficiency of conversion of dietary, digestible protein to milk protein is 70 to 80%. The amount of digestible protein in the feed must also include that required for maintenance.

Minerals. Calcium and phosphorus are the minerals most likely to be deficient in a lactation ration. The high-producing dairy cow secretes more calcium in the milk than she usually consumes in the early part of lactation, and bone stores are depleted. Later in lactation, when milk production has declined, the ration replaces the bone stores. The resorption of calcium from the bones is governed by a parathyroid hormone. Low blood levels of calcium trigger the hormone mechanism. When the drop in calcium blood level is precipitous, as in milk fever, the controlling mechanism is too slow, and the condition must be corrected by injecting calcium gluconate intravenously (see Chapters 7 and 32 for further discussion of this subject).

The calcium and phosphorus content of milk remains relatively constant, that is, if calcium or phosphorus are lacking in the ration, the cow will not produce milk with lower calcium or phosphorus values, but production will be decreased. Phosphorus supplementation has increased production 50 to 146% in phosphorus deficient areas. Calcium supplementation has produced similar results in local areas.

Sodium chloride supplementation has improved production and increased palatability of rations. Iodine, copper, and cobalt deficiencies may be a problem in certain areas, and supplementation required, but the other minerals are usually supplied in normal rations.

Vitamins. Vitamin A supplementation will increase the vitamin A content of the milk but will not increase yield if the ration contains good quality hay or silage. Cows deprived of sunlight require vitamin D for lactation. Massive doses of this vitamin 3 to 7 days prior to parturition have been used to prevent milk fever.

The B-vitamin requirement is a function of the energy metabolism in milk formation plus the amount of vitamins secreted in the milk but it is not of practical importance for cows because of bacterial synthesis of these vitamins in the rumen.

30-6. AVIAN EGG PRODUCTION

The nutrient requirements for egg production are determined by the level of production and composition of the egg. The chemical composition of eggs of the domestic species are not greatly different. Chicken, turkey, duck,

TABLE 30-1. | *Total constituents per egg for the different species of poultry.*

Constituent	Chicken	Turkey	Duck	Goose
		Grams per Egg		
Water	38.0	52.8	46.4	125.0
Solids	13.6	18.8	20.2	52.0
Organic matter	13.2	18.3	19.5	49.9
Proteins	6.60	9.38	9.12	24.8
Fats (lipids)	6.09	8.38	9.59	23.0
Carbohydrates	0.52	0.50	0.80	2.1
Inorganic (ash)	3.35	—	—	—
Calcium	1.98	2.8	2.8	8.7
		Calories per Egg		
	95	161	153	351

Source: Taken from A. L. Romanoff and A. J. Romanoff, 1949. The Avian Egg. John Wiley and Sons, Inc., New York.

and geese eggs contain 70 to 74% water, 12 to 14% protein and 12 to 14% fat, but there is a large difference in the amount of material per egg (Table 30-1).

Energy. The onset of egg production may increase the feed intake by 60 to 95% more than that required for maintenance. This feed increase must supply the energy for producing the egg as well as the energy stored in the egg. Each factor requires about 50% of the energy intake above maintenance.

Feed intake is normally regulated by the energy level of the ration; the hen eats to obtain a certain energy level rather than to consume a specific weight of feed. However, if a feed is very bulky, or low in energy content, the hen might not be able to obtain all the energy she needs because of the capacity limits of her digestive tract. Therefore, it is recommended that rations for laying hens have at least 1,285 metabolizable kcal per pound of feed, as compared to those for growing chicks, which should have 1,095 kcal per pound of feed (Combs, 1962).

Protein. The protein, or amino acid, requirement is determined largely by the rate of egg production and composition of the eggs, but the energy content of the ration is also important. For example, as the energy content of the ration increases, the requirement for the essential amino acid, methionine, increases (Table 30-2).

TABLE 30-2. | *Calculated methionine requirement for 4-lb. laying hens gaining 1 g per hen per day (percent of ration).*

Season	Grams Egg per Hen per Day			
	57	52	47	42
1,260 kcal. of metabolizable energy per lb. of feed*				
Summer	0.295	0.288	0.281 †	0.272
Autumn	0.286	0.280†	0.273	0.263
Spring	0.280†	0.274	0.267	0.258
Winter	0.265	0.259	0.252	0.244
1,370 kcal. of metabolizable energy per lb. of feed**				
Summer	0.321	0.313	0.305	0.295
Autumn	0.311	0.304	0.297	0.286
Spring	0.304	0.298	0.290	0.280
Winter	0.288	0.282	0.274	0.265

Source: Comb, G. F. 1962. Nutrition of Pigs and Poultry. Edited by J. T. Morgan and D. Lewis, Butterworth's Scientific Publications, London.
* 920 kcal. of productive energy per lb.
† NRC requirement is 0.28%
** 1,000 kcal. of productive energy per lb.

The egg protein is very high in quality, so a high-quality protein must be fed. Feeding a low-quality protein does not change the composition of the egg but decreases or stops egg production.

Minerals. Eggs contain appreciable amounts of calcium, phosphorus, sulfur, chlorine, and sodium, and smaller amounts of magnesium, iron, and copper. The dry matter of the egg contains 15% calcium; (the dry matter of milk contains only 1%). This illustrates why the laying hen has such a high requirement for calcium. Calcium and phosphorus supplements are necessary to build up stores prior to onset of egg production because, as with lactation, mobilization of these minerals from bone takes place when the hens are laying.

Vitamins. The vitamin requirements for laying hens have not been extensively studied, but apparently the onset of egg production does not increase the requirement much beyond that required for growth. However, the vitamin content of the egg can be increased by feeding high proportions of vitamins to the hen.

30-7. WORK

The number of draft animals in the United States is decreasing, but knowledge of the nutrient requirements for work are important because of the increasing numbers of pleasure and race horses. Furthermore, knowledge of nutrient expenditures required for walking and grazing are useful in evaluating management practices for cattle and sheep, particularly under range conditions, in which the animals may be required to travel significant distances to feed and water.

Energy. The energy requirement for work is affected by the extent of exertion, the condition of the worker, and, to a small degree, the constituents of the diet. The rate of walking is an example of how speed affects the energy requirement. Energy expenditure by man is linearly proportional to speed that ranges from 3 to 6.5 km per hour, but at higher speeds, energy expenditure increases at faster rates (Passmore and Durnin, 1955). Clapperton (1964) compared the energy cost of sheep walking both on level ground and on gradients at speeds of 24.3 m per/minute and 48.5 m per/minute. The apparent energy cost of level walking was greater at the higher speed, but when the sheep walked on gradients the cost was greater at the lower speed.

The effect of training on work efficiency is illustrated in the data obtained by Nadal'jak (1962). Untrained stallions required 204 kcal per 100 kg of body weight to pull 100 kg for a distance of 1 km; after training only 189 kcal were needed.

A high carbohydrate ration may be a more efficient energy source than a high fat diet for work, but the effect is apparently very small and not of any practical importance, except perhaps under conditions of severe and prolonged muscular effort (Mitchell, 1962).

Protein. Work does not increase the protein requirement directly, that is, protein tissue is not broken down as the result of work. However, prolonged muscular exercise will cause an increase in nitrogen retention and muscle size.

Minerals. Although phosphorus, calcium, and magnesium are associated with muscular contraction, work does not significantly increase the requirement. The loss of sodium and chloride in sweat must be replenished.

Vitamins. The requirement for vitamins, such as thiamine, concerned with energy metabolism increases with work but is not of practical importance. There is no evidence that work increases the requirement for vitamins A, C, D, E, and K.

REFERENCES AND SELECTED READINGS

Blaxter, K. L., 1957. The effects of defective nutrition during pregnancy in farm livestock. *Proc. Nutr. Soc.*, 15:52.

Breirem, K., A. Ekern, and T. Homb. 1961. Relation of nutrition of the young animal to subsequent fertility and lactation. *Fed. Proc.* 20 (Part III) 275.

Brody, Samuel, 1945. *Bioenergetics and Growth*. Reinhold Publishing Corp. New York.

Clapperton, J. L., 1964. The energy metabolism of sheep walking on the level and on gradients. *Brit. J. Nutr.*, 18:47.

Combs, G. F., 1962. The interrelationships of dietary energy and protein in poultry nutrition. In *Nutrition of Pigs and Poultry*. Ed. by J. T. Morgan and D. Lewis. Butterworth's Scientific Publications, London.

Cunha, T. J., 1957. *Swine Feeding and Nutrition*. Interscience Publishers, Inc., New York.

Fowler, S. H., and M. E. Ensminger, 1961. *Relationship of Plane of Nutrition to the Improvement of Swine for Meat Production Through Selection*. Washington Agr. Exp. Sta. Tech. Bull. 34.

Garrett, W. N., J. H. Meyer, and G. P. Lofgreen, 1959. The comparative energy requirements of sheep and cattle for maintenance and gain. *J. Animal Sci.*, 18:528.

Guilbert, H. R., C. E. Howell, and G. H. Hart, 1940. Minimum vitamin A and carotene requirements of mammalian species. *J. Nutr.*, 19:91.

Kielanowski, J., 1964. Estimates of the energy cost of protein deposition in growing animals. In *Proc. 3rd Symposium on Energy Metabolism*. Troon. Scotland. European Assoc. Animal Prod.

Kleiber, Max, 1961. *The Fire of Life, An Introduction to Animal Energetics*. Wiley, New York.

Leitch, I., and J. S. Thomson, 1944. The water economy of farm animals. *Nutr. Abstr. and Revs.*, 14:197.

Lofgreen, G. P., 1965. Personal communication.

Maynard, L. A., and J. K. Loosli, 1962. *Animal Nutrition.* McGraw-Hill, New York.

Mitchell, H. H., 1962. *Comparative Nutrition of Man and Domestic Animals.* Academic Press, New York.

———, W. E. Carroll, T. S. Hamilton, and G. E. Hunt, 1931. *Food Requirements of Pregnancy in Swine.* Illinois Agr. Exp. Sta. Bull. 375.

Moe, P. W., H. F. Tyrrell, and J. T. Reid, 1963. Further observations on energy requirements of milking cows. *In Proc. Cornell Nutr. Conf.,* p. 66.

Nadal'jak, E. A. 1962. Effect of state of training on gaseous exchange and energy expenditure in horses of heavy draught breed. *Nutr. Abstr. and Revs.,* 32:463.

Nehring, K., and R. Schiemann, 1961. Utilization of energy by different species. In *Proc. 8th Int. Cong. Animal Husbandry,* p. 190.

Passmore, R., and J. V. E. A. Durnin, 1955. Human energy expenditure. *Physiol. Revs.,* 35:801.

Livestock Management

Chapter 31

Beef Cattle Management

Be thou diligent to know the state of thy flocks, and look well to thy herds.

Prov. 28

31-1. INTRODUCTION

Sound, modern-day beef cattle management involves the application of the sciences of genetics, physiology, nutrition, and microbiology to the production of beef cattle. The beef animal readily adapts itself to a wide variety of environmental conditions. As will be pointed out later in this chapter, there are many ways in which the environment can be modified to increase the comfort, health, and doing ability of cattle. Management practices will vary, therefore, according to environment. The successful beef cattle producer will use those management practices found to be most desirable and profitable for his area.

31-2. AREAS AND PHASES OF PRODUCTION

Approximately 68 million head of beef cattle (cows, heifers, calves, steers, and bulls) are on farms in the United States. Certain types of production are common to certain areas. The seventeen western states are predominantly range states. Many areas in this section are not suitable for the production of harvested crops, but the native forage available does provide suitable grazing for livestock. Beef cattle production constitutes a major part of the agricultural economy of this section. Many commercial range herds contain 500 to 1000 cows.

The western area produces the necessary young females for herd replacements, and herd bulls are secured from registered herds located in both the western states and other parts of the nation. For many years this area has been the principal source of stocker and feeder cattle for the corn-belt feeder. Recent years have seen the development of highly specialized feeding areas in the West and Southwest, some of which can accommodate 25,000 or more cattle at one time.

In the western states, there are many famous grazing areas, noted for their abundance of high-quality grasses. These include the Sandhills of western Nebraska, the Jackson Hole area of Wyoming, the Flint Hills of south central Kansas, and the Osage in northern Oklahoma. The forage available in these areas is highly prized. In recent years there has been an increase in commercial cow herds in these specialized grazing areas. Feeder calves are often identified according to the area in which they originate, such as Sandhill calves or Flint Hill calves.

The Midwest, or corn belt, has long been famous as a region for fattening beef cattle. Here the most common type of beef production is the farmer-feeder operation. Although the fattening of beef cattle predominates, some commercial cow herds are found. These are usually small herds numbering from 20 to 100 head, depending on the size of the farm operation.

The southern section of the United States has experienced a great change in agricultural production in recent years. With reduced cotton acreage and an increase in grasses and soil-improving crops, beef cattle have played a much larger role in utilizing the forage. As a result, beef cattle have increased very rapidly in the South in the past twenty years, and southern calves compete with feeders and stockers produced in the West. Fattening cattle for slaughter has also increased. One major difference found in the southern section is the widespread use of zebu or Brahman cattle. Brahman bulls crossed on native stock produce offspring that are better adapted to hot and humid climates. Many calves are now sold for slaughter directly off the cow.

While beef cattle are not as important in farming operations in the East and Northeast, certain sections such as Lancaster County, Pennsylvania, are noted for fattening cattle for market.

The raising of registered beef cattle is a specialized phase of beef production. Registered herds are found in all sections of the nation. The primary responsibility for the improvement of a given breed of beef cattle rests with the purebred breeder. Therefore, it is important for the purebred producer of beef cattle to avoid breeds that become temporarily popular but have little if any economic value. He should realize that his primary responsibility lies in the production of sound, useful, and high performing seedstock for use in commercial herds.

Many management practices are common to all areas. The great diversity of climate, soils, and rainfall will, however, necessitate a variety of management practices.

31-3. MEANS OF IMPROVING THE ENVIRONMENT FOR BEEF CATTLE

Beef cattle have the ability to adapt themselves to various climatic conditions. Modern breeds reflect selection that has been based on the responses of animals to specific climatic environments. The Santa Gertrudis breed was developed by the King Ranch, Kingsville, Texas, as a strain of cattle particularly adapted to the climate of the southern part of the United States. The practice of crossing European and Brahman breeds results in animals better adapted to hot climates than those of the European breeds and with better meat qualities than those of the Brahman breed.

Much effort and study has been devoted to developing strains of cattle that are adapted to specific climatic conditions. However, there are many ways to modify the environment and to achieve improved performance of animals not ideally suited to a given region. In hot climates, the environment can be modified to increase heat loss: through convection by the use of fans, through evaporation by a mist spray, through conduction by providing cool drinking water (Fig. 31-1), or through cutting down of radiant heat by means of shades (Ittner et al., 1958). In a cold environment, the pattern of practices to modify the natural environment is reversed: windbreaks reduce loss of body heat by convection, shelters in stormy weather reduce loss of heat by evaporation, and warm drinking water reduces loss of heat by conduction.

Fig. 31-1. Cooling the drinking water by refrigeration is a means of increasing gains in cattle in hot climates. The water in this tank is thus cooled and heating of the water from the sun is reduced by a shade over the tank. [From Ittner *et al.* (1958).]

Tests have shown that use of an overhead shelter will result in a slightly faster rate of gain, and will reduce feed costs per 100 lb. of gain.

Animals seek environments that are most comfortable and that consequently permit them more nearly to express their potential for growth and production. Tests have shown that a good shade will reduce the heat load as much as 50% in areas where animals are kept under very high ambient temperatures.

Artificial shades have proved valuable in reducing heat stress. Fifty to sixty square feet per animal will normally be sufficient to avoid crowding of the animals (Fig. 31-2). Such shades should be oriented in accordance with the prevailing summer winds and should be placed 10 to 12 ft. above the ground. Material for covering shades may vary from galvanized sheet iron to hay or straw (Ittner et al., 1958).

A slight increase in rate of gain can be expected if feedlot animals are maintained in paved lots rather than dirt lots (Self et al., 1964).

The alert progressive beef-cattle producer should develop practical means of environmental control wherever and whenever they are economically justified.

31-4. MANAGEMENT OF THE BREEDING HERD

In commercial range herds a single person may care for a hundred or more brood cows. Additional labor is required only when the entire herd is being sorted or animals are being handled for routine operations, such as vaccination, branding, or castration. In some registered herds a single worker may be responsible for only a few animals, especially when a portion of the herd is being prepared for show and sale. Registered animals must be identified as individuals, but animals produced in most commercial herds are handled as groups.

Growing and Development of Beef Females. The growth and development of the beef heifer is very important. Management practices during the period from birth to first lactation may influence lifetime productivity. The development of the beef heifer is determined by her inherent capacity for growth and by the nutrients provided. Studies by Chambers (1954) indicate that a heifer should weigh approximately 600 lb. before she is exposed to the bull, since heifers of this weight have less trouble at calving time than do smaller heifers. Well-fed heifers reach this desired weight as yearlings, but other factors may make it feasible to delay breeding. Some breed registry associations will not record calves from dams under 2 years of age. Many ranchers believe that a range cow should calve first at 3 years of age. In many parts of the West, this is the most common age for heifers to produce their first calf. Studies by Pinney et al. (1960) have shown that there is

Fig. 31-2. Shade is an important means of reducing solar radiation of heat. Wire corrals (*a*) allow for better movement of air than do wooden corrals (*b*). [From Ittner *et al.* (1958).]

some advantage in having heifers calve first at 2 years of age. Results of one such long range study is shown in Table 31-1.

Care of the Cow Herd. The ratio of the number of calves weaned each season to the number of cows maintained in a herd is one of the most important measures of the economical production of beef. This ratio is usually designated as "percent calf crop." For example, if 100 cows exposed to bulls in a given season wean only 90 calves, this would be expressed as a 90% calf crop. Many commercial beef cattle producers have their cows examined for pregnancy after the breeding season. The "open" cows are then culled, thereby reducing feed costs per calf produced.

Normally a beef cow will produce a calf each 12-month period. Running the bull with the cow herd the year long may result in a slightly higher percent calf crop, but most better commercial herd owners prefer to run their bulls with the cow herd only during a specified season, resulting in calves that are more uniform in age and weight. Such calves will usually sell to a better advantage.

The beef cow may be managed to produce a calf at any season of the year, but spring or fall are more favorable for growth and development of

TABLE 31-1. | *Production records of 11½-year-old cows that calved first as two- or as three-year-olds.*

	Age at First Calving	
Data	*Two-Year-Old*	*Three-Year-Old*
Number of cows at start of experiment (10/48)	60	60
Number of cows remaining (10/59)	41	42
Number of possible calvings	532	480
Number of calves weaned	484	422
Percent calf crop weaned	91.0	87.9
Number of calves weaned per cow	9.10	7.91
Average wean weight (corrected for age and sex)	476	487
Average calving date	3/11	3/9

Source: Oklahoma Agricultural Experiment Station Publication MP-57, 1960.

the calf. In most western range areas, herds are managed so that calves are dropped in the spring months. In these range areas, spring calves are more desirable because of the availability of forage during spring and early summer months. To fit in with an abundant feed supply, fall and winter calves are common throughout the southern and southwestern areas of the country. Many factors—winter and summer forage, labor, market conditions—must be taken into consideration before determining whether fall or spring calving is most desirable.

The beef cow will normally give birth to a live calf without assistance. The cow herd should be observed daily during the calving season, and the herdsman should make sure that all cows are accounted for during this herd check. He should locate calving cows not with the herd in order to check the newborn calf or render assistance if calving is difficult. Heifers calving for the first time should be separated from the older cows and observed at more frequent intervals. This allows the herdsman to assist with abnormal births or other calving difficulties and will result in a higher percent calf crop.

To be a profitable member of the herd, a cow must wean a calf each year. Calf weight at weaning time is usually a reliable measure of a cow's productivity. Botkin and Whatley (1953) have shown that the cow tends to repeat her performance from year to year, as measured by the adjusted weaning weight of her calves. Gifford (1953) has stated that the more milk a cow produces in a given lactation, the heavier will be her calf at weaning. It appears that a moderate flow of milk during the entire lactation period is more desirable than a heavy milk flow in the early stages, followed by a rapid decline in milk production.

During the spring and summer months, most cow herds are maintained on pasture, which, if well managed, will normally provide sufficient feed for the lactating cow (Fig. 31-3). During periods when the forage is dormant, or during droughts, it may be necessary to provide supplemental feed in order to maintain a desirable milk flow for the calf. Cows nursing calves during the winter will normally require some supplemental feed.

A common practice under range conditions is for the dry cow to graze cured native grass throughout the winter. Such range forage is usually supplemented with sufficient protein and energy feeds to meet the needs of the cow (Fig. 31-4). Other systems of winter management involve the use of winter pastures, with a reserve feed supply in case of drought or pasture failure.

The beef cow normally needs little special attention during pregnancy other than routine care and management. It is unwise to handle cows through a squeeze chute or in close quarters during the latter part of pregnancy. Injuries or added stress during this period may result in abortion. The average length of gestation in the beef cow is about 280 days (see Chapter 6). First-calf heifers will sometimes calve earlier than this, while older cows may have a somewhat longer gestation. Beef calves are ordinarily weaned from the cows at 6 to 10 months of age. Calves of about the same age are removed at the time time. A 3- to 5-day "bawling out" period, by both cows and calves, usually follows weaning.

Fig. 31-3. Commercial cow and calf operations in the western United States are best adapted to range conditions. The forage in this California foothill range is still lush but approaching maturity in the latter part of May. The cows received a supplement of cottonseed meal during the previous fall and early winter when forage was scarce. [Courtesy K. A. Wagnon.]

Fig. 31-4. When forage is scarce, supplementation with cottonseed meal is frequently advantageous. Note the thinness of the animals as compared with animals (Fig. 31-3) near the end of the green forage season. [Courtesy K. A. Wagnon.]

Herd owners follow the practice of culling a cow when she ceases to be a profitable member of the herd. Culling is usually based on age, unsoundness or disease, and productivity. Production records, when available, can serve as an effective tool to aid in culling the cow herd. A cow should be left in the herd as long as she weans a calf heavier than that of a first-calf heifer.

Care of Herd Bulls. One of the most important decisions to be made in the management of the cow herd is the selection of the herd bull. The ideal bull would be one that is physically sound, structurally correct, and capable of siring uniform high-quality calves for extended periods of time. He should be active, vigorous, and yet mild-tempered. Consideration should be given to soundness in feet and legs. Fertility and the adaptability of the individual bull to his environment are also important.

Bull calves are normally weaned at 7 to 8 months of age. Individual calves that show promise of developing into high grading bulls are often given special attention. In some registered herds promising calves are "gain tested" in order to select the bulls more effectively. Well-balanced rations, which will permit the calf to express his growth potential, should be used in such gain tests. Following the completion of a feeding test, bulls should be so fed and managed that they may attain their normal growth. To be useful for as many years as possible the herd bull should be maintained in good flesh.

Young bulls can be used for light service as early as 1 year of age, but most ranchers prefer them to be at least 2 years old. When hand mating is practiced, a bull may be used on 50 to 75 cows in a given season. Under pasture mating conditions one bull will be required for each 15 to 30 cows. The number of cows per bull will depend on the age and vigor of the bull, the terrain, and size of the pasture. By the use of artificial insemination it is possible to use a bull on several hundred cows in a given season. In herds where spring-dropped calves are desired, the bull should be mated during late spring and early summer. For fall-dropped calves the breeding season occurs during the winter months.

Herd bulls not in use should be maintained in a good thrifty condition. When forage is inadequate, supplemental feeding should be practiced. Adequate exercise is important and feet should be checked frequently, and trimmed if necessary, to maintain a sound animal. Good herd bulls are normally used until they become unserviceable.

Artificial Insemination and Estrus Control in Beef Cattle. Although artificial insemination is widely and successfully practiced with dairy cattle, it is not used extensively by beef producers. The wide divergence of management practices between dairy and beef herds explains this difference in acceptability; the inconvenience of detecting estrus in cows on the range and then bringing them to the corral for insemination has been a deterrent. Another factor that has tended to reduce widespread use of artificial insemination, especially with purebred beef herds, has been the restrictive registration requirements imposed by breed registry associations. Most beef breed registry associations require that before a breeder can record calves that are the result of artificial insemination from a particular beef sire, he must be the recorded owner of at least one-third interest in the sire. Some breed registry associations have further restrictions. Before a breeder of purebred beef cattle attempts to use artificial insemination in his herd, he should familiarize himself with the rules established by his particular breed registry association.

A renewed interest in the use of artificial insemination in beef herds has resulted from the recent partially successful trials on the synchronization of estrus in beef females; the development of orally effective progestational compounds that inhibit estrus in cows gives promise of widespread application, once techniques for practical use are developed (Zimbelman, 1963).

31-5. MANAGEMENT OF STOCKER AND FEEDER CATTLE

Stockers and feeders are market terms used primarily to describe steers and heifers from weaning age to maturity. Stockers are usually lighter in weight and in thinner condition; such animals are usually maintained on roughage or grazed on pasture prior to fattening. Feeder cattle are usually heavier and in higher condition, and are usually placed in the feedlot and fattened for market. When selecting stocker and feeder cattle, consideration should be given to age, sex, condition, and general health.

Stocker cattle wintered in good thrifty condition and making weight gains of $\frac{1}{2}$ to $\frac{3}{4}$ lb. daily will usually be more profitable than cattle wintered to make greater weight gains. During the summer months stocker cattle are usually maintained on pasture, and will make daily gains of 1 to $2\frac{1}{2}$ lb. per head. Gains will depend on the forage available, age, sex, and condition of animals at the beginning of the grazing season.

31-6. CARE AND MANAGEMENT OF FEEDLOT CATTLE

In recent years cattle-feeding operations have become highly specialized. While much of the full feeding of cattle is still done in the corn belt by farmer feeders, the southern, southwestern, and western areas have developed highly specialized cattle-feeding operations, much of it mechanized. A modern cattle feedlot is truly a beef factory (Fig. 31-5).

When selecting cattle for the feedlot, consideration should be given to grade, sex, age, condition, and freedom from disease and parasites. Disposition of animals placed in the feedlot is very important. Animals that are wild or of a nervous temperament will seldom do well.

Successful feedlot operation requires advanced planning and careful supervision. Lots should be cleaned, water tanks cleaned and filled, and adequate feeding facilities made available. When animals are received—usually following a long haul by truck or rail—they should be given access to water and a limited quantity of grass hay, and then be permitted to rest for a time before being handled or sorted. Cattle that are to be full-fed on grain should be started on feed gradually. Daily grain allowances of 1 lb. for calves and 2 lb. for older animals are considered safe amounts with which to start. These amounts should be increased gradually over a period of 10 to 14 days.

Cattle are assumed to be on full feed when they are consuming approximately 2 lb. of concentrates (or 3 lb. of a mixed ration) for each 100 lb. of body weight. Mixed rations containing 35 to 50% chopped roughage can safely be offered to feedlot animals in greater initial quantities. Rations of this kind can also be increased at a faster rate than high-concentrate rations. A common practice is to allow cattle a self-feeder in the lot.

Initial weight and condition, kind of ration used, desired market weight, and grade are factors that determine the length of the feeding period. When feedlot cattle have reached the desired market weight and grade they should be sold. Since weight gains made in the later stages of the feeding

Fig. 31-5. Mr. Harvey McDougal, owner of the McDougal Livestock Company, Collinsville, Calif., overlooking a portion of a feedlot accommodating 15,000 head. The yard is designed for feeding from an automatic delivery truck. Note the slope of the yards for drainage and the mounds in each pen for loafing areas in wet weather. Concrete or redwood log aprons about 10 feet wide along the feed troughs prevent miring. [Photo by Don Holt.]

period are much more costly than gains made earlier it is important to sort out and sell those cattle that have reached the desired weight and grade. This practice is referred to as "topping-out."

All cattle, when handled or worked—as in dehorning or branding—will lose weight. During shipment, too, whether by truck or by rail, animals will lose weight. This weight loss is referred to as "shrink," and it is affected by length of haul, the weather, condition of animals prior to shipment, type of ration fed, and fill at the market. When animals are to be shipped from feedlot to market they should be handled in the normal or accustomed manner prior to shipping. Any abnormal change in ration or other nonroutine changes will usually result in greater shrink to market.

Feedlot cattle are subject to several familiar ailments. Perhaps most common is going "off feed," usually a result of some digestive disturbance or other organic cause. Many feeders first check the feed supply to make certain that no abrupt changes have been made in the ration or that spoiled feed has not been used. Another common feedlot problem is the occurrence of "bloat" among individual animals, although this seldom occurs in a large number of animals at the same time. "Urinary calculi" results from blocking of the urinary tract by calcium deposits from the bladder. This ailment occurs infrequently, but when it does prompt action must be taken to save the animal. "Foul foot," characterized by severe lameness in advanced stages, can be a serious problem in all classes of cattle. Many experienced feedlot operators have ways of treating minor cases of these ailments. In acute or persistent cases it is wise to seek the service and help of a qualified veterinarian.

Use of Estrogens in Fattening Cattle. The greatest advance in the efficiency of beef production in this century is based on the use of diethylstilbestrol (more commonly known as stilbestrol) to increase the rate of gain of cattle in the feedlot (Dinusson et al., 1950). Stilbestrol is a synthetic estrogen that can be produced cheaply. Other synthetic estrogens, such as dienestrol and hexestrol, or androgens, such as methyl testosterone, are less effective. The response to a combination of progesterone with the natural estrogen, estradiol, has been as good as that to stilbestrol, but the cost is greater.

Stilbestrol can be administered in the feed in the amount of 10 mg per head daily or by implanting a 24 to 36 mg pellet subcutaneously at the base of the ear. Implantation in the ear obviates the possibility of any residue present at slaughter contaminating the meat. There is no appreciable difference in the net effects of the two methods of administering the hormone; the procedure used will depend upon how well the method fits into the general management practices.

The amounts of stilbestrol suggested cause no appreciable effect on carcass grade; there is an increase in rate of gain of 10 to 20% with a concomitant increase in feed efficiency. Larger doses, although resulting in slightly greater increases in rate of gain, cause a reduction in carcass grade, due primarily to decreased marbling; other undesirable side effects, such as udder development and vaginal prolapse in heifers, excessive "riding," and elevation of the tail head, may be manifested.

As mentioned above, cattle on high energy rations respond best to stilbestrol administration. Cattle supplemented with grain on pasture will respond but the response is generally poor on low energy rations.

Cattle gain most rapidly within 60 to 80 days after implantation, but the desirability of reimplantation has not been fully established.

The response to stilbestrol is greater in steers than in heifers or bulls. Increased gains have been obtained in suckling calves, but further study needs to be done on the effect of this treatment on subsequent performance in the feedlot.

31-7. SPECIAL NUTRITIONAL PROBLEMS IN BEEF CATTLE MANAGEMENT

The beef animal is capable of utilizing both roughage and high energy feed grains. Therefore, under normal conditions, most feedstuffs will supply adequate amounts of needed nutrients. Some soils are deficient in certain needed mineral elements, and the forage produced on such soils is also deficient in these minerals. Large losses in grazing animals have occurred in several parts of the world as a result of phosphorus deficiency. Roughages in general are low in phosphorus and high in calcium; grains, high in phosphorus and low in calcium. However, the quantities found in the feed can be markedly influenced by the availability of these minerals in the soil.

Vitamin A needs of the beef animal are dependent on (1) age, (2) life-cycle phase, (3) stability of vitamin A in feeds, (4) factors that influence metabolic conversion of carotene to vitamin A. Normally green pasture or high-quality roughage will supply adequate amounts of carotene for conversion to vitamin A. However, under range conditions, especially in periods of prolonged drought, some form of synthetic vitamin A may be desirable. Neumann et al. (1960) observed that feedlot animals may exhibit vitamin A deficiencies even when a supposedly adequate amount of carotene is present in the ration. It has been suggested that an excess amount of nitrates and nitrites in grains and forage (due to heavy nitrogen fertilization) interferes with the conversion of carotene to vitamin A in the body of the feedlot animal. Under such conditions it is a good management practice to rely on synthetic vitamin A to meet requirements.

Vitamin E is normally present in adequate amounts in natural feedstuffs. Deficiency of this vitamin occasionally occurs in young calves and is characterized by muscular dystrophy or white muscle disease. Although vitamin E is required for normal reproduction in both male and female, beef animals seldom if ever exhibit this deficiency.

Sodium chloride (common salt) is needed by all cattle. Animals maintained in certain geographic areas may require supplemental amounts of minerals such as phosphorus, cobalt, iodine, and copper. There appears to be a special relationship between copper and molybdenum (see Fig. 12-9). In certain parts of California, excess amounts of molybdenum occur in the forage in irrigated pastures and result in an interference with the utilization of copper. Cattle on such pastures have diarrhea and become emaciated and anemic. Becker et al. (1953) have shown the need for specific mineral supplementation in the Florida area. Iodine deficiencies are recognized in the Great Lakes and Pacific Northwest area.

A condition known as "grass tetany" has been attributed to abnormal magnesium blood values. Cattle, especially lactating cows, when grazing winter small grain fields in the southern Great Plains occasionally are subject to grass tetany or wheat pasture poisoning. Certain climatic and temperature conditions must prevail for animals to exhibit the disease. Sound management practices in this area suggest that the cattlemen should observe the cow herd closely when it is grazing such pastures. Animals that exhibit symptoms can usually be saved by intravenous injections of calcium gluconate if they are found in time.

Sound husbandry practices would suggest that, when a herd health problem arises that is not readily attributable to a pathogenic organism, the vitamin and mineral content of the feed should be checked as a possible cause of the disorder. Optimum performance of the herd cannot be expected when vitamin or mineral shortages or imbalances occur.

The daily requirements of energy and protein for growing, pregnant, lactating, and fattening beef cattle are given in Table 31-2. Note that a lactating cow has essentially the same requirement as that of a steer on full feed. The needs of a pregnant cow are relatively low.

31-8. COMMON MANAGEMENT PRACTICES

Marking and Identification. For many years stockmen have made a practice of marking their animals primarily to establish and maintain ownership. The development of our purebred breeds of livestock has made it necessary that individual animals be identified. In fact, purebred registry associations require that each individual animal recorded in the herd books be permanently marked and identified.

The most common means of marking beef cattle for permanent identification is hide-branding. Much of the history and tradition of the western ranges centers around this practice, and it is still widely used in the West and Southwest. Hot iron brands are applied with either a fixed-style brand or a so-called "running iron." Chemical brands are used in a limited way. Other means of marking range cattle include ear notching, ear slitting, and slitting a portion of the dewlap to form wattles.

In purebred herds the most common method of identification is ear tattooing. When properly applied, this provides a lifetime identification and in no way disfigures the animal. Other means used to mark and identify individual animals are horn brands for horned cattle, numbered neck chains or neck straps, and numbered metal or plastic tags. In many experimental herds a hot-iron number brand is used as a permanent mark of identification. While this method may be objectional in herds of registered cattle because it disfigures the hide, it does offer a permanent, easy way of identifying individual animals.

Vaccination. Beef calves are born with—or develop it shortly after birth —immunity to certain diseases. For some diseases this immunity is only

TABLE 31-2. | *Daily nutrient requirements of beef cattle.*

Type of Animal	Body Weight (lb.)	Average Daily Gain (lb.)	Daily Feed per Animal (lb.)	Digestible Protein (lb.)	TDN* (lb.)	Digestible Energy		
						For Maintenance (kcal.)	Per Pound Gain (kcal.)	Total (kcal.)
Growing heifer	1000	1.0	21.1	1.0	10.6	13,300	7,800	21,100
Pregnant heifer	1000	0.5	18.0	0.8	9.0	—	—	18,000
Cow nursing calf	900 to 1100	None	28.0	1.4	16.8	—	—	33,600
Fattening yearling steer	600	2.6	17.5	1.3	11.4	9,000	5,300	22,800
	800	2.7	22.3	1.7	14.5	11,200	6,600	29,000
	1000	2.6	25.8	1.9	16.8	13,300	7,800	33,600

Source: Adapted from Nutrient Requirements of Domestic Animals, No. 4. Nutrient Requirements of Beef Cattle. National Research Council Publication 1137, 1963.
* Total digestible nutrients.

temporary or of limited duration. Vaccines that will supplement this natural immunity have been developed and provide lifetime protection against the occurrence of a few of these diseases. Some of the more common diseases for which vaccines are available are blackleg (*Clostridium feseri*), malignant edema (*Clostridium septicus*), and contagious abortion (*Bacillus abortus*). The advice and service of a well-qualified veterinarian is of utmost importance in maintaining herd health and sanitation.

Parasite Control. Beef cattle are subject to the usual cattle parasites. Two of the most common external parasites are flies and cattle lice. In some sections of the country screwworms are a serious problem during the late summer and early fall months. Certain internal parasites are also common in cattle, perhaps the most common being the cattle grub, in the late winter and early spring months. This parasite may be very costly to the cattle feeder at certain seasons, since buyers will usually discount an animal that is heavily infested with cattle grub. In some areas of the country certain kinds of stomach worms are a serious menace.

The control measures recommended by livestock specialists in the various areas are usually effective in controlling most of the common cattle parasites. When using chemicals to control parasites, it is very important to observe all the precautions and recommendations of the manufacturer, especially when applying the product to feedlot animals. Regulations of the Pure Food and Drug Administration of the federal government require that animals slaughtered for beef contain no chemical residues, which may result from treating them for parasite control.

Castration. Bull calves are castrated for primarily economic reasons. Steers produce a more desirable beef carcass and they are more docile than bulls. Castration may be done at any age, from day-old calves to mature animals, but it is somewhat more hazardous to castrate a calf after 3 months of age. The most common method of castration involves the surgical removal of the testicles from the scrotum. While this is a rather simple operation, an inexperienced person should consult a veterinarian or other qualified person before attempting to castrate a bull calf. During the operation the general rules of animal health and sanitation must, of course, be observed.

Recently Pope et al. (1960) have shown that it is possible to accomplish many of the results of castration by the use of stilbestrol implants. Weight gains and carcass characteristics of intact bull calves implanted with 24 mg stilbestrol at $3\frac{1}{2}$ months of age were intermediate between intact bull and steer calves of the same age.

Dehorning. Some breeds of beef cattle are horned. Although well-trained horns may be considered a desirable asset in a registered herd, they are not desirable in commercial herds. Commercial cattlemen follow a practice of dehorning all but the herd bulls, since dehorning commercial beef animals ordinarily increases their market value. Horned animals may be dehorned at any age; it is much more desirable, however, to dehorn them as calves. Many herd owners follow a practice of dehorning, castrating, and vaccinating at the same time, usually prior to 3 months of age.

There are several common dehorning tools, including a dehorning tube, dehorning spoon, specially designed dehorning pincers or clippers, and a dehorning saw. A bell-shaped hot iron and an electric heated hot iron are also available as dehorning tools. Still other herd owners use a caustic chemical as a means of dehorning very young calves.

Before attempting to dehorn an animal, the inexperienced person should be instructed by an individual familiar with the use of a particular dehorning tool. Some general precautions should be exercised in dehorning. Avoid extremes in weather conditions, and avoid dehorning during fly season. Early spring or late fall are generally considered the most desirable seasons for dehorning animals. The animal's health and customary sanitary precautions must be observed.

Record Keeping. Complete and accurate herd records are a valuable asset to the management of any beef cattle herd. Herd records are essential in the operation of a purebred herd when the management expects to register the animals. Breed registry associations have established certain basic requirements, which must be met before animals can be recorded in the herd books. In addition to making sure that animals meet the basic eligibility requirements for registry, the breeder must provide information on date of birth, sex, color and markings, whether horned or polled, tattoo or other individual identification, name of animal, breeder, first owner, and perhaps a statement relative to whether the calf is the result of artificial insemination. In short, the owners of a registered herd must have a complete set of records.

The more progressive owners—of both registered and commercial herds —maintain records of the productivity of their cow herds. Such records are very valuable for economic reasons. Records may include information on number of calves born, number weaned, weaning weight and grade, and whatever other information is desired by the owner. Such records provide valuable information when selecting herd replacements and aid in culling the cow herd.

In recent years beef cattle producers have shown much interest in records of performance and production of beef cattle, for such records serve to identify the most efficient animals. Data on regular calving, mothering ability of the cow, rate and efficiency of gain, and quality of carcass are all very important in determining net income from a beef cattle enterprise.

In 1955 a national organization was established to encourage the keeping of records on the performance of beef cattle, for the purpose of selecting and producing more productive breeding cattle. Many states have similar organizations administered by the State Agricultural Extension Service or by a state association. Many well-known herds, both registered and commercial, are now keeping and using performance records to aid in the improvement of their cattle.

The importance of keeping accurate records in a selection program is illustrated by the two cows shown in Fig. 31-6; both have raised eleven calves under identical environmental and feed conditions. Calves from the cow shown at left (No. 44) have averaged 521.3 lb. at 210 days of age. Calves from the cow at right (No. 125) have averaged 383.1 lb. at 221 days of age. The difference in lifetime production of these two cows is 1520 lb.; in other words, cow 44 has produced the equivalent of three additional calves, weighing 506 lb. each, with little additional feed or maintenance costs.

Equipment for Beef Cattle. Equipment for handling beef cattle should be practical, safe, and reasonable in cost. Major items of equipment required include adequate shelter, good fencing, paved feedlots in which animals are confined to a small area, adequate water and feeding facilities, corrals, holding pens, some form of restraint equipment, and preferably a scale.

Shelter requirements for beef cattle vary, depending on climatic conditions. In midwestern and northern range sections animals must be provided protection from wind and rain. Beef cattle can withstand severe cold if they are adequately fed and kept dry. Continued exposure to cold winter

Fig. 31-6. A good set of records will help in deciding which of these cows is the more profitable.

rains and wind will result in severe weight loss as well as a much higher body maintenance requirement. Open sheds facing the south generally provide sufficient shelter except in case of severe winter storms.

In much of the West and Southwest, especially the more arid regions, many cow herds are maintained the year long on pasture, the only shelter being that of timber and other natural windbreaks. In areas where shelter must be provided, consideration should be given to the location of sheds and barns. Shelter areas should be well drained and readily accessible from both pasture and service areas. Where winter feeding is done inside barns or sheds the feeding areas should be convenient to both the feeder and the cattle.

Good fences, properly constructed, are one of the most valuable assets on livestock farms and ranches. Fencing need not be expensive, but the useful life of a fence will be largely determined by the manner of its construction.

Corrals or holding pens are essential when groups of animals are to be handled, and should be designed for convenience and ease of handling. They should be strong and made from material requiring a minimum of maintenance. Gates in pasture fences and corrals should be sturdy and located to provide easy access to all areas where animals are to be moved and handled. On many large ranches an auto gate or "cattle guard" is located alongside the gate used to move animals from pasture to pasture. Cattle guards save considerable time and labor when driving about the ranch.

One of the most essential items on cattle farms and ranches, even though not generally included as equipment, is an adequate water supply. In many areas natural streams or lakes provide excellent water sources. Since great numbers of beef cattle are raised in the more arid sections of the country, the problem of an adequate supply becomes a major concern on many large ranches. Farm ponds or earthen tanks are widely used in the range area as a source of stock water. Windmill-operated wells are also commonly used to supply water. On many large ranches one man or a crew of men is detailed to the job of keeping the windmills in good repair.

Cattle-feeding equipment should be convenient, have a low maintenance cost, and remain serviceable over an extended period of time. The amount and kind of feeding equipment required depends on the size and type of cattle operation on a given farm or ranch. Many cow herds are maintained on the range with little or no feeding equipment. During periods when supplemental feeding is necessary, herds may be fed pelleted protein supplement on the ground. Feed bunks are desirable in areas where more frequent feeding is required.

One desirable item of equipment, regardless of the size of operation, is a "squeeze chute," used to restrain animals when they are being handled. A well-designed, sturdy, well-located loading chute is an essential item on farms and ranches where numbers of animals are moved by truck.

Saddle horses are also very necessary on large cattle ranches. This is one phase of agriculture in which the horse is still as valuable as he was 50 years ago. Many of the large cattle ranches in the West still keep a band of mares and a good stallion to grow the saddle stock needed on the ranch.

Other items of needed equipment include balling guns, branding irons, calf puller, castrating knife, dehorning equipment, fencing equipment, flashlights, lariat rope, saddles, syringes, tattoo machine, and a supply of common biologics and medications.

Special Problems in Management of Herds of Registered Beef Cattle. The management of herds of registered beef cattle presents some problems not found in the operation of commercial herds. The detailed records needed to register animals, the merchandising of animals, fitting for showing at livestock expositions, and special herd health consideration are concerns not ordinarily connected with management of commercial cow herds.

One of the major problems in the conduct of a registered herd is the merchandising of cattle. Registered animals can be and sometimes are sold through the normal market channels. In order to provide the operator with a reasonable return on his investment, registered animals must sell for substantially higher prices than commercial animals. Successful selling of registered animals requires salesmanship, advertising, fair dealing, and a reasonable guarantee of the soundness of animals sold. Animals are normally bought and sold by private contract or at public auction.

Fitting and showing of registered animals at livestock shows provide an opportunity to display part of the herd to the public. Many herd owners feel that this is one of the most effective forms of herd advertising, although records of performances are more valuable and may have greater advertising value than show-ring winnings.

Federal and state animal-health agencies regulate the interstate movement of breeding animals. The official state animal-health agency issues health certificates, which permit interstate movement of breeding stock. The prudent cattleman will make sure that breeding animals purchased in another state have been tested for contagious diseases and that an official health certificate—showing them to be free from diseases—accompanies the interstate shipment of such animals. Animals exhibited at livestock shows must be accompanied by an official health certificate.

Sound beef cattle management will always pay good dividends.

REFERENCES AND SELECTED READINGS

References marked with an asterisk are of general interest.

Becker, R. B., P. T. D. Arnold, W. G. Kirk, G. K. Davis, and R. W. Kidder, 1953. *Minerals for Dairy and Beef Cattle.* Fla. Agr. Expt. Sta. Bul. 513.

Botkin, M. P., and J. A. Whatley, Jr., 1953. Repeatability of production in range beef cows. *J. Animal Sci.,* 12:552.

Chambers, D., J. A. Whatley, Jr., and W. D. Campbell, 1954. *A Study of the Calving Performance of Two-Year-Old Hereford Heifers.* Okla. Agr. Expt. Sta. Misc. Publ. MP-34.

Clegg, M. T., and F. D. Carroll, 1957. A comparison of the method of administration of stilbestrol on growth and carcass characteristics of beef cattle. *J. Animal Sci.,* 16:662.

Dinusson, W. E., F. N. Andrews, and W. M. Beeson, 1950. The effects of stilbestrol, testosterone, thyroid alteration and spaying on the growth and fattening of beef heifers. *J. Animal Sci.,* 9:321.

Gifford, W., 1953. *Record of Performance Tests for Beef Cattle in Breeding Herds.* Ark. Agr. Expt. Sta. Bull. 531.

Ittner, N. R., R. E. Bond, and C. F. Kelly, 1958. *Methods of Increasing Beef Production in Hot Climates.* Calif. Agr. Expt. Sta. Bul. 761.

Nelson, A. B., L. R. Kuhlman, and L. S. Pope, 1961. *Stilbestrol for Range Beef Cattle.* Okla. Agr. Expt. Sta. Misc. Publ. 64.

Neumann, A. L., 1960. Another look at vitamin A in cattle feeding. In *Proc. Illinois Nutr. Conf.,* University of Illinois, Urbana.

Pinney, D., L. S. Pope, A. B. Nelson, D. F. Stephens, and G. Waller, Jr., 1960. *Effect of Different Amounts of Winter Supplement on the Performance of Spring Calving Beef Cows.* Okla. Agr. Expt. Sta. Misc. Publ. MP-57(60).

Pope, L. S., E. J. Turman, L. E. Walters, K. Urban, and J. Halbert, 1960. *Implanting Steer, Bull, and Heifer Calves in a Fat Slaughter Calf Program.* Okla. Agr. Expt. Sta. Misc. Publ. MP-57(26).

Self, H. L., C. E. Summes, Dale Hill, Fred Roth, Harold Stockdale, W. G. Smolek, J. B. Herrick, and Robert Rust, 1964. *Shelter and Paving Benefits for Feedlot Cattle.* Iowa State Univ., A. S. Leaflet R 52.

* Snapp, R. R., and A. L. Neumann, 1960. *Beef Cattle.* 5th ed. Wiley, New York.

* Wagnon, K. A., R. Albaugh, and G. H. Hart, 1960. *Beef Cattle Production.* Macmillan, New York.

Zimbelman, R. G., 1963. Determination of the minimal effective dose of 6α-Methyl-17α-Acetoxyprogesterone for control of the estrual cycle of cattle. *J. Animal Sci.,* 22:1051.

Chapter 32

Dairy Cattle Management

There is in all of us a psychological tendency to resist new ideas.
BEVERIDGE, W. I. B., *The Art of Scientific Investigation*

32-1. INTRODUCTION

Since the beginning of history the cow has been useful to man in many ways. Not only has she been a source of food and a beast of burden, but she has even played an important role in his religion, mythology, and political economy. Dairying undoubtedly had its beginning long before the advent of historical writings; the oldest graphic records are pictures discovered in the Libyan caves, believed to have been made no later than 9000 B.C.

The oldest written records, the cuneiform writings of the Sumerians, indicate that dairying existed 6000 years before Christ. Remains of Swiss lake dwellers that date back to about 4000 B.C. include not only the skeletons of cattle but also equipment for cheese-making. Records of an era a thousand years later, discovered in Egypt in excavations of tombs, reveal that milk and milk products were probably widely used by the Egyptians. The Old Testament contains many references to milk and butter.

Yet in spite of man's dependence on the dairy cow during these early historical times, accomplishments in dairying during the thousands of centuries indicated are insignificant in contrast with developments of the last century.

One hundred years ago, dairying was largely a family affair. Even in towns and villages most families kept a cow for their own use; the milk was usually consumed in the raw state, and the surplus was made into butter and cheese in the home. Dairying gradually became more specialized, and people bought milk, butter, and cheese from farmers farther out in the country.

Today, obtaining milk from the cow is only a first step in the very complex process of producing dairy foods. With the gradual development of large centers of concentrated populations, the dairy industry has become divided into three separate and distinct phases—production, processing, and distribution. The producer of milk plays an important role in this extremely complex industry.

Since dairy products constitute a quarter of the human diet, the steady increase in population has required a tremendous increase in annual milk production. During 1964 dairy cows in the United States produced an estimated 126.6 billion pounds of milk, even though the numbers of cows have decreased since 1946 in every year except 1953. In contrast, the average production of milk per cow has increased 26% in the past 9 years (USDA, 1964a).

Increasing costs of production—a result of the rising costs of labor, feed, and modern equipment—have necessitated greater efficiency in production. Dairymen have been forced to improve the methods of feeding, breeding, and caring for the dairy cow—phases that may be combined under the general term "management."

Progress in methods of management has been considerable, but many problems still await solution. Testing associations have published results that suggest what more can be done. During 1964, cows enrolled in the Standard Plan of the Dairy Herd Improvement Association testing program produced at an average yearly rate of 11,685 lb. of milk and 447 lb. of milk fat (USDA 1964b), whereas all cows in the United States averaged only 7,880 lb. of milk (USDA 1964a). This difference reflects improved management and the use of bulls capable of siring daughters that respond to such management.

Consider also that many cows under excellent managerial conditions have produced over 25,000 lb. of milk and 1,000 lb. of milk fat. There are at least two reasons why one could not expect today to obtain a herd average comparable to the recent world record of the Holstein cow Princess Breezewood R. A. Patsy—36,821 lb. of milk and 1,866 lb. of milk fat on twice-daily milking in 365 days. First, we do not have enough information on the mode of inheritance of maximum capacity for milk and fat production, and, second, the care and feeding of this animal was doubtless more elaborate than would be consistent with present commercial practice. Even so, information on such inheritance is increasing, and improved managerial practices will continue to improve average production per cow.

Modern dairy management encompasses a wide field of activities involving a constantly increasing application of many sciences. Only the briefest reference can be made here to its numerous phases. Training in chemistry, genetics, and nutrition, and specialized courses in feeds and feeding methods and dairy cattle management are required for comprehensive understanding of the problems involved and the advances of the last 25 years.

Initial outlay is large in establishing a dairy herd, with its requirements for land, buildings, corrals, and equipment. As a rule dairymen start with milking cows in order to provide an immediate cash income. For the purpose at hand, however, we will start with the calf and continue briefly through the various phases of dairy management.

32-2. CALF RAISING

It is becoming increasingly difficult to maintain and improve a dairy herd by purchasing replacements for the milking herd. With purchased stock the dairyman is unable to follow a breeding program that will improve the productive capacity of his herd. He is also in constant danger of introducing infectious diseases. For the average dairyman, herd improvement involves raising female calves sired by bulls of superior transmitting ability.

The dairy calf is born following a gestation period averaging 278 to 288 days, depending on the breed—shortest for the Jersey, longest for the Brown Swiss. At birth the normal calf has a strong inherent capacity to grow. The dairyman, however, must provide the proper environment (feed and care) for normal development. Successful methods of raising calves are based on fundamental principles of nutrition, physiology, and disease control.

The growth of the calf from birth until it enters the milking herd falls naturally into five stages.

The first stage is the two days following birth, when, starting within an hour of birth, the calf should receive colostrum, the mammary secretion produced by the cow immediately following parturition. Not only is colostrum most nearly perfect nutritionally for the newborn calf, but it contains antibodies that help the calf resist certain infectious diseases encountered during the first few days of life (Smith and Little, 1922).

The second stage extends through the second week of life. At this time the calf should receive whole milk fed at a rate of 7 to 8% of its body weight. If an unlimited milk-feeding method is to be used after two weeks, the daily rate may be increased to 10%.

The third stage is from 2 weeks to 6 months, during which a change is usually made to one of the minimum milk methods, wherein a limited amount of milk is supplemented with a partial milk replacer. Research at many colleges and universities continues to seek more exact information on the specific nutrients required by the young calf. This third stage carries

Fig. 32-1. Calf mortality due to disease may be reduced through use of movable calf pens where climatic conditions permit. [Mead, 1930].

the calf past the period of greatest mortality, averaging about 10 to 12%, largely from infectious diseases. Unsanitary conditions and improper nutrition are important factors in reducing the young calf's resistance to disease. Where climate is favorable, many dairymen use individual pens that can be moved to clean ground each time they start a new calf (Fig. 32-1).

The fourth stage, from 6 months to 1 year, presents a relatively simple problem. During this period calves usually receive either pasture with concentrates or hay.

During the fifth stage, from 1 year to 2 years of age or more, the feeding program is similar, requiring only sufficient feed for normal growth. It is during this period of development that the heifer is bred for first pregnancy. Table 32-1 shows the ages at which heifers may be bred if they have

TABLE 32-1. *Age and weight schedule for breeding dairy heifers.*

Breed	Age to Breed (months)	Normal Weight (pounds)
Holstein	18–20	850–900
Ayrshire	17–19	700–750
Guernsey	16–18	600–660
Jersey	15–17	525–570

made the indicated normal growth. Age alone does not necessarily determine the proper time for breeding, since rate of growth can be materially influenced by feed intake. For example, Jersey heifers fed large amounts of concentrates together with roughage reached normal weight for breeding at 10 to 12 months of age (Mead, 1942).

The advantage of early calving is evident, since the period between birth and first lactation is unproductive. Body size and milk-producing ability are related. After freshening, the undersized heifer is unable to consume enough food for both maximum production and further growth.

In a study of first-calf heifers, sired by the same bull but raised in different herds, the heifers showed marked differences in producing ability (Table 32-2). These differences in milk production could depend upon several factors, such as genotype of the dams, management (including feeding during lactation), and differences in size and finish at time of first calving. The differences in size and condition at first calving are considered by the author, who participated in this study to be largely responsible for the variations between herds.

32-3. DAIRY CATTLE NUTRITION

During the first lactation, the cow needs sufficient feed for continued growth, as well as for milk production. In succeeding lactations, more feed will be needed both for the normally expected increase in milk production

TABLE 32-2. | *Heifers raised in different herds, but sired by the same bull, show marked differences in producing ability.*

Sire Number	Herd Number								
	1	2	3	4	5	6	7	8	9
	Heifer's Producing Ability [*] (*lb. milk fat*)								
62A	316		369	376	386				
333C	277					374		375	
370A	365					456	435	448	
372A	284	312		380				374	
372B		297		347					
372C	302	320		361		384			
374B		280		316					369
375B	297					380			362
375E		296	336					389	
498B		286	308	348					
572B		387	379					441	472

[*] All records are for junior two-year-olds; 305-day lactations; twice daily milking.

and for body maintenance. During the last stages of pregnancy, the cow requires sufficient nutrients, not only for the growth of the fetus, but also for building up reserve body tissue, minerals, and vitamins for use during the next lactation. These nutrient requirements, established through research and feeding trials, are presented in the form of feeding standards. Some differences in requirements dependent upon the physiological state are indicated in Table 32-3. Consideration must be given to the following factors in determining the nutrient requirement: growth, lactation (amount of milk and its fat content, and body size. For details, one should consult a modern text on feeds and feeding of livestock.

Roughages and Concentrates. Feeds used by the dairy cow are divided into two general classes: roughages and concentrates. Roughages contain a relatively high percentage of fiber and have a comparatively low-feeding value; examples are hays, pastures, and silages. Classed as concentrates are grains and oil-bearing seeds and their by-products. Most concentrates contain less fiber and have a higher food value than roughages. By-products of oil-bearing seeds, such as cottonseed, linseed, and soybean meals contain a high percentage of protein. Legumes, such as alfalfa and clover, are roughages with a high percentage of protein. The content of protein depends

TABLE 32-3. | *Daily requirements of dairy cattle observed under several different physiological conditions.*

Physiological State of Animal	Digestible Protein (lb.)	Total Digestible Nutrients (lb.)	Calcium (g)	Phosphorus (g)	Carotene (mg)	Net Energy (therms)
Growing 500-lb. heifer	0.92	8.2	14	10	30	7.4
First-calf heifer giving 50 lb. of 4% milk and weighing 1,000 lb.	3.40	25.7	63	50.5	60	22.8
Mature cow giving 50 lb. of 4% milk and weighing 1,000 lb.	3.10	23.9	60	47.5	60	21.3
Mature dry cow weighing 1,000 lb. one month before freshening	1.25	13.9	23	18	90	11.4
Mature bull weighing 1,600 lb.	1.42	13.5	16	16	96	11.9

Source: Taken by permission of the Morrison Publishing Company, Clinton, Iowa, from the 22nd Edition, third printing, 1959, of Feeds and Feeding, by F. B. Morrison and Associates.

upon the stage of growth, methods of curing, and weather conditions—for example, rain leaches out a high percentage of nutrients. Silages are low in total protein and energy, owing to their high moisture content.

Although the cow can consume large quantities of roughage, this feed will not meet her needs during lactation if she has inherited the capability of high production. In recent years, the amount of roughage fed has decreased and concentrates increased in most areas in order to increase production of milk. Largely because of labor, equipment, and housing costs, the economy of dairying depends to a large degree upon the amount of milk produced.

The composition of the concentrate mixture will depend largely upon the quality of the roughage. If good quality legume hays are fed, or cows are pastured on 50% legume and 50% grass, the concentrate mixture need not contain more than 12% digestible protein. If low-protein roughages are fed, such as timothy or oat hays, the concentrate mixture should contain 17 to 18% digestible protein. If corn silage is fed in large quantities, together with legume hays, then the protein content of the concentrate mixture should be increased by 3 to 4%. Among the "economic" factors that determine the amount of concentrate feeding is the cost of nutrients relative to the selling price of milk.

Ground, Pelleted, and Wafered Hay. Finely ground hay, fed in loose form or compressed into pellets, causes a reduction in percentage of milk fat (Ronning, 1960). Presumably, it passes through the rumen so rapidly that sufficient fermentation does not take place to form enough acetic acid, which is transferred to the mammary gland for the formation of milk fat. However, compression of coarsely chopped hay into wafers results in an increase in consumption of hay and an increase in milk production, but no reduction in percentage of milk fat (Ronning and Dobie, 1962).

Milk Fever (Parturient Paresis). This condition occurs most frequently in high-producing cows, either during parturition or, more commonly, one to three days after parturition. The stricken cow becomes paralyzed and needs immediate treatment—otherwise, she will die. In spite of the name "milk fever," an animal with this disease does not exhibit an elevated body temperature. Parturient paresis seldom occurs in first-calf heifers, but rather in cows in their third lactation or beyond. The parathyroid glands, which have as their sole function the control of blood calcium, are presumably unable, in high-producing cows, to increase rapidly enough their secretory activity in order to mobilize the large amounts of calcium required for milk secretion immediately following parturition. Boda and Cole (1954) suggested that the activity of the parathyroid glands could be stimulated through the use of a low-calcium, high-phosphorus diet during the month before parturition. This practice proved to be highly effective in reducing the incidence of milk fever. Hibbs and Pounden (1955) have shown that this disease can also be prevented by feeding large doses of Vitamin D for 3 to 7 days before calving. The common treatment for milk fever is to inject intravenously a 20% solution of calcium gluconate (250 to 500 cc).

32-4. ECONOMY OF HIGH PRODUCTION

Neither herd size nor total milk production determines profit in the dairy business. Profit is determined by income in excess of cost of production. Feed represents 50 to 60% of the total operating expense. The second-largest cost is labor. Other costs include veterinary fees, depreciation, supplies, repairs, utilities, taxes, and interest on investment. A low-producing cow costs about as much to maintain as a high-producing cow. True, feed needs increase with increases in milk production, but the cash return on milk increases faster than increases in feed expense.

The reason why the high producer is profitable is that she uses a smaller proportion of her feed in maintaining her body, even though her inherited stimulation to produce more milk results in a greater food consumption (Fig. 32-2).

Profit in the dairy business depends both on having cows with inheritance for high production and on providing them an environment that permits full expression of their inherited potential.

Productive Life of the Dairy Cow. Though some cows have lived 17 years or more, the average cow is culled or dies between five and six years of age. Since cows calve, on the average, at approximately two years of age this allows only three to four years of productive life. Each year, 20 to 25% of the animals more than two years old are culled or lost from the milking herd because of low production, infertility, mastitis, miscellaneous diseases, or accidents. Thus, since the productive ability of a cow tends to increase with each succeeding lactation up to maturity (around six years old), the average

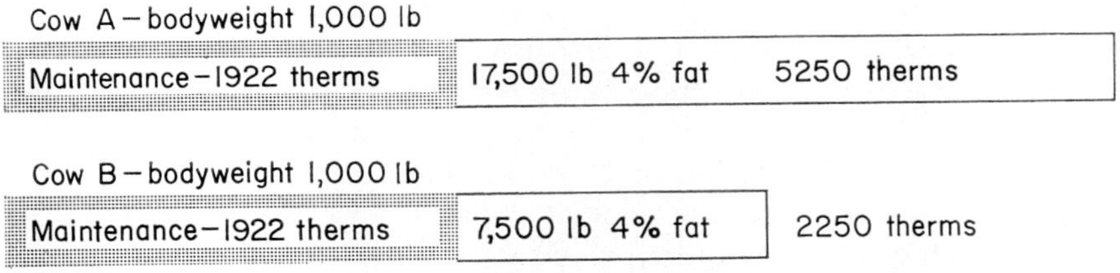

Cow A — bodyweight 1,000 lb

Maintenance – 1922 therms	17,500 lb 4% fat	5250 therms

Cow B — bodyweight 1,000 lb

Maintenance – 1922 therms	7,500 lb 4% fat	2250 therms

Fig. 32-2. Feed for maintenance and milk production. Since both cows use the same amount of energy for maintenance, high-producing cows use less feed per pound of milk. Although the total feed requirement of Cow A was 1.7 times greater than that of Cow B, she required less feed per pound of milk.

cow leaves the herd at or before the age at which she should be returning the greatest profit. The net return from the first lactation of an average producer will scarcely pay for the cost of raising her to first freshening. Improvements in management practices, including disease control, will reduce the excessive cost of replacements, and increase dairy herd profit.

32-5. HERD RECORDS

The modern dairyman cannot operate successfully without complete herd records. The record system must be relatively easy to maintain; otherwise, it will soon be neglected. There are a number of satisfactory methods. One is a file containing a card for each cow in the herd. Identification of all animals in the herd, whether purebred or grade, is very important for intelligent breeding and feeding programs. A simple numbering system is desirable. Each animal is given a number immediately following birth; the animal can be numbered by ear tag, neck chain to which is attached a metal number, ear tattoo, or brand.

Dates of breeding, sire number, due date, calving date and sex, and herd number of calf are recorded on each card during the lifetime of the cow. On the opposite side of the card is recorded any veterinary examinations or treatments. For small herds, a single large card, on which is recorded all data for all cows in the herd will prove satisfactory.

In addition, many dairymen also maintain a herdbook to which they transfer the card records. The same book can contain the monthly milk and fat production and the total lactation for each cow. At present, most production records, following monthly testing, are transferred to a computer center and the results received shortly by the dairyman. He maintains a file for such records or transfers them to his herdbook.

Breeding Dates and Dry Period. Breeding dates should be recorded, in order to predict the expected calving date so that a sufficient dry period will be provided. A cow will produce less, if milked continuously, than if she has a dry period of six to eight weeks between lactations. The proper interval between lactations will depend upon the cow's condition. Milk production is a heavy drain on the cow: she needs a rest period with proper feeding for storing reserves of body flesh, minerals, and certain vitamins for the succeeding lactation. Next to low production, infertility is the reason for culling the largest percentage of cows from many herds. This is a subject on which volumes have been published (for example: Cole and Cupps, 1959; Salisbury and VanDemark, 1961) and on which research is still being conducted at many institutions. Accurate records of breeding dates, followed by a pregnancy diagnosis, not only will indicate the extent of breeding troubles but also will sometimes reveal the cause.

When the cow is pronounced pregnant, the due date should be calculated on the basis of the normal gestation period for her breed, and recorded on the calving schedule.

Many cases of low breeding efficiency have been traced to carelessness in recognizing and recording the estrous periods. The herd should be observed at least twice daily. Cows may not always show the most obvious symptom of "heat"—standing for other cows to mount. All observed periods should be recorded and the cows should be watched closely. The estrous cycle of the normal cow varies between 18 and 24 days. The "heat" period, during which the cow will accept the bull for service, lasts about 18 hours. Ovulation occurs about 10 hours after the end of estrus (heat) (Trimberger, 1948; Moeller and VanDemark, 1951). Variations may indicate abnormal conditions not conducive to efficient reproduction.

The recorded date of calving is important. The normal healthy cow should not be bred for at least 60 days following parturition; the reproductive tract requires that much time in which to regain its normal physiological state. Cows that have had an abnormal gestation period, either too short or too long, or have experienced dystocia or retained placenta, will require a longer rest period. Such cases should be referred to the veterinarian.

Additional Records. The breeder of purebreds must keep additional records. He is responsible to his breed registry association for the accuracy of the ancestry of all animals offered for registration. A registration application form must contain accurate information on date of birth, proposed name, color markings or other identification, and the registration names and numbers of the sire and dam. When the application has been accepted, the breed association issues a certificate carrying the name or names of breeder and owner, registered name and number of the animal, its identification, and the registered names and numbers of its sire and dam.

32-6. MILK AND FAT PRODUCTION AND TESTING PROGRAMS

The milk and fat production of the herd determines its economic value. High production depends on the use of superior sires and efficient feeding and management, including continuous production testing.

A number of production testing programs are available. The National Cooperative Dairy Herd Improvement Association provides three types of programs for unregistered cows, but they are also available to breeders of registered purebreds. In addition, the breed associations provide three programs for registered purebreds.

The National Cooperative Dairy Herd Improvement Association (more commonly known as DHIA) is sponsored by the Agricultural Research Service, USDA, and the extension services of the land-grant colleges. Each local association is operated under the direction of a board of directors consisting of participating dairymen assisted by the College of Agriculture Extension Service. Dairymen may select one of the three programs. The "Standard Plan" requires that a supervisor, hired by the association, must visit the herd each month and weigh the milk produced by each cow during a 24-hour period and determine its milk fat content. The supervisor calculates the monthly production of each cow, and the herd's average for the year. In the "Owner-Sampler" plan the owner (rather than the supervisor) weighs and samples the milk. The samples are tested at a central laboratory. Under the "Weigh-A-Day-A-Month" plan, less information is supplied. The milk of each cow is weighed by the owner; the milk is not sampled. The local extension agent calculates the monthly milk production of each cow and returns the information to the owner. Of the 2,822,522 cows included in all three plans, 2,010,144 are tested under the "Standard Plan"; 752,229 are enrolled in the "Owner-Sampler" program and 60,149 in the "Weigh-A-Day-A-Month" plan, representing a total of 17.1% of all milk cows (USDA, 1964c).

Three programs are provided by the breed registry associations for testing registered cows: "Advanced Registry," "Herd Improvement Registry," and "Dairy Herd Improvement Registry." The rules for conducting these programs have been approved by the Purebred Dairy Cattle Association and the American Dairy Science Association. In the various states, the superintendents of official testing have charge of the tests and appoint the supervisors.

The number of cows in the Advanced Registry program has declined sharply in recent years. This program allows individual cows to be tested, rather than the entire milking herd. At the present time, most dairy breeders are interested in the production of all of their cows and the average annual production of the entire herd. This is accomplished through the selection of either the Herd Improvement Registry or the Dairy Herd Improvement Registry. The Herd Improvement Registry requires that all cows, including those used as nurse cows, be tested, except those that are ten to twelve years old, providing they have a previously completed official record. This testing program requires that the cows be tested monthly by the supervisor, who is required to identify each cow through color markings or by the ear tattoo shown on her registration certificate. During a 24- or 48-hour period, the supervisor weighs the milk of each cow and tests it for fat. When tests have been completed, production certificates are issued to the owners of herds completing a year's record. The Dairy Herd Improvement Registry program is supervised by the State Extension Dairyman and the Breed Superintendent of Official Testing. It provides that records made by registered cows tested under the Standard Plan of the Dairy Herd Improvement Association may be used, if such records meet certain requirements in addition to those provided for in the Standard Plan. At the present time, more cows are tested under this plan than the HIR Program. Detailed rules are in *Unified Rules for Official Testing*, revised in 1960, obtainable from the secretary of the Purebred Dairy Cattle Association, Peterborough, New Hampshire.

32-7. MASTITIS AND MILKING PRACTICES

A disease that results in a great economic loss to dairymen is mastitis, since it causes reduced milk production and is responsible for a high percentage of the cows being culled from the herd. Mastitis is an inflammation of the udder, caused either by infection, or undue stress on the delicate mammary tissues, or a combination of both. In most areas, it is common practice to conduct routine tests for mastitis. Udders or individual quarters of the udder infected with bacteria are treated with certain antibiotics that are effective in eliminating the bacteria.

It is becoming increasingly evident that the most important factors in controlling mastitis are proper milking equipment and good milking practices. Most cows are now milked by machine. To avoid abnormal stress on the mammary tissue, it is important to operate milking machines in accordance with recommendations of the manufacturer, and to keep all equipment in proper operating condition. The teat cup assembly and the milk pipeline must be thoroughly cleaned and sanitized after each milking, both to insure a clean milk supply, and to reduce transmission of mastitis organisms from infected to noninfected cows.

Procedure for the Let-Down of Milk. Prior to attaching the teat cups, the cow must be "primed" for the let-down of milk; if this is not done, less milk will be obtained, and irritation of the teat may result. The udder is gently massaged and one or two streams of milk are drawn from each teat. This massage initiates nerve impulses resulting in the release of the hormone oxytocin from the posterior pituitary gland located at the base of the brain. When this hormone reaches the udder, through the bloodstream, it stimulates contraction of the myoepithelial cells of the alveoli. The milk is then squeezed out into the ducts and, finally, to the teats. This action takes place within one minute following the massage. The teat cups should then be attached or hand milking started. The let-down of milk continues during the milking process, but the release of oxytocin gradually decreases, and milking should take not more than five or six minutes: some cows are completely milked in three minutes. If the cow is frightened, adrenalin, which counteracts the effect of oxytocin, is released, and the milk is not.

32-8. THE HERD SIRE

No other decision is more closely related to the dairyman's success or failure than is selection of the sire. A superior sire will transmit high potential production capabilities to his daughters. Many examples could be cited, one of which is the long-range University of California Jersey Inbreeding Experiment in which sons of University sires with several generations of proved high production were loaned to cooperating dairymen from 1930 to 1959. No other bulls were used during this period. The results (Table 32-4) show a striking increase in milk production over the period of the experiment. Although some of the differences were due to improvements in feeding and management, an important part of the increased production can be attributed to genetic improvement introduced by the superior sires. A few cows were purchased for herds 1 and 3 during the early part of the experiment, in order to enlarge the herds quickly. The increase in size of all the herds was partly a result of genetic improvement, which reduced the need of culling for low production.

Sire Evaluation. Because of the importance of good sire selection, it has been the practice to evaluate or "prove" bulls by use of some sort of standardized rating system. All such systems rate the sire in terms of the production records of his progeny, but differ in the method used to obtain a comparative evaluation of these production records. For many years, sires were evaluated (proved) through daughter-dam comparison. An example of a daughter-dam comparison of a superior sire is shown in Fig. 32-3. Notice that the results of the progeny tests are commonly expressed in terms of twice-daily milking, a lactation period of 305 days, and mature age. Factors have been developed for converting records to this standard basis (USDA, 1955). Limitations of comparing records of daughters with those of their

TABLE 32-4. | *Increase in milk production as a result of genetic improvement and management.*

Herd No.	Before Using University Sires (3-Year Average)		After Using University Sires (Last 3-Year Average)	
	Number of Cows in Herd	Average Yearly Production* (lb. M.F.)	Number of Cows in Herd	Average Yearly Production* (lb. M.F.)
1	19	305	303	439
2	16	354	143	454
3	19	301	208	429
4	17	360	38	448
5	46	368	94	447
6	18	337	44	501
7	13	352	60	417
—	—	—	—	—
Average	21	340	127	448

* Dairy Herd Improvement Association records.

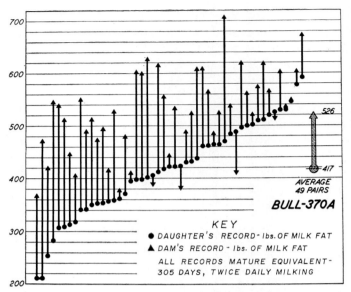

Fig. 32-3. Low production is the major reason for disposal of cows from the milking herd. Superior sires reduce cost of herd replacements.

dams and the advantages of comparing records of daughters with their contemporary herd mates are discussed in Chapter 15.

Management of the Sire. Management of the herd sire is comparatively simple, requiring only good judgment on the part of the caretaker. During early life, male and female calves are fed similarly, except that the milk-feeding period of the young bull should be extended to six or more months of age so as to promote early sexual maturity through maximum growth. Male calves should be separated from the heifers when they are four to five months old. When milk-feeding is discontinued the future herd sire should have free access to best-quality legume hay, if possible, and 4 or more pounds of concentrates daily. The amounts of these feeds must be sufficient to promote rapid growth until the animal becomes three or four years of age. The mature bull, four to five years old and older, should be fed only enough to maintain thrifty conditions. Unlimited amounts of feed, even of good roughage alone, will make him overly fat, and therefore a sluggish and uncertain breeder. The mature bull should receive a limited amount of good-quality roughage, 10 to 15 lb. daily, and 4 to 6 lb. of concentrates, which may be the same as those fed the milking herd. The amounts, naturally, will depend on body size and the extent of use of the bull.

The reason for the interest in early sexual maturity is that a long time is required to determine the true value of the bull. A bull well cared for can be used in natural service once a week, beginning at 1 year of age. Many bulls, however, are not well enough developed for service under 18 to 24 months of age. Even when bulls are used as early as 1 year, it is seldom possible to complete the progeny test before the bull is five to six years old.

32-9. ARTIFICIAL INSEMINATION

Only in dairy cattle has artificial insemination (AI) achieved widespread use in the United States. It is one of the most revolutionary developments in the improvement of dairy cattle. From a small start (initiated by Professor E. J. Perry of New Jersey in 1938) involving about 1,000 animals, the pro gram has expanded to every state; in 1964, 6,500,000 dairy cows were bred to dairy bulls and an additional 1,000,000 were bred with semen from beef bulls. It is reported that 464,959 beef cows were bred with beef semen during the same year. About 41.4% of all dairy cows of breeding age were bred by AI (USDA, 1964e).

Let us briefly review the discoveries that made possible the widespread use of AI in dairy cattle. The first impetus came from a series of important advances made by Russian investigators. They developed an artificial vagina with a water jacket for maintaining the optimal temperature and an attached vial for collection of the semen. They found that if the penis of the bull was directed into the artificial vagina when the bull mounted a restrained cow in estrus, or a dummy, the bull would ejaculate and the uncontaminated semen could be collected in the vial. Next, they found that the semen could be diluted with buffered saline solution and stored for limited periods (24 to 28 hours) before use.

The next significant discovery was made by Phillips and Lardy (1940), who showed that the addition of egg yolk to the semen-diluting medium improved the storage capacity of the semen. Further research by Salisbury et al. (1941) resulted in extensive use of egg yolk and citrate extender.

Finally, Polge and Rowson (1952), in England, found that the addition of glycerol to the diluting medium made it possible to freeze the semen without appreciable loss in viability. This discovery came about when a laboratory technician inadvertently mistook a bottle containing glycerol for the diluting medium bottle; it was subsequently realized that the addition of glycerol enhanced the capacity of the semen to withstand freezing. Frozen semen has greatly improved the flexibility of artificial-insemination operations. Semen from many bulls can be frozen and stored in solidified carbon dioxide or liquid nitrogen for long periods (at least 10 years). Frozen semen of a dairy bull is thus usable long after the bull has died. However, at the present time calves sired by frozen semen from a registered beef bull that has died cannot be registered. Freezing also makes feasible the shipment of semen to distant points.

Advantages of Breeding Dairy Cows by AI. The greatest advantage of AI is that it extends the use of superior sires and makes possible the use of very outstanding sires by dairymen with only a few cows. A bull used in natural service seldom breeds more than 50 cows yearly, whereas each bull used in AI in the United States in 1963 bred an average of 2,999 cows (USDA, 1964e).

In most herds, artificial breeding costs less than natural breeding, if one considers all of the indirect, as well as the direct, costs associated with keeping a bull. Those who run cheap young bulls with cows on pasture can probably beat the cost of AI, but they are ignoring the herd-improvement value of proven sires.

AI is a means of assuring that there will not be long gaps in the breeding program, due to using bulls of low fertility. With artificial breeding, the semen is examined regularly, whereas bulls in natural service may become sterile or low in fertility without being detected for several months.

There is less chance of spreading diseases within a herd when AI is used. The American Veterinary Medical Association has prepared a code of minimum health standards for bulls producing semen for AI that has been adopted by the National Association of Animal Breeders (512 Cherry St., Columbia, Mo.). To meet these standards, the bulls must be tested to make certain that they are negative for tuberculosis, brucellosis, trichomoniasis, and leptospirosis. In addition, the semen is treated with antibiotics for the control of such diseases as vibriosis.

Blood typing of all bulls used in AI helps in assuring the accuracy of the pedigree of an animal. In other words, if a question arises concerning the parentage of a given animal, blood typing is helpful in identifying the sire.

AI facilitates the proving of young bulls. The semen can be used on many cows in several herds. Thus, the time for proving is reduced and the accuracy of the proving is increased because daughters are tested under varying management conditions in different herds.

Artificial breeding eliminates the physical danger involved in handling bulls used in natural service.

Valuable bulls that are not suitable for natural service because of injury or age, or because they are too heavy to use on young heifers, can frequently be used in AI. The semen may be collected either through use of the artificial vagina or, if the bulls will not mount, the semen may be collected by means of an electro-ejaculator.

In herds where natural breeding is preferred, AI can be used during the heavy breeding season to prevent the overuse of herd bulls or to reduce the number of bulls required for the herd.

Disadvantages of Breeding Dairy Cows by AI. Semen from some fertile bulls will not withstand freezing. Inasmuch as the cows of many herds are bred exclusively with frozen semen, certain outstanding sires are excluded from use.

Extensive use of bulls carrying genes for recessive hereditary defects may result in widespread dissemination of these genes. A large number of hereditary defects in dairy cattle have been reported (Mead et al., 1946). The National Association of Animal Breeders has urged all AI studs to request users of their semen to report the occurrence of defects in offspring—such as mule foot, dwarfism, epitheliogenesis imperfecta, cataract, flexed pasterns, sterility, and other hereditary defects.

When fluid semen is used by a stud, it is unusual for the inseminator to have semen from more than two bulls of one breed, and thus the dairyman is limited in his choice of sires. At the present time, however, because most studs are using frozen semen, semen from several bulls of each breed is available. Some purebred breeders prefer to use semen from their own bulls. Many bull studs are custom-freezing semen from such bulls for use by the owner of the bull.

32-10. CONCLUDING REMARKS

The trend in dairying has been toward larger and higher-producing herds, owing largely to the narrowing spread between costs of production and return on investment. To satisfy sanitation requirements, expensive barns and equipment are needed; for this reason, the size of herds has been increased so that these facilities may be used for a greater part of the day, in order to justify the cost of barns and equipment. High labor costs have resulted in the installment of labor-saving equipment and have stimulated effort toward higher production. The cost of maintaining a low producer is nearly equal to that of a high producer. Artificial insemination has simplified the planning of a breeding program and has proved to be the cheapest means by which the dairyman can obtain semen from superior sires to increase production. Knowledge of the nutritive value of feeds and feeding according to production but in line with the cost of the feed and the selling price of the product is an important consideration. Feeds of similar nutritive value often vary in price. The ratio of concentrates to roughage will differ, depending on the costs of the nutrients in these feeds and on the price the dairyman receives for his milk. Proper milking practices can improve production and reduce mastitis, one of the diseases responsible for the heavy loss of cattle from the milking herd. Other diseases, some of which are responsible for low breeding efficiency, must be controlled to reduce culling and thus to lower the cost of production. Dairying has become a highly specialized enterprise requiring a strong background of knowledge in biological sciences as well as training in business management.

REFERENCES AND SELECTED READINGS

References marked with an asterisk are of general interest.

Becker, R. B., P. T. DixArnold, W. G. Kirk, George K. Davis, and R. W. Kidder, 1953. *Minerals for Dairy and Beef Cattle.* Florida Agr. Expt. Sta. Bull. 513.

Boda, J. M., and H. H. Cole, 1954. The influence of dietary calcium and phosphorus on the incidence of mik fever in dairy cattle. *J. Dairy Sci.,* 37:360.

* Cole, H. H., and P. T. Cupps, 1959. *Reproduction in Domestic Animals.* Vols. 1 and 2. Academic, New York.

* Eckles, C. H., and E. L. Anthony, 1960. *Dairy Cattle and Milk Production.* 5th Ed. Macmillan, New York, p. 145.

Gullickson, T. W., L. S. Palmer, W. L. Boyd, J. W. Nelson, F. C. Olson, C. E. Calvery, and P. D. Bayer, 1949. Vitamin E in the nutrition of cattle. I. Effect of feeding vitamin E poor rations on reproduction, health, milk production and growth. *J. Dairy Sci.,* 32:495–508.

Hart, G. H., S. W. Mead, and H. R. Guilbert, 1933. Vitamin A deficiency in cattle under natural conditions. *Proc. Soc. Exptl. Biol. Med.,* XXX:1230–1233.

Hibbs, J. W., and W. D. Pounden, 1955. Studies on milk fever in dairy cows: IV. Prevention by short-time, prepartum feeding of massive doses of vitamin D. *J. Dairy Sci.,* 38:65.

Mead, S. W., 1942. *Raising Dairy Calves in California.* Calif. Agr. Ext. Service Circular 107.

Mead, S. W., P. W. Gregory, and W. M. Regan, 1946. Deleterious recessive genes in dairy bulls selected at random. *Genetics,* 31:574–588.

Moeller, A. N., and N. L. VanDemark, 1951. The relationship of the interval between inseminations to bovine fertility. *J. Animal Sci.,* 10:988.

* Morrison, F. B., 1956. *Feeds and Feeding.* The Morrison Pub. Co., Ithaca, pp. 1087–1088.

* National Research Council, Natl. Acad. of Sci., 1958. Nutrient Requirements of Domestic Animals. No. 111, Nutrient Requirements of Dairy Cattle. Publ. 464, revised.

* Perry, E. J., 1960. *The Artificial Insemination of Farm Animals.* 3rd ed. Rutgers University Press, New Brunswick, New Jersey.

Phillips, P. H., and H. A. Lardy, 1940. A yolk-buffer pablum for the preservation of bull semen. *J. Dairy Sci.,* 23:390.

Polge, C., and L. E. A. Rowson, 1952. Fertilizing Capacity of Bull Spermatozoa After Freezing at —70°C. *Nature,* 169:626.

Ronning, M., 1960. Effect of varying alfalfa hay-concentrate ratios in a pelleted ration for dairy cows. *J. Dairy Sci.,* 43:811.

Ronning, M., and J. B. Dobie, 1962. Wafered versus baled alfalfa hay for milk production. *J. Dairy Sci.,* 45:969.

Salisbury, G. W., H. K. Fuller, and E. L. Willet, 1941. Preservation of bovine spermatozoa in yolk-citrate diluent and field results from its use. *J. Dairy Sci.,* 24:905.

* Salisbury, G. W., and N. L. VanDemark, 1961. *Physiology of Reproduction and Artificial Insemination of Cattle.* W. H. Freeman and Company, San Francisco.

Smith, T., and R. B. Little, 1922. The significance of colostrum to the newborn calf. *J. Expt. Med.,* 36:181–198.

Steenbock, H., 1919. White corn vs. yellow corn and a probable relation between the fat-soluble vitamin and yellow plant pigments. *Science,* 50:352–353 (New series).

Trimberger, C. W., 1948. *Breeding Efficiency in Dairy Cattle from Artificial Insemination at Various Intervals Before and After Ovulation.* Nebr. Agr. Expt. Sta. Research Bull. 153.

USDA, 1955. Agr. Research Service, Dairy Husbandry Research Branch. ARS-52-1.

————, 1961a. Agr. Marketing Service, Crop Reporting Board. *Milk production and dairy products.* Washington, D.C.

————, 1961b. *Economic Research Service,* April 1961.

————, 1961c. National Cooperative Dairy Herd Improvement Program. *Agr. Research Service,* 37:44–97.

————, 1964a. National Cooperative Dairy Herd Improvement Program. *Agr. Research Service,* 40:2.

————, 1964b. National Cooperative Dairy Herd Improvement Program. *Agr. Research Service,* 40:4.

————, 1964c. National Cooperative Dairy Herd Improvement Program. *Agr. Research Service,* 40:8.

————, 1965. *Statistical Reporting Service,* Washington, D.C., Feb., 1965.

Chapter 33

Swine Management

"Swine, women, and bees cannot be turned."
Old English Proverb.

33-1. INTRODUCTION

The concept of swine management that has been developed during the present century in the United States and Canada has carried the industry far from the "pigsty" and "pork barrel" era. There are still pigsties to be found today, but certainly the economic pressures of the modern world are converting the swine industry into one composed of larger units, founded on sound business principles, and operated on a scale that can justify more extensive utilization of scientific knowledge and automation. This improvement is destined to increase as human populations continue to expand and as we move gradually out of periods of grain surpluses and into increasing competition for food between man and his livestock. For a more extensive insight into potential competition for basic food resources, see Leitch and Godden (1941).

One of the outstanding characteristics of swine is their ability to act as a marketing medium for surplus grain or grain damaged by frost, drought, rust, rain, or storage. At times this has been a mixed blessing; when large quantities of feed grain are available, swine marketing increases with resultant price depressions. Hog production has been particularly vulnerable to cyclic fluctuations in selling prices; economists believe, however, that the trend to larger production units and to greater specialization will have a moderating effect on the marketing aspects. The potential effects of expanding human population on the demand for meat has interesting possibilities, too.

Traditionally, the student of animal husbandry has been taught that the three fundamental factors governing success or failure in the livestock business were feeding, breeding, and management. What is management? Is it, as implied, something separate from the other phases of production? Perhaps management should be defined as the responsibility for securing an optimal blend of all the various aspects of production—physiological, nutritional, and genetic—as well as marketing, in order to maximize net profits.

This chapter is designed to focus brief attention on some of the production problems that are not dealt with in detail elsewhere and that have tended, by tradition, to be regarded as management problems. Actually the keen student will find extensive discussion on many of the following points in the literature, and will find also that these areas are not immune to scientific advance, change, or debate.

33-2. SWINE PRODUCTION AS A BUSINESS

Why Raise Pigs? Pigs long ago earned the title of "mortgage lifters," ample testimony that they have generally been profitable to raise. In numbers and in terms of human consumption, swine rank a close second to cattle in both Canada and the United States.

Other factors favoring swine production are the high yields of edible meat from the market hog and the multiplicity of products manufactured by packing houses from hog carcasses.

As long as consumer demand, domestic or export, is high enough to sustain adequate prices in relation to those of other types of products, and particularly when abundant supplies of low-cost grains are available, there will be a thriving swine industry. The operator who stays in business and who is able to survive the recurring periods of low prices will be the man who like pigs and who maintains enough interest in them to keep himself up-to-date on the subject.

Size of Operation and Capitalization. With the day of diversified farming on the decline, economists are suggesting that units marketing less than 200 to 300 hogs per year are likely to be submarginal in nature and not in a position to realize satisfactory returns to labor and capital. Conversely, operations resulting in the sale of 700 to 1,000 hogs or more annually can be considered major enterprises (Figs. 33-1 and 33-2). Only when specialization is quite intensive is there likely to be sufficient economy of overhead to sustain production with narrow operating margins or enough net-income incentive to induce the maximum capitalization and general development of the enterprise.

Statistics on the costs of production of pigs vary with time and location but feed costs usually represent 50 to 75% of the total costs. So important are the feed costs that the difference between low and high feed efficiency (lb.

Fig. 33-1. Interior view of a 500-foot facility for fattening swine that is designed to accomodate 1,000 or more animals. The round self-feeders are filled by means of an automatic unloading truck. Note the tilt-up walls adjoining the alley; they are raised 4 inches above the level of the floor so that the pens can be cleaned by hosing into gutter along side of alley. [Photo by Don Holt.]

feed/lb. gain) often spells the difference between profit and loss. So far as overhead costs are concerned, the allowance for permanent facilities apparently should be in the range of 7 to 19% of the gross returns. Currently this would be equivalent to a new-cost overhead investment of from $250 to $300 per litter. Keeping below this limit constitutes a definite challenge to management.

Production Systems

THE PIG BREEDER. The producer of registered gilts and boars has been the mainstay of the swine industry. Upon him has rested the major responsibility for maintaining and improving the quality of commercial swine, for he has supplied many of the boars and sows used in commercial herds. Traditionally the livestock exposition has been the "testing ground" and show window for his produce but during the last quarter century increasing attention has been paid to progeny and performance testing, including carcass quality appraisal, which provide more useful and direct evaluations.

THE COMMERCIAL PRODUCER. Commercial operation involves the maintenance of sufficient brood sows and boars to produce the feeder pigs needed to fill the available facilities. Ordinarily the operator retains his own female replacements but purchases his sires from purebred breeders.

"PIG HATCHERIES," FEEDER PIG PRODUCERS. Much of the risk of mortality in swine production is associated with the period between conception and the time of weaning. Therefore special attention is required at this stage. Specializing in the production of feeder pigs offers distinct advantages to the industry and a special challenge to enterprising swine breeders—the production of disease-free pigs of outstanding genetic value is an excellent goal. Unfortunately, many such ventures have failed, mostly for reasons associated with inadequate facilities, poor disease control, or unsound financing.

FEEDER OPERATIONS. This type of swine enterprise depends on the purchase of feeder pigs. The operations are usually conducted on a relatively large scale; they often utilize cooked garbage, and consequently may be located near cities.

Feeder units of this type engender a greater degree of risk of diseases and parasites because the previous history of the pigs is often unknown and pigs may have become available as feeders because of their previous unthriftiness. Consequently, special precautions for disease and parasite con-

Fig. 33-2. A portion of facility shown in Fig. 33-1 to show how manure is hosed into gutter along side of alley. [Photo by Don Holt.]

trol must be adopted routinely and the operating margin of profit must be wide on account of the extra risk.

33-3. MAINTAINING THE BREEDING HERD

Assessment of Quality in Swine. Quality must be determined in relation to the important economic traits. These differ somewhat depending on sex and whether the animals are being kept for breeding or for market. Swine shows have played an important role in furthering the production of superior breeding stock but to expand their influence will necessitate the inclusion of carcass and performance data into the judging standards.

Performance Testing and Progeny Testing Programs. Denmark leads the world in hog-testing programs. It has long had a clear objective—to produce lean pork on the most economic basis—and has adopted central station testing and formulated policies whereby the superior strains of breeding stock would be utilized most effectively. Canada has had a testing policy in operation since 1928, patterned to a considerable degree after the Danish system (Fig. 33-3). Both countries were able to organize their schemes

Fig. 33-3. Interior view of a Record of Performance swine-testing station in Canada. Each pen houses 4 litter mates between 50 and 200 lb. Pigs are self-fed balanced rations and slaughtered at 200 lb. for carcass evaluation. [From National Bacon Hog Policy Production Service, Ottawa, Ont.]

on national bases and were perhaps fortunate in being able to concentrate efforts on single breeds.

Somewhat in contrast, the performance testing of swine in the United States did not receive real impetus until after World War II, when great surpluses of animal fat had accumulated. The National Association for Swine Records has now adopted a Production Registry (P.R.) policy and a program for Certified Meat Hogs. It is posssible for a breeder, subject to the regulations of the particular breed involved, to obtain official recognition for prolificness of sows (including their nursing ability, as reflected by 56-day weights of the progeny), for breeding performance of boars that sire 15 or more P.R.–qualifying litters, and for litters (matings) that yield desirable meat-type carcasses within a 180-day age limit.

The Canadian Record of Performance program involves the preweaning inspection of litters on the farm for litter size, disease, abnormalities, breed disqualifications, and so on. Then, if accepted, two gilts and two barrows are station tested. The pigs are fed standard rations to 200 lb. liveweight, are slaughtered, and the carcasses are evaluated. The final report summarizes the performance with respect to litter size, maturity, carcass score, and feed efficiency; it includes similar statistics for the average performance of all pigs tested in that particular station. The breeder can readily determine, if he tests a fair sample of his herd, how his stock rates in any respect with other breeders' pigs, and can, from the published reports, decide what breeders have the quality features that he needs in his next sire. (More details on performance testing are given in Chapter 16.)

Choice of Breed. The motivating force behind the selection of a breed often contains emotional elements and while this bespeaks something of the potential success of a livestock man, one's emotions should not preclude sound judgment based on factual evidence of economic performance. Some breeds are decidedly superior to others in terms of prolificness, vigor, carcass quality, and feed efficiency, and these factors should be considered. Furthermore the producer should not disregard the periodic advances made through animal breeding research and performance-testing programs.

Selection of Female Breeding Stock. Gilts should be from a sow with a superior record in terms of as many important economic traits as possible: soundness and health, prolificness, mothering ability, type, and quality (Fig. 33-4). When performance-testing data are available, it may be possible to select with knowledge of several generations' performance. The particular mating is important. Gilts from tested matings and from strains testing well over several generations are the best selections.

Gilts will breed, if well grown and developed, by the age of 5 to 6 months, but research has proven that larger, stronger litters will be obtained if breeding is delayed until gilts weigh 250 to 280 lb.

Fig. 33-4. An ideal blend of type and performance. Glenafton Duchess 94M-511603-, herself a show winner and from a litter of champions, she scored on R.O.P. 13-156-87-4.25; this Record on Performance is interpreted as 13 pigs weaned, 156 days maturity to 200 lb., 87 carcass score, and 4.25 lb. feed/lb. carcass gain (3.33 lb. feed/lb. live gain). She has 16 functional nipples and a strong family history of both R.O.P. and show winnings. [Courtesy E. F. Richardson & Son, Semans, Sask.]

Selection of Boars. This aspect of management cannot be overemphasized. It is unwise to buy a sire that does not have the genetic material necessary to maintain or improve quality in his offspring. The operator should carefully examine the variations in net profits that might occur on his farm as a result of failure to attain reasonable goals in feed efficiency, market quality, and freedom from abnormalities.

Most breeders buy young unproven boars, largely because of the physical problem of breeding young gilts to mature boars. The use of breeding crates, and perhaps ultimately of artificial insemination, may greatly extend the usefulness of mature boars of proven quality. Far too many boars are recognized as outstanding breeders only after their disposal. The advantages of crossbreeding for the commercial hog raiser are discussed in Chapter 16.

Animal Identification. Any program of selection requires that pigs be identified at least by sire and dam. The easiest method to follow is that of ear notching with special pliers. There are several systems that can be used, depending on the number of swine that have to be identified (Fig. 33-5). Purebred swine must be identified by inked ear tattoos, according to regulations set forth by the appropriate registration authorities.

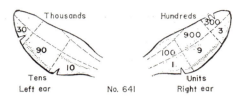

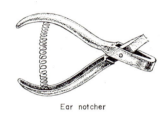

Fig. 33-5. Ear notching is a convenient method of identification. Several numbering systems are available. This one allows numbering about 10,000 pigs without repeating numbers. A maximum of two notches is permitted in the 1 and 3 areas, but only one notch in the central (9) areas.

Ear notcher

33-4. HOUSING AND EQUIPMENT

Swine Performance and Environment. Swine production on this continent embraces a very wide range of climatic conditions and involves extremely important considerations of housing, insulation, shade, cooling, heating, and so on, that vary from region to region. For the first few days of life, pigs have an optimum environmental temperature of about 80°F; growing pigs (of 50 lb. upward) appear to have an optimum temperature of 60 to 75°F. Mature pigs can be maintained satisfactorily during winter temperatures as low as −30° to −40°F if given well-bedded sleeping shelters. Growing and mature pigs begin to exhibit signs of heat stress at about 80°F. Any extended departure from the zone of thermal neutrality involves some reduction in performance. Bearing in mind that swine production is a business, the operator must assess the hazards of unfavorable environment and weigh the risks against the costs of alternative remedial action.

Characteristics of Good Housing. It is difficult to arrive at decisions as to suitable layouts in this field of changing ideas. Great advances have been made in automation, building materials, heating, and ventilation facilities. Regardless of the design and facilities ultimately chosen, however, the costs should be kept in line with regional recommendations of farm economists. As a rule the heaviest investment per unit of floor space will be in the farrowing quarters, where special attention is needed to control temperature, humidity, ventilation, diseases, parasites, and accidental mortality. The amount of space required for swine depends upon the age and reproductive state (Table 33-1).

TABLE | *Floor space requirements for housing swine.*
33-1. |

Class of Swine	Floor Space in Square Feet
Dry sow	20–40
Sow and litter (in farrowing stall)	35–40
Sow and litter (in pen)	50–80
Weaning to 125 lb.	4–10
125 lb. to market weight	8–20

Farrowing Facilities. It is desirable, if not essential, to conduct farrowing operations in quarters separated from other swine (Fig. 33-6). The period from birth to weaning is especially critical. Newborn pigs are easily crushed or trampled; their metabolism is very high and consequently they have

Fig. 33-6. A modern design farrowing house featuring all-metal farrowing stalls, infrared heat lamps, and absence of windows. Windowless barns allow better regulation of temperature and ventilation and facilitate fly control. Each pen is about 5 feet by 7 or 8 feet, with a crate 26 inches wide. [Photo courtesy Beatty Bros., Fergus, Ont., and Shur-Gain Division, Canada Packers Ltd., Toronto, Ont.]

relatively large milk requirements; their temperature regulatory system is slow to adjust to the extra-uterine environment and they are born with virtually no disease antibodies in their blood. Consequently such devices as radiant-heated floors, heat lamps, guard rails in farrowing pens, and farrowing crates or stalls have been widely adopted in efforts to reduce the hazards.

33-5. CARE OF THE PREGNANT SOW

Shelter and Exercise. The preferred accommodation for the bred sow is a pasture lot provided with an inexpensive shelter to protect her from inclement weather. Locating the feed and water supply some distance from the shelter will induce more exercise.

Feeding. During the summer a dry (nonlactating) sow can derive most of her nourishment from green pasture. It is advisable, however, to ensure adequate intakes of essential nutrients by providing supplementary feed, the amount depending on the quality of the pasture and the condition of the sow.

Mature sows given free access to a grain ration may consume up to more than 8 lb. of feed daily and consequently gain excessive weight. In recent studies, sows that were fed balanced diets and limited to intakes of about 5 lb. daily during pregnancy farrowed good litters and nursed them well if they were full fed during lactation.

Nutritional hazards confronting the pregnant sow vary with region and season, but iodine, vitamin A, and calcium are particularly important nutrients. Soils that produce the feed for swine are iodine-deficient in many areas and the relatively high metabolic stress of pregnancy in the sow creates a major demand for iodine. The hormone thyroxine (see Chapter 20) contains iodine and controls metabolic rate. That metabolism is active is apparent when it is seen that a 400 to 500 lb. sow can produce 12 pigs averaging 3 lb. or more at birth in 112 days of gestation.

Iodine is usually supplied by dietary supplements of stabilized potassium or sodium iodide, but some other iodine compounds offering greater stability in rations are being studied. The use of iodized salt (containing 0.007% iodine) as 0.5% of the ration is effective.

Vitamin A (as carotene) is lacking in all feed grains except yellow corn; since this vitamin is so critical in pregnancy, a minimum daily intake of 10,000 I.U., either as carotene from natural feedstuffs or as supplementary vitamin A, should be assured.

None of the cereal grains commonly fed to swine contain more than 0.1% calcium, but the bred sow needs about 0.6%. Some calcium may enter the ration by way of protein supplements like meat or fish meal, but from 0.5 to 1.0% of ground limestone (calcium carbonate), bone meal, or dicalcium phosphate is often needed to provide the necessary minerals.

Prefarrowing Care. Several days prior to the expected farrowing date (112 days gestation period), the pregnant female should be brought to the farrowing quarters, washed off with soap and water, and placed in a clean pen. Operators using farrowing crates or stalls may prefer to delay close confinement until immediately before farrowing in order not to restrict exercise.

The onset of parturition, together with decreased activity and feed changes, cause constipation of the sow. This condition should be avoided by encouraging water consumption by wet feeding, or by including 10 to 25% wheat bran in the ration. There is a definite possibility that toxins absorbed or retained in the body as a result of constipation may adversely affect milk yield and quality and consequently prove harmful to the new-born pigs.

33-6. CARE OF THE LACTATING SOW

Accommodations. Accommodations may vary from somewhat elaborate indoor housing in pens, stalls, or crates to a small summer shelter on fresh pasture. With greater specialization in hog production, the sow and litter will probably be housed indoors, where it is easiest to provide the necessary special care.

Sow and litter pens containing sleeping platforms of concrete surfaced with asphalt have given good service. An adjoining brooder—enabling the piglets to travel directly from platform to brooder—is also advantageous. The brooder arrangement provides opportunity for the young pigs to be fed and handled separately and gives them safe sleeping quarters.

Feeding. The nutrient requirements of the sow during lactation do not differ appreciably in terms of ration composition from those during gestation, but a lactating sow or gilt will consume about twice as much food as a pregnant one. The feed intake is governed largely by the milk yield and there is evidence of great variability in milk production between individuals and breeds. Studies reported in 1935 (Hughes and Hart; Bonsma and Oosthuizen) and more recently in 1960 (Allen and Lasley) indicate 56-day milk yields of about 350 to 400 lb. Lodge (1959) reported an average production in the sows he studied of 993 lb. Contained in these reports are indications of definite breed differences in yield and in time at which the peak of lactation is reached. Obviously there is a challenge to the breeder to select for good milking sows as well as reason to watch feeding methods.

Special care should be taken to encourage high feed consumption by sows. Heavy body-weight losses are not uncommon during lactation and may to some extent be unavoidable, but it is most economical to use food directly for milk production rather than to form body fat as an intermediate. Again, the introduction of substantial quantities of stored fat into the pool of absorbed nutrients may tend to unbalance the nutrient ratios.

A condition known as posterior paralysis sometimes develops during or following lactation. Decalcification of the skeleton occurs, probably as a result of inadequate dietary intakes, which do not meet the animal's demands for calcium and phosphorus during heavy milk production. Weakened lumbar vertebrae collapse under strain, thus damaging the spinal cord and paralyzing nerves controlling the hindquarters.

Posterior paralysis is usually an indication of improper feeding. Lactating sows need rations like those given to growing pigs in the 75 to 100 lb. weight range, except that more vitamin A and D should be used. Rations should contain 13 to 15% crude protein (the higher percentage for gilts), 1,500 kcal. digestible energy (75% TDN), 0.6% calcium, 0.4% phosphorus, 1,500 I.U. vitamin A, and 100 I.U. vitamin D per lb. ration (NRC, 1964).

Rebreeding. Sows may be observed in heat sometime during the first few days after farrowing. If bred at this time, very few conceive. The sow should be bred at the first or second heat period after weaning. Early-weaning practices will enable some acceleration of breeding schedules. Treatment of lactating sows 40 or more days after farrowing with equine gonadotropin will induce estrus and ovulation with consequent conception following mating. This process, however, has not been used commercially to any extent. There is some evidence that "flushing" (breeding during a period of rapid gain) pays dividends; hence delaying breeding until the postweaning recuperative period seems to be advantageous.

Weaning. The preferred practice is to remove the sow from the pen and thus leave the piglets in familiar surroundings. In order to reduce the risks of udder inflammation the sow should be kept on reduced feed for several days after weaning, and some operators who practice hand feeding reduce the allowance a day or two before weaning as well.

33-7. CARE OF THE SIRE

Accommodation. Mature boars can be handled most easily and may retain better temperaments if they are housed with bred sows, but when boars must be isolated they require somewhat stronger and higher partitions or fences than sows. The boar should have plenty of exercise area and should not be penned adjacent to other boars unless separated by a solid fence.

Feeding. The nutritional requirements of the boar vary with his age. Young boars are best maintained on a grower-type ration to ensure normal growth and development. High fit is undesirable. Older boars normally do well on a dry sow-maintenance ration that is either low in energy (maximum 70% TDN—total digestible nutrients) for self-feeding or restricted in amount to about 6 to 8 lb. per day.

Management of the Sire. Young boars in their first season of service should be hand mated to a maximum of about 5 females a week and 20 or 30 in the season. The intensity of service of course varies with vigor, temperament, and other factors. Boars should not be used for breeding until they are about 8 months of age.

Older boars may breed twice daily. Some have been known to handle 100 sows or more in a season, but they were exceptional boars, under excellent management. Boars seldom remain useful after an age of about 5 or 6 years.

De-tusking and Castration. Mature boars should have their tusks removed at least once a year, as a safety measure for the operator and as a precaution against fighting with other swine. This is done by securely snubbing the upper jaw with a rope or cable to a post and clipping the tusks near the gum line. Similar means of restraint can be used for castration of large boars—an operation made necessary by the fact that boars must be castrated a month or two prior to slaughter for human consumption.

33-8. CARE OF YOUNG PIGS

The First Week. The first week of a pig's life is especially critical. Because a pig's temperature-regulating mechanisms are not fully functional during this time, its life is endangered by chills. It is vitally dependent upon getting colostrum in order to obtain disease antibodies, which, in the pig, are not transferred prenatally across the placenta. The small pig is also very subject to mechanical injury. Consequently an unfavorable environment—in terms of physical facilities, temperature, ventilation, or sanitation—may provide sufficient stress to precipitate pneumonia, diarrhea, rhinitis, or some other disorder.

Early versus Late or Normal Weaning. The swine industry has been built upon a program involving weaning at 6 to 8 weeks of age. As a result of phenomenal increase in nutritional knowledge during the past quarter century, however, it is now possible to wean pigs much earlier. In many operations weaning is practiced at 3 weeks, sometimes earlier, with subsequent feeding on rather complex early-weaning formulas. Early weaning requires somewhat more elaborate facilities to care for the pigs and involves judicious use of more costly feeds, but there are compensations: the litters are more uniform because weak pigs get a chance to obtain adequate nourishment before receiving too great a setback, and the sow is ready for rebreeding at least a month earlier.

Opinions vary on the overall advantages of early weaning. Each producer should examine the relative merits of the two systems in light of his own facilities and marketing programs before deciding on his course.

A special problem in the piglet is control of anemia. This calls for the administration of a suitable iron preparation, commencing not later than the second or third day after birth. Either reduced iron, ferrous sulphate, one of a variety of proprietary oral mixtures, or, more recently, injectable iron may be selected. The injections are currently popular because one or two injections suffice, whereas oral preparations must be given at least weekly until substantial amounts of solid feed are being consumed.

Castration and Identification. For the easiest operation, male pigs intended for slaughter should be castrated during the first two months of life. Pigs could be ear notched at this time, or earlier, for purposes of identification. Castrating should not be done coincident with weaning or vaccination because excessive stress may precipitate other problems.

Feeding Formulas and Programs for Pigs up to 3 Months of Age. During this period the pig normally makes the transition from the mother's milk to dry feed. Consequently two factors are of prime importance: (1) the new diet must be palatable, and (2) the diet must be nutritionally adequate. In addition, it is commonly recommended that certain additives like antibiotics and perhaps antihelminthics be included to provide extra protection against diseases and parasites. Because of the complexity of the ration, the producer therefore frequently finds it to his advantage to rely upon commercially prepared rations, often pelleted or crumbled and sweetened. Such rations are rather costly compared to those used for older swine and based on home-grown feeds, but relatively small amounts are needed. It is very important that pigs do not receive a setback at this stage.

33-9. CARE OF GROWING PIGS (50 to 100 lb.)

Accommodation. There is an increasing tendency to raise pigs intended for market entirely indoors (Figs. 33-7 and 33-8). Pigs will do well on pasture and can derive much of their supplementary feed requirements from fresh green vegetation, but pastures tend to give rise to problems—parasites, sunburn (of white breeds), fencing, pasture seeding and maintenance, and others—which some operators prefer to avoid.

When pigs are raised in confinement, the area that is allowed per animal becomes more critical. Economically this is related to overhead costs, but

Fig. 33-7. A modern feeding barn equipped with automatic feeding equipment for floor feeding. Pens are 16 feet long and have slotted floors near the outside walls through which manure discharges into a large gutter. No bedding is required. Floors are flushed with water and the gutters empty periodically into a covered reservoir outside, from which the liquid manure is pumped into a tank truck and spread as fertilizer onto nearby fields. [Photo courtesy Beaty Bros., Fergus, Ont., and Shur-Gain Division, Canada Packers Ltd., Toronto, Ont.]

animal behavior is affected by pen size and shape and by the number of pigs housed per group. Heitman et al. (1961) studied pigs in groups of 3, 6, and 12, with 5, 10, or 20 sq. ft. per pig and found that those pigs having the largest pen areas gained the fastest and were most efficient in converting feed into gains. The larger groups tended to be more efficient in food conversion than the smaller ones. However larger groups and larger areas per pig have not always worked well where slotted floors are used for part of the pen. It appears that some crowding encourages pigs to keep the feeding and sleeping areas cleaner.

Cannibalism occasionally develops in market pigs. Although the causes are unknown, it appears that overcrowding and faulty nutrition may be contributing factors. Once tail chewing commences, steps must be taken

Fig. 33-8. The feed hopper at left is suspended from a track, it is electronically controlled and is on a time clock. It automatically distributes feed and releases water (streaming from pipe) four times daily. The barn has a dunging alley, which is cleaned with a mechanical scraper. [Photo by Don Holt.]

immediately to isolate the offending pigs and to examine management procedures for possible causes.

Feeding Formulas and Objectives. The growing stage of the pig's life involves major physiological emphasis on growth of muscle and skeleton and low priority on fat synthesis. High-quality pork products contain relatively little fat finish and a large proportion of lean meat. One should therefore capitalize on the pig's natural growth potential by providing rations fortified with an adequate amount of good quality protein, minerals, and vitamins. Rapid growth rates are to be desired at this stage. A compensation for adding commercial supplements to farm grains is the high efficiency of feed conversion because of the low fat content of the gains. Good pigs, properly managed and fed, require less than 3 lb. feed per pound of gain at this stage, whereas 4 or 5 lb. of feed will be used per pound of gain during the period just prior to marketing. Studies of growth rates of pigs that were well-fed and well-managed revealed weights of 100 lb. at 100 days of age and 200 lb. at 160 days, with some breed variation, especially in favor of crossbreds.

The feeder ordinarily uses local farm grains as the basis of his feeding program from 50-lb. size on up to market size, but maximum feed efficiency requires adherence to regional recommendations for supplementation. None of the grain species is nutritionally adequate for pigs. Perhaps something less than maximum feed conversion will be desirable during times when the cost ratios between grain and supplements are widely divergent. Interesting publications are available on this challenging border between nutrition and economics.

33-10. CARE OF FINISHING PIGS (100 to 200 lb.)

Objectives. The primary objective in the production of a market pig is to satisfy the consumer demand as reflected through the meat-packing industry. It is difficult, however, to have a clear objective of market demand unless payment is made on a carcass basis and unless reasonable price incentives are paid for quality. In Canada, rail grading of market-weight hogs (140- to 195-lb carcasses) has been in effect since 1935 and has been compulsory since 1940. The ideal carcass weighs between 135 and 170 lb., has a minimum length from forerib to aitchbone of 29 to 30 inches, a maximum shoulder fat of $1\frac{3}{4}$ to 2 inches, and maximum loin fat of $1\frac{1}{4}$ to $1\frac{1}{2}$ inches, each factor varying with carcass weight.

With the demand for more lean and less fat, plus the desire to keep rate of gain high and feed costs down, there are clear challenges to the nutritionist, the geneticist, the practical breeder, and also to the meat industry, which may have to supply material encouragement to the producer in the form of appropriate price differentials. At least some of the answers currently available are (1) to market pigs at weights of 200 lb. or less, in order to produce less fat and to avoid the poor feed efficiency of fattening as such; (2) to use breeds of pigs of lean, meat, or bacon type; (3) to feed low-energy (maximum 70% TDN) rations during the 120 to 200 lb. stage in order to reduce rate of gain and degree of fattening; (4) feed restriction. The third and fourth methods limit the intakes of digestible nutrients (especially energy) thereby leaving smaller margins of nourishment, above the requirements for maintenance and growth, that can be stored as fat. Dilution of the ration with fibrous material such as ground roughage reduces the digestible energy content of the diet. The *weight* of feed eaten per day may or may not be reduced, depending upon the kind of diluent used. Since the daily gain is reduced by dilution of the ration, the total feed required during the finishing period may be increased. Although it has been proved that ration dilution will reduce fat deposition, it is obvious that one must consider carefully the cost of the low-energy ration, the added cost of housing and caring for pigs for an extra week or two, and the prospects of receiving sufficient extra returns to compensate fully for the increased costs involved.

Limited feeding or feed restriction simply means allowing pigs somewhat less than they would voluntarily consume; possibly 80% of *ad libitum* intake. This can be done manually but various automatic devices are available.

33-11. PRODUCTION OF BREEDING STOCK, AND HERD REPLACEMENTS

Since the primary objective in selecting herd replacements is to provide seed stock for commercial hog production, the feeding and management practices should be basically the same. The results of the special attention given to "show pigs" have too often led to serious disappointment when such animals were selected for breeding stock. The factors of prime importance are reliable evidence of superior performance in all important economic traits plus evidence of freedom from physical unsoundness, diseases, and parasites. As mentioned previously, good pastures are desirable for breeding stock, at least by the time they reach 100 to 200 lb. Exercise is beneficial and excessive fattening is highly undesirable.

Commercial production is likely to be increasingly based on crossbreeding in order to capitalize on the advantage of heterosis (Chapter 16). Successful crossbreeding, however, is very dependent on the availability of superior breeds, or strains within breeds, that have been proven out in performance or progeny tests. A measure of the potential benefits of crossbreeding is evident from a recent British survey of about 35,000 litters (Smith and King, 1964) in which it was shown that in crossbred litters, in comparison to purebred litters, there were 2% more pigs at birth, 5% more pigs at weaning, and 10% greater litter weight at weaning. In litters from crossbred *sows,* there were 5% more pigs at birth, 8% more pigs at weaning, and 11% greater litter weight at weaning. Such results justify the current interest in three-way crossing.

Crossbreeding has some disadvantages. The acquisition and housing of sires of two or more breeds and the maintenance of sows of the desired breed or breed cross complicate management. Control of diseases and parasites is made more difficult by introductions of new stock.

An interesting recent development is the production of "pathogen-free" breeding stock. This technique, developed at the Hormel Institute and first applied in practice in Nebraska, involves farrowing pigs aseptically by hysterectomy. The pigs are then reared artificially in an isolated environment and constitute the nucleus of a new pathogen-free breeding herd. In this way the pigs are protected from exposure to the more troublesome diseases, such as virus pneumonia and atrophic rhinitis, for which no good preventative or curative measures exist.

It is important to realize that pathogen-free pigs have not been protected from *all* disease hazards. If they were, they and their progeny would have little chance of survival upon exposure to commercial herds because so much of a pig's resistance to infection depends upon acquired immunity. The Nebraska results to date, however, are encouraging and may well play an increasing role in disease control in swine, at least until more direct and positive countermeasures against certain diseases are available.

33-12. DISEASES AND PARASITES OF SWINE

As testimony to the extent of losses encountered in the United States, the 1956 USDA Yearbook *Animal Diseases* indicates that, of the 40 million pigs lost in 1954, some 50% of these succumbed to diseases and parasites (see Chapter 43).

Swine are subject to a variety of deficiency and nutritional diseases due mainly to faulty feeding (see Chapter 30). These include anemia (iron deficiency); deficiencies of protein, and of vitamins A, B-complex, and D; agalactia or lactation failure; iodine deficiency, hypoglycemia or baby pig disease, and parakeratosis.

Lice and mange are external parasites of swine that can cause considerable irritation of the skin. Both can be controlled by appropriate applications of insecticides, such as lindane, DDT, or methoxychlor. Great care must be taken when using such chemicals to ensure that safety precautions and legal restrictions are observed.

Several internal parasites affect swine. Proper management calls for recognition of the types that occur in each geographic area, since internal parasites impair performance of pigs and may so damage tissues, resulting in their condemnation for human food, and since at least two internal parasites of swine can be transmitted directly to man. These are trichina (produces trichinosis in man: the larvae encyst in muscle tissue), and tapeworm. Probably the most common internal parasite is the large intestinal roundworm, ascarid, which in adulthood may measure 16 inches long by about $\frac{3}{16}$ inch in diameter. This parasite and the nodular worm, which inhabits the colon, can be controlled by any of several antihelminthics, such as piperazine, phenothiazine, or cadmium oxide, properly administered in the feed. Whipworms in the cecum and lungworms occasionally infest swine, but are usually of little economic importance.

Viruses are responsible for a number of serious diseases of swine. It sometimes becomes necessary to impose rigid herd or area quarantines, or to slaughter and bury entire herds. Diseases attributed to viruses are hog cholera (swine fever), foot-and-mouth disease, vesicular exanthema, vesicular stomatitis, rabies, virus pneumonia, swine pox, and swine influenza.

Bacterial infections are involved in pasteurellosis (swine plague or hemorrhagic septicemia), tuberculosis, erysipelas, brucellosis, leptospirosis, and mastitis. Additional diseases, such as edema disease, atrophic rhinitis, bull nose, enteritis, dysentery, transmissible gastro-enteritis, salmonella, serositis, and a few others, are of variable or unknown cause.

To this long list of diseases can be added sunburn, heatstroke, frost-bite, cannibalism, and chemical poisoning (salt, fluorine, arsenic, lead, mercury, insecticides, and so forth), which probably reflect bad management or misfortune, and finally gastric ulcers, the cause of which remains obscure.

The swine breeder would be very much amiss not to be familiar with the symptoms, causes, and control measures for problems likely to be encountered in his area. Full information is available in publications on swine diseases and from qualified veterinarians.

REFERENCES AND SELECTED READINGS

Reference marked with an asterisk is of general interest.

Allen, A. D., and J. F. Lasley, 1960. Milk production of sows. *J. Animal Sci.*, 19:1, 150–155.

Bonsma, F. N., and P. M. Oosthuizen, 1935. Milk production in Large Black sows. *So. African J. Sci.*, 32:360–378.

Heitman, H. Jr., L. Hahn, C. F. Kelly, and T. E. Bond, 1961. Space allotment and performance of growing-finishing swine raised in confinement. *J. Animal Sci.*, 20:3, 543–546.

Hughes, E. H., and H. G. Hart, 1935. Production and composition of sow's milk. *J. Nutrition*, 9:3, 311–321.

*Leitch, I., and W. Godden, 1941. *The Efficiency of Farm Animals in the Conversion of Feedingstuffs to Food for Man.* Imp. Bur. Animal Nutrition. Tech. Comm. Rpt. 14.

Lodge, G. A., 1959. The energy requirements of lactating sows and the influence of level of food intake upon milk production and reproductive performance. *J. Agr. Sci.*, 53:2, 177–191.

National Research Council (NRC). Natl. Acad. of Sci., 1964. *Nutrient Requirements of Domestic Animals. No. 11. Nutrient Requirements for Swine.*

Smith, C., and J. W. B. King. 1964. Crossbreeding and litter production in British pigs. *Animal Production,* 6:3, 265–271.

Chapter 34

Sheep Management

Sir, I am a true laborer: I earn that I eat, get that I wear, owe no man hate, envy no man's happiness, glad of other men's good, content with my harm, and the greatest of my pride is to see my ewes graze and my lambs suck.

SHAKESPEARE, *As You Like It*

34-1. INTRODUCTION

Sheep are unique among domestic animals in their adaptations to adverse conditions. They can convert diverse types of forage, from marginal agricultural land, mountains, hills, moorlands, plains, and deserts, into valuable products for mankind: wool, pelts, lamb, and mutton.

Latitude, altitude, and climatic conditions influence sheep farming. Methods of sheep husbandry may be different on farms or ranches within even the same district. Sheep breeders of the past partly solved their problems by developing breeds to suit the divergent conditions of the various areas in which they lived. Each of the many different breeds found today was developed to suit a certain type of environment. Merinos and Rambouillets are favored for their hardiness in hot desert areas with sparse vegetation and for their gregarious traits. Mutton breeds, such as Suffolk and Hampshire, are popular in cooler climates with abundant vegetation. The increasing importance of lamb as a desirable dinner meat has resulted in the development of Columbias and Targhees, which cross well with Suffolk and Hampshire rams to produce an excellent carcass and a heavy fleece.

Today the importance of wise management, selection within a breed, and crossbreeding, in addition to correct feeding, is widely recognized. One thing is certain, however: breed alone cannot substitute for feed and management. Economical maintenance of breeding animals, a large lamb crop, continuous and rapid growth of lambs, heavy weaning weights, and heavy clean-fleece weights are all based largely on adequate nutrition and manage-ment.

34-2. METHODS OF SHEEP RAISING

Variability of climate, topography, soil, and vegetation, and availability of water has resulted in a diversified agriculture in the United States. Four general systems of commercial sheep production fit into the agricultural pattern: (1) range, (2) farm, (3) purebred breeding, and (4) fat lamb pro-duction.

Range Sheep Production. Range sheep are raised in all of the 17 western-most states. This area constitutes about 40% of the continental United States; approximately 70% of the sheep are located here. The area can be divided into nine range regions (Fig. 34-1): (1) tall grass, (2) short grass, (3) desert grass, (4) bunch or palouse grass, (5) northern or intermountain shrub, (6) southern desert shrub, (7) chaparral, (8) pinion juniper, and (9) coniferous forest. No sharp boundary line exists between many of these regions, and there is justification for further separation of some of them. Other types of sheep raising are also practiced where there are irrigated valleys or abundant feed.

Ranges on the Great Plains are often grazed throughout the year, to their best advantage. However, most ranges of the mountain, desert, and foothill areas are best adapted to grazing in only one season; they may be roughly grouped into winter, spring-fall, and summer ranges. The season in which a range should be grazed is determined by availability of stock water, eleva-tion, depth of accumulated winter snow, season of plant growth, ability of forage to cure, danger from poisonous plants, and the time the range is needed to balance other feed supplies.

Normally, the rangelands on the lower elevations are grazed during the winter (Fig. 34-2). In this area, the annual precipitation is commonly less than 10 inches, soils are often saline, and irrigation water is not available. Desert shrub, sagebrush, and grass-forage types are predominant.

The spring-fall ranges are found in areas where elevation, growing sea-son, rainfall (10 to 18 inches), and topography are intermediate between winter and summer ranges. Vegetation usually consists of sagebrush, ju-niper, mountain brush, and related types, with very little grass.

Summer range areas are in the mountains where precipitation is greatest (18 inches and more), but where low temperatures, short growing season, and steep topography preclude farm-crop production. Ponderosa pine, spruce, fir, aspen, and alpine grasses and sages are found in these areas.

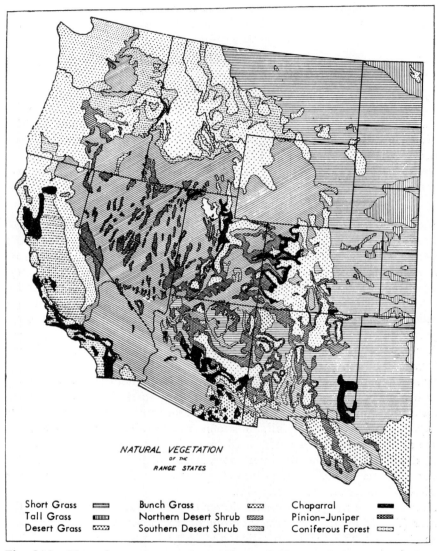

NATURAL VEGETATION
OF THE
RANGE STATES

Short Grass	Bunch Grass	Chaparral
Tall Grass	Northern Desert Shrub	Pinion-Juniper
Desert Grass	Southern Desert Shrub	Coniferous Forest

Fig. 34-1. Natural range vegetation regions of eleven of the western range states. [From Stoddart and Smith (1953.]

With the exception of the coastal region, the western states have hot, dry summers and cold winters with considerable snow.

The breeding flocks in the range areas are largely Rambouillet, Columbia, and Targhee. Purebred rams of these breeds are used to produce replacement ewe lambs, which are kept in the range herds or sold to areas in the Midwest or East as foundation ewes. The ewes are bred largely to Suffolk or Hampshire rams to produce blackface market lambs.

The Farm Flock. Farm flocks are raised throughout the United States; however, they are concentrated in the humid farming areas of the central, southern, and eastern states and on the irrigated farms of the western states. The aim of the farm-flock producer is to grow fat lambs that can be marketed as soon as they are weaned; consequently, most of the farm flocks are of the mutton type.

The Purebred Breeder. Most of the purebred sheep enterprises are located in the areas where farm flocks are produced. Some purebred breeders, however, raise sheep under range conditions part of the year.

The aim of the purebred sheep breeder is to raise rams to sell to commercial sheepmen, and rams and replacement ewes to sell to other purebred breeders. The raising of purebred sheep is a specialized business. Attention must be given to the following: selection and matings of individual parent animals to produce offspring with characteristics that are desired in the breed, feeding and management to develop inherent characters to their

Fig. 34-2. Sheep on the winter range in southern Utah. This range is typical of much of the Intermountain Region. The valleys are broad with low mountains. Note the playa in the background.

maxima, maintenance of accurate records of each animal, proper showing of animals, and building a favorable reputation.

Lamb Feeding

REPLACEMENT LAMBS. Most range operators and farm flock owners in the West select their best ewe lambs and grow them out for replacements for ewes that are culled from the flock each fall. Farm flock owners in the Midwest, East, and South often buy western ewe lambs for replacements.

HOT-HOUSE LAMB PRODUCTION. To produce hot-house lambs, ewes are bred early so that the lambs can be put on the market from December to April. The operator must skillfully feed, house, and manage the ewes and lambs so that the lambs can be marketed at six to twelve weeks of age, when they weigh 30 to 60 lb.

EASTER LAMBS. At Easter time, there is a large demand for lambs that differ widely in quality, age, and weight. Usually, however, the lighter weights are preferred.

SPRING LAMBS. If lambs are marketed after Easter and before July 1, they are referred to as spring lambs. These lambs are marketed directly off their mothers and have been on excellent pasture or they have been creep-fed grain. A lamb creep is a pen with openings large enough so that the lambs can enter but the ewes cannot.

FATTENING LAMBS ON PASTURE OR RANGE. Many lambs are born in the spring and fattened on pasture or range without any grain feeding. In the intermountain region, lambs are usually weaned in August or September. The fat lambs, weighing 85 to 100 lb., are sent directly to market. The thin lambs and those under 85 lb. are put in dry lot and fattened on irrigated or on improved dry-land pasture hay and cropland aftermath and marketed at a weight of 90 to 100 lb.

FATTENING LAMBS IN DRY LOT. Lambs that are from range ewes and that are not fat when weaned are usually shipped to areas where there is an abundant feed supply and are fattened by being fed large amounts of hay and grain in a dry lot (Fig. 34-3). The main areas for dry-lot feeding are eastern Colorado, the corn-belt states, and California; and to some extent, the irrigated valleys of the intermountain states.

34-3. NUTRITION AND MANAGEMENT OF RANGE SHEEP

For adequate nutrition of range sheep, the forage should contain an ample supply of protein, energy, minerals, and vitamins. The animals should also have plenty of water. Under favorable conditions and on well-managed ranges, forage consisting of browse, grass, and forbs frequently supplies all the nutrients necessary. The sheep's diet should be supplemented, however, when there is a scarcity of forage—especially on the winter range, when cli-

Fig. 34-3. (*a*) View of lamb-feeding facility accomodating 30.000 head at Sutter Basin Corporation, Robbins, Calf. Storage facilities for feed are in the background. (*b*) View of same feedlot showing arrangement of mangers for chopped hay or for silage and of self feeders for grain mixtures. Automatic unloading wagons deliver food without entering lots.

matic conditions are unfavorable, or when the forage is deficient in some nutrient. The condition of the sheep, the amount and kind of forage on the range, climatic conditions, and the time of year determine what kinds of supplements to feed and when to feed them.

Nutrient Deficiencies of Range Forage. Usually, nutrient deficiencies occur most frequently on the winter range. Aside from a shortage of water and salt, the deficiencies common among sheep grazing on winter-range forage, particularly on mature, dried forage, are those in phosphorus, protein, and energy. Calcium deficiency is rare.

On some ranges, carotene (provitamin A) may be deficient, especially on a grass-type range or when there is a late spring. However, sheep store large quantities of vitamin A in their livers during the summer when grazing on succulent green forage, and if the winter is open and there is an early spring, animals usually will not suffer from a vitamin A shortage. Plant analyses and supplementary feeding trials indicate that sheep in the intermountain area seldom suffer from a lack of vitamin A.

Data on whether range sheep lack vitamin D are scanty. However, since this vitamin is synthesized in the skin by the ultraviolet rays from the sun, a vitamin D supplement is not recommended in areas of adequate sunlight.

Sheep are able to synthesize in their rumens all the known B-vitamins; therefore supplementary sources of these vitamins are not needed.

Many areas within the United States are deficient in trace minerals. When mineral deficiencies exist, iodine, cobalt, and copper may be supplied by use of trace-mineralized salt.

Feeds to Supplement Range Forage. The classes of feeds for supplements include the following: (1) roughages, consisting of green, leafy, early-bloom alfalfa or clover hay; (2) energy feeds, such as barley, corn, wheat, and molasses (to increase palatability and give adhesive qualities to ingredients in a pellet, (3) protein feeds, such as cottonseed meal (solvent-extracted), linseed meal (solvent-extracted), and soybean meal (solvent-extracted); (4) feeds rich in carotene—such as alfalfa (dehydrated; leafy, green, early bloom; suncured meal)—or vitamin A; (5) mineral feeds high in phosphorus (bone meal, dicalcium phosphate, defluorinated phosphate, or monosodium phosphate), sodium chloride (salt), and, possibly in some areas, cobalt (cobalt sulfate, cobalt chloride, or cobalt carbonate), copper (copper sulfate), and iodine (potassium iodide).

In the western range area, the following feeds will usually be the most economical to use in supplementing range sheep: alfalfa hay, sun-cured; barley grain; cottonseed meal, solvent-extracted; bone meal or dicalcium phosphate; salt; and alfalfa meal. Corn or sorghum grains may replace barley in the states of the Great Plains.

Formulation of a Supplement for Range Ewes. In formulating a supplement for range ewes it is necessary to determine the composition of the forage on which they graze, and then to add sufficient nutrients to make up any deficiencies. In preparing the supplement (see Table 34-1), the following points should be kept in mind:

1. Add enough phosphorus to satisfy the requirement of a ewe weighing about 130 lb.
2. Supply adequate protein to meet the protein requirement.
3. If vitamin A is needed, add sufficient vitamin A or carotene to cover the requirement.
4. Add 1% salt or trace-mineralized salt.
5. Add a sufficient quantity of energy feed to supply the balance of the requirement.
6. Pellet the mixture so it can be fed on the ground.
7. The rate of feeding is important. As the amount of feed is increased the percentage of protein, phosphorus, and vitamin A may be decreased, making it possible to use more of the less-expensive energy feeds, such as barley or corn. It is usually not economical to feed ewes that are in good condition when they go to the winter range enough supplements to keep them gaining or maintaining their weights throughout the winter months.

A pellet containing no more than 12% protein or corn or barley should be fed when sheep graze on sagebrush range, since the diets of sheep grazing there are usually high in protein: the same feeds should also be fed during emergencies, along with alfalfa hay. A 24% protein pellet is more desirable when animals are grazing on mixed browse-grass ranges. A 36% protein pellet or solvent-extracted cottonseed meal should be provided with saltbush type ranges if sufficient grass is available on the range.

Solvent-Extracted Cottonseed Meal, and Salt. In recent years salt has been used to regulate the consumption of solvent-extracted cottonseed meal. This method of feeding saves labor and is recommended where it is not possible to hand feed the animals—that is, where range sheep are handled under fence. It should be kept in mind, however, that salt is used to regulate the amount of meal consumed, and should not be used if the intake of supplement can be controlled by hand feeding.

It is essential that the animals be started on small amounts of solvent-extracted cottonseed meal and salt by hand feeding for a week or more. This will tend to prevent over-consumption of salt, which may be fatal. The salt in the mixture is gradually increased from a sprinkling the first day to the full proportion to be fed in the self-feeders. A mixture of 0.5 lb. of salt and 1.0 lb. of solvent-extracted cottonseed meal is suggested at first. If the animals eat too much of the mixture the salt should be increased; if they eat too little it should be decreased.

Salt-meal mixtures greatly increase water consumption. If this method of feeding is followed, the animals should have access to adequate water every day, since the extra salt must be excreted through the urine. If water is hauled to the animals, an ample supply must be provided. If additional phosphorus or trace minerals are needed, they may be added in the correct proportions to the salt-meal mixture.

Water. It is usually assumed that water is readily available to sheep, but on many range areas it is the most limited nutrient and it may be unpalatable because of dirt, filth, or high mineral content. In other areas, water may be potentially available, but because there is ice, or because facilities are inadequate, the sheep do not obtain an ample supply.

If water or snow is not available, water should be hauled to the sheep. For best production, range sheep should be watered once each day, but the cost of supplying water sometimes makes it necessary to water range sheep every other day. On some winter ranges, reservoirs have been built to provide water for sheep (Fig. 34-4). When soft snow is available, range sheep do not need additional water. When the snow is crusted with ice the crust should be broken so the sheep can eat the snow. When dry feeds, such as alfalfa hay and pellets, are used, sheep may not obtain sufficient water from the amount of snow they are able to consume.

Fig. 34-4.   Sheep being watered from a reservoir on the winter range.

Range Sheep Management. Largely because of differences in environment, management practices vary widely between different areas; they may vary even within a given area, because of such factors as limited availability of summer ranges or because of a unique abundance of by-product feeds to choose from.

Both the physiology of sheep and the environment play important roles in determining management practices. Sheep are seasonal breeders and, until artificial means can be used to control the time of initiation of the sexual season, the lambing season must necessarily be limited according to the season in which ewes will breed. Also, lactation in the ewe requires a plentiful supply of nutrients. Lambing is therefore planned to allow for abundant forage during the period between lambing and weaning. Lambing during midwinter is usually avoided in colder climates because the young lamb is more subject to adverse climatic conditions than is the mature animal.

The following description of range-sheep management is widely applicable in the western United States. The reader should bear in mind, however, that it is merely a general summary of the major factors involved in sheep management.

Lambs are weaned in late August, September, or early October, at which time the ewes are on the high summer range. The heavy lambs are sent directly to market, and the smaller lambs are placed in feed yards. About October 15th, the ewes are trailed to the fall foothill range, where they remain for about two weeks; they are then trailed to the winter range. This trail may be 50 to 250 miles long. When the ewes are at the beginning of the trail, they are in good condition since they have been on lush feed all summer. By the time they arrive at the winter range they have lost some weight. In recent years, many operators have been shipping their animals by truck or rail instead of trailing them.

The winter range forage is dormant and in most areas is low in protein and phosphorus. The sheep, therefore, do not consume sufficient amounts of feed to keep them gaining. They should usually be fed about 0.25 to 0.5 lb. of one of the pellet mixtures outlined in Table 34-1. This extra feed helps to balance the diet and to increase the chances for the ewe to become pregnant.

For range lambing, the rams are usually put with the ewes during November or December, depending on the facilities the operator has for lambing. For shed lambing, the rams are put with the ewes after the lambs are weaned in August.

January, February, and March are critical months on most winter ranges. It is recommended that during this time the ewes be fed 0.25 lb. or more of one of the pellet mixtures of Table 34-1. If the ewes become snowbound during January and February, they should be fed about 2 lb. of alfalfa hay and about 1 lb. of the 12% protein pellet (Table 34-1).

TABLE 34-1. | *Formulas for range supplements for sheep.* *

Feed	Recommended Amount of Protein (%)		
	High	*Medium*	*Low*
Barley, grain, ground	—	32.25	57.0
Corn, grain, ground	5.0	10.0	15.0
Sugarcane, molasses	5.0	5.0	10.0
Beet, pulp, dried	—	—	10.0
Cottonseed meal, solvent-extracted	62.5	32.5	5.0
Soybean meal, solvent-extracted	10.0	10.0	—
Dicalcium phosphate	4.0	3.0	2.0
Salt or trace-mineralized salt	1.0	1.0	1.0
Alfalfa, dehydrated; or suncured meal	12.5	6.25	
Total	100.0	100.0	100.0
Suggested Composition of Supplements			
Protein (N × 6.25) %	36.0	24.0	12.0
Phosphorus %	1.5	1.0	0.5
Carotene mg./lb.	15.6	7.8	—
Rate of feeding ewes (in lb. per day)	0.25	0.33–0.50	0.20–1.0

* These supplements are usually pelleted.

By adequate feeding and good management of the ewes on the winter range, it is possible to increase the lamb crop by 15 to 30%, and the amount of wool produced per ewe by about 1 lb. Death losses of ewes will also be less.

SHEARING. In April the ewes are usually sheared with portable shearing machines (Fig. 34-5). The wool is usually not graded at shearing time. However, black wool is generally separated from the white wool, and the lamb wool (first-year wool) from the rest of the wool. Tags are usually removed from each fleece and packed separately. The wool is put in large bags for shipment to market. Sorting of the wool into grades is most often done in warehouses in the West or in Boston.

After the ewes are sheared they are trailed to the spring range, where they are lambed. There are two types of lambing procedures, range lambing and range shed lambing.

RANGE LAMBING. For range lambing, a protected area is necessary, with sagebrush or scrub oak on the south slopes of the foothills, where there is plenty of water and grass. The ewes are divided into groups of about 500, and each group is bedded in one area. Each morning the ewes that have not lambed are moved to a new area.

Sometimes small portable tents are put over the lambs and ewes at the time of lambing; these are especially useful if it is storming. The ewes that have lambed are kept in the area for two or three days, until their lambs

Fig. 34-5. Sheep being sheared with portable equipment. This procedure is used frequently in the West and protects the range from being trampled and overgrazed, since only one band of sheep is sheared in a given area.

become strong. The more progressive operators feed the ewes a high-protein pellet fortified with salt and bone meal or dicalcium phosphate after lambing. Many shepherds think that this procedure prevents the lambs from eating soil and keeps the ewes from consuming poisonous plants.

As the lambing proceeds, the animals are separated into three groups: those that have lambed (Fig. 34-6), those that are about to lamb, and those in which lambing is not imminent. (For details on how to treat the lambs at birth see Section 34-6.)

RANGE SHED LAMBING. For range shed lambing, feeding yards and pasture may be provided. If possible, ewes are lambed on crested wheatgrass pasture or on a good mixed legume and grass pasture. If pasture is not available, yards are provided with racks for feeding hay. Ewes are commonly confined to the yards one month before lambing. Grain is fed preferably in a reversible bottom trough, which provides a clean surface for the feed each day. Ewes are fed alfalfa hay and about 0.5 to 0.75 lb. of barley. The ewes are given free access to salt and a mixture of one part dicalcium phosphate and one part trace-mineralized salt.

SUMMER RANGE. Soon after lambing is completed the ewes and lambs are trailed to the summer range, where they are divided into bands of about 1,000 ewes with their lambs. On many summer ranges, two men look after one or two bands; one of the men tends camp while the other does the herding. A "home on the range" camp is shown in Fig. 34-7.

The shepherd uses a horse and dogs to control the sheep. At daylight he goes out to see the sheep and to guide them in the direction he wishes them to go, staying with them until about 11 A.M. At this time, he gives them a mixture of three parts salt or trace-mineralized salt and one part of dicalcium phosphate in portable containers. This supplement of minerals makes the sheep easier to handle, and possibly prevents the ewes from eating poisonous plants. The sheep usually have a drink of water and then bed down for a few hours. During this time the shepherd returns to the sheep camp and has his dinner. Late in the afternoon he guides the sheep to the bedding ground on a high ridge, bedding them down just before dark. The shepherd then returns to his camp for the night. During the summer, the ewes are bedded in a different place each night. The shepherd guides the sheep so they will graze the forage to the best advantage. In October, when the lambs are weaned, the ewes are combined into bands of 2,000 to 4,000, to be taken to the fall and winter range.

The range sheep industry produces wool and lambs that are sent directly to market or that are finished on crop aftermath and pasture or in dry lot with the feeding of roughages and concentrates.

Fig. 34-6. Ewes that have just lambed. When ewes are lambed on the range they are distributed over a large area, allotting each ewe and offspring a larger area for grazing.

Fig. 34-7. "Home on the range" for camp tender and shepherd. In some areas tents are used along with pack animals. Dogs are used with horses on most western ranges. Note the mutton hanging on the front of the camp. It is wrapped up and put in the bedroll during the day to keep it cold and retard spoiling.

34-4. FARM SHEEP PRODUCTION

Whereas range sheep are maintained on the range, except for short periods during which they may be fed hay and grain mixtures, farm sheep are usually kept on improved pastures during the summer and are fed hay and possibly grain supplements during the winter.

General Feeding Practices. Farm sheep are fed largely on pasture, crop aftermath, and harvested roughages. At times they need to be given protein supplements, phosphorus, iodine, copper, cobalt, and the vitamins A, D, and E.

Sheep do best if provided with a succulent, fertilized pasture of legumes and grasses. A good irrigated pasture, grazed in rotation, will carry 8 to 12 ewes and their lambs per acre. If the pasture is part of a crop rotation system, the control of sheep parasites is simplified, because the animals do not graze the same areas continuously.

Hay that contains a high leaf-to-stem ratio is preferred. This ratio can be achieved by cutting legumes in the early-bloom stage and grasses just as they begin to head or before they become stemmy. The green color is kept in the hay and the leaves are conserved by harvesting them as quickly as possible and storing them under a shed to protect the hay from the weather. Sheep prefer alfalfa, clover, or lespedeza hay. In the West, the standard grain is barley; in the Midwest and East, it is corn.

FEEDING MINERALS. Sheep should have free access to crushed salt in one side of a self-feeder and one part dicalcium phosphate and one part salt in the other side (Fig. 34-8). If trace minerals are needed, a trace-mineralized salt mixture may be used in place of plain salt. Trace minerals that are usually added to salt include iodine, cobalt, manganese, and copper. One lb. of salt or trace-mineralized salt and 1 lb. of dicalcium phosphate are often added to every 100 lb. of grain or pellet mixture. When sheep are fed roughages composed entirely of corn silage and cereal grains, 0.02 to 0.03 lb. of limestone (38% calcium) should be added to the daily ration to supply required amounts of calcium.

34-5. MANAGEMENT DURING THE BREEDING SEASON

The Sexual Season and Its Control. Ewes have a period of anestrus and a sexual season during which estrus occurs at 17-day intervals. The period of anestrus varies among breeds, but usually occurs between January and June. Ewes with considerable Dorset or Tunis blood may come into estrus in June and July. Some Merino and Rambouillet ewes also come into heat

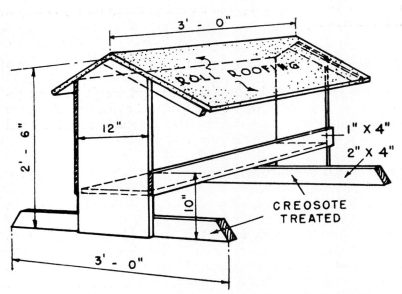

Fig. 34-8. A portable mineral box. Salt or trace-mineralized salt is fed in one side of the self feeder and a phosphorus supplement composed of one part salt and one part dicalcium phosphate is fed in the other side.

in early summer. Other breeds usually come into heat in late summer or fall.

Estrus in ewes ranges from 20 to 42 hours, with an average of 30 hours. Ovulation occurs about 24 to 30 hours after the onset of estrus. If the ewe is not bred, or if she fails to conceive, estrus recurs after an interval of about 17 days.

Certain reproductive processes can be controlled so that estrus and ovulation are synchronized or occur at approximately the same time in a group of breeding females. This synchronization continues within the limits of normal animal-to-animal variation in the length of the estrus cycle. This modification and control of reproductive processes is made possible primarily by the use of progesterone, a female sex hormone. Synthetic progesterone-like hormones, which are similarly effective, have been developed. Many of the synthetic hormones have the added advantage of being orally active (can be administered in the feed), whereas progesterone must be injected. These synthetic compounds are not yet available commercially, however.

One problem that has developed with the use of progesterone or similar synthetic compounds is a frequent decrease in fertility during the first breeding after treatment. Because of the uniformity of the estrous cycles of sheep, good synchronization is maintained through two or more cycles. If sheep are bred at the second estrus after treatment, when fertility is again normal, the advantages of the synchronization can be realized. This procedure is less effective in cattle and swine because of the greater variation in the length of their estrous cycles. Additional research in this field will doubtless solve the present problems so that these hormones can be used for practical and economical control of reproductive processes of farm animals. Such control, combined with the use of artificial insemination, would result in greater uniformity in age of offspring; it could also result in more efficient utilization of labor, facilities, and markets.

Puberty. Puberty marks the beginning of the ability of a young animal to reproduce. Rams reach puberty at the age of 4 to 7 months, but there is considerable variability. Many young ewes will breed at the age of 8 to 10 months.

Ewe lambs that have been born in January and February and that are well matured can be bred in the first fall under farm conditions. If ewe lambs are born later than February, however, they should not be bred until the next breeding season. Ram lambs that have developed well may be used with 15 to 20 ewes the first breeding season, but usually they are not used until after they are one year old.

Time of Year to Breed. The time of year to breed ewes depends on the method of sheep raising, climatic conditions, and the feed and equipment available at lambing time. If range ewes are lambed without sheds it is more desirable to have the lambs born in April and early May. If sheds are available, however, range and farm lambs may be born in February and March. For hothouse, Easter, and early spring lambs, the ewes are bred so that the lambs are born in December, January, and February.

Flushing and Breeding. Flushing consists of supplying the ewe with an abundance of feed to make her gain weight just before and during the breeding season. It is believed that this process increases the number of eggs ovulated.

After the lambs are weaned the ewes are usually kept on pasture. Sometimes they may be fed roughage. They should be managed so they are in thrifty condition—not too thin or too fat. About 3 to 4 weeks before the breeding season the amount of feed is reduced until the ewes are losing weight. Two weeks before the breeding season the amount of feed is increased to permit the ewes to gain weight. This may be done by feeding 0.5 to 0.75 lb. of grain (Table 34-2) or by putting the ewes on fresh succulent pasture. Breeding is begun 2 weeks after the ewes have started to gain. The supplementary feeding is continued during the breeding season, which is usually 6 weeks in length.

In warm climates the flock is sheared and a cool place is provided for the sheep. In cool climates the ewes are crutched and most of the belly of the ram is sheared. If early lambs are desired the flock may be put in a dark building and held there to decrease the amount of light. Decreasing the amount of light causes the ewes to come into heat sooner.

Number of Ewes per Ram. The average number of ewes bred per ram is as follows:

Age	Farm	Range
Ram lamb	20	15
Yearling	35	30
Mature	40	35

If the ewes are hand bred, one ram can serve up to 100 ewes. Under this system it is best to use a teaser ram, with an apron tied under his belly to prevent breeding, or a vasectomized ram. His chest is colored every two days, so that he marks the ewes in heat. At night the marked ewes can be put with a selected ram for breeding. The breeding ram may be painted on the brisket or he may be fitted with a harness that carries a colored crayon. Only paints that can be scoured from the wool should be used.

The ram should be checked for live, viable sperm, blindness, normal tes-

TABLE 34-2.	*Grain mixtures for ewes.*

	Mixtures (lb.)*			
Feed	1	2	3	4
Wheat, bran	10	—	—	—
Barley, corn, or wheat, grain	60	70	75	50
Oats, grain, whole	30	—	20	50
Beets, pulp, dried	—	25	—	—
Cottonseed, soybean, linseed, or peanut meal, solvent-extracted	—	5	5	—
Salt or trace mineralized salt	1	1	1	1
Dicalcium phosphate or equivalent	1	1	1	1

* Use mixtures 1 and 4 with good quality roughage, and mixtures 2 and 3 if the roughage has been rained on, is low in protein, or was cut after full bloom.

ticles, inflamed penis, lameness, and crooked legs before he is used for breeding. It pays to use high-quality purebred rams.

The Pregnant Ewe. The length of the gestation period in ewes is approximately five months. The down breeds, such as the Hampshire and Suffolk, have the shortest gestation period—144 to 148 days. The gestation period of long-wools, including the Romney and Lincoln, is 146 to 149 days. Fine-wools, including the Rambouillet, have the longest gestation period— 148 to 152 days.

In warm climates pregnant ewes can be kept on pasture throughout the winter, but in cool climates they are kept on fall pasture or crop aftermath until snow comes. They are then fed one of the rations outlined in Table 34-3. Usually it is wise to begin feeding 0.5 to 0.75 lb. of one of the concentrate mixtures (Table 34-2) six weeks before lambing.

TABLE 34-3.	*Rations for ewes during the winter or when pasture is not available.*

	Rations (lb.)			
Feed	1	2	3	4
Legume hay, such as alfalfa, clover, or lespedeza	3.0–4.5	1.5–2.0	2.0–2.5	—
Maize or sorghum, ensiled	—	4.0–5.0	—	6.0–8.0
Native hay; or wheat, oat or barley, straw	—	—	1.0–1.5	—
Cottonseed, soybean, linseed, or peanut meal, solvent-extracted, 90%; plus limestone, 10%	—	—	—	0.25

34-6. MANAGEMENT DURING LAMBING

Sheds. A shed for sheep should provide a dry place where the sheep may lie, free from drafts but with adequate ventilation. The shed should be located on well-drained ground with the yards on the south side or away from the prevailing wind. A shed should preferably be 20 to 30 feet deep, with the slope in one direction. With this type of shed all the moisture does not go into the yards, and the sun will help to keep the yards dry. Some lambing sheds consist of a wooden frame with a canvas top. Early lambing necessitates warm quarters, which can be insured by insulating a small area of the shed to be closed off and used for lambing pens. In New Zealand and Australia most of the sheds are equipped with a slatted floor as an aid to keep animals clean and reduce labor costs. In certain areas in the United States, the slatted floor might be advantageous.

Immediately after the lambing season, the lambing shed should be cleaned. Lime should be put on the floor; the canvas should be removed and the interior of the shed should be exposed to the sun. These measures help to control diseases.

Shed Lambing.

PREPARATION FOR LAMBING. Lambing pens are usually set up with panels in a portion of the sheep shed previous to the time the ewes are to lamb, or a special lambing shed with lambing pens is provided. If the weather is cold, some of the pens are equipped with brooders and heat lamps. Pens should receive dry straw daily; if newborn lambs are kept dry and free of drafts, they can stand considerable cold. Legume hay is placed in each pen. Water is provided in each pen by using a bucket or by placing a small V-shaped trough in the back.

Unless the ewes have already been sheared, they should be crutched and the wool removed from around the udder about four weeks before lambing. This prevents the lambs from sucking strands of wool when they nurse.

NORMAL LAMBING. When lambing time approaches, the vulva swells, the udder fills, and the ewe is uneasy. A hollow appears between the ribs and the hip, and the ewe begins to strain. The water bag that has cushioned the unborn lamb bursts with a gush of fluid. Usually the ewe will lamb without help. In a normal birth, both front feet appear first, with the head lying snugly between them.

HELPING THE EWE TO LAMB. The ewe should not be helped unless the lamb is in an abnormal position or is excessively large, or unless the ewe has strained for an hour or more and no part of the lamb has appeared. Before aiding in the birth, the attendant needs to wash his hands and arms in soap and water and be certain that his fingernails are short. He then lubricates his hands and arms with mineral oil or soap. He lays the ewe carefully on her right side by holding her under the jaw with the left hand, reaching under her with the right hand, and grasping the right hind leg well toward the hoof. He pulls gently on the ewe's leg so she will go over on her side. The attendant washes the vagina with soap and water, and then carefully inserts his hand.

The lamb should be born with the front feet first and the head lying between the forelegs. After its position has been determined, it is put in the correct position by pushing it forward to allow room to straighten out all parts into a normal lambing position. The attendant pulls outward and downward as the ewe strains. After the birth the ewe should be examined to make certain there are no other unborn lambs. The attendant inspects the ewe's udder and milks it a little to see that the milk canal is open. If the lamb is weak, he places the ewe's teat in its mouth and milks a small amount of milk into it. In doing this, care must be taken that the milk does not get into the lamb's lungs.

In cold weather a lamb brooder is used to keep the lambs warm. Single lambs are left in the lambing pen for about 12 hours. Twin lambs are left for about 24 to 48 hours or until the ewe claims them.

As each ewe lambs, a number should be marked in paint on her side and the same number marked on the side of her lamb or lambs. This permits immediate identification of lambs that may have become lost or that are not receiving proper care from their mothers. Only paint that will scour out of the wool should be used.

In purebred flocks, ear tags are used showing the name of the owner and the number of the sheep. Sometimes permanent numbers are tattooed in the ears. A good numbering system consists of putting a 0, 1, 2, 3, 4, 5, 6, 7, 8, or 9 in front of the identifying number to indicate year born. For example, all lambs born in 1960 would carry a 0, and all 1961 lambs, a 1. With this system age can be readily determined. With purebred flocks accurate records should be kept of dates of births of lambs and ancestry.

FEEDING LAMBS AND EWES DURING LACTATION. After the ewes lamb, their rations are gradually increased. Ewes may be fed one of the rations given in Table 34-3 plus 0.5 to 1.0 lb. of one of the grain mixtures of Table 34-2. In the West, where excellent quality alfalfa hay is available, it may not be necessary to feed grain. If succulent pasture is available ewes usually do not need any grain.

When it is desired to market the lambs early, they are often fed extra hay and grain in a lamb creep. The lamb creep should be provided with a trough to hold a grain mixture and a rack for leafy, green legume hay.

34-7. LAMB AILMENTS

Chilled Lambs. If a lamb is chilled, it can be rubbed dry with a coarse cloth or a sack. If the lamb continues to shake, it should be wrapped in a blanket or put under a heat lamp placed in the corner of the lambing pen.

Boxes heated with jugs of hot water also make good lamb warmers. A stiff, cold lamb can often be revived by immersing it up to its neck in warm water, as warm as one's elbow will bear, for 2 to 10 minutes. When it revives, it should be rubbed vigorously with a coarse cloth until dry, given some warm milk, and wrapped in a sheepskin or old blanket. At this time it may be well to inject the lamb with penicillin and streptomycin to prevent secondary complications such as pneumonia.

Feeding and Care of Orphan Lambs. If possible, an orphan lamb should be given to another ewe that has lost her lamb. The ewe is tied in a lambing pen; the lamb is smeared with ewe's afterbirth, or some of the ewe's milk is rubbed on the lamb's nose and rump. If this fails, the hide of the ewe's dead lamb can be put on the orphan lamb. Sometimes, to tie a dog near the pen will help; sometimes tranquilizers are helpful.

Orphan lambs should be given colostrum from their mother or from other ewes that have just lambed, or from a cow. A supply of cow's colostrum may be kept frozen for this purpose. The orphans are then transferred to cow's milk with a high-fat content or to artificial milk warmed to about 90 to 100°F. A medium-sized duck-bill nipple is used. The lambs should be fed often (6 to 8 feedings per day to begin with) and in small amounts (1 to 2 oz. per feeding). When the lambs are 6 weeks old the daily feedings are gradually reduced to 3 or 4, and the amount at each feeding is gradually increased to 10 to 16 oz.

34-8. LAMB FEEDING

Fattening in Dry Lot. To make a profit in feeding lambs in dry lot, the feeder must exercise judgment and skill. Feeder lambs are usually purchased from the range operator, from a commission firm, or from an auction. Most range feeder lambs are produced by mating white-faced ewes (Rambouillet, Columbia, or Targhee) to purebred black-faced rams (Suffolk or Hampshire).

Though practices vary widely, depending upon climatic conditions, lambs in most range areas in the United States are usually weaned in August, September, or early October. They are then topped out. Fat lambs weighing more than 85 lb. are sent directly to slaughter; the smaller lambs weighing between 65 and 85 lb. are sold as feeders.

After the lambs have been shipped by truck or by rail from the range to the feed yard, they are rested in dry lot and given free access to water, hay, and salt for a few days. Preferably during this time, they are vaccinated for sore mouth and a disorder from overeating (enterotoxemia) and treated for internal parasites if the need is demonstrated.

If crop aftermath is available, lambs may be turned in to eat the grain stubble, corn, or beet tops. They are then put in dry lot and fed grain and roughage to finish them for market. In commercial feedlots the lambs are put directly on hay and a grain mixture.

Use of Antibiotics. Creep rations for suckling lambs and lamb-finishing rations have improved rate of gain when an antibiotic has been included at the rate of 5 to 10 mg per lb. of feed. The greatest response seems to have occurred under conditions of stress. Chlortetracycline (aureomycin) and oxytetracycline (terramycin) are the antibiotics used most frequently in lamb rations. There is some evidence that the incidence of enterotoxemia in feeder lambs has been reduced when these antibiotics have been included in self-fed finishing rations.

Use of Diethylstilbestrol. Weight gain in finishing lambs may be increased by feeding a supplement containing 2 mg of diethylstilbestrol per head per day. If a complete ration is fed, it is recommended that 1.2 g per ton of diethylstilbestrol be used. If the lambs consume about 3.3 lb. of feed per day they will obtain approximately 2 mg of diethylstilbestrol. Lambs may also be implanted at the base of the ear with 3 mg of diethylstilbestrol.

Diethylstilbestrol will increase weight gains about 10 to 15% and feed efficiency per pound of gain about 8 to 12%.

Hand Feeding. Lambs are separated into weight groups (Fig. 34-9) and started on about 0.1 lb. of grain per head per day and given all the alfalfa or clover hay they can consume. The concentrates are gradually increased and the hay decreased according to the schedule in Table 34-4. Oats are an excellent grain with which to start lambs on feed. Grain mixtures are shown in Table 34-4. Toward the end of the feeding period bulky concentrates, such as oats, beet pulp, or wheat bran, are decreased and high-energy concentrates, such as barley, corn, or wheat, are increased. When the lambs are hand fed, excellent gains can be obtained with either barley or corn as the only grain. Silage from corn, alfalfa, or grass may be used to replace about half the hay. It takes approximately 3 lb. of silage to equal 1 lb. of hay.

If lambs scour, they should be checked for infectious diseases and coccidiosis. The amount of grain allowed should be reduced, and about one-third of the hay replaced with straw.

Self-Feeding. Under commercial conditions, one of the simplest methods to use to feed lambs is that of self-feeding. The hay is put through a hammer mill, and the grains are rolled or ground coarsely and mixed in the proportions given in Table 34-4. Barley, corn, wheat, or oats or mixtures of these grains are satisfactory. If the above grains are used, 0.5 lb. of salt and 0.5 lb. dicalcium phosphate or equivalent are added to each 100 lb. of roughage and grain mixture.

Fig. 34-9. Chute for separating sheep into various weight or age groups. By having cutting gates along the chute, the sheep can be separated more easily into groups.

Pellet Feeding. Until recently, pellets were used largely to supplement range ewes or to feed to lambs, with hay as a creep feed. Now, many large commercial feeders are pelleting the entire diet for fattening lambs.

Several experiments have been conducted to compare pelleted and nonpelleted diets for lambs. Lambs on a pelleted ration ate 6% more feed, gained 23% in weight, and required 19% less feed per lb. of gain than those fed nonpelleted rations. These results are phenomenal and show that lambs can make good use of a completely pelleted diet. When the entire diet is pelleted, more hay can be fed in proportion to concentrates than under other

TABLE 34-4. | *Feeding schedule for fattening lambs.*

	Hand Feeding		Self Feeding	
Days on Feed	Grain Mixture* (%)	Hay (%)	Grain Mixture* (%)	Hay (%)
7 to 14	25	75	25	75
15 to 28	35	65	45	55
29 to 56	45	55	55	45
57 to 100	55	45	55	45

Suggested grain mixtures for the above feeding schedule

Grain	%
Maize or wheat, whole or rolled*	65.0
Beet, pulp with molasses, dried; or oats, grain	33.0
Salt or trace-mineralized salt	1.0
Dicalcium phosphate	1.0
	100.0

* If hay is of poor quality, substitute 5 to 10% of linseed, cottonseed, or soybean meal, solvent-extracted, for part of these grains.

feeding systems. Whether or not it pays to feed pellets to lambs depends largely on the cost of pelleting and processing the feed.

A suggested mixture for a pelleted diet is the following:

Ingredient	Percent
Alfalfa hay	65
Barley grain, ground	9
Wheat grain or corn grain, ground	10
Beet pulp, dried, or oats	10
Sugarcane molasses	5
Salt or trace-mineralized salt	0.5
Dicalcium phosphate	0.5
Total	100.0

Fattening Lambs on Pasture. In the South and in California, many lambs are fattened on pasture. In these areas, alfalfa fields are also used, particularly in the early spring. Later in the season, when losses from bloat may become severe, sheepmen generally prefer native feed, such as bur clover and alfilaria, for finishing spring lambs. Sudan grass is an excellent summer pasture crop under irrigation, and it is a favorite green forage crop among purebred breeders. In the early-lamb districts, the lambs are always marketed by late May, before the green feed dries up.

Rotation of Pastures. Good pasture with plenty of fresh green feed is necessary to produce an ample milk flow in ewes and to develop the lambs quickly. This necessary condition can be provided by dividing pastures into small units by cross fencing.

A 32-inch, eight-bar woven wire fence, with one barbed wire at the bottom and two at the top, is ideal for sheep pastures.

Temporary fences can be made of wood panels. The ewes and lambs are turned into each pasture in rotation and kept there for about two days. This practice insures a fresh feed supply for the ewes and lambs at all times. The lambs develop more rapidly, attain a heavier weight, and a larger number of sheep can be grazed on a given area.

34-9. CONTROL OF PARASITES

The more important parasites of sheep include stomach worms, nodular worms, bankrupt worms, lungworms, tapeworms, liver flukes, coccidia, ticks, and lice.

Phenothiazine controls stomach and nodular worms. Each year, the ewe flock and rams should receive phenothiazine at the following times: two weeks before the breeding season; before beginning any lot-feeding in cold climates; in the fall in warm climates where sheep are kept on pasture; within the week before ewes and lambs go to spring pasture. From June 15 to June 30, all ewes and those lambs weighing over 40 lb. should be treated; both ewes and lambs should be treated at weaning time.

Ticks (keds) can be controlled by dipping or spraying with benzene hexachloride, DDT, or lindane, or by dusting with rotenone.

34-10. HANDLING SHEEP

A sheep should be caught by the flank or by the hind leg (least-preferred method). Never catch or hold a sheep by the wool since this will bruise the flesh just under the skin. If the left hand is under the jaw and the right hand on the sheep's rump, the animal can easily be controlled. To mouth a sheep, straddle the neck, raise the head with the left hand under the jaw, and part the lips with the right hand. To examine the conformation of the sheep, the fingers should be kept close together.

In early days, the blackface sheep of Scotland were termed "Colly" and the sheep dogs were called "Colly dogs." The "Colly dog" became the Border Collie of today. A well-trained Border Collie is a distinct asset to any sheepman, and can gather, drive, or pen sheep.

34-11. PREDATORS

On western ranges, large losses of sheep have occurred from predators. This has resulted in widespread control programs consisting of the use of both government trapping and private trapping and hunting, much of it organized by livestock associations. The coyote, a species of wolf, is by far the most important of the predators, which include bobcat, bear, and cougar (puma or mountain lion). Through the use of poison baits, trapping, and cyanide guns, most of these predators are being controlled. This may have resulted, unfortunately, in an increase in the population of small rodents and rabbits, which consume a great deal of range forage in some areas.

REFERENCES AND SELECTED READINGS

References marked with an asterisk are of general interest.

Alexander, M. A., W. W. Derrick, and K. C. Fouts, 1952. *Farm Sheep Facts.* Neb. Agr. Ext. Serv. Circ. 255.

Austral. Sci. and Indus. Res. Organ., 1951. *Drought Feeding of Sheep.* C.S.I.R.O. Series. Melbourne.

* Belschner, H. G. 1956. *Sheep Management and Diseases.* 4th Ed. Angus and Robertson, London.

Blickle, J. D., D. L. Moe, and J. H. Pedersen, 1961. *Sheep Equipment Plans.* Midwest Plan Service, Iowa State University, Ames, Iowa.

Cook, C. W., L. A. Stoddart, and L. E. Harris, 1954. *The Nutritive Value of Winter Range Plants in the Great Basin.* Utah Agr. Expt. Sta. Bull. 372, pp. 1–56.

Cox, R. F., T. D. Bell, and H. E. Reed, 1951. *Sheep Production in Kansas.* Kansas Agr. Expt. Sta. Bull. 348.

* Harris, L. E., C. W. Cook, and L. A. Stoddart, 1956. *Feeding Phosphorus, Protein and Energy Supplements to Ewes on Winter Ranges of Utah.* Utah Agr. Expt. Sta. Bul. 398, pp. 1–28.

Jordan, R. M., 1950. *Sheep Production in South Dakota.* S. Dak. Agr. Expt. Sta. Circ. 82.

* Kammlade, W. G., Sr., and W. G. Kamm-

lade, Jr., 1955. *Sheep Science.* Revised Ed. Lippincott, New York.

* McKinney, J., 1959. *The Sheep Book.* Wiley, New York.

* Morrison, F. B., 1956. *Feeds and Feeding.* The Morrison Pub. Co., Ithaca.

* National Research Council, Natl. Acad. Sci., Committee on Animal Nutrition, Subcommittee on Sheep Nutrition, 1964. *Nutrient Requirements of Sheep.* Pub. 504. Washington, D.C.

Ringer, Wayne B., 1964. *Plans for Farm Buildings and Equipment.* Utah State University Ext. Circ. 289.

Slen, S. B., F. Whiting, and K. Rasmussen, 1953. *Range Sheep Production in Western Canada.* Canada Dept. of Agr. Pub. 886.

* Stoddart, L. A., and A. D. Smith, 1954. *Range Management.* 2nd Ed. McGraw-Hill, New York.

Wallace, L. R., 1948. The growth of lambs before and after birth in relation to the level of nutrition. *J. Agr. Sci.,* 38:93–153.

Watson, I., 1958. *Range Sheep Production.* New Mexico Agr. Ext. Serv. Circ. 290.

Weir, W. C., and R. Albaugh, 1954. *California Sheep Production.* Calif. Agr. Expt. Sta. Manual 16.

Willman, John P., 1955. *Sheep Production.* Cornell Ext. Serv. Bul. 828.

Chapter 35

Horse Management

Look back at our struggle for freedom,
  Trace our present day's strength to its source;
And you'll find that man's pathway to glory
  Is strewn with the bones of a horse.

ANON.

35-1. INTRODUCTION

Horses and mules in the United States declined steadily in numbers from a high of almost 27,000,000 in 1918 to 3,089,000 on January 1, 1960 (Agricultural Statistics, USDA). Census figures on horses have not been included by the USDA since 1960, but there is reason to believe that their number has increased considerably since that time. Our horse population today is made up almost entirely of light horses and ponies, draft horses and mules having been largely replaced by trucks and tractors. Table 35-1 shows the number and value per head of horses and mules in the United States from 1918 to 1960. Table 35-2 lists the seven leading states in horse and mule population in 1960.

Although breeds and types are widely dispersed throughout the country, the Quarter Horse breed is most popular in the range country of the West and Southwest. The American Saddle Horse is most popular in Kentucky, Tennessee, Missouri, and Virginia, and the Thoroughbred is most numerous in California, Kentucky, Maryland, and New York. Appaloosas, Arabians, palominos, and pintos are found principally in the West, the Tennessee Walking Horse in the South and southeastern states, and ponies, both Welch and Shetland, in all parts of the country.

TABLE 35-1. *Number and value of horses and mules in the United States.*

Year	Thousands	Value per Head ($)
1918	26,723	109
1920	25,742	108
1925	22,569	69
1930	19,124	74
1935	16,683	84
1940	14,478	88
1945	11,950	84
1950	7,781	61
1955	4,309	56
1960	3,089	113

Note: The census of horses was discontinued after 1960.

Most Americans, rural and urban, love horses, and a great revival of interest in them has spread across the nation since World War II. With increased time for recreation, riding clubs, sheriff's posses, trail rides, and horse shows have sprung up in every state. Horses in the United States today provide more pleasure, sport, and recreation for more people than in any other period in the history of our country, and the trend is likely to continue.

35-2. REPRODUCTION

Female

AGE AT PUBERTY. Small, early-maturing breeds of horses and ponies reach sexual maturity as early as 12 to 18 months of age, whereas larger, slower-

TABLE 35-2. *Leading states in numbers of horses and mules in the United States, 1960.*

Leading States	Thousands	Value per Head ($)	Total Value ($1000)
Texas	243	126	30,618
North Carolina	155	122	18,910
Kentucky	155	105	16,275
Mississippi	147	84	12,348
Tennessee	146	112	16,352
Missouri	108	99	10,692
Alabama	100	85	8,500

Source: USDA, 1961. *Agricultural Statistics.*

maturing types do not as a rule reach sexual maturity until they are 24 to 30 months of age. Liberal feeding, producing rapid growth and body development, results in earlier sexual development than scant feeding or a low plane of nutrition. Year-round grazing on the adequate forage available in the southern or semitropic climates produces earlier maturity than northern climates with long severe winters, especially if wintering rations are inadequate.

AGE TO BREED. Size rather than age may be a more practical indication of proper time for first breeding. Small, early-maturing types and breeds may be mature enough to breed as two-year-olds, whereas the larger breeds are normally not bred until they are three, and thus foal when they are four. Although the added nutritional demands of the developing fetus are not great, the demands of lactation are heavy. It is therefore unwise to breed a filly until she has reached nearly mature size. If she is undersize or too immature, her own body needs for maintenance and growth plus the necessary milk for her foal are beyond her capacity. Even with liberal feeding the young lactating mare cannot assimilate enough food to meet her requirements. Accordingly, she becomes thin and her foal fails to make normal growth. The foal may be permanently stunted and the mare fails to come into estrus (heat) or fails to conceive. Thus we have a stunted foal and a barren mare, living testimonials to poor management.

THE ESTROUS CYCLE. Estrus (or heat) is the period during which the mare will accept the stallion. Although the mare may exhibit signs of estrus at any time during the year, the spring months of March through May are the periods of greatest sexual activity. The duration of estrus averages from 5 to 7 days but may vary in extreme cases from 2 to 15 days. The time between the beginning of one estrous period and the next averages about 21 days, and the events occurring during this period are known as the estrous cycle. According to Hammond (1952) the length of the estrous cycle varies directly with the duration of estrus. Mares that remain in estrus 5 to 7 days will have a 21- to 23-day cycle. A longer estrus results in a longer cycle. Hammond also found that very young or very old and thin mares remain in estrus longer than the average, due to the slow ripening of the egg. Mares having an abnormally long estrus are usually poor breeders. Improving the nutritional level of mares in poor condition immediately prior to and during the breeding season improves the regularity of the estrous cycle and increases the chances for conception.

Ovulation generally occurs about 24 hours before the end of estrus in the mare. By daily rectal palpation during estrus an experienced person can determine within a few hours the exact time of ovulation. Knowing when a given mare will ovulate permits mating her to the stallion at the proper time so that many live sperm will be present to fertilize the egg.

Fertilization depends to a great extent on timing of mating during estrus as it relates to ovulation. It has been established by research with other farm animals that the egg remains capable of fertilization only a short time after ovulation, possibly less than 24 hours. Sperm remain alive and capable of fertilizing the egg only a short time inside the female genital tract, possibly 24 to 48 hours. Furthermore, it has been established that 6 to 12 hours may be required for the sperm to travel from the vagina or cervix of the female, where semen is deposited by the stallion during mating, to the upper or ovarian end of the Fallopian tube where the egg is deposited at time of ovulation. To insure fertile mating requires knowledge of the duration of estrus and time of ovulation in the mare so that she may be bred about two days before the end of estrus. Experience has proven this to be the optimum time to breed when only one service is given. Breeding on the third day of estrus and on alternate days thereafter as long as the mare remains in heat is a common practice when the stallion is not over-worked.

GESTATION. Gestation is the time or period from fertile mating to birth of the foal, a period of 340 ± 20 days. Like the duration of estrus, the gestation period is subject to considerable variation. The plane of nutrition affects the gestation period; a low plane tends to lengthen, a high plane shortens gestation. Other factors such as age, season, and individuality may influence the gestation period (Chapter 18).

DIAGNOSIS OF PREGNANCY. Because of the comparatively low conception rate (65–70%) in mares, compared to other farm animals, their value per head, and their long gestation period, early pregnancy diagnosis is important. Failure of the mare to come into estrus on schedule (15 to 16 days) following date of last service is indicative of pregnancy. However, many mares will skip two or more cycles and then return to estrus. Others will show no signs of estrus but fail to produce a foal. Why many mares behave in this manner is not understood, but a disturbance of the delicate hormone balance necessary for implantation and nourishment of the embryo is suspected. Probably fertilization occurs, but the embryo dies and is resorbed.

The blood test. Cole and Hart (1930) and later Catchpole and Lyons (1934) found that between the 45th and 145th days after breeding, the blood serum of a pregnant mare contains enough gonadotropin to give a reliable test when injected into laboratory animals. A small quantity of serum injected into 21-day-old female mice or rats will, if the mare is pregnant, stimulate the ovaries and uterus to mature size and sexual activity in a 48-hour period. This is a reliable and highly useful test because it can be made early in pregnancy.

Rectal palpation. It is also possible to diagnose pregnancy by inserting the hand into the rectum of the mare and palpating the uterine horns through the rectal wall. This requires some skill, acquired only with practice, but it is a reliable technique any time after 60 days from last service.

Male

AGE AT PUBERTY. The colt (young male) normally first exhibits an interest in the female when he is a yearling and will try to mount a mare in estrus. However, except for small, early-maturing types, sexual maturity does not occur until the colt approaches two years of age. As in fillies, plane of nutrition influences growth and sexual development.

BREEDING CAPACITY. Colts that are large and well-developed may be used for light service as two-year-olds, being limited to 10 to 12 mares during a 6- to 8-week season, but individual variation will determine the capacity of a given colt. A three-year-old horse can service 25 to 30 mares in a season, and active stallions 4 years old and more may service 50 to 60 mares in a 90-day season if hand mated with good management.

FERTILITY AND STERILITY. It is not uncommon to find stallions that fail to settle their mares. Examination of the semen usually reveals abnormal sperm. A normal stallion ejaculate will vary from 50 to 100 ml, with a sperm concentration of 100 to 200 millions per ml. Although it is not known exactly what the minimum requirement is for high fertility, experience has demonstrated that a significant reduction below the normal average in volume or sperm count results in lowered fertility. It has also been reported by several research workers that less than 70% normal sperm is an indication of lowered fertility. Frequent microscopic evaluation of semen for sperm numbers, motility, and abnormality is routine procedure in a well-managed stud.

Management at Breeding Time

SYSTEMS OF MATING. Among wild horses stallions fight for their mares, and the victor has his band of mares, which he keeps together. Stallions are still turned out with bands of mares on some ranches in the West and Southwest. Most mares are bred by stallions on halter, however, and artificial insemination is practiced to a limited degree.

THE MARE. Healthy normal mares usually come into estrus between the fifth and seventh day after foaling (foal heat) and remain in estrus until the tenth or thirteenth day. Accordingly, the usual practice, where it is desirable to keep the mare on schedule, is to breed her on the ninth day after foaling. If for any reason the mare has not recovered completely from foaling, or has any type of infection in her reproductive tract, she should not be bred until she is completely sound and free of infection.

For protection of the stallion and for ease in handling, mares should be hobbled and the tail wrapped for breeding. It is also advisable to wash the external genitalia of the mare with a mild soapy water. Following mating, mares should be isolated and kept as quiet as possible for a minimum of 12 hours.

THE STALLION. Only mature stallions (4 years and over) should be permitted to run with a band of mares and 20 mares is about the maximum number if the breeding season is restricted to 60 to 90 days. If hand mated, the stallion should have an individual stall or shed with an adjoining exercise paddock. Exercise and green feed or good-quality hay plus enough grain to maintain weight are essential. Avoid overfeeding and confinement without exercise. Stallions should be trained to mount and dismount properly. If a stallion is being stood for public service, only clean, healthy mares should be accepted. It is advisable to thoroughly wash his penis with a mild soap solution after each service to reduce danger of carrying infection from one mare to the next.

ARTIFICIAL INSEMINATION. Artificial insemination has been practiced in a limited way for many years. It was originally used as a last effort to breed mares that failed to conceive from normal service. More recently it has been used to extend the services from valuable sires. Modern techniques allow the stallion to mount a mare in estrus, and by directing the penis of the stallion into an artificial vagina, a normal semen ejaculate is collected. From this one collection, five to ten mares can be inseminated, depending on the volume and sperm count. The semen may be introduced directly through the cervix into the uterus by means of a sterile syringe and plastic tube, or by means of a gelatin capsule. Sanitation and knowledge of semen characteristics are essential. Satisfactory techniques for dilution and storage have not been developed, as they have for bull semen; hence use of stallion semen is limited to a relatively short time after collection. By reducing the temperature of the semen to 34 or 36°F and excluding oxygen by sealing in small containers, the life and fertilizing capacity of good semen can be extended to 24 to 48 hours. Some breed associations have restrictions concerning the use of artificial insemination.

35-3. THE MARE AND HER FOAL

Care of Mare Before She Foals. As with all pregnant animals, the mare should be maintained in normal flesh and permitted or forced to get regular and moderate exercise. Confinement and overfeeding should never be permitted. Pasture that provides green feed, exercise, and sunshine is an excellent environment. If it is necessary to confine the mare, she should be exercised daily until she foals. If pasture is not available or if it is sparse, the mare should be allowed all of the hay, preferably good-quality grass, or mixed hay plus oats if necessary, to maintain her in good strong flesh, but without fat.

At Foaling Time. The weather will determine the protection the mare needs at foaling time. Clean pasture is a good place for the mare to foal, although this may make it difficult to observe her regularly or to give assistance when needed. If confined, she should be in a clean box stall at least 12 by 12 feet, bedded with clean straw or shavings. If the temperature is low it may be advisable to close the stall, to prevent drafts and chilling. Except in severe weather a dry windbreak is all the protection needed.

Because of the great variation in length of gestation of mares it is difficult to predict foaling date. Frequent observation is essential. The presence of milk in the udder and a waxy (colostrum) exudate from the teats usually indicates foaling within a few days. As labor approaches the mare may exhibit signs of nervousness and fail to eat, or she may lie down and get up frequently. As long as she is making progress she should not be disturbed. But if little or no progress is being made, after a period of two or more hours of hard labor, assistance should be given. Normal presentation of foal is front feet first with head between forelegs. If the position is normal, assistance can be rendered by steady pulling out and down as the mare labors. A soft, cotton rope may be attached to the feet to permit a stronger pull. Should the foal be in malposition—the head or either foot turned back—a veterinarian or an experienced person should be called to correct the position and deliver the foal.

Care of Newborn Foal. Tincture of iodine should be applied to the navel cord. If the foal is normal and strong, it will be on its feet within one-half to one hour after birth. Its legs are long and unsteady and it may be necessary to help it up and assist it to nurse. This is usually all the attention a newborn foal needs. If the weather is severe, however, it is well to dry the foal by rubbing vigorously with a towel or burlap bag. The mare's udder should be checked for milk and the teats squeezed to expel any dried colostrum that might clog the canals. The foal should be observed closely for normal bowel movements and urination. Failure of either within 12 hours after birth is cause for treatment, and a veterinarian or experienced person should be called. Preferably the mare and foal should be turned out to pasture as soon as possible.

35-4. NUTRITIONAL REQUIREMENTS

Specific information on the nutrition of horses is limited, and much of the data available have been collected on draft horses or have been calculated from experimental results obtained with cattle. Little research has been done on the nutritional requirements of horses for maintenance, growth, gestation, lactation, or work. Such data, even if it were available, would at best serve as a rough guide only. Horses vary so widely in temperament, in their response to the environment to which they are subjected and to work, or lack of it, that the use of specific nutritive requirements is much less practical for horses than for other farm animals. Thus, the proper feeding of horses is and will probably continue to be a mixture of art with a general understanding of the principles of nutrition, applied to each individual horse according to his need.

Table 35-3 shows the estimated nutrient requirements of horses. In horses at work, the energy requirement increases as the work increases, all other requirements remaining essentially at the maintenance level. During late pregnancy there is a marked increase in the protein, calcium, phosphorus, and carotene requirements essential for the rapidly growing fetus. The energy requirement is only slightly higher than it is for maintenance of the nonpregnant mare, partly because of her reduced activity during late pregnancy. The energy and protein demands of peak lactation are more than double those of late pregnancy, illustrating the importance of adequate nutrition for the mare and her growing foal.

Special Nutritional Considerations. Good, mixed grass-legume pasture or hay will provide all the nutrients for maintenance or even light work for mature horses. Additional energy can be supplied from oats, corn, milo, or barley, in that order of preference, as the work load increases. Young growing horses, or mares in late pregnancy or in lactation may need additional protein supplement, such as linseed, soybean, or cottonseed meal, unless the roughage portion of the ration is made up of a liberal amount of high-quality legume hay. Green, leafy alfalfa or clover hay, fed with an equal quantity of good grass hay, such as timothy or prairie, makes an ideal roughage for young growing animals.

Horses are more susceptible to digestive disturbances than other farm animals. Because of this, greater care must be exercised in selection of feeds free of mold and dust. Regularity of feeding is important, and changes in kind or amount of feed given should be gradual. The following, in pounds feed per 100 pounds of live weight, will serve as an average guide for mature horses; the amounts can be increased or decreased according to the fatness or thinness of the horse.

Maintenance: $1\frac{1}{2}$ to 2 lb. of grass or grass-legume hay
Light work: $\frac{1}{3}$ to $\frac{1}{2}$ lb. grain plus 1 to $1\frac{1}{2}$ lb. hay
Medium work: $\frac{3}{4}$ to 1 lb. grain plus 1 to $1\frac{1}{4}$ lb. hay
Hard work: 1 to $1\frac{1}{4}$ lb. grain plus 1 to $1\frac{1}{4}$ lb. hay

Total feed intake will average between 2 and 2.5 lb. per 100 lb. body weight, with the grain portion increasing as the work load increases. Horses at hard work should be fed three times daily with most of the roughage fed at night. Working horses confined to their stalls should have the grain portion of the ration reduced to half or less to prevent digestive disturbances when idle for a day or more. Turning out to pasture or to exercise is a recommended practice for idle horses or for horses at work after the evening feed. Iodized block salt, and mineral mixture of two parts steamed bone meal to one part salt should be available to the horse in his stall, paddock, or pasture.

TABLE | *Nutrient requirements of horses. (Based on air-dry feed containing 90% dry matter,*
35-3. | *63% TDN, and 1.25 meg cal of digestible energy per pound.)*

		Percentage, or Amount per Pound, of Feed			
Body Weight (lb.)	Daily Feed (lb.)	Digestible Protein (%)	Ca (%)	P (%)	Carotene (mg per lb.)
Growing Horses, 1,000 Pounds Mature Weight					
200	6.7	12.5	.52	.36	.4
400	9.9	7.6	.33	.27	.6
600	11.4	5.9	.27	.23	.8
800	12.3	5.3	.23	.22	1.0
1,000	10.9	5.5	.22	.22	1.4
Mature Horses at Light Work					
400	8.3	3.6	.16	.16	.7
600	11.2	3.7	.16	.16	.8
800	13.8	3.7	.16	.16	.9
1,000	16.3	3.7	.16	.16	.9
1,200	18.7	3.7	.16	.16	1.0
1,400	21.0	3.7	.16	.16	1.0
Mature Horses at Medium Work					
400	9.6	3.1	.18	.18	.6
600	13.0	3.2	.17	.17	.7
800	16.2	3.2	.16	.16	.7
1,000	19.0	3.2	.16	.16	.8
1,200	21.9	3.2	.16	.16	.8
1,400	24.5	3.2	.16	.16	.9
Mares, Last Quarter of Pregnancy					
400	5.8	6.9	.34	.30	4.8
600	8.0	6.9	.33	.30	5.3
800	9.8	6.8	.31	.29	5.7
1,000	11.7	6.8	.30	.28	6.0
1,200	13.4	6.9	.30	.28	6.3
1,400	15.0	6.9	.29	.28	6.5
Mares, Peak of Lactation					
400	15.4	7.9	.26	.19	1.8
600	17.6	7.8	.29	.23	2.4
800	20.8	7.9	.29	.23	2.7
1,000	23.0	7.8	.29	.23	3.0
1,200	25.4	7.9	.29	.23	3.3
1,400	29.0	7.5	.29	.23	3.4

Source: Adapted from *Nutrient Requirements of Horses*, National Academy of Sciences, National Research Council Pub. 912 (1961).

Nutritional Diseases. So many horses are now confined to their stalls, and fed and cared for by inexperienced owners that nutritional diseases are common. Such diseases are rarely seen when horses have access to green grass, sunshine, and exercise. Only a few of the more common ones will be mentioned here.

Azoturia (Monday Morning Sickness)—faulty carbohydrate metabolism, association with overfeeding during idleness.

Founder—inflammation of laminae, or laminitis, resulting in abnormal hoof growth; it is caused by overeating.

Heaves—abnormal breathing (difficult exhaling) due to damaged lung tissue and aggravated by dusty or moldy feed.

Periodic Ophthalmia (Moonblindness)—periodic cloudy vision in one or both eyes, frequently leading to blindness; it may be associated with riboflavin deficiency in the diet; some evidence indicates a hereditary susceptibility.

Xerophthalmia (Night Blindness)—faulty vision, especially noticeable when animal is forced to move at night or in twilight; such condition indicates advanced stage of vitamin A deficiency.

35-5. MAINTAINING HEALTH AND SOUNDNESS

Importance. Because his usefulness is so dependent on his ability to perform the tasks called for, soundness and health are of first importance. Normal eyesight, sound feet and legs, and good lungs are a must in a top horse. Similarly, freedom from parasites and disease are essential to a healthy, good-working horse. Space does not permit more than a general statement on preventive measures that are consistent with good management practices.

Parasites. According to Dr. A. O. Foster (1942), "Some 150 kinds of internal parasites infest horses and mules, and probably no individual animal is ever entirely free of some of them. Fortunately, comparatively few do real damage—but those few can be extremely harmful and sometimes deadly." Conditions favorable for parasitic worms occur when numbers of horses are maintained on the same restricted pasture acreage for prolonged periods of time. Most actual damage to the horse is done by the immature or larval forms, after they have entered the host and before they reach the adult stage. Almost every tissue in the body of the horse is subject to attack by immature parasitic worms. The alimentary tract, lungs, body cavity, and blood are most frequently parasitized.

Symptoms of parasitism develop gradually and may go unnoticed or be confused with other conditions, until they become pronounced. Young animals are most susceptible; if they are infested, they appear unthrifty, have rough hair coats, are thin in flesh, and may be stunted in growth.

WORMS. The large round worm (ascarid) is most common and especially damaging to foals. Clean stalls, and pastures free of fresh droppings or manure from stalls will prevent heavy infestation. When symptoms of parasitism appear, a veterinarian should be called. He can determine the worm burden by the egg count in the feces and prescribe treatment accordingly.

BOTS. Bot flies are serious pests to horses throughout the United States. They appear in the spring and deposit their eggs on hairs under the throat, or on the neck, legs, or belly. Horses are annoyed and frightened by bot fly strike and may become uncontrollable. Eggs hatch from the heat and moisture of licking and the tiny larvae are taken into the animal's mouth. The larvae are swallowed and eventually attach themselves to the wall of the horse's stomach where they mature 10 to 12 months after being ingested. Hundreds of these larvae the size of the end of a man's finger can do severe harm and produce marked symptoms of parasitism. Clipping the hairs, or rubbing the eggs off the hairs before they hatch by means of a hot wet rag will prevent the bots getting into the stomach of the horse. If such preventive measures have not been taken, routine treatment in the fall by a veterinarian is recommended.

EXTERNAL PARASITES. Lice and mites are the most common external parasites. They occur in greatest numbers in winter when horses have their long hair coat, are closely confined, and may be poorly nourished. Loss of hair, rubbing or scratching, and loss of flesh are commom symptoms. Sucking lice are more harmful than biting lice because of the loss of blood taken from the host. Sucking lice, which are about one-eighth of an inch long, are larger than biting lice. They have pointed heads and bluish bodies; the biting lice have a yellowish body and a brown head.

If the skin on the horse shows bare, rough, and thickened, or wrinkled areas, it indicates mange mites. They can be found only in deep skin scrapings and are barely visible with the naked eye. Complete wetting with high pressure spray, using lindane or benzene hexachloride, is an effective method for ridding the horse of both mites and lice. Methoxychlor and DDT are also effective against lice.

Communicable Diseases

DISTEMPER. This is a widespread, highly communicable disease caused by a bacterium *Streptococcus equi*. The organism is spread through contact at feed bunks or watering troughs, or by inhalation. Loss of appetite, high fever, and a yellowish pus like nasal discharge are common symptoms. Isolation in clean, comfortable quarters and good nursing are important. Severe cases may require veterinary services. Prevention by immunization is partially effective, and is recommended before exposure to other horses.

ENCEPHALOMYELITIS (SLEEPING SICKNESS). This disease occurs throughout the United States, and is believed to be transmitted by blood sucking insects.

It is most prevalent in late summer or early fall. Vaccination is the only safe preventive, and horses should be routinely vaccinated each year in the spring or early summer.

NAVEL INFECTION (JOINT OR NAVEL-ILL). This is a disease of newborn foals. Its symptoms are loss of appetite, a depressed appearance, and a stiffness and swelling of the joints. Good sanitation at breeding and foaling, and subsequent immersion of the navel cord of the foal in tincture of iodine are effective preventive measures. Early treatment by a veterinarian may save the foal but advanced cases are usually fatal.

ABORTION. It is estimated that only half of the mares bred each year have normal, healthy foals. In well-managed studs, 65 to 75% of the mares produce healthy foals. Abortion is responsible for most of the failures to deliver a normal foal. This may occur any time after conception. Death of the embryo during early pregnancy may go unnoticed, the return of the mare to heat being the first indication she has aborted. After the third or fourth month the fetus and membranes are large enough to cause noticeable symptoms, and they may be found in the pasture or stall.

About 50% of abortions in mares are caused by bacterial and viral infections; a few may be due to injuries; the cause of the remainder is unknown. Streptococcal infection is responsible for many abortions in early gestation, and is caused by the bacteria gaining entrance to the uterus after a previous foaling or during mating. Breeding of mares and stallions free of infection, after their external genital organs have been thoroughly washed, is an effective measure against streptococcal infection.

Contagious equine abortion usually occurs during the latter half of gestation and is caused by the bacterium *Salmonella abortive-equinus*. Diagnosis may be made either from cultures from the aborted fetus, or from the fetal membrane. Elimination of infected animals and vaccination of pregnant mares with a specific bacterin are effective control measures.

Viral abortion results from a fever or respiratory infection of pregnant mares, causing them to abort between the fifth and eleventh months. Abortion comes quickly without warning. Mares recover as they do after normal foaling. The disease is highly contagious and extreme care should be exercised to avoid its spreading. Vaccination with virus abortion vaccine in late fall or early winter is recommended for mares bred to foal the following spring in areas where the disease has occurred previously.

Unsoundnesses. Any unsoundness that interferes with action, speed, strength, or endurance is serious in the horse. The feet and legs are the most commonly affected parts, because they support the weight of the horse and rider and provide the means of locomotion. Conformational defects, strain, or injury may be responsible for unsoundness of feet and legs. The old axiom—"No feet, no horse" or "No legs, no horse"—is worth keeping in mind when buying or evaluating a horse.

Many unsoundnesses of the feet and legs can be traced to conformational defects. Straight rather than sloping shoulders, with short, straight pasterns, gives the horse a poor natural shock-absorbing mechanism. This condition is hard on horse and rider, and it is widely believed that such horses are susceptible to stiffness, lameness, ringbones, and sidebones.

The hock joint is one most frequently unsound, partly because of conformation defects, such as sickle hock, cow hock, or too straight a leg. Such defects weaken the joint, and, since all of the driving force of the horse is transmitted through the hock, it is subjected to constant and sometimes great strain in the running horse or by the quick stops, starts, and turns of the cutting horse.

Clean, flat joints and bones are much preferred to round, meaty bones and joints. Figures 35-1 and 35-2 show the parts of the horse and the locations of some of the more common unsoundnesses.

35-6. DETERMINING THE AGE AND HEIGHT OF HORSES

It is extremely important to be able to determine age accurately. Ordinarily horses are not ready for much work until the age of three. Provided they are sound, they will appreciate in value until seven or eight years of

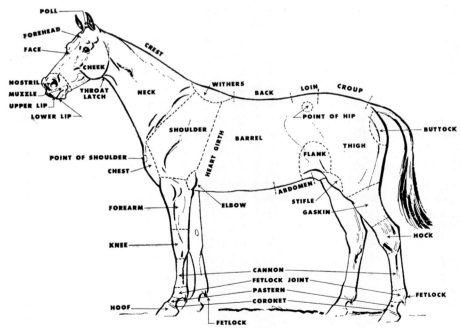

Fig. 35-1. The parts of a horse. [From Oregon 4-H Horse Bulletin H24.]

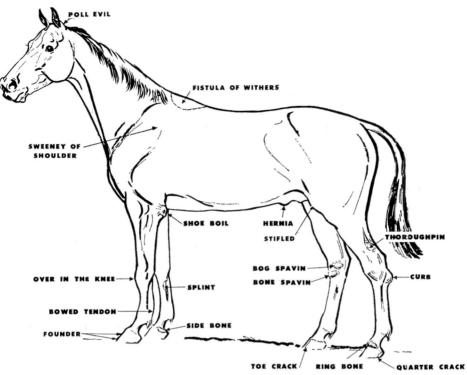

Fig. 35-2. Location of common unsoundnesses. [From Oregon 4-H Horse Bulletin H24.]

age, then depreciate with each additional year. As a matter of policy, all horse breed associations add another year to the age of a horse on January 1. Thus a foal becomes a yearling on January 1, regardless of the month in which it was born.

The young animal has 24 temporary teeth, commonly called milk teeth, consisting of 12 incisors and 12 molars. The milk teeth are shed and replaced by permanent teeth at fairly definite periods, which serve as a guide in determining the age of young horses (Fig. 35-3). At six to ten months of age, the foal will have all 12 incisors or front teeth—3 pairs above and 3 pairs below. The crown on the middle pairs will show wear at one year of age, the intermediate pairs will show wear at one and one-half years, and by two years all teeth will show wear. At two and one-half to three years of age, the middle pairs (above and below) will be replaced by permanent incisors; at four the two pairs of permanent intermediates appear, at five the corner pairs of permanent teeth appear, giving the five-year-old horse a "full mouth" of permanent teeth.

Beyond five years, the age is estimated by the wear of the teeth as in-

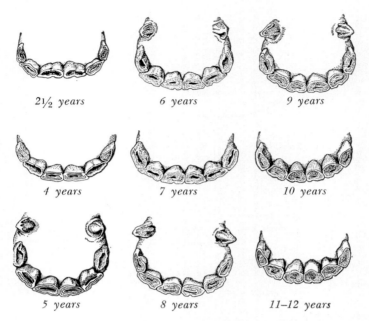

2½ years	6 years	9 years
4 years	7 years	10 years
5 years	8 years	11–12 years

Fig. 35-3. The appearance of the incisors from age 2 to 12 years. [From USDA Farmers Bulletin 2127.]

dicated by the shape and depth of the dental cavity or cup (Fig. 35-3). With wear and age, the cup changes from a long narrow cavity to a round dental star. The wear occurs in the same order as that in which the teeth appeared, and is most pronounced in the lower incisors. Experience is required to estimate accurately the age beyond eight years. The angle (as viewed from the side) at which the upper and lower incisors meet is indicative of age, the angle becoming smaller with increasing age.

The height of a horse is measured in hands, a hand being 4 inches. The measure is the vertical distance from the highest point on the withers to the ground, with the horse standing squarely on level ground. A calibrated, straight bar with a sliding right angle arm is used for determining the height. A horse that is 15-2 hands is 62 inches high.

35-7. CARE AND TRAINING

Housing. Except for newborn foals or extremely old or sick horses, shade and a dry windbreak are all the protection needed from the elements. Too much protection in the form of warm, poorly ventilated, or drafty stalls is worse than not enough. If a horse is adequately fed he can tolerate almost any kind of weather.

Clean stalls and paddocks free from loose wire, nails, glass, or other sharp objects are much more important than the kind of stall or barn. Most injuries to horses can be prevented. Box stalls should measure not less than 12 by 12 feet, with solid walls at least 4 feet high and ceiling 10 feet high or more. Avoid cracks where a foot can get hung. Board fences around paddocks and small lots are much safer than wire. Farm machinery should never be left where horses exercise.

Exercise. All horses need exercise and will get enough when left out on pasture. If confined to a stall, they should be turned out for exercise or ridden daily. Horses left in stalls without exercise frequently get stiff, their lower legs swell, and they go stale. Next to feed and water, exercise is the most important requirement of a horse.

Grooming. A good horseman keeps his horse well groomed. Regular and thorough brushing not only keeps the horse clean, but keeps his coat glossy and his skin healthy, and strengthens his loyalty and devotion to his master. To spare the brush is to spoil the horse. It is especially important to "rub him down" and "cool him out" after a hard and hot workout. He should be wiped dry, rubbed vigorously, blanketed (in cool weather), and walked for 20 to 30 minutes while he cools.

Feet. The hard, horny outer covering of the foot, known as the wall, is like the nails of the fingers (Fig. 35-4). It continues to grow throughout life and requires regular and correct trimming. The frequency of trimming varies with individuals, the kind of terrain, and whether the horse is kept confined or is on pasture. If the horse is on sandy or gravelly pasture, the feet wear enough that little trimming is necessary. If the horse is on wet, soft ground or in confinement, the feet should be checked every six weeks

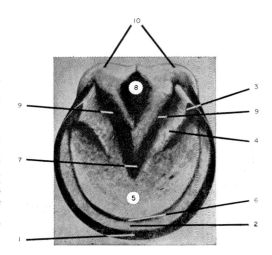

Fig. 35-4. Ground surface of a normal front foot. (1) Basal or ground border of wall; (2) laminae of wall; (3) angle of wall; (4) bar; (5) sole; (6) junction of wall and sole (white line); (7) apex of frog; (8) central sulcus of frog; (9) collateral sole: between frog and bars; (10) bulbs of hoof. [From Sisson, *Anatomy of the Domestic Animals* (4th ed.), Saunders, Philadelphia, 1953.]

and trimmed to shape as needed. Regular cleaning of the feet to remove manure or dead tissue is essential to maintaining sound feet of a horse kept in the stall or in a muddy paddock.

Horses used regularly need shoes, and this is a job for a farrier. Good farriers trim and shape the foot and correct ill-shaped feet before fitting the shoe. Shoes need to be reset about every six weeks, owing to wall growth. Keeping a horse properly shod is expensive, but it is a necessity if he is to travel over rough or gravel roads.

Training and Handling of Foals. A foal should be haltered at a week to ten days of age and handled regularly to gentle him early. When he is two months old, his feet should be trimmed and at regular intervals of four to six weeks thereafter. Keep the foot level with heels down to encourage pressure on the frog.

Foals should be weaned at four to six months of age, and this requires good management. Take the mare from the foal quietly, and take her far enough so that the foal cannot hear its mother. The foal should be left in a clean, tight, and safe stall, fed lightly, kept quiet, and watered by hand. Turn the weanling out to exercise in a tightly fenced paddock as soon as it can be done with safety. As soon as the weanling is eating well, treat him for internal parasites. Start handling him in the stall as soon as he settles down, by brushing, foot inspection, tying up, and teaching him to respond to the halter. Accustom him to discipline and to all routine handling until he no longer shows signs of fear. Allow the weanling all the good mixed legume hay he will eat and gradually increase the oats to 5 or 6 quarts daily, depending on size, appetite, and condition. This routine should be followed until spring, when the weanling is ready to turn to pasture or continue his training.

Colts. Unless they are to be used for breeding, colts should be gelded in the spring when they are yearlings. Normally colts can safely be run in groups of three to five as yearlings, although occasionally it is necessary to separate them.

Yearlings. Breed and future use will determine the schedule of training for the yearling. Racing horses and those being prepared for yearling sales are kept on a regular schedule of schooling, grooming, and feeding for maximum growth and good manners by sale time. They are introduced to the bridle, the saddle, and the exercise boy—in that order—and by the fall of their yearling year they are large for yearlings and well advanced in training.

Many yearlings raised for ranch work or for pleasure are allowed the run of the pasture during the summer and get no further training until they are two-year-olds and big enough to be saddled and broken to ride. This requires good horsemanship. It is much better to make the training from foal to two-year-old a continuous unbroken process, through saddle and rider. The rider should be light in weight (not exceeding 120 lb.) and ride just enough to acquaint the yearling with the saddle and rider. As the yearling approaches two years of age, he can carry a heavier load, stand more work, and graduate into a "man-sized" horse with limited work. From that point, his training continues in reining, cutting, changing leads, learning the gaits, and good manners. The finished product will depend on his breed and what is expected of him, but the basic early training is the same for all horses.

Few persons have the patience and "horse sense" to train a horse properly. Horses are like humans: they require time, kindness, firmness, and understanding. No finer example of affection, loyalty, and understanding can be found between man and beast than that which exists between man and his well-trained horse.

REFERENCES AND SELECTED READINGS

References marked with an asterisk are of general interest.

Catchpole, H. R., and W. R. Lyons, 1934. Gonad Stimulating Hormone of Pregnant Mares. *Am. J. Anat.*, 55:167.

Cole, H. H., and G. H. Hart, 1930. Sex Hormones in the Blood Serum of Mares. *Am. J. Physiol.*, 94:597.

Doll, E. R., 1956. Diseases and Parasites Affecting Horses and Mules, Animal Diseases: In *USDA Yearbook of Agriculture.* p. 534.

* Ensminger, M. E., 1963. *Horses and Horsemanship.* 3rd Ed. The Interstate Printers & Publishers, Danville, Illinois.

* Foster, A. O., 1942. Internal Parasites of Horses and Mules. Keeping Livestock Healthy. In *USDA Yearbook of Agriculture,* p. 459.

* Hammond, J., 1952. *Farm Animals, Their Breeding Growth and Inheritance,* 2nd Ed. Edward Arnold & Co.

Mott, L. O., and H. R. Seibold, 1942. Periodic Ophthalmia of Horses. Keeping Livestock Healthy. In *USDA Yearbook of Agriculture,* p. 402.

* Roby, T. O., and L. O. Mott, 1956. Equine Periodic Ophthalmia. Animal Diseases. In *USDA Yearbook of Agriculture,* p. 531.

Chapter 36

Poultry Management

36-1. INTRODUCTION

In approaching the problems of poultry management one is impressed by the similarity of good management practices for all livestock. Common sense indicates the importance of good nutrition, adequate housing or environmental control, disease prevention, and cost-accounting or business methods. These factors transcend such classifications of management as poultry, swine, horses, or cattle. The distinctive features are to be found when one asks the question, "How is the poultry business different from the others?"

A few examples of such distinctive features may be cited. Poultry females lay eggs and have a short productive life. Since poultry are not ruminants, a concentrated ration is required. Body temperature is high, 106°F or more, and since they are all small with more surface area per unit of weight, nonsweating, and covered with feathers, poultry do not withstand heat as well as larger farm animals. Poultry, being omnivorous, are more cannibalistic than herbivorous animals. Poultry production has a very short economic cycle. Less than a period of two years elapses between peak egg prices. In recent years the variation in seasonality of egg prices received by the producer has been considerably reduced.

Because the poultry industry has a short economic cycle, it reveals changes that have been and are taking place in animal agriculture throughout the United States. Exhibition shows and judging, like cockfighting, are in the historical past of poultry. The farm flock, formerly the backbone of the poultry industry, also will soon be nonexistent. Poultry production is a highly specialized business, and large-sized flocks are the rule rather than the exception. Whereas a man could formerly make a comfortable living with 1,000 laying hens, he now needs 5,000 or more. Twenty-five years ago,

1,000 to 2,000 turkeys raised for meat would produce a living; it now takes 10,000 or more turkeys to provide the same standard of living. These increases in efficiency have resulted from advances in mechanization and efficiency of production. Oddly enough, the prices the consumer pays have decreased. Only when one considers that the value of the dollar is only half of its value a decade ago does the significance of the accomplishment become clearly apparent.

High cash costs, small margins of profit, and larger volume have been associated with the agricultural revolution. Poultry farming today is a modern business requiring accurate records and sound management of both money and poultry. The businessmen have entered the poultry farmer's field. The size of flocks is continually increasing. One economist has calculated that 50,000 flocks of 5,000 layers producing at 70% could supply the entire nation's egg requirement. Some flocks in California now have more than 100,000 hens. One might reverse the numbers above so that 5,000 flocks of 50,000 hens could do the job. Furthermore, fewer hens are now producing more eggs. In the United States during a five-year period, 1950 through 1954, the total number of laying flocks declined nearly one-fifth, from 4.2 to 3.4 million. Yet there was a tremendous increase in flocks having 3,200 or more hens.

The hatchery business changed drastically in the years from 1953 to 1957. The number of hatcheries declined from 7,000 to 5,000—a 27% decrease. It is noteworthy that failure of a franchised hatchery—one under contract with a breeder to merchandise its chicks—rarely occurred. Small breeder-hatcheries were hard-hit, and many became franchise hatcheries in order to survive.

The fryer or broiler business has become localized in most of those states in which it is an important business. Georgia is one of the leading states in the production of fryers.

Turkey production for many years has been localized in areas that have had a climatic advantage in the control of the disease "blackhead." With the discovery of suitable drugs to control the disease, production has increased tremendously in other areas, particularly Minnesota, now a leading state in turkey production.

California ranks first in farm income from eggs, followed by New York, Iowa, and Minnesota. In the future we may expect large mechanized commercial egg plants. Egg production will approach an average of 300 eggs per hen per year. In 1960 California led the nation with an average of 229 eggs per hen; the United States average was 206 eggs. There are fewer egg handlers in the egg business than there were before World War II. Those that are still in the business have increased in size and have become more mechanized. Egg handlers are insisting that their producers follow rather rigid management and handling practices. The candling of eggs by hand will soon disappear as a common practice.

The poultry industry was one of the first agricultural businesses to combine many separate operations formerly handled by independent business agents. Grower, hatcheryman, feed manufacturer, feed salesman, and poultry processor may now all be employees, working for the same management. Many types of partly integrated operations are also possible. The man raising the poultry may be paid wages by the integrated organization, or he may own his own farm and be paid so much per pound, per bird, or per dozen. Often there is a bonus to stimulate efficiency. Production units are large, and by careful planning, spreading the risks, and increasing the volume, this type of operation has been successful in fryer and turkey production. There is growing evidence that in the future egg production will also be integrated. Company-owned flocks may in the future be restricted in growth only by the size of the investment required.

Sources of credit have played a major part in the expansion and integration of the poultry business. Both private sources and cooperatives have played a part in financing producers. Credit in vast amounts is extended by feed companies and hatcheries. The poultry processor also extends credit to assure full use of his equipment. It is now clear that one does not start in the poultry business unless he has a market for his products. One does not make money on poultry every year, or indeed every other year.

36-2. OPTIMUM ENVIRONMENT

The optimum environment for poultry is neither easy to describe nor to attain. On one hand, there are the environmental factors: temperature, light, humidity, air movement, and altitude. On the other hand, we have the various species, breeds, and ages of poultry, all of which have a bearing on the optimum for a particular environmental factor. To make the situation more complex, we have interactions between air temperature, humidity, and air movement. These interactions have much the same effect on the comfort of poultry as on that of man. Although multidimensional graphs are difficult to grasp and are even more difficult to construct, simple graphs can be made to show the optimum for four of the recognized environmental factors on chickens, using maxima, minima, and optima published in the literature. These graphs (Figs. 36-1 to 36-4) indicate a few of the gaps in our knowledge of optimum environment.

The environmental conditions best suited for poultry production dictate to some degree the type of housing required. Poultry may be kept (1) on the range, (2) in litter-floor housing, (3) on wire floors or slats, or (4) in individual cages. The agricultural engineer may design houses without windows for control of light and air movement, reducing to a minimum the hazard of the human factor.

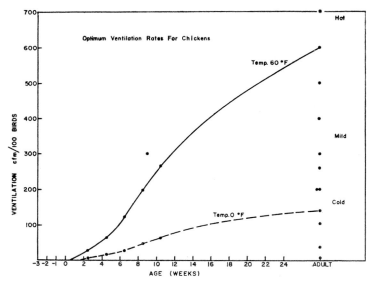

Fig. 36-1. Ventilation rates vary with age of chickens and outside temperatures.

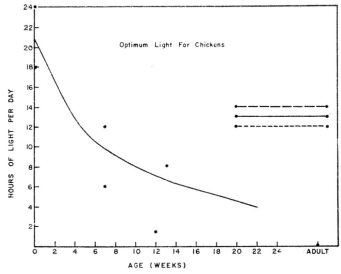

Fig. 36-2. Light requirements are higher for very young chicks and for early sexual maturity.

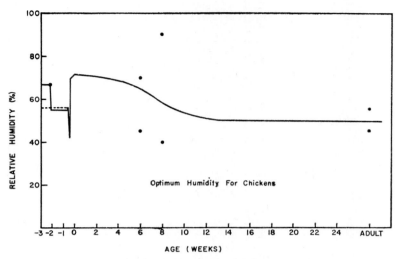

Fig. 36-3. High relative humidity is required for hatching and early feathering.

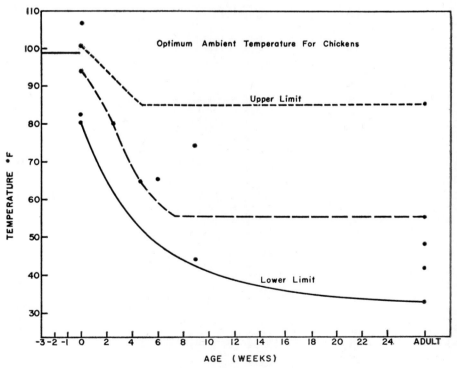

Fig. 36-4. High temperatures are required for hatching and brooding.

Ventilation. Strangely enough the main function of ventilation is not to supply oxygen and remove carbon dioxide. It is to remove ammonia and water and to aid in temperature control. Therefore it is not surprising to see in Fig. 36-1 that air temperature greatly influences the need for air exchange. When the air temperature is 90 to 99°F, a ventilation rate of 2,000 cubic feet of air per minute per 100 birds may be desirable. Tests on efficiency of windbreaks have shown that the velocity of the wind coming through does not vary greatly whether the fence is 25%, 50%, or 65% pervious. Lowered costs would favor the most pervious type of windbreak.

Light. When chicks are young they need more than the normal photoperiods (day lengths); see Fig. 36-2. It has been suggested that this is associated with the size of the crop and the fact that chicks sometimes do not learn to eat enough, at an early age, during the period of normal daylight. After they are approximately four weeks old, chicks actually grow better when not given too much light. As they approach maturity—20 or 22 weeks of age—they may be given the full amount of light.

By reducing the amount of light given in the growing period, egg production can be improved in winter-hatched chickens. The decrease of day lengths will delay sexual maturity. Once the birds are in production the cardinal rule is to avoid exposing the layers to decreasing day lengths.

Recommendations for the use of light for the various species of poultry are shown in Table 36-1.

Humidity. Very little is known about the optimum humidity for poultry, as shown in Fig. 36-3. On one extreme, with low humidity, excessive amounts of dust may be present, and chicks may not feather properly. With high humidity, heat loss becomes a problem and, when birds are on litter, the incidence of coccidiosis may increase, as may the number of dirty eggs.

Altitude. The only information on altitudes available is that on the tolerable altitudes for eggs and chicks at different ages. The effect of high altitudes is mainly one of incubation. Once hatched, chickens and turkeys may be reared successfully at altitudes up to 12,500 feet. The hatchability of turkey eggs is more sensitive to altitude than that of chicken eggs. A decrease in hatchability becomes apparent above 3,000 feet.

Temperature. The ambient or surrounding temperature is usually the major environmental factor involved in heat loss. Its effect can be modified by air movement and humidity. Brooding temperatures of 90 to 95°F under the hover for the first week reflect the inability of very young poultry to maintain homeothermy (constant body temperature). This is partially physiological and partially due to the lack of insulation from feathers. The highest temperature shown in Fig. 36-4 represents the temperature at which

TABLE 36-1. | *Recommendations for the use of light for poultry expressed as photo-periods and intensities.*

Type and Age of Birds (weeks)	Minimum Light (foot-candles)	Light-to-Dark Ratio (hours)	
		Method 1	Method 2
Replacement pullets for egg production			
0–3	1.0	20:4 hours	20:4 hours
3–12	0.1 to 0.5	Seasonal daylight	Decreasing at rate of 15 to 30 minutes per week
12–21	0.5	Seasonal daylight, if decreasing; otherwise, short daylengths 6:18 or 8:16 hours	Continue with simulated decreasing daylengths
Commercial chicken layers			
21 or over	1.0	Seasonal light if increasing, or 16:8 or 15:9 hours, depending on latitude	Increasing at rate of 15 minutes per week
Chicken fryers			
0–3	1.0	20:4 hours	20:4 hours
3 until marketed	0.1 to 0.5	20:4 hours	Seasonal daylight and intermittent night light 1:4 hours
Turkey meat birds			
0-3	2.0	20:4 hours	20:4 hours
3 until marketed	1.0	Seasonal daylight	Seasonal daylight and intermittent night light 1:4 hours
Breeder turkeys			
0-3	2.0	20:4 hours	20:4 hours
3–8	1.0	Seasonal declining daylight	Seasonal daylight
8–28	1.0	Naturally declining day lengths	6:18 or 8:16
28 or over	2.0	14:10	16:8

Source: Calif. Agr. Exp. Sta. Ext. Service Circular 529, 1964.

egg size and shell thickness are reduced; it is not the lethal temperature. Sick birds benefit from brooding temperatures higher than those used for normal chicks. Houses with mechanical ventilation designed to give light as well as temperature control are commonly used. For a more complete discussion of this type of housing, see Hart et al. (1964).

36-3. TYPES OF POULTRY ENTERPRISES

The end products of poultry farms or ranches are either meat or eggs. The production of by-products, including feathers, is not important commercially in any species of poultry. With the development of specialization, a limited number of ranches may specialize in the rearing of "started" pullets for subsequent egg production.

Types of poultry enterprises for chickens include fryer or broiler production, egg production, and rearing started pullets. Turkey enterprises include meat production and hatching egg production. Ducks as a source of meat or eggs have never been popular in the United States, though a notable exception is the Long Island duck farming area. Geese, pigeons, guinea fowl, and other poultry are handicapped by a low rate of egg production, which seriously limits the numbers which may be produced.

The approximate distribution of costs of production of five common types of poultry enterprises are given in Table 36-2.

Meat production of chickens and turkeys is characterized by having a high percentage (60 to 70%) of the total cost represented by feed costs. The costs of production for fryer production may be biased because the sample included a number of growers who raised caponettes (fryers with implants of diethylstilbestrol). In all types of the enterprises, feed costs were approximately half of the cost of production, by far the largest single item. Egg production in both chickens and turkeys shows a relatively high labor cost, since much of the routine work is not mechanized. The percentage cost of the stock is high for pullet rearing because sexed pullet chicks of special-

TABLE 36-2. | *Production costs of poultry enterprises, expressed as percent of total cost.*

	Chickens			Turkeys	
Items of Cost	*Fryers*	*Layers*	*Pullets (16 wk.)*	*Meat*	*Hatching Eggs*
Feed	60.9	59.9	46.9	63.6	50.9
Stock	17.4	8.2	30.8	14.1	17.2
Labor	7.6	14.2	12.3	8.4	16.3
Fuel	2.2		1.4	0.7	
Interest	1.1	4.1	1.0	6.5	1.8
Depreciation	4.3	5.3	3.8	2.5	2.8
Miscellaneous	6.5	8.4	3.8	4.2	11.0

Source: University of California Agricultural Extension Service. The date and California counties represented for the columns in respective order are as follows: 1956–57 Los Angeles, 1960 State wide representing over 1 million layers, 1958 Orange, 1959 Los Angeles and 1960 Fresno, Tulare and Kings. Especial thanks are due W. F. Rooney and A. Shultis.

ized breeding are used and because feed consumption to 16 weeks is low. Other costs of production include interest, depreciation, taxes, fuel, vaccination, and drugs. These constitute only a minor percentage of total cost.

Most fryers are generally reared in floor-type housing and confined so that one bird is allotted one square foot, but this practice may and should be varied, depending on other economic factors of input-output relations.

Laying flocks are kept chiefly in colony cages with wire floors; slat floors and litter floors are less common, since they require more floor area for each bird than do cages. In cages, as small an area as 0.5 square foot per bird may be adequate. The cages may also be multi-decked inside houses. Either wire floors or little floors may be used in pullet production. Turkeys are usually floor brooded but are range reared.

Management standards for five types of poultry operations are given in Table 36-3.

36-4. PRODUCTION ECONOMICS

The poultry producer wishing to make the greatest net return per square foot of housing area soon learns that there is no standard for optimum space allotments, but that these are relative. The relationship between input of factors of production (such as feed, birds, labor, equipment, housing, and other materials of production) and output of product (pounds of poultry produced) varies both with the production costs and with prices received. An understanding of principles is useful in evaluating the more common problems on poultry ranches.

The grower seldom wishes to maximize output, but invariably seeks to maximize profit.

Dr. R. K. Noles of the University of Georgia has prepared guides to the use of resources, using fryer production as an example. Similar input-output data could be prepared for egg production and other phases of the business.

A digest of Dr. Noles' report follows: The maximum-profit solution can

TABLE 36-3. | *Information on amount of feed, mortality, labor, and eggs per bird, and culling rate for several types of poultry operations.*

	Chickens			Turkeys	
	Fryers	Layers	Pullet Rearing	Meat	Hatching Eggs
Feed per bird (lb.)	9	95	12	80	100
Mortality (%)	3	12	5	6	5
Labor per bird (hrs)	0.02	0.5	0.1	0.3	1
Eggs per hen per year		250			65
Culling rate (%)		80			

be obtained by using the equi-marginal principle, since profits are maximized by equating marginal (or added) costs and marginal (added) returns. The calculation is quite simple. If $\Delta Y/\Delta X$ represents the ratio of the change in output of Y (the product) to the change in input of X (the resource) and if Px/Py represents the ratio of resource and product prices, then the relationship for the solution to the problem becomes $\Delta Y/\Delta X = Px/Py$. If the price of a pound of fryer meat was 15 cents and feed 4 cents per pound, the marginal gain in body weight at which profits were maximized would be $\Delta Y/1.0 = .04/.15$: ΔY is therefore 0.2667 pounds. In order to maximize profits, additional pounds of feed should be fed only until the gain in body weight per pound of feed added is 0.2667 pounds. In an example given by Dr. Noles, this occurred between the tenth and eleventh pound of feed (Table 36-4). It should be mentioned that in this example emphasis was placed on developing a technique for maximizing net returns from a single lot of fryers. A more realistic approach is to consider the operator who is attempting to maximize his net return over a given period of time, for example, one year. Time then becomes a crucial factor in the analyses of the solution of this operator's problems.

Several lots of fryers can be grown within 12 months. The producer will need to determine the age and weight at which each lot is to be marketed.

TABLE 36-4.	*Comparison of added returns and costs for a single lot of fryers based on feed at 4¢ and meat value of 15¢. The variable input (feed) and concomitant variable output (meat) indicates that 10 lb. of feed gave the greatest return over feed costs.*

Feed Fed per Bird (*lb.*)	Total Weight per Bird (*lb.*)	Gain in Weight per Unit of Feed Added (*lb.*)	Total Value of Feed at 4¢ per lb. (*¢*)	Total Value of Meat at 15¢ per lb. (*¢*)	Returns Above Feed (*¢*)
0	0.186	—	—	—	—
1	0.698	.512	4.00	10.47	6.47
2	1.185	.487	8.00	17.77	9.77
3	1.648	.463	12.00	24.72	12.72
4	2.085	.437	16.00	31.27	15.27
5	2.496	.411	20.00	37.44	17.44
6	2.883	.387	24.00	43.24	19.24
7	3.245	.362	28.00	48.67	20.67
8	3.582	.337	32.00	53.73	21.73
9	3.895	.313	36.00	58.42	22.42
10	4.181	.286	40.00	62.71	22.71
11	4.442	.261	44.00	66.63	22.63
12	4.679	.237	48.00	70.18	22.18
13	4.891	.212	52.00	73.36	21.36

Source: Personal communication from Dr. R. K. Noles.

This choice influences the number of birds and lots that may be grown in a year. If the fryers are fed to heavier weights, more floor space per bird is needed, and fewer birds will be raised per unit of space. With an increase in the number of input factors that need to be considered and the many possibilities of varying one or more at any given time, the production alternatives seem infinite. The specification of technological conditions can simplify the conditions.

Using published information on body weight and feed consumption, Dr. Noles uses a prediction equation to supply this specification of technological conditions. The number of lots produced per year permits one week for clean-up between lots.

The physical relationships between average live weight of fryers and technological conditions are shown in Table 36-5.

Costs budgets, based on the floor space expressed in terms of a square foot, are given in Table 36-6. The cost per pound of fryer produced under these conditions can be determined by dividing the data in Table 36-5 by the respective total weight produced per square foot per year, as shown on the last line of Table 36-5. The results of these calculations are presented in Table 36-7.

In order to determine the net income per unit of floor space per year, the total live weight per year per square foot from the bottom line in Table 36-5 is multiplied by the selling price per pound of fryers—either 13, 14, 15, or 16 cents. From this is subtracted the total costs per square foot, which is the bottom line of Table 36-6. The remainder of such computations are to be found in Table 36-8.

TABLE 36-5. | *Physical relationships between average live weight of fryers and technological conditions.*

	Body Weight (lb.)			
Conditions	*2.5*	*3.0*	*3.5*	*4.0*
Age at marketing (wks.)	6.96	7.99	9.22	10.34
No. of lots per year	6.55	5.80	5.10	4.60
No. of finished birds per sq. foot	1.399	1.167	1.000	0.875
Mortality rate per lot (%)	.0348	.0399	.0461	.0517
No. of chicks started per sq. foot per year	9.4824	7.0395	5.3351	4.2329
Feed per finished bird (lb.)	5.0140	6.3161	7.7508	9.3608
Feed per sq. foot per year (lb.)	45.9456	42.7512	39.5291	37.6772
Minutes of grow out labor per lb. live weight	.2344	.2240	.2217	.2175
Minutes of total labor per lb. live weight (includes clean out)	.2410	.2308	.2288	.2249
Total live finished weight per year per sq. foot (lb.)	22.9085	20.3058	17.8500	16.1000

Source: Personal communication from Dr. R. K. Noles.

TABLE 36-6. *Allocation of total cost per sq. ft. per year. Most charges are based on number of chicks started.*

| | Body Weight (lb.) | | | |
Items of Cost	2.5	3.0	3.5	4.0
Feed ($.04 per lb.)	183.7824	171.0048	158.1154	150.7088
Chicks ($.10 each)	94.8240	70.3950	53.3510	42.3290
Labor ($1.20 per hr.)	11.0419	9.3732	8.1682	7.2418
Equipment (including depreciation, repairs, insurance, taxes, & interest)	11.4435	11.4435	11.4435	11.4435
Vaccine & drugs ($.015 per chick)	14.2236	10.5592	8.0026	6.3494
Brooding fuel ($.010 per chick)	9.4824	7.0395	5.3351	4.2329
Litter ($.005 per chick)	4.7412	3.5198	2.6676	2.1161
Electricity ($.0004 per sq. foot per wk. of use)	1.8180	1.8480	1.8760	1.8960
Miscellaneous ($.001 per chick)	0.9482	0.7040	0.5235	0.4233
Total	332.3052	285.8870	249.4939	226.7411

From Table 36-8 it may be observed that as the price paid for fryers increases, fryers should be sold at lighter weights and conversely, as the price paid for fryers decreases, fryers should be grown to heavier weights before selling. Adjustments in production, based on changes in the prices of inputs, such as feed or chicks, may also be made. If fryers are grown continuously, as in the above example, the adjustment to changes in the prices of feed or chicks is as follows: As the price of feed or chicks increases, fryers should be grown to heavier weights or conversely as the prices paid for feed or chicks decrease, fryers should be sold at lighter weights. Unfortunately, the

TABLE 36-7. *Allocation of average total cost in cents per pound produced.*

| | Body Weight (lb.) | | | |
Items of Cost	2.5	3.0	3.5	4.0
Feed	8.0225	8.4215	8.8581	9.3608
Chick	4.1392	3.4667	2.9889	2.6291
Labor	0.4820	0.4616	0.4576	0.4498
Buildings & equipment	0.4995	0.5636	0.6411	0.7108
Drugs & vaccine	0.6209	0.5200	0.4483	0.3944
Brooding fuel	0.4139	0.3467	0.2989	0.2629
Litter	0.2070	0.1733	0.1494	0.1315
Electricity	0.0794	0.0910	0.1051	0.1178
Miscellaneous	0.0414	0.0347	0.0299	0.0263
Average total per lb.	14.5058	14.0791	13.9773	14.0833

TABLE   *Net returns per sq. ft. when broilers sell for 13 to 16 cents per pound.*
36-8.

Cost of Broilers per Pound	Body Weight (lb.)			
	2.5	3.0	3.5	4.0
13 cents	−34.4947	−11.5862	11.3223	34.2308
14 cents	−21.9116	−1.6058	18.7000*	39.0058*
15 cents	−17.4439	+0.4061*	18.2561	36.1061
16 cents	−17.4411*	−1.3411	14.7589	30.8589

* Optimum.

producer who is part of an integrated operation is not able to adjust his production to maximize his income.

36-5. BROODING AND REARING

It is not within the scope of this chapter to discuss this important stage in the growth and development of the chick. It is a stage during which proper environmental surroundings are most important. The rearing period begins when brooding ends—on the average, when the birds are 6 to 8 weeks old. Rearing is no longer a seasonal job. To maintain a constant supply of eggs one needs to count on a year-round replacement program. Fryers are produced, processed, and sold every week of the year.

36-6. REPLACEMENT SCHEMES

When the entire flock replacement is made at one time in the spring, wide variation in seasonal egg production is common. Such a procedure was formerly justified because fall and winter egg prices were always the highest of the year. Spring-hatched chicks were the only ones available from many hatcheries. Such a program provided full poultry houses only in the fall months. The procedure changed to hatching all year round—in fact, brooding every three weeks became a common practice for early cage operators. However, such a scheme is hard to maintain in view of the necessity of disease prevention. Production from pullet flocks is more profitable than it is from older hen flocks.

The specialized pullet-rearing farm may be independent, or it may be connected with a hatchery or feed company in order to increase sales and have pullets for an integrated organization. On very large poultry farms, pullet-rearing may be under the management of one man.

In the all-in-all-out program suggested by Arthur Shultis, Extension Economist, University of California, Berkeley, and Stanley Coates, Farm Advisor, Alameda County, California, one specialized pullet-raising enterprise may contract to supply egg farms with pullets for four years. Such a program has certain advantages: operator specialization, production of quality products, use of improved disease-prevention techniques, and cooperative supervision and decision. In addition, the program could increase volume and orderly marketing, which would lead to full use of equipment. It is important that the operators in this program have a written contract in which the terms of operation are spelled out and agreed to by each party. Where people are mutually dependent for success in operation, it is important that these be people who can work together, who can get the job done, and who can solve problems as they arise.

Another plan has been developed for an individual grower, rather than a group, to make efficient use of housing and equipment and to keep egg production on a high year-round level. In this plan the brooder house or area is large enough to brood enough chicks to replace about 20% of the ranch's laying hens. A house for growing pullets from 6 to 16 weeks old has similar capacity but the actual size will be larger because the birds to be accommodated will be larger. Five or preferably six laying houses, each with a similar capacity for laying birds, are used. There is a complete replacement of each laying section approximately every 14 weeks. The hens in each laying house are treated as a one-age-group section until they are sold. The sixth house is desirable because it adds flexibility to the system; for example, if repairs and improvements are needed, they can be made when the house is empty. If prices of eggs are high, additional hens may be kept so that a good market may be taken advantage of.

When economic conditions are unfavorable, as at a time of low egg prices, hens may be kept longer by canceling one or more replacement orders. Layers that are more than 18 months of age should be culled so that 60% rate of lay is maintained. The pullets 6 to 12 months of age should be laying 72% or better, and birds 12 to 18 months of age, 66% or better. The average flock rate of production, from birds that are 6–18 months of age, should be at least 65%. The best way to bring up the flock rate of production is by weekly culling of old hens.

36-7. FEEDING SYSTEMS AND NUTRITIONAL CONSIDERATIONS

When feeding poultry one feeds to suit the needs of the flock rather than the individual. No set of rules can be given as to the best feeding method, but in any system used, the birds usually have some type of feed at all times. With large-sized flocks and hired labor, all-in-one rations are often used to reduce to a minimum the possibility of incorrectly balanced rations.

Mechanical feeders and continuous watering are commonly accepted practices. Bulk handling has replaced most of the sacked feeds. Since it is hard to feed whole grains and mash in the same mechanical feeder, many growers use either pellets, crumbles, or mashes that are complete in all nutrients except (perhaps) calcium, which may be fed as oyster shell or limestone grit. Because of the extra labor involved in feeding, green feeds, wet mash, and other supplemental feeds are generally no longer part of the feeding program. There are good reasons for feeding pellets or crumbles to broilers, but there is added cost, which may or may not be justified.

Although mechanical feeders are popular for use with birds on the floor, birds in cages and colony cages may be fed nearly as fast with the use of a power cart and a traveling feed hopper. In a time-and-travel study made in California in 1957, it was found that feeding required the most travel time. The use of a power cart and mechanical feeding halved the feeding time. The use of large hoppers that hold several days' supply of feed also reduces the feeding time for birds kept on litter-floored houses. Turkeys on range are often fed by a tractor-trailer arrangement whereby feed is augered into feeders.

Although poultry rations are very similar in many respects, rations designed for a given age and type of poultry should be fed as directed. Young turkeys or poults should not be given a chick starter mash. In a given type of ration there is little reason for hesitating to change the grains in a mash. With nutritional consultants to supervise, changes in rations may be safely made, based on the use of computers for poultry feed formulations.

Although water is necessary for all classes of livestock, it was recently discovered that hens given water only three times a day, 15 minutes each time, did as well as birds given water continuously. The hens use their crops as storage, instead of drinking frequently, and the problems of disposal of wet droppings under wire floors is alleviated by this practice.

Feed consumption guides are useful only in helping to set up budgets and measure results (see Table 36-9). Poultry breeders working with broiler stocks have already passed the planning stage of having three-pound males at 6 weeks of age. Poultry nutritionists will soon have developed broiler rations for use until marketing time that will require only 1½ lb. of feed to produce 1 lb. of gain.

Knowledge of poultry nutrition has been advanced considerably by the acceptance of chicks as "laboratory animals." As a result, minimum requirements for a number of vitamins, minerals, and amino acids have been established. Certain natural feedstuffs may be found to contain inhibitors or deleterious substances that limit their use in poultry rations.

Quantity of Feed Consumed. The amount of feed eaten by poultry depends on the species, age, body weight, production, and environmental temperature, as well as the nutritional and energy content of the feed. The nutrient requirements of chickens is shown in Table 36-10. The protein levels shown in this table apply to rations containing approximately 900 kilocalories of productive net energy or 1,300 kilocalories of metabolizable

TABLE 36-9. | *Approximate feed consumption in pounds per 100 replacement pullets.*

Age (in weeks)	Light Breeds		Heavy Breeds	
	Per Day	Cumulative	Per Day	Cumulative
1	2	14	2	14
2	4	42	4	42
3	6	84	6	84
4	7	133	9	147
5	9	196	12	231
6	10	266	13	322
7	11	343	14	428
8	12	427	16	532
9	13	518	17	651
10	14	616	18	777
11	16	728	19	910
12	17	847	20	1050
13	18	973	22	1204
14	19	1106	23	1365
15	20	1246	24	1533
16	20	1386	25	1708
17	20	1526	26	1890
18	21	1673	27	2079
19	21	1820	28	2275
20	21	1967	28	2471

Source: New Hampshire Poultry Management Manual Ext. Circ. 357 (1960).

energy per pound (see Chapter 28 for explanation of these terms). For information on the nutritive requirements of other species, the reader is referred to National Research Council Publication 827 (1960). The feed consumption of different species of poultry is given by Ewing (1951).

Deficiencies in Poultry Nutrition. The intermixing of certain feedstuffs to provide nutritional balance must be done carefully; for example, an excess of one mineral may so bind another mineral nutrient that its value is lost.

The high requirement of the laying hen for calcium is well known, and laying rations must contain at least 3% calcium. Chicks, on the other hand, have a much lower requirement, 0.8%, which needs to be in balance with the content of phosphorus, manganese, and zinc. A deficiency of manganese or an excess of phosphorus may cause a slipped tendon or perosis in growing chicks. Sodium chloride is added to most poultry rations.

Established vitamin requirements for poultry are those for vitamin A,

TABLE 36-10. | *Nutrient requirements of chickens. (In percentage or amount per pound of feed).*

Nutrients	Starting Chickens 0–8 Weeks	Growing Chickens 8–18 Weeks	Laying Hens	Breeding Hens
Total protein (%)	20	16	15	15
Vitamins				
Vitamin A activity (U.S.P. Units)*	1200	1200	2000	2000
Vitamin D (I.C.U.)	90	90	225	225
Vitamin E†	—	—	—	—
Vitamin K_1 (mg)	0.24	?	?	?
Thiamine (mg)	0.8	?	?	?
Riboflavin (mg)	1.3	0.8	1.0	1.7
Pantothenic acid (mg)	4.2	4.2	2.1	4.2
Niacin (mg)	12	5.0	?	?
Pyridoxine (mg)	1.3	?	1.3	1.3
Biotin (mg)	0.04	?	?	?
Choline (mg)	600	?	?	?
Folacin (mg)	0.25	?	0.11	0.16
Vitamin B_{12} (mg)	0.004	?	?	0.002
Minerals (%)				
Calcium (%)	1.0	1.0	2.25	2.25
Phosphorus (%)	0.5	0.6	0.6	0.6
Sodium (%)	0.15	0.15	0.15	0.15
Potassium (%)	0.2	0.16	?	?
Manganese (mg)	25	?	?	15
Iodine (mg)	0.5	0.2	0.2	0.5
Magnesium (mg)	220	?	?	?
Iron (mg)	9.0	?	?	?
Copper (mg)	0.9	?	?	?
Zinc (mg)	20	?	?	?

Source: Adapted from National Academy of Sciences, National Research Council, Publication 827 (1960).
Note: These figures are estimates of requirements and include no margins of safety.
* May be vitamin A or pro-vitamin A.
† Requirement of vitamin E varies greatly, depending upon the nature of the diet.

vitamin D_3, riboflavin, pantothenic acid, niacin, choline, and vitamin B_{12}. In addition to retarded growth, other symptoms, such as general weakness, rickets, curled-toe paralysis, dermatitis, poor feathering, perosis, and low hatchability, may be observed if there are deficiencies of the above mentioned vitamins.

The protein content of the ration supplies the twelve amino acids essential for chicks. The amino acids most likely to be deficient are arginine, lysine, methionine + cystine, and tryptophan. To supply these amino acids, protein-rich materials, such as fish meal, meat scraps, or soybean and cottonseed meal are used. Grains do not contain enough tryptophan and lysine to meet the poultry requirement.

Whether commercial feeds or ranch-mixed feeds are used is most often a matter of economics.

Feed for poultry is usually delivered in bulk, stored in bins, and mechanically fed to the poultry. This makes it difficult to determine closely the amount of feed consumed daily. Since decline in feed consumption often precedes decline in egg production or in rate of growth, it is important to get some measure of daily feed and water consumption. The ratio of water intake to feed consumption varies with the ambient temperature. At moderate temperatures this is approximately 2:1, but as temperature increases it may reach 4:1. The influence of temperature on feed consumption and egg production efficiency is given in Table 36-11.

36-8. DISEASE PREVENTION AND CONTROL

The individual bird loses its identity in poultry management; one is concerned with the health of the flock. The depopulation procedure for controlling poultry diseases is very costly unless it is a planned part of normal management procedure. The all-in-all-out system of fryer- or pullet-raising has the inherent advantage of controlling certain poultry diseases.

Sanitation is always important in disease prevention. Sanitation has been defined as keeping the bird—and everything the bird comes in contact with —clean. The producer can carry on from there.

Vaccination is effective against certain poultry diseases: fowl pox, laryngotracheitis, Newcastle, and bronchitis. Since fowl pox is spread by such insects as mosquitoes, seldom if ever does it pay not to vaccinate against this

TABLE 36-11. | *Influence of temperature (°F) on feed consumption of Leghorns and egg production efficiency.*

Ambient Temperature	Feed per Day per 100 Hens (lb.)	Eggs per Day per 100 Hens (number)	Egg Weight per Dozen (oz.)	Feed per lb. Eggs (lb.)
23	41	26	24.3	12.3
37	35	65	23.9	4.0
45	33	74	23.8	3.5
55	31	78	23.5	3.3
65	29	75	23.2	3.3
75	27	68	22.7	3.4
85	25	56	22.1	3.9

Source: USDA Misc. Publ. No. 728.

disease. Whether or not to vaccinate against the other diseases depends upon the local situation. Only a flock in good health should be vaccinated and, preferably, with only one vaccine at a time. The vaccination program should be completed before the pullets are 4 months old and before they start to lay. It should be done by, or under the supervision of, a qualified specialist, since most vaccines are live viruses, capable of causing disease if improperly used.

Coccidiosis occurs in floor-reared birds. It can be prevented, or at least held in check, by keeping the litter dry and by following such management practices as using wire around the watering and feeding areas. Medication with coccidiostats is effective if an outbreak occurs. Small amounts of such drugs are usually added to fryer rations as a preventive measure. They are often added to the starter and growing mashes of replacement birds reared on litter.

Chronic respiratory disease, associated with *mycoplasma gallisepticum,* is a serious disease of both chickens and turkeys. Control or eradication is vital to the industry. Transmission may be by egg, contact, air, or mechanical carriers.

If a disease is to be properly treated, it must first be identified by a pathological laboratory or a qualified veterinarian. A few diseases such as fowl pox may be recognized by the caretaker.

Among the most frequent parasites is roundworm, which may be a problem if chickens are reared on old litter or contaminated ground. Heavily infested birds should be treated with nicotine or piperazine preparations. Tapeworm control is indirect and can be accomplished by controlling intermediate hosts such as flies and beetles. A major step toward fly control is keeping the droppings dry.

Mites, both red and fowl mites, are often a problem in flocks, especially where wild birds are present. Insecticides approved by the Food and Drug Administration may be used. Red mites are killed by spraying roost, nest, and equipment, whereas lice and fowl mites are killed by spraying or dusting the birds. Insecticides presently used include Sevin (Carbaryl), Malathion, and CoRal (Coumaphos).

36-9. EGG HANDLING

The gathering and processing of eggs for market is one of the most time-consuming chores on a poultry ranch. The production of clean eggs is a goal of every poultryman. The number of birds he may keep in a given area is dependent on the litter conditions. Clean eggs are associated with clean litter or wire floors.

Nests of many shapes and sizes are used for poultry. Up to the present time the limiting factor in litter-floored housing has been the lack of a nest adaptable to mechanical egg gathering. Wire roll-away nests have been used with limited success in litter houses, but too often the hens prefer to lay in the litter. These nests are suitable for cages and colony cages and greatly reduce the time and travel involved in gathering eggs.

Cracked or dirty eggs in excessive numbers may be associated with the season, age of layers, type of nests, nutrition, and gathering procedures. A tremendous amount of time can be saved if it is possible to pack the eggs without having to rehandle them for cleaning.

Many mechanical operations, such as washing and grading for size, are performed in the egg room. The use of bulk candling with an underlit roller-table to remove eggs with defects is less costly than hand candling each egg. Eggs may be packed in cartons on the ranch for marketing through chain stores and other outlets. This practice is usually satisfactory for freshly produced eggs from a ranch, where the care and age of the egg is known.

36-10. DECISION MAKING FOR LAYING HENS

The number of birds to keep per cage or per square foot of floor space will vary with the price received for eggs (the product) and the cost of production. The same principle applies in deciding whether to use replacement pullets or to force molting hens and keep them for several extra months of laying. The purchasing of equipment and of automation is also primarily an economic decision, based on the wish to save labor, or to increase the number of hens and maintain the same hours of labor.

Data from poultry cost-account studies can be applied by the individual poultryman. With the extension of business methods to poultry production it becomes necessary for the poultryman to keep records and to use them, in order to survive in competition with poultrymen who do. In some states keeping cost accounts has been encouraged by the Agricultural Extension Service.

Poultry management studies are conducted in California in the following way. Cooperators send their monthly cost-of-production information to the Agricultural Extension office, which publishes the data in a monthy progress report, that lists each cooperator by code. At the end of the year, a detailed report is published, giving all production and income data. The report, which includes a poultry management supplement describing management practices followed on each ranch, helps the cooperator to analyze his poultry enterprise and to determine the strong and weak spots in his management.

The individual cost-account records of a poultryman, if he keeps them on his own, may be compared with those of the Agricultural Extension Service, random-sample egg-laying tests, and various county and state poultry management studies. The records provide the basis for improvement and set the standards toward which to work. For example, poor egg production quickly shows up in the study but analysis on the ranch is required to pinpoint the cause. High mortality and high feed costs require the same analysis.

When considering published cost-account data it should be remembered that these are not true averages; they are likely to biased in favor of the better poultryman. Poorer poultrymen do not bother to keep records. "They ain't farming as well as they know how already."

Although all data are of interest to the poultryman who wants to know how he stands, there are a number of particularly important production factors in the egg business: number of eggs per hen, pounds of feed per hen, average flock replacement, culling and mortality, labor per hen, and egg size. Economic factors include feed cost, price received for eggs, and miscellaneous costs per hen.

36-11. PROBLEMS OF POULTRY MANAGEMENT

Lest the student think that all problems of poultry management have been solved, current problems are listed here with suggestions for a possible solution to each. Better control methods are urgently needed throughout the industry.

Maintaining Body Weight. The practice of maintaining proper body weight of layers is an essential of good management. This practice, on uniform stock from modern poultry breeders, is very important for management of established strains and crosses. Many managemental factors are reflected in body weight. Feed restriction for more than even one day will result in a loss of egg production that is presumably caused by a lack of sufficient circulating gonadotropins. Layers that have been in production for a considerable length of time are very susceptible to decreases in body weight.

Flies. The use of wire floors on poultry ranches, although it encourages large flocks in a small area, creates a fly problem. When only one bird is kept in a cage, with proper drying conditions the manure can be dried. But such is seldom the case. Some county health officials require that the manure be cleaned out every week, and this imposes an additional cost on the poultryman.

Dust. Turkeys create a dust problem when raised in large numbers on small acreages in a dry climate. Sprinkling the area with water and strip cropping are the two control measures, to date, that work best.

Dirty Eggs. Dust increases the number of dirty eggs in cages and in rollaway nests; removal of this dust from the wire, protection of the wire with covers, or lids over the egg trays reduces the problem. Damp litter also increases the incidence of dirty eggs. Clean nest litter is a partial remedy, but the solution should also include consideration of ample space per hen, good ventilation, suitable waterers, and, possibly, removal of excessive water from the chicken house by pit cleaners.

Cannibalism. Much evidence has been presented to show that nutrition, genetics, and management may all contribute to solving this problem. Cannibalism defies experimentation, because one cannot always get the experimental birds to cannibalize one another. Debeaking or removing one-half or more of the upper beak with a debeaking machine has given the best results so far.

Low Winter Egg Production. Lighting that provides 14 hours of light per day and housing that prevents exposures to temperatures below 35°F will help to maintain winter egg production. Improvements in breeding, feeding, and management—for example, hatching more than once a year— have helped to reduce this problem.

Hot Weather Problems. Every new poultryman must learn that hens in cages are more susceptible to heat prostration than birds on the floor. When temperatures exceed 95°F for four hours or more, some unacclimated birds may die. The new poultryman should remember that heat loss must equal heat production, or fever—and eventually death—will result. One may use "foggers" or mist sprayers to dampen the birds with water, thus cooling them by evaporation. (Wetting the birds with a hose is too time-consuming for a large farm.) Cool drinking water helps to reduce the bird's internal heat.

Predators and Stampeding. Unless kept confined, poultry and their eggs may become prey to predators. With turkeys and other poultry the number of birds killed by piling up or stampeding in fright often exceeds the number killed by the predator. The key word is vigilance.

36-12. WHY POULTRY RANCHES FAIL

When looking for the cause of financial failure, the poultryman may find himself at fault. Failures may be due to various causes.

(1) Lack of fundamental knowledge about poultry, and failure to keep abreast with up-to-date production and marketing methods. There is no substitute for keeping accurate records and using them as bases for management decisions.

(2) Lack of adequate marketing. Many steps are necessary to get eggs from ranch to consumer. Usually it pays to produce a quality product and

obtain the highest price, but producing without an assured market can result in distressingly low prices.

(3) High feed costs. Use the lowest-cost feed that is consistent with good performance, and avoid waste.

(4) Low egg production. Purchase chicks from a proven source.

(5) High mortality. Follow a vaccination or disease-control program tailored to fit the ranch and its conditions. Obtain accurate diagnoses from a qualified veterinarian.

(6) Inadequate financing. Take advantage of discounts, loans, and quantity purchases. Long- and short-time debt loads should not exceed $5.00 and $1.00, respectively, per layer. Maintain a reserve in case of many months of poor prices.

(7) Zoning and population pressure. Flies and odors are considered public nuisances and can force the poultryman from a neighborhood. See that manure and moisture are properly handled.

(8) Location. The operation must not be too distant from centers of population and good transportation must be available.

(9) Lack of volume. High production per bird is not enough. Volume is very important, not only in making the total return to the individual greater but also in finding a market.

(10) Inefficient use of labor and equipment. Keep the ranch filled to capacity. Use as a motto for efficient labor: "Pick up an egg only once." Process it before it is put down. Taxes may be reduced by planning ranch operations and keeping complete accounts. Buildings and equipment should be designed to reduce labor.

REFERENCES AND SELECTED READINGS

Bankowski, R. A., and A. S. Rosenwald, 1956. *Poultry Vaccination—Why and How.* Calif. Agr. Expt. Sta. Circ. 455.

Card, L. E., 1960. *Poultry Production.* 9th Ed. Lea and Febiger, Philadelphia.

Ewing, W. R., 1951. *Poultry Nutrition.* W. R. Ewing Publisher. Pasadena, California.

Hart, S. A., T. Cleaver, W. O. Wilson, and A. E. Woodard, 1958. Housing and operational studies on California egg farms. *Poultry Sci.,* 36:1386–1395.

Hart, S. A., W. O. Wilson, and P. J. Lert, 1964. *Light and Temperature Controlled Housing for Poultry.* Calif. Agr. Expt. Sta. Circ. 526.

Hartman, R. C., and D. F. King, 1956. *Keeping Chickens in Cages.* 4th Ed. Roland C. Hartman, Redlands, Calif.

Longhouse, A. D., H. Ota, and W. Ashby, 1960. Heat and moisture design data for poultry housing. *Agr. Eng.* 41:567–576.

National Research Council, 1960. *Nutrient Requirement of Poultry.* Committee on Animal Nutrition. Natl. Acad. of Sci. Natl. Research Council Publ. 827.

Shultis, A., 1959. *The Egg Production Business in California.* Calif Agr. Expt. Sta. Circ. 483.

Wilson, W. O., W. F. Rooney, R. E. Pfost, and J. R. Tavernetti, 1964. *Artificial Light for Poultry.* Calif. Agr. Expt. Sta. Circ. 529.

Classification, Grading, and Marketing of Livestock and Their Products

Chapter 37

Classification and Grading of Meats

37-1. INTRODUCTION

One of the primary reasons for the development of our present efficient methods for marketing meat has been the widespread understanding and use of a system of identifying its market acceptability. This system involves the process of classification and grading.

The purpose of a classification and grading system for meat is to provide the means for identifying—or grouping together—carcasses that have similar characteristics that are important in determining their market value. The widespread use of such a system by the various market agencies—packers, wholesalers, jobbers, and retailers—materially facilitates trading and reduces marketing costs.

The origin of meat classification and grading probably dates back to the time when men first started trading in this commodity. Even then, in appraising the relative merits of different carcasses, buyers and sellers applied some of the basic principles of the presently used system of classification and grading. Because the early markets in this country were highly localized and almost all trading was based on personal inspection of the meat by the buyer, a distinct terminology was developed in each market area to describe preferences and trade practices. The growth of large urban centers and the improvement in shipping facilities stimulated the rapid growth of large competitive livestock markets and packing companies. Consequently, there also developed a need for a nationally recognized system of classes and grades for livestock and meat, in order that prices among competitive markets could be equitably compared.

Federal class and grade standards for livestock and meat were first formulated and issued in tentative form by the USDA in 1916. These original standards were based on much of the earlier research conducted on market classification and grading of livestock and meat by the University of Illinois.

Since their inception, the federal meat grade standards have been used by the USDA as the basis for the National Meat Market Reporting Service. These standards also serve as the basis upon which the Federal Meat Grading Service grades beef, veal, calf, lamb, mutton, and pork.

The federal class and grade standards for meat are published and thus generally available to all segments of the industry. However, their use by the industry has always been strictly voluntary, except for two periods of national emergency, during World War II and the Korean conflict, when the federal grading of beef, veal, calf, lamb, and mutton was compulsory under federal maximum price regulations.

The federal grading of meat grew out of a request by a producer group known as the Better Beef Association, which organized in 1926 for the primary purpose of establishing a federal program for the grading of the higher qualities of beef. It was the contention of the Better Beef Association that if a reliable identification of excellence were placed on these better grades of beef, consumers would buy them with greater confidence and that this would stimulate the production of better beef cattle. Accordingly, in May, 1927, the USDA inaugurated the federal grading of beef on a voluntary, experimental basis for one year. At the end of this first year, the program was continued on the same voluntary basis, but the users of the service were charged fees to cover the operating costs of the service. The service was later expanded to include the grading of other meats. Federal meat grading is now available in all of the principal marketing centers of the country. The federal class and grade standards have been revised from time to time as new research information has become available, for purposes of clarification of the standards, to meet changes in consumer preferences or to meet changes in production and marketing practices.

The marketing and distribution of meats in the United States is a complex process, and since classification and grading are very specialized functions in this process, only a very brief treatment of these can be included in this chapter.

Individual meat packers and retailers may, and often do, use their own brand names for identifying the meat they sell. Although these private branding systems generally include many of the same factors as those upon which the federal standards are based, the standards used by private companies are not published. Since there are no controls over the use of brands by private companies, these may vary from one company to another and they may also differ from federal grades. Therefore, the following discussion of classification and grading will be limited to the federal standards as published and used by the USDA.

37-2. BASIS FOR CLASS

Class, as used in a broad sense, is a designation that identifies or groups together carcasses or cuts that have a similar commercial use and that come from the same species or kind of animal. For example, lamb and mutton are both derived from the ovine species. However, since their commercial uses are not interchangeable, they are considered as two distinct classes of meat. Subclasses of carcasses within a specific class or kind of meat usually refer to the sex condition of the animal from which the meat was derived.

37-3. BASIS FOR GRADE

For beef, calf, veal, lamb and mutton, quality evaluation and conformation evaluation are combined into a single "quality" grade. The estimated yield of retail cuts of beef is indicated by a separate yield grade. Beef may be graded, therefore, for quality alone or for quality and yield. On the other hand, grades of pork carcasses combine both aspects of quality and yield into a single grade identification.

Quality Grades. Quality grades are based on two criteria: (1) factors that contribute to the palatability of the meat, frequently referred to as "quality" factors, and (2) conformation, from which the percent of the carcass consisting of the more valuable cuts and the ratio of meat to bone in the carcass can be estimated.

Grade is a designation that identifies carcasses or cuts falling within the same group on the basis of certain utility or value-determining characteristics. A grade includes a range of grade-determining factors narrow enough that carcasses or cuts of the same grade have a high degree of interchangeability; that is, though two carcasses may vary in characteristics such as maturity or fatness, the various factors balance each other to result in comparable acceptability.

The two basic objectives in grading meat are these: (1) to reflect differences in the proportion of the more desirable to the less desirable parts of the carcass and in the ratio of meat to bone; (2) to evaluate the characteristics of the lean meat that are associated with its ultimate consumer acceptability.

CONFORMATION. Conformation refers to the proportionate development of the various parts of the carcass and to the ratio of meat to bone. It is primarily a function of the relative development of the muscular and skeletal systems. However, since overall thickness and fullness of a carcass is a part of the conformation evaluation, the extent to which the conformation has been improved by the quantity of external fat on the carcass is also a factor that must be considered. Superior conformation implies a high proportion of meat to bone and a high proportion of the weight of the carcass in the more valuable parts. It is reflected in carcasses which are very thickly muscled, are very full and thick in proportion to their length, and have a very plump, full, and well-rounded appearance.

QUALITY. Quality,[1] as used in grading meat, refers to its expected palatability—its tenderness, juiciness, and flavor. The most important factors considered in evaluating the quality of a carcass are these: (1) the firmness of the lean and its degree of marbling or intramuscular fat (interspersion of fat within the muscles); (2) the indications of the maturity of the animal from which it was produced. Differences in maturity are evaluated from the color and texture of the lean; the color, size, and shape of the rib bones; the ossification of the cartilages, particularly those on the ends of the split thoracic and lumbar vertebrae and the cartilaginous connections of the sacral vertebrae.

Since increasing firmness and marbling and advancing maturity have opposite effects on quality, the federal standards permit increased marbling and firmness to compensate, within certain limits, for advancing maturity to maintain the same degree of quality. Excellent quality in meat, as evidenced in the cut surface, usually implies a full, well-developed, firm muscle of fine texture and bright color containing a liberal amount of marbling and a minimum of connective tissue.

Illustrations of the degrees of marbling used in the federal grade standards for beef are available from the USDA.

The USDA beef grade standards that became effective in June, 1965, required, for the first time, that all carcasses be ribbed, that is, the hindquarter separated from the forequarter between the twelfth and thirteenth ribs to expose the large rib-eye muscle, which lies adjacent to the backbone. This permits a direct observation of the cut surface of the largest muscle in the carcass. However, most ovine and veal and calf carcasses are graded as intact (unribbed) carcasses, without the benefit of observation of a cut surface of the lean. Therefore, when grading these carcasses, it is necessary to evaluate their quality-indicating characteristics indirectly, on the basis of the development of other closely related characteristics. The development of certain interior fats—feathering between the ribs, fat streaking in the inside flank muscles, fat covering over the diaphragm, and overflow fat over the ribs—is one of the most important indications of quality in unribbed carcasses. Although the quantity of external fat is positively associated with quality, this association is quite low. Therefore, the quantity of external fat is given practically no consideration in determining quality. Color of fat is a price-determining factor in some instances, but since it has not

---

[1] The term quality is also used, at times, to refer to the general excellence or acceptability of meat for a specific purpose. When used in this less restrictive sense, "quality" may also include such factors as proportion of fat and lean, color of fat, and so forth.

proven to be directly related to quality, it is not used as a factor in grading. However, the character of fat is a factor associated with quality in grading unribbed carcasses. Firm, brittle fats indicate a higher degree of quality than do soft, oily, or powdery fats.

Yield Grades. The yield grades for beef are predicated on an entirely different basis. These were adopted in June, 1965 and identify carcasses for differences in yields of closely trimmed, boneless retail cuts. These grades will be discussed in Section 37-4, Classes and Grades of Bovine Carcasses.

USDA Meat Grades. Table 37-1 summarizes the classes and grades of meat for which the USDA has issued official grade standards. (These official grade standards are listed as references at the end of this chapter.)

37-4. CLASSES AND GRADES OF BOVINE CARCASSES

Classification. Carcasses and cuts produced from animals of the bovine species are grouped into three classes based essentially on the evidences of maturity present in the carcass. These three classes—from youngest to oldest

TABLE 37-1. | *Classes and grades of meat.*

Class and Subclass	Grades
Beef	
Steer, heifer, and cow*	Quality Grades—Prime, Choice, Good, Standard, Commercial, Utility, Cutter, and Canner
	Yield Grades—1, 2, 3, 4, 5
Bull	Quality Grades—Choice, Good, Commercial, Utility, Cutter, and Canner
	Yield Grades—same as steer, heifer, and cow
Stag	Quality Grades—same as bull
	Yield Grades—same as steer, heifer, and cow
Calf	Prime, Choice, Good, Standard, Utility, Cull
Veal	Same as Calf
Lamb	Prime, Choice, Good, Utility, Cull
Yearling mutton	Same as Lamb
Mutton	Choice, Good, Utility, Cull
Pork carcasses	U.S. No. 1, U.S. No. 2, U.S. No. 3, Medium, Cull

* Cow carcasses not eligible for the Prime grade.

—are veal, calf, and beef. The primary carcass characteristics used to differentiate between these classes include the color and texture of the lean, the character of the fat, the size, shape, and color of the bones, and the degree of ossification of the cartilages. Although typical carcasses of each class have distinctive characteristics, carcasses near the borderline between two classes seldom show an equal development of each characteristic. Hence, there may be some overlapping of these characteristics between classes, and the final determination must represent a composite evaluation of all characteristics.

The color of lean is the most important characteristic used in differentiating veal from calf. The lean of typical veal carcasses is grayish-pink and very smooth and velvety. Such carcasses also have slightly soft, pliable, fat and round, red rib bones. Most veal carcasses weigh less than 150 lb., but some, particularly in the higher grades, exceed that weight. Most veal carcasses are produced from vealers less than three months old.

The lean of typical calf carcasses is grayish-red and is usually somewhat firmer than that of veal. The fat is drier and flakier and the rib bones are flatter and lack some of the redness characteristic of veal carcasses. Calf carcasses usually weigh from 150 to 275 lb. and rarely exceed 350 lb. even in the higher grades. Most calf carcasses are from calves three to eight months old.

All carcasses with evidences of more advanced maturity than those in the calf class are considered as beef.

In beef carcasses the sex condition of the animal at time of slaughter is an important economic consideration and affects the trade use of the beef. Therefore, there are five subclasses of beef—steers, heifers, cows, bulls, and stags.

Steer carcasses are identified by the rough, rather irregular fat in the region of the cod, the small pelvic cavity, the small "pizzle eye," the curved aitchbone, and the small area of lean posterior to the aitchbone.

Heifer carcasses are identified by the smooth udder fat, by the slightly larger pelvic cavity and straighter aitchbone than those of steers, and by the much larger area of lean posterior to the aitchbone.

Cow carcasses are characterized by a large pelvic cavity and a nearly straight aitchbone. The udder is usually removed, but if a cow is not lactating at the time of slaughter, the udder may be left on the carcass. Cow carcasses usually have at least slightly prominent hips, and since most cows are rather advanced in age when marketed, the bones and cartilages are usually hard and white.

Bull carcasses are identified by their disproportionately heavy muscling in the round, the heavy, crested neck, and the large, prominent "pizzle." The cut surface of the meat is usually dark and coarse. Stag carcasses exhibit characteristics somewhat intermediate between those of steers and bulls.

In the federal grading system, steer, heifer and cow carcasses are graded on the same standards and the grade stamp does not include an identity of the sex condition. In trading, however, steer beef of a given weight and grade usually sells for a higher price than does heifer or cow beef. In the past, there was rather widespread belief that steer beef was more palatable than heifer beef of the same grade, but this has not been supported by research. At the same quality grade and weight, however, steer carcasses usually have less waste fat than heifer or cow carcasses. The resulting higher average yield of trimmed retail cuts or "cutability" of steers is one reason for their continued trade preference.

Bull and stag carcasses are graded on separate standards and the word "bull" or "stag" is included with the grade when such carcasses are graded by a federal grader. Most bulls marketed are rather advanced in age and usually have very little fat and quite dark and coarse lean. Bull beef is generally considered inferior to beef of the other subclasses for sale as fresh or "block" beef through retail outlets. However, it is very desirable for use in frankfurters, bologna, and similar sausage products. Because of the increasing recognition of the importance of "cutability" as a factor affecting the value of beef carcasses and because bull carcasses usually will have considerably less fat and therefore a higher degree of cutability than either steers or heifers of the same weight and age, there is increasing interest in the production of young bulls. A limited amount of research indicates that beef from young bulls is not materially inferior in palatability to steer beef with the same indications of quality.

In veal and calf carcasses the sex condition does not materially affect the quality of the flesh. Therefore, all sexes are graded on the same standards.

Quality Grades. Although the federal standards for beef, calf, and veal are contained in three separate standards, it is intended that they be considered a continuous series. Therefore, the grade of a carcass that is near the borderline in maturity between two of these groups would be essentially the same, regardless of the class under which it was graded.

The quality grade standards for beef describe the quality requirements in terms of firmness and marbling of the rib-eye muscle for different degrees of maturity. The Prime, Choice, Good, and Standard grades are restricted to carcasses from young cattle and involve two maturity groups (Fig. 37-1). Although no references are made in the standards to chronological age, the maturity groups described are designed to correspond, in general, to certain chronological ages. In the Prime and Choice grades, the "younger" group is intended to include beef from cattle about 9 to 30 months of age. The "older" group for these same grades is designed to include beef from animals about 30 to 42 months old. In the Good and Standard grades the "younger" group includes beef from cattle about 9 to 30 months old, and the "older" group includes those from approximately 30 to 48 months of

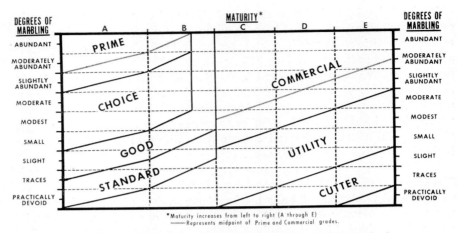

Fig. 37-1. Marbling-maturity chart. Determination of quality grade from degree of marbling and maturity.

age. The Commercial grade is restricted to carcasses older than those eligible for Good and Standard and includes three general maturity groupings. Carcasses within the full range of maturity may qualify for the Utility, Cutter, and Canner grades; in these grades, therefore, five maturity groups are recognized. For each grade, the degree of marbling and firmness of lean are specified for each maturity group. Since marbling and maturity are the most important factors used in the determination of quality, Figure 37-1 shows how these two factors are combined into a quality evaluation, provided the firmness of lean is proportionately developed with the marbling.

The standards for each quality grade also give detailed descriptions of the conformation requirements. By evaluating the conformation of the various parts, an overall evaluation of the conformation of the carcass can be made.

Balancing Quality and Conformation Grade Factors. To arrive at the final grade, the quality and conformation evaluations must be combined. The relative importance of each of these factors varies somewhat with grade. For example, in the Prime, Choice, and Commercial grades, a superior

Figs. 37-2 and 37-3. U.S. Prime beef carcass. This young carcass, as indicated by the red chine bones, the cartilaginous "buttons" on the thoracic vertebrae, and the distinct separation of the sacral vertebrae, tends to be blocky and compact and is thickly fleshed and plump throughout, thus slightly exceeding minimum Prime grade conformation. The slightly abundant degree of marbling exhibited by the ribeye qualifies this carcass for the lower limits of the Prime grade.

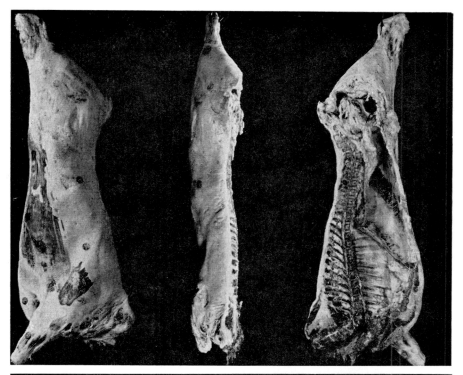

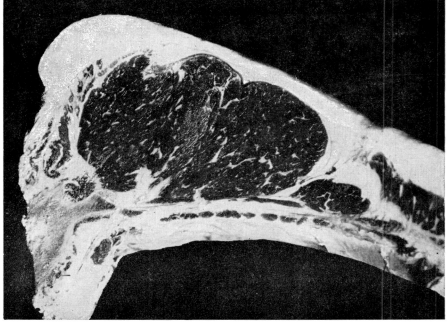

development of conformation is not permitted to compensate for a deficient development of quality. In the other grades, this type of compensation is permitted but is limited to one-third of a grade of deficient quality. This means that in the Prime, Choice, and Commercial grades, carcasses must have the quality requirements specified regardless of how superior their conformation may be, whereas, in the other grades, carcasses with conformation at least one-third of a grade superior to that specified for the particular grade may have quality equivalent to the upper one-third of the next lower grade. In all grades, however, a superior development of quality is permitted to compensate for an inferior development of conformation on an equal basis, and to an unlimited extent through the upper limit of quality. For example, a carcass with average Choice quality would qualify for the Choice grade with only average Good conformation. By the same token, a carcass with low Prime quality and low Good conformation would also qualify for Choice.

The underlying theory behind these differences in rate and extent of compensation between different developments of conformation and quality is based on the use that is made of the different grades. Since the eating quality of the lean is the primary consideration in grading beef, the standards for all grades permit a superior development of quality to compensate, without limit, for a deficient development of conformation. The reverse type of conformation, that is, a superior development of conformation for a deficient development of quality, is not permitted in the Prime, Choice, or Commercial (the highest grade of mature beef) grades because beef of these grades is utilized largely by hotels, restaurants, and individual consumers whose major interest is the quality of the lean. For such users, any extra increase in the ratio of edible to inedible portions that might result from a superior development of conformation would not compensate for a lowering of the eating quality of the lean. On the other hand, to consumer-buyers who normally use the lower grades of beef, the ratio of edible to inedible portions is important enough that in these grades a superior development of conformation is permitted to compensate for a deficient development of quality but only to the extent of one-third of a grade.

Figures 37-2 through 37-5 illustrate Prime and Standard grade carcasses. Their legends briefly describe the characteristics of these carcasses in relation to the official standards.

Yield Grades. In the early 1950's, the USDA recognized the need for a change in the beef grade standards that would precisely distinguish carcasses within the same quality grade by differences that existed in their yields of usable meat. During the next ten years, studies involving measurements, evaluations, and cutting data on hundreds of carcasses served as the basis for the development of such standards. These studies also indicated that variations in yields of usable meat among carcasses probably have a greater effect on value than do variations in the quality of their lean.

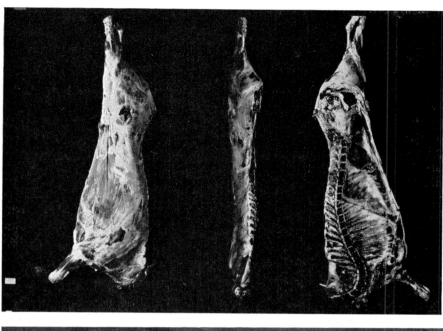

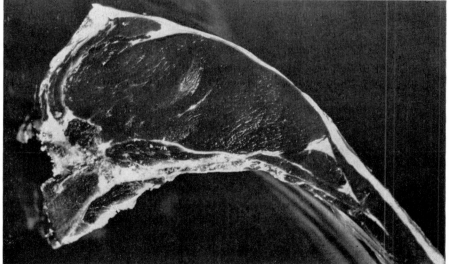

Figs. 37-4 and 37-5. U.S. Standard beef carcass. This young carcass, as indicated by the red chine bones and the cartilaginous "buttons" on the thoracic vertabrae, is rangy and angular and tends to be slightly thin fleshed, thus qualifying for average Standard grade conformation. Because the ribeye is practically devoid of marbling, this carcass is restricted to the lower third of the Standard grade.

The adoption of yield grade standards on June 1, 1965, was perhaps the most significant change made in the grading of beef since its inception in 1927. This is because (1) the yield grades are based largely on factors that can be measured objectively, (2) they were formulated on more sound research than any previous change, and (3) they identify important differences in value among carcasses of the same quality grade, differences that had previously received only token recognition in trading. Price differentials based on yield grades could provide the entire industry with the much needed guide and incentive to increase the production of truly meat-type cattle—those that combine thick muscling with a high quality of lean and a minimum of excess fat.

Specifically, these grades identify carcasses for differences in "cutability" or yield of boneless, closely-trimmed retail cuts from the round, loin, rib, and chuck. In a more general sense, however, they identify differences in yields of all retail cuts from a carcass (Fig. 37-8). There are five yield grades, numbered from one to five. Carcasses qualifying for yield grade one have the highest degree of cutability and those qualifying for yield grade five have the lowest. Among steer, heifer, and cow carcasses each yield grade reflects a specific range in yields of retail cuts regardless of sex, weight, quality grade, or other consideration (Fig. 37-9).

A limited amount of research indicates that bull carcasses of a specific yield grade will have a higher degree of cutability than a steer, heifer, or cow carcass with the same development of yield grade factors. The same is probably true for stag carcasses.

The yield grade of a carcass is determined by considering four characteristics: (1) the amount of external fat, (2) the amount of kidney, pelvic, and heart fat, (3) the area of the ribeye muscle, and (4) the carcass weight.

The amount of external fat on a carcass is evaluated in terms of the thickness of fat over the ribeye muscle at a point three-fourths of the length of the cross section of this muscle from its chine bone end, where this

---

Figs. 37-6 and 37-7. U.S. Commercial beef carcass. This hard-boned carcass with cartilages on the ends of the chine bones, which are completely ossified and barely visible, has rather typical, minimum Commercial grade conformation with its slightly thick but slightly concave round, its prominent hips, sunken loin, slightly thin chuck, and rough, irregular contour. However, despite the apparent rather firm, high quality exterior fat covering, the interior fats—feathering, overflow fat, and fat streaking in the inside flank muscles—are very scanty and insufficient to indicate that the rib eye will have the required moderate degree of marbling to qualify it for the Commercial grade. Therefore, the carcass is eligible only for the Utility grade before ribbing. After ribbing, however, the rib eye (Fig. 37-7) exhibits a moderate amount of marbling and the carcass thereby qualifies for the lower limits of the Commercial grade.

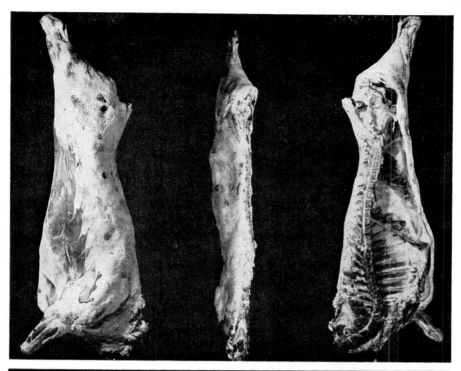

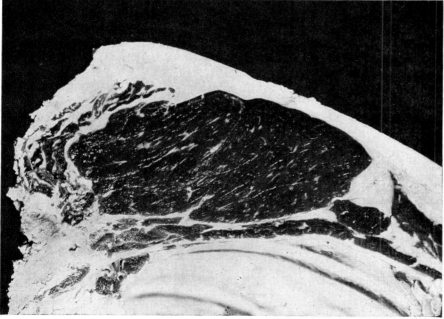

muscle is exposed by ribbing. This measurement may be adjusted, as necessary, to reflect unusual amounts of fat on other parts of the carcass. The quantity of external fat is the most important factor affecting cutability. With the other three yield factors remaining constant, as the external fat increases, the cutability decreases; each four-tenths of an inch change in fat thickness over the ribeye changes the yield grade by one full grade.

The amount of kidney, pelvic, and heart fat considered in determining the yield grade is expressed as a percent of the carcass weight. It is evaluated subjectively. As the amount of these fats increases, the cutability decreases; a change of 5 percent of the carcass weight in these fats is required to change the yield grade by one full grade.

The area of the ribeye muscle is determined where this muscle is exposed by ribbing between the twelfth and thirteenth ribs. In normal grading operations, the area usually is estimated subjectively; however, it may be measured. When making ribeye measurements, federal graders use transparent grids calibrated in tenths of a square inch. An increase in the area

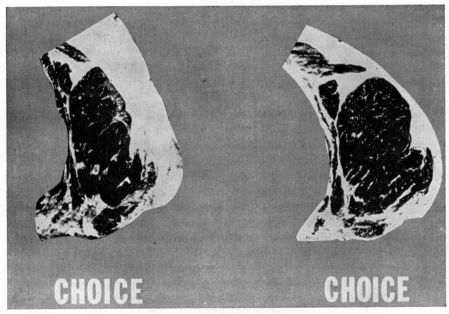

Fig. 37-8. Variation in Cutability. These two ribs illustrate wide variation in factors determining yield grades. The rib on the right has only .4 inch of fat over the ribeye and a 13 square inch ribeye and was produced from a Choice 2 carcass. The rib on the left has about 1.2 inches of fat over the ribeye, a 9.5 square inch ribeye and was produced from a Choice 5 carcass. The difference in value between carcasses represented by these ribs would be about $10.00 per hundredweight.

PERCENT OF BONELESS RETAIL CUTS

FROM ROUND,LOIN,RIB AND CHUCK

YIELD GRADES

1	2	3	4	5
52.4 AND ABOVE	52.3 TO 50.1	50.0 TO 47.8	47.7 TO 45.5	45.4 AND BELOW

———— PERCENTAGE ————

Fig. 37-9. Yields of closely trimmed, boneless retail beef cuts from the round, loin, rib, and chuck associated with each yield grade. These yields are based on a specific method of cutting and trimming used by the Livestock Division, USDA, and illustrate the style of cutting used by many retailers making boneless cuts.

of ribeye increases the cutability—a change of 3 square inches in area of ribeye changes the yield grade by approximately one full grade.

The hot carcass weight is the weight used in determining the yield grade. As carcass weight increases, the cutability decreases slightly—a change of 250 lb. in hot carcass weight is required to change the yield grade by approximately one full grade.

The yield grade standards actually are based on a mathematical equation derived from research involving the very detailed measurement and cutting of a large number of carcasses (Murphey et al., 1960). That equation is as follows:

Yield grade = 2.50 + (2.50 × adjusted fat thickness, inches)
+ (0.20 × percent kidney, pelvic, and heart
fat) + (0.0038 × hot carcass weight, pounds)
− (0.32 × area of ribeye, square inches)

In addition to the equation, the yield grade standards for each grade also describe the development of the yield grade factors for two weights of carcasses—500 and 800 lb. These descriptions are not specific requirements but are included as general illustrations of the development of the yield grade factors in carcasses of these weights. An adaptation of the yield grade equation to provide a simplified, short-cut method of determining yield grade also has been developed by the USDA. Copies of this are available upon request.

Figure 37-10. Yield grade stamp.   **Fig. 37-11.** Quality grade stamp.

Figure 37-8 illustrates Choice grade ribs from carcasses of yield grades two and five. Such carcasses have a difference in value to a retailer of about $10.00 per hundredweight when the average price of Choice grade carcasses is about $40.00 per hundredweight. A further discussion of yield grades for beef will be found in Section 37-7, Developing Improved Grade Standards.

Under the USDA grades, beef carcasses may now be identified for quality grade, yield grade, or both, at the option of the packer. The stamps used to identify beef for these grades are illustrated in Figs. 37-10 and 37-11. The quality grade is applied in a ribbon-like manner so that the grade name appears on most retail cuts. The yield grade is applied once on each quarter.

37-5. CLASSES AND GRADES OF OVINE CARCASSES

Classification. Carcasses and cuts produced from animals of the ovine species are grouped into three classes, essentially on the basis of evidences of maturity present in the carcass. These three classes—from youngest to oldest—are lamb, yearling mutton, and mutton.

Lamb carcasses always have "break joints" on their front shanks; these are moist and fairly red and show well-defined ridges. The rib bones are moderately flat and slightly wide; the lean is light red and has a fine texture.

Yearling mutton may have either break joints or spool joints on their front shanks; the rib bones tend to be flat and moderately wide and the lean is slightly dark red and is slightly coarse in texture.

Mutton carcasses always have spool joints on the front legs and have wide, flat rib bones. The lean is dark red in color and coarse in texture. Figure 37-12 illustrates a break joint and a spool joint.

In the United States, there is a very strong preference for lamb over yearling mutton or mutton because of the progressively stronger flavor that develops in ovine carcasses as they advance in maturity. Price differentials between these classes of carcasses are substantial. For this reason, a high percentage of ovines are marketed as lambs. Mutton carcasses are largely from ewes that are culled from breeding flocks. Mutton is used largely in manufactured meat products.

Grading. The grade standards apply to all carcasses without regard to their sex. However, carcasses that show heavy shoulders and thick necks typical of uncastrated males are discounted in grade from less than one-half a grade to two full grades, depending on the extent to which these secondary sexual characteristics are developed. Grade is determined on the same general basis as for beef—by considering variations in conformation and quality.

Except for one important difference, the combination of quality grade and conformation grade into a final grade is much the same as for beef. Lamb and mutton carcasses that have at least midpoint Choice conformation are permitted to have a development of quality equivalent to the minimum of the upper third of the Good grade and still be graded Choice. This compensation is not permitted in Choice beef.

Another difference between the standards for these two species relates to the extent to which a superior development of quality can compensate for a deficiency in conformation. The lamb and mutton standards specify that, regardless of how high a degree of quality a carcass may have, it will not be considered for a given grade if it does not have a development of conformation equivalent to at least that of the next lower grade. For example, a carcass could not be graded Choice if it had less than Good grade conformation. There is no such minimum conformation specified for beef. Also, the lamb and mutton standards provide for a minimum external fat covering for carcasses in the Prime and Choice grade. Figures 37-13 through 37-16 illustrate differences in two grades of lamb carcasses; the accompanying legends describe the characteristics of the carcasses in relation to the official standards.

37-6. CLASSES AND GRADES OF PORK CARCASSES

Classification. Carcasses produced from swine are grouped into five classes —barrows, gilts, sows, boars, and stags—based on their sex condition.

Barrow carcasses have the typical pocket in the split edge of the belly (where the sheath has been removed) and have a small "pizzle eye." Gilt carcasses have a smooth split edge of the belly but do not show any development of mammary tissue. Sow carcasses have a smooth belly edge similar to gilts but show a rather pronounced development of mammary tissue as a result of advanced pregnancy or lactation.

Boar carcasses have a somewhat larger belly pocket than barrows and have a large, coarse "pizzle eye." The shoulders are heavy and the skin and joints are coarse. The lean is dark red and coarse. Stag carcasses, depending on the age at which the stag was castrated, have characteristics somewhat intermediate between barrows and boars.

Barrow and gilt carcasses are considered interchangeable in trading, since they both produce pork of acceptable quality. Barrow and gilt carcasses are graded on the same standards. Sow carcasses are graded on separate stand-

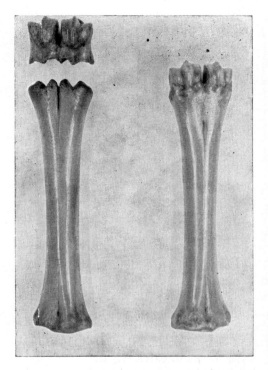

Fig. 37-12. Ovine break joint and spool joint.

ards, but some young, lightweight sows also produce cuts that are considered interchangeable with those from barrows and gilts.

Carcasses from most boars have a strong "sex" odor and, when this is present, such carcasses are not passed for use as food in federally inspected packing plants. There are no Federal grade standards for boar or stag carcasses.

Grading. In general, carcasses in U.S. No. 1, U.S. No. 2, and U.S. No. 3 grades all have an acceptable degree of quality of lean. Differentiation between these grades is based solely on characteristics associated with differences in the yield of lean cuts. This is primarily a function of fatness in relation to weight or length. Thus, a U.S. No. 1 carcass has about the minimum degree of fatness required to produce an acceptable quality of pork. Carcasses of U.S. No. 2 grade are overfinished; U.S. No. 3 carcasses are decidedly overfinished. There is a progressive reduction in the yield of lean cuts between U.S. No. 1, U.S. No. 2, and U.S. No. 3 carcasses. Medium grade carcasses are underfinished, and, although the ratio of lean to fat is higher than in the U.S. No. 1 grade, the quality of the lean is not generally acceptable. Cull grade carcasses are decidedly underfinished and have a correspondingly lower quality of lean.

Objective guides are incorporated in the standards for grades of pork carcasses. Those applicable to barrow and gilt carcasses are presented in Table 37-2. This approach to grading pork carcasses is in contrast to the grading of other meats, in which the standards are presently subjective in nature. Studies of measurement and cutting data for barrow and gilt carcasses have established that the average thickness of backfat in relation to either carcass length or weight is a reliable guide to the yield of cuts and quality of meat.

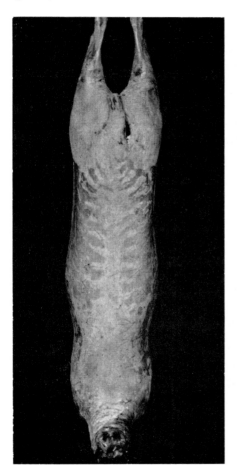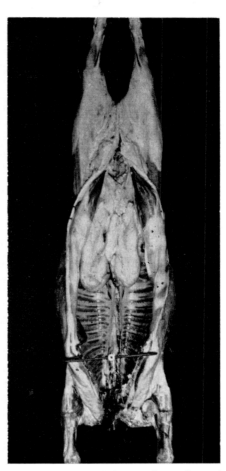

Figs. 37-13 and 37-14. U.S. Prime lamb carcass. The conformation of this young lamb carcass, especially through the back, exceeds the minimum specified for the Prime grade. The moderate amount of feathering between the ribs, the modest streaking of fat on the inside flank muscles, and the moderately thick and full flanks and firm lean also slightly exceed the minimum quality requirements for Prime and qualify the carcass for the lower third of the Prime grade.

TABLE 37-2. | *Weight and measurement guides to grades for barrow and gilt carcasses.*

Carcass Weight or Carcass Length*	Average Backfat Thickness (in inches)† by Grade				
	U.S. No. 1	*U.S. No. 2*	*U.S. No. 3*	*Medium*	*Cull*
Under 120 lb. or under 27 in.	1.2–1.5	1.5–1.8	1.8 or more	0.9–1.2	Less than 0.9
120 to 164 lb. or 27 to 29.9 in.	1.3–1.6	1.6–1.9	1.9 or more	1.0–1.3	Less than 1.0
165 to 209 lb. or 30 to 32.9 in.	1.4–1.7	1.7–2.0	2.0 or more	1.1–1.4	Less than 1.1
210 or more lb. or 33 or more in.	1.5–1.8	1.8–2.1	2.1 or more	1.2–1.5	Less than 1.2

* Either carcass weight or length may be used with backfat thickness as a reliable guide to grade. The table shows the normal length range for given weights. In extreme cases, where the use of length with backfat thickness indicates a different grade than by using weight, final grade is determined subjectively as provided in the standards. Carcass weight is based on a chilled, packer-style carcass. Carcass length is measured from the forward point of the aitch bone to the forward edge of the first rib.

† Average of measurements made opposite the first and last ribs and last lumbar vertebra.

However, other characteristics are also considered in grading to achieve a more accurate evaluation of yields and quality. The use of these other characteristics is restricted to the grading of carcasses which, on the basis of their fatness and their length or weight, are near the borderline between two grades. In no instance may these other characteristics be used to change the grade of a carcass more than one-half a grade from that indicated by its thickness, backfat, and weight or length.

The U.S. No. 1, U.S. No. 2, and U.S. No. 3 grades differ only in their yield of cuts. Therefore, the only factors other than backfat thickness in relation to length or weight used in grading carcasses near these borderlines are quantitative characteristics, such as meatiness, conformation, and fat distribution. On the other hand, the U.S. No. 1, Medium, and Cull grades differ only in their quality. For this reason, the only factors other than backfat thickness in relation to length or weight used in grading carcasses near these borderlines include only quality characteristics such as firmness, quantity and distribution of internal fats, and belly thickness.

Firmness of fat, as related to the degree of finish, is also considered in grading. However, carcasses whose fat is soft or oily, owing to the type of feed consumed, are graded without consideration of this characteristic, but are identified as soft or oily along with the grade. Differences between U.S. No. 1 and U.S. No. 3 pork carcasses are shown in Figs. 37-17 and 37-18. The legend accompanying each picture indicates the average thickness of back‑ fat and length of body.

Sow carcasses are graded on exactly the same basis as described for barrow and gilt carcasses, except that the objective guides to grades are based entirely on thickness of backfat without reference to length or weight. This relationship was established through measurement and cutting data on sow carcasses. Table 37-3 shows the range in average thickness of backfat for each of the grades of sow carcasses.

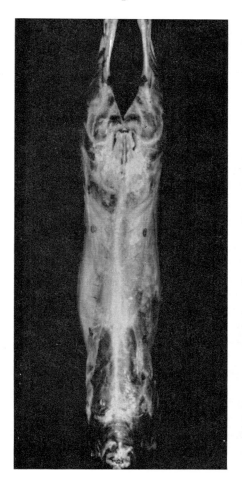 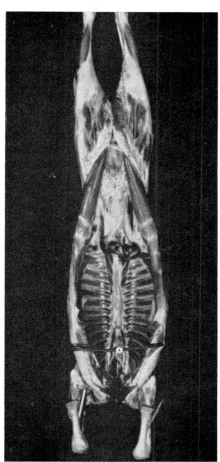

Figs. 37-15 and 37-16. U.S. Utility lamb carcass. The conformation of this lamb carcass exceeds the requirements for the midpoint of the Utility grade. However, its development of quality—the lack of feathering and fat streakings, and the very soft, watery lean—is equivalent to only the Cull grade. The superior development of conformation is sufficient to qualify this carcass for the lower third of the Utility grade.

Grade	Average Backfat Thickness*
U.S. No. 1	1.5–1.9
U.S. No. 2	1.9–2.3
U.S. No. 3	2.3 or more
Medium	1.1–1.5
Cull	Less than 1.1

* Average of three measurements, skin included; made opposite first and lastribs and last lumbar vertebra.

37-7. DEVELOPING IMPROVED GRADE STANDARDS

The Administrative Procedure Act requires that changes in federal grade standards be made only after the changes are published as a proposal in the Federal Register and interested persons given an opportunity to express their views on the proposed changes. The valid points thus expressed are given careful consideration in arriving at a decision relative to adopting, rejecting, or modifying the proposal.

The original federal standards for grades of meat were based largely on trade practices and beliefs. Limited research results were available at that time on the factors responsible for differences in carcass merit. The USDA has long recognized the desirability of having grade standards based on sound research and has continually supported and encouraged research in this area. In the past few years, considerable research has been conducted that has been useful in developing grade standards on a more scientific and objective basis.

Much of this research has been concerned with beef. In one general area of research on beef carcass merit, studies by the USDA and by various colleges and universities have indicated that each of the quality grades of beef includes carcasses with a very wide range in cutability or yield of trimmed major retail cuts and that these differences have a very important effect on carcass value (Pierce et al., 1956, and Murphey et al., 1960). They also indicated that these differences can be estimated with a high degree of accuracy by considering only four carcass characteristics—the thickness of fat over the ribeye, the quantity of kidney and pelvic fat, tht area of the ribeye muscle, and the carcass weight. Pilot studies involving application of a cutability grading system also demonstrated that this could be done with a very satisfactory degree of speed and accuracy.

Based on these findings, in April, 1962 the USDA proposed a "dual-grading" system for beef carcasses. This system provided that carcasses would be separately identified for two different factors, (1) cutability and

(2) quality. The cutability grades had a numerical designation from one to five, with one representing the highest yield of cuts. The same eight grade names currently used in beef grading were applied to the quality grades. In this proposal, the quality grade did not include consideration of conformation since there is no scientific evidence that indicates that variations in conformation alone have any effect on differences in quality (Tyler et al., 1964). This proposed system of grading was used on a trial basis for one year. Although the system proved quite successful in some areas where a

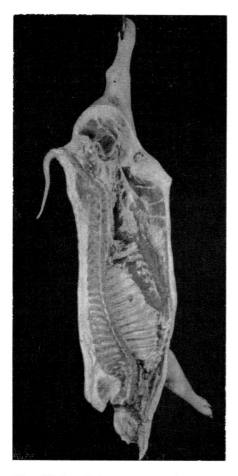 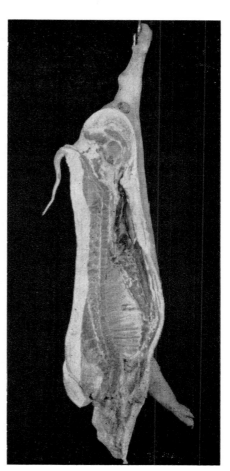

Fig. 37-17. U.S. No. 1 pork carcass. Average thickness of backfat, 1.4 inches; length of body, 29.8 inches; estimated yield of ham, loin, picnic, and Boston butt, 52.6%.

Fig. 37-18. U.S. No. 3 pork carcass. Average thickness of backfat, 2.1 inches; length of body, 27.5 inches; estimated yield of ham, loin, picnic, and Boston butt, 45.1%.

sufficient volume of beef was graded to develop a competitive price pattern between the various grades, the proposed dual-grading system was not adopted. The comments, however, provided the basis for a second proposal containing the cutability concept. In this proposal of September 18, 1963, yield grades were proposed for optional use, either separately or in addition to the conventional beef grade standards. This approach had both strong support and strong opposition from practically all segments of the industry. After full consideration of all relative material the yield grades were adopted concurrently with a revision of the quality grades, effective June 1, 1965.

Similar studies of cutability have been made with lamb carcasses (Hoke et al., 1961). These studies are being further expanded, refined, and tested.

Much research also has been conducted on the subject of beef quality. However, identifying and determining the relative influence of various factors on palatability has proven to be a very complex problem. A major study conducted by the American Meat Institute Foundation under contract with USDA (Doty et al., 1961) showed measurable differences in palatability of the grades tested and revealed significant relationships between certain basic factors such as marbling and palatability. Many other studies have been aimed at evaluating marbling differences and their influence on palatability. These studies have provided conflicting results, but, for the most part, they have shown a positive correlation between marbling and tenderness, juiciness, and general palatability. Although these relationships have often been disappointingly low, the degree of marbling, in relation to the degree of maturity, is still considered the most important market criterion for evaluating quality; a more reliable approach has not been identified by research to date. Research has substantiated the general relationship of maturity to palatability. With other factors remaining constant, palatability decreases with increasing maturity.

In recent years, research results have become available that indicate that in beef from animals up to about 30 months of age, the standards required too rapid an increase in marbling to compensate for the increase in maturity. Based on this research, in July 1964, the USDA proposed a change in the quality grade standards to implement these findings. This proposal evoked both strong support and strong opposition from various segments of the industry but, as indicated above, it was adopted concurrently with the yield grades on June 1, 1965.

The foregoing emphasizes the point that, regardless of the merit of a change in standards, it is most difficult to establish the understanding and acceptance desired to make a change in standards completely effective. Because most changes have different immediate and long-range effects on different segments of the trade, it is also almost inevitable that any meaningful change in standards that may be proposed will become a controversial issue.

Therefore, although there is a real need to emphasize research efforts to establish a more objective and factual basis for grade standards, the adoption and application of such information into a grading system will undoubtedly continue to lag well behind the basic research findings.

In order to be of maximum use to the industry, grades must be understood and used in some manner. Consequently, there is a very practical limitation on the extent to which grade standards can be more advanced than the industry they serve.

REFERENCES AND SELECTED READINGS

References marked with an asterisk are of general interest.

Doty, D. M., and J. C. Pierce, 1961. *Beef Muscle Characteristics as Related to Carcass Grade, Carcass Weight, and Degree of Aging.* USDA Technical Bulletin No. 1231.

Hoke, K. E., C. E. Murphey, J. C. Pierce, and M. M. Van Zandt, 1961. *Factors Affecting Yield of Cuts in Lamb Carcasses.* Proceedings of the 14th Reciprocal Meat Conference.

Murphey, C. E., D. K. Hallett, J. C. Pierce, and W. E. Tyler, 1960. *Estimating Yields of Retail Cuts from Beef Carcasses.* J. Animal Sci. 19:1240 (abstract).

Pierce, J. C., C. E. Murphey, C. L. Strong, and M. M. Van Zandt, 1956. *Some Factors Influencing Yields of Wholesale and Retail Cuts from Beef Carcasses.* J. Animal Sci. 15:1267 (abstract).

Tyler, W. E., B. C. Breidenstein, D. K. Hallett, K. E. Hoke, and C. E. Murphey,

1964. *Effects of Variation in Conformation on Cutability and Palatability on Beef.* J. Animal Sci. 23:864 (abstract).

* USDA, 1926. *Official United States Standards for Grades of Carcass Beef.* Service and Regulatory Announcement No. 99. Reprinted June 1965.

* ———, 1928. *Official United States Standards for Grades of Veal and Calf Carcasses.* Service and Regulatory Announcement No. 114. Reprinted September 1956.

* ———, 1931. *Official United States Standards for Grades of Lamb, Yearling Mutton, and Mutton Carcasses.* Service and Regulatory Announcement No. 123. Reprinted April 1960.

* ———, 1952. *Official United States Standards for Grades of Pork Carcasses (Barrow and Gilt; Sow).* Service and Regulatory Announcement No. 171. Reprinted April 1958.

Chapter 38

Classification and Grading of Livestock

38-1. INTRODUCTION

As pointed out in Chapter 37, during the early growth of our country the main agricultural production areas moved progressively farther away from the large urban areas. Concurrent with this relocation, large, highly competitive central livestock markets and packing plants were developed adjacent to these areas. This was made possible largely by the improvement in transportation facilities. However, it was also materially facilitated by the development and widespread use of a nationally recognized and understood system of classes and grades of livestock.

To a somewhat greater extent than has been the case with meat, the livestock industry has adopted and used the system of classification and grading of livestock developed and published by the U.S. Department of Agriculture. Although there has never been an official federal livestock grading program, the federal standards are used by the USDA as the basis for its nationwide livestock market news service. Some states, however, conduct official livestock grading programs in which state employees officially grade livestock at the time it is offered for sale—usually at auctions. In most such programs, the federal standards are used as the basis for the grading.

The trend toward increased selling of livestock through auctions and by direct sales at the ranch, farm, or feedlot means that some producers are assuming the responsibility for marketing their livestock, rather than delegating it to commission agents as they did when livestock was sold on a central market. Because of this trend, it becomes increasingly important that livestock producers have as much understanding as possible of the market desirability of their livestock—primarily their class and grade. Federal class and grade standards have been developed for slaughter animals of all species; standards have also been developed for feeder animals of some species.

The class and grade standards for slaughter livestock are based on the class and grade of carcass they will produce. The class and grade standards of feeder animals are based on the class and grade of slaughter animals they will produce under normal feeding and management practices. This closely interrelated system of standards thus provides the means of communication whereby consumers' wants and preferences for meat can be relayed back through the entire marketing channel to the producers. This enables producers to plan their production and marketing program more intelligently.

Class, as used in its broader sense, refers to the segregation of animals into groups on the basis of their commercial use. Thus, there are two general classes of livestock—slaughter and feeder. Subclasses of livestock are usually based on their sex condition. Thus, steers, heifers, cows, bulls, stags are considered subclasses of cattle.

In formulating standards for these various classes of livestock, some of the subclasses may be grouped together. For instance, grade standards for slaughter steers, heifers, and cows are combined in the same manner as are the standards for grades of slaughter barrows and gilts.

38-2. GRADES OF SLAUGHTER LIVESTOCK

Basis for Grade. Since the grade of a slaughter animal is the estimated grade of carcass it will produce, the characteristics used in grading are the same as used in grading carcasses—(1) conformation and (2) quality of meat.

In evaluating conformation, the primary emphasis is placed on estimating the ratio of lean meat or muscle to bone and the ratio of the more demanded to the less preferred parts. Animals that have thick, plump muscles in relation to their skeletal structure are considered to have good conformation, particularly if this muscling is well developed in the region of the most valuable parts. Since the actual thickness and plumpness of the animal is influenced greatly by the fatness of the animal, plumpness and thickness of muscling can best be appraised from those parts of the animal on which relatively little fat is deposited. Thus, the round in cattle, the ham in hogs, and the leg in sheep are particularly useful parts to observe closely in evaluating conformation. Other parts of the carcass should not be ignored, but their evaluation—particularly in highly finished animals—should be made with appropriate adjustments for the fat that may be present. The loin and back are two parts of the animal over which fat is deposited at a relatively fast rate; in rapidly fattening animals, the contour of these parts changes very rapidly. Since an animal's carcass is made up only of muscle, fat, and bone, two animals that are of the same weight and scale or skeletal structure but that differ considerably in fatness will necessarily show a difference in muscular development. The fatter of the two will normally have the least thickness of muscling and thus the poorest conformation.

An accurate appraisal of a live animal for the quality of meat that will be contained in its carcass is usually more difficult to make than is an appraisal of its conformation. This results from the fact that the criteria that are useful for estimating marbling and other quality characteristics of the lean are, at best, only moderately accurate indicators of these characteristics. The amount of finish carried by an animal is, of course, the best single criterion of the quality of the meat in its carcass. However, the distribution and firmness of the finish are probably equally as important as the amount. Maturity is also a very important consideration in grading slaughter livestock. Among the criteria that are useful in evaluating maturity are these: (1) the size of the animal, (2) the width of the muzzle (this reflects differences in the number of small, temporary teeth and of the larger, permanent teeth), (3) the length of the tail (in vealers and calves the hair on the tail is relatively short, since it has not had time to grow long), (4) the general symmetry and smoothness of outline (as animals advance in maturity they frequently develop more prominent hips and become irregular in contour) and, (5) in horned cattle, the size of the horns. As animals advance in maturity, they are required to have greater amounts of finish to qualify for a given grade.

USDA Grades for Slaughter Livestock. The following tabulation summarizes the applicable grades for the various kinds of slaughter livestock for which the USDA has issued official standards. (The official United States standards for grades of livestock are listed as references at the end of this chapter.)

38-3. GRADING SLAUGHTER CATTLE

The various sex conditions of slaughter cattle are defined as follows: *bull* —an uncastrated male; *steer*—a male castrated when young, prior to development of the secondary sexual characteristics associated with a bull; *stag*—a male castrated after it has developed, or has begun to develop, the secondary sexual characteristics associated with a bull; *cow*—a female that has developed, through reproduction or with age, the relatively prominent hips, the large middle, and other physical characteristics typical of a mature female; *heifer*—an immature female that has not developed the physical characteristics typical of a cow.

The determination of the quality grade of the live animal—that is, the

TABLE | *Classes and grades of slaughter animals.*
38-1. |

Slaughter Class and Subclass	Grade
Cattle	
Steer, heifer, and cow	Quality Grades—Prime, Choice, Good, Standard, Commercial, Utility, Cutter, and Canner
	Yield Grades—1, 2, 3, 4, 5
Bull	Quality Grades—Choice, Good, Commercial, Utility, Cutter, and Canner
	Yield Grades—same as steer, heifer, and cow
Stag	Quality Grades—same as bull
	Yield Grades—same as steer, heifer, and cow
Calf	Prime, Choice, Good, Standard, Utility, and Cull
Veal	Same as calf
Lamb	Prime, Choice, Good, Utility, and Cull
Yearling mutton	Same as lamb
Mutton	Choice, Good, Utility, and Cull
Hogs	
Barrows and gilts	U.S. No. 1, U.S. No. 2, U.S. No. 3, Medium, and Cull
Sows	Same as barrows and gilts

quality grade of its carcass—requires well-regulated judgment. Each animal that is graded is likely to present a somewhat different combination of grade-determining characteristics. It is not unusual for an animal of one grade to have some characteristics that are associated with animals of one or more other grades. A composite evaluation of the total inherent characteristics of the animal is essential for accuracy in determining grade.

As previously indicated, maturity is one of the important factors considered in grading slaughter cattle; there are maximum maturity limitations for each of the four higher quality grades. Since evidences of maturity in beef carcasses vary among animals of the same approximate age, only general maximum age limitations can be indicated for these grades. These are as follows: Prime, 36 months; Choice, 42 months; Good, 48 months; and Standard, 48 months. The Commercial grade is restricted to steers, heifers, and cows that are over 48 months. There are no age limitations for the Utility, Cutter, and Canner grades.

The general fatness of the animal is a very important consideration in evaluating the expected quality of its carcass. However, distribution of fat is probably equally as important as the total fatness. For instance, a steer with considerably more fat in the cod, flank, and brisket than another steer will frequently have a higher quality of meat, even though the two steers may have a very similar degree of fatness over their loin and back. Animals do differ markedly in the relative amounts of fat they deposit as marbling and in the amounts they deposit over the outside of their carcasses and around their kidneys. This is a characteristic that may be influenced by feeding, but it is probably an inherited trait. The variable distribution of fat is of tremendous economic importance, and the identification of factors responsible for it is worthy of much more research effort than has been directed to it in the past.

Excellent conformation in slaughter cattle is evidenced by a wide, thick back, full square rump, and a deep, wide, full quarter. Fullness and thickness should be especially evident in the portions of the body producing the more desirable cuts—loin, rib, and round. Inferior conformation is evidenced by a decided lack of fullness and thickness, and the various parts are frequently angular, sunken, or even concave. Since conformation relates to proportions of the more valuable and less valuable parts and to the ratio of meat to bone, it cannot be appraised directly from the actual width, depth, and thickness of an animal. Rather, these factors must be appraised in accordance with the size or scale of the animal.

The most logical and accurate way to determine the quality grade of a slaughter steer is to evaluate separately its conformation and its expected quality of carcass in terms of grade. These then can be combined into a final quality grade in the same manner as indicated for beef carcasses in Chapter 37. Unless this procedure is followed, there is a distinct tendency to permit conformation to exert a greater influence on grade than it should. This is particularly true in grading thick muscled, high-quality cattle whose conformation is relatively better developed than their carcass quality. Figure 38-1 illustrates slaughter steers that qualify for the indicated grades.

Vealers and slaughter calves are young bovine animals. They are differentiated primarily on the basis of age and evidences of types of feeding. Typical vealers are less than three months old and have subsisted largely on milk. Since they have consumed little or no roughage, they have the characteristic trimness of middle that is associated with limited paunch development. Calves are usually three to eight months of age, have subsisted partially or entirely on feeds other than milk for a considerable time, and have begun to develop the larger middles and other physical characteristics associated with maturity. Most bovines that are nine months of age or older are considered to be cattle. Differences in sex condition are usually unimportant in vealers and calves; for this reason, the federal grade-standards for all types are combined. The factors considered in grading vealers and slaughter calves, and the emphasis placed on variations in development of conformation and quality, are identical with those for slaughter cattle.

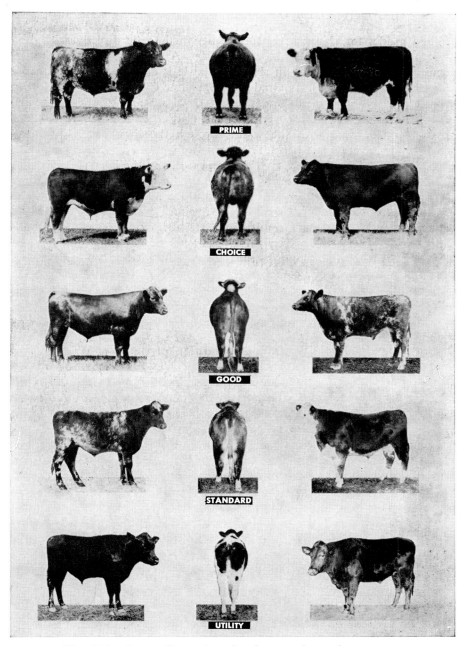

Fig. 38-1. Steers illustrating slaughter cattle grades.

38-4. YIELD GRADES OF SLAUGHTER CATTLE

Yield grades for slaughter cattle are designed to be directly related to corresponding yield grades of beef carcasses. A full discussion of the determination of yield grades in slaughter cattle is contained in the Official United States Standards for Grades of Slaughter Cattle. In general, cattle typical of the higher yield grades (Yield Grades 1 and 2) combine thick muscling with a minimum covering of fat. Cattle typical of Yield Grade 1 are moderately wide with the width through the shoulders and rounds greater than through the back. The top is well rounded with no evidence of flatness. The rounds are deep, thick, and full, and the back and loins are full and thick as are the forearms. These cattle have only a thin covering of fat over the back and rump and have little evidence of fullness in the brisket and cod or udder. Cattle qualifying for Yield Grade 1 will differ widely in quality grade as a result of variations in conformation and indications of quality. Usually only thickly muscled carcasses qualify for Yield Grade 1. Cattle qualifying for the lower yield grades have either less muscling, more finish, or a combination of the two.

38-5. GRADING SLAUGHTER LAMBS, YEARLINGS, AND SHEEP

The various sex conditions and age groups of slaughter ovines are as follows: *ram*—an uncastrated male ovine; *wether*—a male ovine castrated when young, prior to development of the secondary sexual characteristics of a ram; *ewe*—a female ovine; *lamb*—an immature ovine, usually under 14 months of age, that has not cut its first pair of permanent incisor teeth; *yearling*—an ovine, usually between one and two years of age, that has cut its first pair of permanent incisor teeth, but that has not cut the second pair; *sheep*—an ovine, usually more than 24 months of age, that has cut its second pair of permanent incisor teeth.

Since the grade of slaughter ovines is based on the grade of their carcasses, the same principles must apply to grading the animals prior to slaughter as apply to grading carcasses. These are discussed in detail in Chapter 37.

The accurate determination of the grade of a slaughter lamb or sheep requires handling in addition to visual observation. This is necessary because the length and density of the wool is quite variable among individuals, making it almost impossible to appraise the conformation, and the thickness and firmness of finish accurately without handling. Experienced graders may find one quick handling satisfactory. This usually is done by placing one open hand over the back and ribs with the thumb extended just over the backbone. The fingers are held close together and cover the rib section. Applying pressure very lightly, the hand is moved with a slight lateral and forward–backward motion.

If time permits, it is usually desirable to handle forward along the back from the dock to the neck. In doing this, the hand should be open and nearly flat and the fingers should be together. Pressure of the hand should be very light and the motion should be slightly lateral. Both hands may then be used in a similar manner along each side to determine the fleshing over the shoulders, ribs, and hips. Figure 38-2 shows slaughter lambs that qualify for the indicated grades.

38-6. GRADING SLAUGHTER SWINE

The various sex conditions of slaughter swine are defined as follows: *barrow*—a male swine castrated when young, before development of the secondary sexual characteristics associated with a boar; *gilt*—a young female swine that has not produced young or has not reached an advanced stage of pregnancy; *sow*—a mature female swine that has reproduced or has reached an advanced stage of pregnancy; *boar*—an uncastrated male swine; *stag*—a male swine castrated after development of the secondary sexual characteristics of a boar.

Grade standards for slaughter swine have been developed only for barrows, gilts, and sows. Those for barrows and gilts are combined into a single standard, since, when they are marketed, sex condition has not yet exerted an influence on the market acceptability of their carcasses.

The grades of slaughter swine are based on the corresponding grades of carcasses. Therefore, the same principles govern the grading of slaughter swine as govern the grading of pork carcasses. Since thickness of back fat is such an important factor in grading carcasses, this becomes the most important factor in grading slaughter hogs. Swine are much fatter when marketed than are other species of meat animals. Therefore, the variable rate of deposition of fat on various parts of the body is more evident than in other animals, and this can be used to good advantage in grading. As hogs fatten they tend to deposit fat at a faster rate over their back than they do over the lower part of their body or over the lower part of their hams or shoulders. As a result, hogs with a high degree of finish are usually wider over the loin and back than through the underline, and they are wider through the loin than through the center or lower part of the ham. They also tend to taper slightly down their backs from their shoulders to their hams. In many such hogs, the contour of the back from side to side is nearly flat, and there may be a decided break into the sides. The flanks are very deep and full, and the jowls are usually very thick and full.

By contrast, hogs with relatively little fat will appear rather peaked along the top and usually will be wider through the underline than through the back. The hams will be wider than either the shoulders or the loin; the flanks will be thin and the jowls thin and flat. Some of these characteristics

Fig. 38-2. Lambs illustrating the slaughter grades.

are evident in Fig. 38-3, which illustrates the various grades of slaughter barrows and gilts.

38-7. GRADES OF FEEDER LIVESTOCK

Whether a particular animal is considered a feeder or slaughter animal is determined not by his characteristics but, rather, by the use for which he is purchased. If he is purchased for immediate slaughter, he is considered a slaughter animal. If he is purchased for further feeding, he is considered a feeder animal. Although certain combinations of characteristics in animals almost always result in their being purchased either for slaughter or for further feeding, there are also many animals whose characteristics make them suited for feeding under certain economic conditions, whereas under other economic conditions they would be purchased for immediate slaughter. Livestock whose value is almost the same to either slaughterers or feeders are frequently referred to as "two-way" or "warmed-up" animals.

Animals that are in thin condition—particularly those that have a good meat type—will practically always be sold as feeders, since under most economic conditions they will make profitable gains and thereby return a profit to their feeder. Because animals make their most economical gains during the early stages of fattening, it is usually most advantageous to purchase for further feeding only animals that have a limited degree of finish.

The grade of a feeder animal is based upon the grade of slaughter animal it will produce at its "logical slaughter potential" and upon its thriftiness or its likely efficiency of gain. The "logical slaughter potential" is the stage in an animal's development at which its conformation and carcass quality have an equivalent degree of development.

When an animal is fed beyond its "logical slaughter potential" its carcass quality becomes relatively more highly developed than its conformation. This means that when an animal is fed beyond its "logical slaughter potential" it takes a considerably greater increase in quality to make a given increase in the slaughter grade than it does to make a similar increase in slaughter grade prior to the time the "logical slaughter potential" is reached. For example, it is entirely possible for an animal with Good grade conformation to be fed until it has developed Prime grade quality. The slaughter or carcass grade of such an animal would be Choice. However, this would normally require a longer feeding period and would require a more expensive feeding regime, whereas feeding an animal with Choice grade conformation to produce a Choice grade slaughter animal would require feeding to develop only Choice—instead of Prime—grade quality. Thus the feeder grade of the first animal is Good, and the feeder grade of the second is Choice.

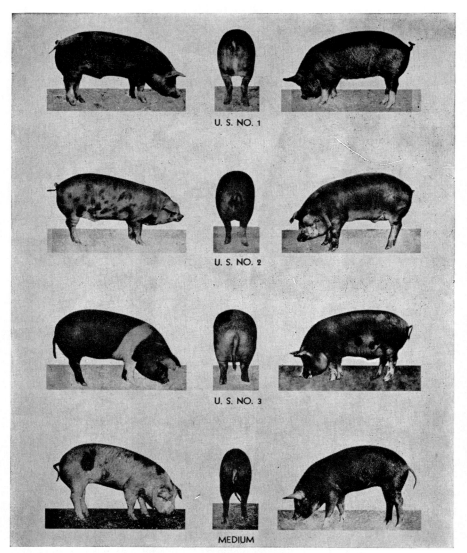

U. S. NO. 1

U. S. NO. 2

U. S. NO. 3

MEDIUM

Fig. 38-3. Animals illustrating the grades of slaughter barrows or gilts.

Conformation in feeder cattle is evaluated in terms of the conformation of the slaughter animal at its "logical slaughter potential." Since most feeder animals are in thin condition, they do not have as thick or as plump an appearance as they do at their "logical slaughter potential." For this reason, the conformation grade descriptions of a particular grade of feeder

PRIME

CHOICE

GOOD

STANDARD

UTILITY

Fig. 38-4. Steers illustrating the feeder cattle grades.

animal require less thickness, fullness, and so forth than the conformation descriptions of the corresponding grade of slaughter animal. This is especially true in the higher grades of feeder animals.

Thriftiness, or the likely efficiency of gain by an animal, is considered in grading feeder animals, but it is not used to raise the feeder grade beyond the animal's grade at its "logical slaughter potential." For example, a feeder animal with a "logical slaughter potential" of Good, but that gave the appearance of being unusually thrifty, would not be graded Choice. On the other hand, a feeder animal with a "logical slaughter potential" of Choice might be graded Good if it appeared to be unthrifty and likely to have a low efficiency of gain. Figure 38-4 illustrates the standards for grades of feeder steers. Standards for grades of feeder pigs were adopted in 1966. These provide for four grades—U.S. No. 1, U.S. No. 2, U.S. No. 3, and Medium.

REFERENCES AND SELECTED READINGS

USDA, 1928. *Official United States Standards for Grades of Slaughter Cattle.* Service and Regulatory Announcement No. 112. Reprinted April 1966.

——, 1928. *Official United States Standards for Grades of Vealers and Slaughter Calves.* Service and Regulatory Announcement No. 113. Reprinted May 1957.

——, 1964. *Official United States Standards for Grades of Feeder Cattle.* Service and Regulatory Announcements No. 183. Issued March 1965.

——, 1951. *Official United States Standards for Grades of Slaughter Lambs, Yearlings and Sheep.* Service and Regulatory Announcement No. 168. Reprinted November 1960.

——, 1952. *Official United States Standards for Grades of Slaughter Swine (Barrows and Gifts; Sows).* Service and Regulatory Announcement No. 172. Reprinted April 1958.

——, 1966. *Official United States Standards for Grades of Feeder Pigs.* Service and Regulatory Announcements No. 189. Issued February 1966.

Chapter 39

Marketing of Livestock and Meats

Livestock is a national commodity. It is produced and marketed in every state of the Union. The distribution of livestock and meats takes place on a national scale; cattle raised on the plains of Texas may be fattened in an Iowa feedlot, slaughtered in Chicago, and the steaks therefrom served in a New York restaurant. Spring lambs produced in California may be slaughtered kosher style in Brooklyn and consumed in Boston. Meat and lard from hogs raised in Illinois may be sold in Philadelphia, San Francisco, and New Orleans. Livestock marketing must therefore be viewed as a national industry rather than as a purely local business.

This huge national industry developed from small beginnings. When the first settlers arrived, no livestock existed in this country. All of our domestic livestock, as we know it today, was originally imported. George Washington, as well as many other famous Americans, was an importer of livestock. In an average year, our annual slaughter of meat animals now totals about 27 million head of cattle, 10 million head of calves, 80 million head of hogs, and 15 million head of sheep and lambs. If this volume of meat is divided among all the people of the country, there are about 95 lb. of beef, 6 lb. of veal, 65 lb. of pork and $4\frac{1}{2}$ lb. of lamb and mutton per year for every man, woman, and child in the United States.

39-1. THE MARKETING PROCESS

Marketing may be defined as the performance of services necessary to move goods, after they have been physically produced, into the possession and ownership of final consumers. The services usually required in the successful marketing of agricultural products are: (1) assembling, (2) grading and standardizing, (3) packaging, (4) processing, (5) storing, (6) financing, (7) transporting, (8) selling, and (9) risk-bearing. Some of these services are performed more than once in the marketing process; others, such as packaging or processing, are not involved for certain products. In the livestock industry, marketing the live animals, slaughtering and processing operations, and moving the resulting meat products through trade channels to consumers are as essential to the completion of the production process as is the work of the primary producer.

Marketing Stages. Marketing of agricultural products can logically be grouped into three major stages: (1) concentration of supplies at shipping points, (2) equalization of supply and demand in wholesale transactions, and (3) distribution to consumers through retailers. Livestock generally passes from the hands of a large number of producers into the hands of a much smaller number of buyers; the slaughter animals are then sent to packers for slaughter and processing, and on to retailers; from the retailer, the final product goes to the consumer.

Sales Channels for Livestock and Meats. The customers or buyers to whom a livestockman or middleman sells his product are his sales outlets. The term sales outlet may be used to indicate either different kinds or classes of buyers, or a series of specific persons or firms.

A sales channel is a series of marketing agencies through which a product moves from the producer to the consumer. Sometimes sales channels are called trade or marketing channels. Figure 39-1 shows the principal sales channels for livestock and meats. When these sales channels are traced in reverse, they indicate the various sources of supply used by each class of middleman.

Direct Marketing Increasing. During the first quarter of this century, terminal livestock markets at Chicago, Kansas City, Omaha, Denver, and other railroad centers dominated the marketing of livestock. In 1923, the earliest year for which data are available from the USDA Agricultural Marketing Service, the following proportions of federally inspected slaughter were purchased at terminal markets: cattle 90%, calves 86%, sheep and lambs 86%, hogs 77% (USDA, 1923).

In recent years, an increasing volume of livestock has moved directly from producers, country dealers, and auctions to packers. The proportion marketed through the terminals has declined. Several terminal markets, such as those at Los Angeles and San Francisco, have ceased operation entirely. The volume of livestock moving through the Chicago terminal market has been reduced materially. In California, for example, it is estimated that in 1964 approximately 90% of the cattle fed for slaughter moved directly from

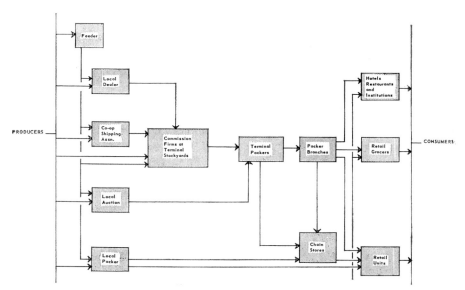

Fig. 39-1. Principal sales channels for livestock and meats.

the feedlot to the packer, bypassing the terminal market. Indications are that the trend toward more direct marketing is increasing in all parts of the United States.

Among the more important changes that have contributed to this trend are improved motor truck transportation, improved market-news facilities, and increased development of large feedlots, which feed and market animals to meet the specifications of packers, who in turn sell meat to fit the more discriminating specifications of large-volume retailers. Also, there has been a decentralization of packing plant facilities away from the terminal markets to minimize the high costs associated with central-terminal traffic congestion, high land values, labor problems, and so forth. New and more specialized and efficient interior packing plants, located closer to sources of livestock supply, have received an increasing proportion of the livestock marketed by producers. Country livestock auctions have increased in recent years to approximately 2,300, whereas terminal markets have declined from more than 65 in number to 45 in 1965.

39-2. PRESENT METHODS OF SELLING LIVESTOCK

Selection of Sales Outlets. One of the major marketing problems is to decide which outlet or combination of outlets best fits the needs of the individual producer. The following are the most important factors to be considered in making such a choice: (1) sales volume handled, (2) quality of livestock usually preferred, (3) kind of livestock usually handled, (4) net returns to producer, (5) location or distance and transportation facilities available, (6) financial resources of seller and/or buyers, (7) personal abilities and preferences, (8) established trade connections.

Production and marketing situations differ widely. The relative importance of the above factors are thus likely to differ considerably in importance from one producer to another.

Local Markets, Country Dealers or Traders. Local markets may also be called assembly yards, buying points, or reload stations. They may be operated by a local dealer, a cooperative, or a packer.

Country dealers or traders are independent operators who buy and sell livestock on a full or part time basis for profit. They may vary considerably in their scales of operation. They usually buy from producers at the farm, but many also buy at auctions or concentration yards. Country dealers or traders conduct their activities in all livestock areas of the country.

Auction Markets. The auction is one of the oldest systems of selling. An auction may be defined as a public market at which an article is offered for sale simultaneously to several prospective buyers and is sold by the auctioneer to the highest bidder.

A livestock auction may be conducted by a proprietorship, partnership, general business corporation, or a cooperative association. The rapid rise in the volume of business handled by auctions indicates that farmers have found that the advantages of this method of sale outweigh its disadvantages.

Central or Terminal Markets. Terminal markets (also referred to as central markets) differ from other livestock markets principally in the scope and type of activity conducted. However, it is difficult to draw a clear-cut distinction between terminal markets and all other livestock markets.

Originally, our seaboard livestock markets were terminal markets in the true sense of the term. These markets were located on the east coast, in such cities as Baltimore, Philadelphia, and Jersey City, and were truly the end of the line at that time. Once livestock reached the coast it was rarely shipped to any other point.

Now, however, the term "terminal market" is only relative. In general, the term implies a market about the size of the one in Chicago, where there is a concentration of livestock for sale, a concentration of buyers, and a concentration of meat plants and other related services. Not all livestock shipped to terminal markets, however, is sold or processed there; a sizeable volume is shipped to other markets for sale. Another portion may be purchased and shipped to meat-packing plants located in other cities.

It is commonly thought that market transactions at terminal markets are the greatest single influence in determining livestock prices from day to day. Market-news agencies usually base their reports on the terminal markets.

In recent years, however, with a declining volume moving through central markets, more information is needed about direct and final transactions in the country and in the feedlots to obtain a more complete picture of prices and movements.

Commission Firms at Central Markets. Commission firms at central markets receive livestock on consignment from producers, local dealers, or cooperative associations, and act as the shipper's agent in selling the livestock to buyers, chiefly packing companies. When a consignment of livestock has been sold, the yardage, feed, selling commission, and any other marketing charges are deducted from the total receipts for the animals, and the net receipts are forwarded to the shipper. The basic service of the commission firm is to supply highly skilled specialists in the sale of each class of livestock in order that the actual selling may be done by a person whose technical competence is on a par with the buyer's experience.

Commission firms usually assume full charge of the livestock from the time of delivery by the consignor. The animals are sorted and penned by species, and sometimes by class. Sorting may also be done by weight and grade to provide for more effective sale. Commission firms also keep their patrons informed on the market situation and assist in the collection of claims against carriers for losses in transit. Commission firms are bonded under supervision of the Packers and Stockyards Division of the USDA.

Selling on Contract for Future Delivery. Contracting for purchase or sale of livestock in advance is commonly practiced in the western livestock country. This system offers one means of insuring against possible price changes.

There are times when the producer gains by contracting to sell his livestock in advance or to purchase feeders or replacements, especially if he is able to interpret market trends correctly. On the other hand, if the producer misjudges market conditions, sale by contract for future delivery may have a distinct disadvantage.

Studies made of contract selling of livestock seem to show that the advantages of contracting are usually more with the buyer than with the producer. The buyer usually has more complete information and is able to interpret market trends more accurately than the average producer.

Stockmen who sell their animals on contract should be thoroughly familiar with the fundamentals of good livestock purchase contracts. Taylor (1953) and Stucky (1952), who have made separate studies of selling cattle on contract, point out the advantages and disadvantages of this type of selling and some of the pitfalls that might be avoided.

Cooperative Marketing of Livestock. In their day-to-day marketing operations, livestock cooperatives function within the same competitive framework as all other segments of the livestock industry. The cooperatives deal with the same livestock interests, trade with the same buyers, and are subject to the same general customs and practices of the industry. On the surface their operations do not vary greatly from those of noncooperative livestock marketing agencies.

Cooperative associations are owned and controlled by producers. Control is exercised through a board of directors elected by the membership. To become a member and to be eligible to vote for directors, a producer must patronize the association and apply for membership. Membership is restricted to bona fide livestock producers. The primary purpose of the cooperative is to market livestock for members in the most effective way possible and to serve the members' interest in particular. Those livestock cooperatives that handle sufficient volume to set the competitive pace are instrumental in bringing about improvements in practices and in pricing for the benefit of producers. Capable management, adequate volume, and loyal membership support are vital to their success.

39-3. HOW LIVESTOCK PRICES ARE DETERMINED

Stated in simple terms, the prices of meat (and therefore of livestock) are determined by supply and demand—supply, the number of animals raised and sent to market and the yield of meat from these animals; demand, the number of dollars that consumers are able and willing to spend for beef, pork, lamb, and veal. It must be realized further that a great many factors—all of them constantly changing—affect both supply and demand. It is the complex interplay of all those factors that determines the price that livestock producers get for cattle, calves, hogs, or lambs on the day, and at the place, they are sold.

Illustrations of the Price-Making Process. The biggest single reason for the fluctuation in livestock prices is that the amount of total meat slaughtered and offered for sale by the packers varies from day to day or from week to week.

Meat is a perishable product that cannot be stored economically. Hence, there is strong pressure for it to be sold each day. There is really no such thing as a stored surplus of meat. Most of it is moved into the channels of trade within two weeks of the time it is slaughtered. Of course, small amounts may be stored as canned meat or in some other processed form, but only the lower grades of meat are handled this way.

If, during a particular week, supplies of a certain class and grade of livestock are large and haven't been sold by the end of the week, the holdover causes that particular class of livestock to sell for less when the market opens the following week. This will be especially true if the current market run turns out to be heavy.

Changes in the meat prices that the packers get in the wholesale market are, in turn, almost immediately reflected in the prices that packers will bid for livestock. The prices that packers can pay for livestock are also greatly dependent on the prices they receive for by-products, such as tallow, hides, tankage, lamb pelts, and the like. Changes in the market price of these products can make a sizable difference in the price a packer can pay for livestock.

The amount of money spent by the consumer for meat depends mostly on the number of dollars available for all kinds of consumer goods. As disposable income goes up, the consumer tends to spend more money on meat. The livestock producer should recognize, however, that his product is competing with other meats, such as poultry and fish, as well as with plant foods. Thus, to maintain his market, he must present a superior product to the consumer at an attractive price. Furthermore, promotion is essential in order that meat can compete successfully with other consumer expenditures.

Consumer income changes very gradually. It does not increase or decrease nearly as often or as much as does the number of livestock raised and marketed. Whatever their incomes may be, consumers tend to spend, within rather narrow limits, about the same percentage of their income for meat. Accordingly, about the same amount of money is available for buying meat from one day, week, or month to the next. This means that when livestock marketings increase significantly, prices must go down in order to stretch the same number of consumer dollars over a larger meat supply. And when marketings are lighter, consumers will pay higher prices for the smaller supply of meat. This emphasizes the importance of orderly livestock marketing in helping to prevent violent price changes.

Competition from poultry, fish, eggs, cheese and many other products also has a great influence on meat prices. An increasing supply and variety of such items are available in convenient form at reasonable prices, and the consumer will continue to buy red meat only if she can buy the quality she desires at what she considers to be the right price.

Such factors as weather, seasonal habits, and religious beliefs also affect the demand for meats. Weather conditions, which may either cause drought or increase grass supplies, can also greatly influence the movement of livestock to slaughter. Each of these factors has its particular influence on livestock and meat prices and must be considered by the alert producer.

Seasonal Marketing Patterns and Price Trends. Seasonal trends in prices of meat animals reflect the seasonality of production and of market sales. To a large extent, cattle and sheep are born in the spring and marketed in the fall, although there are many variations in different parts of the country. Their prices are usually lowest in the fall, when supplies are abundant, and climb to a spring high, when supplies are scarce.

Farrowing of pigs is concentrated in two seasons, spring and fall. Marketings and prices of hogs accordingly have two up-and-down swings each year. A major price peak occurs in late summer; a secondary peak in late winter.

Figure 39-2 shows how marketings are highly concentrated at the end of the grazing season for all meat animals except hogs. A November–January high in hog marketings reflects large March–April farrowings.

Within these broad trends are many separate seasonal patterns for individual grades and classes of livestock. Differences are especially great for the various kinds of cattle. Those that have not been specially fed—lower grade steers and cows, as well as all stocker and feeder cattle and calves for slaughter—conform to the pattern of peak supply and lowest prices in the fall. Feeding raises the grade of cattle and delays the peak supply for the higher grades until progressively later seasons.

Feeding of lambs evens out the yearly slaughter supply of lambs, but not as much as cattle feeding does for the beef supply. Lamb prices retain the same general pattern as do lower grade cattle prices—lowest in the fall, highest in the spring.

Seasonal patterns change with time, as innovations are made in livestock production and marketing. Seasonal trends in prices of veal calves were markedly different in the 1950's than they were in the 1920's because since 1920 marketings have been affected by a shift from spring calving of milk cows to year-round calving. Similarly, seasonal marketings of fed cattle have been smoothed out, particularly in California, where cattle feeding has expanded more than 500% since 1945 and is now done on a year-round basis. As a result of such developments, seasonal fluctuations in slaughter cattle prices are not nearly so great because marketings are more uniform from month to month.

39-4. VERTICAL INTEGRATION IN THE LIVESTOCK INDUSTRY

In recent years, the subject of integration in its various forms has been one of growing interest and concern in livestock circles.

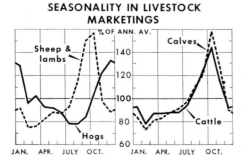

SEASONALITY IN LIVESTOCK MARKETINGS

Fig. 39-2. Marketings are highly concentrated at the end of the grazing season for all meat animals except hogs. A November–January high in hog marketings reflects large March–April farrowings.

Fig. 39-3. Sheep offered for sale at a typical country auction market.

Definition and Trends. There are two general types of integration—horizontal and vertical. Horizontal integration is the combination of businesses that are alike into one larger business concern, for example, the merger of two or more ranches under one management. Vertical integration means the control of two or more stages in the chain of production, processing, and distribution under one single management or decision-making unit. It may come about either through cooperative arrangements, through contracts, or through ownership of the various steps in the process.

Integration in the livestock industry is not new. At one time, livestock production and meat consumption were completely integrated. All of the stages in the chain of production were under the control of one man, namely, the farmer. This chain included growing the feed, producing and butchering the livestock, and consuming the meat on the farm or selling it directly to consumers. Since those early days, however, we have had much specialization in various stages of production, processing, marketing, and distribution.

In recent years, many significant changes have occurred in food marketing that have also influenced trends toward integration. For example, food retailing is becoming controlled by an increasingly smaller number of larger firms. These large supermarkets depend upon uniform quality and continuous supply. To be assured of the desired uniformity and quantity of meat, they are interested in integration or in contracting with suppliers in advance for future delivery. Specifications of final products can be more closely controlled through some form of integration.

Forces Influencing Integration. Kramer (1958) indicates that the following forces have been important causes of the trend toward integration in the livestock and meat industry: (1) efforts of slaughterers and distributors to assure themselves of an adequate and stable supply of livestock, poultry, or meat, (2) efforts of slaughterers and distributors to control the quality of livestock, poultry, and meat, (3) efforts of farmers to spread or reduce risks and to reduce production costs, (4) efforts of farmers to obtain additional production capital, (5) efforts of feed companies and other farm suppliers to expand the market for their supplies, (6) efforts of slaughterers and distributors to reduce dependence upon others, (7) efforts of slaughterers and distributors to reduce transaction costs.

In summary, it might be said that today's emphasis on vertical integration has grown out of the struggle of the agencies engaged in production and marketing to shorten the gap between the producer and the consumer. The primary ends sought are cost reduction and more market power of influence over supply, quality, and price. This is the basis for the growth of integration. For society in general, it means better quality, more uniform products at lower prices, more stability of supply and price throughout the year, and a smoother seasonal and geographical flow of product. It can be said that the success of integrated programs will be determined largely by the quality and effectiveness of management and by efficiency in the use of capital. Integration poses a threat to the inefficient, high-cost producer because he can't sell and make a profit on the same market as can the more efficient, integrated operation. On the other hand, it holds promise for many capable producers who will find opportunity to expand their business, reduce their risk, and increase their profits.

39-5. LIVESTOCK MARKET NEWS AND OUTLOOK INFORMATION

Market News. Market news, in a strict sense, consists of market supply, demand, and price information pertaining to daily trading in livestock and livestock products. This information is assembled from daily coverage of principal shipping points and terminal markets. Timeliness is the essence of market news. To be of most use to the producer, the information must be gathered during the most active trading period of the day and be made available almost immediately.

Accuracy of market news is always important, but there are often conflicts between timeliness and completeness. For this reason, some kinds of market news can be assembled and distributed daily, whereas other information cannot be complete enough, except on a weekly or monthly basis.

Owing to the different needs of farmers, packers, wholesale handlers, and retailers, several kinds of prices and price reports are issued.

Federal-State Market News Service. The Market News Service of the USDA, in cooperation with the states, is the largest and most important livestock market-news-gathering agency in the country. It operates on all the important livestock markets through a system of leased wires covering thousands of miles of teletype lines, which link scores of field and terminal markets. Reports are made available through radio, television, daily newspapers, wire news services, and the mail. Most of these reports are available to producers free of charge upon written request to the offices that issue them. For all publications issued at Washington, requests should be sent to the Office of Information, Agricultural Marketing Service, United States Department of Agriculture, Washington 25, D.C. For publications issued elsewhere, address the request to the originating office. These offices will also furnish a list of the various types of reports released as well as the date of release. In addition to market news issued by government agencies, some of the most timely and accurate information regarding transactions is often disseminated to clients by commission firms and other market-agency personnel actively engaged in market trading.

Outlook Information. Market outlook projection begins where fact collecting leaves off. Whereas market news is often thought of as information that needs to be available on an hourly or daily basis, outlook information generally relates to the next month, quarter, or year. It contributes to a better understanding of current market news. Basic facts are necessary for an understanding of the economic structure of the livestock industry, but facts alone are not enough. They must be interpreted, analyzed, and related before they have true meaning.

Farmers and ranchers must constantly make decisions concerning production and marketing practices. However, the results of most decisions will affect income at some future date rather than at the time the decision is made. Therefore, the outcome of the decision often depends not upon the prices and market conditions prevailing at the time the decision is made but rather upon prices and market conditions at some future date. Thus, reliable outlook information is very important in assisting the livestock producer to arrive at sound conclusions. By studying various outlook reports, the livestock man should be able to arrive at a conclusion as to what the general level of prices and the price risks or advantages will be. By weighing these against his costs, he can select the course of action that will be most likely to provide him with the greatest net income.

Livestock outlook reports are issued periodically from nearly all land-grant colleges and extension services as well as from livestock organizations, trade journals, and business sources. Producers may receive such reports by requesting them from their nearest Land-Grant College Department of Agricultural Economics or by contacting their favorite livestock journal or reporting service.

39-6. FOREIGN TRADE IN MEAT

Meat is produced and consumed mainly in developed countries. For several reasons, particularly those of low incomes and less favorable natural conditions, meat production is low in less highly developed countries, with the exception of Argentina and Uruguay.

The volume of world meat trade and of exports from meat exporting countries depends largely on the large consumer demand and the trade policies of the four largest meat importing nations—Britain, the United States, Federal Republic of Germany, Italy. In 1963, the imports of these four countries accounted for 71% of world imports of all meats and 76% of world imports of beef.

United States Beef Imports. In recent years, the United States has emerged as the major importing nation for beef and veal, importing 35% of the total amount exported in 1963. Britain was second with 24%, and Italy third with 12%. Table 39-1 shows the trend in imports since 1940. In 1963, imports amounted to 10.7% of total United States beef and veal production.

The exports of beef and veal from Argentina, Australia, and New Zealand account for about 56% of the total entering world trade. Imports of beef and veal from Argentina into the United States are limited to canned and processed meats because of dangers of foot-and-mouth disease. Australia is, by far, the most important exporter to the United States, followed by New Zealand, Ireland, and Mexico. Shipments from Australia to the United States rose from 16 million lb. in 1958 to 509 million lb. (product weight) in 1963, accounting for 81% of total Australian exports. During the same period, imports from New Zealand increased from 152 million to 235 million lb., accounting for about 85% of New Zealand's total exports.

The bulk of the recent imports of beef has consisted of frozen, boneless lean meat (derived from grass-fed cattle), used primarily in the production of manufactured beef products, including hamburger, frankfurters, bologna, canned preparations, and frozen prepared dinners. Nearly all such imported beef has been comparable in quality to United States grades of cutter or canner. Small quantities of beef cuts have been imported for consumption as table beef.

Importers of beef and veal include brokers, jobbers, domestic processors, packers (including domestic branches of international companies), and United States agents of foreign packinghouses Relatively few concerns handle the bulk of the imports.

TABLE
39-1.
United States imports of cattle and beef compared with meat production, 1940–1963

	Imports					
Year	Dutiable Cattle* (1,000 head)	Beef Equivalent of Live Cattle† (mil. lb.)	Beef & Veal** Carcass Weight (mil. lb.)	Total Cattle & Beef Carcass Weight (mil. lb.)	Beef and Veal Production (mil. lb.)	Imports as Percentage of Production
1940	630	173	168	341	8,156	4.2
1941	733	204	257	461	9,118	5.1
1942	653	188	212	400	9,994	4.0
1943	630	156	226	382	9,738	3.9
1944	341	90	190	280	10,850	2.6
1945	489	131	128	259	11,940	2.2
1946	516	141	20	161	10,816	1.5
1947	54	24	64	88	12,037	0.7
1948	419	173	356	529	10,498	5.0
1949	412	153	254	407	10,773	3.8
1950	438	157	348	505	10,764	4.7
1951	220	91	484	575	9,896	5.8
1952	138	47	429	476	10,819	4.4
1953	177	62	271	333	13,953	2.4
1954	71	35	232	267	14,610	1.8
1955	296	93	229	322	15,147	2.1
1956	141	43	211	254	16,094	1.5
1957	703	221	395	616	15,728	3.9
1958	1,126	340	909	1,249	14,516	8.6
1959	688	191	1,063	1,254	14,588	8.6
1960	645	163	775	938	15,835	5.9
1961	1,023	250	1,037	1,287	16,341	7.9
1962	1,232	280	1,440	1,720	16,311	10.5
1963‡	834	180	1,679	1,859	17,330	10.7

Source: USDA. *Livestock and Meat Situation.*
* Includes milk cows.
† Estimated at 53% of the live weight of all dutiable imports of cattle.
** Canned and other processed meats have been converted to their dressed weight equivalent.
‡ Preliminary estimates.

Factors Contributing to Increased United States Beef Imports. A number of concurrent developments in the United States and in the principal foreign cattle producing areas, particularly Australia and New Zealand, contributed to the marked rise in United States imports of beef and veal from 1958 to 1963. The most important of developments were these:

1) Increasing demand of United States consumers for frankfurters, ham-

burger, sausages, television dinners, and a great variety of luncheon meats caused manufacturers of these products to look for foreign and cheaper sources of supply.

2) The United States with its high consumer demand for beef offered higher prices than most importing countries. Also, United States tariff rates (3 cents per pound) and other nontariff barriers are lower than any other major meat importing nation. Under such conditions, meat imports naturally increase.

3) More cows were held back to increase herds, resulting in a decline in slaughter of the canner and cutter cows that normally move into processing meat channels. The volume of "two-way" cattle marketed also declined. The reduced supply of domestic processing beef, coupled with rising consumer demand for manufactured items, resulted in the United States market drawing upon large supplies of foreign beef.

4) No less important, however, was the loosening of trade ties among Britain's Commonwealth countries. Australia, in particular, began looking toward the United States market in 1958 when the United Kingdom–Australian Meat Agreement, which had bound Australia to ship most of her meat to Great Britain, was modified. Thereafter, Australian beef shipments to the United States rose very rapidly.

5) Coupled with the change in the United Kingdom–Australian Meat Agreement was the large increase in beef and other meat production in Great Britain, as well as expanded production in Australia, New Zealand, and the Central American countries, most of which increased export shipments to the United States.

6) The protectionist policies of the European Common Market (EEC) in 1962–63 reduced shipments of meat to that area by nonmember exporting countries, thus placing more pressure on the nonmember countries to ship meats to the United States.

Voluntary Import Agreements. The developments cited above coupled with declining United States prices for fat cattle in 1963–64, led cattlemen to press for legislation to restrict the volume of foreign beef imports. While cattlemen, through their organizations, were working with congressmen to obtain "beef quota" legislation based on some reasonable figure (1959–63 five-year average), the governments of Australia, New Zealand, and Ireland, in February, 1964, entered into voluntary agreements with the United States government to limit their annual exports of beef, veal, and mutton to the United States. In May, 1964, Mexico signed a similar agreement. Each of these agreements, which may be canceled by either party on 180 days' notice, is subject to renegotiation after three years. The export quotas to United States from each country during the three years from 1964 through 1966 are shown in Table 39-2.

TABLE 39-2. | *Export quotas of meat to the United States from four principal meat exporting countries.*

Country and Type of Meat*	Million Pounds Product Weight		
	1964	1965	1966
Australia: beef, veal, mutton	542	562	582
New Zealand: beef and veal	231	240	249
Ireland: beef and veal	76	79	82
Mexico: beef and veal	66	69	71

Source: U.S. Tariff Commission, TC Publication 128, Report of Investigation No. 332–44 under Section 332 of the Tariff Act of 1930, Beef and Beef Products, Washington, D.C., June 1964.
* Includes all forms except canned, cured, and cooked meat and live animals.

The quota limitations specified in Table 39-2 allowed for a growth factor of 3.7% per year, which was estimated to be the average growth of the United States market.

Meat Import Quota Bill P. L. 88–482. The United States cattle industry was not satisfied with the voluntary agreements to restrict imports. They continued to press for quota legislation, which culminated in the signing of the "Compromise Meat Import Quota Bill" (HR-1839) by President Johnson on August 22, 1964.

This bill authorizes import quotas for fresh, chilled, or frozen beef, veal and mutton only. It does not include lamb, nor does it include canned or processed meats. The most important provisions are these:

1) The imposition of quotas whenever the imports of beef, veal, and mutton exceed 725,400,000 lb., plus an allowance for the growth of the market. (The amount of 725,400,000 lb., which is equal to the average imports of these meats during the five-year period from 1959 through 1963, becomes the base.)

2) The "growth factor," as defined in the bill, is the percentage by which estimated average annual, domestic, commercial production of these meats in the current calendar year and the two preceding years (a three-year average) exceeds the average annual domestic, commercial, production of these meats during the years 1959 through 1963.

3) The Secretary of Agriculture is required to make quarterly estimates of anticipated imports. The bill allows a leeway of 10% above the base figure plus the growth factor before quotas will be imposed. Thus quotas on imports would not be imposed in 1965 unless the Secretary estimated that 1965 imports would be 10 percent above the allowed quota level. (That is, imports for 1965 would have to be estimated at approximately 930,000,000 lb. or more before quotas would be imposed.)

4) If quotas were imposed in accordance with the formula, the total import quota would be allocated among the supplying countries on the basis of the shares such countries supplied to the United States market during a "representative period."

5) The President of the United States has certain discretionary powers that authorize him to suspend quotas or refrain from imposing them at all if he determines that (*a*) such action is required by "overriding economic" or national security interests, (*b*) the supply of meat items described in the bill will be inadequate to meet the domestic demand at reasonable prices, or (*c*) trade agreements entered into after date of enactment of this act ensure that the policy set forth will be carried out. All determinations by the President and the Secretary of Agriculture shall be final.

The bill may insure the domestic industry against a runaway import situation, in which the United States market might possibly be used as a dumping ground for surplus world meat.

The passage of this bill indicates to foreign producers that the United States livestock industry, working with its Congress, will be alert to major changes in world trading patterns for beef that would threaten the domestic industry.

United States Foreign Trade in Other Meats. Imports of pork and pork products consist primarily of canned hams and shoulders and some canned bacon. Total United States pork imports on a carcass weight equivalent basis are larger than exports. Since 1958, imports have ranged from 1.6 to 1.8% of domestic production, and exports from 0.6 to 0.7%, except in 1963 when exports equaled 1.1% of production.

United States exports of lamb and mutton are very small, amounting to 1.5 to 2.6 million lb. annually since 1958. Both lamb and mutton imports, on the other hand, have been several times larger than total United States lamb and mutton exports in recent years.

Mutton imports are several times as large as lamb imports; lamb imports have averaged between 1 and 2% of United States production in recent years, and mutton imports have approximated 2.5 times United States mutton production. Practically all mutton imports arrive in boneless, frozen form and are used in soups and processed items. The bulk of the mutton comes from Australia; New Zealand supplies most of the lamb, as either frozen lamb carcasses or frozen cuts.

United States Exports of Meat By-products. The high level of United States meat production yields larger quantities of by-products than can be consumed in this country at profitable prices. An important part of United States output of products, therefore, is sold abroad. The principal by-products entering significantly into United States export trade are variety meats, lard, tallow and grease, hides and skins. The value of United States exports of all meat and by-products in 1963 was $364,200,000. Of these exports, more than one-third were tallow and greases, followed by hides and skins. By comparison, United States imports of all meats and by-products amounted to $870,700,000 in 1963. The major items imported were beef and veal, valued at $361,800,000, followed by wool, valued at $210,600,000.[1]

39-7. GOVERNMENT PURCHASE PROGRAMS FOR MEAT

In 1935, Congress enacted legislation to widen market outlets for agricultural products being produced in surplus, as one means of strengthening prices received by farmers. This special legislation, known as Section 32 of the Act of 1935, contains a clause authorizing government purchases of surplus commodities "to encourage the domestic consumption of . . . commodities or products by diverting them from the normal channels of trade and commerce or by increasing their utilization among persons in low-income groups." [2]

This program is initiated only when prices of particular commodities need bolstering. On March 2, 1964, the USDA started purchasing frozen and canned beef of the choice grade to assist in strengthening beef prices. Between March 2 and December 11, 1964, a total of 404,000,000 lb. were purchased at a cost of $222,000,000. This is equivalent to approximately one million head of cattle, or 3% of commercial production.

The meat so purchased is used in school lunch programs, charitable institutions, such as hospitals, homes for the aged, correctional institutions, and child care centers. It is difficult to measure the impact of such a program on beef prices, but it undoubtedly aids in strengthening them somewhat.

Section 32 programs are financed by a continuing appropriation equal to 30% of the import duties collected on all commodities entering the United States under the customs laws—nonagricultural commodities, as well as agricultural, plus unused balances to the extent of $370 million. Funds used to purchase any single commodity are limited to 25% of the total.

39-8. DISTRIBUTION OF MEATS AND MEAT PRODUCTS

The meat distribution system in the United States has been changing rapidly. Changes in the meat industry include development of large-volume retail firms, mass buying of meat by retailers on a specification basis, widespread use of the federal meat trade standards, and adjustments in the distribution system. Packers using national systems of distribution have

---

[1] Livestock and Meat Products Division, FAS, USDA.
[2] Agricultural Information Bulletin 135, Price Programs, USDA.

declined relatively in volume of meat sales. On the other hand, the number of independent packers and independent wholesale distributors has increased. At the same time, the specifications of the large-volume retail buyers have brought about changes in the methods of operation and in the principal types and qualities of meat handled by packers and wholesalers. The increased use of federal grade standards for beef, veal, and lamb, especially by large food chains, has also affected the entire meat industry.

Trends in Retailing and Merchandising. The big changes in meat marketing and merchandising have come in the organization of the marketing system and in the introduction of packaging and selling technologies. The development of self-service meat merchandising accelerated the move from the small grocery store to the giant supermarket. Self-service merchandising has enabled processors and retailers to take advantage of advances in packaging, such as heat-shrunk, transparent film and nitrogen-gas-filled film containers for packaging cured beef and pork products. Today, nearly 50% of the meat sold in the United States is purchased at a self-service counter. This trend is likely to continue.

Through prepackaging and maintaining high quality, processors and retailers have been able to carry product branding through to the consumer level. These new packages permit visual inspection of the product and enable the processor to identify his brand at the consumer level. Because of these developments in packaging methods, meat processors have been increasing the amount of product-branding at the retail level, the goal of each packer being, of course, to capture as much of the consumer market for his individual product as possible.

Retailers have found, since the advent of self-service and centralized packaging, that they are able to stock a greater variety of meat and processed products. This means that consumers have a greater variety of meat cuts and meat products available to them. They also have more built-in food services available than ever before. This wide differentiation of products has helped to broaden the market for meat.

One of the impacts of self-service on meat merchandising has been to increase the demand for higher quality. This development, of course, has had its influence on the producer and his management practices. Self-service merchandising has also led to closer trimming standards for retail cuts. Consumers are becoming better judges of values as a result of their self-service experience. As they shop in different stores, they recognize when the retailer is not doing as good a job of trimming as his competitor.

In recent years, several major meat processors have made attempts to process and package meat cuts at the plant and distribute them as packaged, boneless, defatted, frozen meat cuts. In general, however, consumers have not been willing to accept such meat cuts at current prices. For the same amounts of edible meat, the cost per pound of boneless frozen fresh meat cuts may not be any higher than the cost per pound of traditional un-boned, untrimmed, fresh meat cuts. However, the cost per pound is considerably higher than that of an equal number of pounds of regular fresh meat cuts—thus, more consumer resistance.

Consumer Preferences. The introduction in recent years of these many new products in new forms has coincided with significant changes in consumer income and consumption patterns for meat. The combination of these and other changes has tended to increase the complexity of the choice-making process facing the consumer.

It is important for any industry to gear itself to produce what the customer wants. Although the average housewife has little concept of what makes high quality meats, she does know that the family likes tender, flavorful meats. Furthermore, she is very conscious of excessive fat on the meat. First, from the standpoint of efficient buying, she desires to cut down on waste; second, from the standpoint of health, she is extremely sensitive to obesity. Usually housewives want 'thin-skinned steaks' and light, meaty roasts. They want some marbling in the beef, and they like a bright red color in the lean meat. The key to the future is to give the consumers the kind of product they prefer in the form, quality, and place preferred. These points are significant to producers, handlers, and merchandisers throughout the trade. Each must keep abreast of new developments and run his operations efficiently or competition will take its toll.

REFERENCES AND SELECTED READINGS

DeGraff, Herrell, 1960. *Beef Production and Distribution.* University of Oklahoma Press, Norman, Oklahoma. 252 pp.

Kramer, R. C., 1958. What is integration and its current status? Proceedings, Eighth National Institute of Animal Agriculture, Purdue University, April.

Livestock Marketing Journal, 1960. Swan, Texas. July 18.

Stucky, H. R., 1952. *Contracting vs. Selling at Delivery Time.* Mont. Agr. Ext. Service, Bozeman, Montana. Mimeographed report.

Taylor, M. H., 1953. *Livestock Purchase and Sale Contracts.* Utah Agr. Ext. Bull. 211, Logan, Utah.

Chapter 40

Classification, Grading, and Marketing of Wool and Mohair

40-1. INTRODUCTION

The production of wool on a world-wide basis (Table 40-1) fluctuates from year to year as a result of periodic droughts in some of the major sheep-producing areas, notably Australia and South Africa. Approximately one-half of the 270 million lb. of United States wool production comes from the eleven western states. Texas, Wyoming, California, Montana, Colorado, South Dakota, Utah, and Idaho rank in descending order in wool production.

There are many different breeds and crosses of sheep in the world. Each produces wool of certain specific characteristics. In addition, fleeces from animals of a single breed may be quite variable. Moreover, each fleece does not contain uniform fiber throughout, but varies in length, fineness, color, purity, and shrinkage according to the region of the body from which it comes. Raw wool, then, is not a uniform product.

On the other hand, the manufacturer must have uniformity in order to insure that, after processing, his textile will have the necessary appearance, handle, drape, and wearing and insulating properties characteristic of the particular fabric. The ultimate uses of a specific type of wool are dependent to a large degree on its properties. For this reason, raw wool is divided into uniform lots at some step in the marketing process, before it reaches the manufacturer.

TABLE 40-1. | *World wool production, 1964–65.*

Area	Grease Basis (millions of lb.)
Australia	1,813
New Zealand	625
United Kingdom	133
Argentina	415
South Africa	320
United States	270
Uruguay	198
Other	895
Total Free World	4,669
Soviet Bloc	1,170
World Total	5,839

Source: Estimated world wool production, *Wool Situation*, TWS-69. USDA, Economic Research Service. Oct. 1964, p. 29.

40-2. FACTORS DETERMINING THE VALUE AND UTILITY OF WOOL

Grading Systems and Standards Based on Fiber Diameter. Wool is divided into different grades on the basis of the diameter of its fiber. In the United States, two different grading systems are used. One of these is the American blood system. This system arose from the introduction of Merino sheep, which produced fine wool, and from their crossing with the common, coarse-wooled, somewhat hairy sheep of the colonies, which were imported earlier from the British Isles. The wool from this first cross with the common sheep was coarser than that from Merinos and was termed half blood. Wools of descending degrees of fineness were designated as ⅜ blood, ¼ blood, common, or braid, the latter being the coarsest. Ultimately, the origin of these terms was forgotten, and they came to mean wools of a certain fiber size.

The other grading system, which is used throughout the world, is the counts, or Bradford, system. This system was originally based on the number of "hanks" of yarn that could be spun from one pound of clean wool. A hank of yarn is 560 yards long. Thus, from a pound of clean "60's" wool the manufacturer can spin 60 hanks of yarn—33,600 yards, or slightly more than 19 miles, of yarn.

This counts system is now used more specifically to designate wools of a

TABLE | *Specifications for counts system grades of wool as related to blood system.*
40-2.

Blood System	Counts Grades	Range for Average Fiber Diameter (microns)	Standard Deviation* (microns)
Fine	Finer than 80's	Less than 17.70	3.59
	80's	17.70 to 19.14	4.09
	70's	19.15 to 20.59	4.59
	64's	20.60 to 22.04	5.19
Half blood	62's	22.05 to 23.49	5.89
	60's	23.50 to 24.94	6.49
Three-eighths	58's	24.95 to 26.39	7.09
	56's	26.40 to 27.84	7.59
Quarter blood	54's	27.85 to 29.29	8.19
	50's	29.30 to 30.99	8.69
	48's	31.00 to 32.69	9.09
Low quarter	46's	32.70 to 34.39	9.59
Common	44's	34.40 to 36.19	10.09
Braid	40's	36.20 to 38.09	10.69
	36's	38.10 to 40.20	11.19
	Coarser than 36's	Over 40.20	—

Source: Adapted from American Society for Testing Materials, *Textile materials—Fibers and Products; Leather.* (ASTM Standards Part 25) Philadelphia, 1964. p. 102.

* The standard deviation given here is the maximum allowable for the grade; those samples with standard deviations exceeding this are assigned to the next coarser grade. As indicated in the table, the allowable standard deviation is greater for the coarser than for the fine wools. This is because there is greater variation in fiber diameter in the coarse wools.

definite range in fiber diameter. Table 40-2 shows the blood and counts grades and their specifications in terms of average fiber diameter.

Carpet wool is not a grade of wool in the strict sense of the word, but refers to wools that are highly variable in fiber length and diameter. Practically all carpet wools used in the United States are imported from other countries and are shorn mostly from unimproved hairy breeds of sheep.

Commercial wool grading is done by visual inspection. Wool graders are highly trained individuals whose experience enables them to grade wools rapidly and accurately. They check their visual standards periodically by inspecting standard sets of samples from wools that have been sectioned and measured for fiber diameter.

Production of Grades by Breeds. A fairly wide range of wool grades is commonly found among fleeces from any single breed of sheep, though

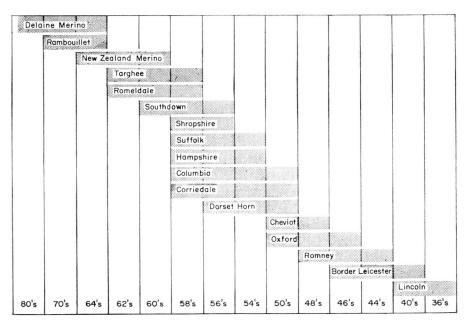

Fig. 40-1. Grades of wool produced by breeds of sheep common in the United States.

most fleeces from a given breed will fall within a narrower range. Figure 40-1 lists common American sheep breeds and the range of grades commonly found among their fleeces.

Length Classification. In addition to being graded on the basis of fiber diameter, wool is classified on the basis of staple length. The staple or combing length refers to the necessary length that permits a particular grade to be combed on the English or Noble comb. This length standard varies with wool grade. Shorter wools must be combed on the French or Heilman comb; wools that are too short to comb are known as clothing wools and go into woolen rather than worsted manufacture. Longer wools are more valuable on a per pound basis. Table 40-3 shows the length classification of wools according to grade.

Shrinkage. Raw wool consists of wool fibers, suint from the suint (sweat) glands in the skin, wool grease (the waxy secretion from the sebaceous glands in the skin), as well as moisture, dirt, vegetable matter, and other foreign material. Wool grease protects the growing fiber from wear and weathering. In purified form it is known as lanolin, a common component of salves, cosmetics, and skin lotions. The weight of the clean wool fiber content of the fleece, calculated as a percent of the total fleece weight, is

TABLE | *Staple length (in inches) designation by grade.*
40-3. |

Commercial Length Class	Fine	½-Blood	⅜-Blood	¼-Blood	Low ¼-Blood	Common
Staple	2.5 and longer	3.0 and longer	3.5 and longer	4 and longer	4.5 and longer	5.0 and longer
Good French combing	2.0	2.5	3.0	3.5	—	—
Average French	1.5	2.0	2.0	2.5	—	—
Short French	1.0	1.5				
Clothing and stubby	Under 1.0	Under 1.5	Under 2.0	Under 2.5	Under 4.5	Under 5.0

Source: E. M. Pohle *et al.*, 1953, USDA Oct. Mimeo., Washington, D.C.
Note: The length designations are based on unstretched staple length and represent a minimum length for the bulk of the staples in a sample.

known as the clean yield. This represents the proportional weight of clean wool fiber weighed at standard conditions (65% relative humidity and $70 \pm 2°F$). These conditions have been set up as standard for wool by the American Society for Testing Materials, because wool changes in weight, diameter, and other properties with small changes in temperature and humidity.

In commercial practice the wool buyer usually estimates shrinkage by manually hefting representative fleeces from a given lot and by inspecting them for dirt, grease, and vegetable content. A recently developed apparatus is used to remove a series of sample cores from bags or bales of wool of a given lot. These cores are then enclosed in plastic bags and sent to a testing laboratory to ascertain the clean yield of the samples. This procedure permits the buyer or seller to evaluate the shrinkage of a certain lot of wool on a more accurate basis. Today, many large lots of wool are purchased at a price that is determined by core-test results.

Purity. Purity refers to the relative amount of foreign material present in the wool. This term is useful in describing wool as to estimated shrinkage or to damage to the clip due to the presence of such substances as certain seeds and unscourable branding paint, which lowers the wool value.

Soundness. Healthy animals will produce wool fiber that is sound throughout its length. When animals have been suffering from senility, prolonged malnutrition, or chronic disease, the wool will be fragile throughout its length. This is known as tender wool. Some wool will show a line of weakness at a specific point in its staple, such that if a lock of wool is pulled it will break at this point. In other wools, the break occurs spontaneously, and the animal may shed portions of its fleece. This shedding usually indicates that either severe starvation or high fever has recently affected the animal. Tender or broken wools are of low value, since such wool will break apart when carded or combed, with the result that the final processed fiber will be shorter than desired.

Color. Pigmented wool, as well as hair fibers, are common in some breeds of sheep and are found in certain individuals of other breeds. Since anything other than white wool is undesirable for manufacturing, the inclusion of pigmented fiber limits the use of the wool. A white fabric cannot be made from brown or black fleeces. Even minor variations in color cannot be tolerated in the manufacture of white or pastel-shaded garments.

Some wools will show permanent staining from a variety of causes, such as bacterial or fungal damage to the fleece when on the animal, sacking wet wool, or drugs spilled on the animal during treatment for disease. The most common color defect in unpigmented wools is known in the trade as "canary yellow"—a bright yellow stain thought to be due to a combination of moisture and heat or light on the suint in the fleece.

Character. Grease wool that has a bright lustrous appearance, a desirable color (usually white to creamy), crimp pattern characteristic of the grade, and large well-defined locks is said to have excellent character. The term is used to denote all of the recognized characteristics that contribute to the general attractiveness of a fleece or clip.

Crimp. Wool from the improved breeds of sheep shows a wavy pattern. The distinctness of this waviness or crimp pattern varies somewhat with animal and breed and is greatly influenced by nutrition and health. This is a useful characteristic of wool and contributes to the springiness of the finished fabric. Menkart and Detenbeck (1957) noted that wools of identical fiber diameter, but that have more crimps per inch of unstretched length, can be made into a fabric with more desirable drape, handle, and shape retention characteristics than can wools with less crimp.

Uniformity of Product. Wool from some animals varies in fiber diameter and in length from one part of the fleece to another. Furthermore, such animals commonly show a wide variation in length and fineness of fiber within the lock of a given area. This is primarily a hereditary defect.

The fleeces of primitive sheep and of occasional animals in improved breeds may contain dead-white, opaque fibers that are hollow (medullated), uneven in diameter, and very weak in tensile strength. These are called kemp fibers. Other undesirable fibers are either coarse and hair-like, white, or colored. Animals whose fleeces contain such fibers should be culled from the flock. Lack of uniformity decreases desirability of the fleece for manufacturing.

40-3. PROCESSING AND HANDLING BETWEEN PRODUCER AND CONSUMER

If the sheep breeder knows the uses to which his clip may be put, he will understand the reasons behind his management procedures and will obtain a clearer idea of the value of his wool. Following shearing, fleeces are tied individually and sacked in large burlap bags, which may then be sold directly off the ranch or taken to a warehouse for storage before sale. The following paragraphs outline the general steps involved in processing wool from the time it leaves the sheep until it is made into a fabric.

Grading (Classing). At a wool warehouse belonging to a wool coopera-tive, wool merchant, or commission merchant, the wool is commonly classed by a reputable wool grader. As the fleeces are examined, they are thrown one at a time into large baskets or bins marked for individual grades, lengths and types of wool. This is ordinarily the first step in making up lots, or lines, of wool. There is usually a black (or colored) line, a tender (or broken) line, and, if wools from different areas are being graded, there may be a high-shrinking as well as a low- or medium-shrinking line. Divi-sion, however, is primarily made on the basis of fiber diameter, length, color, and soundness.

Graded lines of wool are resold from the warehouse to mills. The classing process permits a particular mill to buy just the grade and length of fleece required for its particular operations. Some lots of wool from large, uniform flocks are highly desirable in that they do not require grading. These are called original-bag wools and may be sold off the ranch directly to a mill.

Manufacturing. The following is a résumé of the general steps involved in wool manufacturing. Subsidiary steps, such as oiling, fulling, or the many finishing operations, have been left out in the interest of brevity.

SORTING. Graded lines and original-bag wools are still not uniform enough for a certain specific mill use. At the mill each fleece must be untied, tossed flesh side up onto a special table, and divided into its component types of wool. The sorting process involves skirting or removing the soiled, heavy-shrinking edge of the fleece, removing the seedy, high-shrinking belly, and, finally, sorting the principal grade areas of the fleece. The fleece from the legs and britch area is commonly the coarsest. Each type of wool goes into a different sorting line destined for a particular use.

SCOURING. Wools are usually run through an opener, duster, and burr picker, which serves to tear open the locks, remove some of the dust, and pick out larger burrs and seeds. In scouring, wool is passed through succes-sive baths of solution to remove dirt, suint, and wool grease. The solution may be a mixture of detergent, sal soda, and warm water; in a few scouring plants, fat solvents are used. Efforts are made to disturb or agitate the wool as little as possible to avoid cotting or felting.

Scoured wool comes out of the final rinse as a white product in contrast to the dirty appearance of raw wool. Particularly seedy lots may be subjected to strong sulfuric acid (carbonized), which attacks materials of plant origin and breaks them down. The remaining residue crumbles easily and can be dusted from the wool. Since the wool is slightly affected by the acid, the reaction to dyes may be altered somewhat. Carbonization may be done on scoured wools or on rolls of finished fabric. In recent years, most seed removal has been accomplished through the use of Peralta rolls—heavy steel cylinders through which the wool is passed subsequent to carding. The rolls crush and powder the plant material, but allow the wool to pass through virtually undamaged.

CARDING. A wool card is a machine that consists of a succession of cylinders studded with needle-like points between which the wool is passed. The card serves to homogenize the wool fibers and to distribute them in a smooth strand of carded wool called card sliver.

WORSTED, WOOLEN, AND FELT MANUFACTURE. There are three principal methods of wool manufacture, each of which includes a great variety of specialized procedures and utilizes different wools or blends of wool in the finished product. Wools for these manufacturing systems all go through the carding process.

In worsted manufacturing, the card sliver is combed on a special machine that arranges the fibers parallel to each other in the resulting strand of sliver, which is called wool top. In the combing process, short fibers that are somewhat smaller in average diameter than others in the wool are combed out. This material is called noil and is used in either felt or woolen manufacture. Some manufacturers stop their processing at this point and sell wool top in the ball to other mills. Such manufacturers are called topmakers. Wool top is spun into yarn, woven, and dyed to make the finished fabric. Worsteds are usually tightly woven, hard-finished fabrics that are made into such garments as men's suits. Fine worsteds, used in sheer, summer-weight fabrics, are made from long wools that are extremely fine in diameter.

Woolen yarns are spun directly from the card sliver and do not go through the combing process. The wools used in the woolen process are shorter and are usually coarser than those used in worsted processing, although the principal difference is in length rather than in diameter of fiber. Because lambs' wool is short and has a soft feel, it is highly desired for the manufacture of soft woolens. Woolen yarns are larger in diameter than worsted yarns and are rougher to the feel. Woolen fibers do not lie parallel to each other as do those of worsted yarn. This is because the wool has not been combed into top before spinning. After having been spun into yarn, the wool is woven into fabric and dyed. Woolens are usually bulkier, more loosely woven, and rougher to the feel than worsteds. A wide variety of fabrics are woven; these, in turn, are subjected to many different finishing processes. In the end, it may be difficult to distinguish some finished woolen garments from worsteds. Blankets, sweaters, some types of suiting, coat and dress materials, and socks are typical examples of woolen products.

Felt making is the oldest type of wool manufacture known to man. This process rests on the well-known ability of the wool fibers to mat or pack together into an unwoven fabric when subjected to heat, friction, and moisture. In the process, the fibers become so tightly interlocked that separation is difficult.

There are two principal types of modern felts. One is manufactured from carded wool, which is often mixed with cotton or other plant fibers, hair, or artificial fibers. Large amounts of reprocessed wool and noil are used for some felts. Wool up to $1\frac{1}{2}$ inches long is used, including all grades from 64's down to coarse carpet wool.

In processing, the fibers are mixed, thoroughly carded, and made into batting, which in turn is steamed and subjected to friction and heat. Fulling agents are worked into the batting, which is hammered repeatedly, after which the felt is washed to remove the fulling agent. Finally, this fabric is dyed and subjected to a variety of finishing processes. Felts are most familiar to us in the form of hats, felt washers, and rug pads.

Woven felts are made from wool and other fibers that are carded, combed, spun, and woven into heavy fabrics that are in turn felted and finished to make the final product. Some of our finest wool felts, such as billiard cloth, are made in this way.

KNITWARE. Knitting yarns may be either worsted or woolen yarns. The knitting industry is large and turns out a great variety of materials. Knitting is accomplished on intricate machines, which, depending on the model, can manufacture a tubular or flat fabric in a variety of stitches. Common examples are stockings, knit dresses, and sweaters.

40-4. MARKETING CHANNELS FOR WOOL

Some clips are sold by the grower directly to the manufacturer. These clips consist mostly of large, original-bag lots, which require relatively little classing or grading before mill use. Other clips are sold to local buyers engaged to buy wool for a particular wool merchant. Wool merchants grade or class wools and resell the graded lines to manufacturers. Wool merchants also act as commission merchants and may have wool consigned to them for grading and even for scouring, thus acting as wool handlers for the original growers. From the final sales' receipts they deduct a charge for their commission as well as the cost of the operations they perform.

In recent years cooperative wool marketing has grown rapidly. In a wool cooperative, growers combine and pool their clips for sale and handling purposes. In some instances nothing is done with the wool except that it is all for sale at one location; individual owners of the cooperative merely have the advantage of volume sales for buyer attraction.

Some cooperatives, however, maintain a complete warehousing and grading facility where wools are classed and sold to mills in uniform graded lots. Here each grower's wool may be weighed into the different lots. He is paid for each lot according to the amount of wool he has in it. The cooperative method works well if the management is efficient, honest, and well-acquainted with the needs of different mills in order that grading can be tailored to meet specific mill needs. Such a system is a particular advantage to the small grower, whose clip is not large enough to attract bids from several buyers. Large growers benefit by having their wool graded on a cost basis.

40-5. COMPETITION BETWEEN WOOL AND SYNTHETIC FIBERS

Wool producers face a difficult problem in the competition between wool and synthetic fibers. Thus far, no one synthetic fiber has all the desirable properties characteristic of wool. Some of them, however, do have certain properties similar to those of wool and can be made into fabrics simulating wool. A number of synthetics are used in wool blends.

Synthetic fibers do have some advantages over wool. Since they are manufactured by extrusion from small openings, synthetic fibers are exceedingly uniform in diameter. This is not true of wool. Furthermore, synthetic fibers can be made in any length, whereas wool fibers are limited to the growth provided by the sheep in 12 months. Some sheep have difficulty in producing wool of combing length (that is, fine wool, $2\frac{1}{2}$ inches long) in 12 months. Moreover, synthetics can have a crimp or wavy pattern imparted to them mechanically and can be made in a variety of cross-sectional forms. Fibers can be extruded that are oval, round, square, W-shaped, or U-shaped in cross section. Each one of these cross-sectional shapes will impart a slightly different handling characteristic to the fabric. Modern wool research is currently changing the wool fiber to compete with synthetics more efficiently (Chapter 4). Wool growers can help meet competition from synthetics by selection for uniformity of wool growth in their sheep, by selection for increased length of staple, and by growing and marketing their product in the best way possible.

40-6. MOHAIR

Origin of Mohair. The mohair goat was originally from Angora, Turkey. It was first introduced to this country in 1849, in South Carolina; with continued importation, it gradually spread to the South and West, where it has proved most popular. Texas furnishes approximately 95% of the mohair produced in the United States.

Grades of Mohair. Specifications and the total number of grades of mohair have not been firmly established. Figure 40-2 illustrates the series of grades of mohair based on graded samples from the spring clip. The fall clip has a slightly different appearance, owing to long exposure to sun, and the presence of dirt and seed defect accumulated during the summer season.

Table 40-4 shows the commercial grade names and specifications in common use. Some trade sources include an additional grade, 26's.

Characteristics of Mohair. Mohair is a beautiful, smooth, very lustrous fiber that is somewhat coarser than wool and that has a less pronounced waviness or crimp. Mohair tends to be more uniform in diameter than

Fig. 40-2. The USDA proposed counts system grades of mohair as shown from samples of the spring clip. See Table 40-4 for the commercial names and diameter measures of the grades.

TABLE | *Fineness grade guide for mohair.*
40-4.

Commercial Name	Counts Grades	Average Fiber Diameter (microns)
No. 1 Kid	40's	24
	Bulk 36's	26
No. 2 Kid	Bulk 32's	28
No. 1 Grown	Bulk 28's	30
No. 2 Grown	Bulk 24's	34
No. 3 Grown	Bulk 20's	38
No. 4 Grown	Bulk 16's	41

Source: Courtesy of Elroy M. Pohle, Livestock Div., Wool Laboratory, USDA, Agricultural Marketing Service, Denver, Colorado.

wool within the lock. By comparison with wool, the fiber feels slick, owing to its smooth scale pattern.

The goats are usually shorn twice a year (spring and fall); shedding may occur if this is not done. In addition, animals thrive better in shorter fleece. Improved strains are able to produce clips averaging one inch per month in growth, and have grease fleece weights up to 12 lb. per year (two shearings). In 1959, the yearly average United States grease fleece weight was 6.4 lb. per animal. Mohair goat breeders recognize three general kinds of fleece on the basis of lock type (Gray, 1958). The ringlet (C-type) sometimes shows some twist or spiral but typically hangs in long, tight locks or curls about the diameter of a pencil. This type of fleece usually has sufficient length, but may lack density, thus allowing excessive penetration of dirt into the fleece. A second category is the flat lock (B-type), which hangs in flat, wavy locks. According to Gray, long-range selection for this type of fleece may result in a shortening of staple length and, therefore, lighter fleece weights.

The third lock type, called the web lock, or round lock, is intermediate between the other two. Goats having web lock fleeces are said to produce heavier fleeces of good staple length for a greater number of years than goats having the other two fleece types. Since light and heavy or coarse and fine fleeces are found within each of these three fleece types, some growers are still not convinced of the superiority of any one type, although the web lock type seems to be most favored.

Marketing Mohair. Mohair is sold in much the same way as wool, usually by companies that act as warehousers, buyers, and handlers for both products. Some producers sell the clip on a graded basis; for this reason, a handling company often maintains a grading or classing service. The sale of graded lots from the clip is credited to the account of the consigning

owner. Relatively little mohair is marketed through grower-owned cooperatives.

The most valuable raw mohair is that shorn from the kids in fall or
spring. Kid hair shrinks less and is softer and finer than that from mature
goats. Yearling mohair clipped at about 18 months of age is more desirable
than adult mohair.

Sixty percent or more of the United States mohair clip is exported annually, principally to the United Kingdom; some is sent to other European
countries and Japan.

Use of Mohair in Textiles. A résumé of the characteristics and uses of
mohair has been given by Marincowitz (1959). Mohair is used in the production of fine velvets and in the manufacture of plush upholstery fabrics
for cars and furniture. It is highly desired for such uses because it is stiff
and lustrous, holds up well under long continued use, and shows off color
to advantage, owing to its glossy appearance. Since the fiber is stiff, it is
noted for its role in shape retention; in some uses, it imparts a slick, hard
finish to the fabric. Finer grades of mohair are used in the manufacture of
suits, gowns, robes, and ties, for which they furnish crease resistance, durability, and excellent drape characteristics. The slickness of the fiber requires that spinning be performed carefully, since fibers tend to slide past
each other quite easily. Large quantities of mohair are used in rugs and
carpets.

REFERENCES AND SELECTED READINGS

Gray, J. A., 1958. Texas Angora goat production. *Sheep and Goat Raiser.* Dec., p.
  10.

Menkart, J. and J. C. Detenbeck, 1957. The
  significance of wool fiber crimp. Part I.
  A study of the worsted system. *Textile
  Research J.*, 27:665–689.

Marincowitz, G., 1959. High prices have
  been paid for mohair. *Sheep and Goat
  Raiser.* Nov., p. 30.

USDA, Agr. Marketing Service, 1950. *Notice
  of Proposed U.S. Standards for Grades of
  Mohair.* Federal Register, Oct. 18, p.
  6969.

Chapter 41

Marketing of Milk and Dairy Products

Many there be who from their mothers keep
The new-born kids, and straightway bind their mouths
With iron tipped muzzles. What they milk at dawn,
Or in the daylight hours, at night they press;
What darkling or at sunset, this ere morn
They bear away in baskets—for to town
The shepherd hies him—or with dash of salt
Just sprinkle, and lay by for winter use.

VIRGIL, *Georgics, III*

Compared to the marketing of livestock and meats or poultry and eggs, the marketing of milk and dairy products is an extremely complex subject. The dairy industry may be considered as a collection of industries, each one having its own unique marketing problems and systems. The differences depend upon the type of product and geographical and time factors.

First we will consider in a general way the nature of dairy products as a part of the food industry. Milk, because of its universal use by infants, has been made the object of legislation to control its quality. State, municipal, and federal laws have been devised to control its composition and bacteriological quality at both producer and consumer levels. Quality of product and quality of package and knowledge of the importance of milk in nutrition have made milk and its products highly acceptable to the public. It is recognized generally in the industry that it is impossible to get a high-

TABLE | *Per capita consumption of dairy products* (1963).
41-1. |

Dairy Products	Pounds of Product
Fresh whole milk	275.0
Skim or low fat milk	29.0
Cream	8.5
Sweetened condensed milk	0.4
Unsweetened condensed milk	1.8
Evaporated milk (whole)	9.4
Evaporated and condensed skim milk	4.6
Butter	6.8
American cheese (cheddar)	6.2
Other type cheese	3.1
Cottage cheese	4.4
Dry whole milk	0.3
Non-fat dry milk	5.4
Dry whey	0.3
Malted milk	0.1
Dry buttermilk	0.4
Frozen desserts (net milk used)	51.9
Ice cream (net weight)	18.0
Ice milk (net weight)	6.0
Sherbet (net weight)	1.5
Other frozen dairy products (net weight)	0.2
Mellorine (net weight)	1.3

Source: USDA, Dairy Situation DS 302, August 1964.

quality product from low-grade raw milk. Hence, giving high-quality milk to the processor of fluid milk or of manufactured dairy products is the first step in the successful marketing of dairy products.

The dairy industry may be considered to be made up of two major divisions—fluid milk, including fluid by-products, and manufactured products. Table 41-1 lists the per capita consumption of the principal dairy products in pounds of finished products, such as butter and cheese. Fluid whole milk can yield a number of products; for example, 100 lb. of 3.7% fat milk with 8.8% milk-solids-not-fat can yield 4.6 lb. of butter (80% fat), and 9.1 lb. of nonfat dry milk and dry buttermilk (8.6 lb. of nonfat dry milk and 0.5 lb. of dry buttermilk at 96% solids). Expressed in other terms, 1 lb. of butter can be made from 21.8 lb. of milk, and 1 lb. of nonfat dry milk requires the milk-solids-not-fat from 11 lb. of milk.

Fluid milk and cream utilize slightly more than 50% of the total milk production. In 1963 the total per capita consumption of milk in all forms was 628 lb., of which 308 lb. was in the form of fluid milk and cream. Table 41-1 lists the 18 manufactured products that accounted for 320 lb. per capita.

41-1. FLUID MILK MARKETING

Fluid-milk marketing will be considered under two subdivisions: marketing of the raw milk to the processors and the marketing of the finished products to consumers via retail and wholesale channels. Before discussing these two aspects of the problem, we will characterize the nature of this market and differentiate it from the marketing of manufactured milk products.

Fluid milk is bulky in comparison to its solids content, and it is more perishable than manufactured products. Because of these two factors, it must be produced fairly close to the point of consumption. However, in recent years improved refrigeration, better transportation equipment, and faster highways have extended the area of production of fluid milk for a given market. The extension of the production area has contributed to the complexity of the marketing problem in terms of state and federal regulations of fluid milk marketing. In a review of barriers to trade in fluid milk, Hardin (1955) has stated the problem well: "Economics and politics meet in the production and distribution of fluid milk."

Producer's Markets. The milk producer must find a market for fluid milk in his area. Since processors have increased in size and decreased in number, the bargaining power of individual dairymen has decreased. In order to compensate for the loss of bargaining power, most dairymen belong to a cooperative marketing association, which we may classify into three separate categories, on the basis of the activities of each:

1) An association strictly for bargaining with no processing facilities.

2) A bargaining association, with processing facilities for handling surplus.

3) A cooperative distributing association, usually competing with private processors.

PRICE PLANS. The milk marketing economists refer to the different systems of buying milk by processors as "price plans."

The present complex price structures used in the purchasing of milk in the large markets have been gradually evolved from the experience of the past forty years. Because of the large fluid-milk surpluses at certain seasons, producers have sought a price plan that would permit payment of milk on a use basis and yet reward those who attempted to limit the seasonal surplus by maintaining a fairly even production throughout the year.

It is impossible to discuss each price plan in detail, but a list and brief description of the principal plans that have been used will indicate the many solutions tried for special situations (Roadhouse and Henderson, 1950; Spencer, 1956).

The flat rate plan. This plan was common before 1918. The producer was paid a uniform price for all milk purchased, regardless of use.

The use classification plan. Milk was assigned a class according to its use: Class I (fluid milk), Class II (cream), Class III (ice cream and other products). Class I, the highest in price, is usually the negotiated price. A classification of milk according to use requires a "pooling system." The pooling system refers to the method whereby the market is divided among the producers on an equable basis. In this plan the producer receives a blended price, depending on the use of the milk. It involves the use of a dealer pool or a market pool. The dealer pool is operated in the absence of a cooperative association or a federal or state board, functioning to operate a market-wide pool. The use classification plan is most frequently combined with the two plans that follow.

The basic surplus or quota plan. This plan is a variation of the pooling system of the use classification plan. The purpose of the plan is to attempt to favor producers who will regulate their production to accord with sales requirements. Many variations of the plan are used.

The formula plan. This plan is based on cost of feed, labor, and other costs (Spencer, 1956).

Federal and State Control of Price to Dairymen. Both state and federal laws have been established to regulate the price, and in some states, the market area, for Class I milk.

FEDERAL MARKET ORDERS. The federal laws are known as Federal Market Orders. As of January, 1963, there were 82 market orders in operation. About two-thirds of the nonfarm population of the United States now reside in marketing areas defined by Federal Market Orders, operating under the minimum price terms to dairymen. One of the basic functions of Federal Milk Orders is to encourage the production of milk adequate to supply the needs of a given market (Bartlett, 1956). The orders define marketing areas, classify milk sold in such areas according to its use, fix prices for each classification, and provide for pooling of receipts and the ascertainment of blend prices so that producers will be paid equally (except for the differentials for quality and transportation). The plan is operated by a federal official. The processors pay the entire cost of the Federal Market Orders program.

STATE AND MUNICIPAL CONTROL. During the period from 1933 to 1935, 17 states passed laws that attempted to control wholesale, retail, and producer prices of milk. Since 1935, 28 states have introduced similar legislation. Today 19 states fix the minimum prices to be paid to producers; 11 of these states also fix minimum retail and wholesale prices (USDA, 1964). **Mu-**

nicipal control has largely been related to inspection of sanitary requirements. It appears that many sanitary codes are needlessly detailed and serve
to "build a wall" around their local markets in favor of their own dairymen and processors.

Trends in Dairy Production. Dairy farming in recent years has been following the trends of all types of production activities; that is, securing more
product per man hour. This trend has resulted in fewer but larger dairy
farms with increased production per cow, a development that has been
necessary to maintain the income of the dairymen who remain in business
More and more of the total milk production is used for fluid (Class I) milk,
and this type of operation requires more investment in building and equipment and more care in handling the cows and the milk.

The relationship of the producer to the processor is also changing. Federal Market Orders now cover areas including approximately 190,000 dairymen and supply markets furnishing fluid milk to approximately 100,000,000
people. The Federal Orders Program tends to equalize the competitive position of all dairymen in the market area. The requirement that class prices,
location differentials, and butterfat differentials be uniform is a major
benefit to processors as well as to producers. The pooling provisions of the
Federal Orders make it possible for each producer to find a market on
equal terms with other producers. It is thus not necessary for individual
producers to undercut each other in price in order to retain a market for
their milk. The orders do not require processors, however, to purchase
milk from a particular group of producers or to purchase milk in any particular amount and consequently cannot guarantee a market for producers'
milk. The orders provide only for fixing prices on an equitable basis where
the order is in force. The cooperative dairy associations play an important
role in marketing in Federal Orders areas.

In those states where the Federal Orders do not operate and state control
laws function, the relationship of the producer to the processor is usually
more direct. Contracts are made either with the individual producer or
with a cooperative marketing association if one exists in the producer's
area and he is a member. The contracts usually specify a minimum amount
of Class I milk that must be supplied. The price received by the dairyman
is a blended price, depending on the amounts of Classes I, II, III purchased.
Changes in producers furnishing a particular processor may take place.
There are a number of causes of these changes: unsatisfactory quality is
the most common reason for a change; unsatisfactory financial arrangements for loans may be a reason for cancellation of contracts.

Processing. Processing facilities and methods are important in considering the marketing of fluid milk. Changes in the market milk industry in

the past 50 years have converted the basic unit of the industry from a simple, nearly cottage-type operation to an industrial factory operation (Henderson, 1956). The following types of operations now exist.

PRODUCER-DISTRIBUTOR. As a result of increased labor and other costs, this type of operation is decreasing in importance. In certain isolated areas, however, the producer-distributor is able to compete with larger operators.

SINGLE INDEPENDENT PLANTS. Before 1925 the single plant was the most important form of organization. Many single-plant operations are too small for economical processing unless they have unique operating methods or market outlets. Consolidation of production facilities may be expected to continue to favor multiple-plant operations.

MULTIPLE-PLANT OPERATIONS. The trend is toward consolidation of plants into local or regional chains and to national chains. The increased costs of processing and distribution favor this type of operation since it contributes to better utilization of labor and capital. Large plants have automation in many departments, which reduces processing costs. The largest eight dairy chains of the United States, regional and national, probably handle 40% of the fluid milk and cream.

Distribution. Distribution of fluid milk is classified as retail or wholesale.

RETAIL DISTRIBUTION. This is effected in three ways: company-owned trucks, driver-owned trucks, or distributors who may have a fleet of trucks serving a specific area. Retail distribution is the most expensive type of delivery since it does not lend itself to an efficient use of labor and equipment. In recent years, however, certain procedures have been introduced to increase the output of labor and equipment. Some of these practices are delivery every other day or three times a week, discounts for increased quantity of product delivered at one time, and additional products handled —eggs, ice cream, and other food items. In many large cities approximately 30 to 35% of milk is sold from retail trucks. Most of the retail milk is packaged in glass bottles.

Drive-in dairies. In recent years the drive-in dairy has become a factor in milk sales in certain areas. The drive-in dairies are of three types: (1) Ranch drive-ins, which combine milk production, processing, and sales in one location; (2) dock drive-ins, which combine only processing facilities and sales; (3) drive-ins that sell milk processed elsewhere at the same prices as grocery stores. The type of agriculture and the market areas determine to a great degree the type of drive-in. If dairy farming is adjacent to large urban areas with convenient major highways, the ranch type drive-in predominates, and if dairy farming facilities are not close to urban populations, the dock type drive-in or the drive-in selling milk processed elsewhere tends to predominate.

The drive-in type of operation thrives best where minimum prices to producers and consumers are under state control. California has perhaps the largest percentage of milk sold in drive-ins. A price differential of 1½ to 2 cents per quart doubtless stimulated this growth. At the present time the differential is 1 cent per quart. The peak in sales was reached in central California in 1962, when 6.4% of milk sales were from drive-ins. In 1963 the total sales in the same market was 5.8%.

WHOLESALE DISTRIBUTION. Approximately 65% of the milk sold in many urban areas is distributed from food stores, some types of which are discussed here.

Independent or chain dairies—local, regional, or national in scope. The fluid milk and by-products may be packaged in the brand of the dairy or in the private brand of the retailer. Smaller stores are usually served every other day; supermarkets, at least once a day and often several times. Wholesale delivery is usually limited to a 5-day operation to conform with the 40-hour week required in most union contracts. Quantity discounts are usually on the basis of volume delivered at one time. Nearly 100% of the milk sold in stores is packaged in single-service fiber cartons.

Captive dairies. A group of small supermarket chains or a voluntary group of single supermarkets may control a dairy that processes and distributes exclusively to the members of the chains. The business is conducted like that of the independent chains, except that a sales force is not needed and advertising is usually included in the store advertising.

Dairy chains owned by supermarket chains. A number of large national grocery chains operate a chain of dairy plants in their principal marketing areas. Their operation is similar to that of the captive dairy, with the difference being that the ownership is in the control of one large chain rather than being the joint venture of a group of smaller chains or a voluntary group of single large supermarkets.

Trends in Fluid-milk Marketing. Fresh fluid milk will doubtlessly continue to dominate the market but technological developments are making other products acceptable as substitutes and introducing new forms of milk that tend to increase the total consumption. Some of these developments are instant nonfat dry milk, instant powdered whole milk, canned sterilized whole milk, aseptic concentrated milk, frozen and fresh 3:1 concentrated milk, recombined milk made with low-heat nonfat dry milk, and anhydrous (pure) milk fat. Efforts to cut costs of processing and delivery of fluid milk or acceptable alternates will dictate the future trends in this segment of the dairy industry. We can expect increases in the size of producer dairies, but fewer of them, and parallel changes in the processing plants. Automated operations of many parts of the process will further reduce labor costs. Five-day processing in plants will become a general practice and equipment will be designed to permit rapid handling of large volumes of milk. Retail delivery will be less frequent as quantity discounts become more general. Household dispensers will doubtless increase in importance, and it is probable that delivery will be reduced to once or twice a week. The 10-quart polyethylene bag in a fiberboard container is the most recent development in the home delivery of milk. Single service plastic bottles blown in the processors plant are now available in half-gallon and gallon sizes and may soon be in general use, especially for wholesale milk distribution.

Vending machines for $\frac{1}{2}$ pint, $\frac{1}{3}$ pint, quart, and $\frac{1}{2}$ gallon—in factories, apartment areas, depots, and other central locations—will make milk more available. Many surveys have shown also that increased total consumption of milk results from such distribution.

41-2. MANUFACTURED PRODUCTS

With the exception of cottage cheese and frozen desserts, most manufactured dairy products can be produced in areas remote from consumption centers, are reduced in H_2O content as compared to whole milk, and can be held for relatively long periods of time without deterioration. Table 41-1 lists 22 manufactured products, which will be discussed under the classifications of butter, cheese, dried milk, condensed and evaporated milk, and frozen desserts.

Butter. The production and marketing of butter has undergone many changes in the past 50 years. Before the introduction of cream separators in 1878, much of the butter was made on the farm, where cream was secured by gravity separation. The separators made it possible for the dairyman to sell cream to a centralized butter plant and to retain the skim milk at home for feeding to pigs, chickens, and calves. Cream was often delivered after it had undergone some deterioration, and the resultant "centralizer" butter was of poor quality. With the development of evaporators and driers to utilize skim milk for human food, creameries were established to receive whole milk from farms and to convert it into butter and nonfat dry milk. This development, which resulted in an improvement in the quality of butter and in better utilization of the skim milk, was accelerated during World War II.

The annual per capita consumption of butter is approximately 7 lb. per person. The consumption of oleomargarine is slightly higher. This relationship has been maintained for a number of years and it is likely that the improved quality of butter and advertising supported by the dairy industry will maintain consumption at the current level, although the public's concern with fat in the diet tends to restrict the per capita consumption of all fats in major markets.

Cheese. Many types of cheese are manufactured and sold in the United States. Some types of cheese are imported, principally from Switzerland, Holland, France, Italy, and Denmark.

AMERICAN CHEESE (CHEDDAR TYPE). This type accounts for approximately 65% of the cheese sold in the United States. The per capita consumption is 6.2 lb. a year. In recent years considerable improvements have been made in the packaging and the merchandising of American cheese. Large wheels of cheese bandaged with cheesecloth and treated with paraffin are being replaced by rindless blocks. This type of package lends itself to cutting into wedges, slices, or bricks; the blocks are more attractive in appearance and aid in better merchandising.

OTHER TYPES OF CHEESE. These include all other varieties except cottage cheese. The annual per capita consumption is 3.1 lb., or approximately one-half that of Cheddar cheese. Swiss type, blue cheese, Monterey, Gorgonzola, and many other interesting cheeses can be effectively merchandised and increase the utilization of milk.

COTTAGE CHEESE. Although cottage cheese is a manufactured product, it is usually sold along with fluid milk in both retail and wholesale outlets. The annual cottage cheese consumption, as shown in Table 41-1, is 4.4 lb. per capita, but in one state it is nearly 10 lb. This high rate of consumption is due to a high-quality product, effectively merchandised for a period of years.

Dried Milk. The development of the evaporator and drier for dairy products has resulted in increased use of milk products for human food and has made it possible to provide milk in many parts of the world where it would not otherwise be available. Some types of dry milk products have been designed for special purposes.

NONFAT DRY MILK. Nonfat dry milk is made by either the spray or roller process. The spray process is by far the most common. The USDA has established commercial grades for both spray and roller powder. Spray powder is available in three classifications, depending upon the heat treatment used in processing. High-heat powder is used in making bread and in producing many other foods; low-heat powder is used for beverages and for making cottage cheese, recombined milk, and ice cream; intermediate-heat powder is used wherever either the high- or low-heat product is not indicated. The American Dry Milk Institute has been instrumental in developing standards for dry milk and in preparing educational material to promote its use in foods.

The per capita consumption of nonfat dry milk is 5.4 lb. annually, and consumption is increasing rapidly. The growth has been accelerated by the introduction of instant nonfat dry milk in consumer packages.

DRY WHOLE MILK. Dry whole milk shows a low per capita consumption

—only 0.3 lb. The recent development and introduction of instant whole milk will doubtless result in a significant per capita increase in the dry product. It doubtless will be used to augment and supplement fluid milk in many areas, especially in those where a dependable quality of fluid milk is not available at a competitive price. Foreign markets will be an important factor in utilizing milk in the form of instant dry whole milk.

MISCELLANEOUS DRY MILKS. The per capita consumption of dry whey is now 0.3 lb. The utilization of edible dry whey in many products will increase this consumption. The lactose (milk sugar) and protein of whey are valuable food constituents and many applications for their use for human food are being developed.

Dry buttermilk is a by-product of butter manufacture. The per capita consumption of 0.4 lb. is likely to remain constant, since butter production is not expected to increase. Dry buttermilk is used in ice cream and in many types of packaged food.

Condensed and Evaporated Milk. Evaporated and condensed milk plants are generally located in manufacturing milk-production areas rather than in the milk shed of a large urban marketing area. Freight rates, however, make it essential that evaporated milk plants be not too far removed from the areas of consumption.

EVAPORATED MILK. The evaporated milk industry is in the hands of a few large companies because a factory requires a large investment and the product is an item of large volume and low margin of profit. The per capita consumption of evaporated milk is 9.4 lb., but it has been declining. A new development in the field, an aseptic product, may reverse the decline in consumption. Evaporated milk is sold exclusively through grocery stores.

CONDENSED MILK. Sweetened condensed milk in cans is no longer an important form of manufactured milk in the United States. Considerable amounts, however, are still manufactured for sale in the Orient. Sweetened condensed whole milk in bulk is manufactured for use in candy and to a lesser extent for use in ice cream. Condensed skim milk is used for milk solids in ice cream and other foods.

Frozen Desserts. Frozen desserts, such as ice cream and ice milk, as Table 41-1 shows, accounts for a per capita consumption of 52.4 lb., in terms of milk. The per capita consumptions of ice cream and ice milk are 4 and 0.9 gallons, respectively. Ice cream is no longer regarded as a luxury—it is a stable food item. The advent of the half-gallon container, the increase in home freezers, and the wide distribution of ice cream at attractive prices in stores and especially in supermarkets have accelerated the per capita consumption of ice cream and other frozen desserts. The increase in the sale of ice milk in recent years has accompanied the interest of the general public in foods that are low in fat, high in protein, and low in total caloric value. This trend was noted in the discussion of fluid products, which include nonfat milk fortified with milk solids and in connection with the "900 diet foods," which use milk solids-not-fat.

Imitation ice cream of the mellorine type continues to grow and depress the margins on all frozen desserts. This type of product can now be legally manufactured in 12 states.

41-3. BASIC FACTORS INFLUENCING THE MARKETING OF MILK AND DAIRY PRODUCTS

Quality. Dairy products, like all foods, must have quality appeal if they are to be successfully merchandised. The dairy industry is the most strictly regulated of any food industry with respect to sanitation of producers' and processors' premises and to the bacterial quality and composition of the finished products. Most dairy companies attempt to control the quality of the product from the cow to the consumer. Plant laboratories of most companies maintain constant checks on incoming milk and finished products.

Knowledge of Nutritive Value of Milk and Dairy Products. Scientific interest in milk as a food began with the discovery of vitamin A by Dr. E. V. McCollum in 1913. This study stimulated research on the other nutritional factors in milk. The work of Dr. H. C. Sherman on the nutritional factors required to achieve "buoyant" health dramatically illustrated the benefits of drinking milk regularly in adequate amounts.

The National Dairy Council program brought the nutritional benefits of milk to the attention of doctors, dentists, nurses, nutritionists, and teachers. In 1924 the council initiated the program for serving milk in schools. Today about $2\frac{1}{2}\%$ of the total milk production—154 half pints per child in the participating schools—is utilized in the school milk program. The school milk and lunch programs have made teachers and children acquainted with the benefits of milk drinking and have undoubtedly done much to stimulate the use of milk.

Advertising and Sales Promotion. A third factor in marketing dairy products is advertising and sales promotion. All food products are competing for a market limited by the amount of food a person will consume. It is essential that the story of milk and its advantages be continually kept before the potential customer in order that the dairy industry will receive its "fair share" of the food dollar.

The American Dairy Association was organized to advertise and promote the sale of all types of dairy products. Millions of dollars are collected from the dairy producers each year to finance the program of the ADA. All types of media are used—TV, radio, newspapers, billboards, point-of-sale material, and printed booklets on many dairy product subjects. Such advertising is for the benefit of the whole industry, increasing the sale of all types of dairy products. Many dairy companies, however, use the ADA material and tie it in with their own advertising programs to take advantage of the well-developed ADA programs.

Independent companies also spend millions of dollars a year in advertising their own brand names and the merits of their products. Attractive cartons and point-of-sale material, in addition to a quality product, are necessary to make an advertising program a continuing success.

Special Problems of the Dairy Industry. The dairy industry in recent years has been plagued by a number of developments largely outside of its control. Nuclear fallout, pesticide residues, and cholesterol have occupied the attention of the public, because of news releases suggesting that the public health was in danger from strontium 90, iodine 131, DDT and other pesticides, and from the possible relation of cholesterol to atherosclerosis.

These factors are now being largely resolved in favor of the dairy industry. Since the banning of nuclear tests, strontium 90 levels are gradually decreasing. If tests are resumed and the strontium levels become a public health problem, equipment and processes are available to remove most of the contamination. The amounts of pesticide residue in milk have been greatly reduced since 1960, owing to efforts of dairymen, processors, and regulatory agencies, and are much lower than those permitted in other foods.

Finally, the cholesterol problem is now considered to be only one factor in lipid metabolism and not the primary cause of atherosclerosis. Glycerides, phospholipides, sugars, and perhaps other factors, such as stress and lack of physical activity, are apparently involved.

41-4. PARTICIPATION OF UNITED STATES GOVERNMENT IN DAIRY PRODUCTS' MARKETING

Marketing. A discussion of the marketing of dairy products should take into account the part that government plays. The federal government is committed to support the price of milk at not less than 75% or more than 90% of parity. In order to implement this law, milk products are purchased rather than milk. Before April of each year an estimate of the price at which manufacturing milk and milk fat must be sold to reach 75 to 90% of parity is made. The Secretary of Agriculture determines the percent of parity to use, depending upon supplies of milk and other factors. The products purchased are butter, Cheddar cheese, and nonfat dry milk. The prices paid for the products are determined by the price of milk per 100 lb. that would attain the desired parity. The Department of Agriculture must buy all milk products offered that will meet their specifications. The program is intended to support the market, but its net effect is to stimulate production artificially. As support prices drop, marginal producers leave the market and production more nearly equals supply.

The disposal of the products purchased is accomplished in a number of ways. (1) The products can be sold in the domestic market if the price paid (plus storage and handling costs) is recovered. (2) The school lunch program receives some of the supported products. (3) The Army and Navy can use the products if the use will not interfere with commercial operation. (4) The products can be given to charities. (5) The mutual aid program for foreign countries disposes of large amounts of the products, especially nonfat dry milk. (6) The products may be sold to commercial interests at reduced prices if they are exported from the United States and reduced in price to equal world market prices.

REFERENCES AND SELECTED READINGS

References marked with an asterisk are of general interest.

Bartlett, R. W., 1956. Government control of the dairy industry other than health measures and standards. *J. Dairy Sci.*, 39:892–899.

* Dairy statistics-supplement for 1959 to statistical bulletin No. 218, June, 1960. Agricultural Marketing Service, USDA, Washington, D.C., p. 91.

Hardin, C. M., 1955. *Political and Institutional Barriers to Trade in Fluid Milk. Marketing Efficiency in a Changing Economy.* A report of the National Workshop on Agricultural Marketing, University of Kentucky. Agricultural Marketing Service, USDA, Washington, D.C.

Henderson, J. L., 1956. Market milk operations, 1906 vs. 1956. *J. Dairy Sci.*, 39:812–818.

* Roadhouse, C. L., and J. L. Henderson, 1950. *The Market Milk Industry.* 2nd Ed. McGraw-Hill, New York, pp. 534–547.

Spencer, L., 1956. Significant developments in the distribution and pricing of market milk in the United States, 1900–1956. *J. Dairy Sci.*, 39:884–891.

USDA, 1964. Dairy Situation: Economic Research Service, DS 302. Washington, D.C. August, 1964.

Chapter 42

Marketing of Poultry and Eggs

What's the use? Yesterday an egg, tomorrow a feather duster.
MARK FENDERSON, *"The Dejected Rooster"*

42-1. INTRODUCTION

Marketing of poultry and eggs has changed rapidly during the past few years. From a pattern of institutions, agencies, and procedures seemingly well established, a nearly complete alteration has occurred or is in the process of taking place in the production and distribution of poultry and egg products, including the channels of distribution and the marketing agencies involved in moving the product to market.

There are a number of reasons for this change. First, the average size of broiler and egg-producing flocks has increased. Small producers have been declining in numbers for some time, but the rate of withdrawal has increased materially during recent years. Not only have many small producers gone out of business, but many of those who remained substantially increased the average size of their flocks. The increased size of production units has led to a shift in the markets and market agencies used and to a more widespread adoption of grades and standards for poultry and egg products. The large producers are considerably more quality-conscious than were the earlier small producers.

The second factor that has been instrumental in changing the poultry and egg marketing pattern has been a shift in the location of the production areas, the most noteworthy of which has been the development of broiler meat and egg production in the southern states. These states, formerly a deficit area, are now producing a tremendous surplus of poultry meat and are rapidly reaching the point of surplus egg production. As this area further develops its production facilities, and as other major production areas attempt to gain premium outlets for their products, through the reorientation of their marketing programs, increased attention will be given to the production of quality products and to the wider use of grades and standards in marketing poultry and eggs.

A third factor is the change in size and location of marketing firms handling poultry and eggs. These firms were historically small, and situated in large cities. In recent years, however, there has been a marked shift to large firms with headquarters in the production rather than the consumption areas. This particular shift—in size and location—means that marketing firms must emphasize quality and grade standards to facilitate the marketing of large quantities of uniform products in distant areas and in premium markets.

A fourth factor in the changing marketing program is the pronounced shift to more direct marketing in many areas. Producers follow this practice by selling direct to retail stores or to chain store warehouses. Formerly they sold to country buyers, who in turn sold to larger egg handlers; these men sold to city wholesalers, who eventually sold to the chain store warehouses or to retail stores. Marketing firms also are engaging in more direct marketing through the packaging of poultry and eggs in consumer packages at country processing points and shipping them directly to either stores or chain store warehouses. In direct marketing, the wholesale market facilities, formerly used to a large degree for all poultry and eggs, have been bypassed.

Direct marketing is of significant economic advantage to producers and processors. However, it presents the industry with a lack of large central wholesale markets wherein the movement and pricing of poultry and eggs can be measured effectively, as was done in earlier years. The net result is that the wholesale markets have declined substantially in volume and many producers have become extremely critical of the validity of prices established in these markets when the bulk of the product moves direct to stores or chain warehouses. Two requirements for successful direct marketing of poultry and egg products are the maintenance of a high standard of uniform quality and the availability of a sufficient volume of the product to meet the needs of the buyers throughout the year.

Contract farming has made its most noticeable progress in the poultry field, in which the bulk of the broilers and turkeys are now produced on a contract basis, and increasing attention is being given to contract production of eggs. Although contract farming is essentially a means of financing production, it has led to the more uniform flow of production and to improved quality of the products. The more uniform flow results because producers are essentially working on a salary basis; they tend to keep their facilities at full operation rather than change the volume to follow the market prices. Moreover, in typical contracts, specifications are included concerning the management of the flock and the handling of the products. This in turn has encouraged and facilitated the development of uniform, high-quality products.

Finally, throughout the poultry and egg industry, renewed attention has been given to developing new poultry and egg products. The industry has for years relied upon the sale of shell eggs and of broilers, fryers, roasters, and stewing chickens, but more recently the types of products sold have been reduced to shell eggs, fryers, and occasionally stewing chicken. The widespread trend toward the production of convenience food items for modern housewives developed at a time when the poultry industry was lagging seriously. Chances are good that the next few years will see the production of many new products, which will broaden the market for poultry meat and eggs.

TABLE 42-1. | *Average civilian per capita consumption of eggs, chicken and turkey, United States, 1910–1964.*

Year	Eggs	Chicken* (lb.)	Turkey* (lb.)
1910–14	309	15.0	
1915–19	296	13.8	
1920–24	313	13.9	
1925–29	334	14.5	
1930–34	312	14.5	1.7
1935–39	300	13.4	2.2
1940–44	326	18.1	2.8
1945–49	387	19.4	3.4
1950–54	385	21.8	4.7
1955	371	21.3	5.0
1956	369	24.4	5.2
1957	362	25.5	5.9
1958	354	28.2	5.9
1959	352	28.9	6.3
1960	334	28.2	6.2
1961	326	30.3	7.5
1962	323	30.0	7.0
1963	315	30.8	6.7
1964†	313	31.4	7 3

Source: USDA. *Egg and Poultry Statistics Through 1957*, Statistical Bulletin No. 249, May 1959; *Poultry and Egg Situation*, Nov. 1963; *The National Food Situation* Aug. 1964.
  * Ready-to-cook.
  † Preliminary.

42-2. CONSUMPTION

Consumption data are available for eggs, chickens, and turkeys (Table 42-1). Data for the first two items are available from 1909; those for turkeys first became available in 1929.

Eggs. The consumption of eggs has varied widely during the past 50 years. Peak consumption—402 eggs per capita—was reached in 1945. The lowest rate of consumption occurred in 1935, at 280 eggs per capita. From 1910 to 1941, except for World War I and the Great Depression, egg consumption has averaged slightly more than 300 eggs per capita. With the outbreak of World War II, consumption rose to its peak level, held relatively high during the late 1940's and early 1950's, then declined. Estimated consumption in 1964 was 313 eggs per capita, the lowest in more than 20 years.

Although the demand for eggs has declined in recent years, during the 1940's there was a rising demand for eggs (Fig. 42-1). During the 1950's, however, not only were consumers using fewer eggs, but, relative to their incomes, they were paying less for them, a trend that continued in the

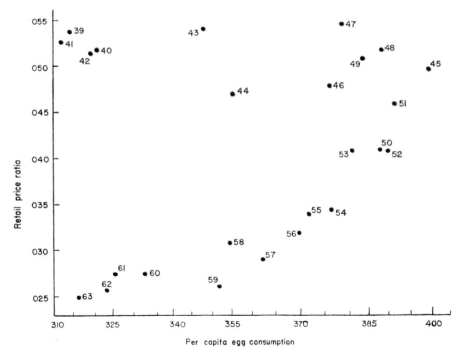

Fig. 42-1. Changes in the demand for eggs, 1939–1963. The number at each point refers to the year.

Retail price ratio: retail price divided by per capita disposable income.

1960's. Also, the data indicate a less elastic demand for eggs. This means that consumption remains relatively steady even with major price concessions, and that prices respond sharply to important changes in the actual supply of eggs.

Of the total egg consumption, about 7% is consumed as frozen and dried eggs, and the remainder as shell eggs. According to data available, approximately three-fourths of the shell eggs purchased by consumers are used as table eggs and the rest are used in cooking (Pincock).

Chickens. From 1910 to the beginning of World War II the consumption of chicken remained at about 13 to 15 lb. per capita (Table 42-1). During World War II consumption rose to a peak of 23 lb. per capita (1943) and then declined somewhat after the war ended. During recent years consumption rose to a new all-time high level of more than 31 lb. The development of the present-day commercial broiler has been important in increasing the consumption of chicken. Certainly there was little or no indication of an increase in chicken consumption during the time when cull hens made up the bulk of the chicken meat supply. With the advent of the broiler, which now constitutes about 70% of the poultry meat supply, plus favorable price considerations, consumption of poultry meat has been stimulated and may be expected to increase still further in the years ahead. Unlike the consumption of beef and pork, which tends to fluctuate cyclically, the consumption of chicken has shown a rather steady upward trend since the late 1930's (Fig. 42-2).

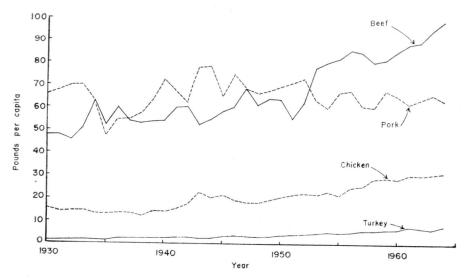

Fig. 42-2. Consumption of chicken, turkey, beef, and pork in the United States, 1930–1964.

A study conducted by the USDA (1957) showed that 93% of the home-makers served or bought broilers or fryers during the preceding year and nearly 60% served broilers or fryers during the previous week. This indicates the present widespread use of broilers. The most popular way of using broilers or fryers is to serve them fried, as reported by 94% of the housewives who had served this item during the previous year. Other popular ways of preparing this type of chicken meat are baked or roasted, broiled, and barbecued. Broilers and fryers are usually purchased either whole or cut up. In the purchase of broilers and fryers housewives consider such factors as plumpness, fat covering, and lack of bruises, discoloration, pinfeathers, and skin tears.

Turkeys. The consumption of turkeys in the 1930's averaged about 2 lb. per capita. Since the late 1930's, the consumption of turkey meat has been increasing steadily (Fig. 42-2). Estimated consumption for 1964—7.3 lb.— is about the highest on record. Here again the development of a lighter-weight bird, the marketing of turkeys at lighter weights, and the recognition that turkey meat is good at times other than the fall holiday season, have encouraged consumption.

42-3. CLASSES AND GRADES

The reasons for classifying and grading poultry and eggs are numerous. Some of the major reasons are listed here. (1) Classifying and grading facilitate exchange of products, especially under circumstances where the product is offered for sale by distant sellers. (2) Products can be grouped according to their best and most efficient use. (3) Premium markets can be found for best quality. (4) Price comparisons are made easy. (5) Buying and selling on description are possible. (6) Risk of fraudulent market practices is eliminated. (7) Pooling of shipments is made possible.

In general the grades for poultry and eggs are established by visual inspection, or candling. Candling involves twirling the egg in front of, or rolling it over, a strong light in such a way that the light rays penetrate and illuminate the interior of the egg. In some of the modern plants, machines to detect eggs with cracked shells or blood spots are used to supplement the candling process.

Eggs. The following information on the parts of an egg and the basic egg grades has been adapted largely from USDA (1958, 1960).

MAJOR PARTS OF AN EGG. To understand the basis of our egg grades and to develop a limited knowledge of egg grading, one must be at least reasonably familiar with the four major parts of an egg: shell, yolk, albumen or white, and air cell. (A more detailed discussion of these parts is given in Chapter 5.)

The shell is the outer covering of the egg. It is a calcareous layer deposited around the outer shell membrane. In egg grading, the shell is considered for cleanliness, shape, and texture.

The air cell, usually at the large end of the egg, is the space that results when the contents of the shell shrink upon cooling after being laid. Air passes through the shell and fills a space between the inner and outer shell membranes. When the egg has just been laid and cooled, the space is quite small. As the egg ages and the water evaporates, the air cell becomes larger.

The albumen or white of an egg is the nearly colorless material surrounding the yolk. It consists, in order, of a thin watery outer layer, a thick layer, a thin watery inner layer, and finally a thin layer of thick albumen surrounding the yolk, which branches out on two sides of the yolk to form the cordlike chalaza. When the egg is laid, the albumen is generally firm, with enough of the thick albumen to hold the yolk in the center of the egg. With time, the albumen becomes thinner or more watery, thus permitting the yolk to move freely from its normal center position in the egg. In grading eggs, albumen characteristics considered are freedom from foreign bodies, freedom from blood, and firmness.

The yolk is normally in the center of the egg, held there by the thick albumen. Because of its relatively high fat content, it moves upward and closer to the shell and becomes flatter as the albumen becomes thinner and weaker—just as cream rises to the top in milk. The yolk also enlarges because of the absorption of moisture from the albumen. Thus the size, shape, and position of the yolk in the egg are important indices of egg quality. Yolk factors considered in grading eggs are position in the egg, shape of yolk, outline of yolk, freedom from defects, and germ development.

GRADES. The basis for egg grades is resemblance to normal new-laid eggs. For this reason, care has been taken to explain briefly the characteristics of new-laid eggs and the changes occurring with age. Quality standards have been established for individual eggs; they provide for four main classes: AA quality, A quality, B quality, C quality. In addition, three other quality standards—dirty, check, and leaker—are established for eggs with dirty or broken shells. A summary of the standards for individual shell eggs is given in Table 42-2. The four major grades are illustrated in Fig. 42-3.

Three sets of grades, based on the quality standards for individual shell eggs, are used in this country: consumer grades, used in the sale of eggs to individual consumers; wholesale grades, used in the wholesale channels of trade; and United States procurement grades, used for institutional buying and Armed Forces purchases. Attention in this chapter will be focused on the consumer and wholesale grades.

Consumer grades. These are used for lots of eggs that have been carefully candled and graded for retail sale. The four consumer grades of eggs are Grade AA or Fresh Fancy Quality, Grade A, Grade B, and Grade C; see Table 42-3.

Wholesale grades. These differ from consumer grades in the tolerance

TABLE 42-2. | *Summary of United States standards for quality of individual shell eggs. (Standards effective Dec. 1, 1946; amended Mar. 1, 1955).*

Quality Factor	AA Quality	A Quality	B Quality	C Quality
Shell	Clean, unbroken, practically normal	Clean, unbroken, practically normal	Clean to slightly stained; unbroken; may be slightly abnormal	Clean to moderately stained; unbroken; may be abnormal
Air cell	$\frac{1}{8}$ inch or less in depth; practically regular	$\frac{2}{8}$ inch or less in depth; practically regular	$\frac{3}{8}$ inch or less in depth; may be free but not bubbly	May be over $\frac{3}{8}$ inch in depth; may be free or bubbly
White	Clear, firm	Clear; may be reasonably firm	Clear; may be slightly weak	May be weak and watery; small blood clots or spots may be present
Yolk	Well centered; outline slightly defined; free from defects	May be fairly well centered; outline may be fairly well defined; practically free from defects	May be off center; outline may be well defined; may be slightly enlarged and flattened; may show definite but not serious defects	May be off center; outline may be plainly visible; may be enlarged and flattened; may show clearly visible germ development but no blood; may show other serious defects

of lower quality eggs permitted, and in possible inclusion of some "loss" or inedible eggs. The grade designation given wholesale eggs are Specials, Extras, Standards, Trades, Dirties, and Checks. In general, Specials consist of 20% AA quality eggs and 80% A quality eggs; Extras, 20% A quality eggs and 80% B quality eggs; Standards, 20% B quality eggs and 80% C quality eggs or better; Trades, 83% C quality eggs.

Weight classes. In the marketing of eggs, weight and quality are often confused. Weight is separate and distinct from quality, however. A large egg may be either Grade AA, A, B, or C, as may a jumbo or a small-sized egg. The weight classes established for consumer grades of eggs are those shown in Table 42-4. Weight classes for wholesale grades of eggs have been established by the USDA.

Producing quality eggs.[1] Egg quality is at its peak at the time the egg is

[1] Adapted largely from L. B. Darrah, *Business Aspects of Commercial Poultry Farming.* Ronald, 1952.

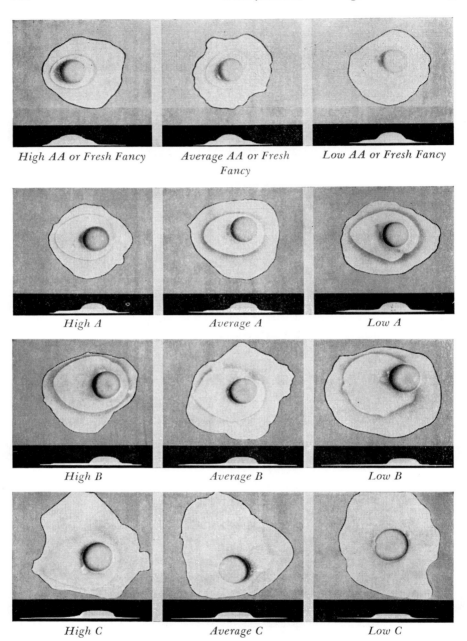

High AA or Fresh Fancy	Average AA or Fresh Fancy	Low AA or Fresh Fancy
High A	Average A	Low A
High B	Average B	Low B
High C	Average C	Low C

Fig. 42-3. The interior quality of eggs that meet the specifications of the U.S. Standards for Quality of Individual Shell Eggs with respect to albumen and yolk quality. [From Poultry Division, Agricultural Marketing Service, USDA.]

TABLE 42-3. | *Summary of United States consumer grades for shell eggs.*

U.S. Consumer Grade	At Least 80% (*Lot Average*)* Must Be:	Tolerance Permitted† Percent	Tolerance Permitted† Quality
Grade AA or fresh fancy quality	AA Quality	15 to 20 Not over 5**	A B, C, or Check
Grade A	A quality or better	15 to 20 Not over 5**	B C, or Check
Grade B	B quality or better	10 to 20 Not over 10**	C Dirty or Check
Grade C	C quality or better	Not over 20	Dirty or Check

* In lots of two or more cases, no individual case may fall below 70% of the specified quality, and no individual case may contain more than double the tolerance specified for the respective grade (that is, in lots of Grade A, not more than 10% of the qualities in individual cases within the sample may be C or Check, provided the average is not over 5%).

† Within tolerance permitted, an allowance will be made at receiving points or shipping destination for ½% leakers in Grades AA, A, and B and 1% in Grade C.

** Substitution of higher qualities for the lower qualities specified is permitted.

laid. It is not possible to improve the original quality. The most that can be done is to produce a high-quality product and then exercise the necessary precautions through proper practices to maintain as much of the original quality as possible. In general, the major considerations in producing and marketing high quality eggs are as follows:

1) Management practices:
   (a) Whether males are in the flock.
   (b) The number of times eggs are gathered each day.
   (c) Whether the laying flock is confined.
   (d) The condition of the nest and floor litter.

TABLE 42-4. | *United States weight classes for consumer grades for shell eggs.*

Size or Weight Class	Minimum Net Weight per Dozen (oz.)	Minimum Net Weight per 30 Dozen (lb.)	Minimum Weight for Individual Eggs at Rate per Dozen (oz.)
Jumbo	30	56	29
Extra large	27	50½	26
Large	24	45	23
Medium	21	39½	20
Small	18	34	17
Peewee	15	28	

2) Holding conditions:
  (a) Temperature in the egg room.
  (b) Humidity in the egg room.

Flocks with males generally market lower-quality eggs than those without males. Males should be in the flock for hatching egg production only. The removal of males is especially important on farms where eggs are gathered only once each day, or where the egg-laying flock runs loose in the barnyard. Confinement of the flock when there is one daily gathering of eggs helps quality but it is even more helpful to gather eggs twice or, better still, three times daily.

Frequent gathering of eggs is essential in maintaining quality under the usual types of nesting arrangements. Gathering eggs twice each day rather than once gives a substantial boost in quality. Gathering them three times a day rather than twice further improves quality, but the gain is not as great as that obtained by twice-daily gathering rather than once-daily gathering.

Holding conditions of temperature and humidity are always important, but they are especially so when other practices are poor. In general, egg rooms should be maintained at 45° to 55°F. The rooms should be kept moist; 80% or higher relative humidity is recommended.

While not listed as a factor, the temperature at which the eggs are candled or graded is important in the interpretation of egg quality. Under warm temperatures, eggs appear worse than they are; under cool conditions, better.

Poultry. Poultry, like eggs, has many classes and grades. Although this may seem unnecessary and confusing, it is necessitated by the wide variety of poultry meat items offered for sale (USDA, 1960).

CLASSES. The classes of chicken commonly used are the following:

Broiler or fryer. A young chicken (usually under 16 weeks of age) of either sex, tender-meated, with soft, pliable, smooth-textured skin and flexible breastbone cartilage.

Roaster. A young chicken (usually under 8 months of age) of either sex, tender-meated, with soft, pliable, smooth-textured skin and breastbone cartilage that is somewhat less flexible than that of a broiler or fryer.

Capon. An unsexed male chicken (usually under 10 months of age), tender-meated, with soft, pliable, smooth-textured skin.

Stag. A male chicken (usually under 10 months of age), with darkened flesh and considerable hardening of the breastbone cartilage. Stags show a condition of fleshing and a degree of maturity intermediate between that of a roaster and a cock or old rooster.

Hen or stewing chicken or fowl. A mature female chicken (usually more than 10 months old), with meat less tender than that of a roaster, and with nonflexible breastbone.

Cock or old rooster. A mature male chicken, with coarse skin, toughened and darkened meat, and hardened breastbone.

GRADES. The standards of quality used with poultry are applicable to individual birds. Factors considered in establishing standards for live poultry are these: health and vigor, feathering, conformation, fleshing, fat covering, and degree of freedom from defects. The grades used for live poultry are Grade A or No. 1 Quality, Grade B or No. 2 Quality, and Grade C or No. 3 Quality.

In general, Grade A birds are alert, vigorous, well-feathered, well-fleshed, well-covered with fat, free from skin tears, and virtually free of bruises, and have a normal conformation. Grade B birds are of good health and vigor, fairly well-feathered, fairly well-fleshed with a modest fat covering, and free of skin tears, and have a practically normal conformation and a moderate extent of bruises. All other birds suitable for food would be classed as Grade C.

Grade names for dressed and ready-to-cook chicken are A Quality, B Quality, and C Quality. The three grades of dressed chicken are illustrated in Fig. 42-4.

In general, Grade A birds have a normal conformation and are well fleshed, well covered with fat, and practically free of pinfeathers, hair, cuts, tears, bruises, and discoloration. Grade B birds have a practically normal conformation and are fairly well fleshed, moderately fat, and reasonably free of physical and processing defects. All other birds considered suitable for food would be classed as Grade C.

The wholesale grades for dressed and ready-to-cook poultry are Extras, Standards, and Trades. Extras consist of at least 90% A Quality birds, with

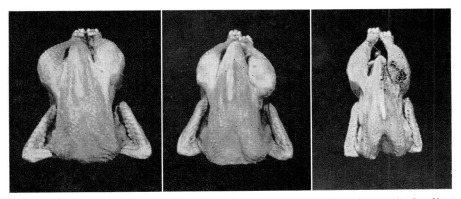

Fig. 42-4. Grades of dressed chickens. *Left,* A Quality. *Center,* B Quality. *Right,* C Quality. [From Poultry Division, Agricultural Marketing Service, USDA.]

the remainder of B Quality; Standards, 90% B Quality, the remainder C Quality; Trades, all C Quality.

Production of quality poultry meat is largely the result of starting with good stock, proper feeding, proper housing and sanitation, and attention to other phases of good management.

42-4. MARKETING CHANNELS

The channels used in the marketing of eggs and poultry have changed rapidly during the past few years. Twenty years ago most eggs moved from producers to country buying stations, or to hucksters and peddlers, then to central assembling plants and shippers, from which they were shipped in large quantities to city wholesalers and jobbers. These wholesalers and jobbers sold to chain store warehouses or retail stores. In recent years the desire to sell to premium outlets has had a major impact on the channels used in the areas close to market. In areas close to the market, a substantial proportion of the eggs are moving from producers either direct to consumers or to retail stores. Because of this manner of marketing, country buying stations and city wholesalers are gradually being eliminated. In the surplus-egg-producing states located long distances from market, direct marketing is relatively less important, but selling to large assembling plants rather than to country buying stations is growing in importance. These large plants in turn ship to wholesalers in distant cities or direct to chain store warehouses. A few firms are making arrangements for sales direct to retail stores, even in distant areas.

Poultry marketing, too, has changed rapidly in recent years. This is the result of a substantial increase in commercial broiler production and a change in the way of preparing poultry for the retail market—from New York dressed (blood and feathers removed only) to ready-to-cook birds or parts. Formerly, poultry moved through country buying stations and then largely to country shippers, to be sold to wholesalers of live chickens. These wholesalers moved the chickens to city processors and then either to wholesalers of dressed chickens or direct to retail stores. The wholesalers, of course, sold to retail stores. Today, two major patterns exist. In the heavy-producing areas birds are moved directly from farms to the processing plants and then to chain store warehouses or direct to retail stores. In other areas the buying station type of agency still plays an important role in assembling the small quantities of poultry from individual producers for sale in larger quantities to city or country processing plants.

In summary, one of the most striking developments in poultry and egg marketing is the simplification of the marketing channels. Formerly, many agencies were necessary because the poultry and eggs had to be collected from the millions of small farms that produced these items, and then assembled in large quantities for processing and for sale in distant markets.

Today, with fewer but larger flocks, sales are becoming much more direct. Many small country buyers as well as numerous city wholesalers and jobbers are no longer needed or used.

42-5. MARKETING COSTS

The cost of marketing eggs has fluctuated widely over the years. In general, though, costs declined during the 1920's and 1930's and reached the lowest point on record in the period from 1935 to 1939, when the average marketing margin for eggs was 10.4 cents per dozen (Table 42-5). There-

TABLE 42-5. | *Marketing margins for eggs and chickens, 1920–1963.*

Year	Eggs				Chickens (Broilers and Fryers, Ready-to-Cook)			
	Retail Price per Dozen (cents)	Farm Value* (cents)	Farm-Retail Spread (cents)	Farmer's Share of Consumer's Dollar (%)	Retail Price per Pound (cents)	Farm Value† (cents)	Farm-Retail Spread (cents)	Farmer's Share of Consumer's Dollar (%)
1920–24	49.4	34.8	14.6	70				
1925–29	46.3	32.1	14.2	69				
1930–34	30.5	19.2	11.3	63				
1935–39	32.8	22.4	10.4	68				
1940–44	43.1	30.5	12.6	71				
1945–49	61.8	44.9	16.9	73				
1950	57.1	38.0	19.1	67	57.0	37.4	19.6	66
1951	69.7	49.4	20.3	71	59.7	39.0	20.7	65
1952	63.6	43.2	20.4	68	60.0	39.7	20.3	66
1953	66.8	49.0	17.8	73	58.5	37.0	21.5	63
1954	56.2	37.5	18.7	67	52.8	31.6	21.2	60
1955	58.1	40.1	18.0	69	54.8	34.6	20.2	63
1956	57.7	39.8	17.9	69	47.8	26.9	20.9	56
1957	54.9	36.6	18.3	67	46.7	25.9	20.8	55
1958	57.9	39.5	18.4	68	46.1	25.4	20.7	55
1959	50.8	32.1	18.7	63	42.0	22.0	20.0	52
1960	54.9	37.3	17.6	68	42.7	23.1	19.6	54
1961	54.9	36.5	18.4	66	38.5	19.3	19.2	50
1962	51.8	34.7	17.1	67	40.7	20.9	19.8	51
1963	52.8	35.0	17.8	66	40.1	20.0	20.1	50

Source: USDA, Agricultural Marketing Service. *Farm-Retail Spreads for Food Products.* Misc. Pub. 741. *The Marketing and Transportation Situation.* Jan. 1961, Feb. 1962, Feb. 1963, and Feb. 1964.
* Payment to farmers for 1.03 dozen eggs.
† Payment to farmers for 1.37 pounds live chicken.

after, marketing costs rose and reached a high of more than 20 cents per dozen in 1951 and 1952. Since then costs have declined somewhat, but remain at about 17 to 18 cents per dozen.

The cost of marketing eggs, as with other farm products, is "sticky"; that is, it changes slowly with price changes. As a result, marketing margins tend to remain relatively steady over short periods of time, even though farm prices and retail prices fluctuate greatly. In 1959, for example, farm and retail prices were considerably lower than they had been before, or than they have been since, but marketing costs actually increased.

Because eggs require relatively little processing and relatively simple packaging, the farmer's share of the consumer's dollar spent for eggs has always been relatively high. During the period for which data are available, the farmer's share has ranged from a high of 73% of the consumer's dollar in the period from 1945 to 1949, and in 1953 to a low of about 63% during the period from 1930 to 1934. In periods of prosperity the farmer's share is relatively high; in periods of depression, relatively low.

The biggest single cost item in marketing eggs is the retail operation. Although the retail margin varies widely, it generally averages close to 10 cents per dozen, which represents about half of the total marketing margin. Other important cost items are labor, cartons, and transportation from the farm to the retail market.

Marketing margins for ready-to-cook chicken are available for the period beginning in 1950 (Table 42-5). During the 1950's, marketing margins averaged slightly more than 20 cents per pound but have declined somewhat since that time. The farmer's share of the consumer's dollar, however, has been declining. In the early 1950's the farmer's share averaged about two-thirds of the consumer's dollar. In the most recent year for which data are available, the farmer's share averaged half of the consumer's dollar spent for chicken.

The makeup of the marketing margin for chicken is difficult to assess because of the tremendous impact that loss-leader sales have on the marketing of the chicken. Evidence available indicates that a substantial share of the chicken is sold during special sale periods of chain stores and that a relatively small part is sold at other times. Thus margins are appreciable when sales are low, but practically nonexistent when sales are high. In general, however, about half of the marketing margin, as with eggs, consists of the gross markup at retail. Operations of processing plants, including assembling of chickens and distribution of processed items to consuming centers, averages 7 to 8 cents per pound; wholesale operations account for the remainder of the margin.

In general, assembling costs in the marketing of poultry and eggs have declined as production has become more concentrated in certain areas and as flocks have become larger. On the other hand, the rising cost of labor, and the demand by the consumer for new and better services and products, have more than offset the reduction in assembling costs. Thus, total marketing costs have tended to rise over the years.

REFERENCES AND SELECTED READINGS

References marked with an asterisk are of general interest.

*Darrah, L. B., 1965. *Food Marketing.* Cornell Campus Store, Ithaca.

* Darrah, L. B., 1952. *Business Aspects of Commercial Poultry Farming.* Ronald, New York.

* Darrah, L. B., and Reid, R. G., 1963. *Barriers to Egg Consumption.* A.E. Res. 113. Cornell Univ. Agr. Exp. Sta., Ithaca.

* Marshall, Joseph H., 1964. *Expanding the Market for Fowl Through New Products.* Bull. 998. Cornell Univ. Agr. Exp. Sta., Ithaca.

Pincock, M. G., 1952 and 1957. *Consumer Purchasing Practices and Quality Recognition for Eggs.* A.E. Res. 1073. Cornell Univ. Agr. Exp. Sta., Ithaca.

* Robertson, J. O., 1964. *Characteristics of and Barriers to Chicken Consumption.*
A.E. Res. 152. Cornell Univ. Agr. Exp. Sta., Ithaca.

USDA, 1957. *Selected Highlights from a Study of Consumer's Use of and Opinions About Poultry.* AMS-159.

————, 1958. *Regulations Governing the Grading and Inspection of Shell Eggs and United States Standards, Grades, and Weight Classes for Shell Eggs.* Poultry Division, AMS.

————, 1960a. *Regulations Governing the Grading and Inspection of Poultry and Edible Products Thereof and United States Classes, Standards, and Grades with Respect Thereto.* Poultry Division, AMS.

————, 1960b. *Farm-Retail Spreads for Food Products.* Misc. Publ. 741, AMS.

————, 1960c. *The Marketing and Transportation Situation.* AMS.

Livestock Diseases

Chapter 43

Infection and Its Effects on Livestock Production

There is nothing, Sir, too little for so little a creature as man. It is by studying little things that we attain the great art of having as little misery and as much happiness as possible.

JOHNSON *to Boswell for his Private Journal*

Infectious diseases have plagued mankind and his domestic animals throughout the ages. History is replete with accounts of mysterious sicknesses that affected large populations of people or animals. These maladies changed history by influencing the success or failure of wars and the material progress of nations. The results of infection are multiple: the human host may be directly attacked or infectious diseases may indirectly affect man by attack on animals. When animals are attacked, the general plane of human nutrition is lowered by loss of meat, milk, and eggs, and the economy is harmed by inefficiency in livestock production. When disease is prevalent, the yields of animal products may be far too small in relation to the effort expended. Animal losses by disease in the United States have been estimated at a billion dollars annually. Consider the importance of these problems in less prosperous parts of the world where efforts at control are minimal and the population involved must bear the full impact of disease attacks.

Though much remains to be done in infectious disease control, we can look back with satisfaction at what has been accomplished. It was only near the end of the last century and the beginning of our twentieth century that man began to understand the nature of these problems. The scientific giants of the early developments were people such as Pasteur, Koch, and Ehrlich. These are familiar names to all of us because their accomplishments opened the way for a great period of progress in improving the health of man and animals.

In this chapter we shall explore the problem of the infectious diseases of domestic animals and see how various areas of science have contributed in the past and presumably will contribute in the future to their prevention and cure. The interrelationship of livestock production, veterinary medicine, and public health will become obvious as we proceed.

43-1. PARASITISM

Infection involves a host and a parasite. The hosts that concern us here are the domestic mammals and birds that man has selectively bred and trained for various specialized purposes. The parasites are creatures who have become dependent on the host for their existence. They have lost the capacity to survive as independent forms of life. Therefore they spend all or part of their life cycles either on the surfaces or in the tissues of the host. There are many forms of parasites. A partial list includes, in order of increasing structural complexity, the following: viruses, bacteria, fungi, protozoa, helminths (worms), and arthropods (insects). It must be remembered that the parasitic forms constitute a very small part of the total number of species in a given biological group. For example, only a very small percentage of fungi found on earth are parasitic. The viruses are an exceptional group, of which all forms are parasitic.

These various forms of life have given rise to several fields of science— virology, bacteriology, mycology (study of fungi), and parasitology (this last term includes the study of protozoa, helminths, and insects). When dealing with infectious diseases, physicians and veterinarians may contribute both as professional men trained in the healing arts and as scientific specialists in a particular field of knowledge. Other scientists prepare themselves by intensive basic study in one of these sciences. The best progress in research is made when knowledge from disease-oriented educations (M.D. or D.V.M.) and scientific specialty educations (Ph.D.) are combined among investigators working together.

To exist, parasitic forms of life have restrictions on their freedom of development, determined by four requirements for perpetuation.

The Parasite Must Gain Entrance to the Host. The portal of entry, which is limited to body orifices and the skin, varies for different parasites. The mouth and nose are probably the most important since they provide approaches to the respiratory and gastrointestinal tracts, and thus indirectly to the blood. The tissues of mouth and nose are constantly exposed to contamination from feed, water, and inhaled air. An important aspect of good livestock management, therefore, is to prevent contamination of feed and water with animal excretions. Certain disease organisms gain access to the host during the act of breeding; these cause the venereal diseases. Although the skin is usually an effective barrier against the entry of parasites, some organisms do penetrate the skin and others enter at sites of skin abrasions. Biting insects may serve as vectors for disease organisms by penetrating the skin and thus depositing the parasite in the tissues.

The Parasite Must Multiply and Adapt to the Environment. Following entry, the contest between host defenses and parasitic adaptation begins. It is in this phase that injury to the host is created—the period during which a disease is in progress. It has been the usual center of interest in veterinary medicine.

The Parasite Must Have a Satisfactory Portal of Exit. The parasite must have a means for leaving the host because each individual host will eventually die—either because of the parasitism or from other causes. The organisms affecting the respiratory tract usually escape in respiratory excretions. This may occur from coughing or sneezing. Parasites entering by way of the digestive tract are frequently shed from the host in the droppings or feces. Those entering by insect bite usually escape by gaining access to the tissues of a new insect when the host is bitten.

The Parasite Must Have an Effective Mechanism for Transmission to a New Host. Many disease agents pass constantly from the infected excretions of one individual to another individual of the same species. In their evolutionary development, such parasites frequently become limited to a single host. Numerous parasites have more involved mechanisms of transmission—for example, the formation of resistant spores. This is a dormant physical state of the parasite that does not require prompt transmission to a new host. The spore can lie in soil and other places for prolonged periods, awaiting the arrival of a new host animal. Some worms and protozoa have very complex life cycles, which may depend upon development in one or more intermediate hosts before gaining access again to the domestic animal host.

43-2. DEFENSE MECHANISMS OF THE HOST

The science of immunology concerns the study of means by which the host may protect itself from parasites. Immunologists continuously search for additional ways to aid the host defense mechanisms. Certain infectious diseases affect only one host; hog cholera in swine and diphtheria in man are examples. Man has high, natural resistance to hog cholera virus but is susceptible to the diphtheria organism; the hog has high natural resistance to diphtheria but a susceptibility to hog cholera. Therefore, we see that some parasites have a high degree of host specificity; some hosts, on the other hand, have high levels of natural resistance to certain parasites. The natural resistance of the host manifests itself at the time the parasite is attempting to multiply and adapt to the host environment. Precise knowledge of the mechanisms of natural resistance is lacking, and the refractoriness of the host to a given parasite is simply considered a genetic endowment. Some parasites infect several species and are spoken of as organisms with a wide host range. Examples of diseases affecting both man and several domestic animals are rabies, psittacosis, tuberculosis, tetanus, and trichinosis. Not only are variations in resistance to parasites found between species but, furthermore, within a given species of susceptible host, one can find wide variations of susceptibility. This has been shown to be inherited resistance or susceptibility. Genetic selection of breeds or families of domestic animals demonstrating increased resistance to certain infections has been accomplished. It is one of the important hopes for the future that this work may be expanded to assist livestock production in this important way.

Normal Defenses of the Host. Many structures and body fluids of the host are ingeniously designed for the prevention of parasitic invasion (Raffel, 1953). The skin itself is not only a mechanical barrier but its secretions also have the property of killing some bacteria. The mucous membranes such as those lining the nasal chambers, respiratory tract, and digestive tract secrete a mucous film entrapping many parasitic organisms and mechanically sweeping them away. In addition, the secretions of such membranes may contain substances that act directly to destroy the parasite. The animal body has a remarkable series of cells—phagocytes—that are of central importance in preventing disease. These cells are widely distributed in such places as the liver, spleen, blood, and bone marrow. They have amoebalike behavior and appearance in that they are capable of taking particles into their cell substance. If the particle is an invading bacterium or some other parasite, it will very likely be destroyed within the phagocytic cell. Inflammation in the tissues is another important means for defending the host. Everyone has experienced the infection of a finger following a skin abrasion or the troubles associated with a "boil" in the skin. In these situations the affected tissue is swollen, painful, red, and warm; it is inflamed. The pathologist studies such affected tissues and knows that the inflammatory response of the host is walling off the invading microorganism and assisting in the destruction of the invader.

Acquired Immunity. Although the natural defense mechanisms play a great role in the prevention of disease among the domestic animals, it **is**

common knowledge that these natural defenses may be overwhelmed. In speaking of acquired immunity, we are interested in knowing what can be added to the natural defenses to protect the host. It is convenient to divide acquired immunity into active and passive categories.

ACTIVE ACQUIRED IMMUNITY. In this condition the tissues of the host play an active part in bringing about an increase in resistance. Children recovering from measles or chickens recovering from an attack of coccidiosis are examples wherein the live disease organism grows in the tissues and is eventually destroyed, leaving the host with a higher degree of resistance than it had naturally. Since the natural disease frequently kills or leaves the animal in a useless physical condition for commercial purposes, this is not a satisfactory way to gain acquired immunity. Consequently, artificial means are sought for inducing the desired changes in the host that will not harm the health of the animal. This procedure is called immunization or vaccination. By vaccination procedures, the veterinarian induces the body to form certain kinds of antibodies and perhaps enhances the capacity of certain cells to combat the parasite. The antibodies are protein molecules formed by lymphoid tissues and discharged into the lymph and blood. We test for their presence by various procedures referred to as serological tests.

PASSIVE ACQUIRED IMMUNITY. The accumulation of protective antibodies for a specific organism in the serum means that a transfer of serum from an actively immunized animal to a normal animal may afford a degree of added protection. The recipient of the serum is passive in the sense that it did not form the antibodies. Passive immunity occurs naturally in the newborn animal because antibodies are received from the dam. In some species antibodies from the maternal serum pass across the placenta and reach the blood of the developing fetus in the uterus. In other species such as the horse, cow, and sheep, the maternal antibodies are stored in the colostrum and are taken up by the newborn when nursing begins. Only in the newborn animal do these antibodies escape destruction by the digestive fluids and enter the blood stream in appreciable amounts. Resistance thus acquired is lost in a matter of weeks or in a few months at the most. Passive immunity is induced artificially by the injection of antiserum, a procedure affording immediate increased resistance to diseases such as tetanus, botulism, swine erysipelas, and canine distemper. However, the passive protection afforded by serum injection is dissipated in a matter of weeks.

In the remainder of this chapter some of the important groups of parasitic organisms will be described and one or two examples from each group will be considered. The examples demonstrate the diversity of approaches required for infectious disease control, but the reader should realize that there are numerous infectious disease problems that must await future research before they can be brought to the advanced state of control depicted in the examples.

43-3. THE VIRUSES AND THEIR CONTROL

The viruses are a large group of very small microorganisms that can only multiply when inside a host cell (Pelczar and Reid, 1958). The unit of virus measurement is the millimicron (mμ). This is 1/1,000,000 of a millimeter or 1/25,400,000 of an inch. The viruses range in size from about 5 to 300 millimicrons. Many of them are within the size range of large molecules. Such structural simplicity does not permit them to live independently because they lack vital parts needed for growth and multiplication. Consequently, they are the most dependent and burdensome of all the world's creatures. A virus in a susceptible host will seek out certain cells for completion of its life processes. The cells chosen by the virus may then be injured, resulting in disease in the host. The nature of the disease depends upon the particular cells involved. Thus the clinical signs and the damaged tissue (lesions) can indicate to the veterinarian which virus is responsible.

The scientist working with viruses rarely sees the microorganism he is studying. How does he know what he is doing in such a situation? He can determine this by knowing the nature of the disease produced in experimental animals by the virus under study and by the use of serological procedures. The many serological techniques cannot be described here, but some procedures will be considered briefly in a subsequent section of the chapter.

We said that viruses can only grow in living cells and this means that the virologist must cultivate them in living cells in the laboratory. Two main methods are used for this. One procedure is to grow the virus in the living tissues of the developing chicken embryo. Another method of great current importance is virus cultivation in tissue culture. In this technique, cells from an animal are grown on a glass surface in bottles! If the virus grows in the kidney cells of the calf, then we may grow kidney cells of the calf in tissue culture and use these cells as hosts for virus.

Viral Encephalitis. Throughout the world there are several different viruses that cause inflammation of the brain (encephalitis) in vertebrate animals. However, our attention will focus on one of these, the western equine encephalitis virus (W.E.E.). In the United States this virus has infected horses in every state west of the Appalachian Mountains. The virus was named after the horse (equine), but it was soon realized that man also suffered from this disease.

According to figures compiled by the USDA, around 2,000 cases of equine encephalitis have been reported annually in horses in recent years. At certain times this reaches far greater proportions, as in 1938, when 184,662 cases occurred. The stricken horse first goes through a brief period when it has a fever and mild depression (Hagan and Bruner, 1961). The virus is in the blood of the horse at this time. Many infections are arrested at this early stage and go completely undetected. However, if brain invasion occurs, the animal shows many peculiar forms of behavior. It may shy from objects that ordinarily cause no concern, or it may walk in circles or force its way through fences. At these stages it will refuse to take feed and water and may develop a sleepy attitude, in which case it will rest its head on any convenient structure such as the manger. Paralysis follows and the horse eventually collapses and remains down until death ensues. Similar symptoms occur in man.

What is the source of the virus that causes this bizarre and frequently fatal disease? In search of an answer to this question, scientists have uncovered many interesting facts that provide a basis both for understanding the infection and for attacking the control problem (Fig. 43-1). The greatest

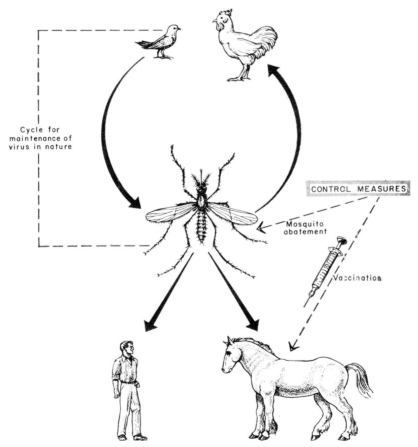

Fig. 43-1. The parasitic cycle of the western equine encephalitis virus and the places where control efforts are directed. Transmission from horse to man is not considered an important part of the infection chain.

number of cases occur in August and September, and this information led workers to look for an insect vector. We now know that certain mosquitoes play a key role in transmitting this virus among vertebrates. But where does the mosquito get the virus? Does the horse serve as the reservoir and, therefore, the source of infection for man? Research has indicated that wild and domestic birds are the probable sources of infection for the mosquito vector. High percentages of domestic chickens have been shown to carry antibodies against the virus in places where epidemics occur among horses and man (Hammon et al., 1945). Wild migratory birds may also carry antibodies. Experimental infection of these bird hosts demonstrates that they develop high yields of virus in their blood for brief periods but show no signs of sickness. This, therefore, is infection without disease! Mosquitoes biting birds when virus is in their blood can readily pass it on to man or to the horse.

A two-pronged effort at control is made. The first control effort is to block the infection chain at the point of transmission from the wild bird host to the horse or man. This means control of the mosquito vector. In several parts of the United States mosquito abatement programs are in effect for control of viral encephalitis, and for pest control. Though costly, this procedure will undoubtedly become more common as additional lands go under irrigation and create environments for the breeding of mosquitoes. Many problems concerning the operation of large mosquito abatement programs remain unsolved.

The other control measure is directed toward raising the resistance of the susceptible host. In the case of the horse, this is done by annual vaccination with killed encephalitis virus. Virus is grown in embryonating chicken eggs. The chicken embryos die within 24 hours of the virus inoculation, at which time the embryo tissues are heavily saturated with virus. The embryo is ground into a paste and the virus is killed with formalin. This suspension is then injected between the layers of the skin in the horse. Vaccination of horses is performed in the late spring or early summer so that a high level of resistance will be in effect during the peak of the mosquito season.

In summary it can be said that as more lands are irrigated the control of viral encephalitis will become an increasingly important problem. Better methods of mosquito control and improved methods for the immunization of horses and possibly man will be needed.

Hog Cholera. Hog cholera is the most important disease of swine in the United States. It occurs throughout the country but is most prevalent in the concentrated swine raising areas of the Midwest. This very costly and destructive disease can kill nearly 100% of susceptible animals in a given outbreak (Hagan and Bruner, 1961). The pigs suffer from degeneration in the walls of the small blood vessels in numerous parts of the body and a decrease in the number of phagocytic cells. This latter effect lowers the resistance of the host to a variety of other parasites. Death from hog cholera is therefore often associated not only with the virus infection itself but with pneumonia and inflammation of the intestinal tract, secondary processes caused by other organisms.

How does this parasite persist in nature and continue to cause its great losses to livestock production? To perpetuate itself in nature, the virus must be protected from the deleterious effects of drying and sunlight. Characteristically in a hog cholera outbreak, a few animals in the drove sicken initially. In a matter of a few weeks nearly all the swine in the immediate environment become ill. The virus leaves the infected host in a variety of secretions and excretions. Once the first pig in a drove of swine develops hog cholera, the remaining pigs will become infected by direct transfer of virus to their digestive tracts. But how did the first pig contact the virus? Frequently, the virus may be tracked from farm to farm on contaminated footwear and other objects. It is unfortunate that swine are often slaughtered for food purposes at the beginning of a hog cholera outbreak to limit economic loss. Virus-laden pork trimmings from infected carcasses may thus be present in garbage that is fed to susceptible swine. This can initiate a new outbreak of the disease. These methods all explain how the disease outbreaks may spread from one farm to the next during an epidemic season, but there has long been a question of how the virus persists in the relatively long periods that often intervene between outbreaks. Shope (1958) demonstrated that the swine lungworm may serve as a reservoir and intermediate host for the virus of hog cholera. Lungworms in virus-infected swine may take up the virus and pass it on to a new generation of lungworms through the eggs. The newly hatched larval lungworms may in turn survive for many months in the soil as parasites in the common earthworm. Thus, months later, completing this rather complicated cycle, rooting swine may ingest earthworms and take up the virus.

With this knowledge, what means are available for controlling hog cholera (Fig. 43-2)? Certainly rigid use of quarantine to prevent direct transfer of virus from farm to farm is an obvious step. The cooking of garbage to prevent the ingestion of live virus in pork scraps is another important control procedure, now required throughout the United States. Laws requiring the cooking of garbage also have beneficial effect for the control of several other diseases.

Paramount in importance for control is the immunization of swine. The passive immunity afforded by antiserum is very valuable at the beginning of hog cholera outbreaks to block the entrance of virus into the cells of susceptible swine. The development of means for inducing active immunity is an important chapter in veterinary medicine which goes back to the work of Dorset, McBride, and Niles (1908). These investigators showed that virus injected simultaneously with antiserum would cause the swine to develop a long lasting immunity. This procedure required a delicate balance be-

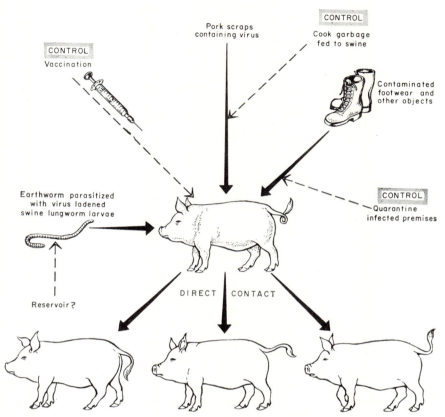

Fig. 43-2. Different ways in which hog cholera may start on a farm and the control measures used for its prevention.

tween the killing power of the virus, the protective capacities of the serum, and the general state of health of the host animals. The procedure has been a tremendous help to the swine industry but it has not always been free of difficulties. So-called "virus breaks" and "serum breaks" have resulted in the loss of animals from hog cholera following the immunization procedure. This kind of vaccination procedure requires a good deal of judgment on the part of the veterinarian in deciding when and how to proceed. Unfortunately, the vaccination has frequently been performed without the aid of expert opinion, and losses have, not uncommonly, been extensive. The goal of the veterinary profession has been to eradicate this disease, and it was long recognized that this could not be done when the immunization procedure required the use of live virus, because vaccination kept virus in the environment. Work was undertaken to prepare vaccines made with killed viruses, and two vaccines of this kind are now used in the United States. They have not replaced the simultaneous serum and virus vaccination procedure because they do not stimulate immunity for as long a period; 2 to 3 weeks are required following vaccination for immunity to develop fully.

Work continued in search of yet a better means for immunizing swine. Adaptation of the hog cholera virus to multiply in the domestic rabbit was an important development (Baker, 1947). By alternately passing the virus between the rabbit and the pig, it was eventually trained to grow and multiply in the tissues of the rabbit. This rabbit-adapted virus produced little more than fever in the rabbit host, and likewise it lost its disease-producing capacity in the pig. This paved the way for a new chapter in hog cholera control. Tamed hog cholera virus can be used to immunize swine and induce a resistance greater than that obtained with killed virus. The tamed or modified virus can be used with simultaneous injection of anti-serum. This means not only that protection is afforded immediately—rather than waiting the 2- to 3-week period necessary with killed virus vaccination—but, more importantly, it obviates the use of the live, dangerous, disease-producing virus. The prospects for complete control of hog cholera are now very good. The significant discoveries leading up to the tamed hog cholera virus vaccine afford an excellent example of how advances in disease control are achieved.

43-4. THE BACTERIA AND THEIR CONTROL

The bacteria constitute a large group of single-celled organisms considered to be the smallest form of plant life. In size the bacteria may be a little smaller or a little larger than a micron; that is, they are only about 1/1,000 of a millimeter in a given dimension. Bacteria usually reproduce by binary fission, the formation of two cells from a single cell (Pelczar and Reid, 1958). They may take numerous shapes: spheres, rods, or spirals. Extracts or broths made from the tissues of animals support the growth of bacteria very well. Thus, the bacteria thrive on the same nutrients required for the maintenance of higher animals. Nutrients prepared in the laboratory for growing bacteria are referred to as media, many of which are in the liquid state and are called broths. Substances that cause gel formation may be added to broths; these media are referred to as solid media. The use of media for the cultivation of bacteria permits diagnosis of diseases by growing the organism from infected tissues and then recognizing the cultivated organism as a disease agent. Of the many thousands of bacterial species on earth, only a few cause disease. Therefore, special knowledge is required to detect a disease-producing organism since nearly all bacteria will grow quite readily on media.

Brucellosis. Brucellosis is an important disease affecting both man and animals on a world-wide scale. There are three main species of bacteria that cause brucellosis. These closely related organisms incite a similar disease process but have different species of animals as principal hosts. Thus *Brucella abortus* affects primarily cattle, *Brucella suis* affects primarily swine, and *Brucella melitensis* affects primarily goats. However, all three species cause brucellosis in man; it is frequently called undulant fever because the human host is likely to have a series of daily fever peaks over an extended period.

Let us consider in some detail the problem of *Br. abortus* infection in cattle. This disease has been given several different names—contagious abortion, infectious abortion, Bang's disease, and brucellosis. The main clinical effect is indicated by the term contagious abortion (Hagan and Bruner, 1961). The cattleman may suffer extensive calf crop losses due to abortions late in gestation. Some weak calves are born alive, but die during the first few days of life. The reason for early expulsion of the calf from the cow's uterus can be explained by considering the habitat of this bacterial parasite. The organism multiplies within the cells of the placental membranes surrounding the calf in the uterus. The membranes become inflamed and undergo degenerative changes. This unhealthy state of the tissues interferes with the nourishment of the calf and finally results in the act of abortion. After aborting once or twice, infected animals may carry a normal calf to full term even though the uterine fluids are contaminated with the organism at the time of calving. Though the host has developed some increased resistance, it is still unable to rid its tissues of the parasite. The udder constitutes the other important site for long-term survival of the *Brucella* organism. The important aspect of infection in this tissue is the frequency with which the bacteria are shed in the milk. The implication of contaminated milk for transmission of the disease to man is obvious.

The *Brucella* organism has become a well-adapted parasite in the vertebrate host, persisting for years in an infected animal. How can we know when an animal is infected? To detect the organism in milk, we can inject milk into guinea pigs and see if disease develops in that experimental animal. We can examine the membranes and fluids from the calf by staining them and studying the cells under the microscope, or by culturing these materials on media in an attempt to grow the *Brucella* organism. These methods are useful for study of the disease but they are too laborious and time-consuming for large-scale diagnosis. Therefore, we use serology as the major diagnostic procedure, utilizing our knowledge that the serum of infected animals contains antibodies. There are many different serological procedures for the diagnosis of infection by the detection of antibodies. The one used for detecting *Brucella* antibodies is called agglutination. It is the least complex serological test to perform and it has been the cornerstone for brucellosis diagnosis. The essential features of such a test will be described.

Blood is taken from the animal to be tested for brucellosis and allowed to clot. As the clot contracts, serum will separate from it and this serum is then tested to determine the amount of antibody activity present. To do this we dilute the serum out through a series of tubes (Fig. 43-3); the fluid used for diluting is a special salt solution. The *Brucella* antibodies are protein molecules whose presence cannot be detected unless we put them in contact with the organism. A suspension of killed *Br. abortus* is placed in each tube containing the serially diluted serum. The surface of the bacterial cell is not smooth, but has a pattern like the tread on an automobile tire. Even though the parts of the pattern are very tiny, they are repeated several times over the surface of the organism. This pattern is characteristic and specific for the organism under study. When antibodies are in the serum, they attach to a portion of the surface pattern because they have the mirror image of that pattern on a portion of the antibody surface. You could compare this to the perfect fit of a lock and key. The antibody has two places on its surface that can attach to the bacterial cell; therefore each

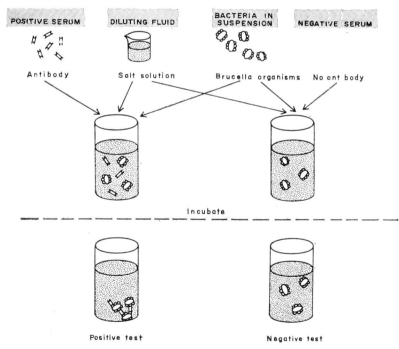

Fig. 43-3. The tube agglutination test as used in the diagnosis of brucellosis. All reacting components are represented diagrammatically.

antibody can link two bacteria. This makes it possible to form a network if one mixes large numbers of bacteria with large numbers of antibody molecules. As this linking occurs, the masses of bacteria become large enough to settle out of suspension and appear as granular masses in the bottom of the tube. In a negative test, this does not occur and the cells remain in suspension giving an opaque appearance to the solution. The agglutination test may be modified so that it can be performed on the surface of a glass plate. In this case, a few drops of serum and bacteria are mixed and the test is read within ten minutes. Visible clumping occurs in a positive reaction. This rapid plate test is commonly used in large disease-control programs.

The reaction of agglutination is also used when seeking antibodies in secretions from the mammary gland. This may be done by separating the casein from the whey and testing the whey in a manner similar to serum testing. Another procedure, called the milk ring test (MRT) is very useful for detecting antibodies in bulk milk. Whole milk is mixed in a tube with *Brucella* organisms that have been stained with a purple dye. If the milk contains antibodies, agglutination will begin, and as the cream layer rises the entrapped masses of bacteria will be carried with it. Thus, a positive test demonstrates a purple cream layer with white milk below. A negative test is exactly the opposite; the cream layer is white and the milk is diffusely purple because the stained bacterial cells are still in suspension.

Brucellosis is controlled for two important reasons. One is for the prevention of economic loss to the livestock producer from abortions and infertility in his cattle. The other reason relates to the public health problem of brucellosis in man. Great numbers of *Brucella* organisms are shed at the time of calving. When susceptible cattle eat or drink in an area contaminated by infected fluids or lick an aborted fetus, they in turn are infected. Transmission among cattle commonly occurs in this way, although venereal transmission is also possible. Man may contract brucellosis by ingestion of contaminated milk, by working around infected animals at the time of calving or abortion, or by handling parts of infected carcasses in slaughterhouses.

In 1935 the United States began a disease-control program aimed at the eventual eradication of bovine brucellosis. This is a cooperative federal-state program, which has two approaches—vaccination and test-and-slaughter. Vaccination of young cattle is performed by inoculating a vaccine containing live *Br. abortus* organisms. The vaccine is prepared from a particular strain, chosen because it produces an increased level of resistance in cattle but has no ill effect on the animals, when properly used. This is a tamed *Brucella* organism (Buck, 1930). Cattle are ordinarily vaccinated as calves, and by the time they reach sexual maturity (time for the first calf) they may be resistant to the disease. Vaccination has been of great assistance in reducing economic loss to the livestock industry from brucellosis. However, vaccination alone will not eradicate the disease, and when the level of infection reaches a low percentage, the test-and-slaughter program is used. In this situation, cattle are diagnosed as having brucellosis on the basis of agglutination tests. Animals reacting positively to the test are removed from the herd. As compensation for loss of the animal, the owner receives the full carcass value from slaughter plus an additional sum (indemnity) from the government. The brucellosis eradication program has made extensive progress. When begun in 1935, more than 11% of the cattle in the United States were infected, whereas the figure today is less than 1%. This decline has been paralleled by a similar reduction in the number of human cases of brucellosis.

Tuberculosis. Tuberculosis is well recognized as one of the most serious diseases affecting man. However, it is less commonly known that various forms of tuberculosis plague nearly all vertebrate animals. Even reptiles may suffer from tuberculosis, but we shall devote our attention to the important problem of this disease in cattle. The bacterium responsible is named *Mycobacterium bovis,* a different species from the so-called human tubercle bacillus, *Mycobacterium tuberculosis.* These organisms are so closely related that a dispute existed in the past as to whether they were separate species. The bovine tubercle bacillus will cause tuberculosis in man just as readily as the human tubercle bacillus. The reverse is not true; cattle are quite resistant to the human tubercle bacillus.

Tuberculosis is a chronic or long-lasting disease resulting in weight loss, general lack of vigor, and eventually death. Similarity of the control problems to those in brucellosis are readily seen. Infection causes extensive loss to the livestock producer and is also an important threat to human health. Man may contract this infection by drinking the milk of the cow or by handling infected animals.

Diagnosis of bovine tuberculosis is difficult because animals appear quite normal until they are in advanced stages of the disease. Therefore, detection of the infected individual in an early stage of the disease process is the key problem. The agglutination test, as used in brucellosis, is unsatisfactory for diagnosing this disease; an entirely different phenomenon is used as the basis for testing. Certain diseases, such as tuberculosis, induce a form of allergy or hypersensitivity in the infected individual. The host becomes allergic to the protein portions of the bacterial cell and will react quite violently to the proteins. The veterinarian uses this allergic state as the basis for diagnosis. Tubercle bacilli are grown in liquid media, and, as they grow, protein from the cells is extruded into the liquid. This liquid, after processing, is called tuberculin. If a small amount is injected between the layers of the skin of an infected individual, the allergic reaction manifests itself as an inflammatory lesion showing swelling, redness, and sometimes necrosis of the cells in the skin. Such a reaction would be called positive and would indicate infection by the *Mycobacteria*.

In 1917 a federal-state cooperative program for the eradication of tuberculosis in cattle was begun. The program is based on a test-and-slaughter procedure. The tuberculin test is used for diagnosis and is usually performed on cattle by injecting tuberculin into the skin near the base of the tail or the side of the neck. Animals that are allergic to the tuberculin are removed from the herd. The livestock producer is compensated by receiving the salvage value of the carcass and an indemnity payment, as described for brucellosis control. In 1917, about 5% of the cattle in the United States were tubercular; by 1940 this figure was reduced to less than 0.5%. This important accomplishment has been described by Myers (1940). Nevertheless, the battle with tuberculosis in cattle has not ended because scattered herds with active infection continue to appear. If care is not exercised, these foci of infection can readily spread to other herds. As the number of tuberculous herds has been reduced, the problem of accurate diagnosis among the remaining infected animals becomes more critical. Additional procedures for refining the diagnosis are being sought in the research laboratories.

A vaccine, which is called BCG, is available for immunization against tuberculosis. This vaccine consists of live tubercle bacilli that do not cause the progressive fatal disease. This is a tamed organism similar to the one used for brucellosis vaccination. In many parts of the world BCG vaccination is used on both man and animals for the control of tuberculosis. The question may be asked, why not vaccinate for tuberculosis in the United States? The answer is very simple: vaccination makes the individual allergic to tuberculin. Obviously, vaccination would make our diagnostic test with tuberculin useless. The decision to vaccinate or to use a test-and-slaughter procedure depends upon the density of tuberculous animals and the economic status of an individual country. The decision to use test-and-slaughter in the United States was wise since cases of human infection from the bovine tubercle bacillus are now very rare, and we are well along the way toward abolishing livestock loss from tuberculosis.

43-5. THE PROTOZOA AND THEIR CONTROL

The protozoa constitute the phylum of the simplest animals. The name explains this since the Greek word for first is *protos* and animal is *zoon*. These so-called single-celled animals are often complex in structure; they vary greatly in size, shape, and physiological characteristics (Pelczar and Reid, 1958). The protozoa are divided into four classes and parasitic forms are found among all classes. As an example, we shall consider the class Sporozoa, which contains several parasitic forms, such as the agent of ma-

laria in man, the cause of Texas Cattle Fever, the organism causing toxo-
plasmosis in man and animals, and the agents of coccidiosis.

Coccidiosis. The coccidia are an important group of protozoa; all are
parasitic and produce disease in cattle, sheep, goats, swine, dogs, cats, rabbits, ducks, turkeys, chickens, and other animals. Because these parasites are
of great economic importance to the poultry industry, we will examine the
problem of coccidiosis in chickens.

Many protozoa have complex life cycles. To understand coccidiosis one
must know the developmental stages (Fig. 43-4). The infectious form of the
parasite is called an oocyst. When the chicken swallows a mature oocyst,
eight sporelike structures escape from it and enter the cells lining the intestinal tract. Inside these host cells the parasite begins to round up and

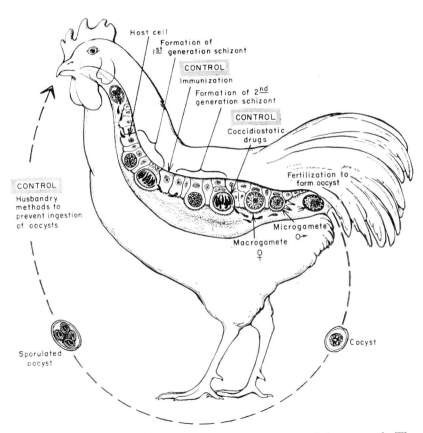

Fig. 43-4. The coccidial life cycle and measures used for control. The
digestive tract of the chicken and infected host cells are represented diagrammatically.

become larger. This is followed by division of the larger structure into numerous, small banana-shaped bodies. The infected cell containing these organisms is very swollen and eventually ruptures, releasing the parasites. This increase in numbers is by asexual reproduction, in which the parasite simply divides many times. The process of increase is called shizogony. The released parasites enter new cells and repeat the cycle. Thus, there may be repeated generations of shizogony; the number varies with the species of coccidium concerned. This phase of the life cycle involves many cells in the lining of the intestinal tract and is, therefore, very destructive to the tissues. When this stage is completed, the parasite gives rise to male and female sexual forms called microgametes and macrogametes, respectively. As these forms are freed from their host cells, they unite in the intestinal canal to form a fertilized structure, an oocyst. The oocyst is passed from the bird in the droppings and represents the resistant form of the parasite that can serve to infect a new host. To become infective, the oocyst must undergo a period of maturation called sporulation. Under the proper conditions of moisture, air, and temperature, sporulation can take place in about 24 to 48 hours. Obviously, much research was necessary to understand the numerous parts of this life cycle. The problem was further confused by finding that eight different species of coccidia affect the chicken. Each one of these produces its own characteristic disease depending upon the specific part of the intestinal tract that the parasite attacks, and the character of the damage thus caused. One of the more severe forms of coccidiosis in the chicken is called cecal coccidiosis because the organism causing it (*Eimeria tenella*) attacks the cells lining the ceca, a portion of the large intestine. The disruption to the lining membrane of the ceca is so severe that massive hemorrhage occurs, with consequent blood loss and local necrosis of tissue. On the other hand, a few species of coccidia affecting chickens produce a very mild disease, which goes undetected unless one looks for the infection by microscopic examination of droppings.

Because of the importance of coccidiosis in poultry, knowledge regarding the life cycle of the parasite has modified husbandry methods used in raising birds for commercial purposes (Biester and Schwarte, 1959). If one prevents the swallowing of sporulated oocysts, the disease is prevented. Efforts to accomplish this are the basis for the common practice of raising birds on wire. Wire mesh or hardware cloth is used in various poultry houses, thus permitting the droppings to fall through the wire and prevent the ingestion of fecal material by the birds. Management programs for poultry ranges may use a three-year rotation system wherein birds are kept on the range for a season and the land is then used to grow a cereal or grass crop for the following two years. This rotation plan permits the oocysts to die in the interim, before birds again occupy that section. Research has shown us that oocysts, under proper conditions for survival, may live longer than a year in the soil, making a long dormant period necessary. These methods are useful for controlling coccidiosis, but absolute prevention of infection is not possible even under these strict circumstances. Because of this limitation, investigators have continued looking for additional ways to control the disease. Development of the sulfonamide drugs and certain other nitrogen-containing compounds opened a new chapter in the control of coccidiosis. These drugs are referred to as coccidiostats. Most of them do not prevent the beginning of the life cycle stages of parasitic growth, but they stop the process after the initial shizogony cycle has been completed. This is important for two reasons. Interruption of the life cycle at the second generation of shizogony prevents the highly destructive effects of the disease on the individual bird. The second important effect is that the drugs allow the infection cycle to begin, thus stimulating a response in the bird to form resistance or immunity.

That birds can develop immunity to coccidia, if they are able to survive an initial infection, has been known for some time. In a haphazard way, this can occur during the usual husbandry methods wherein some individuals in the flock will receive a light, natural infection and thus develop resistance to a larger dose of challenging organisms which may reach them later. However, by leaving this to chance the poultryman is never sure when he will unintentionally create the circumstances for a massive build-up of coccidial organisms that can be highly destructive to the flock. Another factor in coccidial immunity is that it is highly specific. This means that immunity to one of the eight species of coccidia affecting the chicken does nothing to induce resistance to the other seven species. Investigators have tried to use all of this knowledge for the practical immunization of poultry. The procedure is to make a preparation containing known numbers of oocysts of the important species affecting the health of the bird. Birds receive this dose of organisms and are then placed on a coccidiostatic drug, which will prevent the immunizing organisms from causing disease but permit immunity to develop (Dickinson et al., 1951). Immunizing procedures of this kind are expected to have an important place in coccidiosis control in the future.

In summary, coccidiosis is controlled by efforts from three directions; (1) efforts to prevent the swallowing of sporulated oocysts by various husbandry methods, (2) the use of coccidiostatic drugs for the treatment of birds to prevent overwhelming losses during an outbreak, and (3) procedures to immunize the birds.

43-6. WORMS (HELMINTHS) AND THEIR CONTROL

There are many free-living forms of worms, such as the common earthworm, but our concern is with the parasitic worms, which we call helminths. There are numerous species of helminths, and it is certain that all of the earth's vertebrates are subject to some form of helminth infection. The helminths occur as flatworms and roundworms. The flatworms include the tapeworms and flukes, and the roundworms include many cylindrical forms, such as the common roundworms (ascarids) of the dog and pig.

Trichostrongylidosis. The trichostrongylids are a family of roundworms (Lapage, 1956). Five genera from this family contain species that inhabit the stomach and intestinal tract of sheep and cattle, producing a disease condition called trichostrongylidosis. This disease, which may masquerade in many forms, causes extensive losses in the livestock industry. The extremes of parasitism by these worms may cause intestinal irritation with diarrhea, anemia due to blood-sucking worms, great loss of general condition, and eventually death. Many milder patterns of disease are seen, including infections with no apparent damage to the host. These variations in the disease pattern are governed by the size of the worm burden and the resistance status of the animal. By far the greater number of trichostrongyle infections do not kill the host, and one might ask what their importance is in this case. It must be remembered that not only is the livestock producer concerned with the survival of his animals, but he is obliged to have healthy animals capable of making economic weight gains as he prepares them for a competitive market. The greatest economic loss from trichostrongyle infections is the result of their effect on the general thriftiness of the animals.

The life cycle of a typical trichostrongyle is both interesting and involved (Fig. 43-5). The male and female adult worms mate within the intestinal tract of the host, and the female then begins laying eggs. The eggs are passed from the host in the feces or droppings. The parasitologist can determine which of the numerous trichostrongyles are affecting the host by microscopically studying the eggs. Eggs on the pasture hatch into larvae and then begin a period of nonparasitic existence. The larvae eat bacteria found in their environment and live as free organisms on the grass and soil. Stages of development follow wherein the larvae become somewhat larger and more complex in structure. They go through a molting process at each developmental stage. When molting, the outer skin or sheath is usually cast off in a manner similar to the shedding of the outer skin by a snake. Third-stage larvae that retain the skin of the second stage as a protective sheath are infective to sheep and cattle. These larvae are located on grass and soil and are ingested by grazing animals. The larvae emerge from the protective sheath, burrow into the mucous membrane lining the intestinal tract, and undergo another molt to become fourth-stage larvae. Fourth-stage larvae molt once more to a fifth stage, and in this last form they increase in size to become adult worms. Even as adults these worms are only about $\frac{1}{2}$ to 1 inch in length; thus an observer must use care to avoid overlooking them when examining tissues with the naked eye. They make up in numbers what they lack in size; a lamb showing signs of disease may be carrying 60,000 parasites.

As in many kinds of infectious disease, much can be accomplished in terms

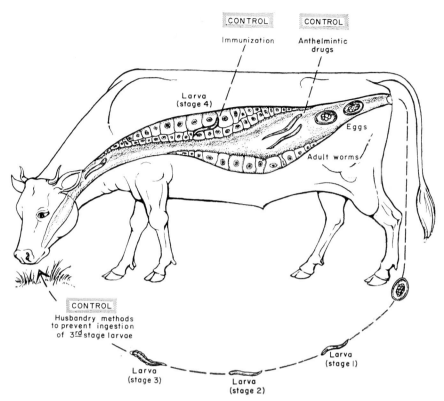

CONTROL
Immunization

CONTROL
Anthelmintic drugs

Larva (stage 4)

Eggs

Adult worms

CONTROL
Husbandry methods to prevent ingestion of 3rd stage larvae

Larva (stage 3)

Larva (stage 2)

Larva (stage 1)

Fig. 43-5. The life cycle of a trichostrongylid worm and measures used for control. The digestive tract of the calf is represented diagrammatically.

of minimizing trichostrongyle infections by the use of wise management practices. Eggs passed in the feces may give rise to infective larvae in 5 to 7 days, and presumably the larvae may survive for 6 months. Obviously the livestock producer should avoid overstocking and continuous use of pasture lands; these practices permit a build-up of infective larvae, particularly around places of animal concentration. A wise procedure, if it can be adapted to practical circumstances, is to keep animals moving frequently to new pastures. Studies of the host indicate that this group of worms infect cattle, sheep, goats, and deer, but they do not usually affect species such as the horse, pig, or chicken. The contaminated land may be used for grazing animals that are not susceptible to this form of parasitism or for growing another crop while waiting for the larvae to die.

It is common knowledge that success in avoiding parasitism is highly variable and we must deal with great numbers of animals carrying worm burdens that affect their health. In these situations one must resort to other means of control—for example, the use of drugs to kill the worms. Such drugs are called anthelmintics, and currently the most valuable one for the control of the trichostrongyles is phenothiazine. Since this drug can have toxic effects on sheep and cattle, the achievement of useful results from its administration requires skill on the part of the veterinarian. Phenothiazine is given orally to the animals so that the drug will contact the cuticle (outer skin surfaces) of the parasite within the digestive tract of the host. Research on anthelmintics is a continuing effort and recent findings have brought new light on the use of phenothiazine. It has been shown that small particle size and purity of the drug determine the efficiency with which worms can be killed (Douglas et al., 1959).

New drugs, such as the organophosphate compounds, hold great promise as anthelmintics. Preliminary studies have indicated that these preparations have outstanding properties for the treatment of trichostrongylidosis, particularly in cattle, for which phenothiazine is often of limited value (Baker et al., 1959). This is a new area in anthelmintic research and caution is required before application becomes general, because these compounds have toxicity for the host animal. As research progresses with the organophosphates and phenothiazine, we will continue to have better ways for controlling these parasites by reducing the worm burden in apparently normal animals, and by saving the lives of visibly sick sheep and cattle.

Young lambs and calves are particularly prone to develop severe effects from the trichostrongyles. After they reach the age of four months and older they become increasingly resistant. They eventually acquire a resistance status wherein many thousands of infective larvae may be ingested, but only a limited number will develop into adult worms. The greater number of infective larvae are repelled and destroyed by the host. The mechanisms of immunity to helminth infections have been difficult to study and up until very recently they have been beyond the control of man. It has long been recognized that a small dose of infective larvae could induce the animal to undergo these changes toward a resistance status. This means of immunizing by natural infection is untrustworthy because of the lack of control over the numbers of ingested larvae. Efforts made to immunize animals by injecting extracts prepared from dead worms have not been successful. The evidence suggests that certain substances released from the worm while it is viable and growing in the tissues are necessary to induce resistance. As the larvae go from one stage to another during the molt, they release certain fluids, which are apparently important for inducing immunity. Research workers have conceived the idea of inducing the immune response by infecting animals with irradiated larvae (Soulsby, 1961). Such larvae burrow into the mucous membrane and go through the molt but do not complete their life cycles. During the period of viability, they would excrete the substances necessary for immunization. These research findings are very encouraging and indicate that we will one day artificially immunize against helminth infections.

43-7. INSECTS AND THEIR CONTROL

The insects belong to an extensive class of organisms, including lice, gnats, bugs, mosquitoes, fleas, and flies. Most of these are known to the reader as insect pests and for their important role in disease transmission. Our attention will be focused on the important insect problem of myiasis, the infestation of host animal tissues by fly maggots.

The Screwworm Problem. Several species of flies deposit their eggs in animal wounds as well as decaying vegetable or animal material (Herms, 1950). Larvae hatch from the eggs and, when the deposition is on a live animal, the resultant myiasis can be quite troublesome. A special form of myiasis, however, is caused by the fly *Callitroga americana*. This is the screwworm fly, whose larvae must hatch on the tissues of a warm-blooded animal and obtain their nutrition by eating the living flesh. This characteristic of the screwworm fly has made it a persistent threat to economical livestock production. Any time a break occurs on the body surface of an animal it becomes subject to attack by the screwworm. Common sites for screwworm infestations include castration wounds, docking wounds, shear cuts, dog bites, dehorning wounds, the navels of newborn animals, and all manner of snags and scratches.

The female fly, which is a bit larger than the common housefly, attacks and deposits as many as 400 eggs along the edges of the wound (Fig. 43-6). In less than one day the eggs hatch into larvae which begin to eat host tissues. The larvae have rings of spines encircling the body which give them the appearance of a wood screw—thus the name screwworm. Extensive damage can be done, as the larvae feed on the tissues for a period of 5 to 6 days. After this feeding period, they drop from the wound and burrow into the ground, where they take a new form, called the pupae. After about 8 days they emerge as flies and in an additional 6 days are ready to lay eggs. The whole life cycle is completed in about 24 days.

Efforts to prevent extensive losses from screwworms have required continuous vigilance on the part of livestock men for many years. Good management practices will do much to limit the problem. Screwworms are especially active during the warm months and this period requires frequent inspection and prompt treatment of all wounds. Entomologists have developed preparations such as smear 335 and smear EQ-62, which are useful for killing screwworms in wounds and repelling flies to avoid reinfestation. The organophosphate compounds, mentioned previously for the control of helminths, are an additional weapon in the fight against the screwworm fly. Sprays containing these compounds may be used at the time animals receive wounds or skin abrasions. Sprays have certain advantages over the smear preparations: (1) longer-lasting action, (2) sufficient power of penetration to reach and kill larvae that may be deep within the wound, and

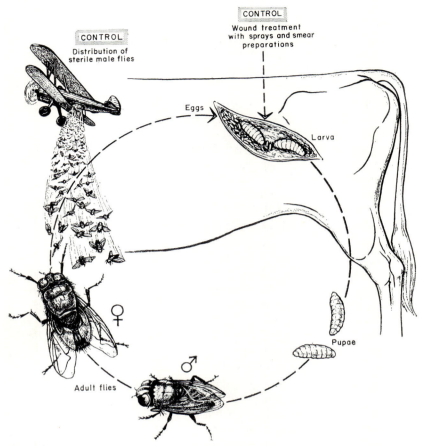

Fig. 43-6. The life cycle of the screwworm fly and methods used for its control.

(3) no interference with normal wound healing. Thus, we do have means for combating this problem, but the control measures require extensive effort and vigilance on the part of the livestock producer.

The Agricultural Research Service of the USDA estimates the annual loss from the screwworm in the southwestern states at approximately 20 to 25 million dollars. The southeastern United States was not concerned with this problem until 1933, when infested animals were brought in from the drought-stricken southwest. Estimated yearly losses in the southeast soon amounted to about 20 million dollars. These extensive losses further stimulated research on the part of entomologists, and eventually an important new control procedure for parasite eradication was developed.

The screwworm fly is a semitropical insect. This means that it is particularly important in states that have winters mild enough to permit year-round activity. During the warm months this parasite ranges into the central and northern states, but such far-ranging insects are invariably killed during the colder months of the year. The over-wintering areas, those areas in which the screwworm can survive the winter, constitute a far more restricted region of the United States than is apparent from the distribution of the organism at certain times. Observation of the life cycle revealed that the female fly mates only once. Early work had shown that X-radiation of fruit flies would cause a change in the sperm of the male so that their progeny, or offspring, would not survive. Studies by Bushland and Hopkins (1953) showed that radioactive cobalt-60, which emits gamma radiation, would sterilize the insects—eggs fertilized by an irradiated male failed to hatch.

With this background the parts of the puzzle could now be put together. In the overwintering areas sterilized male flies were to be turned loose in great numbers. These flies would seek out and mate with fertile females but the deposited eggs would not hatch. Competition by the sterile males would, after a period of time, tend to overwhelm fertile males found in the environment. Eventually this could cause the eradication of the screwworm. This novel eradication plan was first tried on a small island in the Carribbean Sea, the Island of Curaçao. The island was geographically isolated by water, which prevented reinfestation, and it was soon demonstrated that eradication could be accomplished. The southeastern United States was thought to be a favorable place for an eradication program because the overwintering region was limited to the peninsular part of Florida. This meant that the focus of infestation was limited by water on three sides and by colder winters in the region to the north. Early in 1958 a federal-state cooperative program was begun for the extensive release of sterile flies in that region (Agricultural Research Service Report, 1959). A laboratory was set up in which great numbers of flies were produced under artificial conditions and subjected to gamma radiation in the pupal stage. The sterile flies were then carried to various parts of the eradication region by airplane and distributed from the air. The control program in that region has been eminently successful, and in November of 1959 the program in Florida was discontinued because eradication had been accomplished. This was a remarkable achievement in livestock disease control and one wonders whether the same procedure can be followed in the extensive areas of the southwest, which are similarly plagued by the screwworm fly. Unfortunately, the problem is more complicated there because the great land mass is contiguous with Mexico. The cost of such a program would be staggering. It would require maintenance of a large buffer zone across the whole width of Mexico, in which sterile male flies would have to be continuously present. In Florida there were only 50,000 square miles of over-wintering area, but the southwestern area requiring eradication is about 145,000 square miles. At present it does not appear feasible to attack the problem in this way. More research and new ingenious ideas will be required before complete eradication of this disease can be accomplished.

REFERENCES AND SELECTED READINGS

References marked with an asterisk are of general interest.

Agricultural Research Service, USDA, October, 1959. *Report on the Screwworm Problem in the Southwestern United States and Northeastern Mexico.*

Baker, J. A., 1947. Attenuation of hog-cholera virus by serial passage in rabbits. *J. Am. Vet. Med. Assoc.,* 111:503–505.

Baker, N. F., P. H. Allen, and J. R. Douglas, 1959. Trial with a new organic phosphate as an anthelmintic in cattle. *Am. J. Vet. Research,* 20:278–280.

*Biester, H. E., and L. H. Schwarte, 1959. *Diseases of Poultry.* 4th Ed. Iowa State University Press, Ames, Iowa.

Buck, J. M., 1930. Studies of vaccination during calfhood to prevent bovine infectious abortion. *J. Agr. Research,* 41:667–689.

Bushland, R. C., and D. E. Hopkins, 1953. Sterilization of screw-worm flies with X-rays and gamma rays. *J. Econ. Entomol.,* 46:648–656.

Dickinson, E. M., W. E. Babcock, and J. W. Osebold, 1951. Coccidial immunity studies in chickens I. *Poul. Sci.,* 30:76–80.

* Dorset, M., C. N. McBryde, and W. B. Niles, 1908. *Further Experiments Concerning the Production of Immunity from Hog Cholera.* U.S. Bur. Animal Indus. Bull. 102.

Douglas, J. R., N. F. Baker, and W. M. Longhurst, 1959. Further studies on the relationship between particle size and anthelmintic efficiency of phenothiazine. *Am. J. Vet. Research,* 20:201–205.

Graham, O. H., B. Moore, M. J. Wrick, S. Kunz, J. W. Warren, and R. O. Drummond, 1959. A comparison of Ronnel and Co-Ral sprays for screw-worm control. *J. Econ. Entomol.,* 52:1217–1218.

Hagan, W. A., and D. W. Bruner, 1961. *The Infectious Diseases of Domestic Animals.* 4th Ed. Comstock Publishing Associates, Ithaca, N.Y.

Hammon, W. McD., W. C. Reeves, S. R. Benner, and B. Brookman, 1945. Human encephalitis in the Yakima Valley, Washington, 1942. *J. Am. Med. Assoc.,* 128:1133–1139.

Herms, W. B., 1950. *Medical Entomology.* 4th Ed. Macmillan, New York.

Lapage, G., 1956. *Veterinary Parasitology.* Charles C. Thomas, Springfield, Ill.

Myers, J. A., 1940. *Man's Greatest Victory Over Tuberculosis.* Charles C. Thomas. Springfield, Ill.

Pelczar, M. J., and R. D. Reid, 1958. *Microbiology.* McGraw-Hill, New York.

Raffel, S., 1953. *Immunity-hypersensitivity-serology.* Appleton-Century-Crofts, New York.

Shope, R. E., 1958. The swine lungworm as a reservoir and intermediate host for hog cholera virus. I. The provocation of masked hog cholera virus in lungworm-infested swine by ascaris larvae. *J. Expt. Med.,* 107:609–622.

Soulsby, E. J. L., 1961. Mechanism of immunity to helminths. *J. Am. Vet. Med. Assoc.,* 138:355–362.

Index